PEPTIDES

Structure and Function

Proceedings of the Ninth
American Peptide Symposium

Edited by

Charles M. Deber
Department of Biochemistry
University of Toronto
Toronto, Ontario

Victor J. Hruby
Department of Chemistry
University of Arizona
Tucson, Arizona

Kenneth D. Kopple
Department of Chemistry
Illinois Institute of Technology
Chicago, Illinois

PIERCE CHEMICAL COMPANY
ROCKFORD, ILLINOIS

PIERCE CHEMICAL COMPANY
3747 N. Meridian Road
P.O. Box 117
Rockford, Illinois 61105 U.S.A.

Library of Congress Cataloging in Publication Data

American Peptide Symposium (9th: 1985: Toronto, Ont.)
Peptides: structure and function.

Meeting held in June 1985 in Toronto, Ont.
Includes bibliographies and index.
1. Peptides–Congresses. I. Deber, Charles M.
II. Hruby, Victor J., III. Kopple, Kenneth D.
IV. Title [DNLM: 1. Peptides--congresses.
W3 AM595 9th 1985p / QU 68 A512 1985p]
QO552.P4A43 1985 574.19′2456 85-28500
ISBN 0-935940-04-9

Printed in the United States of America

PREFACE

"Toronto" is reputedly derived from an Indian word for "meeting place." Appropriately, the Ninth American Peptide Symposium was held in Toronto, Ontario, Canada, on the campus of the University of Toronto, June 23-28, 1985. About 700 participants, more than 200 of them visitors to North America, met to exchange research results in peptide science. The 60 talks and 250 posters reflected current trends in peptide research, and over 200 of these have been collected into this Proceedings Volume.

The topics discussed at the meeting ranged from new physical methods, such as vibrational circular dichroism, through asymmetric syntheses of non-protein amino acids, protein semisynthesis and protein modification, to the identification of antigenic sites, schemes for producing synthetic vaccines, and the stimulation of the immune system by peptides and peptide conjugates. Considerable attention was also directed to peptide-lipid interactions and amphiphilic secondary structures, and to structure-activity relations in peptide neurotoxins.

Continuing intense interest in peptide hormones was apparent. The design, synthesis and testing of analogs of the enkephalins, vasopressin, LHRH, cholecystokinin, somatostatin, and other peptide hormones were the subjects of almost a quarter of the papers. Another major thread was the further development of peptide analog inhibitors of enzymes, especially of renin and the angiotensin converting enzyme. Conformational considerations are now an integral part of analog design for both hormones and enzyme inhibitors. The use of non-peptide amino acid residues, and of isosteric replacements for the peptide bond is clearly increasing, and a particularly exciting paper described screening fermentation broths for non-peptide ligands for peptide receptors. A session organized by Ralph Hirschmann presented the state of clinical, physiological and synthetic research on an important new class of hormones, the atrial natriuretic factors.

In addition to numerous papers in which the emphasis was on synthetic methods and synthesis of longer peptides, the symposium featured a session dealing with problems in peptide coupling, organized by Bruce Merrifield. Speakers at this session presented evidence that the frequently observed instances of hindered coupling at intermediate chain lengths, both in solution and on solid supports, may result from the formation of beta structures.

Professor Robert Schwyzer, of the Eidgenössiche Technische Hochschule, Zürich, received the fifth Alan E. Pierce Award in recognition of his contributions to the techniques and strategies of peptide synthesis, to the understanding of peptide conformation and the development of research using cyclic peptides, and for influential concepts of the mechanism by which peptide hormones carry their messages. In his award address he presented new ideas about the thermodynamic basis of high affinity binding of hormones to membrane-bound receptors.

Peptide chemistry and biology clearly continue their growth in scientific, clinical and commercial interest, and it was a pleasure for us to have been able to organize a meeting in which the scientific participation was as enthusiastic as it was at the Ninth American Peptide Symposium. Many people contributed to bringing this Symposium off successfully, and we are pleased to thank them here. The secretarial assistance of Homer Harwood and Steve Cap (to KDK) and Christie Chantler (to CMD) was outstanding. The students of the "Phi Psi Omega" fraternity helped the organizers and participants in a thousand ways. We thank Queenie Teichman, Irene Derrett, and the staff of Insight Planners, Inc., for a fine job with the registration, accommodations, tours and the "Canada Sea to Sea" banquet. Maurice Jones, our liason officer with the University of Toronto, smoothed operations from early planning to final accounting. We are grateful to Tiiu Kask, Instructional Media Services, for ceaseless attention to detail in getting the Program and Abstract books set up and printed, to Padam Shukla for arranging the closed circuit TV, and to Jack Riley and his staff for the audio-visual services. Eva McGrath of the Word Processing Center at the Hospital for Sick Children most efficiently got our names and peptides straight. Thanks to all of them, to the friendly weatherman, and to the Toronto setting, the "peptide dreams" of the organizers became reality. Finally, for this permanent scientific record of the Symposium, we want to thank Steve Seely and Melba Rinaldo and their associates at Pierce Chemical, our publishers.

Charles M. Deber
Kenneth D. Kopple

Toronto, Ontario
Chicago, Illinois

PEPTIDE DREAMS

Theme from "Ninth American Peptide Symposium"

Words and music © by C. M. Deber, Toronto, Canada 1985

NINTH AMERICAN PEPTIDE SYMPOSIUM

University of Toronto
June 23-28, 1985

PLANNING COMMITTEE

Charles M. Deber
Lila M. Gierasch
Victor J. Hruby
Kenneth D. Kopple
Garland R. Marshall
Miguel A. Ondetti
Daniel H. Rich
Jean E. Rivier
Michael Rosenblatt
Peter W. Schiller
Clark Smith
John M. Stewart
Daniel F. Veber

PROGRAM COMMITTEE

Kenneth D. Kopple, Chairman

Michael J. Brownstein
Charles M. Deber
Bruce W. Erickson
Victor J. Hruby
Danute E. Nitecki
Jean E. Rivier
Peter W. Schiller
Daniel F. Veber

The Symposium was Sponsored by:

The University of Toronto

The Symposium Acknowledges Grants from:

National Institutes of Health
U.S. Public Health Service

National Institute of Medical Sciences

National Institute of Arthritis, Diabetes
and Digestive and Kidney Disease

and from

The U.S. Army Medical Research and Development Command

Generous Support in Service and Released Time was Provided by:

The Hospital for Sick Children, Toronto
The Illinois Institute of Technology, Chicago

Industrial Contibutors to the Ninth American Peptide Symposium

Abbott Laboratories
Bachem Inc.
Beckman Instruments, Inc.
Bio-Mega, Inc.
Boehringer Ingelheim Ltd.
Bristol-Myers Co.
Burroughs Wellcome Co.
Chemical Dynamics Corp.
CIBA - GEIGY Corp.
Du Pont NEN Products
E.I. du Pont de Nemours and Co.
Farmitalia Carlo Erba
Ferring AB
Hoechst-Roussell Pharmaceuticals, Inc.
Imperial Chemical Industries PLC
Institut Armand-Frappier
Lilly Research Laboratories
McNeil Pharmaceutical
Merck Sharp and Dohme Research Labs
Monsanto Company
Organogen Gmbh.
Organon
Ortho Pharmaceutical Corp.
Peninsula Laboratories, Inc.
Peptides International
Pfizer, Inc.
Pharmacia, Inc.
Pharmacia (Canada), Inc.
Pierce Chemical Co.
Sandoz, Inc.
G.D. Searle & Co.
Smith Kline & French Laboratories
The Squibb Institute for Medical Research
Syntex Research
UCB Bioproducts, S.A.
The Upjohn Co.
Vega Biotechnologies, Inc.

The Symposium wishes to acknowledge the generous support of these companies. Their contributions and interest helped greatly to make a successful meeting.

TABLE OF CONTENTS

PEPTIDES: Structure and Function

Proceedings of the Ninth American Peptide Symposium

FIFTH ALAN E. PIERCE AWARD LECTURE

ANTIGENIC SITES AND IMMUNOPEPTIDES

Peptide Antigens

CONTENTS

Physical Methods and Conformation: Calculation

Physical Methods and Conformation: X-Ray

Physical Methods and Conformation

Conformation: Cyclic Peptides

Conformation: Secondary Structure

CONTENTS

Metal Binding Peptides

SYNTHESIS

Coupling and Protection

Solid Phase Methods

Syntheses of Larger Peptides

Semisynthesis and Protein Modification

Glycopeptides

PEPTIDE HORMONES

Opioid Peptides

CONTENTS

Gonadotropin Releasing Hormone

Cholecystokinin, Glucagon, Secretin

Oxytocin, Vasopressin

Somatostatin

Neuropeptides

CONTENTS

Peptide Hormones

CONTENTS

Peptide-Lipid Interactions

ATRIAL NATRIURETIC FACTORS

Atrial Natriuretic Factors

FIFTH ALAN E. PIERCE AWARD LECTURE

ROBERT SCHWYZER

PEPTIDE CHEMISTRY AS A TOOL IN MOLECULAR BIOLOGY: DOES THE MEMBRANE LIPID PHASE CATALYZE PEPTIDE-RECEPTOR INTERACTIONS?

ROBERT SCHWYZER

PEPTIDE CHEMISTRY AS A TOOL IN MOLECULAR BIOLOGY: DOES THE MEMBRANE LIPID PHASE CATALYZE PEPTIDE-RECEPTOR INTERACTIONS ?

Robert Schwyzer

Laboratory of Molecular Biology and Biophysics, Swiss Federal Institute of Technology (ETH), CH-8093 Zürich, Switzerland

My sincere and heartfelt thanks go to my colleagues in the United States and Canada and to Mr. Pierce for electing me as the 5th recipient of the prestigious Alan E. Pierce Award. In accepting it, I feel greatly honored. However, the distinction also honors my parents, my long-gone teachers, the many persons who assisted me, the institutions that have furthered my efforts, and particularly my dear wife who provided me the atmosphere for unhampered investigation.

Molecular biology is the magic word of modern science. It is often used to designate its presently most spectacular branch: molecular genetics. However, a widely accepted definition runs as follows:[1]

"Recent technical advances in physics and chemistry have made possible quantitative studies of the molecular structures and events that underlie all biological processes. This facet of the biological sciences is known as molecular biology."

In this fundamental, unrestricted sense, all of us are contributing to molecular biology in our effort to advance peptide chemistry and its application to biological problems.

Before turning to the main subject of this presentation, a chapter from *molecular pharmacology*, I wish to mention two other recent contributions which fascinate me. One is the development, as a sequel to peptide synthesis, of the CAMET-CASET family of phosphate protecting groups that may prove to be the most versatile links known at present for the preparation of oligodeoxynucleotide synthons on solid or soluble carriers.[2 3 4] The other is the use of tobacco mosaic virus-neuropeptide conjugates as cooperative affinity labels for receptor localization and isolation.[5 6]

The Lipid Membrane as a Catalyst of Peptide-Receptor Interactions

Since many years I have been interested in structure-activity relationships and molecular mechanisms of peptide-receptor interactions.[7 8] Using the methods of membrane capacitance minimization (CM),[9] infrared attenuated total reflection spectroscopy (IR-ATR),[10-12] vesicle-mediated hydrophobic photolabeling (VMHL), and equilibrium dialysis (ED),[13] we have recently discovered that structure-activity relationships of adrenocorticotropin and dynorphin peptides are reflected in their physical and chemical behavior towards model lipid membranes.[14-17] This section reviews some of our findings, examines their thermodynamic plausibility, and deals with the possible intermediacy and catalytic activity of cell membranes in the reactions of peptides with their receptors.[18]

Specific interactions between peptides and artificial lipid membranes.- Dynorphin-(1-13)-tridecapeptide (dynorphin$_{1-13}$) and adrenocorticotropin--(1-24)-tetracosapeptide (ACTH$_{1-24}$) become reversibly adsorbed from aqueous solutions onto neutral, planar bilayer membranes prepared from 1-palmitoyl--2-oleoyl-sn-glycero-3-phosphocholine (POPC). The associated Gibbs free energy changes are approximately $\Delta G° = -6$ to -7 kcal/mol and the lipid areas necessary for binding one peptide molecule are about 75 and 110 nm^2,

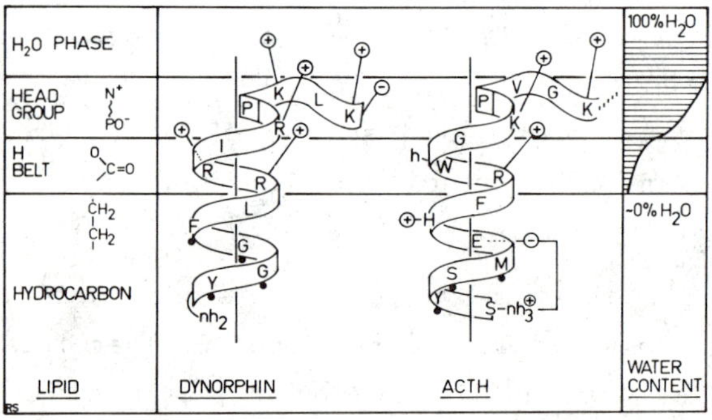

Fig. 1. Schematic Representation of the Interactions of Dynorphin$_{1-13}$ and ACTH$_{1-24}$ with Neutral Lecithin Membranes. The N-terminal amino groups and the indole hydrogen of Trp are indicated with lower-case letters. Unsatisfied H bonds of the helix end are indicated by dark spots. Positive and negative charges are shown in their relative positions. ACTH segment 14-24 is omitted (.....). The C-terminal segments probably extend into the water phase and do not remain in the head group layer.[12]

respectively.[12 19] During adsorption, preferred conformations and orientations are imposed upon the otherwise flexible, randomly structured molecules as shown in Figure 1. The N-terminal segments contact the hydrophobic membrane layers as perpendicular helical structures, whereas the C-terminal segments remain in contact with the aqueous layers as random coils.[10-14]

Beta-endorphin, mixed with POPC, yields IR-ATR spectra in the amide I and II regions that suggest a mixture of α-helical and random structures (amide I at 1658 cm^{-1}). After equilibration with liquid H_2O or vapor, the lipid becomes ordered and the amide I band shows two maxima at 1665 and 1631 cm^{-1} that clearly indicate the presence of short α-helices and of β-structures. Chou-Fasman interpretation of the amino acid sequence suggested about equal probability of α- and β-structures if β-endorphin were a crystalline protein. However, in the water-lipid interphase of an ordered membrane such predictions need not hold (see[12]). We are not yet sure whether the observed α- and β-structures are part of the same or of different molecules and how they are oriented with respect to the membrane. Thus, an interesting conformationselective interaction of beta-endorphin and POPC membranes is found (report in preparation).

Energy considerations.- Vertical intrusion of peptide helical segments into the hydrophobic phase of a membrane appears at first sight to be quite impossible, because of the unfavorable free energy changes caused by the charged α-amino group and by unsatisfied H bonds at the helix end. Similar destabilizations are caused by other residues such as Glu, Tyr, Ser, His, and Trp. Furthermore, upon binding, the peptides lose one degree of translational and two degrees of rotational freedom. These energy and entropy losses, as well as the observed free energy of binding, must be accounted for by the free energy differences for the transfer of $ACTH_{1-24}$ and $dynorphin_{1-13}$ from their random coil conformations in water to their membrane-bound conformations in the membrane.

Intramolecular side chain interactions[20] and interactions of charges and H bonds with water molecules may stabilize the helices as shown in Figure 1 and Table I. Yet the free energy of transfer remains difficult to assess, because the interphase boundary between bulk water and the hydrocarbon layer is not sharp. In this intermediate region containing the lipid head groups and the ester links (H belt[21]), concentration and activity coefficient of dissolved water are expected gradually to decrease to their low values in the hydrocarbon layer. Variations of the depth of helix insertion may modify the interaction energy in an unpredictable manner. As a first approximation, I ignored the influence of the diffuse layer and calculated the free energy differences for the transfer from an aqueous into a hydrophobic bulk phase, using known parameters.[22-24]

Table I. Possible Side Chain Interactions in the α-Helix Models of $ACTH_{1-24}$ (a) and $Dynorphin_{1-13}$ (b).

```
                                  H2O H2O      H2O
         (h) (h)        (h)+       +   h        +
          |   |          | |       |   |        |
nh3-Ser°Tyr°Ser°Met-Glu-His-Phe-Arg°Trp°Gly°Lys-Pro-Val-Gly-...    (a)
 |  |________h_______| |   |
 +                     -  (h)
 |_____________________|

                       H2O H2O     H2O     H2O     H2O H2O
   (h)                  +   +       +       +       +   -
    |                   |   |       |       |       |   |
nh2-Tyr°Gly°Gly°Phe-Leu-Arg°Arg°Ile°Arg-Pro-Lys-Leu-Lys-┘          (b)
```

Functional groups of interest are indicated with lower-case letters (e.g., h means hydrogen in an H-bond, whereas (h) is an 'unsatisfied' H-bond). Helix end groups are indicated with °. H_2O shows accessibility of groups to water. Residues with hydrophobic contacts are underscored.

Table II summarizes the results. The interacting segments were assumed to adopt an α-helical conformation in the hydrophobic phase, the rest of the molecules to remain exposed to water as random coils. Binding of $ACTH_{1-24}$ becomes possible if residues 1-8 interact; the experimentally observed binding energy is reached in the nonapeptide. Dynorphin residues 1-5 (enkephalin) are insufficient to cause membrane binding. However binding becomes feasible through interaction of the heptapeptide and reaches its experimentally observed strength with the octapeptide. Thus the calculations agree excellently with our proposed models, Figure 1.

Membrane binding energy, amphiphilicity, and biologic activity.- Given the necessary amphiphilic structure,[10,13] peptide-membrane interactions are governed by the free energy of transfer of the interacting segment from water to a hydrophobic phase. Biologic activity appears to correlate with the calculated binding energies of Table II. The shortest peptide with typical dynorphin-like activity on κ-receptors is the heptapeptide, and arginine in position 7 is important.[25] At this stage, membrane interactions become possible because of the binding energy and amphiphilicity introduced by the two arginines, 6 and 7. The binding is strong enough to convert a 3-dimensional to a 2-dimensional receptor search by the agonist.[26] The binding of $ACTH_{1-24}$ through the N-terminal octa- and nonapeptides is made possible by a very specific 'Maxfield-Scheraga' contact[20] between Ser^1 and Glu^5 that compensates the unfavorable contributions of the charges and H bonds (Table I). This may be the reason why Ser^1 is essential for ACTH activity.[27]

Table II. Estimated Free Energy Differences (ΔG_{ass}, kcal/mol) for the Transfer of $ACTH_{1-24}$ and $Dynorphin_{1-13}$ from Random Coil Conformations in Water to their Membrane-bound Conformations (Figure 1) in the Aqueous--hydrophobic Interphase of a Neutral Lipid Bilayer Membrane (entropy loss of about 10 kcal/mol taken into account).

$ACTH_{1-24}$ Interacting segment	ΔG_{ass}	$Dynorphin_{1-13}$ Interacting segment	ΔG_{ass}
SYSME...	+4.2	YGGFL...	+5.0
SYSMEH...	+5.7	YGGFLR...	+1.1
SYSMEHF...	+0.5	YGGFLRR...	-2.7
SYSMEHFR...	-3.4	YGGFLRRI...	-7.1
SYSMEHFRW...	-9.8	YGGFLRRIR...	-7.1

Association of the two isolated fragments, $ACTH_{1-10}$ and $dynorphin_{1-5}$ (enkephalin), with membranes is very weak: the amphiphilicity introduced with the hydrophilic tail segments is necessary for strong, specific, and oriented interaction with membranes.[10 13] This is reflected in the biologic potency: $ACTH_{1-10}$ is practically devoid of steroidogenic activity and enkephalin is a very poor κ-agonist compared to $dynorphin_{1-13}$. The observed, very weak enkephalin-membrane interaction probably involves lipid head groups[14] as also evidenced by enantioselective interactions of D and

L enkephalins with cerebroside sulfate liposomes.[28] Lack of amphiphilicity may also explain why increased hydrophobicity of enkephalin analogs does not markedly increase their membrane interactions[14] or alter their biologic activity.[29] Amphiphilicity, either primary or secondary, is obviously a prerequisite for strong peptide-membrane associations and connected biologic activities.

$ACTH_{1-24}$ contains three residues that are destabilized by being forced into a hydrophobic environment: Tyr^2 and Ser^3 (hydroxy) and His^6 (charge and 2 H bonds), Table I. These residues are essential elements of the potentiator and message segments that appear to interact directly with the receptor.[7] In dynorphin, the amino and hydroxy functions of Tyr^1 are destabilized: these groups are believed to be essential for opiate receptor interaction (in analogy to morphine). Interactions of such destabilized groups to form H bonds or salt bridges with receptor recognition sites exposed in the hydrophobic membrane layer[10] will be rapid and strong.

Is the peptide-membrane interaction physiologically significant at low agonist concentrations ? The apparent affinity constants for peptide--receptor interaction, measured pharmacologically or by binding, are in the nanomolar range ($\Delta G°$ about -12 to -14 kcal/mol). Does the low affinity of peptides for neutral membranes have any physiologic significance? I propose that low affinity paired with a large number of binding sites (larger than that of receptor sites) will amplify receptor occupancy at physiologic agonist concentrations.

Peptide-membrane association and catalysis of chemical reactions by detergent micelles.- The just described regio-, conformation-, and orientationselective binding of $ACTH_{1-24}$ and $dynorphin_{1-13}$ to model lipid bilayer membranes displays structural characteristics closely resembling those held responsible for the catalysis of bimolecular chemical reactions by detergent micelles. Thus the binding energy between the peptide and the membrane may be utilized to overcome the entropy requirements involved in bringing the reacting groups, peptide and receptor, together (see micelle catalysis by induced association[30]). Furthermore, the free energy from the binding interaction between the membrane and part of the peptide may be utilized to force the reacting portion of the peptide into an environment in which it is destabilized and can reach the transition state more easily (see catalysis by substrate destabilization[30]). Both mechanisms have been extensively studied in micelle-catalyzed reactions and are known to influence both the reaction rates and the equilibrium constants of chemical reactions.

The multistep model of peptide-receptor interaction.- Membrane-mediated peptide-receptor interactions involve at least two sequential steps: binding of the peptide to the membrane, followed by binding of the peptide to the receptor in the membrane. The total free energy of the reaction is thus divided between at least two steps. The activation energy of a single-step 'lock-and-key' process is also partitioned between the steps. This model has been described and its influence on receptor association and dissociation rates discussed.[18 31] Particularly interesting consequences of the multistep model emerge if the numbers of binding sites of the different steps are unequal.

The effect of successive binding steps on macroscopic ligand binding characteristics was calculated by Dr. David F. Sargent in my laboratory.[32] Peptide binding to membranes is saturable.[12 19] We use:

$$[A][B] = K_d[AB] \tag{1}$$

$$\Delta G° = - RT\ln K_d. \tag{2}$$

For a macromolecule M (or any large binding unit, such as a unit of area of membrane surface) containing n binding sites of which on the average $\nu = L_b/M$ are occupied by ligand L, eq. (1), after rearrangement, yields $L_b = \nu M = nML_f/(K_d+L_f)$, where L_f, L_b are the free and bound concentrations.

For each step in a sequence, each with n_i sites, K_{di}, ν_i bound ligand molecules, we have:

$$L_i = \nu_i M = n_i ML_{(i-1)}/(K_{di} + L_{(i-1)}), \tag{3}$$

thus

$$L_1 = n_1 ML_0/(K_{d1} + L_0); \quad L_2 = n_2 ML_1/(K_{d2} + L_1); \quad .....$$

For low ligand concentrations such that $L_{(i-1)} << K_{di}$ for each step:

$$L_n = L_0 M^n \prod n_i / \prod K_{di}.$$

Setting M = 1 for illustrative purposes, as in Figure 2, and using $\prod K_{di} = K'_d$ for the same total $\Delta G°$, we find:

$$L_n/L' = (L_0 \Pi n_i / \Pi K_{di}) / (L_0 n'/K'_d) = \Pi n_i/n' \qquad (4)$$

where ' indicates the corresponding value for a single-step reaction.

The binding curves of Figure 2 show the effects of intermediate binding steps on the saturation at the final site. If each stage has the same number of binding sites (n_i) and the total change in free energy, $\Delta G° = \Sigma \Delta G_i$ (= -12 kcal/mol, chosen for illustrative purposes) is the same, then the binding curve is not affected and coincides with that of the one--step reaction (curve A). If, however, an early stage has a higher number of sites, the subsequent stages saturate at agonist concentrations lower by a factor $\Pi n_i/n'$ where n' is the number of sites in the final stage (curves B, B'). Thus at physiological agonist concentrations, receptor occupancy may be modified in a wide range by the multiple sequential binding steps of our model of membrane catalysis.

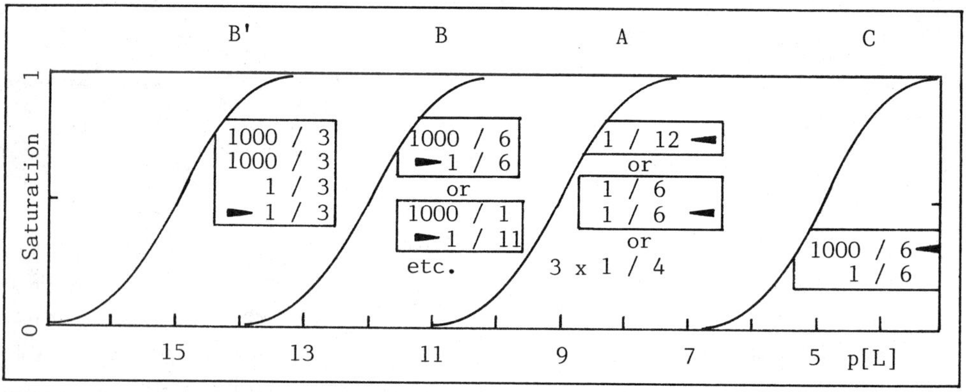

Fig. 2. Effect of Successive Binding Steps on Macroscopic Ligand Binding Characteistics. Binding curves show the saturation at the site indicated as a function of the decadic logarithm of free ligand concentration for various combinations of successive steps. The number pairs give the number of binding sites and $\Delta G°$ value per step (n_i; $\Delta G°_i$). $\Sigma \Delta G°_i$ is constant (-12 kcal/mol) for all curves.

The example of Figure 2 is quite realistic: in a target cell with a surface area of about 1000 μm^2 in which only 25% of the surface is occupied by lipid, the number of binding sites may exceed that of receptor binding sites by three orders of magnitude. With agonist concentrations in the physiologic pM to fM range, the condition $L_o << K_d$ is met. Curve C shows membrane binding under these conditions and curve B the saturation at the final stage, irrespective of the true intrinsic peptide-receptor association constant.

Conclusion.- Insertion of the N-terminal message segments of dynorphin$_{1-13}$ and ACTH$_{1-24}$ is energetically possible through hydrophobic interaction of the amphiphilic molecules with lipid membranes. A new function of the lipid bilayer cell membrane is proposed: catalysis of peptide-receptor interactions. The catalytic action shares two classical features with micellar catalysis of chemical reactions: catalysis by induced association and by substrate (agonist) destabilization. Two features are non-classical: rate increases by reduction of dimensionality and shifts of receptor occupancy by the effect of preceding binding stages with a higher number of binding sites. 'High-affinity low-capacity' binding need not be due to an exceptionally strong intrinsic affinity of the receptor for its agonist, but may merely be the consequence of preceding 'low-affinity high--capacity' sites (lipid membrane sites).

References

1. Villee, C.A. (1971) in Collier's Encyclopedia, Vol. 4, p. 172c.
2. Schwyzer, R., Felder, E., and Failli, P. (1984) Helv. Chim. Acta 67, 1316-1327.
3. Felder, E., Schwyzer, R., Charubala, R., Pfleiderer, W., and Schulz, B. (1984) Tetrahedron Lett. 25, 3967-3970.
4. Felder, E. and Schwyzer, R. (1985) in Peptides 1984, Proc. 18th Eur. Peptide Symp., Ragnarsson, U., Ed., Uppsala, pp. 157-160.
5. Schwyzer, R. and Kriwaczek, V.M. (1981) Biopolymers 20, 2011-2020.
6. Wunderlin, R., Sharma, S.D., Minakakis, P., and Schwyzer, R. (1985) Helv. Chim. Acta 68, 12-22.
7. Schwyzer, R. (1977) Ann. N.Y. Acad. Sci. 297, 3-26.
8. Schwyzer, R. (1980) Proc. R. Soc. Lond. B210, 5-20.
9. Schoch, P., Sargent, D.F., and Schwyzer, R. (1979) Biochem. Soc. Trans. 7, 846-849.
10. Gremlich, H.-U., Fringeli, U.-P., and Schwyzer, R. (1983) Biochem. 22, 4257-4263.
11. Gremlich, H.-U., Fringeli, U.-P., and Schwyzer, R. (1984) Biochem. 23, 1808-1810.

12. Erne, D., Sargent, D.F., and Schwyzer, R. (1985) Biochem. 24, in press.

13. Gysin, B. and Schwyzer, R. (1984) Biochem. 23, 1811-1818.

14. Gysin, B. and Schwyzer, R. (1983) Arch. Biochem. Biophys. 225, 467--474.

15. Gysin, B. and Schwyzer, R. (1983) FEBS Lett. 158, 12-16.

16. Schwyzer, R., Gremlich, H.-U., Gysin, B., and Fringeli, U.-P. (1983) in Peptides 1982, Blaha, K. and Malon, P., Eds., Walter de Gruyter & Co., Berlin, N.Y.,pp. 55-71.

17. Schwyzer, R., Gremlich, H.-U., Gysin, B., Sargent, D.F., and Fringeli, H.-P. (1983) in Peptides: Structure and Function, Hruby, V.J. and Rich, D.H., Eds., Pierce Chem. Co., pp. 657-664.

18. Schwyzer, R. (1985) in Natural Products Chemistry 1984, Zalewski, R.I. and Skolik, J.J., Eds., Elsevier Science Publishers B.V., Amsterdam, pp. 591-598.

19. Schoch, P. (1980) Thesis No. 6699, Swiss Federal Institute of Technology, Zürich.

20. Maxfield, F.R. and Scheraga, H.A. (1975) Macromolecules 8, 491-493.

21. Chatterjie, N. and Brockerhoff, H. (1978) Biochim. Biophys. Acta 511, 116-119.

22. Chothia, C. (1976) J. Mol. Biol. 105, 1-14.

23. Von Heijne, G. and Blomberg, C. (1979) Eur. J. Biochem. 97, 175-181.

24. Von Heijne, G. (1981) Eur. J. Biochem. 116, 419-422.

25. Chavkin, C. and Goldstein, A. (1981) Proc. Natl. Acad. Sci. USA 78, 6543-6547.

26. Berg, H.C. and Purcell, E.M. (1977) Biophys. J. 20, 193-219.

27. White, W.F. (1955) J. Am. Chem. Soc. 77, 4691-4692.

28. Schwyzer, R. (1983) Proc. 18th Solvay Conf. on Specific Reversible Interactions in Life Processes, Katchalski, E., Chairman, Bruxelles, in press.

29. Do, K.Q., Fauchère, J.-L., Schwyzer, R., Schiller, P.W., and Lemieux, C. (1981) Hoppe-Seyler's Z. Physiol. Chem. 362, 601-610.

30. Jencks, W.P. (1975) Adv. Enzymol. 43, 219-410.

31. Schwyzer, R. (1984) Naito Found. Internatl. Symp. Nat. Prod. Biol. Act., Tokyo, in press.

32. Sargent, D.F. and Schwyzer, R., in preparation.

ALGORITHMIC IDENTIFICATION OF ANTIGENIC DETERMINANTS IN PROTEINS OF KNOWN STRUCTURE

David W. Fanning and George D. Rose

Department of Biological Chemistry, Hershey Medical Center, Pennsylvania State University, Hershey, PA 17033

John A. Smith

Departments of Molecular Biology and Pathology, Massachusetts General Hospital and Harvard Medical School, Boston, MA 02114

Introduction

The surface topography of protein molecules is of key importance in intermolecular interactions such as the binding of antibodies. The topographic features of the molecular surface are not defined intrinsically; instead, they are dependent upon the way the surface is mapped. For example, when the protein surface is surveyed by a small, sub-atomic probe, it will have many subterranean crevices and valleys that are inaccessible to a larger, protein-sized probe.

Frequently, the protein surface is taken to be synonomous with the solvent-accessible surface. Lee and Richards[1] developed a widely applied algorithm to measure solvent-accessible surface in proteins of known structure. In this algorithm, the surface is probed uniformly by a water-sized sphere (radius = 1.4 Angstroms); regions of the molecule that come into contact with the probe sphere comprise the surface, while remaining regions are said to be buried within the interior.

In the case of antigen:antibody interactions, however, it is the antibody-accessible surface that is relevant to intermolecular association. Two approaches to mapping the antibody-accessible surface have been described recently. Novotny _et al._[2] use the original Lee and Richards algorithm[1] with a larger probe (radius = 10 Angstroms) in order to better approximate the size of an antibody binding domain. In addition, Fanning _et al._[3] and Novotny _et al._[2] contour the protein surface to reveal its topography in a manner that is analogous to the familiar topographic maps of the earth issued by the U.S. Geological Survey.

In the following sections, the construction of protein topographic maps is described and then used to examine antigenic sites on myoglobin and lysozyme. Such sites are seen to correlate well with regions of the surface that are globally exposed -- the high mountains and unencumbered ridges of the protein.

Contour Maps

Contouring is a common method used to graphically represent topographic surface features of irregular objects. Each contour line is a trace of constant altitude above an assumed starting elevation. When the interval between contour lines is constant, the patterns formed by the lines quantitate surface features such as peaks, valleys, and steepness of slope. A significant advantage of a contour map is that global features such as the highest portions of the surface can be easily and quantitatively assessed.

To contour a protein, a starting elevation or sea-level must be defined, the equivalent of a standard state elevation. We have defined this sea-level to be proportional to the ellipsoid of intertia calculated for all protein atoms. From classical mechanics[4], the ellipsoid of inertia of a set of atoms is a smooth, convex body superimposed upon the atoms and having the same rigid-body dymamics. That is, the ellipsoid has the same principal moments of inertia as the atoms themselves and is, therefore, uniquely defined. The sea-level ellipsoid is defined as a tri-axial ellipsoid with axes scaled to be 0.5 times the protein's principal moments.

The molecular surface is generated for a protein using an algorithm due to Connolly[5]. Points of this surface that are below sea-level are eliminated, while remaining points are assigned an altitude. These altitudes are then contoured in intervals of 1 Angstrom using a standard program[6]. Details of the method are available in Fanning _et al_.[3]

Figures 1 and 2 depict contour maps of lysozyme and myoglobin. Only contours corresponding to the most highly exposed regions of the surface, those with contours in excess of 8 Angstroms, are shown. These have been extracted from the set of all contours and annotated by residue number. When the curved surface of the protein is projected onto a plane, as it is in these figures, systematic distortions are introduced, similar to the familiar distortion of Greenland seen on cylindrical map projections of the earth. Nevertheless, topographic features on the map have a one-to-one correspondance with physical features on the protein's molecular surface.

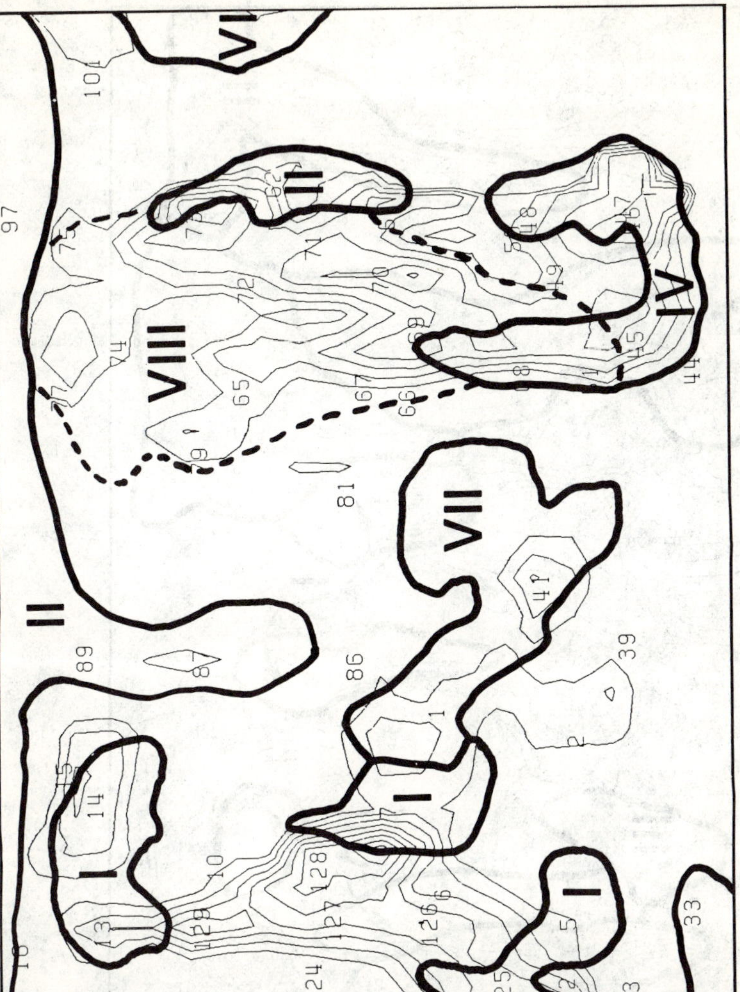

Figure 1. Contour Map of Lysozyme

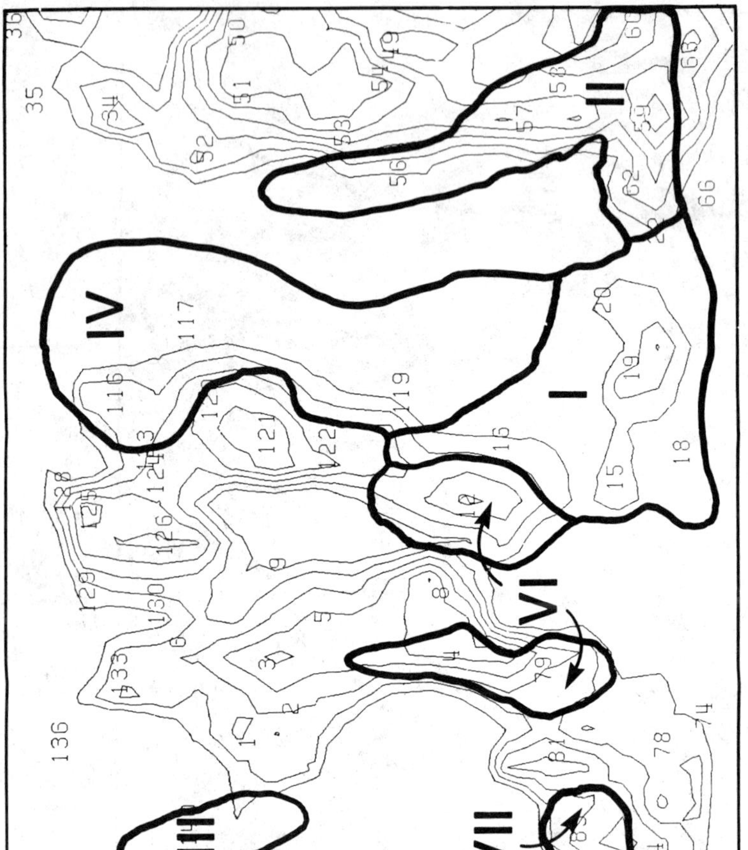

Figure 2. Contour Map of Myoglobin

Antigenic sites and accessible surface

Antigenic sites for lysozyme and myoglobin were culled from the literature and listed in table 1.

TABLE 1

Lysozyme

Site No	Residues	Reference
1	5,7,13,14,125	Atassi (1978)[7]
2	62,87,89,93,96,97	Atassi (1978)[7]
3	33,34,113,114,116	Atassi (1978)[7]
4	45-48,68	Smith-Gill *et al*. (1982)[8]
5	19,21	Smith-Gill *et al*. (1984)[9]
6	102,103	Smith-Gill *et al*. (1984)[9]
7	1,41,84	Smith-Gill *et al*. (1984)[9]
8	64-80 (loop)	Arnon (1977)[10]

Myoglobin

Site No	Residues	Reference
1	15-22	Atassi (1975)[11]
2	56-62	Atassi (1975)[11]
3	94-99	Atassi (1975)[11]
4	113-119	Atassi (1975)[11]
5	146-151	Atassi (1975)[11]
6	4,12,79	Berzofsky *et al*. (1982)[12]
7	83,144,145	Berzofsky *et al*. (1982)[12]
8	140	Berzofsky *et al*. (1982)[12]

These sites are also shown superimposed upon the protein's contoured surface in figures 1 and 2.

It can be seen from the figures that known antigenic

sites correspond generally, though not perfectly, to highly exposed regions of the molecular surface. However, there remain regions of this surface that are not covered by any known antigenic site. Nevertheless, such negative evidence is not conclusive because additional antigenic sites may yet be found. In fact, very recent work of Amit *et al.*[13] on the structure of a lysozyme:anti-lysozyme complex implicates additional regions of the remaining highly exposed surface in figure 1 (*viz.* residues 22, 117, 119, and 121).

Only one antigenic site fails to correspond to a highly exposed region, site IV in myoglobin (figure 2). However, this site does occupy the flank of a highly exposed region composed of residues 120-130, and it continues into a broad groove on the protein surface. The fold on the protein is wide enough (approximately 15-20 Angstroms), and the curvature of the protein is great enough, that the site may still remain accessible to an antibody molecule.

The correlation between antigenic sites and highly exposed surface prompts the conjecture that globally exposed residues play a primary role in determining the antigenic structure of a globular protein. These correlations are consistent with the view of Benjamin *et al.*[14] in which the entire protein surface is thought to be potentially capable of binding antibodies. For a protein of known structure, the method presented here provides a quantitative way to elucidate the relevant surface.

References

1. Lee, B.K. and Richards F.M (1971) J. Mol. Biol. 55, 379-400.

2. Novotny, J., Handschumacher, M., Haber, E., Bruccoleri, R.E., Carlson, W.B., Fanning, D.W., Smith, J.A., and Rose, G.D. (1985) Submitted.

3. Fanning, D.W., Smith, J.A., and Rose, G.D. (1985) Submitted.

4. Goldstein, H. (1950) Classical Mechanics (Addison-Wesley, Reading, MA).

5. Connolly, M.L. (1983) Science 221, 709-713.

6. National Center for Atmospheric Research. (1980) Fortran Subroutine SRFACE, Boulder, Colorado.

7. Atassi, M.Z. (1978) Immunochem 15, 909-936.

8. Smith-Gill, S.J., Wilson, A.C., Potter, M., Prager, E.M., Feldman, R.J., and Mainhart, C.R. (1982) J. Immunol. 128, 314-322.

9. Smith-Gill, S.J., Lovoie, T.B., and Mainhart, C.R. (1984) J. Immunol. 133, 384-393.

10. Arnon, R. (1977) Immunochemistry of Enzymes and Their Antibodies., (M.R.J. Salton, Editor), (Wiley, New York).

11. Atassi, M.Z. (1975) Immunochem. 12, 423-438.

12. Berzofsky, J.A., Buckenmeyer, G.K., Hicks, G., Gurd, F.R.N., Feldman, R.J., and Minna, J. (1982) J. Biol. Chem. 257, 3189-3198.

13. Amit, A.G., Mariuzza, R.A., Phillips, S.E.V., and Poljak, R.J. (1985) Nature 313, 156-158.

14. Benjamin, D.C., Berzofsky, J.A., East, I.J., Gurd, F.N.R., Hannum, C., Leach, S.J., Margoliash, E., Michael, J.G., Miller, A., Prager, E.M., Reichlin, M., Sercarz, E.E., Smith-Gill, S.J., Todd, P.E., and Smith, A.C. (1984) Ann. Rev. Immunol. 2, 67-101.

HYDROPHILICITY PARAMETERS IN PEPTIDES. 1. PREDICTION OF PEPTIDE RETENTION IN REVERSED-PHASE HPLC (RPC). 2. PREDICTION OF HYDROPHILIC REGIONS ON THE SURFACE OF PROTEINS.

D. Guo, C.T. Mant, J.M.R. Parker and R.S. Hodges
Department of Biochemistry and Medical Research Council of Canada Group in Protein Structure and Function, University of Alberta, Edmonton, Alberta, Canada T6G 2H7

Introduction

The ability to predict the elution profiles of peptides of known composition would greatly enhance the utility of RPC. Also we reasoned that the interaction of hydrophilic or hydrophobic amino acid side chains in a peptide with a hydrophobic support as in RPC would provide a more accurate measure of the hydrophilicity of amino acid residues in a protein. We have determined retention coefficients and hydrophilicity parameters for each amino acid residue by measuring their contributions to the retention of a model synthetic peptide: Ac-Gly-X-X-(Leu)$_3$-(Lys)$_2$-amide, where X was substituted by the 20 amino acids found in proteins.

Results and Discussion

The excellent resolving power and selectivity of the TFA/H_2O -TFA/acetonitrile (pH 2) gradient system is demonstrated by the representative mixture of model peptides separated on a C-18 column (Fig. 1, top). The amino acid coefficients determined from subsequent HPLC of all our model peptides (Table 1) may be used to predict the retention time of any peptide of known composition provided a standard peptide is always run to correct the change in retention due to column aging or supports of var-

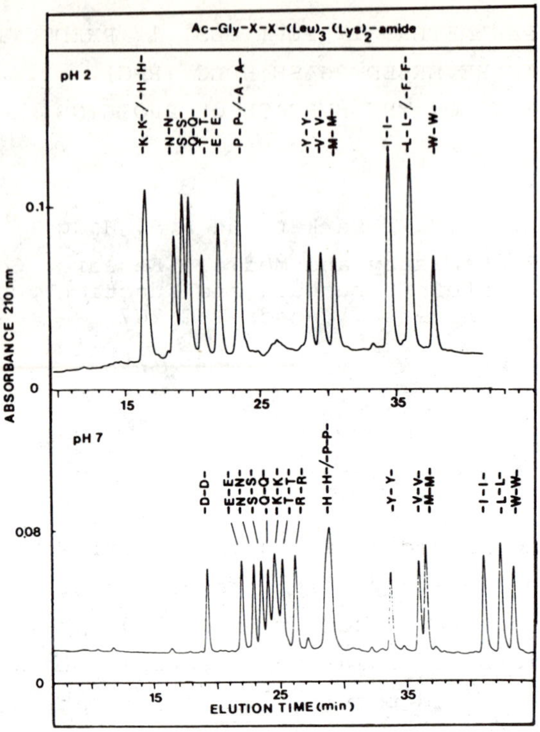

Figure 1. Reversed-phase HPLC of synthetic peptides having the sequence Ac-Gly-X-X-$(Leu)_3$-$(Lys)_2$-amide where position X is substituted by the 20 amino acids found in proteins. The top panel was performed at pH 2.0 and the bottom panel at pH 7.0. Conditions: AB gradient; at pH 2.0, solvent A consisted of 0.1% TFA-H_2O and solvent B of 0.1% TFA in acetonitrile, gradient rate 1% B/min at a flow rate of 1 ml/min; at pH 7.0 solvent A consisted of 10 mM $(NH_4)_2HPO_4$ - 0.1M $NaClO_4$ buffer and solvent B consisted of 40% (0.1 M $NaClO_4$-H_2O) and 60% acetonitrile, gradient rate 1.67% B/min which is equivalent to 1% acetonitrile/min, flow rate 1 ml/min. The elution profiles shown are representative runs from a C-18 Synchropak RP-P C-18 column, 250 mm x 4.1 mm I.D., 6.5 μm particle size, 300 Å pore size, carbon loading ∿10%.

ying n-alkyl chain length and ligand density as well as instrumentation variations. When chromatographing the peptides at pH 7 the addition of $NaClO_4$ (7) to the gradient buffers was discovered to be essential for producing the resolution shown in Fig. 1 (bottom).

Despite the excellent resolution of peptide mixtures at either pH, the volatility of the solvents used at pH 2 makes this the more desirable system for most purposes. Side chain retention coefficients for pH 7 were calculated and used to derive a new set of hydrophilicity parameters (Table 1). These parameters were compared with eight sets of hydrophilicity parameters for three proteins, myoglobin, lysozyme and influenza to

TABLE 1

RETENTION COEFFICIENTS AND HYDROPHILICITY PARAMETERS OF AMINO ACID RESIDUES

Amino Acid Residue	Retention Coefficients[a] pH 2.0 (min)	pH 7.0 (min)	Relative Hydrophilicity Parameters[b]
Trp	8.8	9.5	-10.0
Phe	8.1	9.0	- 9.2
Leu	8.1	9.0	- 9.2
Ile	7.4	8.3	- 8.0
Met	5.5	6.0	- 4.2
Val	5.0	5.7	- 3.7
Tyr	4.5	4.6	- 1.9
Cys	2.6	2.6	1.4
Pro	2.0	2.2	2.1
Ala	2.0	2.2	2.1
Glu	1.1	-1.3	7.8
Thr	0.6	+0.3	5.2
Asp	0.2	-2.6	10.0
Gln	0.0	0.0	5.7
Ser	-0.2	-0.5	6.5
Gly	-0.2	-0.2	6.0
Arg	-0.6	+0.9	4.2
Asn	-0.6	-0.8	7.0
His	-2.1	+2.2	2.1
Lys	-2.1	-0.2	6.0
α-Amino	-6.9		
α-COOH	-0.8		

[a]The retention coefficients (in min.) were determined from retention times on reversed-phase HPLC as shown in Fig. 1. The predicted retention time for a peptide equals the sum of the retention coefficient for the amino acid residues plus t_o (the time for elution of unretained compounds). In these experiments, t_o = 2.0. All parameters are calculated for N^{α}-acetylated and C-terminal amide peptides and only the values of the end groups shown need be considered.
[b]Hydrophilicity parameters were determined from the HPLC retention coefficients at pH 7.0. A standard scale was used where the maximum negative retention coefficient was assigned a value of +10 for the maximum hydrophilic parameter and the maximum positive retention coefficient was assigned a value of -10 for the maximum hydrophobic parameter. The remaining parameters were scaled relative to these maximum and minimum values.

determine if the predicted hydrophilic segments matched the antigenic sites proposed by several workers (2,3,4). In Fig. 2, we have shown a hydrophilicity profile for myoglobin calculated from the most widely used Hopp and Woods parameters (1) and our own HPLC parameters. Although all parameters were able to predict some of the antigenic sites, the HPLC parameters in this report, the average surrounding hydrophobicity (5) and accessibility (6) parameters scored highest.

Our retention coefficients have worked successfully in retention prediction with many peptides in our laboratory and work is in progress to determine the predictive accuracy with a further 50 known peptides (up to 40 residues). The effect of gradient rate, flow rate and molecular weight will also be examined.

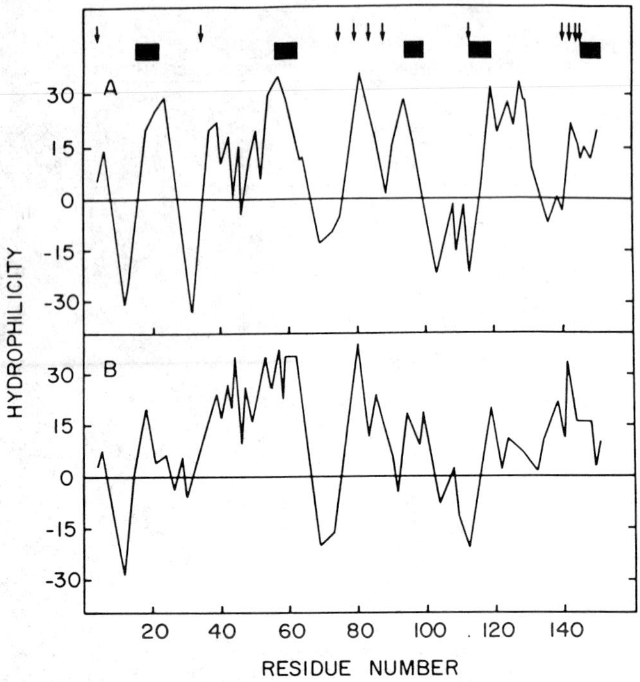

Figure 2. Hydrophilicity profiles of myoglobin. The summed parameters for each seven residue segment was assigned to the fourth residue and this segment was shifted from the N to C-terminus of the myoglobin sequence by one residue each time. The results were plotted as a function of the amino acid residue in the sequence. Panel A is the myoglobin profile using the parameters listed in Table 1 and panel B is the profile obtained with the parameters of Hopp and Woods (1). The solid bars indicate the proposed antigenic sites from Atassi (2) while the arrows show the proposed sites from Benjamin et al. (3).

References

1. Hopp, T.P., and Woods, K.R. (1981) Proc. Natl. Acad. Sci. USA 78, 3824-3828.

2. Atassi, M.Z. (1984) Eur. J. Biochem. 145, 1-20.

3. Benjamin, D.C., Berzofsky, J.A., East, I.J., Gurd, F.R.N., Hannum, C., Leach, S.J., Margoliash, E., Michael, J.G., Miller, A., Prager, E.M., Reichlin, M., Sercarz, E.E., Smith-Gill, S.J., Todd, P.E., and Wilson, A.C. (1984) Ann. Rev. Immunol. 2, 67-101.

4. Webster, R.G., Laver, W.G., Air, G.M., and Schild, G.C. (1982) Nature 296, 115-121.

5. Manavalan, P., and Ponnuswamy, P.K. (1978) Nature 275, 673-674.

6. Janin, J. (1979) Nature 277, 491-492.

7. Meek, J.L. (1980) Proc. Natl. Acad. Sci. USA 77, 1632-1636.

ENHANCEMENT OF IMMUNE RESPONSE USING B-LYMPHOCYTE MITOGENS COVALENTLY LINKED TO ANTIGENS

Günther Jung*, Karl-Heinz Wiesmüller and Jörg Metzger
Institut für Organische Chemie, Universität Tübingen

Hans-Jörg Bühring and Claude P. Muller
Medizinische Universitätsklinik, Universität Tübingen

Lothar Biesert and Wolfgang G. Bessler
Arbeitsbereich Mikrobiologie und Immunologie, Universität Tübingen
D-7400 Tübingen, FRG

The synthetic lipopeptide N-palmitoyl-S-(2,3-bis(palmitoyloxy) -(2RS)-propyl)-(R)-Cys-(S)-Ser-Ser-Asn-Ala (Pam_3Cys-Ser-Ser-Asn-Ala)[1,2] is a polyclonal B-lymphocyte activator as potent as the native lipoprotein from the outer membrane of E.coli[3]. Recently we have shown that the even shorter lipopeptide segment Pam_3Cys-Ser[4] is a fully active mitogen stimulating proliferation and immunoglobulin synthesis in B-lymphocytes in vivo and in vitro, whereas Pam_3Cys and lipopeptide derivatives carrying only one fatty acid, e.g. Pam-Cys-Ser-Ser-Asn-Ala, are considerably less active. Pam_3Cys-Ser-Ser-Asn-Ala induces the production of interleukin 1 in human mononuclear leukocytes and the release of prostaglandin (PGE_2) from mouse peritoneal macrophages (ref[5] and to be published). Binding proteins for the lipopentapeptide could be isolated by affinity chromatography[6], which may represent specific recognition structures. A potent adjuvant effect of the substances was observed in vitro and in vivo: The synthetic lipopeptide derivatives markedly increased the immune response against TNP-SRBC in mouse phagocytes (ref[7] and unpublished results).

Among the various new immunologically active analogues prepared, the mitogenic fluorescent cell markers Pam_3Cys-Ser-Lys_4 (FITC and RITC) were found to be useful for cell sorter experiments and immune fluorescence microscopy.

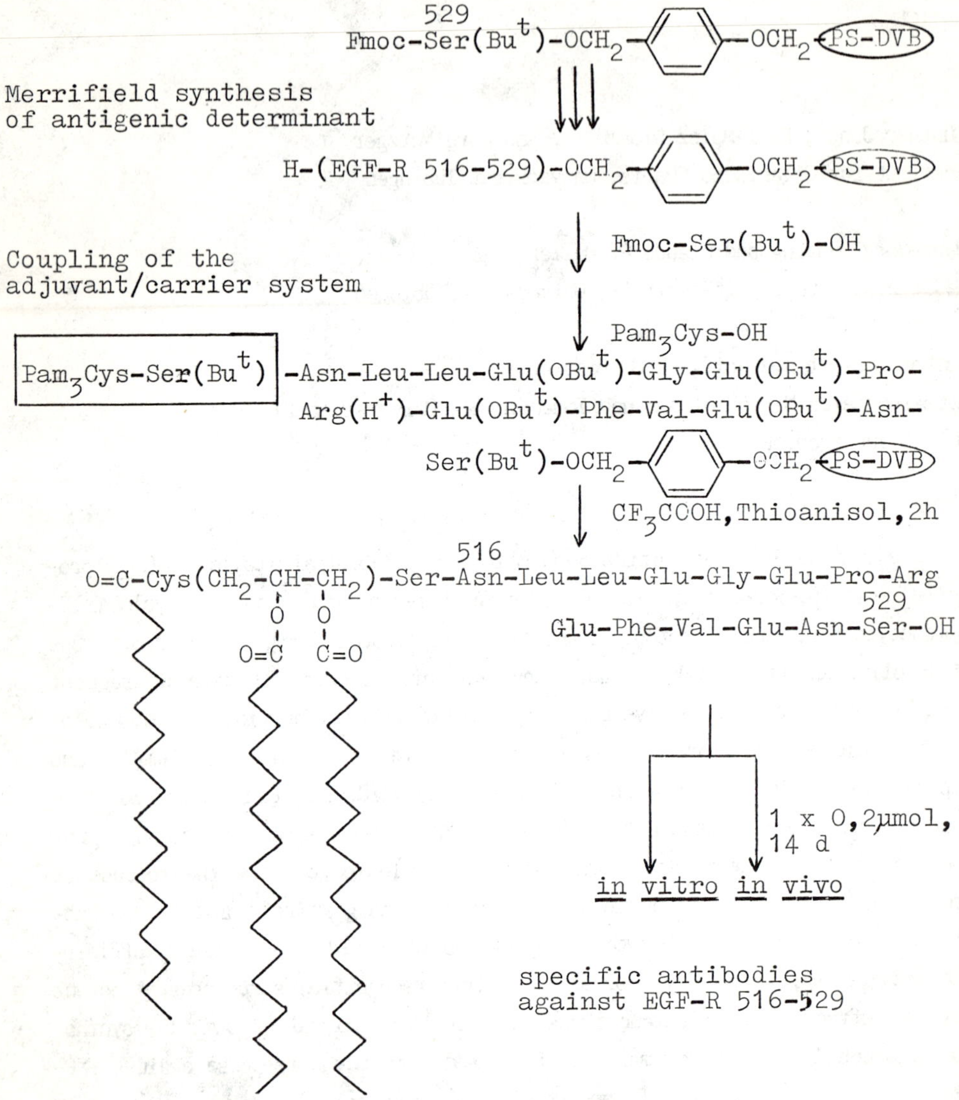

Fig. 1. Routine incorporation of the lipopeptide carrier and adjuvant Pam_3Cys-Ser during Merrifield synthesis of the antigenic determinant EGF-R 516-529.

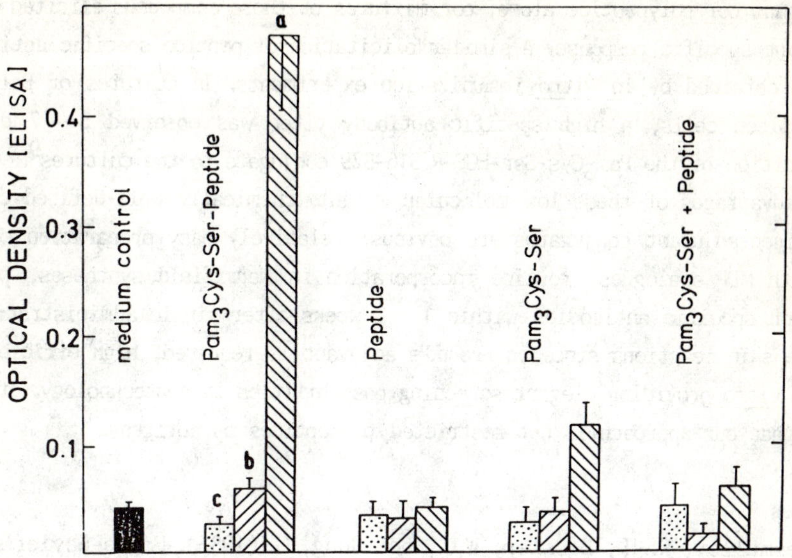

Fig. 2. Titers of peptide specific antibodies elicited by in vivo immunization with the conjugate and with controls. Each experiment was done with groups of 3-5 mice injected each with 0.2 ml of 1 mM (a), 0,1 mM (b), and 0.01 mM (c) solutions. Antibody titers were determined by ELISA[9].

Here we report most promising experimental results using tripalmitoyl-S-glyceryl-cysteinyl-serine as a mitogenic carrier for antigens. The covalent linkage of antigens to the mitogenic lipopeptide and the use of conformationally stabilized lipophilic α-helices for antigen presentation on the plasma membrane of lymphocytes has been discussed in context with the design of the "Pam_3Cys-Antigen-Helix System"[8] as a synthetic vaccine, as demonstrated in Fig. 1. We recommend the routine incorporation of Pam_3Cys as the last residue in the solid-phase peptide synthesis of antigens. The resulting lipopeptides are highly active mitogens, and they induce antigen-specific immunoglobulin production (IgG and IgM) in vivo and in vitro without addition of any further adjuvants.

Using the in vivo approach (Fig.1) we obtained, via a single i.p. administration, specific antibodies against e.g. the non-immunogenic tetradecapeptide EGF-R 516-529 of the extracytoplasmatic region of the epidermal growth factor receptor, as measured by ELISA (Fig.2). In contrast to the conjugate, neither the adjuvant nor polypeptide alone, nor mixtures of these compounds elicited a significant specific response. A similar elicitation of peptide specific antibodies was obtained by in vitro immunization experiments. In cultures of Balb/c murine spleen cells, a high specific antibody titer was observed 5 - 7 days after addition of the Pam_3Cys-Ser-EGF-R 516-529 conjugate to the cultures[9].

The advantages of these low molecular weight, chemically well-defined carrier-antigen-adjuvant conjugates are obvious: relatively easy preparation compared with MDP analogues, routine incorporation in Merrifield syntheses, production of specific antibodies within 1 - 2 weeks after single administration, no severe skin reactions since no Freund's adjuvant is required, high efficiency also in vitro providing elegant screening possibilities in gentechnology. It is obvious that our approach is not restricted to peptides as antigens.

References

1. Wiesmüller, K.-H., Bessler, W.G., and Jung, G. (1983). Hoppe-Seyler's Z. Physiol. Chem. 364 593-606.
2. Bessler, W.G., Wiesmüller, K.-H., Scheuer, W., Johnson, R.B., Fouad, H. and Jung,G. (1984). in Chemistry of Peptides and Proteins, Vol. 2, Voelter, W., Bayer, E., Ovchinnikov, Y.A., Wünsch, E. (eds.) De Gruyter, Berlin. 87-96.
3. Johnson, R.B., Köhl, S., Wiesmüller, K.-H., Jung, G., and Bessler, W.G. (1983). Immunobiol. 165 27-35.
4. Bessler, W.G., Cox, M., Lex, A., Suhr, B., Wiesmüller, K.-H., and Jung, G. (1985). J. Immunol., in press.
5. Bessler, W.G., Flad, H.-D., Schade, U., and Rietschel, E. (1984). in Lymphokine Research, Vol. 3, Nr.4, 234.
6. Bessler, W.G., Kustermann-Kuhn, B., and Scheuer, W.V. (1981). In: Mechanisms of Lymphocyte Activation, H.Kirchner and K. Resch (eds.), Elsevier, North Holland, 476-479.
7. Lex, A., Vordermeier, H.-M., Sprenger, R., Suhr, B., and Bessler, W.G. (1984). Immunobiol. 168, 71.
8. Jung, G., Proc.Int.Cong. on Synthetic Antigens, Siena, Oct.29-30, Ann. Sclavo 1984, n. 2, 191-208.
9. Bessler, W.G., Suhr, B., Bühring, H.-J., Müller, C.P., Wiesmüller, K.-H., Becker, G. and Jung, G. Immunobiol, in press.

A NOVEL CROSSLINKING AGENT AND ITS APPLICATION

Lois Aldwin and Danute E. Nitecki
Cetus Corporation, 1400 Fifty-Third Street, Emeryville, CA 94608

Chemical crosslinking of polyfunctional molecules such as proteins to proteins or peptides to proteins presents a very special challenge, yet it is very much required in preparation of peptide immunogens or assay reagents in ELISA procedures. Use of homobifunctional reagents, such as glutaraldehyde, or general reagents, such as carbodiimide, results in an extensive intramolecular crosslinking leading to considerable modification of the starting materials with concomitant general damage to bioactivity or epitope integrity. It is therefore desirable to use crosslinking agents which act through well defined and heterologous foci in the two molecules to be crosslinked. Hence, heterobifunctional agents, reacting usually with amino groups in the first molecule and with sulfhydryl group in the second, are the reagents of choice.

There are other desirable properties in a crosslinking reagent. It should be water soluble; it should produce a quantitative signal for monitoring the reaction progress; it should react rapidly and cleanly with the defined first focus, and be stable enough for removal of excess reagent before the second reaction; it should not react with the solvent-water; it should be easy to handle and to store.

We have achieved some of these properties in crosslinking agents based on active esters formed with 1-hydroxy-2-nitrobenzene-4-sulfonic acid sodium salt (HNSA)[1]. The esters of this phenol have been used before in designing peptide syntheses in water [2,3,4].

The esters are easily prepared via carbodiimide activation in DMF or DMSO, followed by ether precipitation. The esters can be purified by chromatography on LH-20 Sephadex columns in DMF. When suitably desiccated, these esters are stable indefinitely. Frozen aqueous solutions of these esters have been stored at -20°C for nine months with <5% loss upon thawing. The esters are freely water soluble, and solutions of 1.0 M have been prepared in 0.1 M phosphate buffers. The solubility allows for a more extensive and controllable reaction with proteins than can be obtained with sparingly soluble esters, such as those based on N-hydroxysuccinimide[1].

HNSA esters exhibit no absorption above 350 nm. After reaction with nucleophiles, such as the epsilon amino groups of lysine, the liberated dianion has a molar extinction coefficient of 4600 at 406 nm. The free HNSA has a pK_a of 5.5 and therefore obeys Beer's law at pH above 7. The liberation of free HNSA dianion allows monitoring the labeling reaction with proteins as the reaction occurs, and thus controlling the degree of labeling. The HNSA esters react rapidly with amines, yet are fairly resistant to water hydrolysis. The half life for water hydrolysis at pH 7.5 is 408 minutes, whereas the half life for aminolysis in 0.1 M bovine serum albumin is 36 minutes. Practically, under the conditions of 10 to 30 minutes of reaction time, the hydrolysis is negligible.

For crosslinking of peptides to carrier proteins for immunogenic purposes, we have extensively employed N-maleimido - 6 - aminocaproyl - HNSA ester (Figure 1)., as a crosslinking agent between amino groups of the carrier protein and cysteinyl sulfhydryl group in the peptide.

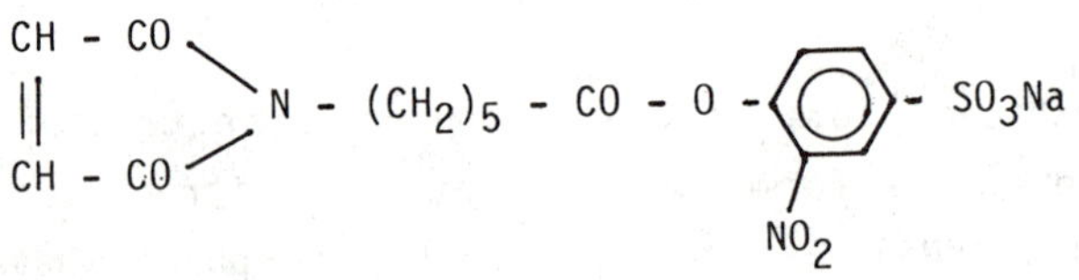

Fig. 1 N-maleimido-6-aminocaproyl-HNSA ester.

Efficient, rapid protein labeling can be achieved with this reagent at pH's ranging from 7.0 to 8.0. This pH range is preferred to protect the maleimide group from hydrolysis, although aliphatic maleimides are substantially more stable than aromatic maleimides[5]. The maleimide-labeled protein is separated from excess reagent in pH 6-7 buffer by chromatography on a Sephadex G-25 column. The sulfhydryl-containing peptide is added as rapidly as possible to the pooled protein fractions.

Since the coupled peptide contains only one cysteine, a well-defined immunogen is obtained which retains all the functional side chains intact. These immunogens invariably have elicited antibodies that recognize the peptide. Frequently, but not always, the antibodies were found to bind the proteins from which the peptide sequence was selected.

An application of this crosslinking reagent was used for the study of point mutations in the ras p21 oncogene protein family. A synthetic peptide that substituted serine for glycine at a position corresponding to amino acid 12 of the p21 protein was coupled to KLH with mal-sac-HNSA. Antibodies elicited by the serine peptide conjugate were capable of distinguishing between the normal p21 protein and the serine p21 variant found in Kirsten sarcoma virus transformed cells.

References

1. Bhatnagar, P.K., Nitecki, D.E., and Raubitschek, A. (1981) in Peptides: Synthesis-Structure-Function. Proceedings of the Seventh American Peptide Symposium, Rich, D.H. and Gross, E. Eds., Pierce Chemical Co., Rockford, Illinois, pp. 97-100.

2. Gershkovich, A.A. and Serebryany, S.B. (1978) Bioorganicheskaya Khimiya 4, 1129-1131.

3. Gershkovich, A.A. and Serebryany, S.B. (1979) Bioorganicheskaya Khimiya 5, 1125-1132.

4. Klausner, Y.S., Meiri, T.H., and Schneider, E. (1977) in Peptides, Proceedings of the Fifth American Peptide Symposium, Goodman, M., and Meienhofer, J., Eds., John Wiley & Sons, New York, NY, pp. 536-538.

5. Kitagawa, T., Shimozono, T., Aikawa, T., Yoshida, T., Nishimura, H. (1981). Chem. Pharm. Bull. 29,1130-1135.

THE C-TERMINAL HOMOCYSTEINE THIOLACTONE AS A PRECURSOR FOR THE SYNTHESIS OF IMMUNOGENS. APPLICATION TO THE TACHYKININS

Solange Lavielle, Gérard Chassaing, Sylviane Julien and Andrée Marquet.
Laboratoire de Chimie Organique Biologique, CNRS 040493, 4, Place Jussieu 75230 Paris Cedex 05, France.

Introduction

Besides substance P (SP), two tachykinins, neurokinin A (NKA), and neurokinin B (NKB), have been recently isolated from mammals[1].

SP	Arg-Pro-Lys-Pro-Gln-Gln-Phe-Phe-Gly-Leu-Met-NH_2
NKA	His-Lys-Thr-Asp-Ser-Phe-Val-Gly-Leu-Met-NH_2
NKB	Asp-Met-His-Asp-Phe-Phe-Val-Gly-Leu-Met-NH_2

Specific antisera against these three peptides are required. Unfortunately, the conventional preparations of the immunogens involved the coupling (carbodiimide, glutaraldehyde, bis-diazotized benzidine) of the hapten to the protein carrier via the aminoacids from the variable N-terminal part of the tachykinins. The antibodies thus obtained recognize the common C-terminal sequence and present a high cross-reactivity[2,3].

We have developed a new methodology for coupling specifically a peptide to the bovine serum albumine,(BSA), by the C-terminal aminoacid. The reaction sequence takes advantage of the C-terminal homocysteine thiolactone analogs we have previously described[4]. These $[Hcy^{11}]$SP, $[Hcy^{10}]$NKA, $[Hcy^{10}]$NKB thiolactones may be used, after adequate modifications as a reagent for coupling to the thiol-modified BSA and as a precursor for the synthesis of iodolabeled tracers.

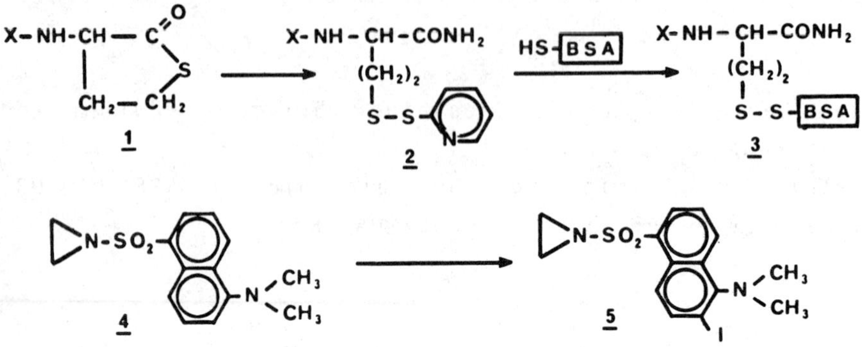

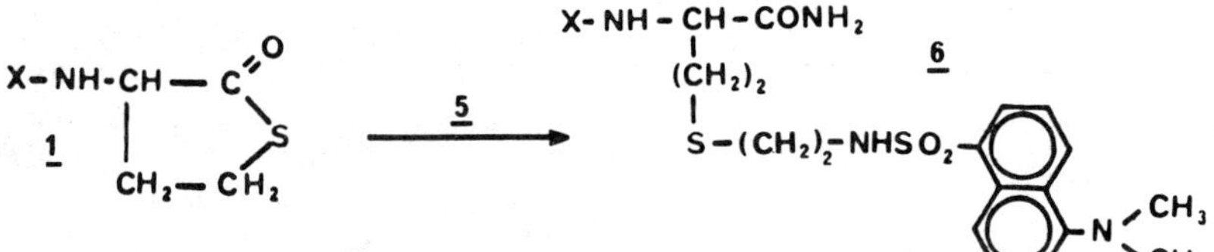

X_a = CH_3-CO- X_b = CT_3-CO-

X_c = Arg-Pro-Lys-Pro-Gln-Gln-Phe-Phe-Gly-Leu-

X_d = His-Lys-Thr-Asp-Ser-Phe-Val-Gly-Leu-

X_e = Asp-Met-His-Asp-Phe-Phe-Val-Gly-Leu-

Results and discussion

Synthesis of the hapten-coupling reaction to the carrier protein

The thiolactones 1 are easily opened in liquid ammonia and alkylated by 5 equiv. of 2-pyridine thiol in the presence of 5 equiv. of $K_3Fe(CN)_6$ for one hour at -55°C yielding the analogs 2. Then, the coupling efficiency of 2 to the thiol-modified BSA[5] has been monitored whith the tritiated 2b analog (0.36 mCi/mmoles). The reaction is quantitative using a 5-fold excess of 2b in 0.1M NH_4OA_c pH 7.5 in one hour at 25°C. SP,NKA and NKB have been conjugated with BSA via the 2c, 2d and 2e analogs following the same procedure. After a last dialysis step against 0.08M NaCl at 4°C, the antigenic conjugate solutions have been emulsifiec with an equal volume of Freund's complete adjuvant and used for immunizing male rabbits.

Synthesis of the iodolabeled analogs

We have previously established[6] that the specific thiol alkylating reagent 4 reacted in liquid ammonia with 1d. Thus, we have studied the iodination conditions of this p-dimethylamino substituted compound. As shown by NMR, this dansyl aziridine is quantitatively iodinated at 4°C, in 30 s by 1.1 equiv. NaI, 10 to 20 equiv. chloramine T in CH_2Cl_2-MeOH, 1:5. The iodinated analog 5 presents a different UV spectrum from that of the starting compound and a slight decrease, 1/9th of the emission maximum at 540 nm. (The excitation maximum for both is 345 nm). Alkylation of the thiolactones 1 in liquid ammonia by 1.5 equiv. of 5, for one hour at -55°C, leads to the iodolabeled coumpounds 6.

Table I. Characterization of Compounds 2 and 6

	2a	2c	2d	2e	6a	6c	6d	6e
TLC	0.45(A)	0.13(B) 0.63(C)	0.15(B) 0.62(C)	0.26(B) 0.67(C)	0.32(D)	0.18(B) 0.70(C)	0.21(B) 0.71(C)	0.33(B) 0.83(C)
D 22	-	-88°$^+$	-28.8°$^+$	-26.8°$^=$	-	-	-	-
HPLC		17min. (24%)	21.2min. (21.6%)	17min. (30%)	-	14.3min. (34.2%)	11.5min. (34.2%)	12.1min. (40.2%)
NMR*/FAB	NMR	MH^+	MH^+	MH^+	NMR	-	-	-

(A): AcOEt-hexane, 2:8; (B): n-butanol-AcOH-H_2O, 4:1:5; (C): n-butanol-pyridine-AcOH-H_2O, 5:5:1:4; (D): MeOH-$CHCl_3$, 1:9.

\+ (c 0.25, 10%AcOH) ; = (c 0.25, 50%AcOH)

μ-Bondapak C-18, isocratic 1.5mL/min, percent CH_3CN in 0.25 M TEAP buffer pH 3.0

* 250 MHz ^{1}H-NMR in $CDCl_3$

This method seems to be more attractive than those involving the coupling of a cysteine containing peptides since the thiolactone precursors are stable and therefore might be easily purified. Thus, as an alternative to the cysteine, we proposed the introduction of a C-terminal homocysteine derivative with either a S-t-butyl protecting group or the HF stable S-Npys group[7].

Acknowledgments

We are indebted to Dr. J. Bolard for carrying out the fluorescence experiments and to Dr. D. Godin for the FAB Mass Spectral Analysis.

References

1. Kimura, S., Okada, M., Sugita, Y., Kanazawa, I. and Munekata E. (1983) Proc. Natl. Japan Acad. 59, Ser. B, 101-104.
   In accordance to the recommendation of the committe from the IUPHAR Satellite Symposium on Substance P (August 1984, Maidstone, England), Neurokinin α, identical to Substance K and Neuromedin L, and Neurokininβ, identical to Neuromedin K, have been named NKA and NKB respectively.
2. a) Kimura, S., Ogawa, T., Sugita, Y., Munekata, E. and Kanasawa, I. and b) Shults, C.W., Yajima, H., Buck, S., Gullner, H., Burcher, E., Chase, T.N. and Donohue, T.L. (1984) IUPHAR 9th International Congress of Pharmacology, Satellite Symposium, to be published.
3. Minamino, N., Masuda, H., Kangawa, K. and Matsuo, H. (1984) Biochem. Biophys. Res. Comm., 124, 731-738.
4. Chassaing, G., Lavielle, S., Julien, S. and Marquet, A. (1985) Tetrahedron Letters, 623-626.
5. Carlsson, J., Drevin, H. and Axen, R. (1978) Biochem. J. 173, 723-737.
6. Chassaing, G., Lavielle, S., Marquet, A. (1983) J. Org. Chem., 48, 1757-1760.
7. Chassaing, G., Lavielle, S., Marquet, A., Morgat, J-L, Genet, R. and Fromageot P. 9th American Peptide Symposium, this issue.

ANTAGONISTIC ANALOGS OF HUMAN TRANSFORMING GROWTH FACTOR ALPHA

John J. Nestor, Jr., Sherry R. Newman, Barbara M. DeLustro, and Alain B. Schreiber

Institutes of Bio-Organic Chemistry and Biological Sciences, Syntex Research, Palo Alto, California 94304

Introduction

Two families of transforming growth factors (TGF), mitogenic proteins which induce a transformed phenotype in normal cells, have been identified.[1,2] TGFα's are structurally homologous to epidermal growth factor (EGF), bind to EGF receptors, and induce all of the known EGF-mediated biological actions. TGFβ does not bind to EGF receptors but potentiates the action of TGFα. The sequences of TGFα's from rat[2] and human[3] exhibit homology to EGF, particularly in the arrangement of the three disulfide bridges.

We have searched for antigenic or receptor binding regions of proteins by focusing on predicted β-turn regions[4,5] (Figure 1). A synthetic replicate of region II of rTGFα (the third disulfide loop, residues 34-43) linked to KLH elicited specific antibodies which recognized rTGFα but which did not cross-react with EGF. This synthetic peptide also bound weakly to the EGF receptor.[4] We now report studies with synthetic analogs of the corresponding region of hTGF which were designed for increased receptor affinity.

Fig. 1. Predicted β-turns in rTGFα ($V^{41} \rightarrow A^{41}$ for hTGFα).

Analog Design

The parent structure 1 (C-H-S-G-Y-V-G-V-R-C), with free amino- and carboxy-termini, exhibited weak but detectable receptor binding (Figure 2), and we sought an approach for increasing the receptor affinity. Ligand-membrane interaction was hypothesized to be an important first step in ligand-receptor binding,[6] and we reasoned that facilitation of this interaction might lead to increased apparent receptor binding.

Our first approach was to make the loop more hydrophobic by capping the ends, e.g. Ac-TGFα(34-43)-NH_2 (2). Although 2 exhibited a lower IC_{50} (i.e. increased receptor affinity), this might be due to removal of charge and addition of H-bonding sites found in hTGF. When the C-terminus is blocked as a methyl ester, a more hydrophobic modification, there is a further increase in apparent receptor affinity (3).

In LHRH agonists, substitution at the C-terminus with an NHEt function gave increased biological activity, thought to be due to protection from proteolysis. However, LHRH analogs containing NHEt are also more hydrophobic than their parent structures and, where known, their receptor affinities are higher. This hydrophobic substitution (4) caused a

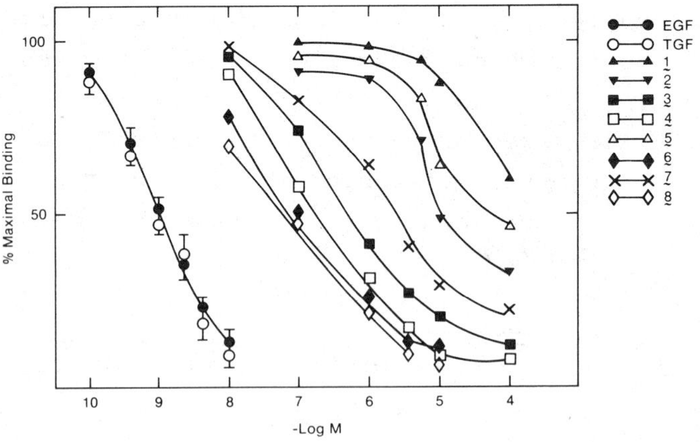

Fig. 2. Binding of analogs to EGF receptors on A431 cells.

substantial increase in apparent receptor affinity.

We recently described a new class of amino acids, the $N^G,N^{G'}$-dialkylhomoarginines [hArg(R_2)], designed for interaction with phospholipid membranes.[7] This design rationale involves an ionic interaction between the guanidinium ion and the phosphate head group of the phospholipid, which is further stabilized by hydrophobic interactions of the alkyl chains on the guanidine with the membrane. This approach to increased membrane affinity also results in greater receptor affinity.

Addition of D-hArg(Et_2) to the TGFα loop C-terminus (5) caused decreased receptor affinity compared to (2). However, when added with a spacer to the N-terminus (6), a 100-fold increase in receptor binding affinity resulted. The corresponding L-hArg(Et_2) analog (7) was not as effective a substitution, but this may be due to proteolysis in this intact cell binding assay. Capping of the N-terminus of the most potent analog provided a small increase in binding (8).

Although these analogs bound to the EGF receptor, they did not induce mitogenesis at concentrations up to 10^{-5} M on HFF. Furthermore, when present in HFF cell cultures containing a mitogenic concentration of EGF or TGF, these analogs resulted in antagonism (Table I) of induced cell growth (DNA synthesis). The antagonism was correlated with the receptor binding affinity of the analogs and was dose related (Table I).

Conclusions

Residues 34-43 (the third disulfide loop) comprise a useful antigenic and an important receptor binding region of hTGFα. Modifications of this structure designed to increase membrane interaction resulted in increased apparent receptor affinity. These analogs demonstrated effective antagonism of the mitogenic effect of TGFα in cell culture.

Table I. Biological Activities of TGFα Analogs

#	Compound	Binding[a] IC_{50} (μM) A431	HFF	Mitogenesis Inhibition[b] IC_{50} (μM) EGF	TGF
1	H-hTGFα(34-43)-OH	60	80	80	>100
2	Ac-[1]-NH_2	10	5	6	NT
3	Ac-[1]-OMe	4	2	2	2
4	Ac-[1]-NHEt	0.7	0.8	0.5	NT
5	Ac-rTGFα(34-43)-D-hArg(Et_2)-NH_2	40	NT	30	NT
6	H-D-hArg(Et_2)-$(Gly)_2$-[1]-OH	0.2	0.4	0.2	0.5
7	H-hArg(Et_2)-$(Gly)_2$-[1]-OH	3	5	2	NT
8	Ac-D-hArg(Et_2)-$(Gly)_2$-[1]-NH_2	0.1	0.2	0.2	NT

[a]IC_{50} calculated from inhibition of binding of 1 nM [^{125}I]EGF to A431 cells or to human foreskin fibroblasts (HFF). [b]IC_{50} is conc. which halves the enhancement of [^{3}H]methyl-thymidine incorporation in HFF by either 1 nM EGF or TGFα.[4]

References

1. DeLarco, J.E. and Todaro, G.J. (1978) Proc. Natl. Acad. Sci. USA 75, 4001-4005.
2. Marquardt, H., Hunkapiller, M.W., Hood, L.E., and Todaro, G.J. (1984) Science 223, 1079-1082.
3. Derynck, R., Roberts, A.B., Winkler, M.E., Chen, E.Y., and Goeddel, D. (1984) Cell 38, 287-297.
4. Nestor, J.J. Jr., Newman, S.R., DeLustro, B.M., Todaro, G.J., and Schreiber, A.B. (1985) Biochem. Biophys. Res. Commun. 129, 226-232.
5. Nestor, J.J. Jr., Moffatt, J.G., and Chan, H.W. (1985) U.S. Patent 4,493,795.
6. Schwyzer, R., Gremlich, H.-U., Gysin, B., and Sargent, D.F. (1983) Peptides: Structure and Function, Hruby, V. and Rich, D.H., Eds., Pierce Chemical Company, Rockford, IL pp. 657-664.
7. Nestor, J.J. Jr., Ho, T.L., Tahilramani, R., Horner, B.L., Simpson, R.A., Jones, G.H., McRae, G.I., and Vickery, B.H. (1984) LHRH and Its Analogs: Contraceptive and Therapeutic Applications, Vickery, B.H., Nestor, J.J. Jr., and Hafez, E.S.E., Eds., MTP Press Limited, Lancaster pp. 23-33.

SYNTHETIC 30-50 PEPTIDE FRAGMENT OF CHOLERA TOXIN B CHAIN BINDS TO GM_1 GANGLIOSIDE

Robert H. Reid, Charles A. Hooper, Roberta L. Richards, Carl R. Alving and Robert C. Seid
Walter Reed Army Institute of Research, Washington, D.C. 20307-5100

Robert H. Boykins
Office of Biologics, Research and Review, Food and Drug Administration, Bethesda, Maryland 20205

Introduction

The synthetic peptide Ser-Leu-Ala-Gly-Lys-Arg-Glu-Met-Ala-Ile-Ile-Thr-Phe-Lys-Asn-Gly-Ala-Thr-Phe-Glu-Val representing the 30-50 amino acid peptide fragment of cholera toxin B chain has been found to inhibit the binding of cholera toxin to the gangliosides of human and rat erythrocytes and to suppress the increase of corticoid secretion in mice surrenalin Y_1 tumor cells challenged with cholera toxin suggesting that the peptide probably binds to GM_1 ganglioside as reported by Rivaille at the Eighth American Peptide Symposium. The purpose of this study was to determine if the 30-50 amino acid peptide fragment of B chain could bind to GM_1 ganglioside. Moss, Richards, Alving and Fishman[1] found that cholera toxin and its B protomer also called B subunit made up of 5B chains would bind to multi-lamellar liposomes containing GM_1. The 30-50 amino acid peptide was synthesized and tested for its ability to bind to multi-lamellar liposomes containing GM_1. An additional octapeptide (unrelated to cholera toxin) Gly-Asn-Thr-Ile-Val-Ala-Val-Glu was added to the 30-

50 amino acid peptide by extended n-terminal synthesis. This 29 amino acid peptide was also tested for GM_1 binding.

Methods

The peptides were synthetized by the Merrifield solid phase method using a Vega Biochemical automated peptide synthesizer. The peptides were cleaved from the polystyrene beads by hydrofluoric acid hydrolysis and purified using preparative high pressure liquid chromatography. The longer 29 amino acid peptide was synthesized by continuing the solid phase synthesis while the 30-50 amino acid peptide was still attached to the beads and fully protected.

The multi-lamellar liposomes contained dimyristoyl phosphatidylcholine:cholesterol:dicetyl phosphate in molar ratios of 1:0.75:0.11 with 10 nmol of phospholipid and 0.5 nmol GM_1 ganglioside per microliter of liposomes.[2]

Fifty microliters of liposomes with and without GM_1 were each reacted with one microgram of peptide in a milliliter of normal saline for two hours at room temperature. The liposomes were then spun down with 20,000 G force by centrifugation at 25 °C and washed twice with normal saline. The liposomes were then disrupted by chloroform extraction using one half milliliter and an equal volume of normal saline. The phospholipid and GM_1 remained in the chloroform and the peptides in the saline which was collected and dried. Following acid hydrolysis with 6 normal hydrochloric acid under vacuum for 18 hours, quantitative amino acid analysis was done using the Beckman 6300 instrument. The results were expressed as pmol of glutamic acid and converted to pmol of peptide.

Results and Discussion

Results of duplicate experiments indicated that 344 pmol of the synthetic 30-50 amino acid peptide (21 amino acid peptide) bound to liposomes containing GM_1, minus the 150 pmol peptide non-specifically binding to liposomes without GM_1, gave 194 pmol of 21 amino acid peptide specifically binding to liposomes containing GM_1. Since the total amount of peptide added to the liposomes was 440 pmol, 194 divided by 440 times 100 equals 44% as the percent specific binding.

Similar results were found using the 29 amino acid peptide. Results of duplicate experiments found that 245 pmol of the 29 amino acid peptide bound to liposomes containing GM_1, minus the 117 pmol of peptide non-specifically binding to liposomes without GM_1, gave 128 pmol of 29 amino acid peptide specifically binding to liposomes containing GM_1. Since the total amount of peptide added to the liposomes was 326 pmol, 128 divided by 326 times 100 equals 39% as the percent specific binding.

In summary, there appears to be little difference in the percent specific binding between the two peptides - 44% with the 21 amino acid peptide and 39% with the 29 amino acid peptide. The GM_1 binding suggests that the synthetic 30-50 amino acid peptide might serve as a carrier for peptide antigens for intestinal immunization as the five B chains of B subunit or protomer of cholera toxin do for proteins. McKensie and Halsey[3] found that B subunit served as an effective carrier for covalently coupled horseradish peroxidase for intestinal immunization. Mice immunized with the horseradish peroxidase-B subunit conjugate showed significantly higher levels of IgA antihorseradish peroxidase in intestinal washes and bile, as well as increased levels of serum IgG antihorseradish peroxidase.

The IgA antihorseradish peroxidase responses in gut washes were 33 to 120-fold higher when the conjugate was used as when the B subunit plus horseradish peroxidase or horseradish peroxidase alone were used. These results suggest that vaccines to stimulate mucosal immunity to any antigenic determinant might now be prepared by covalent conjugation to effective mucosal immunogens such as cholera toxin B subunit or possibly its synthetic 30-50 amino acid peptide GM_1 binding fragment. It has been postulated by Pierce[4] that the GM_1 binding confers the mucosal immunogenicity by aiding uptake by M cells or trapping mucosal lymphocytes or macrophages or both.

In conclusion, a synthetic 21 amino acid peptide representing the 30-50 amino acid peptide fragment of cholera toxin B chain binds to multi-lamellar liposomes containing GM_1 ganglioside. Also, a synthetic octapeptide can be added by n-terminal extended synthesis without interfering with the GM_1 binding property.

References

1. Moss, J., Richards, R.L., Alving, C.R. and Fishman, P.H. (1977) J. of Bio. Chem. 252, 797-798.

2. Alving, C.R., Shichijo, S. and Mattsby-Baltzer, I. (1984) in Liposome Technology, Vol. II, Gregorviadis, G., Ed., CRC Press, Cleveland, OH, pp 157-175.

3. McKenzie, S.J. and Halsey, J.F. (1984) J. of Imm. 133, 1818-1824.

4. Pierce, N.F. and Gowans, J.L. (1975) J. Exp. Med. 142, 1550-1563.

NEUTRALIZING ANTIBODIES AND LOCAL PROTECTION INDUCED BY A SYNTHETIC PEPTIDE OF CHOLERA TOXIN B SUBUNIT

Anne Guyon-Gruaz, Agnès Delmas, S. Pedoussaut, G. Milhaud, Daniel Raulais, Pierre Rivaille.

LA 163 CNRS, CHU Saint-Antoine, 27, rue Chaligny, 75012 Paris, France.

Introduction

Synthetic fragments of protein sequences are a successful approach for the production of antibodies able to neutralize these proteins. Immunization with such peptides can even lead to neutralization of infectious agents and induction of protective immunity. It would be of interest to develop protection at the mucosal level, using synthetic peptides, since the initial invasion of many pathogens occurs via mucosal surfaces. As stimulation of local immunity is necessary for adequate protection against cholera infection, we chose cholera toxin (CT) as a model to test the feasibility of developing local protection with synthetic peptides. CT is a multimeric protein containing one A subunit and one B subunit. The B subunit composed of five non covalently linked β chains binds the toxin molecule to a specific receptor of intestinal mucosal cells, the ganglioside GM_1. The toxic A subunit penetrates the cell and stimulates adenylate cyclase, which promotes the secretion of water into the lumen of the small intestin. As the local immune response obtained after oral immunization with CT or its B subunit is explained by their binding to the GM_1 receptor, and as neutralizing antibodies are principally directed against the B subunit, we decided to synthesize selected peptides from within the predicted binding area of the β chain of the toxin.

Results and discussion

- Choice and synthesis of the peptides :

Since the result of Dufy et al (1) suggested that ^{35}Arg and probably ^{73}Arg in the CT β chain or the regions nearby are involved in the binding process of the toxin, we decided to synthesize two peptides, each containing one of these arginines and corresponding to the following criteria : It should occur on the protein surface. This localization has been predicted by the methods of Hopp (2) and of Fraga (3), taking into account the hydrophilicity and the recognition factor of each constitutive amino acid.

The synthesized sequences have been chosen to contain the maximum number of tetrapeptides with the fewest homologies with the sequences contained in the NBRF sequence database. Effectively, a unique sequence is more likely to be immunogenic than an ubiquitous one. As the number of tetrapeptides unique to a protein increases, the more epitopes specific to this protein it will contain, because a tetrapeptide is the minimal number of aminoacids for an epitope.

The two peptides P30-50 and P50-75 were synthesized by the solid phase method (4).

- Immunological reactivity : both peptides raised antibodies in mice even when administered per os (Table 1). The cross-reactivity of the antipeptide antisera with CT was tested in an ELISA. Titers are expressed as total immunoglobulins reacting with CT.

Anti CT antibody titers (ELISA)

	I.P		P.O	
	30-50	50-75	30-50	50-75
Priming	30	500	640	1460
Booster 1	70	580	450	560
2	100	580	200	260
3	160	600	250	260

Table 1

Immunization every 14 days (2 μg peptide)

The highest serum titers were obtained with P50-75 (max titer: 5120). P30-50 immunized mice produced much lower antisera titers (max titer : 1280). In contrast to the response seen after i.p. injection, oral immunization produced antibody titers which were maximal after the initial immunization but decreased after several boosters (Table 1). The CT anti CT reaction was not inhibited by either of the two peptides : These results show that the specificities of the antibodies raised against the intact molecule are different from those obtained with P30-50 and P50-75. The anti P30-50 and P50-75 antibodies also cross-react with E. Coli LT enterotoxin (LT), as expected because there are sequence homologies between the CT and LT β chains.

- Neutralization of biological activity : the CT neutralizing potency of the antisera was determined by their ability to neutralize the increase of vascular permeability of rabbit dorsal skin caused by CT (5). The anti P50-75 serum (ELISA titer : 2560) was able to neutralize the CT activity at a dilution of 1/20 while anti P30-50 serum had no neutralizing effect at the same dilution.

LOCAL PROTECTION AGAINST CT ACTIVITY (ligated intestinal loop test)

Synthetic immunogen	weight/length g/cm	n protected / n challenged	mean antibody titer
30-50	0.201±0.010	0/9	0
50-75	0.128±0.012	8/14	24
50-75-NH2	0.159±0.014	3/10	0
PCA-50-75-NH2	0.132±0.019	6/10	8
PCA-CYS(Acm)-50-75-NH2	0.167±0.013	2/10	4
PCA-CYS-50-75-NH2 PCA-CYS-50-75-NH2	0.139±0.010	4/10	4
BUFFER	0.176±0.008		

Table 2 : *Immunization (5 μg peptide) on days D, D + 10, D + 16, D + 22, ligated loop test on day D + 28.*
Animals were considered as protected when the ratio W/L was less than the mean control values minus 2 standard deviations : 0.124

- Local protection : The ability of the peptides to protect animals against CT was evaluated with the ligated loop test (6) (7). Groups of mice were orally immunized with each peptide or with buffer (controls). Only mice treated with P50-75 or its analogues were protected against a further challenge with CT (Table 2). Some sera from protected mice contained no detectable seric anti-CT antibodies, suggesting that this protection is not related to seric antibodies, but, probably, to local antibodies.

Conclusion

Thus anti P30-50 and anti P50-75 antibodies cross-react with CT and LT and anti P50-75 antibodies are capable of neutralizing CT activity. This is in agreement with the results of Jacob et al. who reported that anti P50-64 antibodies neutralized CT and LT activity (8). The P50-75 peptide when administered orally has the capacity to raise serum CT-neutralizing antibodies and is furthermore able to elicit a local protection in gut. The 50-75 peptide might be a good candidate for oral synthetic vaccination against cholera.

References

1. Duffy, L.K., and Lai, C.Y. (1979) Biochem. Biophys. Res. Comm. 91 (3), 1005-1010.
2. Hopp, T.P., and Woods, K.R. (1981) Proc. Natl. Acad. Sci. USA 78, 3824-3828.
3. Fraga, S. (1982) Can. J. Chem. 60, 2606-2610.
4. Delmas, A., Guyon-Gruaz, A., Halimi, H., Raulais, D., and Rivaille, P., (1985) Int. J. Pep. Prot. Res. 25, 421.
5. Craig, J.P., Cray, W.C. Jr., and Sacci, J.B. Jr. (1982) Infect. Immunity 37 (2), 687-694.
6. Fujita, K., and Finkelstein, R.A. (1972) J. Infect. Dis. 125, 647-655.
7. Lange, S., and Holmgren, J. (1978) J. Acta Pathol. Microbiol. Scand. Sect. C 86, 145-152.
8. Jacob, C.O., Pines, M. and Arnon, R. (1984) EMBO Journal 3, 12, 2889-2893.

DEVELOPMENT OF SITE-SPECIFIC ANTIBODIES FROM SYNTHETIC PEPTIDES OF THE C-MYC ONCOGENE PROTEIN: STUDIES TOWARD STRUCTURE-FUNCTION RELATIONSHIPS

David R. Bolin, Iou-Iou Sytwu, Johannes Meienhofer and Richard Chizzonite*

Peptide Research Dept. and *Molecular Genetics Dept., Roche Research Center, Hoffmann-La Roche Inc., Nutley, N.J. 07110

Introduction

The viral myc oncogene, of the avian myelocytomatosis retrovirus, is structurally related to a highly conserved gene in normal eukaryotic cells.[1] The avian MC29 virus has been found to induce a variety of tumors in chickens, including sarcomas, renal and hepatic carcinomas, and myelocytomatoses.[2] Expression of the c-myc gene in avian cells results in production of a gag-myc fusion protein, possessing DNA binding properties. The human c-myc gene, identified as a possible element of malignancy in Burkitt lymphomas, encodes a 439 residue, 65Kd, protein.[3] Immunological studies have shown that the human c-myc protein also occurs predominately within the cell nucleus and associated with DNA. In leukemic cells, the apparent cause of cellular transformation has been the translocation of the myc gene to a position on chromosome 14, close to the transcriptionally active immunoglobulin heavy chain locus.[4] Several mechanisms for activation of the c-myc gene have been proposed with enhanced c-myc protein expression resulting in transforming activity.

Methods

Synthetic peptides were prepared by standard solid phase methodology on benzhydrylamine resin with appropriately protected amino acids, using a Vega Model 250A Synthesizer. After construction of the desired sequence, cleavage was performed by treatment with HF under a modification of the Tam procedure.[5] Crude peptides were extracted from the resin with 10% AcOH and lyophilized. Purification was achieved by gel filtration on G-10 and/or preparative reversed phase HPLC. All purified compounds were analytically pure by HPLC and were characterized by amino acid analysis and FAB-MS.

Synthetic peptide segments were conjugated to thyroglobulin with glutaraldehyde and gel filtered on G-75 in phosphate buffered saline. Aliquots of the conjugate were mixed with an equal volume of Freunds complete adjuvant and used for primary immunization of mice and rabbits. Boosting with conjugate and Freunds incomplete adjuvant occurred on Day 14 and at subsequent 4 week intervals.

Peptide	Region[a]	Mouse[b]	Rabbit[b]
SER-PRO-SER-TYR-VAL-ALA-VAL-THR-PRO-PHE-SER-LEU-NH_2	71-82	185(10-14)	35(5-7)
ASP-SER-GLY-SER-PRO-ASN-PRO-ALA-ARG-GLY-HIS-SER-VAL-NH_2	158-170	<1	nd
ALA-SER-GLN-ASP-SER-SER-ALA-PHE-SER-PRO-SER-SER-NH_2	209-220	31(5-7)	110(5-7)
LYS-ARG-GLN-ALA-PRO-GLY-LYS-ARG-SER-GLU-SER-GLY-SER-PRO-SER-NH_2	269-283	900(18-20)	204(5-7)
ILE-SER-ASN-ASN-ARG-LYS-ALA-THR-SER-PRO-ARG-SER-SER-NH_2	336-348	480(10-14)	204(5-7)

[a]From c-myc protein sequence as in Ref. 2. [b]The ELISA titer is expressed as the dilution (x1000) where binding of the primary antisera falls to 50% of the maximum value. The amount of peptide bound to the microtiter wells is held constant at 100 ng and the amount of primary antibody bound to the peptide is determined with an anti-antibody coupled to horseradish peroxidase. Weeks after primary immunization are in parenthesis.

Table I. Synthetic Peptide Segments and Representative Antisera Titers

Results and Discussion

A combination of Chou-Fasman secondary structural prediction[6] and Hopp-Woods hydropathicity analysis[7] were utilized in order to identify peptide segments residing on the surface of the protein. Five segments were chosen for synthesis as shown in Table I.

Rabbits and mice developed high anti-peptide titers after immunization with peptide conjugates (Table I). The high titers could be maintained over an extended period when periodic boosts of the peptide conjugate were administered, but the titers eventually decreased to low levels when the boosts were halted.

Rabbit and mouse antisera identified c-myc protein in COS cells transfected with c-myc DNA (Figure 1). The immunofluorescence pattern localized the c-myc protein within the nuclei of COS cells (Figure 1). Both rabbit and mouse antisera immunoprecipitated a 62-65Kd protein from an *in vitro* transcription-translation coupled system when c-myc DNA was

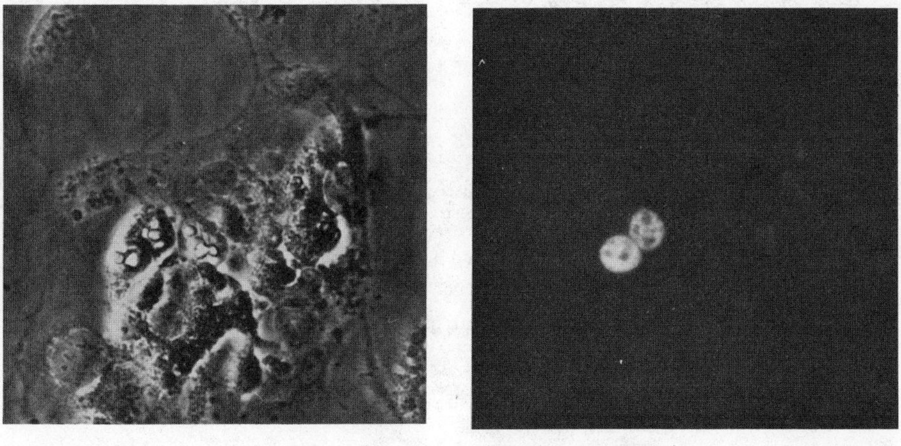

Fig. 1. Phase contrast of COS cells (l) and immunofluorescence detection of c-myc protein (r).

used as the priming DNA. Three or four other labelled proteins which are immunoprecipitated by these antisera may represent smaller c-myc proteins which result from internal initiation sites within the coding region of the c-myc DNA.

"Western" blotting[8] of protein extracts from E. coli transformed with an expression vector containing the c-myc plasmid detected a 60Kd protein in addition to two smaller molecular weight bands. Specific rabbit and mouse antibodies were found to differentiate between these fragments as being derived from either exon 2 or 3. Because these antibodies appear to distinguish separate regions of the myc protein, it may be possible to determine the DNA binding site by antibody neutralization techniques.

References

1. Watson, D., Psallidopoulos, M., Samuel, K., Dalla-Favera, R., and Papas, T. (1983) Proc. Natl. Acad. Sci. USA 80, 3642-3645.

2. Colby, W., Chen, E., Smith, D., and Levinson, A. (1983) Nature 301, 722-725.

3. Pachl, C., Schubach, W., Eisenman, R., and Linial, M. (1983) Cell 33, 335-344.

4. Hamlyn, P. and Rabbitts, T. (1983) Nature 304, 135-139.

5. Tam, J., Heath, W., and Merrifield, R. (1982) Tet. Letters 23, 2939-2942.

6. Chou, P. and Fasman, G. (1978) Ann. Rev. Biochem. 47, 251-276.

7. Hopp, T. and Woods, K. (1981) Proc. Natl. Acad. Sci. USA 78, 3824.

8. Burnette, W. (1981) Analyt. Biochem. 112, 195-203.

USE OF SYNTHETIC PEPTIDES TO EVALUATE CROSS-REACTIVITY OF MONOCLONAL ANTIBODIES RAISED AGAINST FROG RHODOPSIN

Grazyna Adamus, Anatol Arendt, Paul A. Hargrave, Robert Jackson, J. Hugh McDowell and Agnieszka Szary

Departments of Medical Biochemistry and Medical Microbiology, Southern Illinois University School of Medicine, Carbondale, Illinois 62901

Heidi E. Hamm

Department of Physiology and Biophysics, University of Illinois, Chicago, Illinois 60680

Introduction

Monoclonal antibodies (MAbs) have been raised against the photoreceptor protein rhodopsin to use as tools in the study of its structure and function.[1] In order to determine the site on rhodopsin to which each antibody binds, synthetic peptides have been used as competitors for antibody binding to rhodopsin in an immunoassay.[2] For the present study six MAbs were raised in mice using frog rhodopsin as antigen. Five of these MAbs were found by immunocytochemical methods to recognize rhodopsin in rod cells, as expected, and to also recognize two 80-85,000 MW proteins in Müller cells (which was unexpected).[4] In order to understand the biological significance of this finding it was important to determine the binding specificity of the MAbs. Use of synthetic peptides enabled elucidation of the amino acid sequences to which the MAbs bound.

Methods and Results

Soluble peptides formed by the cleavage of bovine rhodopsin with cyanogen bromide,[5] were used as competitors for binding of antibody to rhodopsin in an enzyme-linked immunosorbent assay (ELISA). Competition was observed only with the CNBr peptide CB-1 which contains the rhodopsin sequence 2-39, indicating that all MAbs bound to rhodopsin's amino-terminal region. Synthetic peptides 1-12 and 2-32 were effective competitors (Fig. 1). We were surprised to find that peptide 24-34 was also a competitor. For the six MAbs tested peptide 1-12 was from 1.1 x 10^4 to 1.4 x 10^5 more effective than peptide 24-34. Based upon these results we looked for and observed homology between the two peptides: the rhodopsin sequence 5-11 is homologous to the rhodopsin sequence 25-31.

5-11	Glu-Gly-Pro-Asn-Phe-Tyr-Val
25-31	Glu-Ala-Pro-Gln-Tyr-Tyr-Leu

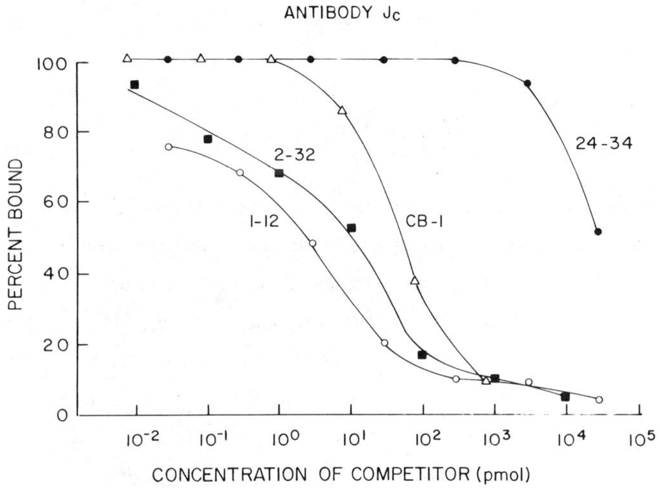

Fig. 1 Peptides as competitors in an ELISA for rhodopsin. One of the anti-rhodopsin MAbs, J_c, was tested for binding specificity. Peptide CB-1 represents the native glycosylated peptide, 2-39 in the sequence. No competition was observed for peptide 13-23 (not shown). Peptides were synthesized and purified essentially as described previously.[2] Bovine rhodopsin's amino terminal sequence 1-39 is Met-Asn-Gly-Thr-Glu-Gly-Pro-Asn-Phe-Tyr-Val-Pro-Phe-Ser-Asn-Lys-Thr-Gly-Val-Val-Arg-Ser-Pro-Phe-Glu-Ala-Pro-Gln-Tyr-Tyr-Leu-Ala-Glu-Pro-Trp-Gln-Phe-Ser-Met.

In order to assess which amino acids were most important for antibody binding, we synthesized the following peptides to use as competitors:

Peptide name	Peptide Sequence
5-11	Ala-Ala-Glu-Gly-Pro-Asn-Phe-Tyr-Val-Ala-Ala
5-11 (Tyr^9)	Ala-Ala-Glu-Gly-Pro-Asn-Tyr-Tyr-Val-Ala-Ala
5-11 (Gln^8)	Ala-Ala-Glu-Gly-Pro-Gln-Phe-Tyr-Val-Ala-Ala
5-11 (Leu^{11})	Ala-Ala-Glu-Gly-Pro-Asn-Phe-Tyr-Leu-Ala-Ala
5-11 (Ala^6)	Ala-Ala-Glu-Ala-Pro-Asn-Phe-Tyr-Val-Ala-Ala
25-31	Ala-Ala-Glu-Ala-Pro-Gln-Tyr-Tyr-Leu-Ala-Ala

Results from the use of these peptides is illustrated for antibody I_c (Fig. 2). The results were qualitatively the same with the other MAbs tested. Substitution of Tyr^9 for Phe^9 causes the most severe loss of antibody binding (a factor of 1×10^4), and substitution of Gln^8 for Asn^8 also has a substantial effect (1.7×10^3). Substitution of Leu^{11} for Val^{11} is less important (1.5×10^2), and the antibody shows no preference for Gly over Ala in position 6. These results suggested to us that the MAbs might be able to show a considerable degree of cross-reactivity with proteins containing sequences similar to but not identical with rhodopsin's 5-11 sequence.

A computer search showed that the amino acid sequence Glu-Gly-Pro-Asn-Phe-Leu-Val was present in the unrelated protein, ferredoxin from *Azotobacter vinelandii*. This sequence is identical in 6 of its 7 amino acids with amino acids 5-11 in the bovine rhodopsin sequence. The native protein ferredoxin was found to compete in the ELISA with MAb I_c as effectively as peptide 5-11 [Gln^8].

Although the sequences of the Müller cell proteins which cross-react with the anti-rhodopsin MAbs are not known, we feel that it is likely that the proteins contain a region of sequence which is homologous to the rhodopsin sequence 5-11. The results of this study suggest that it is important to know the binding specificities of MAbs and to use several MAbs of different specificity in order to ensure that the immunological reactions observed are biologically significant.

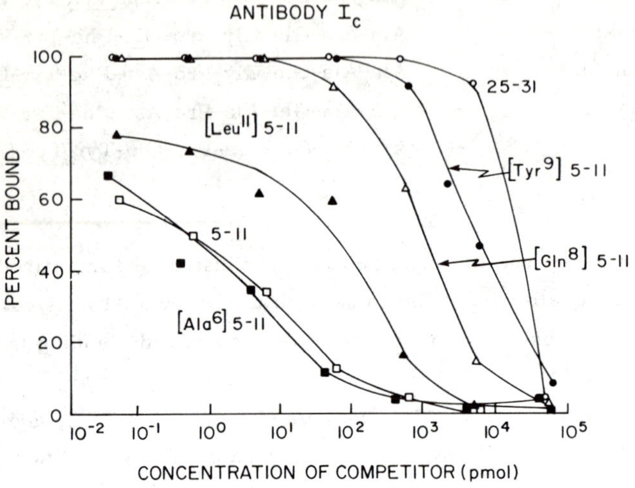

Fig. 2. Use of synthetic peptides to determine important amino acids in antibody binding to rhodopsin. One of the antirhodopsin MAbs, I_C, was tested for binding specificity. Peptides represent the rhodopsin sequences 5-11, 25-31 and their analogues, with dialanyl sequences at each end.

References

1. De Grip, W.J. (1985) in Progress in Retinal Research, Vol. 4, Osborne, N.N. and Chader, G.J., Eds., Pergamon Press, New York, pp. 137-180.

2. MacKenzie, D., Arendt, A., Hargrave, P., McDowell, J.H. and Molday, R.S. (1984) Biochemistry 23, 6544-6549.

3. Witt, P.L., Hamm, H.E. and Bownds, M.D. (1984) J. Gen. Physiol. 84, 251-263.

4. Takahashi, J.S., Mezey, E. and Hamm, H.E. (1984) Invest. Ophthalmol. Vis. Sci. 25 Suppl., 63.

5. Hargrave, P.A., McDowell, J.H., Curtis, D.R., Fong, S.-L. and Juszczak, E. (1982) Methods in Enzymology 81, 251-256.

USE OF SITE-DIRECTED-POLYCLONAL ANTIBODIES AS IMMUNO-TOPOLOGICAL PROBES FOR THE *lac* PERMEASE OF *ESCHERICHIA COLI*.

Waleed Danho, Raymond Makofske, Frank Humiec and Thomas F. Gabriel

Peptide Research Department, Roche Research Center, Hoffmann-La Roche Inc., Nutley, N.J. 07110

Nancy Carrasco and H. Ronald Kaback

Roche Institute of Molecular Biology, Nutley, N.J. 07110

Introduction

The *lac* carrier protein (i.e. *lac* permease) in *Escherichia coli* is an intrinsic membrane protein, the product of the *lac* Y gene, that catalyzes the translocation of β-galactosides with hydrogen ion in a symport (i.e. co-transport) reaction.[1] The *lac* permease has been solubilized from the membrane, purified to homogeneity, reconstituted into phospholipid vesicles and shown to catalyze all of the transport reactions typical of the native membrane with comparable activities and kinetic properties.[2] The permease is a 46.5 KDa polypeptide containing 417 amino acid residues of known sequence.[3] Based on circular dichroic measurements indicating that the protein has an exceptionally high helical content and on an analysis of the sequential hydropathic character of the protein, a secondary structure model has been proposed.[4] The model suggests that the protein consists of 12 hydrophobic α-helical segments that traverse the membrane in zig-zag fashion connected by 11 shorter hydrophilic segments (Fig. 1). The model makes explicit predictions regarding those portions of the molecule that should be accessible to the solvent at the surfaces of the membrane. To test these predictions and obtain more information about structure/function relationships, peptides corresponding to each hydrophilic segment, the N- and C-termini have been synthesized and used to generate site-directed polyclonal antibodies in rabbits.

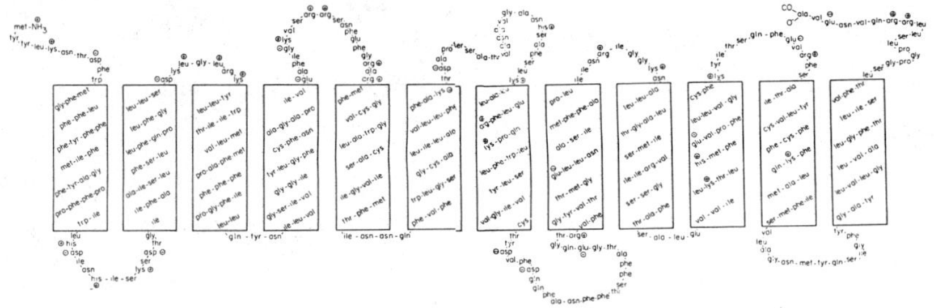

Fig. 1. Secondary structure model of the <u>lac</u> permease

Results and Discussion

The peptides were synthesized by solid-phase peptide methodology using p-hydroxymethyl polystyrene resin, the symmetrical anhydride procedure and a Vega 250 chemistry module peptide synthesizer. The synthesizer was controlled by a Model 300 microprocessor from Vega Biochemicals. Deprotection and cleavage from the resin were achieved by treatment with anhydrous liquid HF using the modified procedure of Tam et.al. The peptides were purified by preparative high performance liquid chromatography using a μBondapak C_{18} column. The purity of the peptides was ascertained by analytical HPLC and amino acid analysis. Table I shows the peptides synthesized with their corresponding sequence.

The peptides were coupled to bovine thyroglobulin with glutaraldehyde and the conjugate was used to produce site-directed-polyclonal antibodies in rabbits. The IgG fractions were purified from serum by affinity chromatography on protein A/sepharose followed by affinity chromatography on a column containing the peptide conjugated to sepharose. Antibodies were shown to be specific for the lac permease protein by immunoelectroblotting.[5]

Amino Acid Sequence	Sequence	Hydrophilic Segment
H-Met-Tyr-Tyr-Leu-Lys-Asn-Thr-Asn-Phe-OH	1-9	1
H-Leu-His-Asp-Ile-Asn-His-Ile-Ser-Lys-Ser-Asp-Thr-Gly-OH	34-46	2
H-Leu-Ser-Asp-Lys-Leu-Gly-Leu-Arg-Lys-Tyr-Leu-Leu-Trp-Ile-OH	66-79	3
H-Ile-Phe-Gly-Pro-Leu-Leu-Gln-Tyr-Asn-Ile-Leu-Val-Gly-Ser-OH	94-107	4
H-Glu-Lys-Val-Ser-Arg-Arg-Ser-Asn-Phe-Glu-Phe-Gly-Arg-Ala-Arg-Met-Phe-OH	130-146	5
H-Thr-Ile-Asn-Asn-Gln-Phe-OH	163-168	6
H-Phe-Phe-Ala-Lys-Thr-Asp-Ala-Pro-ser-Ser-Ala-Thr-Val-Ala-OH	185-198	7a
H-His-Ser-Ala-Phe-Ser-Leu-Lys-Leu-Ala-Leu-Glu-Leu-Phe-Arg-Gln-Pro-Lys-OH	205-221	7b
H-Thr-Tyr-Asp-Val-Phe-Asp-Gln-Gln-Phe-Ala-Asn-Phe-Phe-Thr-Ser-Phe-Phe-Ala-Thr-Gly-Glu-Gln-Gly-Thr-Arg-OH	235-259	8
H-Asn-Arg-Ile-Gly-Gly-Lys-Asn-Ala-Leu-OH	284-292	9
H-Arg-Ile-Ile-Gly-Ser-Ser-Phe-Ala-Thr-Ala-Leu-Glu-Val-OH	302-315	10
H-Phe-Lys-Tyr-Ile-Thr-Ser-Gln-Phe-Glu-Val-Arg-Phe-OH	334-345	11
H-Gly-Asn-Met-Tyr-Glu-Ser-Ile-Gly-Phe-Gln-Gly-Ala-Tyr-Leu-OH	370-383	12
H-Leu-Ser-Leu-Leu-Arg-Arg-Gln-Val-Asn-Glu-Val-Ala-OH	406-417	13

Table I. Summary of the peptides synthesized

Binding studies with radiolabeled antibodies in right-side-out and inside-out membrane vesicles demonstrate that hydrophilic segments 1 (the N-terminus), 5, 7a and 13 (the

C-terminus) are more accessible from the cytoplasmic surface of the membrane (Table II). On the other hand, hydrophilic segments 2, 6, 8, 10, 11 and 12 are inaccessible from either surface suggesting that these segments either do not protrude from the membrane or are buried within the tertiary structure of the polypeptide.

Sequence	Hydrophilic Segment	Ab vs Peptide	Ab vs lac carrier protein	Accessibility of peptides in the membrane
1-9	1	+++	++	+ cytoplasmic domain
34-46	2	++	+	-
66-79	5	+++	++	+ cytoplasmic domain
163-168	6	+++	++	-
185-198	7a	+++	++	+ cytoplasmic domain
235-259	8	+++	++	-
302-315	10	+++	++	-
334-345	11	+	+	-
370-383	12	+++	+	-
406-417	13	+++	+++	+ cytoplasmic domain

Table II. Accessibility of peptides in the membrane

References

1. Kaback, H.R. (1983) J. Membr. Biol. 76, 95-112.
2. Matsushita, K., Patel, L., Gennis, R.B. and Kaback, H.R. (1983) Proc. Natl. Acad. Sci. USA 80, 4889-4893.
3. Büchel, D.E., Sonenberg, B.M. and Müller-Hill, B. (1980) Nature (London) 283, 541-545.
4. Foster, D.L., Boublik, M., and Kaback, H.R. (1983) J. Biol. Chem. 258, 31-34.
5. Carrasco, N., Herzlinger, D., Mitchell, R., De Chiara, S., Danho, W., Gabriel, T.F., and Kaback, H.R. (1984) Proc. Natl. Acad. Sci. USA 81, 4672-4676.

NEW APPROACHES TO A SAFE FOOT AND MOUTH DISEASE VIRUS VACCINE

Johann J. Leban

Biogen Research Corp., 14 Cambridge Center, Cambridge, MA 02142, U.S.A.

Hans G. Boehm and Juergen H. Thiel

Federal Research Institute for Animal Virus Diseases, Paul-Ehrlich-Str. 28, 7400 Tuebingen, FRG.

Ewald Beck and Eberhard Pfaff

Microbiology, Univ. Heidelberg and ZMBH, 6900 Heidelberg, FRG.

Introduction

Foot and Mouth disease viruses (FDMV) are the causative agents of an economically important disease afflicting primarily cloven-hooved animals. Presently FMDV vaccines are produced by inactivation of the live virus obtained from tissue culture. Even though the vaccine is effective, outbreaks of the disease have been linked to incomplete inactivation.

One of the four capsid proteins (VPl) or a chemically cleaved fragment of it was shown to contain a neutralizing determinant.[1]

A synthetic hexadecapeptide derived from VPl-O_1K sequence 144 to 159 has been used to raise neutralizing antibodies.[2]

Our work was directed toward the development of an effective formulation containing such a peptide, testing its efficiency in small animals and cattle, and studying its efficacy in cross neutralization between subtypes.

Results and Discussion

Four peptides derived form the sequence of VP1-O_1K were synthesized by an automated Merrifield solid-phase method[3]: VP1-O_1K (144-159), VP1-O_1K (140-160), VP1-O_1K (129-159), VP1-O_1K (110-159), in the text referred to as 16 aa, 21 aa, 31 aa, and 50 aa peptides. In the 31 aa and 50 aa peptides, Cys in 134 was exchanged for L-α-aminobutyric acid. All peptides except the 50 aa peptide were conjugated to KLH by the glutaraldehyde method and used to immunize guinea pigs with one injection. After four weeks, sera were analyzed by the baby mouse neutralization test[4] and viral challenge. Results are shown in Table I.

There was no difference in immunological response to the 16 aa, 21 aa and 31 aa peptide when used as KLH-conjugates. However, when the peptides were used without a carrier the 31 aa peptide was effective while the 16 aa and 21 aa peptides were not.

These results show that the region preceding amino acid 144 is not needed to induce neutralizing antibodies. When no carrier is used, however, this region makes the molecules

Table I. Neutralizing Antibody Response to Synthetic FMDV-O_1K Peptides.

Antigen	Dose	Titers (pooled sera)	No. of animals Immunized/protected
16 aa-KLH	200 μg	1.6	4/4
21 aa-KLH	200 μg	1.6	4/4
31 aa-KLH	200 μg	1.1	4/3
16 aa	200 μg	0	2/4
21 aa	200 μg	1.4	1/4
31 aa	200 μg	1.6	4/4
50 aa	200 μg	1.6	4/4
KLH	200 μg	0	4/0

The titer is given as $\log_{10}$ serum dilution that protects 50% of suckling mice.

more stable against enzymatic degradation and the peptide may be better utilized by antigen presenting cells. A 50 aa peptide was equally effective. However, the antibodies did not show cross-reactivity with other O subtypes while the antibodies elicited by the 31 aa peptide did. (Table II). The failure of the 50 aa peptide to cross-react with other subtypes reflects its hydrophobicity and lower conformational flexibility. The cross-neutralization by anti-31 aa peptide antibodies of other FMDV subtypes differing by one or more amino acids (E. Beck, unpublished) within the antigenic region may be due to the conformational flexibility of short peptides and their ability to raise antibodies against different conformations of the peptide.[5]

These observations were supported by the fact that a nonoclonal antibody raised against FMDV subtype O_1K did not neutralize other isolates. Even the primary amino acid sequence within the antibody binding site is identical (Table II). To test the hypothesis that recognition by antibodies is strongly influenced by the secondary structure of the antigenic determinant, we decided to raise monoclonal antibodies against the peptides. The 21 aa-KLH conjugate was

Table II. Serological Analysis of O Subtypes.

Titers	Virus monclonal antibody	31 aa polyclonal antibody (guinea pig)	50 aa polyclonal antibody (guinea pig)	31 aa polyclonal antibody (cattle)
O_1Kaufbeuren	3.0	3.6	2.2	2.9
O_1Wuppertal	0	0	0	2.4
O_1Israel	1.9	1.6	0	2.3
O_1Lausanne	0	1.5	0	1.7
O_2Normandie	0	2.0	0	weak neut.

Cross-neutralization is given as $\log_{10}$ of dilution at 50% reduction in a plaque neutralization assay.

used to immunize mice and after the usual procedure two clones were isolated. One of these predominantly neutralizes O_1L, while another one neutralizes only the O_1K isolate (E. Pfaff, unpublished).

In a preliminary experiment, five head of cattle were immunized with different peptide preparations and VCA and after two weeks again with peptide and FIA. Only the one immunized with the uncoupled 31 aa peptide (1 mg per immunization) had high neutralizing antibody titers and was completely protected even after being in close contact with the other 3 head (data not shown).

In conclusion, we have shown that the region preceeding amino acid 144 in FMV, VP1O_1K is not needed to induce neutralizing antibodies, but a peptide extended into this region is effective without a carrier and shows good neutralization of subtypes with different amino acid sequences. Our reasoning for this phenomenon is supported by cross-neutralization experiments with monoclonal antipeptide antibodies. We could demonstrate that protection in cattle is possible but have not yet demonstrated complete protection of a statistically significant number of cattle.

References

1. K. H. Adam, O. R. Kaaden and K. Strohmaier (1978) Biochem. and Biophys. Res. Comm. Vol. 84, No. 3, 677.

2. E. Pfaff, M. Mussgay, H. O. Boehm, G. E. Schulz, H. Schaller, (1982) EMBO 1, 868-874.

3. R. B. Merrifield, (1963) J. Am. Chem. Soc. 85, 2149.

4. H. H. Skinner, (1952) Proc. 15th Int. Vet. Congr., Vol. 1, part 1, pg. 195.

5. H. Alexander, D. Johnson, J. Rosen, L. Jerabek, N. Green, J. L. Weissman and R. A. Lerner, (1983) Nature 306, 697-699.

IDENTIFICATION OF ANTIGENIC DETERMINANTS IN PROTEINS BY ANALYSIS OF PEPTIDES FROM PROTEOLYSED ANTIGEN-ANTIBODY COMPLEXES.

Ronald Jemmerson and Yvonne Paterson
Department of Immunology, Scripps Clinic and Research Foundation, La Jolla, California 92037

Introduction

Existing methods to study the sites on proteins that interact with antibodies raised against them employ either panels of evolutionarily related proteins in fine specificity studies or synthetic and/or proteolytically or chemically cleaved peptides[1]. Both are limited since only one or two dominant residues in an epitope may be determined from the former approach and non-contiguous or conformational-dependent epitopes can not be examined using the latter. We have developed a new method to analyze antigenic sites on proteins which uses the observations that many protein antigens are susceptible to proteolysis whereas the antigen binding domains of immunoglobulins are not, and that regions of an antigen involved in antibody binding are more protected from proteolytic attack than those located at a distance from these sites. The method is demonstrated by determining the antigenic sites on horse cytochrome c (cyt c) for two anti-horse cyt c monoclonal antibodies (mAbs).

Results and Discussion

Analyses of mAb binding to evolutionarily related cyts c (see Table I) show that mAb C3 binds to cyts c that have a

Table I Reactivity[a] of MAbs to Heterologous Cyts c and to Peptide Fragments

Antigen	Sequence Difference 44	47	60	62	89	92	MAb E8	C3
Horse	P	T	K	E	T	E	80	55
Donkey	-	S	-	-	-	-	85	60
Cow	-	S	G	-	G	-	>1000	60
Rat	A	S	G	D	G	A	>1000	60
Rabbit	V	S	G	D	D	A	>1000	>1000
Guanaco	V	S	G	D	G	A	>1000	>1000
CNBr 1-65 (Horse)							>1000	>1000
41-49 (Horse)							>1000	>1000

[a]Measured as the nmoles of heterologous cyt c required to give 50% inhibition of mAb binding to horse cyt c by RIA

proline or alanine at position 44, but not to cyts c that have valine at that position. MAb E8 binds to horse and donkey cyts c but not to cow cyt c which differs from these two proteins at residues 60 and 89.

Analysis of mAb binding using peptides is also shown in Table I. A synthetic peptide representing the sequence 41-49 in horse cyt c and a CNBr-cleaved fragment representing 1-65 do not hind these mAbs although these peptides have been shown to bind polyclonal antisera raised in rabbits to the region containing residue 44 in horse cyt c[2]. This suggests that the determinants on cyt c for E8 and C3 involve residue interactions from non-contiguous regions of the molecule and/or have conformational restrictions.

The effects of mAbs on the tryptic digestion of horse cyt c after 30-35 minutes are shown in Table II as the HPLC peak height ratios observed for mAb-free compared to mAb-bound cyt c. High peak height ratios indicate protection of those segments of the polypeptide chain from cleavage. The positions of the peptides in the horse cyt c sequence were identified by amino acid composition from an HPLC profile of tryptic peptides obtained from a complete digestion of horse cyt c.

Table II Peak Height Ratios (Free Cyt c:MAb Bound Cyt c) of HPLC Fractionated Peptides of Horse Cyt c after Tryptic Digestion

Monoclonal Antibody	E8	C3
Sequence Identification	Peak height ratio Average ± S.D.[a]	
1-5	4.0 ± 1.7	4.8 ± 1.1
9-13	4.7 ± 2.8	4.8 ± 0.4
14-22	3.8 ± 0.8	3.1 ± 0.3
26-27	5.1 ± 2.5	6.7 ± 0.6
28-33	2.2 ± 0.9	3.0 ± 1.2
28-38	1.7 ± 0.7	5.4 ± 0.2
39-53	5.1 ± 3.3	9.1 ± 1.8
40-53	4.5 ± 3.1	7.5 ± 1.0
56-60/92-97	3.1 ± 1.7	4.8 ± 1.8
56-73	15.3 ± 4.2	4.6 ± 0.8
61-73	>20	4.8 ± 0.5
73-79	2.6 ± 1.5	4.6 ± 0.3
74-79	2.3 ± 1.9	4.3 ± 0.5
80-86	2.6 ± 1.1	2.8 ± 0.3
89-91	4.7 ± 2.7	3.8 ± 0.3

[a]The SD for E8 refers to data from experiments performed at three different times and for C3 to triplicate values obtained from one experiment

It can be seen from Table II that each mAb affects different peptides. The peptides affected by E8 are 56-73 which is an intermediate product in the tryptic degradation not observed in a complete digest, and 61-73 which is its breakdown product. The other breakdown product of 56-73 is 56-60 which overlaps with 92-97 and could not be independently quantified. Thus, residues within the region 56-73, which include residue 60, one of the two residues implicated as antigenic from the fine specificity analysis shown in Table I are clearly involved in the binding site for this mAb. This region folds around one side of the molecule with a distorted helix beginning at residue 62[3]. If residues within this helix are also involved in the epitope as well as residue 60 then the conformational integrity of this helix may be required in stabilizing the antibody/antigen interaction. Under these circumstances CNBr fragment 1-65, which contains only a few of the residues in the helix, would not be expected to bind E8.

C3 affects regions of the cyt c molecule containing residues 26-27 and 40-53 suggesting that these peptides are involved in its binding site. Immunodominant residue 44 which was identified from fine specificity analysis (see Table I) is within 40-53. These two regions of cyt c are spatially close in the X-ray crystal structure of the molecule. In fact the carbonyl oxygen atom of residue 44 is hydrogen bonded to the N-ε-hydrogen of the side chain of histidine 26[3]. Thus, C3 binds to a non-contiguous epitope of cyt c.

Conclusions

Frequently used methods to study antigenic sites on proteins may not map an entire epitope. The approach described here offers an advantage to identifying both non-contiguous and conformational-dependent epitopes and therefore will be useful in the study of protein antigens which are susceptible to proteolysis.

References

1. Benjamin D.C., et al. (1984) Ann.Rev.Immunol. 2:67-101.
2. Paterson, Y. (1985) Biochemistry 24: 1048-1054.
3. Takano, T., and Dickerson, R. E. (1981) J. Mol. Biol. 153: 95-115.

IDENTIFICATION OF IMMUNODOMINANT SITES ON C-TERMINAL REGION OF CYTOCHROME C BY USING SYNTHETIC PEPTIDES

Bhagirath Singh, Patrick D. Kilgannon and E. Fraga

Department of Immunology and MRC Group on Immunoregulation, University of Alberta, Edmonton, Alberta, Canada T6G 2H7

Introduction

Synthetic peptides with amino acid sequences deviating to varying degrees from the original immunogen have been used to study the highly specific interaction between the antigen and the receptors on B and T lymphocytes[1,2]. However, direct comparisons between the fine specificities of B lymphocytes (that produce antibodies) and T lymphocytes (which mediate cellular immunity) have not been made. Cytochrome c offers an excellent model system to explore this issue because of its known primary, secondary and tertiary structure[3] and high immunogenicity[4]. CNBr cleaved 81-104 fragment of pigeon- and synthetic (86-90)-(94-103) fragment of moth-cytochrome c were utilized in our study. These fragments induce excellent T cell responses[2,4]. A series of peptides related to these fragments were synthesized (Table I) to map various residues involved in antibody binding and compare them to the ones involved in recognition by T lymphocytes. Synthesis of the peptides was carried out by liquid phase peptide synthesis as described elsewhere[2]. For the synthesis of 90-(94-104)99Q,103A,104K and 90-(94-104)100K,103A,104K peptides, the 90-(94-96) fragment was assembled by the classical method and was coupled to 97-104 fragments on a solid support[6]. 81-104 fragment of pigeon cytochrome c was a gift from Dr. Ron Schwartz N.I.H. All peptides were purified by HPLC on a C_{18} reversed phase column with a 0 to 30% gradient of CH_3CN-0.1% TFA in water.

Table I. Antigens Used For Inhibition Studies.

		Amount(µg)[1]	
	81 ... 104	A	B
81-104 (Mouse)	I F A G I K K K G E R A D L I A Y L K K A T N E		
81-104 (Pigeon)	I F A G I K K K A E R A D L I A Y L K Q A T A K	1	0.03
<u>Moth</u>			
81-103	I F A G L K K A N E R A D L I A Y L K Q A T K		
(86-90)-(94-103)	K K A N E - - - L I A Y L K Q A T K	0.008	>200
90-(94-103)	E - - - L I A Y L K Q A T K	0.20	>200
94- 103	L I A Y L K Q A T K	>200	
90-(94-102)	E - - - L I A Y L K Q A T	3	
90-(94-103)99Q	E - - - L I A Y L Q Q A T K	>200	
90-(94-103)103A	E - - - L I A Y L K Q A T A	3	>200
90-(94-104)103A,104K	E - - - L I A Y L K Q A T A K	4	0.50
90-(94-104)103A,104E	E - - - L I A Y L K Q A T A E	5	>200
90-(94-104)99Q,103A,104K	E - - - L I A Y L Q Q A T A K		>200
90-(94-104)100K,103A,104K	E - - - L I A Y L K K A T A K		0.50

[1]Amounts of various peptides required for 50% inhibition of the binding of anti moth (86-90)-(94-103) (A) or anti pigeon 81-104 (B) antibody to immunizing antigens in solid phase radioimmunoassay.

Results and Discussion

B10.A mice immunized with whole pigeon cytochrome c, 81-104 fragment of pigeon or (86-90)-(94-103) fragment of moth cytochrome c produced excellent antibody responses as measured by solid phase radioimmunoassay (SPRIA)[7]. The anti-pigeon 81-104 antisera bound 81-104 fragment and the whole pigeon cytochrome c but failed to bind moth (86-90)-(94-103) fragment (Table II). Therefore, this antisera may see residues on 81-104 fragment that are modified from the moth sequence at positions 81-85, 88, 89, 91-93, 103 and 104 (Table I). Competitive inhibition studies performed by pre-incubating the anti-pigeon fragment antibody with various synthetic analogs (Table I) prior to SPRIA indicate that residues 81-85, 88, 89 and 91-93 do not contribute to the binding. Both moth 90-(94-103) and (86-90)-(94-103) peptides failed to bind to the antibody. Similarly, substitution/deletions of residues in the C-terminal end of 90-(94-103) peptide i.e. 90-(94-102); 90-(94-103)103A; 90-(94-104)103A,104E also failed to bind to the anti-pigeon fragment antibody. However, 90-(94-104)103A,104K peptide showed excellent binding to the anti-pigeon antibody. We therefore conclude that Ala-103 and Lys-104 are critical residues in the recognition of the pigeon 81-104 peptide. Gln-100 was not part of the epitope because its replacement by Lys, i.e. 90-(94-104)100K,103A,104K, did not abrogate binding to the anti-pigeon fragment antibody. Peptide 90-(94-104)99Q, 103A, 104K that has a substitution of Gln for Lys at position 99 also does not bind to the anti-pigeon antibody in spite of the presence of 103A, 104K residues. Thus residues Lys-99, Ala-103 and Lys-104 on pigeon cytochrome c are directly involved in recognition by the anti-pigeon cytochrome c antibody.

Anti-moth (86-90)-(94-103) fragment antibodies bound both pigeon and moth fragments as well as the whole pigeon cytochrome c (Table II). These antibodies therefore see residues that are present in all these antigens. The residues in the

Table II. Cross-reactivity of Antibodies[1]

	^{125}I-Goat anti mouse IgG bound (C.P.M.)			
Antigen	Anti Pigeon cytochrome c	Anti Pigeon 81-104	Anti moth (86-90)-(94-103)	Normal Mouse Serum
Pigeon cyt.c	23,802	21,150	12,530	763
Pigeon 81-104	4,124	4,579	4,968	427
Moth (86-90)-(94-103)	831	986	4,347	450

[1]Antigens were coated on PVC microtiter plates at 20 μg/ml. Amount of antibody (1:125 dilution) bound to antigens was determined by using ^{125}I-labelled second antibody in a solid phase radioimmunossay.

86-90 region of the moth fragment appear to stabilize overall α-helical conformation of the 94-103 region[5] and their deletion one at a time gradually reduced the capacity of resulting peptide to inhibit the binding of anti moth antibodies to (86-90)-(94-103) sequence (Table I). Removal of Glu-90 almost completely abolishes the antigenicity of the remaining 94-103 peptide. Substitutions at positions 102, 103 or 104 have little effect on the binding of this antibody. Lys-99 appears to play a significant role in the recognition of moth (86-90)-(94-103) peptide by the antibody. All substitutions at this site i.e. Ser, Glu and Gln in 90-(94-103) fragment reduced their binding to the anti-moth antibody.

Therefore, both anti-moth and anti-pigeon fragment antibodies seem to contact Lys-99 which is present in almost all cytochrome c and has been identified as a critical residue that contacts the T cell receptor[4]. Our results suggest that it is also a contact residue for antibodies. These studies provide direct evidence that the same amino acid residues are seen by both T and B lymphocyte receptors on the C-terminal of cytochrome c.

References

1. Geysen, H.M., Meloen, R.H., Barteling, J. (1984). Proc. Natl. Acad. Sci. U.S.A. 81, 3998-4002.

2. Schwartz, R.H., Fox, B.S., Fraga, E., Chen, C. and Singh, B. (1985). J. Immunol. (In Press).

3. Dickerson, R.E. (1972). Sci. American 226(4), 58-72.

4. Schwartz, R.H. (1985). Ann. Rev. Immunol. 3, 237-261.

5. Pincus, M.R., Gerewitz, F., Schwartz, R.H. and Scheraga, H.A. (1983) Proc. Natl. Acad. Sci. U.S.A. 80, 3297-3300.

6. Kilgannon, P.D., Fraga, E., Singh, B. (1985). Mol. Immunol. (Submitted).

7. Singh, B., Raghupathy, R., Shaw, A.R.E., Tews, D.G., Hamilton, M.S. and Wegmann, T.G. (1982). Transplantation 33, 156-162.

STRUCTURAL STUDIES OF INTERLEUKIN 2

Thomas L. Ciardelli[*+], Kendall A. Smith[*], Robert Gadski[**], Larry Butler[**], Joann Strand[**], and Fred E. Cohen[#]

* Dartmouth Medical School, Hanover, NH 03756
\+ Veterans Administration Hospital, White River Jct., VT 05001
** The Lilly Research Laboratories, Indianapolis, IN 46285
\# University of California, San Francisco, CA 94143

Interleukin 2 (IL-2, formerly T-Cell Growth Factor or TCGF) is a lymphocytotrophic hormone which mediates T-cell clonal proliferation and is therefore, fundamental to T-cell related immunologic response.[1] Antigen receptor triggering activates the expression of two specific genes, one that encodes IL-2 and another that encodes the IL-2 receptor. The magnitude of the resulting response then becomes dependent upon the concentration of IL-2, the density of receptors on the cell surface and the duration of the hormone receptor interaction.

In 1983, the primary amino acid sequence of human IL-2 was first predicted from the nucleotide sequence of an IL-2 cDNA clone derived from a T-cell leukemia cell line.[2] Complete homology with the sequence from normal T-cells was confirmed as well as the organization of the gene[3]. Interleukin-2 is a glycoprotein of M.W. 15,000 containing 133 amino acid residues and one disulfide bridge between Cys-58 and Cys-105 (Figure 1). Glycosylation occurs at Thr-3 but has no effect on <u>in vitro</u> biologic activity. We have used a combination of monoclonal antibodies, molecular modeling, recombinant DNA derived analogs and peptide synthesis to provide further insight into IL-2 structure-activity relationships.

Ala Pro Thr Ser Ser Ser Thr Lys Lys Thr Gln Leu Gln Leu Glu His
Leu Leu Leu Asp Leu Gln Met Ile Leu Asn Gly Ile Asn ↓ Asn Tyr
Lys Asn Pro-Lys Leu Thr Arg Met Leu Thr Phe Lys Phe Tyr Met Pro
Lys-Lys ↓ Ala Thr Glu Leu Lys His Leu Gln **Cys** Leu Glu Glu Glu
Leu Lys Pro Leu Glu Glu Val Leu Asn Leu Ala Gln Ser Lys Asn Phe
His Leu Arg Pro Arg Asp Leu Ile Ser Asn Ile Asn Val Ile Val Leu
Glu Leu Lys ↓ Gly Ser Glu Thr Thr Phe Met **Cys** Gly Tyr Ala
Asp Glu Thr Ala Thr Ile Val Glu Phe Leu Asn Arg Trp Ile Thr Phe
Cys Gln Ser Ile Ile Ser Thr Leu Thr

Underlined regions correspond to hydrophilicity maxima
Cys denotes position of the disulfide bond
↓ indicates separation of exon coding regions

Fig 1. Interleukin 2 primary structure

Three IL-2 related proteins: IL-2-I (residues 30-133, missing the region encoded by the first exon); IL-2-II (residues 1-29 + 50-133, missing the region encoded by the second exon); and IL-2-IV (residues 1-99, missing 34 of the 36 residues encoded by the fourth exon) were prepared by transfection and expression of the appropriate cDNA clone in E.Coli. Isolation was performed by a combination of conventional gel filtration and C-3 reverse phase HPLC. The interaction of these proteins with neutralizing IL-2 monoclonal antibodies DMS-1 and DMS-5 was then determined using the "western" type immunoimaging procedure of binding of radioiodinated antibody to proteins immobilized on nitrocellulose membranes followed by autoradiography. IL-2-I and IL-2-IV both bound DMS-1 while IL-2-II and IL-2-IV bound DMS-5. This suggests that the DMS-5 epitope lies within the first 29 residues while the DMS-1 epitope resides in residues

30-49. The biologic activity of each protein was determined in the IL-2 quantitative bioassay.[4] All three were found to be completely devoid of bioactivity. Noteably, IL-2-IV, which binds both DMS-1 and DMS-5, was found inactive, indicating that the antibodies although capable of neutralizing activity, bind, at best, at or near only part of the active site.

A synthetic peptide C-terminal segment, IL-2 (105-133), was then prepared using manual solid phase peptide synthesis techiques according to Merrifield[5] and isolated using reverse phase HPLC. This peptide was also found to be devoid of biologic activity and bound neither DMS-1 nor DMS-5. However, when Cys-105 of the synthetic segment was allowed to form a disulfide bond with Cys-58 of recombinant IL-2-IV via air oxidation (note: Cys-125 on the synthetic segment was left protected as the acetamidomethyl derivative to avoid incorrect disulfide formation), a disulfide linked synthetic-recombinant hybrid protein (Hybrid I) could be separated from the homo dimers formed via reverse phase HPLC, which possessed distinct biologic activity (Table I).

Table I. Biologic Activity of the Synthetic-Recombinant Hybrid

	3H Thymidine Incorp.*	% IL-2 STD (10 ng/ml)
Hybrid I (2μg/ml)	6965 cpm	40.7
N^1-Formyl Hybrid I (2μg/ml)	97	0.8
Recombinant IL-2-IV (2μg/ml)	71	0.6
Synthetic IL-2 (105-133, 12 μg/ml)	73	0.6

* Determined in CTLL murine cell line pulsed between 20 and 24h according to ref 4.

This indicates that at least two regions of the protein are brought into proximity by the disulfide bond to participate in the active site. A second hybrid protein was prepared using a 29 residue C-terminal synthetic segment with Trp-121 present as the N-indole formyl derivative as the only modification

(N^i-Formyl hybrid I) This protein behaved identically on reverse phase HPLC but repeatedly failed to express biologic activity indicating that this residue participates in important hydrophobic interactions either at the active site or internally in the protein to stabilize vital tertiary structure.

A preliminary 3-dimensional model IL-2 has been generated from a consideration of the primary sequence using a combinatorial approach[6]. Secondary structure is assigned and then hydrophobic surfaces are packed to form a four fold 2-helical bundle. The model suggests that the N-terminal helix (17-31) associates with the C-terminal helix (116-132) through hydrophobic interactions possibly involving Trp 121. This association is further stabilized by the disulfide bridge between Cys 58 and Cys 105. We are investigating the implication of these interactions for bioactivity.

1. Smith, K.A. (1980) Immunological Rev. **51**, 337-357.

2. Taniguchi,T, Matsui, H., Fujita, T., Takaoka, C., Kashima, N., Yoshimoto, R. and Hamuro, J. (1985) Nature **302**, 305.

3. Holbrook, N.J., Smith, K.A., Frunace, A.J., Comeau, C.M., Wiskocil, R.L. and Crabtree, G.R. (1984) Proc. Natl. Acad. Sci., USA **81** 1634.

4. Gillis, S., Ferm, M.M., Ou, W. and Smith, K.A. (1978) J. Exp. Med. **148**,1093.

5. Merrifield, R.B., Vizioli, L.D. and Bowman, H.G. (1982) Biochemistry **21**, 5021.

6. Sternberg, M.J.E. and Cohen, F.E. (1982) Int. J. Biol. Macromol. **4**, 137.

TWIN-α_1, THE HEAD-TO-HEAD ANALOG OF THYMOSIN-α_1, IS A MORE POTENT IMMUNOREGULATORY POLYPEPTIDE DRUG THAN THE PARENT.

Christian Birr[1,2], Thomas L. Ciardelli[3], Thomas Nebe[2], Wolfgang Heinzel[2], Michael Young[4], Anthony D. Ho[5] and Bernd Stehle[5].

1) Max-Planck-Institute for Medical Research, Heidelberg, 2) ORGANOGEN GmbH., Heidelberg (FRG), 3) Veterans Administration Hospital, White River Junction, NH (USA), 4) VENTREX Inc., Portland, ME (USA), 5) Med. Univ. Poliklinik, Heidelberg.

Introduction

In the thymus proteins are processed resulting immunoregulatory peptides which are released into the lymphoid fluid and blood stream. These thymic peptides seem to be further processed in the secondary immune organs yielding small signal peptides, which balance the proliferation of T cell subpopulations[1]. Some structural elements of thymosin-α_1 were found to be immunologically active in the induction of T helper cell proliferation, others induce T suppressor lymphocytes. Aiming at an increased concentration at the receptor site of those processed small signal peptides we have synthesized an exact head-to-head twin thymosin-α_1 from a fully protected synthetic precursor.

Materials

The synthesis of thymosin-α_1 was performed as described earlier[2] starting from Ddz-amino acid mixed anhydrides[3]. The efficiency of the total synthesis from five fragments was increased as previously described for the Gla^{24,25}-thymosin-α_1 [4]. After cleavage of the N-terminal Ddz group with 5% trifluoroacetic acid in dichloromethane the polypeptide was divided into two portions. The first one was reacted with 1 equiv. of succinoyl anhydride. Then the free carboxyl of the fully protected semi-succinoyl thymosin-α_1 was activated with CDI plus 2 equiv. HOBt for the transformation with the second portion of the fully protected N-des-Ddz thymosin-α_1. This, after entire deprotection as described[2] resulted N,N'-succinoyl-bis-thymosin-α_1, which we named "Twin-α_1". After gel chromatographic purification in an aqueous system Twin-α_1 was obtained lyophilized with 43% yield, $[\alpha]^{20}_{D}$ -87.5°, (c 0.078 in H_2O). This preparation was investigated for immunological activity in 4 different assays as described in the following Tables I-IV.

Results

Table I. The Effect of Twin-α_1 as Compared to Thymosin-α_1 in the Mixed Lymphocyte Culture Assay (MLC) on Human Peripheral T Cells.

µg/ml cell culture	5.0	2.5	1.0	0.5	
Twin-α_1 a)	104 b)	85	81	162	156 g)
Twin-α_1	103	89	84	150	
Thymosin-α_1 c)	86	80	85	141	133
Thymosin-α_1 d)	83	80	88	126	

a) purified Twin-α_1, b) % stimulation of responder T cells as compared to blank MLC responses (100 %) of identical cultures, n = 6; c) own synthetic product[2], d) foreign source, g) mean.

Table II. The Effect of Twin-α_1 as Compared to Thymosin-α_1 on Human Peripheral T Cells in the E-rosette Assay with Pre-Inhibition of the E-receptor Expression[5] by 0.2 µg α-Amanitin/ ml Cell Culture.

µg/ml cell culture	5.0	2.5	1.0	0.5
Twin-α_1 a)	95 b)	5	-	29
Twin-α_1	81	71	-	-
Thymosin-α_1 c)	81	67	-	10
Thymosin-α_1 d)	90	-	40	-

a) purified Twin-α_1, b) % restoration to full E-rosette count (100 %, relative scale [5]) without pre-inhibition by α-amanitin. c) own synthetic product [2], d) foreign source. - All cultures n = 5.

Table III. Effect of Twin-α_1 and Thymosin-α_1 on Ecto-5'-Nucleotidase in Human Cord Blood Lymphocytes[6]; p 0.05

Sample	Control	0.2 a)	2.0	4.0	8.0
1	20 b)	24	29(29) c)	24	19
2	5	9	8(6)	6	6
3	15	10	17(-)	16	8
4	5	7	12(6)	10	6
mean	11.25	12.50	16.5(13.7)	14	12.30

a) µg/ml cell culture, b) units/ 10^6 cells, c) in parenthesis Thymosin-α_1

Table IV. Effect of Gel Chromatographic Fractions from an Original Twin-α_1 Preparation in the Mitogen Co-Stimulation Assay (5µg PHA/ml cell culture) on the T Helper/Suppressor Ratio in Peripheral Blood from a Single Allergy Patient.

Fraction of Twin-α_1	A		B		C		D		Control
µg/ml cell culture	1	10	1	10	1	10	1	10	
OKT 8^+	15 a)	15	10.5	7.6	8.7	18	9.7	8.9	9.4
OKT 4 / OKT 8	3.6	3.6	4.7	6.6	5.8	3.0	5.2	5.5	5.7

a) % of total T lymphocytes.

Discussion

The excess response caused by Twin-α_1 in the MLC indicates an increased stimulatory potency of Twin-α_1 (mean +56 % excess stimulation over control) as compared to thymosin-α_1 (mean +33 % excess stimulation) at a concentration as low as 0.5 µg/ml cell culture. Because doubled local concentration at the receptor site, Twin-α_1 is significantly more potent than tymosin-α_1, though the former has only have of the molar concentration in the cell culture. At a concentration 1 order of magnitude higher Twin-α_1 is also superior in the restoration of the E-receptor expression after pre-inhibition of the cell culture by the RNA polymerase B inhibitor α-amanitin. This can be interpreted to be a kinetic necessity, because of an enzymic degradation

of the active species during internalization of the signal peptide through the cell membrane for re-start of the E-receptor biosynthesis.
As indicated by other studies[6], cord blood lymphocytes show a lower level of ecto-5'-nucleotidase activity (14.5 ± 0.7 units/10^6 cells) than those from normal adults (25.4 ± 1.2 units/10^6 cells). T lymphocyte maturation caused by immunoregulatory peptides can therefore be investigated in this functional assay system. The effect of Twin-α_1 at an optimal dose of 2 µg/ml culture medium (mean 16.5 units/10^6 cells) again indicates the superiority of this potential peptide drug for immunotherapeutic treatments.
A surprising effect we have observed with Twin-α_1 on a blood sample from a patient suffering from acute pollen allergy. In a mitogen assay (PHA) co-stimulation of the culture with a chromatographic fraction enriched in Twin-α_1 caused a significant decrease of T suppressor cells (OKT 8^+) so that an uncommon T helper/suppressor ratio became rebalanced (fraction B), whereas fraction C (enriched in N-terminally free thymosin-α_1) at 10 µg/ml culture caused a significant doubling of the T suppressor count, as determined in our laboratory by flow cytometry.

References

1. Birr, Chr. (1984) in THYMIC HORMONES AND LYMPHOKINES, Goldstein, A.L., Ed., Plenum Press, Washington D.C., pp 97-109.
2. Birr, Chr. and Stollenwerk, U. (1979), Angew. Chem. 91, 422-423.
3. Birr, Chr., Nassal, M., and Pipkorn, R. (1979), Int.J.Pept.Protein Res. 13, 287-295.
4. Birr, Chr. and Krueck, I. (1985) in PEPTIDES 1984, Ragnarsson, U., Ed., Almquist & Wiksell, Uppsala, Sweden, pp. 447-450.
5. Birr, Chr., Krueck, I., Stollenwerk, U., Ciardelli, T.C., and Brodner, O. (1981) in PEPTIDES, Rich, D., Gross, E., Eds., Pierce Chem.Comp., Rockford, USA, pp. 545-548.
6. Ho, A.D., Stehle, B., Dietz, G., Hunstein, W. and Hoffbrand, A.V. (1985) Scand.J.Immunol. 21, 221-225.

DESIGN, SYNTHESIS, AND CONFORMATION OF SUPERACTIVE THYMOPOIETIN-ANALOGUES

Horst Kessler, Bernhard Kutscher, Rainer Kerssebaum, Armin Klein, and Jörg Lautz
Institut für Organische Chemie, Johann Wolfgang Goethe-Universität Frankfurt, D-6000 Frankfurt, Germany

Rainer Obermeier and Hubert Müllner
Hoechst AG, D-6000 Frankfurt, Germany

Wilfried F. van Gunsteren, Rolf Boelens, and Robert Kaptein
Laboratory of Physical Chemistry, University of Groningen
NL-9747 AG Groningen, The Netherlands

Introduction

Thymopoietins I and II are 49 amino acid peptide hormones isolated from bovine thymus[1]. They contribute to the selective induction of early T lymphocyte differentiation. The biologically active site, common in both, is the fragment Arg-Lys-Asp-Val-Tyr, TP-5 or thymopentin (positions 32-36). Later a closely homologue peptide, thymopoietin III or splenin[2] was isolated from bovine spleen. Its biologically active site (SP-5) contains Glu in position 34.

Thymopentin has immunoregulatory function in several assays and is therapeutically used as immunostimulant in clinical tests. In contrast to the parent peptide it is rapidly decomposed by proteolytic enzymes[3]. This could be avoided by using cyclic peptides. Furthermore higher activity and better receptor selectivity is expected in using cyclized structures[4,5]. Reduction of conformational freedom enables conformational studies in solution with higher reliability than for those of linear peptides, and there is a higher probability to be close to the bioactive conformation[4] when a cyclic compound is biologically active. The aim of this work was the synthesis of cyclic TP-5 or SP-5 analogues, the study of their conformation in solution by NMR spectroscopy, and the search for conformation - activity relationships.

Synthesis

The syntheses of the linear precursors of the cyclic thymopentin analogues have been accomplished by conventional peptide synthesis. For fragment condensation we used PPA[6] as coupling reagent; hydrogenolytically removable protecting groups were used for side chain protection. Cyclisation of the linear peptide H-Tyr-Arg(NO_2)-Lys(Z)-Asp(OBzl)-Val-X using the azide method[7] did not yield any cyclic product, but using the EDCI/DMAP method[8,9] we obtained a cyclic product in about 30% yield. Amino acid analysis on chiral support showed complete inversion of the C-terminal Val. We conclude that the conformation of an all-L-pentapeptide is unfavourable for cyclisation. We furtheron synthesized TP-5 analogues containing at least one D-amino acid with terminal L- or D-amino acid in the linear precursor. The synthesized cyclic pentapeptides are listed in Table IV (below).

Conformational analysis

The backbone conformation of all cyclic peptides is determined by NMR-spectroscopy on the side chain blocked compounds in DMSO solution. The cyclic structure is proved by one- and two-dimensional NOE-measurements[10] and by H,C-COLOC-spectra[11] (Figure 1) via the long range couplings between the carbonyl-carbons and the amide protons along the peptide bonds. The

^{1}H-NMR-spectra are assigned by using normal[12] and phase-sensitive[13] two-dimensional techniques (H,H-COSY, H-relayed-H,H-COSY and H,H-relayed-H,H-COSY[14]). The combination of these homonuclear techniques and heteronuclear correlation spectroscopy[15] enables an unequivocal assignment of all proton, carbon, and nitrogen resonance signals. Their NMR parameters are used for the discussion of conformation and dynamics.

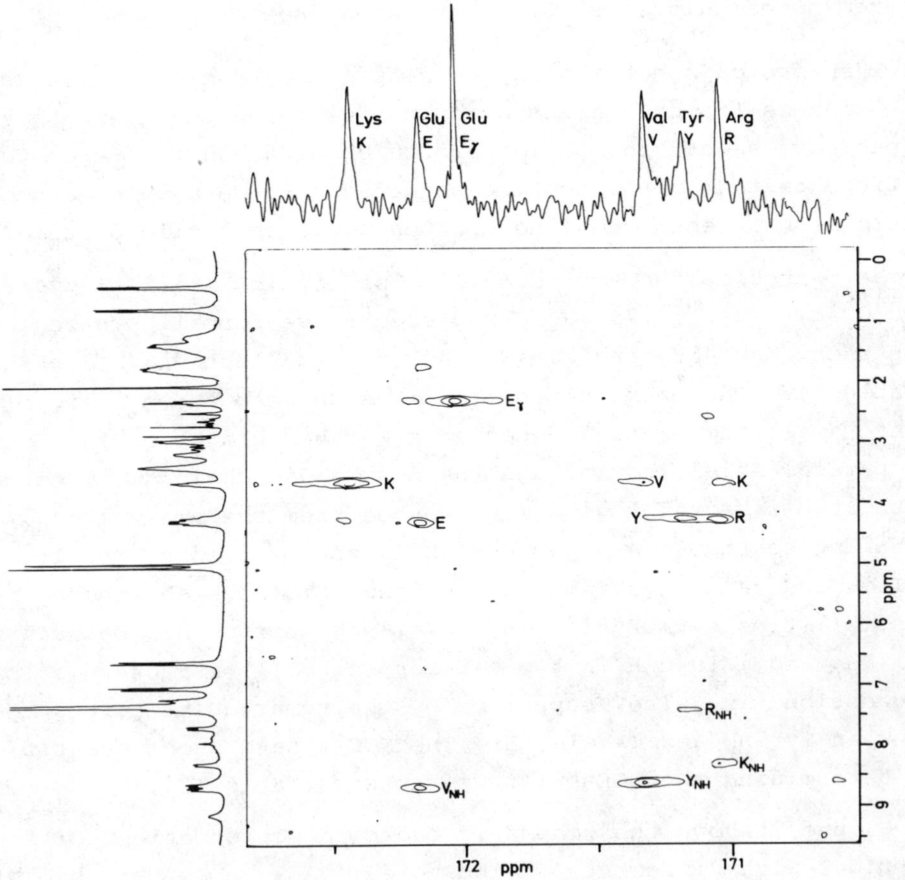

Fig. 1. Heteronuclear correlation via long range couplings (COLOC) for protected **3** (see Table IV) in DMSO (270 MHz ^{1}H). Cross-peaks indicate two- and three bond couplings from carbonyl carbons to NH and $C^{\alpha}H$ protons.

Table I. NMR and MD Parameters

		Arg	Lys	Glu	D-Val	Tyr
δ_{NH}	[ppm]	7.38	8.29	7.68	8.70	8.62
$\Delta\delta/\Delta T$	[ppb/K]	0.7	1.9	-0.7	6.5	6.7
$^3J_{HNC^{\alpha}H}$	[Hz]	9.9	7.0	8.6	7.1	8.5
Φ	[°]	-102	+81	-92	+85	-91
ϕ^a	[°]	-91±16	+68±9	-96±15	+92±16	-94±20

^{a} from MD calculations, 20 ps averaged (see Table III)

The proton spectra of all cyclopeptides show only one set of signals for each amino acid residue, and the temperature dependencies of the amide proton chemical shifts are well differentiated in the peptide sequence. So we assume conformational homogeneity of the backbone of each cyclic peptide[4].

We discuss the conformation of cyclo(-Arg(NO_2)-Lys(Z)-Glu(OBzl)-D-Val-Tyr-) as a representative for all synthesized peptides. Arg-NH and Glu-NH show the smallest $\Delta\delta/\Delta T$ values (Table I). The chemical shift of Lys-NH also shows a significant small temperature dependence which could be explained by a partial shielding through the Lys side chain[16]. Together with the dihedral angles φ, deduced from the vicinal $^3J_{HNC^{\alpha}H}$ coupling constants[17], two possible conformations result: a βII´γ and a $\gamma_1\gamma_2^i$ structure. We can distinguish those two possibilities by NOE effects; e.g. the strong NOE between D-Val-NH and Glu-$C^{\alpha}H$ is typical for a βII´γ-structure. This suggestion is also supported by the remarkable strong NOE between Tyr-NH and D-Val-$C^{\alpha}H$. The NOE´s have been quantified by determining the NOE-built-up-rates[18] (Table II).

Figure 2 shows the schematic backbone conformations of four peptides with different D-amino acids. As can be seen by this comparison, the D-amino acid always prefers the i+1 position of the β-turn in the β,γ conformation. In several cases coupling constants between α and β protons, diagnostic for the side chain orientation, can be extracted by the DISCO-

Table II. NOE - Distance Table[a]

NOE observed		calculated distance
D-Val-NH	- D-Val-$C^{\alpha}H$	2.33
D-Val-NH	- D-Val-$C^{\beta}H$	2.60
D-Val-NH	- Glu-$C^{\alpha}H$	2.34
D-Val-$C^{\alpha}H$	- D-Val-$C^{\beta}H$	2.43
Tyr-NH	- Tyr-$C^{\beta}H$ (pro-R)	2.46
Tyr-NH	- Arg-NH	2.70
Tyr-NH	- D-Val-$C^{\alpha}H$	2.33
Tyr-$C^{\alpha}H$	- Tyr-$C^{\beta}H$ (pro-S)	2.68
Arg-NH	- Arg-$C^{\alpha}H$	2.77
Lys-NH	- Lys-$C^{\alpha}H$	2.51
Lys-NH	- Arg-$C^{\alpha}H$	2.75
Glu-NH	- Glu-$C^{\alpha}H$	2.85

[a] Reference distance between Tyr-β-protons 1.75 Å

technique[19], in which phase-sensitive H,H-COSY-spectra are evaluated. Because of strong overlap this was not possible in the D-Val containing cyclic SP-5 analogue.

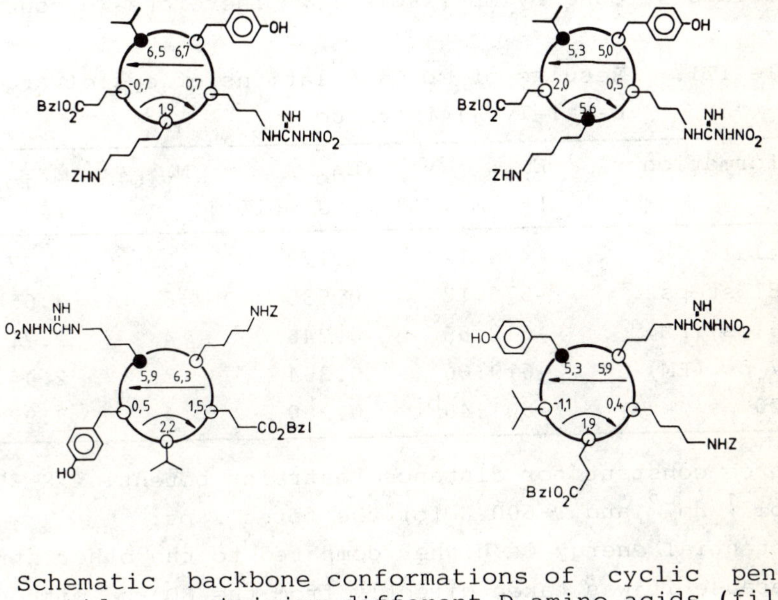

Fig. 2. Schematic backbone conformations of cyclic pentapeptides containing different D-amino acids (filled circles). The numbers are the temperature dependencies of the chemical shifts of the amide protons (ppb/K).

Molecular dynamics

The refinement of the structure information extracted from NMR-data was done by molecular dynamics (MD) methods[20], using the usual types of potential functions[21]. An additional term allows to use distance constraints[20], extracted from 2D-NOE-measurements. As a model for the above discussed cyclic peptide without protecting groups we used cyclo(-Arg-Lys-Gln-D-Val-Tyr-) to check the possibility of hydrogen bonding of the Gln side chain to the backbone. The number of violations between the 12 distances (Table II) extracted from NOE-data and calculated distances is a criterion for the quality of the structure refinement (Table III). The three calculated MD-structures belong to the same configuration space. They differ only slightly in their potential energy, restraint energy, and number of violations. Further we can find a good agreement between the experimental $^3J_{HNC^{\alpha}H}$ coupling constants and the dihedrals Φ found by MD (Table I). Therefore we conclude that

Table III. Results of MD Calculations of cyclo(-Arg-Lys-Gln-D-Val-Tyr-) in vacuo

Conformation	E_{pot} [kJ mol^{-1}]	E_{dc} [kJ mol^{-1}]	N_{viol}	Σ_{viol} [Å]	Δr_{max} [Å]
Initial	-424.12	1.178	8	5.27	1.07
EM 81 steps	-576.12	0.535	6	3.65	0.77
MD 6 ps (EM)[a]	-612.65	0.246	4	2.71	0.44
MD 7 ps (EM)	-619.66	0.354	2	2.44	0.47
MD 20 ps	-501.40[b]	0.289	4	2.92	0.48

[a] Force constant for distance restraint potential K_c=500 kJ mol^{-1} nm^{-2} and T=600 K for the first 2 ps.

[b] Potential energy is higher compared to the other structures, because it was taken directly from the MD run without minimization.

the MD results establish the experimental predicted βII´γ-structure very well[16].

Table IV. Biological Activity of Thymopoietin-Analogues[a]

Substance		PFC-Assay	PHA-Assay
control		-+	1.00
1	Arg-Lys-Asp-Val-Tyr (TP-5)	-+	1.05
2	cyclo(-Arg-Lys-Asp-D-Val-Tyr-)	++	1.69
3	cyclo(-Arg-Lys-Glu-D-Val-Tyr-)	+++	1.92
4	cyclo(-Arg-Lys-Aad-D-Val-Tyr-)	-+	
5	cyclo(-Arg-Orn-Glu-D-Val-Tyr-)	-+	
6	cyclo(-Arg-Lys-Asp-D-Val-Trp-)	-+	
7	cyclo(-Arg-Lys-Glu-D-Val-Trp-)	+	1.25
8	cyclo(-Arg-Lys-Asp-Val-D-Tyr-)	+	
9	cyclo(-Arg-Lys-Glu-Val-D-Tyr-)	-+	
10	cyclo(-Arg-Lys-Glu-Val-D-Phe-)	-+	
11	cyclo(-D-Arg-Lys-Glu-Val-Tyr-)	-+	
12	cyclo(-Arg-D-Lys-Asp-D-Val-Tyr-)	-+	
13	cyclo(-Arg-D-Lys-Glu-D-Val-Tyr-)	-+	

[a] cyclic peptides as diacetats

Biological activity

The deprotected peptides have been tested in the phytohemagglutinine (PHA) and the plaque forming cell (PFC) assay[22]. It is obvious that at least the cyclic thymopoietin analogues containing D-Val exhibit a high biological activity (Table IV). To our knowledge, compound **3** is the analogue which exhibits highest activity in the PFC and PHA assay.

Discussion

The conformation of linear TP-5 and SP-5 has been calculated[23a] and studied in solution by NMR spectroscopy previously[23b,c]. An equilibrium of several conformations has been found. One of these *extended* structures was assumed to be the bioactive one. However, this is in contrast to the activity of the *cyclic* compounds presented here. High activity so far is only observed in cyclic pentapeptides containing a D-Val. The exchange of any other amino acid by its D-enantiomere induces a complete change of backbone conformation accompanied by a reduction of biological activity. The side chain length of Lys (compare **3** and **5** in Table IV) and Glu (compare **3**, **2** and **4**) is optimal for high biological activity.

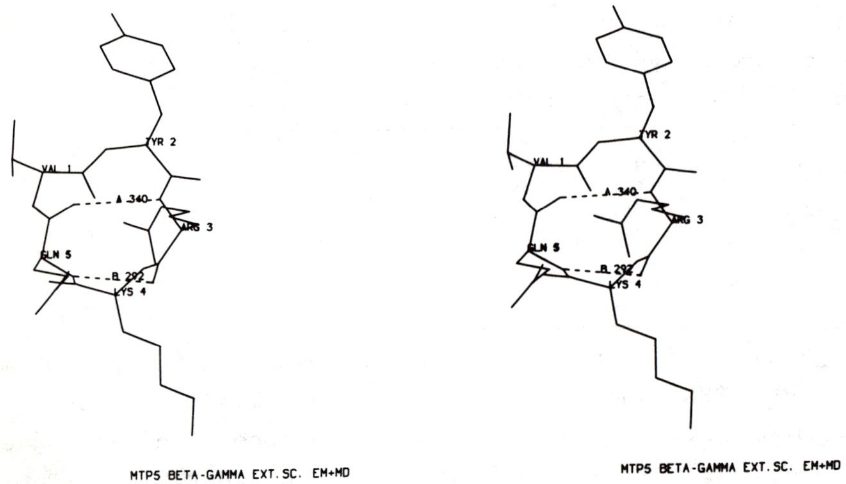

Fig. 3. Conformation of cyclo(-Arg-Lys-Gln-D-Val-Tyr-) (7ps MD after energy minimization).

Acknowledgement

We gratefully acknowledge financial support by the Deutsche Forschungsgemeinschaft and the Fonds der Chemischen Industrie. A.K. thanks the Hermann-Schlosser-Stiftung for a grant.

References

1. Goldstein, G. (1974) Nature **247**, 11-14.

2. Audhya, T., Schlesinger, D.H., and Goldstein, G. (1981) Biochemistry **20**, 6195-6200.

3. Tischio, J.P., Patrick, J.E., Weintraub, H.S., Chasin, M., and Goldstein, G. (1979) Int.J.Peptide Protein Res. **14**, 479-484.

4. Kessler, H. (1982) Angew.Chem.Int.Ed.Engl. **21**, 512-523.

5. Hruby, V.J. (1982) Life Sciences **31**, 189-199.

6. Wissmann, H., and Kleiner, H.J. (1980) Angew.Chem.Int. Ed.Engl. **19**, 133-134.

7. Medzihradsky, V. (1960) Comm. 3rd Europ.Symp.Pept.Synth., Basel.

8. a) Kessler, H., and Kutscher, B. (1985) Tetrahedron Lett. **26**, 177-180,
b) Kutscher, B. (1984) Doctoral thesis, Frankfurt.

9. Höfle, G., Steglich, W., and Vorbrüggen, H. (1978) Angew. Chem.Int.Ed.Engl. **17**, 569-582.

10. Jeener, J., Meier, B.H., Bachmann, P., and Ernst, R.R. (1979) J.Chem.Phys. **71**, 4546-4553.

11. Kessler, H., Griesinger, C., Zarbock, J., and Loosli, H.R. (1984) J.Magn.Reson. **57**, 331-336.

12. Bax, A., and Freeman, R. (1981) J.Magn.Reson. **44**, 542-561.

13. States, D.J., Haberkorn, R.A., and Ruben, D.J. (1982) J.Magn.Reson. **48**, 286-292.

14. Wagner, G. (1983) J.Magn.Reson. **55**, 151-156.

15. Kessler, H., Hehlein, W., and Schuck, R. (1982) J.Am.Chem.Soc. **104**, 4534-4540.

16. Experimental studies of the influence of side chain protecting groups to the backbone conformation are in progress.

17. Bystrov, V.F. (1976) Progress in NMR Spectroscopy, **10**, 41-81.

18. Wagner, G., and Wüthrich, K. (1979) J.Magn.Reson. **33**, 675-680.

19. a) Kessler, H., and Oschkinat, H. (1985) Angew.Chem.Int. Ed.Engl. August issue,
b) Kessler, H., Müller, A., and Oschkinat, H. (1985) Magn.Reson.Chem. in press

20. Gunsteren, W.F.van, Kaptein, R., and Zuiderweg, E.R.P. (1983) in Nucleic Acid Conformation and Dynamics, Report of NATO/CECAM Workshop, Orsay, Ed. Olsen, W.K., pp 79-92.

21. Gunsteren, W.F.van, and Berendsen, H.J.C. (1982) Biochem. Soc.Trans. **10**, 301-305.

22. a) Schmidtke, J.R., and Dixon, F.J. (1972) J.Exp.Med. **136**, 392-397,
b) Stewart, C.C., Cramer, S.F., and Steward, P.G. (1975) Cell Immunol. **16**, 237-250.

23. a) Tourwe, D., Coen, J.L.de, Hallenga, K., and Binst G.van (1984) Int.J.Peptide Protein Res. **23**, 84-93,
b) Krishna, N.R., Huang, D.-H., Chen, D.M., and Goldstein, G. (1980) Biochemistry **19**, 5557-5563,
c) Krishna, N.R., Heavner, G.A., and Vaughn, J.B. (1983) J.Am.Chem.Soc. **105**, 6930-6934.

PROTON NMR ASSIGNMENTS AND SOME SOLUTION CONFORMATION FEATURES OF SERUM THYMIC FACTOR

Joseph B. Vaughn, Jr.

Department of Bioorganic Chemistry and Biochemistry
The Rockefeller University
1230 York Avenue, New York, New York 10021

Introduction

Serum thymic factor (STF), pyroGlu-Ala-Lys-Ser-Gln-Gly-Gly-Ser-Asn, is a thymic hormone initially isolated from porcine serum.[1] The biological activity of the synthetic hormone is secondary to the binding of Zn(II) ions.[2] It is involved in T-cell differentiation and T-cell activity, and is implicated in a regulatory role in antibody production.[3] There is evidence the biological activity of the STF-Zn(II) complex (thymulin) is dependent on the positive charge of Lys^3 in the sequence.[4] Additionally, Asn^9 may also be necessary for activity.[5] In an effort to define the conformation-activity relationships of STF and thymulin, this laboratory is investigating their solution conformations by NMR spectroscopy.

Results and Discussion

The proton NMR chemical shift assignments of STF in D_2O, pH=7.5 are listed in Table I. These assignments are based on typical chemical shifts and coupling patterns of the amino acids, together with 2D J resolved and COSY spectra. The assignments of Ser α and β to Ser^4 or Ser^8 and of Gly α to Gly^6 or Gly^7 remains ambiguous. The chemical shifts listed in

Table I. STF Proton Chemical Shifts

Fig. 1 Label	Proton	δ (ppm)[a]	Fig. 1 Label	Proton	δ (ppm)[a]
A	Ser α	4.51	K	Asn $\beta 1$	2.78
B	Asn α	4.49		Asn $\beta 2$	2.66
C	Ser α	4.44	L	<Glu $\beta 1$	2.54
D	Gln α	4.40	M	<Glu γ	2.42
E	<Glu α	4.36	N	Gln γ	2.39
F	Lys α	4.34	O	Gln $\beta 1$	2.16
G	Ala α	4.32	P	<Glu $\beta 2$	2.05
H	Gly α	4.02	Q	Gln $\beta 2$	2.02
	Gly α	4.00	R	Lys $\beta 1$	1.85
I	Ser β	3.87	S	Lys $\beta 2$	1.77
	Ser β	3.86	T	Lys δ	1.70
J	Lys ϵ	3.00	U	Lys γ	1.46
			V	Ala β	1.39

a Chemical shifts relative to external DSS/D_2O, pH=7.5 at 25°C. β_1 and β_2 are downfield and upfield multiplets respectively.

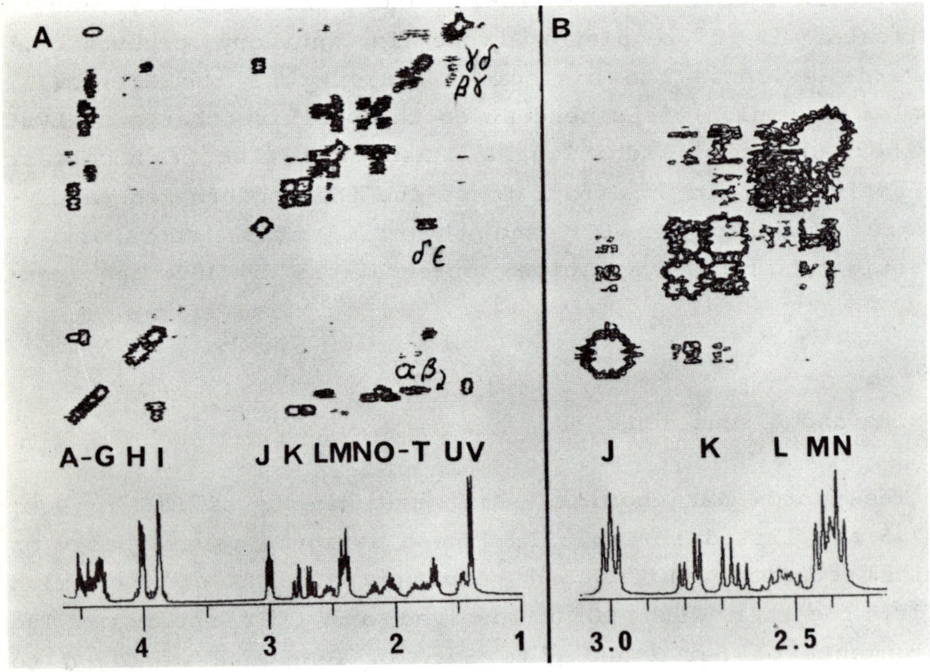

Fig. 1. (A) COSY and 1D and (B) NOE and 1D (2.25-3.08 ppm) NMR Spectra of STF/D_2O pH=7.5 at 25°C.

Table I are determined from the projection onto the chemical shift axis of the 2D J resolved spectrum (not shown) after application of a 45° tilt. The contour plot of the COSY spectrum of 10 mM STF/D_2O, pH=7.5 is shown in Figure 1A. The off diagonal peaks are due to J couplings between nuclei resulting in the spin multiplets in the 1D NMR spectrum in the figure. For example, the contour peaks indicating J coupling between the various protons of Lys^3 have been identified.

The 2.25 to 3.08 ppm region of the NOESY spectrum of STF/D_2O, pH=7.5 is presented in Figure 1B. The spectrum was acquired with a mixing time, τm, of 200 ms. This τm permits observation of NOE between nuclei < 5 Å apart.[6] To insure against the possibility of spin diffusion peaks, it was also acquired with τm=150 ms. The spectrum was essentially unchanged in this region. Additionally, the full NOESY spectrum (not shown) reveals only specific off diagonal contour peaks indicating there are probably no contour peaks due to spin diffusion. An examination of Figure 1B reveals distinct off diagonal contours indicating dipolar relaxation between $Asn^9\beta_1$, and $Asn^9\beta_2$, $Lys^3\epsilon$ and ($Asn^9\beta_1$, $Asn^9\beta_2$), $Asn^9\beta_1$ and (<$Glu^1\gamma$ $Gln^5\gamma$), and $Asn^9\beta_2$ and (<$Glu^1\beta$, <$Glu^1\gamma$, $Gln^5\gamma$) protons. Of course the $Asn^9\beta_1$-$Asn^9\beta_2$ NOE is expected and provides no conformational information. However, the remaining off diagonal peaks do provide relevant information. These peaks indicate that in a significantly populated family of solution conformations the $Lys^3\epsilon$ and $Asn^9\beta_1$ and $Asn^9\beta_2$ are in close proximity. Additionally, $Asn^9\beta_1$ is in close proximity to <$Glu^1\gamma$ and to $Gln^5\gamma$. Also, $Asn^9\beta_2$ must be in close proximity to <$Glu^1\beta$, <$Glu^1\gamma$, and $Gln^5\gamma$. Due to the hydrophilicity of STF it is unlikely, but it is possible these associations are due to dimerization. Inspection of a Kendrew molecular model of STF reveals that in a monomer these spatial associations are not possible in any extended random conformation. However, they are all compatible with a family of bent conformations made possible via a β or γ turn in the Gly^6-Gly^7

region. Additional stabilization of such a family of conformations may be provided by the possible electrostatic interaction between the Lys^3 NH_3^+ and Asn^9 COO^- groups.

These results provide some initial boundary conditions necessary to construct a detailed molecular model of STF. Additional investigations, now in progress, will provide more detailed conformation information about STF and thymulin.

Acknowledgements

The NMR experiments were performed on a Nicolet NT-360 spectrometer in the NMR facility in Professor E. T. Kaiser's laboratory, and has been supported by The Rockefeller University. I also wish to thank Dr. David Cowburn and Mr. Francis Picart for their valuable discussions.

References

1. Bach, J.F. and Dardenne, M. (1973) Immunology 25, 353-366.
2. Dardenne, M., Pleau, J.M., Nabarra, B., Lefrancier, P., Derrien, M., Choay, J., and Bach, J.F. (1982) Proc. Natl. Acad. Sci. USA 79, 5370-5373.
3. Lenfant, M., Millerioux, L., Blazsek, I., and Duchange, N. (1983) Immunology 48, 635-645.
4. Gyotoku, J., Imaizumi, A., Terada, S., and Kimoto, E. (1983) Int. J. Peptide Protein Res. 21, 135-144.
5. Bach, J.F., Bach, M.A., Blout, D., Bricas, E., Charreire, J., Dardenne, M., Fournier, C., and Pleau, J.M. (1978) Bull. Inst. Pasteur 76, 325-398.
6. Dubs, A., Wagner, G., and Wuthrich, K. (1979) Biochim. Biophys. Acta, 577, 177-194.

STRUCTURAL REQUIREMENTS FOR THE BIOLOGICAL ACTIVITY OF THYMOPENTIN ANALOGS

George A. Heavner, Tapan Audhya, Daniel Kroon and Gideon Goldstein

Immunobiology Division, Ortho Pharmaceutical Corporation
Raritan, New Jersey 08869-0602

Introduction

Thymopentin, Arg-Lys-Asp-Val-Tyr, the peptide corresponding to amino acids 32-36 of thymopoietin, the 49 amino acid polypeptide hormone of the thymus, has been shown to reproduce the biological effect of the native hormone in early T cell differentiation and normalization of immune dysbalanced animals, where dysbalance is induced by adult thymectomy, age involution or various experimental procedures[1-5]. Various animal models as well as clinical efficacy in immune dependent disease states, indicate that thymopentin exerts its homeostatic immunonormalizing activity by direct action on mature T cells[6-7]. In order to gain insight into the requirement of each amino acid in thymopentin for activity, 29 analogs and a control peptide were synthesized and tested for binding to a receptor isolated from a human T cell line and for their ability to induce elevation of intracellular c-GMP in the same cell line.

Results and Discussion

Peptides were synthesized by appropriate solid phase or solution techniques and purified by ion exchange chromatography or preparative reverse phase HPLC. Homogeneity was checked by HPLC and TLC and characterization was by enzymatic hydrolysis, amino acid analysis, and by chiroptical and spectroscopic techniques. The radioreceptor and cyclic nucleotide induction assays were run as previously described[8,9]. The results of the assays for thymopentin and the thirty peptides are shown in Table I. For all analogs giving significant displacement in the radioreceptor assay, positive enhancement of intracellular c-GMP levels was also observed, suggesting that binding to the receptor is a sufficient as well as necessary conditions for biological activity.

As can be seen from the data in Table I, substitution in positions 1 and 3 of thymopentin are most constrained with modifications in stereochemistry (D-Arg^1 or D-Asp^3) but not functionality or side chain length being accepted. At position 2, in addition to D-Lys, both the L-Pro and Aib analogs were active. Positions 4 and 5 were the least stringent, accepting a variety of substitutions, some of which gave analogs more potent than thymopentin in both assays. There appears to be no clear correlation between the potency in the receptor assay and the minimum dose necessary to induce intracellular c-GMP elevation.

Conclusions

For the human phenotypic cell line CEM, there is an absolute requirement for the arginine and aspartic acid functionality but not stereochemistry in thymopentin in positions 1 and 3 respectively. Various other substitutions are accepted in the other positions. While there is no requirement for side chains functionality at any of the remaining positions as illustrated by the Pro^2, Ala^4 or Val^5 analogs, the inactivity of the des-Tyr^5, Ala^5, 3-OH-Tyr^5 and the lower activity of the His^5 suggest a pocket in the CEM thymopoietin receptor for the c-terminal portion of thymopentin, requiring a molecule of minimum size and hydrophobicity but accepting variations in stereochemistry. Although these studies have not identified a compound with an antagonisic profile (positive receptor interactions but negative

Table I. Potency in Thymopoietin Radioreceptor Assay and c-GMP Assay in CEM Cells

Structure	c-GMP Assay Threshold Activity î (μg/ml)	Thymopoietin Radioreceptor Assay, Percent Potency*
Thymopoietin	0.01 (2)	
Arg---Lys---Asp---Val---Tyr	1 (17), 10 (6)	95 (5)
Lys---Asp---Val---Tyr	NR+ (2)	2 (2)
D-Arg---Lys---Asp---Val---Tyr	10 (2)	4 (2)
Ala---Lys---Asp---Val---Tyr	NR (2)	2 (2)
Lys---Lys---Asp---Val---Tyr	NR (2)	2 (2)
HomoArg---Lys---Asp---Val---Tyr	NR (2)	2 (2)
His---Lys---Asp---Val---Tyr	NR (2)	2 (2)
Arg---Pro---Asp---Val---Tyr	10 (2)	323 (2)
Arg-D-Lys---Asp---Val---Tyr	10 (2)	17 (2)
Arg---Ala---Asp---Val---Tyr	NR (2)	2 (2)
Arg---Aib---Asp---Val---Tyr	1 (2)	56 (2)
Arg---Arg---Asp---Val---Tyr	NR (2)	2 (2)
Arg---Lys-D-Asp---Val---Tyr	1 (2)	36 (2)
Arg---Lys---Asn---Val---Tyr	NR (2)	2 (2)
Arg---Lys-β-Asp---Val---Tyr	NR (2)	2 (2)
Arg---Lys---Glu---Val---Tyr	NR (3)	2 (2)
Arg---Lys---Ala---Val---Tyr	NR (2)	2 (2)
Arg---Lys---Asp-D-Val---Tyr	1 (2)	155 (2)
Arg---Lys---Asp---Ala---Tyr	1 (3)	210 (2)
Arg---Lys---Asp---Sar---Tyr	10 (2)	370 (2)
Arg---Lys---Asp---Val	NR (2)	2 (2)
Arg---Lys---Asp---Val---Ala	NR (2)	2 (2)
Arg---Lys---Asp---Val---Val	1 (2)	200 (2)
Arg---Lys---Asp---Val---Phe	0.1 (2)	220 (3)
Arg---Lys---Asp---Val-D-Tyr	1 (2)	140 (2)
Arg---Lys---Asp---Val-3-NO_2-Tyr	1 (2)	300 (2)
Arg---Lys---Asp---Val-3-Cl-Tyr	1 (2)	191 (2)
Arg---Lys---Asp---Val-3-OH-Tyr	NR (2)	2 (2)
Arg---Lys---Asp---Val---His	100 (2)	7 (3)
Arg---Lys---Asp---Val---Trp	1 (3)	50 (2)
Asp---Arg---Tyr---Lys---Val	NR (2)	2 (3)

î lowest active concentration (number of determinations)
* percent potency relative to thymopoietin (number of determinations)
\+ NR = No response

cyclic nucleotide induction) the existence of such analogs cannot be definitely excluded.

The information gained from this study is a prerequisite to the design of more complex modification aimed at producing more potent and stable analogs of thymopentin and of conformationally restricted analogs to probe the thymopentin conformation at the receptor. These latter studies may determine whether the predominate Arg-Asp hydrogen bond deduced from NMR solution conformations[10] is responsible for stabilizing the biologically active conformation or is a solution artifact.

References

1. Goldstein, G., Scheid, M.P., Boyse, E.A., Schlesinger, D.H. and Van Wauwe, J. (1979) Science **204**, 1309-1310.

2. Goldberg, E.H., Goldstein, G., Boyse, E.A. and Scheid, M.P. (1981) Immunogenet. **13**, 201-204.

3. Weksler, M.E., Innes, J.B. and Goldstein, G. (1978) J. Exp. Med. **148**, 996-1006.

4. Lau, C.Y., Freestone, J.A. and Goldstein, G. (1980) J. Immunol. **125**, 1634-1638.

5. Goldberg, E.H., Goldstein, G., Harman, D.B. and Boyse, E.A. (1984) Transplantation **38**, 52-55.

6. Malaise, M.G., Franchimont, P., Bach-Andersen, R., Gerber, H., Stocker, H., Hauwaert, C., Danneskiold-Samsol, B., Gross, D., Gerschpacher, H. and Bolla, K. (1985) Lancet **1**, 832-836.

7. Aiuti, F., Fiorilli, M., Quinti, I., Seminara, R., Businco, L., Galli, E., Rossi, P. and Goldstein, G. (1983) Lancet **1**, 551-555.

8. Audhya, T., Talle, M.A. and Goldstein G. (1984) Archiv. Biochem. Biophys. **234**, 167-177.

9. Audhya, T., Heavner, G.A., Kroon, D.J. and Goldstein, G. (1984) Regul. Pept. **9**, 155-164.

10. Krishna, N.R., Huang, D-H., Chen, D.M. and Goldstein, G. (1980) Biochemistry **19**, 5557-5563.

CONFORMATIONAL ANALYSIS OF CONOTOXIN AND ITS ANALOGUE BY NMR MEASUREMENTS AND DISTANCE GEOMETRY ALGORITHM

Yuji Kobayashi, Tadayasu Ohkubo and Yoshimasa Kyogoku
Institute for Protein Research, Osaka University
Suita, Osaka, 565, Japan

Yuji Nishiuchi and Shumpei Sakakibara
Peptide Institute, Protein Research Foundation
Ina, Minoh, Osaka, 562, Japan

Werner Braun and Nobuhiro Go
Department of Physics, Faculty of Science, Kyushu University
Hakozaki, Fukuoka, 812, Japan

Introduction

We are developing a new technique to determine the conformation of polypeptide in solution. The strategy involves so-called "distance-geometry algorithm" which is a mathematical treatment to obtain a three dimensional representation of molecular structure from the information of interatomic distances. The algorithm is used to require a long calculation time and a large memory space of computor. We have resolved this problem by (i) assuming fixed standard bond lengths and bond angles and

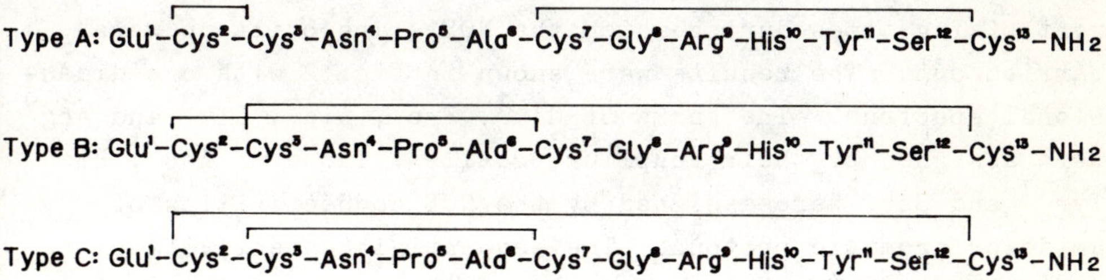

Fig. 1 Amino acid sequence of conotoxin GI

treating only dihedral angles as independent variables, and (ii) determining and minimizing a proper target function which consists of a sum of differences between an experimentally observed interatomic distance and a distance between the same pairs of atoms obtained in a computor generated conformation. Two dimensional nuclear Overhauser enhancement spectroscopy (NOESY) provides experimentally the informations of distances. The solution conformations of conotoxin GI and its analogue of which amino acid sequences are given in Fig. 1 are elucidated by the application of this method.

Experimentals

Materials; Conotoxin GI is a peptide neurotoxin contained in the venom of the marine snail *Conus geographus* which causes postsynaptic inhibition at the vertebrate neuromuscular junction and displays as high a potency as that of tetrodotoxin. Three analogues with different types of disulfide bonds were synthesized and found that Type B among them shows the highest toxicity comparable to that of the native conotoxin GI. Types A and C show one tenth or less toxicities. Types A and B dissolved in deuterated dymethyl sulfoxide were used here as samples.

NMR measurements; The proton NMR measurements were carried out at 20°C by Jeol GX-500 spectrometer performing at 500 MHz. Two dimensional shift correlation spectroscopy (COSY) and nuclear Overhauser enhancement spectroscopy (NOESY) with three different mixing times during which the NOE's are built up were carried out. The results were shown in Fig. 2 with one dimensional spectrum. The peaks of Glu^1, Pro^5, Ala^6, Gly^8 and Arg^9 were assigned by their coupling profiles. Those of Asn^4, His^{10} Tyr^{11} and Ser^{12} were assigned by the NOE connectivities of amide or aromatic protons. The sequential connectivities along the back bone of the peptide chain; spin coupling Between N^iH and $C^i_\alpha H$ and NOE connectivity between $C^i_\alpha H$ and $N^{i+1}H$

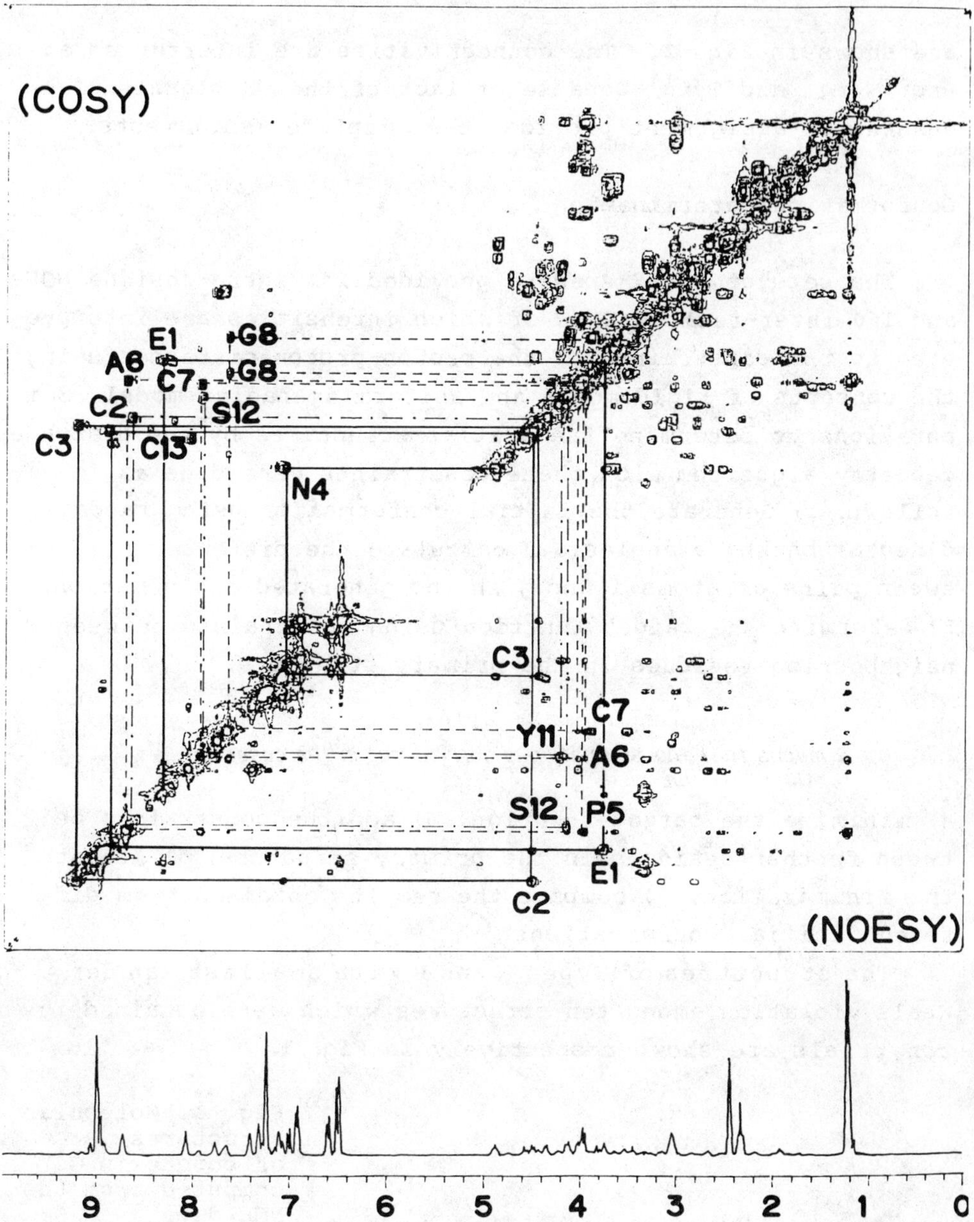

Fig. 2. Two dimensional spectra of conotoxin GI in DMSO-d_6. In this contour plot the upper triangle is COSY the lower triangles is NOESY with a mixing time of 160 ms. COSY-NOESY connectivity diagram for sequential resonance assignment is shown. Cross peaks are identified by the one-letter symbol and the location of the residue in the sequence.

are shown in Fig. 2. The connectivities are interrupted at Pro^5, Arg^9 and Tyr^{11} because of lack of the NH signals. The sequential assignment provided the complete assignments.

Conformation Determination

The sets of NOESY spectra provided 152 intra-residue NOEs and 160 inter-residue NOEs of which intensities are interpre-eted to the constraints of the proton-proton distances using the concepts of rigid model and uniform averaging model. Computations to determine the spatial structures by the distance geometry algorithm with these constraints were done as follows. 1) Generate the initial conformation using random dihedral backbone angles, 2) calculate the distances r_{ij} between pairs of atoms i and j in the generated conformation, 3) determine the target function using constraints between neighbouring residues in the primary structure,

$$T = \sum_{i<j}{}' (r_{ij} - U_{ij})^2 + \sum_{i<j}{}' (L_{ij} - r_{ij})^2$$

Σ' *means that the summation is carried out in the cases of* $r_{ij} > U_{ij}$ *or* $r_{ij} < L_{ij}$.

4) minimize the target function, 5) add the constraints between further residues in the primary structure, 6) iterate the minimization, 7) compare the results obtained from different initial conformations.

The structures of Types A and B with smallest van der Waals violation among ten structures which were obtained in ten trials are shown respectively in Fig. 3.

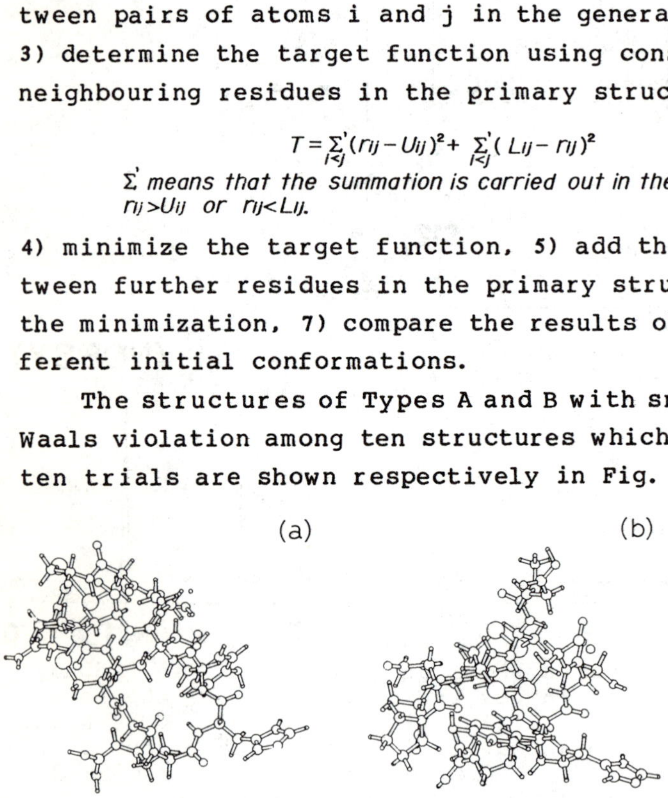

Fig. 3. Molecular structures of conotoxin GI computed from the NMR data
(a) Type A
(b) Type B

THE STRUCTURE OF ARIDICIN A. AN INTEGRATED APPROACH EMPLOYING 2D NMR, ENERGY MINIMIZATION AND DISTANCE GEOMETRY CONSTRAINTS FOR THE DERIVATION OF 3-DIMENSIONAL STRUCTURES IN SOLUTION

Peter W. Jeffs, Luciano Müller, Charles W. DeBrosse, Sarah L. Heald and Richard Fisher

Smith Kline and French Laboratories, 1500 Spring Garden Street, Philadelphia, Pennsylvania 19101.

Introduction

In the present study we describe the elucidation of the full 3-D structure of the glycopeptide antibiotic aridicin A and its aglycone. This glycopeptide, as part of a complex consisting of the aridicin A, B and C from Kibdelosporangium aridium, has been reported recently from our laboratory.[1]

We based our structure elucidation on an interactive approach using estimates of interatomic distances and torsion angles derived primarily from two-dimensional NMR spectra in conjunction with computer assisted modeling to provide a complete 3-D structure of aridicin A. Structural studies of peptides using NMR data has been done successfully in the past. However, these studies were facilitated by the knowledge of structure and stereochemistry of the amino acid residues. The building blocks of new glycopeptides usually are unknown quantities.

The initial goal was to explore the application of 2D NMR methods to derive the primary structure of the aglycone in conjunction with mass spectrometric data. Secondly, we investigated the application of distance geometry constraints obtained from 2D-NOE spectra to derive both configurational and conformational features of the aglycone in conjunction with computer assisted modeling. Following this we proceeded with the elucidation of the 3-D-structure of the intact glycopeptide, Aridicin A.

Aglycone:

First, the primary structure (as depicted in Figure 1a) was elucidated by applying various NMR techniques as well as mass spectrometry to the aglycone and its degradation products. Our analysis heavily depends on the information extracted from proton NMR experiments. The delayed COSY experiment (extra delays are inserted before and after the mixing rf pulse to magnify effect of small long range J-couplings) turned out to be a great value, since it allowed to establish J-connectivity maps between all protons of a given amino acid residue (including connectivities between aliphatic- and aromatic protons). Furthermore, the delayed COSY spectra revealed some unusual J-connectivities between protons that are located on adjacent amino acid residues, e.g. between the protons ANH and E1' across the cis-peptide bond (Figure 1a,b). The 3-D structure of the aglycone was elucidated by building molecular models on our SKF Molecular Modeling System (Evans and Sutherland PS330 graphics system linked to a VAX 11-785) which then were energy-minimized with distance constraints derived from 2D NOE data. Structures were built by systematically modifying all chiral centers with the exception of E1' which was arbitrarily assumed to be R. The so-derived structure is in good agreement with the predicted NH-CH bond angles derived from ^{1}H coupling constants. Also all computed distances agree with our NOE data.

Intact Glycopeptide:

It appeared that the two sugar residues (D-mannose, N-acyl-2-amino-2-deoxyglucuronic acid) do not affect the structure of the peptide core. This was easily verified from the highly resolved ($\Delta\nu$ = 1 Hz) proton spectra recorded in DMSO-d_6/D(H)$_2$O mixtures. The delayed COSY spectra again provided a wealth of cross-peaks, in particular a cross-peak was found between the anomeric proton of mannose and the aromatic proton D4 indicating a long range J-coupling across the glycosidic linkage. This finding and 2D NOE data clearly indicated the site of attachment of D-mannose. The site of

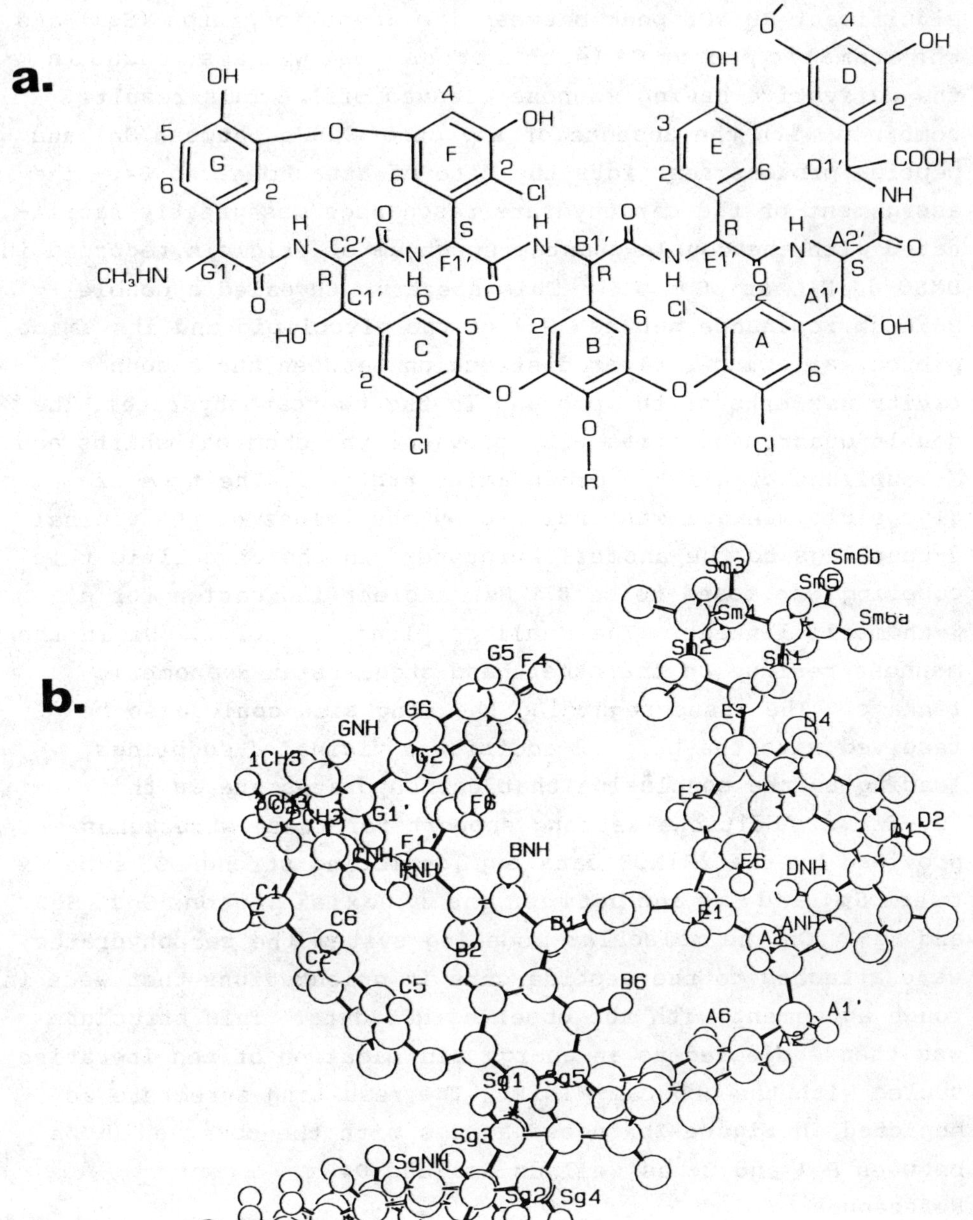
a.
b.
OH
CH₃HN
G1′
C2′
F1′
B1′
E1′
A2′
D1′
COOH
NH
C1′
A1′
HO
OH
Cl
O
R
R′
S
H
G
F
E
D
C
B
A
BNH
DNH
ANH
GNH
1CH3
2CH3
Sm6b
Sm5
Sm3
Sm4
Sm2
Sm1
Sg1
Sg5
Sg3
SgNH
Sg2
Sg4

attachment of the glycolipid is pointed out by a small but significant 2D NOE peak between its anomeric proton (Sg1) and the aromatic proton C5 (such a cross-peak was also found in the derivative having mannose cleaved off). This result combined with the absence of any other NOE's between Sg1 and peptide protons restricts the site of attachment to B4. The assignment of the carbohydrate resonances was greatly facilitated with the double quantum spectrum of aridicin recorded in DMSO-d_6/H_2O at pH = 5.6. This spectrum revealed a double quantum resonance between Sg2 of the glycolipid and its amide proton, allowing a clear distinction between the J-connectivity networks of the protons in the two carbohydrates. The double quantum spectrum also provided the chemical shifts and J-couplings of all the other amide protons. The type of glycosidic linkage was indicatd by the values of the vicinal J-couplings to the anomeric protons. In the glycolipid this coupling was found to be 8.5 Hz, a clear indication for a β-anomeric linkage. The small coupling $J_{1,2}$ of 1.5 Hz in the mannose residue on the other hand suggests an α-anomeric linkage. The issue regarding the ring size could also be resolved with the help of additional vicinal J-couplings leading to the conclusion that carbohydrates are in the $^4C_1$-form. Additional strong support for these structures were provided by the 2D NOE data e.g. depicting strong NOE's between Sm1 and Sm2 and between the syn axial protons Sg1, Sg3 and Sg5. On the molecular modeling system the carbohydrates were attached to the peptide core in orientations that were in rough agreement with the observed NOE data. This structure was then subjected to an energy minimization of ten iterative cycles with the program MINIM. The resulting structure as depicted in Figure 1b indeed agrees with the observed NOE's between Sg1 and C5 as well as Sm1 and D4.

References:

1. Sitrin, R.D., Chan, G., Digerdissen, J., Holl, W., Hoover, J.R.E., Valenta, J., Webb, L. Snader, K.M., Antibiot. 1985, 0000.

DETECTION AND SEQUENTIAL CHARACTERIZATION OF AIB-CONTAINING POLYPEPTIDES ("PEPTAIBOLS") AND UTILIZATION OF THEIR NATURAL MICROHETEROGENEITY FOR AN ASSIGNMENT OF ^{13}C NMR SIGNALS

Hans Brückner
Institute of Food Technology, University of Hohenheim,
D-7000 Stuttgart (F.R.G.)

Michael Bokel
Institute of Organic Chemistry, University of Hohenheim,
D-7000 Stuttgart (F.R.G.)

Michael Przybylski
Institute of Organic Chemistry, University of Mainz,
D-6500 Mainz (F.R.G.)

The production of paracelsin[1] (PC) as a secondary metabolite of the mold *Trichoderma reesei* - a cellulolytically highly active supermutant of great biotechnological importance for the saccharification of waste cellulose - caused us to develop methods allowing the rapid detection, isolation and structural characterization of any of these polypeptide (peptaibol) mycotoxins[2]. These procedures are exemplified with alamethicin (AM) F-50. In analogy to PC and gliodeliquescin[2] (GD), AM F-50 is easily detected and quantitatively determined in the culture broth of fermentations by employing octadecylsilyl bonded silica (e.g. Sep-Pak® cartridges) and using α-aminoisobutyric acid (Aib) and amino alcohols (e.g. phenylalaninol, Pheol) as specific marker constituents[2]. Medium pressure (MP) liquid chromatography (LC) using XAD-2 or 7 adsorber resins (typically 4.5 x 35 cm columns for about 10 l culture broth) followed by silica gel MPLC and/or Sephadex LH-20 LC are used for preparative peptaibol isolations. However, HPLC is indispensible in obtaining uniform components from very microheterogeneous peptaibol mixtures (Figure 1).

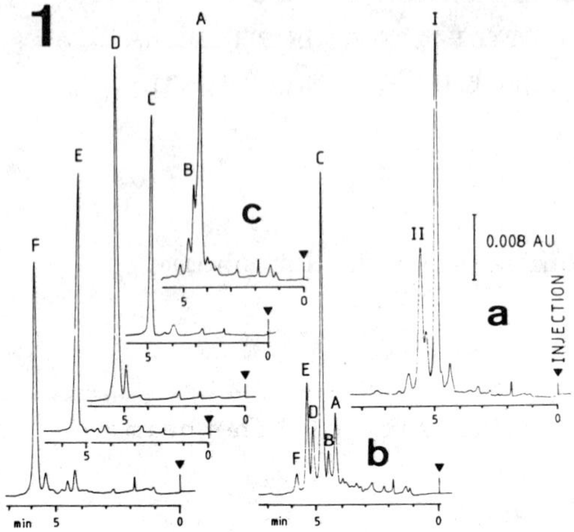

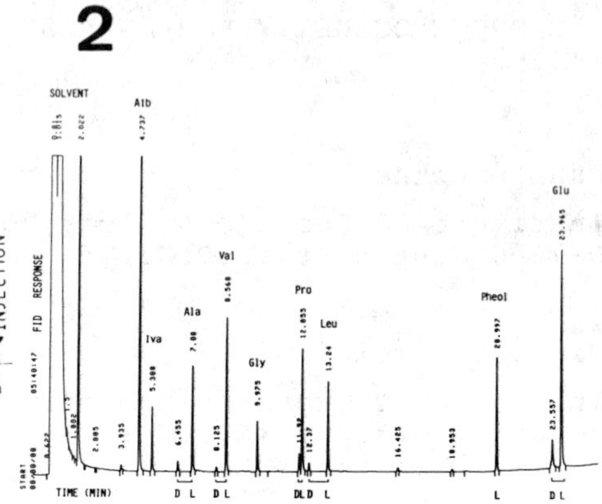

Fig. 1. HPLC of AM F-30 (The Upjohn Company) (a), AM F-50 (b) and isolated components A-F (c) used for sequencing. Column, 250 mm x 4.6 mm; nucleosil C-18, 3 μm, eluent, 85% methanol; flow rate, 0.8 ml/min; absorbance, 200 nm.

Fig. 2. GLC of N-pentafluoropropionyl amino acid n-propyl esters of AM F-50/D. Conditions: glass capillary, 0.25 mm i.d. x 25 m length coated with Chirasil-Val; 3 min 80°C, 4°C/min (80-210°C); carrier gas, hydrogen; p, 50 kPa; split ratio 1:45; temperature injector, 250°C.

In this way AM F-50 (as predominately obtained by our fermentation conditions) could be separated into components A-F. Hereby GLC revealed the presence of isovaline (Iva) in components D and F (Fig. 2). Fast atom bombardment mass spectrometry (FABMS) allowed the sequence determination of components C-E, proofing the identity of C and E to AM F-30 I and II, with the restriction of F-50 having Gln in position 18 (Fig. 3). Most abundant sequence specific fragment and molecular mass ions could be obtained by using tetraethylene glycol (TEG) as liquid matrix and performing a selective trifluoroacetolytic procedure[3,4] (Fig. 4). Furthermore, repetitive HPLC[2] enabled the isolation of up to 100 mg amounts of homogeneous peptaibols to compare them by ^{13}C NMR spectroscopy, thus allowing an unambiguous and direct assignment of resonances of amino acids replaced by others (Figs. 3, 5).

```
        1  [2]  3  [4]  5  [6]  7   8  [9]  10  11 [12]  13 [14] [15]  16 [17]  18  19  20
(a) Ac-Aib-Pro-Aib-Ala-Aib-Ala-Gln-Aib-Val-Aib-Gly-Leu-Aib-Pro-Val-Aib-Aib-Gln-Gln-Pheol (C)
    Ac-Aib-Pro-Aib-Ala-Aib-Ala-Gln-Aib-Val-Aib-Gly-Leu-Aib-Pro-Val-Aib-Iva-Gln-Gln-Pheol (D)
    Ac-Aib-Pro-Aib-Ala-Aib-Aib-Gln-Aib-Val-Aib-Gly-Leu-Aib-Pro-Val-Aib-Aib-Gln-Gln-Pheol (E)

(b) Ac-Aib-Ala-Aib-Ala-Aib-Ala-Gln-Aib-Val-Aib-Gly-Aib-Aib-Pro-Val-Aib-Aib-Gln-Gln-Pheol (A)
    Ac-Aib-Ala-Aib-Ala-Aib-Ala-Gln-Aib-Leu-Aib-Gly-Aib-Aib-Pro-Val-Aib-Aib-Gln-Gln-Pheol (B)
C* {Ac-Aib-Ala-Aib-Ala-Aib-Aib-Gln-Aib-Val-Aib-Gly-Aib-Aib-Pro-Val-Aib-Aib-Gln-Gln-Pheol (C)
   {Ac-Aib-Ala-Aib-Ala-Aib-Aib-Gln-Aib-Leu-Aib-Gly-Aib-Aib-Pro-Val-Aib-Aib-Gln-Gln-Pheol (D)

(c) Ac-Aib-Ala-Aib-Ala-Aib-Ala-Gln-Aib-Val-Aib-Gly-Leu-Aib-Pro-Val-Aib-Aib-Gln-Gln-Pheol (A)
```

Fig. 3. Sequences of alamethicin F-50/C-E (a) compared to paracelsin A-D (b) and gliodeliquescin A (c). Positions assignable by comparative NMR spectroscopy are framed.

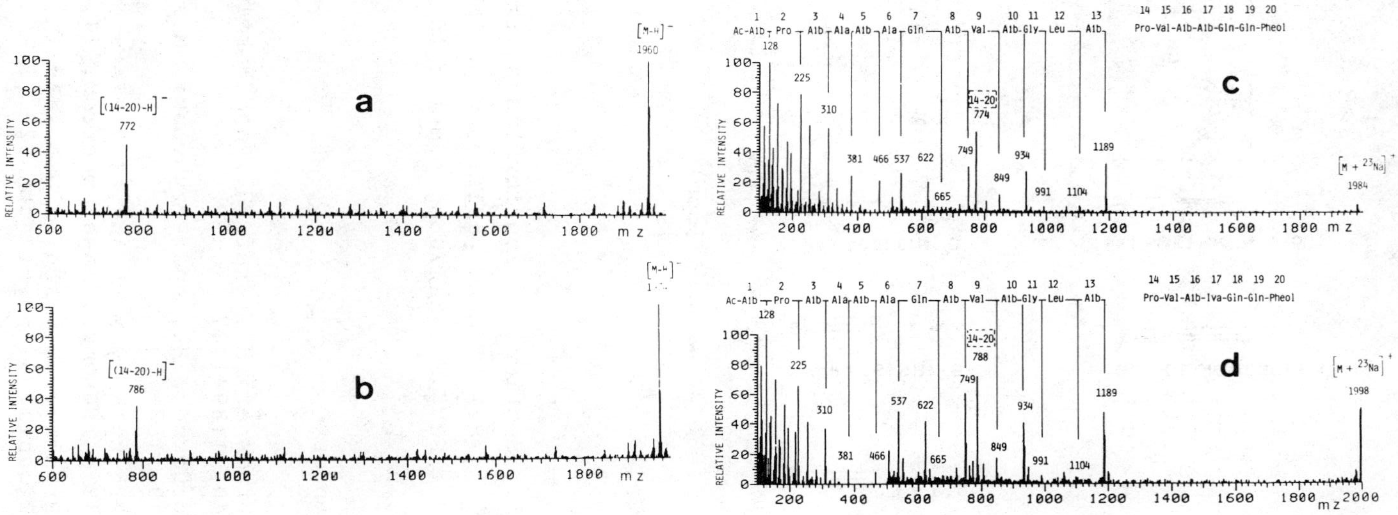

Fig. 4. Negative ion (a), (b) and positive ion (c), (d) FABMS of AM F-50 components C (a), (c) and D (b), (d) taken in TEG. For details of recording conditions see lit.[2,3].

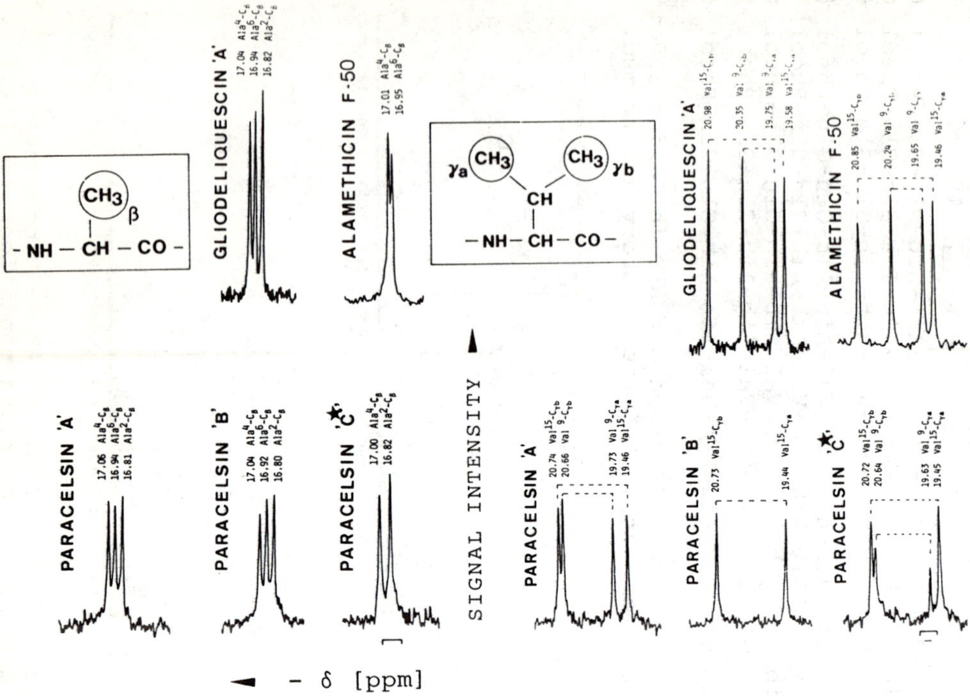

Fig. 5. Sections of Fourier transform ^{13}C NMR spectra (62.89 MHz) of HPLC purified peptaibols GD A, AM F-50/C and PC C★. Ca. 70 mg peptide in 0.5 ml ^{12}C,^{2}H-methanol.

References

1. Brückner, H., Graf, H., and Bokel, M. (1984) Experientia 40, 1189-1197.

2. Brückner, H., and Przybylski, M. (1984) Chromatographia 19, 188-199.

3. Przybylski, M., Manz, I., Dietrich, I., and Brückner, H. (1984) Biomed. Mass Spectrom. 11, 569-582.

4. Brückner, H., König, W.A., Aydin, M., and Jung, G. (1985) Biochim. Biophys. Acta 827, 51-62.

CONFORMATION OF THE PROLINE-RICH FRAGMENT PEPTIDE 97-104 OF MYELIN BASIC PROTEIN

Paul E. Fraser and Charles M. Deber
Research Institute, Hospital for Sick Children, Toronto, M5G 1X8, and Department of Biochemistry, University of Toronto, Toronto, M5S 1A8, Ontario, Canada

Myelin basic protein (MBP) is a highly charged (19 arginine; 13 lysine), 170-residue extrinsic membrane protein found in the oligodendroglial cells of central nervous system myelin (Carnegie and Moore, 1980). The ability of MBP to bind negative lipids may provide the cohesive link necessary for the compaction and maintenance of the multi-bilayer myelin structure. These functional considerations have prompted speculation of MBP's involvement in such demyelinating diseases as multiple sclerosis. As the details of its membrane interactions are being defined on the molecular level (Fraser and Deber, 1984 and references therein) interest has heightened in determining the biologically-functional conformation of membrane-bound MBP.

Comparison of MBP primary sequences from several species reveals a conserved sequence of three consecutive prolines (residues 99-101) which could have important conformational consequences by inducing a reversal in protein chain direction, as suggested earlier by Eylar and Thompson, (1969). Two possibilities by which such a structure could arise involve either a cis X-Pro peptide bond or a β-turn conformation (Martenson, 1981). We have been evaluating these possibilities using ^{13}C and ^{1}H nuclear magnetic resonance to examine the properties of a synthetic octapeptide, Arg_1-Thr_2-Pro_3-Pro_4-Pro_5-Ser_6-Gln_7-Gly_8, corresponding to residues 97-104 of intact MBP, and containing the critical tri-proline region (Fraser and Deber, 1985). Conceptually-analogous studies have been carried out on a 22-residue peptic peptide (91-112) of rabbit MBP (Nygaard et al., 1984).

RESULTS AND DISCUSSION

Cis and *trans* forms of X-Pro resonances are readily discerned in ^{13}C NMR spectra, with Pro-γ-carbons being the most diagnostic (Deber et al., 1976). The sequential nature of our synthesis (Arg + Thr + Pro-Pro-Pro + Ser-Gln-Gly) allowed for unequivocal peak assignments through stepwise comparisons of spectra. Incorporation of Pro-Pro-Pro into the hexapeptide H-Pro-Pro-Pro-Ser-Gln-Gly-OH, first resulted in a marked reduction in *cis* peptide bonds from 45% in the Pro-Pro-Pro tripeptide to 6% in the hexapeptide). Further elongation through addition of the Thr and Arg components produced hepta- and octapeptide MBP fragments for which no *cis* resonances could be detected in aqueous solution. Since the residues of greatest interest - the three prolines - retained their essentially all-*trans* conformational integrity through three chain elongation steps, the tri-proline region was considered to have become structurally stable (in effect, as a turn of a poly-pro II-type helix). The octapeptide 97-104 was thus judged to be of sufficient length to allow valid conformational comparisons with the corresponding region in intact MBP. Natural abundance ^{13}C NMR spectra (125 MHz) of bovine myelin basic protein in aqueous solution, studied in conjunction with the correspondingly simpler spectra of the octapeptide fragment 97-104, then demonstrated that the vast majority of *all* proline residues in MBP (which contains 12 proline residues) were also in the *trans* orientation (Fraser and Deber, 1985).

This lack of *cis* X-Pro bonds did not, however, exclude the possibility of a 'reverse turn' conformation in the tri-proline region, as a bend could arise through a β-turn involving the Pro residues (e.g., Pro-Pro-Ser-Gln). To

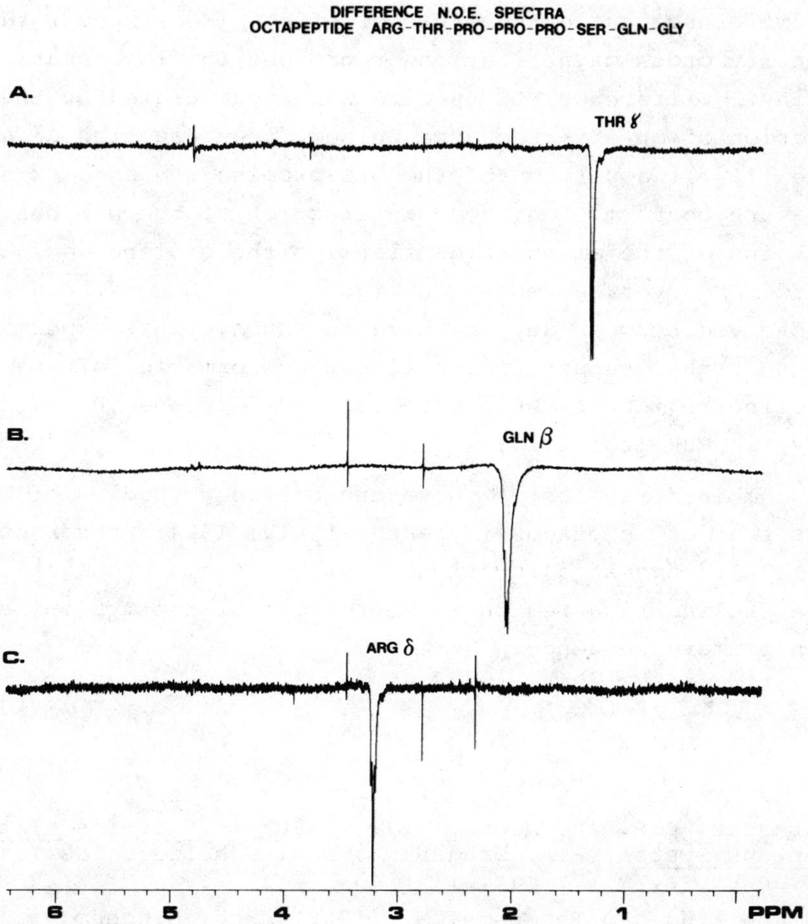

Fig. 1. Difference NOE spectra of the synthetic octapeptide Arg-Thr-Pro-Pro-Pro-Ser-Gln-Gly (D_2O; pH 6) were obtained for (A) Thr-γ; (B) Gln-β; and (C) Arg-δ on a Nicolet NT-360 spectrometer using preprogrammed Nicolet software. Spectra were accumulated in 512 scans with a sweep width of 2400 Hz and a pulse width of 5.0 usec. NOE's were generated with a 43 dB pulse at the indicated specific frequencies. See text for further details.

investigate this possibility, evidence of intramolecular association for the amino acid side chains surrounding the Pro-rich region in octapeptide fragment 97-104 was sought using difference nuclear Overhauser effect (NOE) spectroscopy to identify cross magnetization of protons in close spatial proximity. Difference NOE spectra were accumulated by the subtraction of on/off-resonance pulsed free induction decays (Figure 1). If a portion of the tri-proline sequence exists in a β-turn conformation, some interaction of the protons in side chains of the amino acids flanking the proline would be expected. However, as shown in Figure 1, no NOE correlations were observed between any of these residues. This experiment provides further support for a linear conformation of the tri-proline region of MBP octapeptide 97-104, and, by analogy, in MBP itself.

The combined results of these and previous studies of MBP and its fragment peptides (Nygaard et al., 1984; Fraser and Deber, 1985) suggest a functional role of chain propagation for the proline-rich region in the middle of the myelin basic protein primary sequence.

REFERENCES

1. Carnegie, P.R. and Moore, W.J. (1980) in Proteins of the Nervous System (R.A. Bradshaw & D.M. Scheider, Eds.), Raven Press, N.Y., pgs. 119-143.
2. Eylar, E.H. and Thompson, M. (1969) Arch. Biochem. Biophys. 129, 468-479.
3. Fraser, P.E. and Deber, C.M. (1984) J. Biol. Chem. 259, 8689-8692.
4. Fraser, P.E. and Deber, C.M. (1985) Biochemistry, In Press.
5. Martenson, R.E. (1981) J. Neurochem. 36, 1543-1560.
6. Nygaard, E., Mendz, G.L., Moore, W.J. and Martenson, R.E. (1984) Biochemistry 23, 4003-4010.

SYNTHESIS AND CONFORMATIONAL COMPARISON OF PROLINE- AND HYDROXYPROLINE-CONTAINING PEPTIDES

V.S. Ananthanarayanan and S.K. Attah-Poku

Department of Biochemistry, Memorial University of Newfoundland, St. John's, Newfoundland, A1B 3X9, Canada

In an effort to understand the conformational aspects of the enzymatic proline hydroxylation in collagen, we have been studying the conformation of proline-containing peptides and polypeptides[1,2]. Extension of these studies to peptides containing hydroxyproline (Hyp) should be useful in delineating the conformational consequence of proline hydroxylation. It is known that the hydroxylation results in a conformational change of the polypeptide backbone and renders it thermodynamically more stable[3]. Very few studies[4] are however available on the conformation of Hyp-containing peptides. We present here the results of our studies on such peptides and compare them with those on corresponding proline-containing peptides.

Results and Discussion

The following peptides were synthesized by step-wise condensation using standard procedures[2]: Boc-Pro-Gly-OH (-OMe), Boc-Pro-Gly-X-OH, Boc-Hyp-Gly-OH and Boc-Hyp-Gly-X-OH, where X = Ala, Val or Leu. Their purity was checked by HPLC and elemental analysis. The conformational data were obtained using CD, IR and NMR.

NMR Data: ^{13}C and ^{1}H-NMR spectra of the peptides in $CDCl_3$ and DMSO-d_6 were obtained to (a) estimate the ratio of the cis and trans isomers resulting from rotation around the Boc-Pro and Boc-Hyp bonds and (b) delineate the intramolecular hydrogen bonding pattern in the peptides. A quantitative estimate of the cis/trans ratio, R, was obtained by taking the average of the intensities of the cis and trans ^{13}C signals of all the assignable carbon atoms in each compound. This was done to correct for the possible differential saturation effects arising from the variable spin-lattice relaxation times and the nuclear Overhauser enhancements. An estimate of R within $\pm$ 1% was obtained using this method. The ^{13}C data on Boc-Pro-OH indicated the presence of 63% cis isomer in $CDCl_3$ and 59% in DMSO-d_6. On going to the dipeptide, the population of the cis conformer reduced to 59% in Boc-Pro-Gly-OH and 50% in Boc-Pro-Gly-OMe in $CDCl_3$. The population of the trans isomer (in $CDCl_3$) increases considerably on going from the di- to the tripeptides, Boc-Pro-Gly-X-OH. The Hyp-containing peptides show similar trends in the cis and trans populations (Figure 1). In DMSO-d_6, the cis isomer (69%) predominates over the trans isomer (31%) in Boc-Hyp-Gly-OH. On going from the dipeptide to the tripeptide, Boc-Hyp-Gly-X-OH, there is a large increase in the trans isomer (43-48%). In pure DMSO-d_6, the tripeptides, Boc-Hyp-Gly-Ala-OH and Boc-Hyp-Gly-Leu-OH show a small but significant increase (2%) in the trans isomer when compared with their proline-containing counterparts. This trend is found to persist in solvents consisting of mixtures of 20 to 80% (v/v) $CDCl_3$ in DMSO-d_6.

In the ^{1}H NMR, by following changes in the carboxyl -OH and amide -NH chemical shifts of Boc-Pro-OH, Boc-Pro-Gly-OH and Boc-Pro-Gly-OMe we established that, in the cis isomer, the dipeptide exist in a C_5 conformation and the trans

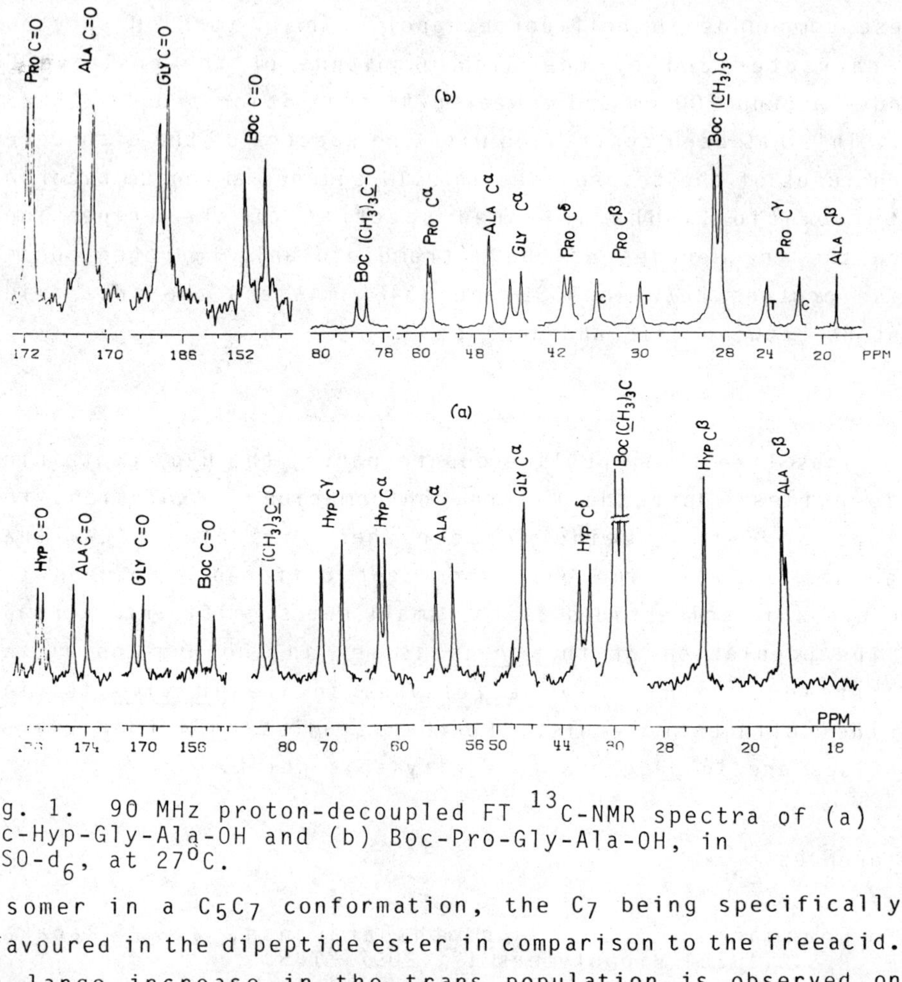

Fig. 1. 90 MHz proton-decoupled FT ^{13}C-NMR spectra of (a) Boc-Hyp-Gly-Ala-OH and (b) Boc-Pro-Gly-Ala-OH, in DMSO-d_6, at 27^oC.

isomer in a C_5C_7 conformation, the C_7 being specifically favoured in the dipeptide ester in comparison to the freeacid. A large increase in the trans population is observed on going from the dipeptides to the tripeptides in the Pro- and Hyp-containing peptides. This is attributed to the formation of a 4 → 1 intramolecular hydrogen bonding between the amide proton of the X-residue and the urethane carbonyl resulting in β-turn conformations[6], as judged from the solvent- and temperature-dependence of the chemical shifts of the amide protons and from IR and CD spectral data. The presence of the β-turn conformers in the Pro- and Hyp- tripeptides is supported by the CD spectra of these compounds in trifluoroethanol. The β-turn CD spectrum is characterized by the high magnitude of the positive CD bands[6] around 200 nm and a weak n-π* transition around 225-230 nm. This has been contrasted with the spectra of the disordered structures of the tripeptides in 0.1M aqueous ammonium bicarbonate solution. The infrared spectra[7] of the tripeptides show the frequencies of the intramolecularly hydrogen bonded amide protons between 3332 and 3348 cm^{-1} and the free amide protons between 3418 and 3434 cm^{-1}.

Thus like their proline counterparts, the Hyp- containing tripeptides adopt the β-turn conformation in solution, the extent of β-turn being greater when X is Ala or Leu than when it is Val. The only important difference between the two types of tripeptides is the small but significant increase in the population of the *trans* isomer in the Hyp-containing tripeptides. This may be relevant in the *in vivo* folding of the collagen molecule. Further studies on larger oligopeptides are in progress to verify this point.

References

1. Brahmachari, S.K., Ananthanarayanan, V.S., Rapaka, R.S. (1978) Biopolymers 17, 2097-2105.
2. Brahamacnari, S.K., Rapaka, R.S., Bhatnagar, R.S. and Ananthanarayanan (1982) Biopolymers 21, 1107-1125.
3. Chopra, R.K. and Ananthanarayanan, V.S. (1982) Proc. Natl. Acad Sci. USA 79, 7180-7184.
4. Roques, B.P., Garbay-Jaureguiberry, C., Combrisson S. and Oberlin, R. (1977) Biopolymers 16, 937-944.
5. Benedetti, V., Pavone, V., Toniolo, C., Banora, G.M. and Palumbo, M. (1977) Macromolecules 10, 1350-1356.
6. Smith, J.A. and Pease, L.G. (1980) CRC Critical Reviews in Biochemistry, 316-400.
7. Boussard, G. and Marraud, M. (1979) Biopolymers 18, 1297-1331.

PROLINE RING MOTION IN SOLID STATE: A ^2H NMR STUDY

Susanta K. Sarkar

Department of Radiology, The Johns Hopkins Medical Institutions Baltimore, MD 21205

P. E. Young

Department of Chemistry, York College, C.U.N.Y. Jamaica, NY 11451

C. E. Sullivan, and D. A. Torchia

Mineralized Tissue Research Branch, NIDR, National Institutes of Health, Bethesda, MD 20205

There is considerable evidence that the proline ring is flexible in solution. The degree of flexibility of different carbons in the ring vary according to $C^{\gamma} > C^{\beta} > C^{\delta} > C^{\alpha}$.[1] We have recently reported the presence of proline ring motion in a cyclic hexapeptide, cyclo(Gly-Pro-D-Ala)$_2$, in the crystalline powder form [2]. In order to study the details of proline ring motion we have synthesized D,L-proline labeled with deuterium at the γ position.[6] ^2H nmr spectroscopy is well-suited to study the molecular dynamics of crystalline amino acids and peptides and can be used to investigate molecular motions over a wide range of correlation times (τ_c)[3-5].

We have measured ^2H powder lineshapes and orientation dependent T_1's of D,L-[γ-$^2H_2$] proline over a wide temperature range. The correlation times for the C^{γ}-^2H bond motion were obtained using a two-site jump model to analyze the T_1 data. The activation energy is determined from the temperature dependence of τ_c and is discussed in terms of the reported crystal structure of D,L-proline.hydrochloride.

Solid state ^2H nmr spectra were measured at 6T(38.45 MHz) on a home-built spectrometer as described[7]. 76.76 MHz spectra were recorded using a home-built probe on a NTC-500 spectrometer after modification for solid state work[8]. ^2H spectra were observed using the solid echo pulse sequence [9], $90^o_{\pm x}-\tau-90^o_y-\tau$, with a 90° pulse of 2.5μs and τ=30μs. Inversion recovery spectra were obtained by applying the echo sequence at a time T after the 180° pulse ($180^o-T-90^o_{\pm x}-\tau-90^o_y-\tau$).

We have measured the inversion recovery spectra of D-L[γ-2^2H_2] proline from 49°C to -156°C, at 38.45 MHz and at 20°C and -35°C at 76.76 MHz. Figure 1 shows the

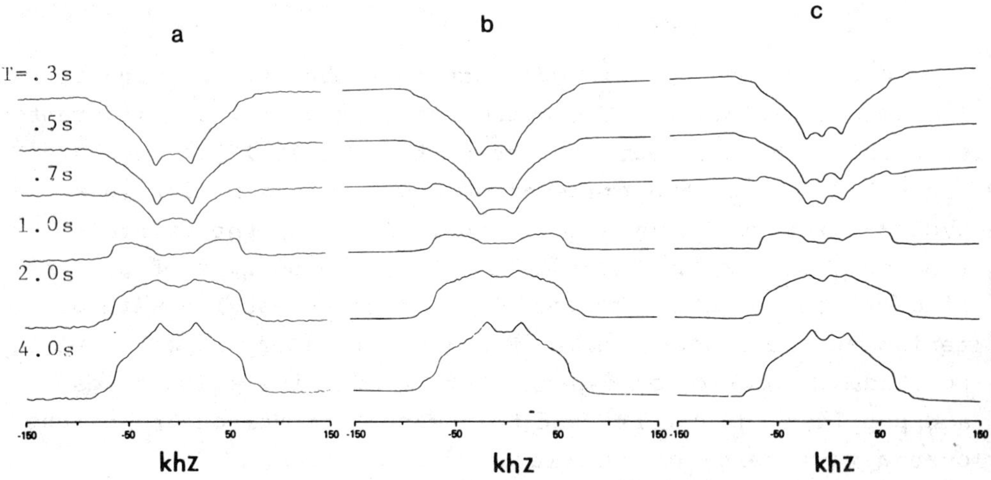

Fig. 1. a and c) Partially relaxed spectra taken at 38.45MHz(a) and at 76.76MHz(c) at -35°C using the pulse sequence $180^o-T-90^o_{\pm x}-\tau-90^o_y-\tau$. b) Partially relaxed spectra calculated assuming a two-site jump model as explained in the text. Calculated spectra included factors that account for spectral distortions due to finite pulse power.

partially relaxed ^2H nmr spectra of D,L[γ-$^2H_2$] proline at -35°C recorded at 38.45 MHz. As is seen in the figure the ^2H lineshape is that of an axially asymmetric $\eta = 0.6$ powder pattern with a splitting of 36.7 KHz between the two peaks of maximum intensity. Similar lineshape is observed for D,L[γ-$^2H_2$] proline at temperatures ranging from 49°C to -85°C. We have assumed a two-site jump model for the motion of the C^γ-^2H bond axis. This model is consistent with the reported crystal structure of D,L-proline.HCl[10]. Using this model, lineshapes were calculated as a function of Θ, the angle made by the C^γ-^2H bond axis in the two orientations. The best fit to the experimental lineshape (from 49°C to -85°C) is obtained using Θ=60°, which corresponds to an r.m.s. angle of 30°.

An examination of the inversion recovery spectra of D,L-[γ-$^2H_2$ proline reveals that T_1 is anisotropic i.e. it depends upon frequency offset and therefore upon orientation of the C^γ-^2H bond axis relative to the field direction. We have determined the correlation time at each temperature using the mathematical expression for angular dependent T_1 for a two-site jump model[11] to simulate the measured inversion recovery spectra. A computer simulation of the partially relaxed spectra at -35°C is shown in Figure 1b. The correlation times calculated in this manner at each temperature is shown in Table I. We find correlation time, τ_c, for

Table I. Correlation Times, τ_c, for the Proline C^γ-^2H Bond Motion from Analysis of T_1 Data using a Two-site Jump Model to Simulate the Inversion Recovery Spectra

T°C	τ_c, ps
49	1.0
20	1.3
-11	2.0
-35	2.0
-85	3.9

the proline $C^{\gamma}-{}^{2}H$ bond jump motion in the picoseconds range above - $100^{o}C$. The activation energy from these data is 1.2 Kcal/mole. This low activation energy can be rationalized in terms of the reported crystal structure of D,L-proline.HCl[10]. In the crystalline amino acid, the packing of the proline molecules is loose with the smallest intermolecular distance involving C^{γ} atom greater than 4 Å. The X-ray study also reports a large temperature factor for the C^{γ} atom. Therefore there are no intermolecular interactions which hinder motion of the C^{γ} atom.

These results demonstrate that the temperature factors in the crystal structure are due to the proline ring motion. We are currently extending these studies to D,L-[γ-${}^{2}H_{2}$] Pro.HCl, D,L-[β,γ, δ-${}^{2}H_{6}$] proline and D,L-[β,γ, δ-${}^{2}H_{6}$] Pro.HCl.

References

1. D. A. Torchia and J. R. Lyerla, Jr. (1974) Biopolymers, 13, 97-114.
2. S. K. Sarkar, D. A. Torchia, K. D. Kopple and D. L. VanderHart (1984) J. Am. Chem. Soc., 106, 3328-3331.
3. D. A. Torchia (1984) in Ann. Rev. Biophys. Bioeng., 13, 125-144.
4. L. S. Batchelder, C. H. Niu and D. A. Torchia (1983) J. Am. Chem. Soc. 105, 2228-2231.
5. H. W. Spiess, (1978) ''NMR Basic Principles and Progress'', Springer Verlag, Berlin/Heidelberg/New York, Vol.15.
6. P. E. Young, and D. A. Torchia (1983) in Peptides: Structure and Function. V. J. Hruby, D. H. Rich., Eds, Pierce, Rockford, Il, pp.155-158.
7. S. K. Sarkar, C. E. Sullivan, D. A. Torchia (1983) J. Biol. Chem., 16, 9762-9767.
8. Y. Hiyama, J. V. Silverton, D. A. Torchia, J. T. Gerig and S. J. Hammond. Submitted to J. Am. Chem. Soc.
9. J. H. Davis, K. R. Jeffrey, M. Bloom, M. I. Valic and J. P. Higgs (1976) Chem. Phys. Lett., 42, 390-394.
10. V. Mitsui, M. Tsuboi, Y. Iitaka (1969) Acta. Crysta., B25, 2182-2192.
11. D. A. Torchia and A. Szabo (1982) J. Magn. Reson., 49, 107-121.

2D-NMR STUDY OF PRO-PRO SEQUENCES. A PROPOSED NMR CORRELATION CONCERNING THE ψ ANGLE

Manh Thong Cung, Bernard Vitoux and Michel Marraud
Cnrs-Ua 494, Ensic-Inpl, 54052 Nancy, France

Claude Giessner-Prettre
Cnrs-Ua 77, Institute of Biological Chemistry,
13, P.M. Curie, Paris, France

André Aubry
Cnrs-Ua 809, University of Nancy I, 54506 Vandoeuvre, France

The Pro residue exhibits particular conformational properties due to _cis-trans_ isomerism of the X-Pro bond and to different puckering modes of the five-membered ring. A conformational analysis by ir, nmr and x-ray diffraction studies was carried out on two dipeptides tBuCO-X^1-L-Pro^2-NHMe (X^1 = L-Pro _1a_, D-Pro _2a_), with reference to the homologous dipeptides with X^1 = L-Ala _1b_, D-Ala _2b_.[1-3] The Pro seven-proton system was delineated by COSY and J-Resolved 2D-nmr experiments followed by PANIC refinements. The correlation between the Pro ψ angle and the Pro H^β chemical shifts was studied by _ab initio_ calculations[4], and compared to experimental data.

Solid state

Dipeptides _1a_ and _1b_ accommodate open-conformations stabilized by intermolecular hydrogen bonds involving a water molecule[2,5], whereas _2a_ and _2b_ are βII'-folded.[3,5]

In all cases, the X^1-L-Pro^2 bond is trans and the L-Pro^2 cycle is C^β-exo with $\phi_2 \simeq -80°$, $\chi_2^1 \simeq 30°$.

Solute state

All four dipeptides experience a cis-trans isomerism for the X^1-Pro^2 bond but the cis and trans conformers are particularly favored for 1a and 2a respectively.

The low stretching frequency of the C-terminal NH bond in chlorinated solvents is indicative of a bonded vibrator for both 1a and 2a. The sharp and strong absorption band at 3367 cm^{-1} for 2a in CCl_4 is typical of the βII'-turn[2]

Table I. ^{1}H-nmr (ppm) and Conformational Data for some C^β-exo Pro-Containing Peptides

Peptide[a]	H^β-exo	H^β-endo	Δδ	Conformer	ψ(°)[b]	Ref.
1a cis	1.97	2.57	0.60	βVI-turn	60	This work
1b cis	2.09	2.56	0.47	βVI-turn	60	
1a trans	1.86	2.28	0.42	γ-turn	75	"
1b trans	1.91	2.38	0.47	γ-turn	75	"
2a trans	2.04	2.35	0.31	βII'-turn	10	"
2b trans	2.01	2.36	0.35	βII'-turn	10	"
3	1.69	2.31	0.62	γ-turn	75	"
4	1.75	2.44	0.69	γ-turn	70	9
5	1.72	2.44	0.72	γ-turn	60	9
6	1.58	2.66	1.08	boat	50	10
7	1.91	2.37	0.46	crown	90	11

[a]tBuCO-L-Pro-NHMe 3, cyclo(Gly-Pro-Gly-D-Ala-Pro) 5, cyclo (D-Pro-D-Pro-Pro) 6, cyclo(Pro_3) 7, [b]rounded values from x-ray data for the derivative itself (2a trans, 2b trans, 6 and 7) or for similar derivatives with the same conformation.

whereas the wide and medium absorption band at 3334 cm^{-1} for 1a is compatible with both βVI and γ-turns.[1,6]

Solvent effects on the C-terminal NH proton chemical shift ($CHCl_3$/MeCN mixtures) reveal flat variations for both trans 2a and cis 1a, indicating a βII' and βVI-folded form respectively. Trans 1a exhibits a decreasing curve with the MeCN content, parallel to that for trans MeCO-L-Pro-NHMe, which can be assigned to the γ-turn conformation.

Correlation ψ angle - H^{β} chemical shift

The delineation of the C^{β}-exo Pro^{2} seven-spin systems in all four dipeptides, and in other peptides containing Pro residues with similar conformation of the C^{α}-C^{β} bond ($\chi^{1} \simeq 30°$) reveals a wide distribution of the Pro H^{β} chemical shifts (Table I) which is probably due to the different rotational states of the Pro C^{α}-C' bond. This was verified by ab initio calculations[4] of the Pro H^{β} shielding constants σ for trans HCO-L-Pro-NH_2 with a C^{β}-exo Pro cycle and for different values of ψ. σ (H^{β}-endo) is much more sensible to the ψ angle than σ(H^{β}-exo) (Figure 1a).

Theoretical shielding constants and experimental chemical shifts are not directly comparable quantities. However, the relative invariance of σ(H^{β}-exo) allows us to compare the differences Δσ and Δδ for the two H^{β}-endo and H^{β}-exo protons (Δσ = -Δδ). Excepting the βVI-folded cis conformers of 1a and 1b, the experimental points from Table I fairly agree with the theoretical curve (Figure 1b), suggesting that the chemical shift difference Δδ could be used for the estimation of the Pro ψ angle. Work is in progress in order to determine the influence of the Pro puckering mode on the theoretical curve ψ vs. Δδ.

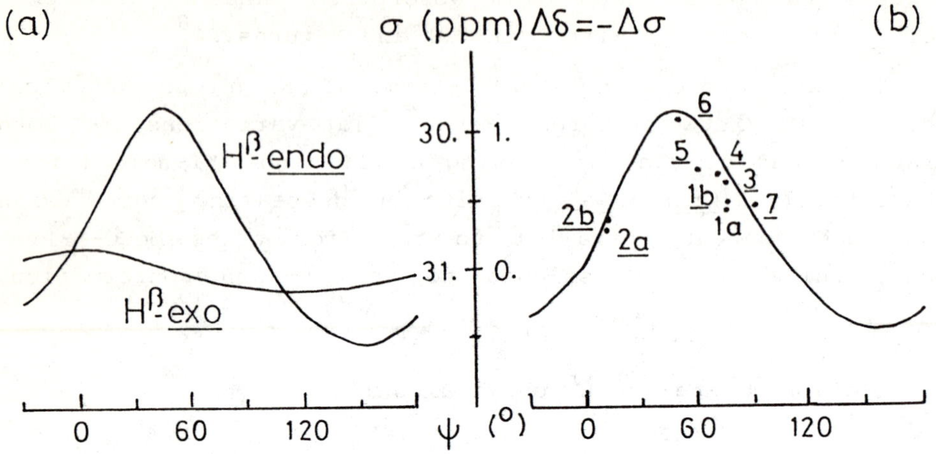

Fig. 1. Variation versus ψ of the theoretical shielding constant σ for the Pro H^{β}-exo and H^{β}-endo protons for HCO-Pro-NH_2 (a). Comparison between the differences of the theoretical shielding constants Δσ and of the experimental chemical shifts Δδ (b).

References

1. Boussard, G., Marraud, M., and Aubry, A. (1979) Biopolymers 18, 1297-1331.

2. Aubry, A., Protas, J., Boussard, G., and Marraud, M. (1980) Acta Crystallogr. B 36, 321-326.

3. Aubry, A., Protas, J., Boussard, G., and Marraud, M. (1979) Acta Crystallogr. B 36, 694-699.

4. Ditchfield, R. (1974) Mol. Phys. 27, 789-807.

5. Aubry, A., Vitoux, B., and Marraud, M. (1985) Biopolymers (in press).

6. Vitoux, B., Aubry, A., Cung, M.T., Boussard, G., and Marraud, M. (1981) Int. J. Peptide Protein Res. 17, 469-479.

7. Bach II, A.C., Bothner-By, A.A., and Gierasch, L.M. (1982) J. Am. Chem. Soc. 104, 572-576.

8. Kessler, H., Bermel, W., Friedrich, A., Krack, G., and Hull, W.E. (1982) 104, 6297-6304.

9. Kessler, H., Friedrich, A., and Hull, W.E. (1981) J. Org. Chem. 46, 3892-3895.

THE CONFORMATION OF N-ACETYL-N-PHENYLGLYCINE: CORRELATION BETWEEN NMR, SINGLE CRYSTAL X-RAY, AND MOLECULAR MECHANICS CALCULATIONS

David E. Portlock, Edward G. Soyke, Jr., Stephen B. Ashton, and Thomas P. Demuth, Jr.
Research and Development Department, Norwich Eaton Pharmaceuticals, Inc., Norwich, NY 13815

Joel D. Oliver, Lawrence C. Strickland, and Fouad S. Ezra
The Procter & Gamble Company, Miami Valley Laboratories, P.O. Box 39175, Cincinnati, Ohio 45247

Introduction

In connection with our studies of angiotensin converting enzyme (ACE) inhibitors, we have studied the solution, solid state, and computer-generated conformation of N-acetyl-N-phenylglycine (**1**), a model of the captopril analogue **2**. In particular, we wished to determine: (a) the relationship between the planes of the phenyl ring and the peptide bond and (b) the preferred amide rotamer population. These two conformational parameters are defined by torsion angles Θ = C(7) - N(1) - C(1) - C(6) and ω = C(8) - C(7) - N(1) - C(9), respectively (Figure 1).

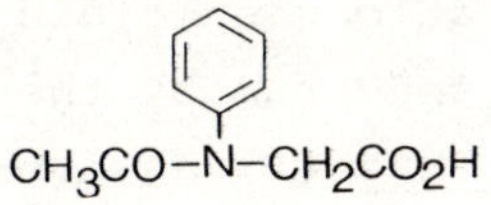

1

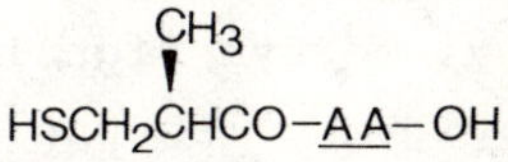

2, AA = N-phenylglycine
captopril, AA = L-Pro

Results and Discussion

The conformation of acetanilide (**3**) is well established both in the crystal and in solution.[1] In the crystal structure the ring is nearly coplanar with the amide ($\Theta = 18^{o}$) and is s-cis (Z) to the oxygen atom ($\omega = 0^{o}$). In solution the molecule assumes a similar conformation, which has been principally supported by H-1 NMR spectroscopy. In contrast to **3** the conformation of N-methylacetanilide (**4**) is notably dissimilar.[1] In the crystal and in solution the ring occupies a position orthogonal to the amide plane ($\Theta = 87^{o}$) and s-trans (E) to the oxygen atom ($\omega = 180^{o}$). We have found that C-13 NMR, UV, and IR spectroscopy provide additional evidence for the reported solution conformations of acetanilide (**3**) and N-methylacetanilide (**4**) (Table I).

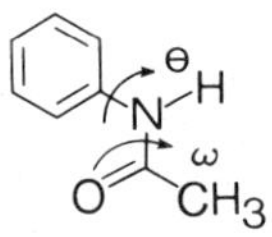

3 $\Theta = 18^{o}$, $\omega = 0^{o}$

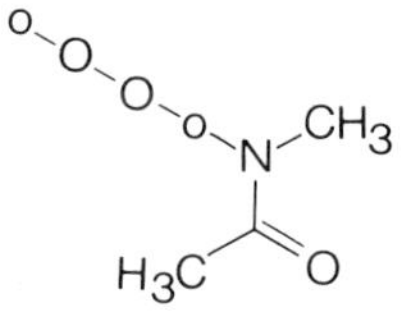

4 $\Theta = 87^{o}$, $\omega = 180^{o}$

To this data base for **3** and **4**, we compared spectroscopic data generated for N-acetyl-N-phenylglycine (**1**) (Table I). The C-13 NMR resonances for the ortho, meta and para phenyl carbons were similar to those observed for **4** and indicate that the phenyl ring and the amide are not conjugated. This conclusion was further verified by the UV and IR spectra of **1** which again closely matched the spectra of **4**.

Regarding the amide rotamers for **1**, the strong shielding of the acetyl methyl signal in the C-13 NMR spectrum by the nearby phenyl ring is consistent with an s-trans (E) assignment. Thus, the solution conformation of N-acetyl-N-phenylglycine **1** appears to be similar to N-methylacetanilide (**4**).

In DMSO, $CHCl_3$, MeOH, and MeOH:H_2O (1:1) only the s-trans (E) rotamer for **1** was observed by H-1 NMR. In each of these solvents the spectra revealed sharp single resonances for the methylene and methyl protons. However in H_2O, two sets of signals appeared for each due to the presence of both the s-trans (E) and s-cis (Z) rotamers in a ratio of 80:20. At coalescence the rate of exchange was calculated[2]: k_{cis} = 4 x k_{trans} = 13 (75°) and 43 (90°) sec^{-1}. The energy barrier to rotation around the amide bond was calculated[2] to be 19 kcal/mol. The phenyl ring makes a negligible contribution to this barrier since the same value was reported for the equally populated rotamers of N,N-dimethylacetamide.

Table I. C-13 NMR[a], UV[b] and IR[c] Spectroscopic Data

Compd	Phenyl Carbons 1	2[d]	3[d]	4	$COCH_3$	λ_{max} (nm)	ε	$\nu_{C=O}$ (cm^{-1})
3	139.2	119.0	128.4	122.8	23.8	240	14300	1688
4	144.3	126.8	129.4	127.1	22.0	220	5700	1647
1	143.3	127.5	129.3	127.5	21.8	220	5600	1657

[a] NMR spectra recorded in DMSO-d_6 at a concentration of 100 mg/ml, chemical shifts in ppm relative to TMS, NMR peak assignments confirmed by the APT technique; [b] UV spectra recorded in absolute ethanol; [c] IR spectra recorded as 0.05M solutions in $CHCl_3$; [d] double intensity

Since the molecular conformation in the solid state will generally correspond to one of the low energy conformations in solution, a single crystal x-ray study of **1** was carried out. A perspective drawing of the crystal structure analysis of **1**, along with its atomic numbering system, is shown in Figure 1. The solid state conformation is consistent with the H-1 and C-13 NMR solution results. The phenyl ring is oriented perpendicular to the amide plane (Θ = -89.2°) and the amide bond is s-trans (ω = 176.8°).

Molecular mechanics parameters were compiled to reproduce the spectroscopic and x-ray derived conformations of **1**, **3**, and **4**. Calculations using MM2P predicted a low energy

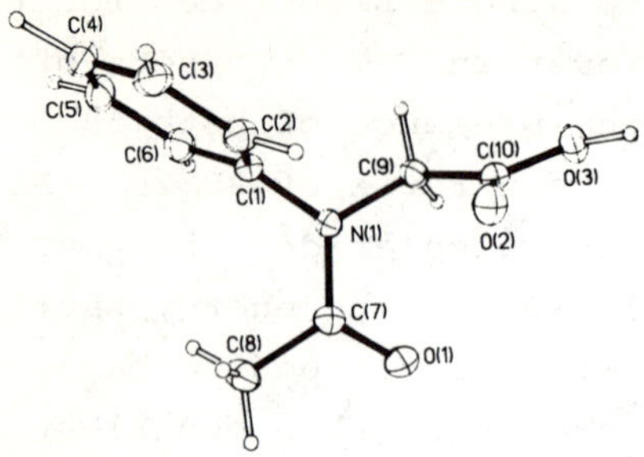

Fig. 1
Solid state conformation of N-acetyl-N-phenylglycine (**1**) showing the endo orientation of the methyl group to the phenyl ring.

conformation of N-acetyl-N-phenylglycine (**1**) that closely matched the conformations observed in solution and the crystal. The amide linkage was calculated to exist in a nearly planar s-trans (E) orientation (ω = 180°). The plane of the phenyl ring was determied to be perpendicular to the amide plane (Θ = 88°). The calculated bond lengths and bond angles for the MM2P minimized structure deviated $\pm$0.01 Å and $\pm$1.3°, on the average, when compared to the x-ray structure. The corresponding s-cis (Z) conformer (ω = 4°, Θ = 87°) was also identified by MM2P calculation as a local energy minimum approximately 2.2 kcal/mole less stable than the s-trans (E) conformer.

In conclusion, a correlation between the solution, solid-state, and molecular modelled conformations of N-acetyl-N-phenylglycine has been obtained. The relationship between the planes of the phenyl ring and the amide is orthogonal. In DMSO, $CHCl_3$, and MeOH the amide rotamer population is exclusively s-trans (E). However in H_2O, both the s-trans and s-cis rotamers were observed in a ratio of 80:20. The phenyl ring has a negligible effect on the barrier to rotation about the amide bond.

References

1. Carter, R.E. (1967) Acta Chem. Scand. **21**, 75-86.
2. Shanan-Atidi, H. and Bar-Eli, K.H. (1970) J. Phys. Chem. **74**, 961-963.

INTRAMOLECULAR MOBILITY IN TWO CYCLIC OCTAPEPTIDES, EFFECT ON SPIN-LATTICE RELAXATION IN THE ROTATING FRAME

Kenneth D. Kopple and Yu-Sen Wang
Department of Chemistry, Illinois Institute of Technology, Chicago, IL 60616, USA

Cyclo-(D-Ala-Gly-L-Pro-D-Phe)$_2$ appears to have in DMSO solution a stable, narrowly defined conformation[1] that differs only slightly from a crystal structure,[2] while NMR data for cyclo-(D-Ala-Gly-L-Pro-L-Phe)$_2$ do not clearly define a solution conformation. Two crystal conformations, differing in the rotation of one Ala-Gly CONH plane relative to the local ring plane, have been found for the latter.[3] We have compared proton rotating frame spin lattice relaxation rates to find differences in flexibility of the two peptides.

Spectra for the D-Ala,L-Phe analog lack features to suggest a single narrowly defined backbone conformation. For example, the nonequivalence of the glycine α-protons is only .21 ppm, compared to .46 ppm in the D-Ala,D-Phe peptide, which suggests more comprehensive averaging of the position of the glycine methylene relative to the magnetically anisotropic CONH groups flanking it. Also, the D,L compound lacks the wide range of N-H chemical shifts found for the D,D-compound (0.65 vs 1.93 ppm in DMSO). Finally, the H-N-C-H coupling constants and the solvent exposure data (N-H temperature coefficients and nitroxyl-catalyzed relaxation rates) for the D,L peptide are equivocal as regards conformation. However, both peptides give spectra with lines of the usual width, so that whatever conformational mobility they possess is fast on the chemical shift time scale.

With the expectation that different internal mobilities of the two peptides might be apparent on a shorter time scale, we compared spin lattice relaxation in the rotating frame ($T_{1\rho}$) for the backbone N-H protons. $T_{1\rho}$ as a probe of peptide mobility has been discussed by Bleich and Glasel.[4] Exchange between conformations that results in fluctuations in chemical shift[5] or coupling constants[6] can result in contributions to the $T_{1\rho}$ relaxation rate. These will be maximal when the exchange occurs near the frequency, γH_1, corresponding to the spin locking field used in the $T_{1\rho}$ experiment. We used a field of 8333 Hz, well into the fast exchange region on the chemical shift scale. Because of the sensitivity of their chemical shifts to hydrogen bonding and so to conformation, the peptide N-H proton resonances are most appropriate for $T_{1\rho}$ examination.

The total $T_{1\rho}$ relaxation rate of the NH protons contains contributions not only from exchange, but from dipolar coupling to nearby protons and adjacent nitrogen, and from scalar relaxation through J coupling to the nitrogen. The dipolar and scalar terms may be evaluated from knowledge of the proton T_1's, the quadrupolar coupling constant of the amide nitrogen, the rotational correlation time τ_C of the backbone and other data. τ_C may be estimated from the proton-decoupled T_1 relaxation rate of the α-carbons. For the two peptides in DMSO at 20°, τ_C is near 4×10^{-10} s, and τ_N, the quadrupolar relaxation time of N, is about 1.3×10^{-3} s. With γH_1 = 8333 Hz, the non-exchange contributions to the N-H proton $T_{1\rho}$ rates were 3.4-4.8 s^{-1} by the proton-proton dipolar mechanism, ~0.95 s^{-1} from the nitrogen-proton dipole coupling, and ~0.0007 s^{-1} from scalar relaxation. The exchange contribution was obtained by difference.

Proton Relaxation Rates, R = 1/T, s^{-1}, DMSO, 20°.

	D-Ala, L-Phe			D-Ala, D-Phe		
	$R_{1\rho}$	R_1	$R_{1\rho}$(exch)	$R_{1\rho}$	R_1	$R_{1\rho}$(exch)
Ala N-H	10.6	2.9	5.9	8.1	3.1	2.9
Gly N-H	11.1	3.0	6.2	8.3	2.7	3.8
Phe N-H	12.5	3.1	7.5	8.8	3.4	3.2

The Table above gives T_1 and $T_{1\rho}$ rates and the exchange contribution to the latter for the N-H protons. As might have been expected, the T_1 relaxation rates, dependent on the rotational correlation time and interproton distances, thus on the general size and shape of the peptides, are similar for the two diastereomers. However, the chemical exchange contributions to $T_{1\rho}$ are distinctly different. It should be noted that this chemical exchange contribution arises from conformational interconversions, not from proton exchange, which is very slow in these solutions. The exchange contribution is greater in the D-Ala,L-Phe peptide, which on the grounds already given seemed to be the more conformationally mobile. However, since there is an exchange contribution to the $T_{1\rho}$ rate for both peptides, both must have internal mobility that is not otherwise detected.

The difference in the exchange contributions may arise from greater chemical shift differences between conformational states in the D,L peptide, which may be associated with larger conformational variations. A second possibility is that internal motion in the D,L peptide could have a larger low-frequency component. This also might correlate with larger excursions of atomic groups.

We have made measurements at varying spin locking fields. For the D,D peptide the exchange contribution is not H_1 dependent, indicating a characteristic time for conformation exchange on the high frequency side of our highest γH_1, $5.2 \times 10^4\ s^{-1}$. A slight field dependence for the D,L peptide suggests that it may have some lower frequency components.

In the absence of a specific model for the conformational exchange, including N-H proton chemical shifts and populations of the conformations involved, these residual relaxation rates cannot be interpreted in detail. Nonetheless, the result is consistent with the indications from chemical shift, solvent exposure and coupling constant data. $T_{1\rho}$ measurements may thus allow investigation of an additional dimension in peptide conformation, internal mobility.

This work was supported by a grant from the National Institutes of Health, GM26071.

References

1. Kopple, K.D., Parameswaran, K.N., and Yonan, J.P. (1984) J.Am.Chem.Soc. **106**, 7212-7.

2. Kopple, K.D., Kartha, G., Bhandary, K.K., and Romanowska, K. (1985) J.Am.Chem.Soc. **107**, in press.

3. Bhandary, K., Kartha, G., and Kopple, K.D. (1983,4) Abstracts Amer. Cryst. Assoc. meeting, Aug. 1983, PD14, p.32, and May 1984, PC13 p.53.

4. Bleich, H.E. and Glasel, J.A. (1978) Biopolymers **17**, 2445-2457.

5. Deverell, C., Morgan, R.E. and Strange, J.H. (1970) Mol. Phys. **18**, 553-559.

6. Strange, J.H. and Morgan, R.E. (1970) J. Phys. C. Solid St. Phys. **3**, 1999-2011.

ENERGETICS AND DYNAMICS OF A PEPTIDE CRYSTAL

David H. Kitson and Arnold T. Hagler
The Agouron Institute, 505 Coast Boulevard South, La Jolla, CA 92037

Introduction

We have undertaken an extensive theoretical computer simulation study of a crystal of the peptide cyclo-(Ala-Pro-D-Phe)$_2$ with two objectives in mind. Firstly, by comparison with the experimental system, we have an excellent opportunity to test the theoretical methodology which we employ. Secondly, through such studies we can investigate several interesting unanswered questions in the field of peptide biophysics such as: what are the effects of crystal environment on peptide conformation and dynamics; what is the relationship between static and dynamic properties; can we account for vibrational and NMR spectroscopic properties by simulating time-averaged structures?

Methods

Cyclo(Ala-Pro-D-Phe)$_2$ crystallises in the orthorhombic space group $P2_12_12$ with crystallographic twofold axes relating the halves of single peptide molecules and with 2 peptide molecules and 16 water molecules per unit cell.[1] The peptides are in a double type II β turn conformation, with L-Pro and D-Phe residues at the corners of the turns. The starting system for our studies was constructed by taking the peptide coordinates from the crystal structure and adding the waters, which the X-ray study indicated range from fully ordered to fully disordered,[1] in random, but sterically reasonable, locations. The system was then minimised and this minimised system was used as the starting system for a 10 picosecond (ps.) dynamics simulation, run at 300K. One of the peptides from the minimised crystal system was minimised *in vacuo*, and was then used as the starting structure for a 60 ps. simulation, again at 300K.

Results and Discussion

Fit to experiment. Table I compares the backbone torsion angles of the experimental and minimised crystal systems (and an isolated molecule, see below). It is reassuring to see that the minimised conformations are similar to the experimental peptides. The biggest difference in torsion angle is for the ψ of the proline residues (average calculated value $= 105°$ *versus* experimental $= 122°$),and the overall shape of the peptides, including the type II β turns, has been maintained. Thus the peptide, in its crystal environment, is reproduced extremely well.

Effect of environment on structure, energetics and dynamics. This study has allowed us to investigate interesting questions related to the effect of the crystal environment on the peptide, for example: is the peptide in the crystal constrained to a conformation which would be unfavourable *in vacuo*, and, if so, what intermolecular interactions stabilise it in this conformation? Table I compares the torsion angles of the peptides from the minimised crystal system with those of the lowest energy conformation found by minimising structures chosen at 3 ps. intervals throughout the 60 ps. *in vacuo* simulation. If we examine this table we can see that the crystal forces have a considerable effect on the molecular conformation. For example, Ala^1 exists in an extended conformation ($\phi,\psi \approx -150,170°$) in the crystal, whilst in the isolated state it is in the C_7 (equatorial) region ($\phi,\psi = -82,116°$). Similarly, the D-Phe residues, which are in left-handed alpha helical (α_L) conformations ($\phi,\psi \approx 80,10°$) in the crystal undergo a transition to C_7 (equatorial) ($\phi,\psi \approx 120,-80°$) *in vacuo*.

We have calculated the intramolecular energies of the peptides in the crystal and isolated states and have found that the energies of the peptides in the crystal conformation (228.3 and 228.5 kcal/mole) are higher than that of the isolated molecule (222.8 kcal/mole). This indicates that the crystal provides an environment favouring an energetically strained conformation of the peptide, which is stabilised in this conformation by intermolecular interactions.

Figure 1 illustrates the effects of the crystal environment on the dynamic properties of the peptide. This figure shows the variation in the ϕ and ψ torsion angles of the D–Phe^6 residue for both the crystal and *in vacuo* simulations and demonstrates clearly that the crystal forces act to damp out the torsional fluctuations in the molecule. An interesting conformational change can be seen in the case of the isolated molecule, in which the D–Phe^6 residue begins in the α_L region of conformational space. During the first 3 ps. this residue undergoes a conformational transition to the

Table I. Torsion Angles of Experimental, Minimised Crystal and Minimised Isolated Structures of Cyclo–(Ala–Pro–D–Phe)$_2$

Residue	Torsion Angle	Experimental[a]	Minimised Crystal Molecule 1	Minimised Crystal Molecule 2	Minimised Isolated
Ala[1]	ϕ	-156.6	-150.4	-152.7	-81.9
	ψ	171.7	169.9	169.1	116.0
Pro[2]	ϕ	-60.4	-59.8	-59.1	-44.2
	ψ	122.5	105.8	107.7	116.3
Phe[3]	ϕ	78.7	83.9	80.7	121.3
	ψ	9.0	4.8	12.7	-68.8
Ala[4]	ϕ	-156.6	-146.8	-152.7	-79.0
	ψ	171.7	167.6	166.7	165.1
Pro[5]	ϕ	-60.4	-55.9	-56.4	-72.3
	ψ	122.5	101.8	108.9	96.1
Phe[6]	ϕ	78.7	85.1	77.8	132.8
	ψ	9.0	11.8	9.9	-76.9

[a] The conformations of the two molecules in the experimental crystal structure are identical.

C_7 (equatorial) region, around which it oscillates for the remaining 57 ps., with the exception of a brief return at about 40 ps. to the α_L state observed in the crystal conformation. This excursion demonstrates the accessibility of the crystal conformation which is visited through dynamic fluctuations by the isolated molecule, and the ability of dynamics to take molecules into higher energy conformations, which may be stabilised by external forces and observed in other states, such as in the crystal or bound to a receptor.

We have also carried out an investigation of the effects of the crystal environment on the vibrational spectrum of the peptide. We have calculated NMR coupling constants from the time-averaged and minimised structures of the isolated peptide and have compared these with the experimental values. This work is described in more detail elsewhere.[2]

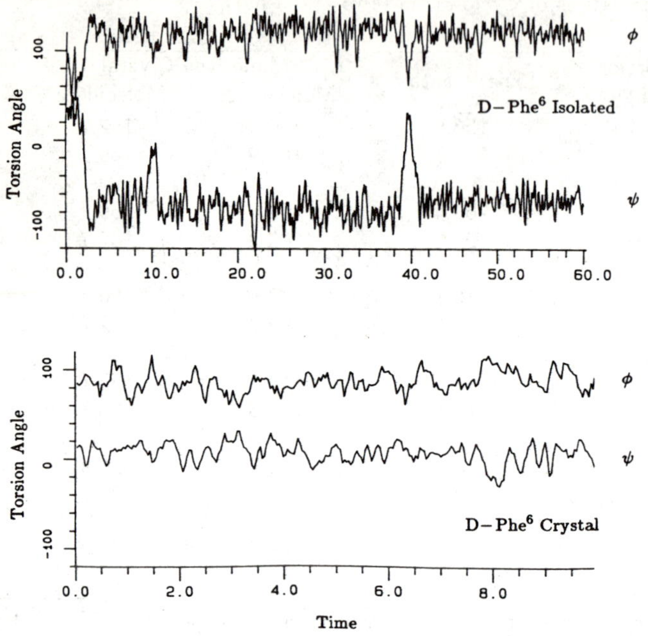

Fig. 1. Variation of D−Phe^6 ϕ and ψ torsion angles with time during a 60 picosecond molecular dynamics simulation of cyclo-(Ala−Pro−D−Phe)$_2$ *in vacuo* and a 10 picosecond simulation in the crystal environment.

Acknowledgements

This work was supported by grant number PCM 8204908 from the National Science Foundation. We wish to acknowledge the provision of computing facilities by Abbott Laboratories.

References

1. J. N. Brown and R. G. Teller, *J. Am. Chem. Soc.*, **98**, 7565-7569 (1976).
2. D.H. Kitson and A.T. Hagler (1985). in preparation

A METRIC ANALYSIS OF MINIMUM ENERGY CONFORMATIONS OF LYSINE VASOPRESSIN GENERATED USING MOLECULAR DYNAMICS SIMULATION

Judith C. Hempel

Department of Medicinal Chemistry, Smith Kline and French Laboratories, Philadelphia, Pennsylvania 19101

Introduction

Questions which arise in the analysis of conformations of peptides include: How comprehensively have conformational search algorithms searched the conformation space of these flexible molecules? How do you come to terms with large numbers of conformations? Metrics (molecular dimensions) and the vector sense of these metrics are used in this study to classify conformations of lysine vasopressin (LVP) generated via molecular dynamics simulation.[1] Families of conformers of LVP generated in the molecular dynamics simulation, i.e. sets of conformers defined by RMS comparisons, examination of stereo plots and of hydrogen bonding patterns, and analysis of the conformational states of residues have been determined previously.[2] These analyses provide specifics of conformational similarity and difference within the data set. Metric analysis represents a new approach to the analysis of peptide conformation which provides an overview both of this conformational data set and of the conformation space of LVP.

Analysis and Results

The peptide ring of VP and oxytocin is critical for binding to antiduretic and uteronic receptors. The two dimensional metric analysis reported here is a conformational analysis of this twenty membered ring, which is a hexapeptide ring closed by a disulfide bridge from Cys^1 to Cys^6. The distance from the midpoint of the disulfide bond to the midpoint of the amide bond linking residues 3 and 4 is a uniquely defined metric for this class of peptides. This metric provides an estimate of ring "length". The distance from the alpha carbon of residue 2 to the alpha carbon of residue 5 is the second metric of the analysis presented here. This metric provides an estimate of ring "width". Data for 95 minimium energy conformations of LVP (derived by minimizing the energy of instantaneous configurations of LVP along dynamics trajectories [2]) has been analyzed. Previously reported conformations for the peptide ring of oxytocin [3,4,5] have also been analyzed [6] and a two dimensional conformational analysis of this data set is presented graphically in Figure 1.

We see from Figure 1 that the metric analysis groups ring conformations into neighborhoods. Not all regions of the ring map are occupied although they are geometrically accessible to the ring. The molecular dynamics data set defines conformational classes not reported in previous conformational searches for this twenty membered ring while a conformational class previously defined by DeCoen et.al.[3] did not arise in the LVP simulation. The DMSO solution conformation model proposed for oxytocin [7,8] is an effectively planar "beta sheet" ring conformation with a beta turn at the residue 3 to residue 4 region of the peptide ring. As illustrated in Figure 1, this class of conformations is "longer" than it is "wide". Planar conformers which are "wider" than they are "long" no longer have this beta turn and the region of the metric map occupied only by DeCoen conformations represents this class of ring conformation.

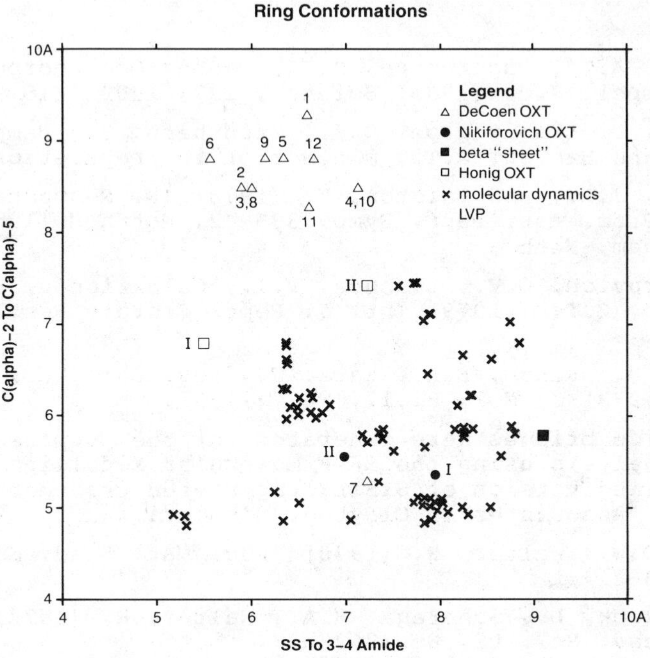

Fig.1. Vasopressin ring map.

The three dimensional sense of ring conformers is conveyed by the vector sense of their metric descriptors. Conformers which are "longer" than 8.5 A (disulfide to 34 amide distance) or "wider" than 7 A (Ca^2 to Ca^5 distance) are effectively planar. That is to say the distance between parallel planes, each containing one metric descriptor vector, is less than 1 A. The vectors are within fifteen degrees of perpendicular for all conformers in the data set analyzed. The maximum deviation from planarity for the LVP minimum energy conformations is 3.3 A and these conformers are "shorter" than 7 A and "narrower" than 5.5A. It is proposed that energy can be incorporated into a metric map concept, and that metrics derived from an analysis of molecular topography can be used to drive conformational searching algorithms.

References

1. Hagler, A.T., Osguthorpe, D.J., Dauber-Osguthorpe, P. and Hempel, J.C. (1985) Science, 227, 1309-1315.
2. Kitson, D., Osguthorpe, D.J., Struthers, S., Hempel, J.C., and Hagler, A.T., manuscript in preparation.
3. DeCoen, J.L., and Ralston, E. (1973) (H. Nesvadba, ed). Pept. Proc. Eur. Pept. Symp. 335-42. North-Holland, Amsterdam, Neth.
4. Nikiforovich, G.V., Leonova, V.I., Galaktionov, S.G., Chipens, G.I. (1979) Int. J. Pept. Protein Res., 13, 363-73.
5. Honig, B., Kabat, E.A., Katz, C., Levinthal, C., Wu, T.T. (1973) J. Mol. Biol., 80, 227-95.
6. Disulfide bridges were generated for the purposes of this analysis using the SK&F Molecular Modelling System, a modified version of SYBYL, registered trademark of TRIPOS Associates in Clayton, Missouri.
7. Urry, D.W., Walter, R. (1970) Proc. Nat. Acad. Sci. US, 68, 956-8.
8. Kotelchuck, D., Scheraga, H.A., Walter, R. (1972) Proc. Nat. Acad. Sci. US, 69, 3629-33.

CRYSTAL STRUCTURE OF OXYTOCIN

Jim E. Pitts, Steve Wood, Ian J. Tickle, Anne-Marie Treharne, Yvonne Mascarenhas, J. Y. Li, Jasmine Husain, Sue Cooper and Tom Blundell

Laboratory of Molecular Biology, Department of Crystallography, Malet Street, Birkbeck College, London WC1E 7HX, UK

Victor J. Hruby

Department of Chemistry, The University of Arizona, Tucson, Arizona 85721, USA

Herman R. Wyssbrod

Department of Physiology and Biophysics, Mount Sinai Medical Center, City University of New York, 1 Gustave Levy Place, New York, NY 10029, USA

Introduction

The neurohypophysial hormone oxytocin is a nonapeptide which elicits smooth muscle contraction causing milk ejection and uterine contractions in mammals;

H_2N Cys Tyr Ile Gln Asn Cys Pro Leu Gly NH_2
S——————————S

Here we present the three dimensional structures of two crystal forms of deamino-oxytocin and a further related form of 6-seleno deamino-oxytocin defined by X-ray analysis at resolutions between 1.0 and 1.9 A.

Results and Discussion

The C2 form of deamino-oxytocin was solved using a "best" phased Fourier map at 2.1 A resolution calculated

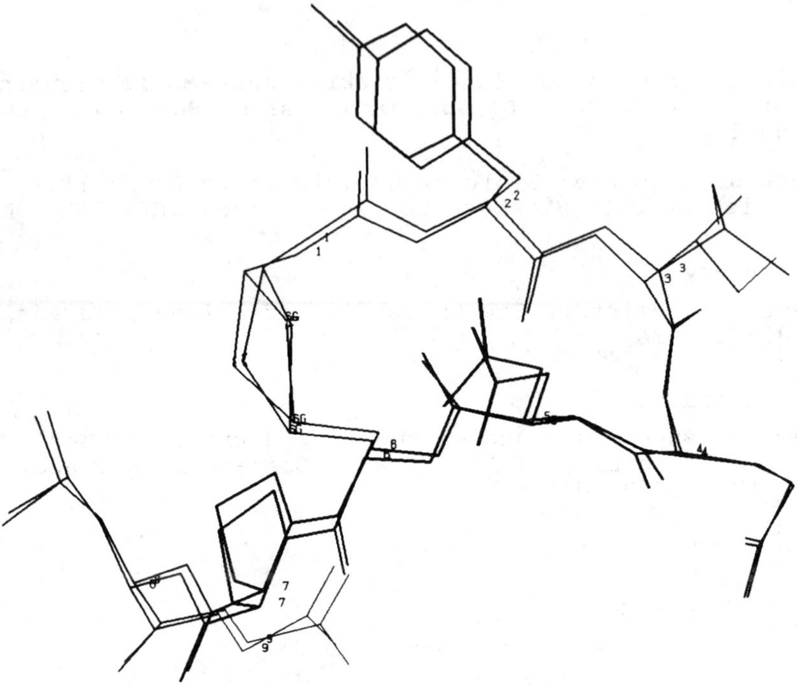

Fig. 1. The two molecules of deamino-oxytocin in the $P2_1$ asymmetric unit are shown superimposed to demonstrate the two conformers in the S1 region and other conformational differences.

using the selenium single isomorphous replacement with anomalous differences (SIRAS) combined with anomalous differences due to sulphur in the native structure. Both unrestrained and restrained refinement of deamino-oxytocin was employed in conjunction with difference maps to define non-hydrogens anisotropically and hydrogens isotropically with R=10% at 1.2A for the C2 crystal form and R=9% for the $P2_1$ form at 1A. The general features of the oxytocin molecules are similar in all three structures examined although the disorder observed in the C2 cell is partially resolved in the $P2_1$ cell.

The crystal structure of deamino-oxytocin (Fig. 1.) has two beta-turns, a type II turn between Tyr 2 and Asn 5 stabilised by two hydrogen bonds between the amide and carbonyl of both residues and a type III turn between Cys 6 CO and Gly 9 NH. The disulphide bridge region exhibits at least two possible conformations which differ in S1 position. In the C2 cell one conformation has a disulphide torsion angle of -101° with a left handed chirality, and the other has $+76^{o}$ and right handed chirality. They are equally populated. In the $P2_1$ cell the S1 disorder is predominantly populated (70/30) in each molecule.

All potential hydrogen bonds are satisfied either by intramolecular interactions or bonding with water and other oxytocin molecules. Of the seven water molecules found in the C2 cell, two are disordered about a crystallographic two-fold axis. Intermolecular hydrogen bonds lie between Ile 3 NH and Pro 7 CO, Asn 5 ND2 and Leu 3 CO, Cys 6 NH and Gln 4 OE1, Cys 1 CO and Asn 5 ND2, Ile 3 CO and the C-terminal amide. The close approach of 2-fold related pairs at the Cys 1 ends may explain why oxytocin itself is not crystallised in this form.

The spectroscopic data taken together with the crystal structures indicate that (i) the two beta turns are stable features of oxytocin and its analogues and (ii) the disulphide conformers exist in solution but in different proportions in changing environments. There may be considerable flexibility in the spatial relationship between the tocin ring and acyclic tail, but it is unlikely in aqueous conditions that there are intramolecular hydrogen bonds between the two parts. Similar conformational preferences, although perhaps more flexibility, are evident in the vasopressins. To understand these motions we have recently carried out normal mode analysis for deamino-oxytocin and calculated 1000 picoseconds of molecular dynamics simulation in vacuo and in the crystal lattice (Treharne et al, unpublished) which will be described elsewhere. Similar work has recently been presented by Hagler et al.

In the absence of specific receptor affinity data, antagonists have played a major role in understanding the binding process. Using such an approach Hruby and coworkers have proposed a general "dynamic" model of peptide hormone activity which complements the ideas of the "cooperative" model of Walter.[2 3] The existence of two conformers in the crystal structures are evidence of conformational flexibility. We propose that these conformers are required for receptor recognition and trandsuction of the biological response and that the receptor is flexible and can adopt two conformations. In the absence of hormone, its conformation can bind the right handed disulphide chiral oxytocin conformers. This inactive state is thus stabilised by rigid, right-handed disulphide containing analogues such as 1-penicillamine derivatives, which act as antagonists.[4] Strong agonists stabilise the receptor in a conformation complementary to the left-handed disulphide oxytocin conformer. A low energy barrier between conformers as in deamino-oxytocin and a strong interaction with the receptor in the active form would both be required for an efficient transduction of the biological response.

Circulating in the blood stream, oxytocin will spend much time in an aqueous environment, but at the receptor water will be excluded from a large part of the hormone. Experimental results from analogues indicate that hydrogen bonded and hydrophobic interactions are likely to dominate receptor binding. We suggest a neutral species binds so that the deprotonated amino terminus can form a hydrogen bond to the receptor which would explain the enhanced activity of analogues which have a hydroxyl group in place of the amino terminus. All of these observations indicate that the conformation of the neutral deamino-oxytocin in an environment which excludes water, in a manner as found in the crystal, may reveal important aspects of the conformation at the receptor.

References

1. Hagler et al., (1985) Science 227, 1309-1315.

2. Hruby, V. J. (1981) in Topics in Molecular Pharmacology (eds. Burgen, A.V.S. and Roberts, G.C.K) 100-126 (North Holland, Amsterdam).

3. Walter, R. (1977) Fed. Proc. 36, 1872-1878.

4. Hruby, V. J. et al., (1983) Int. J. Pept. and Protein Res. 21, 24-34.

4-ZINC HUMAN INSULIN: CONFORMATIONAL FLEXIBILITY AND ITS EFFECT ON STABILITY AS DETERMINED BY X-RAY CRYSTALLOGRAPHY

G. D. Smith and D. C. Swenson
Medical Foundation of Buffalo, Inc., Buffalo, NY 14203

E. J. Dodson, G. G. Dodson and C. D. Reynolds
University of York, Heslington, York, England Y01 5DD.

Introduction

Only monomeric insulin is bound by its receptor. Insulin binding depends upon a number of factors and includes the shape and flexibility of the insulin molecule, the orientations of those residues which interact with the receptor, and the amount of insulin that is present as the free monomer. The balance between the biologically active free monomer and insulin in its hexameric storage form depends upon interactions between amino acids that are responsible for holding the monomers together. One way of obtaining information concerning the flexibility of the insulin molecule and how this affects its aggregation is by comparing the crystal structures of different crystalline forms of insulin from various species.

Results

The structure of 4-zinc human insulin has been refined in our laboratories[1], using 1.85Å resolution data, to a residual of 0.173. The unit cell is rhombohedral, space group R3, with hexagonal cell constants a = 80.953 and c = 37.636Å. As a result of a conformational change, the extended conformation observed in residues B1 through B8 in one of the molecules in the 2-zinc structure[2] is transformed into an α-helical conformation in the 4-zinc structure. This conformational change, illustrated in Figure 1, produces a

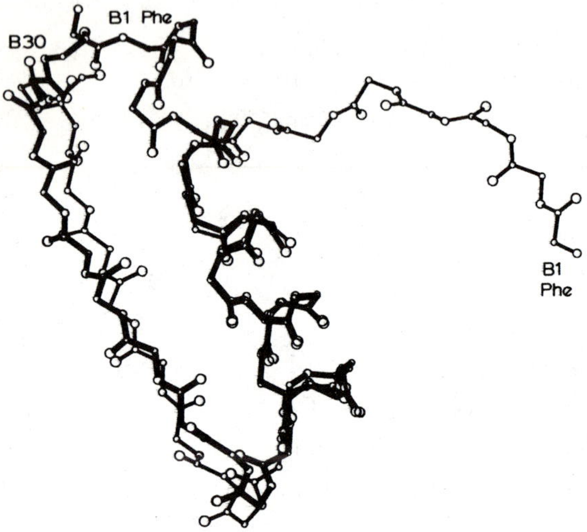

Fig. 1. A comparison of the conformation of the backbone atoms of the B-chain of molecule 1 of 4-zinc human insulin with that of molecule 2 of 2-zinc porcine insulin. The 4-zinc structure is drawn with wide bonds while the 2-zinc structure is drawn with narrow bonds. Backbone atoms of residues B9 through B18 were used in the optimization.

continuous α-helix from B1 Phe through B19 Cys and involves a shift of over 30 Å of the C-alpha carbon of B1 Phe. This phenomenon is of particular interest because the 4-zinc crystalline modification is used in insulin preparations for diabetes therapy and is called a "slow-acting" insulin as it is slow to dissociate to the biologically active monomer. Our results show that, in addition to the change in conformation observed in the B-chain, four additional intermonomer hydrogen bonds result which are not possible in the 2-zinc structure. These hydrogen bonds provide additional stabilization for the dimer and the hexamer.

Another dramatic change which occurs as a result of the shift of the N-terminus of the B-chain is the nature of the surface of the entire hexamer. In 2-zinc porcine insulin, each zinc ion is coordinated by three B10 histidine side chains and three water molecules and lies in a shallow depression on the surface of the hexamer. In the 4-zinc structure, this shallow depression is transformed into a narrow channel bounded by the three newly formed α-helical segments of the B-chain and contains a zinc and a chloride ion at the bottom. This is illustrated in Figure 2.

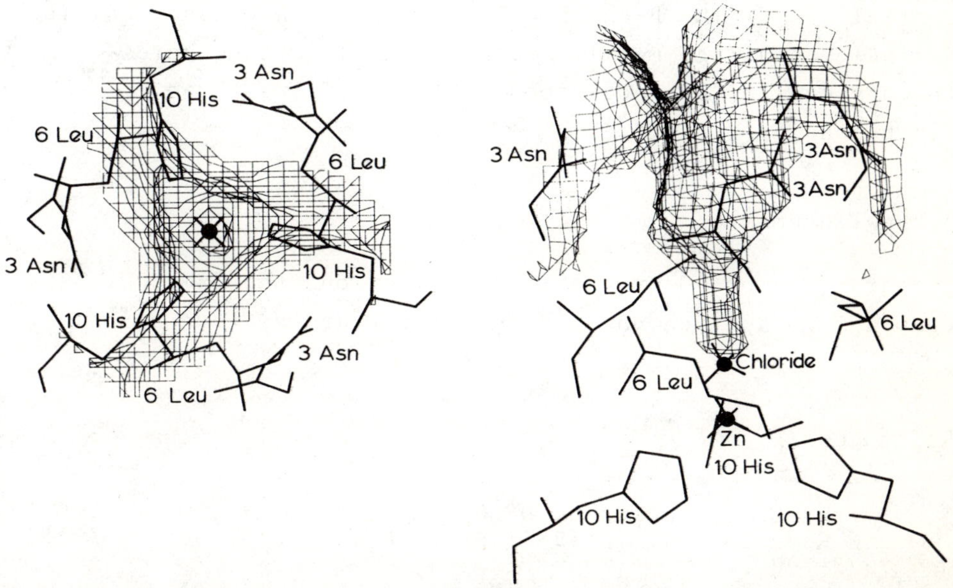

(a) (b)

Fig. 2. Extra radius plot[3] showing access to the zinc and chloride ion in the "4-zinc face" (a) viewed along the 3-fold axis and (b) perpendicular to the 3-fold axis. In this technique, a surface is constructed at a distance from the insulin surface equal to the sum of the van der Waals radii of individual atoms in the hexamer and that of a water molecule and therefore bounds the volume in space from which the center of a water molecule is excluded. The three sets of B3 Asn, B6 Leu and B10 His residues are related by the crystallographic 3-fold axis which passes through the length of the tunnel. The B3 Asn residues lie on the surface of the hexamer while the B10 His residues are at the interior.

Therefore, the net effect of the conformational change is to isolate the chloride ion, and hence the zinc ion, from the environment. Since the zinc ions contribute to the stability of the hexamer, the rate of dissolution is decreased if this zinc ion cannot be easily removed from the hexamer. Therefore, prior to the dissociation of the hexamer into monomers, the complicated changes in conformation from extended to α-helical must be reversed. This includes not only the N-terminus of the B-chain, but also portions of the A-chains since the side chain of A10 isoleucine lies partially over the B-chain. The "slow-acting" aspect of the 4-zinc crystalline modification of insulin in diabetes therapy can now be understood in terms of its three dimensional structure.

Acknowledgements

Research supported by the Kroc Foundation, the James H. Cummings Foundation and the Juvenile Diabetes Foundation.

References

1. Smith, G. D., Swenson, D. C., Dodson, E. J., Dodson, G. G. and Reynolds, C.D. (1984) Proc. Natl. Acad. Sci. USA, 81, 7093-7097.

2. Dodson, E. J., Dodson, G. G., Hodgkin, D. C. and Reynolds, C. D. (1979) Can. J. Biochem., 57, 469-479.

3. Barry, C. D., Beyond the Space-Filling Molecular Model, manuscript to be submitted.

HYDROPHOBIC PEPTIDE INHIBITORS OF SICKLE HEMOGLOBIN AGGREGATION HAVE SIMILAR STRUCTURAL PROPERTIES

Stephen K. Burley

Department of Chemistry,
Massachusetts Institute of Technology,
Cambridge, MA 02139 and
Harvard Medical School,
(Health Sciences and Technology Division)
Boston, MA 02115

Andrew H.-J. Wang, Joseph R. Votano and Alexander Rich
Department of Biology,
Massachusetts Institute of Technology,
Cambridge, MA 02139

Introduction

Previous studies of inhibitors of deoxy-sickle cell hemoglobin (HbS) gelation have revealed that biaromatic compounds are potent non-covalently acting agents[1 2 3 4].

In this investigation we have used single crystal X-ray diffraction to solve the three-dimensional structures (3D) of two modifications of L-Phe and two biphenyl peptides. L-phenylalanyl benzoyl ester (FOB) and N-phenylacetyl-L-Phe (PAF) are antisickling agents[5 6], and L-Lys-L-Phe-L-Phe (LFF) and L-Phe-Gly-Gly-D-Phe (FGGF) are antigelling agents, which cannot pass the red cell membrane[1 7]. Our aims were: (i) to determine the structure of these dissimilar compounds, (ii) to correlate their structures with their antigelling action, and (iii) to establish what structural criteria allow FOB and PAF access to the red cell's interior.

Standard crystallographic methods were used and full details will be published elsewhere. Space groups: $P2_12_12_1$ (FOB and PAF), $P2_1$ (FGGF), and C2 (LFF).[8 9 10 11]

Results and Discussion

Figure 1 illustrates the molecular structure of each of the four biphenyl compounds. The most striking features of these structures are (i) the proximity of the two phenyl rings ($<\delta>\simeq 5Å$), and (ii) their near perpendicular orientation ($<\theta>\simeq 77°$). This side-to-edge interaction, also seen in crystalline benzene[12], brings a C-atom with -ve partial charge near to an H-atom with +ve partial charge, and is energetically favourable. The resulting compact conformation is amphipathic, and is responsible for these compounds' antigelling activity (FAP, LFF, and succinyl-FGGF are 3, 1.5, and 1.3 times more effective than L-Phe, respectively, data not shown). Binding studies of succinylated bi-aromatic molecules (Phe-Phe, Trp-Trp, and Phe-Gly-Phe) have shown that they bind at a preferred site on Hbs ($\Delta G \simeq -3$kcal/mol at 23°C)[13], and prevent gelation. We believe that the four antigelling compounds act by binding at the same preferred site on HbS, and thereby increasing the macromolecule's solubility.

The structural data we report also provide some insight into the problem of designing biaromatic compounds capable of passing the red cell membrane. Positively charged FOB and PAF are readily taken up by the erythrocyte; whereas, +vely charged LFF and neutral FGGF cannot enter the red cell. Therefore, neither the interlocking phenyl ring system nor the compound's charge are limitimg factors for membrane passage. Instead, it appears that the size and hydrophobic content of the agent play important roles in cellular permeability. FOB and PAF, which are similar in size, are substantially smaller than LFF and FGGF, and have quite different hydrophobic densities (FOB and PAP are about 80% C-atoms by volume, and LFF and FGGF are about 66% C-atoms by volume).

Efficient development of any therapeutic agent relies heavily on a molecular framework from which to design and synthesize compounds intended for a specific function. The

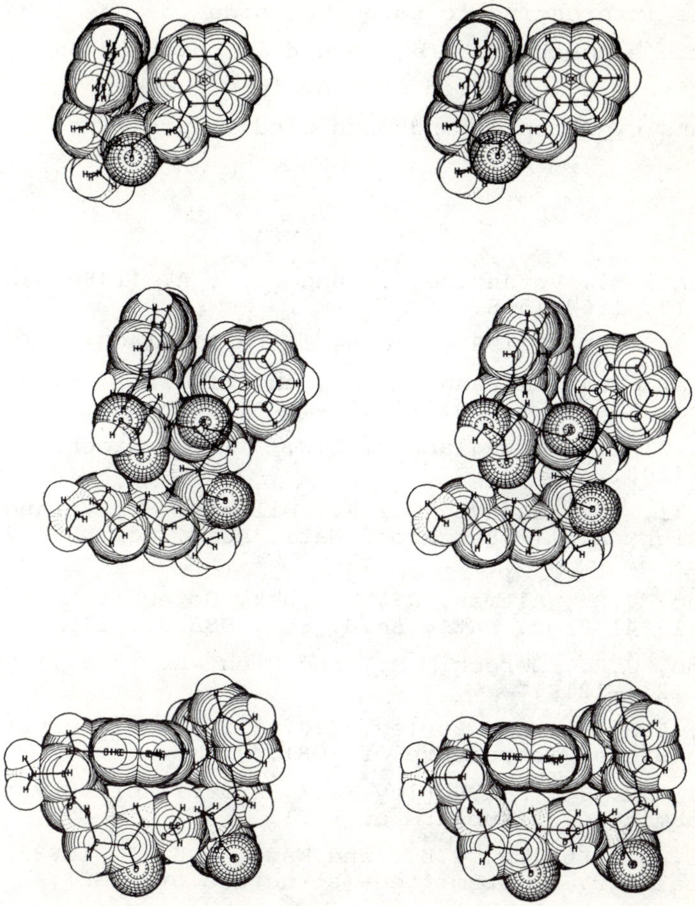

Fig. 1. Van der Waals stereodrawing of the molecular structure of FOB, LFF, and FGGF. PAF is depicted in the abstract book. Both counter-ions and water molecules were omitted for clarity.

conformation of antisickling and antigelling agents we report represents such a framework for future development of potential therapeutic agents for sickle cell disease.

It is interesting to note that side-to-edge interactions of aromatic side-chains are frequently observed in proteins, and constitute a mechanism of protein structure stabilization and a means by which ligands can bind to proteins[14].

References

1. Gorecki, M., Votano, J.R., and Rich, A. (1980) Biochemistry 19, 1564-1568.
2. Poillon, W.N. (1982) Biochemistry 21, 1400-1406.
3. Abraham, D.J., Mehanna, A.S., and Williams, F.L. (1982) J. Med. Chem. 25, 1015-1020.
4. Behe, M.J., and Englander, S.W. (1979) Biochemistry 18, 4196-4200.
5. Gorecki, M., Acquaye, C.T.A., Wilchek, M., Votano, J.R., and Rich, A. (1980) Proc. Natl. Acad. Sci. USA 77, 181-185.
6. Votano, J.R., Altman, J., Wilchek, Gorecki, M., and Rich, A. (1984) Proc. Natl. Acad. Sci. USA 81, 3190-3194.
7. Votano, J.R., Gorecki, M., and Rich, A. (1977) Science 196, 1216-1219.
8. Wang, A.H.-J., and Burley, S.K. (1985) Acta Crystallogr. C, submitted (structure of FOB).
9. Wang, A.H.-J., and Burley, S.K. (1985) Acta Crystallogr. C, submitted (structure of LFF).
10. Fujii, S., Burley, S.K., and Wang, A.H.-J. (1985) Acta Crystallogr. C, submitted (structure of FGGF).
11. Burley, S.K., and Wang, A.H.-J. (1985) Acta Crystallogr. C, submitted (structure of PAF).
12. Wyckoff, R.W.G. (1969) Crystal Structures: Structure of Benzene Derivatives, Interscience Publishers, New York.
13. Votano, J.R., and Rich, A. (1985) Biochemistry 24, 1966-1970.
14. Burley, S.K., and Petsko, G.A. (1985) Science 228, in the press.

AN EXPOSED BACKBONE IN Li^+ PERHYDROANTAMANIDE, AN INACTIVE ANALOG OF THE CYCLIC DECAPEPTIDE ANTAMANIDE

Isabella L. Karle

Laboratory for the Structure of Matter, Naval Research Laboratory, Washington, D. C. 20375-5000

Introduction

The cyclic decapeptide antamanide,[1] Pro^8-Phe^9-Phe^{10}-Val^1-Pro^2
Pro^7-Phe^6-Phe^5-Ala^4-Pro^3,
isolated from the deadly poisonous mushroom <u>Amanita phalloides</u>, acts as an antitoxin to the toxin phalloidin found in the same mushroom in much larger quantities. Antamanide and <u>active</u> synthetic analogs all form <u>ion complexes</u> with Li^+ and Na^+, e.g. <u>Inactive</u> analogs do not make ion complexes except for the perhydro derivative,[1] Pro^8-Cha^9-Cha^{10}-Val^1-Pro^2
Pro^8-Cha^6-Cha^5-Ala^4-Pro^3,
where Cha = cyclohexylalanyl.

X-ray diffraction analyses of many different single crystal forms have already been made of uncomplexed and complexed antamanide and active analogs.[2-6] In this paper are shown the structure and conformation established by x-ray diffraction analysis of the Li^+ complex of the <u>inactive</u> perhydroantamanide. A comparison is made with the Li^+ complex of natural antamanide.[2] Both complexes were made and crystallized in acetonitrile solution by an identical procedure.

Results

The effects of perhydrogenation of Phe residues on antamanide are the following: (1) complete loss of antitoxic potency;[1] (2) Complex formation unaffected; (3) Backbone conformation in complex unaffected; (4) Rotations of 120°-125° about the C^{α}-C^{β} bonds of the four Cha residues (5,6,9,10) as compared to Phe residues, Figures 1 and 2; (5) Exposure of polar backbone to exterior environment with the consequences, shown in Figure 2, that there is ligand formation of carbonyl oxygen O(10) with an extra Li^+ ion on the exterior of the complex, that four amide groups participate in hydrogen bonds of the type NH···Br^-, and that four hydrogen bonds between CO groups and water molecules are formed; (6) Shielding carbonyl oxygen O(4) and O(9) from the exterior environment (compare "bottom" views in Figure 3); (7) Water molecule fifth ligand to internal Li^+ instead of CH_3CN; and (8) Change in hydrophobic areas of surface of complex.

Discussion

Is the exposed backbone in the perhydro complex responsible for the antitoxic inactivity? Antitoxic activity of antamanide appears to occur on or near the surface of liver cell membranes.[1] The binding of the antamanide prevents access by the phalloidin toxin. The ability of antamanide and analogs to complex with ions such as Li^+, Na^+ or Ca^{++} is a necessary, but not sufficient condition for activity. Perhaps the chief role of complexatiion is to fold the backbone into a conformation so that carbonyl oxygens O(4) and O(9) are exposed in an otherwise globular hydrophobic shell, Figure 3. It may be that these carbonyl oxygens bind to the OH groups in the cholesterol that is present in high concentrations in cell membranes. The possibility of such an action is suggested by the inactivity of the perhydro complex where residues 5, 9 and 10 bury O(4) and O(9) into the interior.

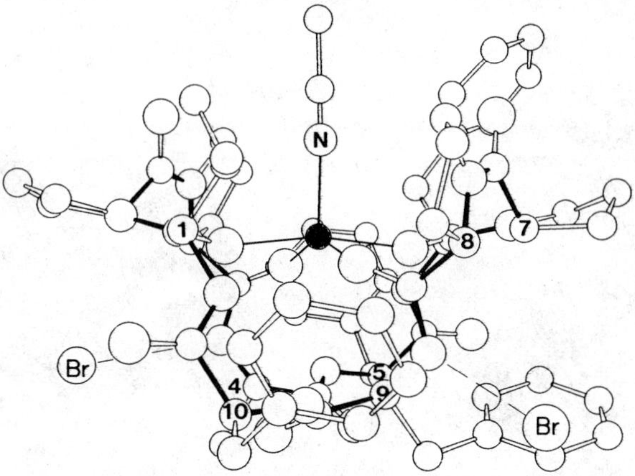

Fig. 1. Side view of Li^{+} antamanide$\cdot CH_3CN \cdot Br^{-}$. The numbered atoms are the C^{α} atoms in each residue. The phenyl groups in residues 5, 9 and 10 fold against the backbone to form part of the hydrophobic exterior shell of the complex.

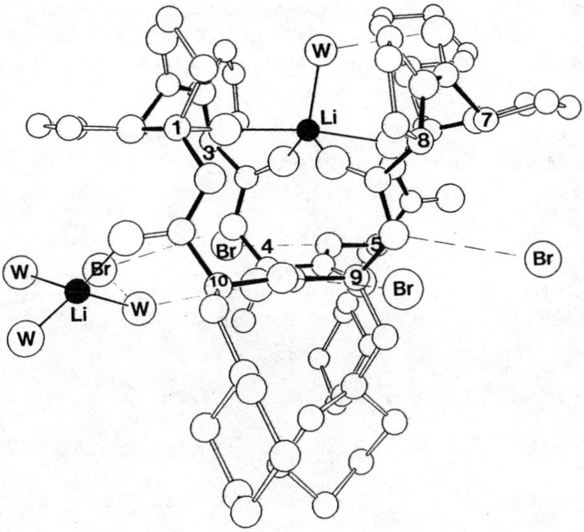

Fig. 2. Side view of $2Li^{+}$perhydroantamanide$\cdot 4H_2O \cdot 2Br^{-}$ in an orientation similar to that in Figure 1. The cyclohexyl groups in residues 5, 9 and 10 are suspended below the folded backbone resulting in exposure of the backbone to the environment. An additional external Li^{+} forms a ligand to O(10).

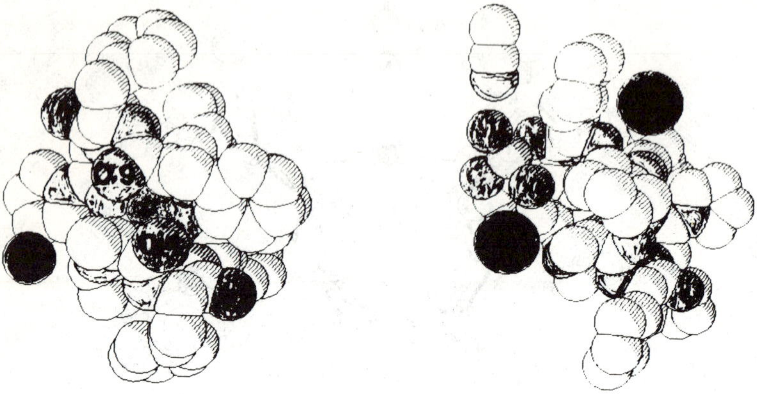

Fig. 3. Bottom views of complexes in Fig. 1 (left) and in Fig. 2 (right) represented by space-filling drawings of atoms. Oxygen atoms from residues 4 and 9 are exposed to the exterior in the left complex, but are shielded by the cyclohexyl groups in the right complex.

This work has been supported in part by NIH grant GM30902.

References

1. Wieland, Th. (1972) in Chemistry and Biology of Peptides, Meienhofer, J., Ed., Ann Arbor Sci., Ann Arbor, MI, pp. 377-396.

2. Karle, I. L. (1974). J. Amer. Chem. Soc. 96, 4000-4006.

3. Karle, I. L. (1974). Biochemistry 13, 2155-2162

4. Karle, I. L. (1977). J. Amer. Chem. Soc. 99, 5152-5157.

5. Karle, I. L. and Duesler, E. (1977). Proc. Natl. Acad. Sci. USA 74, 2602-2606.

6. Karle, I. L., Wieland, T., Schermer, D. and Ottenheym, H. C. J. (1979). Proc. Natl. Acad. Sci. USA 76, 1532-1536.

CONFORMATION OF CYCLO-$(PRO\text{-}GLY)_3$ IN ITS COMPLEX WITH SODIUM AND CALCIUM IONS.

K. Krishna Bhandary and Gopinath Kartha†
Biophysics Department, Roswell Park Memorial Institute,
666 Elm St., Buffalo, New York 14263.

Introduction

The synthetic cyclic hexapeptide cyclo-$(Pro\text{-}Gly)_3$ (cPG3) has been synthesised and studied extensively by Blout and co-workers[1-2]. From these studies they have concluded that the conformation of cPG3 varies significantly with the nature of the medium. In polar solvents cPG3 takes up an asymmetric conformation with one Gly-Pro bond cis and in non-polar solvents and when binding cation it takes up a C_3 symmetric structure. We have studied[3-7] using X-ray crystallographic techniques the conformation of cPG3 in the uncomplexed and complexed state in crystalline complexes with various metal ions. In these studies we have found that cPG3 forms sandwich complexes of varying stoichiometries. It also forms mixed ion complexes with calcium and sodium ions in which the calcium ion coordinates the glycyl carbonyls while the prolyl carbonyls coordinate with the sodium ions. In this paper we report the crystal structure of a new crystalline form of the mixed ion complex of cPG3 with calcium and sodium ions crystallised from a solution containing copper chloride.

Experimental

The mixed ion complex of cPG3 was crystallised from a solution containing calcium complex of the cyclic hexapep-

† Deceased

tide cPG3, sodium thiocyanate and copper chloride in water/methanol. Crystals are monoclinic, space group $P2_1$ with unit cell dimensions a=12.902(2)Å, b=12.616(2)Å, c=19.954(3)Å, β=96.25°(2). The cell dimensions and intensity data were measured on an Enraf-Nonius CAD-4 diffractometer using Cukα radiation. Out of 5055 unique reflections measured by the ω-2θ scan technique 3036 reflections had intensities greater than 2σ(I) and hence used in structure determination & refinement.

Structure was determined using a combination of MULTAN and Fourier techniques. The asymmetric unit contained two cPG3 molecules, one Ca^{2+} ion, two Na^+ ions, one Cu^+ ion, five SCN^- ions, four water molecules and one methanol for a total of 91 non-hydrogen atoms. The structure was refined by block-diagonal least squares method of an R-factor of 0.11 for 3036 reflections.

Results and Discussions.

There are two hexapeptide molecules in the asymmetric unit. Both the molecules have similar conformations with approximate three-fold symmetry. The conformation of one of the molecules is shown with that of cPG3 in the uncomplexed state crystallised from polar solvents in Figure 1. As predicted from solution studies, cPG3 crystallised from polar solvents exhibit an asymmetric conformation with one Gly-Pro

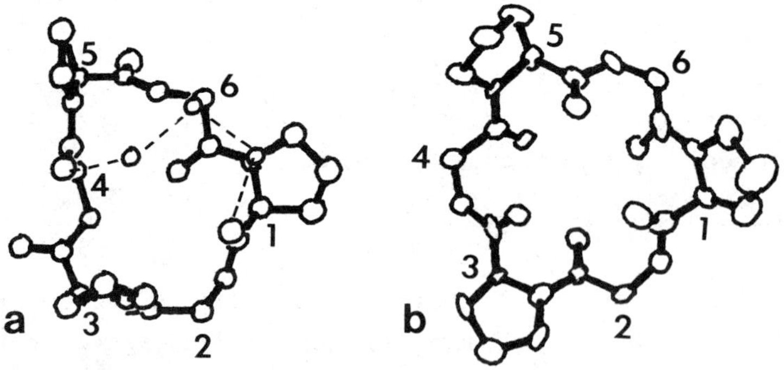

Fig. 1. Conformation of cPG3 in a) uncomplexed and b) metal ion complex. The α-carbon atoms are numbered.

Table 1. Main chain torsion angles for cPG3. The values in the parenthesis are those observed for free molecule.

Residue		φ	ψ	ω	Residue		φ	ψ	ω
Pro	1	-60 (-56)	158 (153)	176 (173)	Gly	2	68 (91)	-168 (-125)	174 (-174)
	3	-54 (-78)	158 (-8)	177 (-170)		4	69 (90)	-176 (159)	-180 (2)
	5	-58 (-77)	159 (-10)	-180 (-170)		6	70 (-96)	-167 (-163)	-177 (-178)

bond cis and the other five peptide links trans. The asymmetric conformer is further stabilised by hydrogen bonds involving three water molecules linking the prolyl carbonyl O1 and the glycyl carbonyl O4. The main chain torsion angles are listed in Table 1. It can be seen that the major changes occur in the peptide backbone upon complexing a metal ion. The hexapeptide attains a C_3 symmetric conformation in the complex with all peptide links trans and the glycyl carbonyls & the prolyl carbonyls pointing towards opposide sides of the peptide ring. Three carbonyls on either side of the ring form 'cups' of 3 oxygen to which a metal ion can bind.

In the crystalline state cPG3 forms different types of sandwich complexes with various metal ions[3-7]. The infinite sandwich formed in the present structure is shown in Figure 2. The repeating unit of the infinite sandwich has the sequence $(solvent\text{-}Na\text{-}cPG3\text{-}Ca\text{-}cPG3\text{-}Na)^{4+}$. The calcium ion is sandwiched

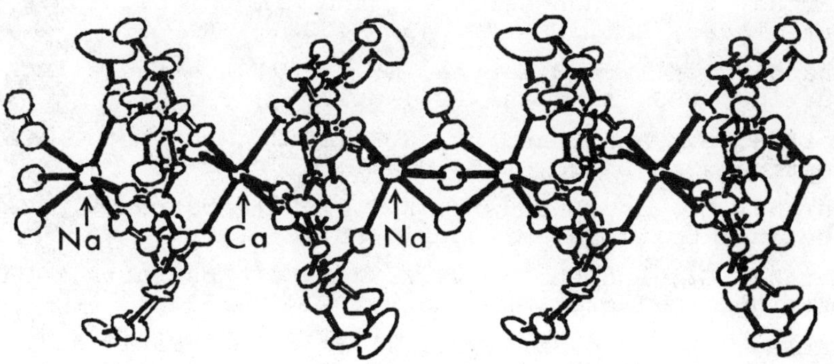

Fig. 2. The conformation of the mixed ion complex.

between two peptide molecules and is coordinated by 6 glycyl carbonyls with an average Ca-O distance of 2.30Å. On either side of the sandwich the prolyl carbonyls are coordinated to two sodium ions at an average Na-O (Prolyl) distance of 2.35Å. The sodium ions complete the octahedral coordination by coordinating to two water molecules (Na-O:2.54Å) and one methanol oxygen (Na-O:2.40Å). The solvent layer is shared by two sodium ions. In all the metal ions complexes of cPG3 when there is a divalent cation the glycyl carbonyls coordinate the metal ion. This is different from the 1:2 complex[8] of Mg^{2+} with cyclo$(Gly\text{-}Pro\text{-}Pro)_2$ in which 4 oxygens from the gly residues and 2 oxygens from the Pro residues are used to complete the octahedral coordination. In the crystal the metal ion sandwiches form columns along the b axis. The voids between these columns are filled by Cu^+.3SCN^- units, (Cu-N:1.96 A), solvent molecules and SCN^- counter ions.

Acknowledgement: This work is supported by NIH grant GM 22490 and the New York State department of Health.

References:

1. Madison,V., Atreyi, M., Deber, C.M., and Blout, E.R. (1974) J. Amer. Chem. Soc. 96, 6725-6734.

2. Deber, C.M., Madison, V., and Blout, E.R. (1976) Acc. Chem. Res. 9, 106-113.

3. Kartha, G., Varughese, K.I., and Aimoto, S., (1982) Proc. Natl. Acad. Sci. USA 79, 4519-4522.

4. Bhandary, K.K. and Kartha, G. (1982) Abstracts, ACA meeting, NBS, Maryland PB4, p.19.

5. Kartha, G., and Bhandary, K.K.(1982) Abstracts, ACA meeting, San Diego, PG5, p.44.

6. Kartha, G., and Bhandary, K.K.(1983) Abstracts, ACA meeting, Columbia, Missouri, PB17, p.31.

7. Kartha, G., and Bhandary, K.K.(1984) Abstracts, ACA meeting, Lexington, Kentucky, PC14, p.53.

8. Karle, I.L., and Karle, J. (1981) Proc. Natl. Acad. Sci. USA 78,681-685.

VIBRATIONAL CIRCULAR DICHROISM OF POLYPEPTIDES

T. A. Keiderling, S. C. Yasui, and A. C. Sen,
Department of Chemistry, University of Illinois at Chicago, Chicago, Illinois 60680

C. Toniolo and G. M. Bonora
Biopolymer Research Center, C.N.R., Institute of Organic Chemistry, University of Padova, 35131 Padova, Italy

Introduction

We and others have shown that vibrational circular dichroism (VCD), the CD of infrared vibrational transitions, can be used to gain insight into the stereochemical nature of molecular structure.[1] Recently it has been shown that the Amide A, I, and II bands of α-helical polypeptides yield VCD patterns characteristic of the helix sense.[2] In subsequent studies, we have shown that VCD for β-sheet and 3_{10}-helical oligopeptides may also be characteristic.

Here we will illustrate some of our solution phase VCD studies with results for polylysine in random-coil and β-sheet conformations, polytyrosine in random-coil and α-helical conformations,[3] and two globular proteins. A final example will be given demonstrating the minimum length of an Aib containing oligopeptide needed to establish characteristic 3_{10}-helical VCD spectra.

Experimental

Our methods of obtaining VCD of solution polypeptides

have been described elsewhere in detail.[1-3] In brief, we now have ability to measure VCD down in frequency to ~900 cm^{-1} with an adequate signal-to-noise ratio (S/N). For measurements on aqueous solutions, VCD must be obtained on very thin, concentrated, deuterium exchanged samples in D_2O. Even then, only the amide I band can be studied with our present instrumentation, due to interference by solvent bands. These samples were held between two clamped CaF_2 windows separated by a teflon spacer. The pH values are uncorrected for deuterium isotope effects. In non-aqueous environments, we can also study Amide A and II bands. We have separately discussed the limitations of using films for VCD.[4]

Results and Discussion

In Figure 1a are shown the VCD and absorption spectra of poly-L-lysine at pH ~7.3 (random-coil) and ~11.2 after heating (antiparallel β-sheet). Similarly, in Figure 1b are shown the Amide I and II VCD of poly-L-tyrosine in DMSO (random-coil) and DMSO:DCA, 80:20 (α-helical, right-handed). These exemplify three basic structural types for polypeptides and their characteristic Amide I VCD. Both the α-helical and random-coil Amide I bands exhibit VCD couplets at ~1650 cm^{-1} but of opposite sense. The antiparallel β-sheet has two negative VCD bands correlating to the absorption maxima near 1613 and 1680 cm^{-1}. In comparison to the near-uv CD, the random-coil results have a surprisingly large magnitude, being within a factor of two, in terms of $\Delta A/A$, of the α-helical result. In fact, the random-coil results are quite similar to the previously reported left-handed α-helical Amide I VCD.[2] Differentiation between these two structures is possible in the Amide A region

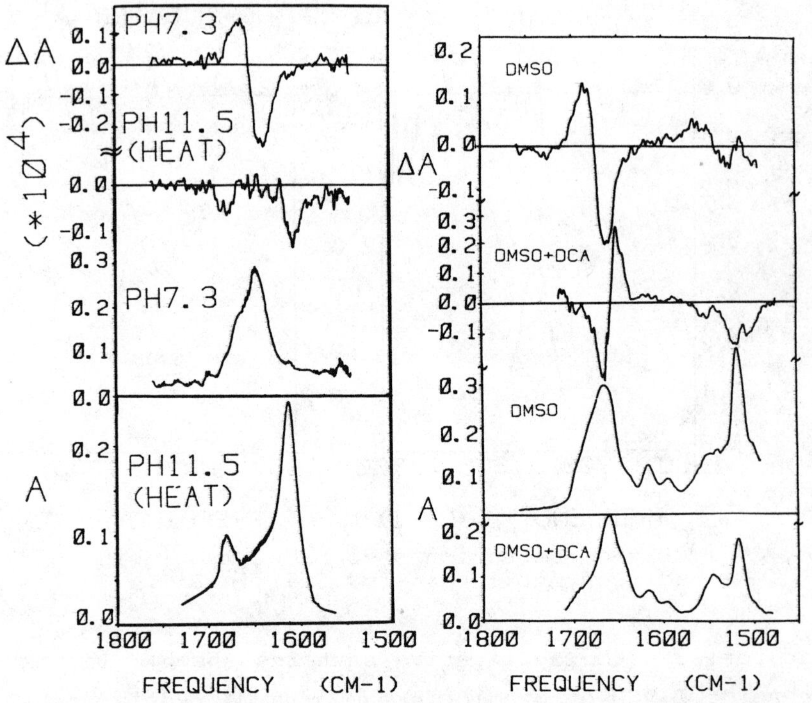

Fig. 1. VCD and absorption spectra of (a) poly-L-lysine in D_2O at pH 7.3 (random) and pH 11.5 (β-sheet) and (b) poly-L-tyrosine in DMSO (random) and DMSO:DCA, 80:20 (α-helical) in the Amide I and II regions.

where a broad, ill-defined band occurs for the random-coil.[3] These results indicate that VCD, while sensing the secondary structure, is of a more local nature than electronic CD and may offer complementary information. It also implies that the variety of well-resolved chromophores accessible in VCD have a distributed information content.

The VCD structural models above are analogous to those used early in the development of electronic CD for characterization of protein secondary structure. While, at present, our data base is insufficient to make a significant impact upon such studies, we can obtain globular protein VCD

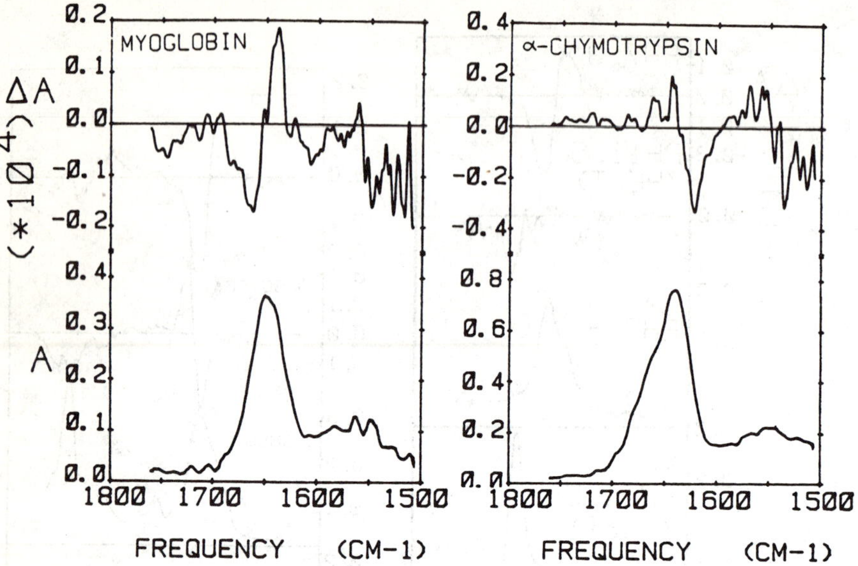

Fig. 2. VCD and absorption spectra of myoglobin (pH 7.6) and α-chymotriypsin (pH 3.9) in D_2O.

that reflects the qualitative patterns shown. In Figure 2 are Amide I VCD of myoglobin and α-chymotrypsin in D_2O . The region below 1550 cm^{-1} is obscured by noise due to solvent absorption. Here the predominantly α-helical myoglobin clearly shows a VCD couplet of the same sign as the α-helical polytyrosine (Figure 1) but of smaller magnitude. A negative band lying lower in frequency, characteristic of a deuterated right-handed α-helix,[2,5] is also evident. The α-chrymotrypsin, having a high β-sheet component, has a predominantly negative band, paralleling the high pH polylysine results. This qualitative success suggests a role for VCD in protein stereochemical studies if improved S/N and data for other amide bands can be obtained.

The dependence of our polypeptide results on polymer length and end effects is important to their ultimate applicability to protein secondary structure analysis. We have made studies of β-sheet[4] and 3_{10}-helical[5] forming oligopep-

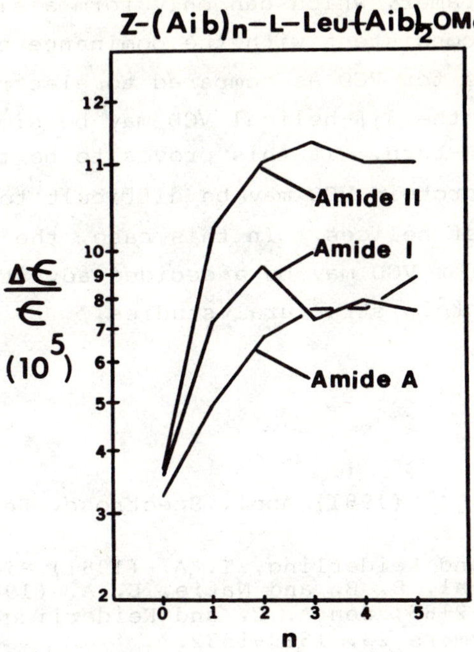

Fig. 3. Plot of ΔA/A vs. n for A_nLA_2 Aib containing oligopeptides. VCD was measured in $CHCl_3$.

tides. For the latter studies, several oligopeptides were synthesized[5] with the general formula A_nLA_2 where A is Aib, aminoisobutyric acid, and L is L-leucine. The octapeptide, n = 5, has previously been established as having a right-handed 3_{10}-helical structure by analogy to the A_8 and A_3VGLA_2 (V = L-Val, G = Gly) oligopeptides using nmr, x-ray and VCD studies.[5,6]

We measured the VCD of the Amide A, I, and II bands of A_nLA_2 as a function of n to determine when the characteristic 3_{10}-helical VCD developed. As can be seen in Figure 3, for all 3 bands the characteristic magnitudes of ΔA/A are fully developed by n = 3 which corresponds to almost two turns (3 H-bonds) if an ideal 3_{10} helix were formed. The spectra, in fact, evidence the same qualitative VCD as early as n = 1, the tetramer, which can only form a single turn. Such a result is consistent with the dominance of relatively local interactions for VCD as compared to electronic CD. It thus appears that the 3_{10}-helical VCD may be similar to that of the analogous β-turn. If this proves to be true, β-turn contributions to protein VCD may be difficult to differentiate from those of helices. In this case, the variety of bands accessible for VCD may be a decided advantage for application to protein structural studies.

References

1. Keiderling, T. A. (1981) Appl. Spectrosc. Rev. 17, 189-226.

2. Singh, R. D. and Keiderling, T. A. (1981) Biopolymers 20, 237-240; Lal, B. B. and Nafie, L. A. (1982) Biopolymers 21, 2161-2183; Sen A. C. and Keiderling, T. A. (1984) Biopolymers 23, 1519-1532.

3. Yasui, S. C. and Keiderling, T. A. (1985) Biopolymers 24, (in press).

4. Narayanan, U.; Keiderling, T. A.; Bonora, G. M. and Toniolo, C. (1985) Biopolymers 24, 1257-1263, and additional results to be submitted.

5. Yasui, S. C.; Keiderling, T. A.; Bonora, G. M. and Toniolo, C. (1985) Biopolymers 24, (in press); and references therein.

6. Toniolo, C.; Bonora, G. M.; Barone, V.; Baroso, A.; Benedetti, E.; Di Blasio, B.; Grimaldi, P.; Lelj, F. and Pedone C. (1985) Macromolecules 18, 895-902; Toniolo, C.; Bonora, G. M.; Benedetti, E.; Bavoso, A.; Di Blasio, B.; Pavone, V. and Pedone, C. (1985) Int. J. Biol. Macromol. (in press).

INTRAMOLECULAR DISTANCE DISTRIBUTIONS IN GLOBULAR PROTEINS DETERMINED BY EXCITATION ENERGY TRANSFER MEASUREMENTS

Elisha Haas

Department of Life Sciences, Bar-Ilan University, Ramat-Gan 52100, Israel

Introduction

The flexibility of peptide and protein structures and the dynamics of their structural transitions are essential components of the development and activities of any living organism. Understanding the mechanism of the folding/unfolding transitions of globular proteins depends on our ability to characterize the unfolded state and the partially folded intermediate structures which compose the pathway of folding. These structures are, by definition, short-lived and unordered and, hence, should be characterized by distributions of intramolecular distances rather than by single distances or coordinates. Measurements of nonradiative excitation energy transfer[1] has been shown to be one of the most suitable approaches for the study of flexible conformations in polypeptides.[2] It was shown that distributions of intramolecular distances between labeled sites,[2] as well as Brownian motions of labeled segments,[3] can be determined by the analysis of donor fluorescence decay curves.

We have recently extended this experimental approach to the study of the unfolding transition of bovine pancreatic trypsin inhibitor (BPTI).[4] BPTI derivatives labeled by donor and acceptor chromophores were prepared and their conformations studied by nanosecond pulse fluorometry.

Results and Discussion

BPTI was labeled at the N-terminal amino group by a 2-methoxy-naphthyl-1-methylenyl residue (MNA) (a donor of excitation energy), via a reductive alkylation reaction.[5] The pure labeled MNA-N- $-Arg^1$-BPTI (abbreviated

MNA-BPTI) was labeled by a second chromophore, 7-dimethyl-amino-coumarin-4-acetyl residue (DACA) (an acceptor). It was attached to the ε amino groups of lysine side-chains in a non-selective partial acylation reaction of its N-hydroxy-succinimide ester.[6] The products of this reaction were resolved by affinity chromatography and by reversed phase HPLC. Four doubly labeled (donor and acceptor) derivatives, MNA-N--Arg^{1}-DACA-N- -lys^{n}-BPTI (n=15,26,41,46) (abbreviated (1-n)BPTI) were thus prepared. HPLC tryptic peptide mapping was used to determine the labeled sites in each derivative and its purity. All the labeled derivatives were active inhibitors of trypsin and regained their inhibitory capacity after denaturation/reduction and renaturation/oxidation cycle. The excitation and emission spectra showed contributions by both the MNA and DACA chromophores with relative intensities which depended on the site of labelling and the conformational state of the protein.[4]

The fluorescence decay of the excitation energy donor, MNA, attached to the protein in the absence of an acceptor is monoexponential with a lifetime of 6.7 nsec (at room temperature in neutral aqueous solution). It has been shown[2] that the distance distributions functions between labeled sites, f(r), can be obtained from donor fluorescence decay curves by a least-squares curve fitting procedure, using Eq. 1 for the calculated decay curve $I_C(t)$:

$$(1) \qquad I_C(t) = m\int_0^{\infty} f(r)\exp - t/\tau\{1 + (\frac{R_o}{r})^6\}dr$$

In Eq. 1, R_o is the Förster critical distance[1] and τ is the fluorescence lifetime of the donor in the absence of an acceptor. We have used a skewed Gaussian expression for f(r): $f(r) = 4\pi r^2 e^{-a(r-b)^2}$, where a and b are free parameters.

The kinetics of the decay of the fluorescence of MNA in each one of the four doubly labeled derivatives (1-n)BPTI (n=15,26,41,46) was faster than that of MNA-BPTI and not monoexponential. Analysis of the experimental curves using Eq. 1 gave the donor acceptor distribution functions shown in Fig. 1. The average values and widths at half height of these distributions are given in Table I, which also include the distances between the labeled amino groups calculated from their coordinates obtained from x-ray crystallography.[7] The averages of the distributions follow the crystal data quite closely, which shows that we have a reliable method for determining distributions of intramolecular distances in globular proteins. The width of the calculated distributions includes the experimental noise, the flexibility of probes and the protein-carrying segments. In the native state, shown in Fig. 1, the three contributions cannot presently be

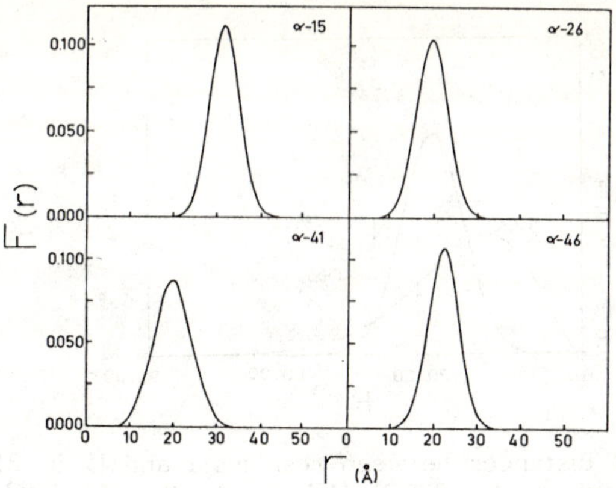

Fig. 1. Interchromophoric distance distributions calculated from donor fluorescence decay curves of (1-n)BPTI (n=15,26,41,46) in aqueous solution at room temperature. The derivative corresponding to each curve are indicated as "α-n".

factored out, but any increase caused by conformational change should be interpreted as an increase in the flexibility of the protein molecules. It is pertinent to note that these measurements are carried out in very dilute solutions under a wide range of conditions, solvents, temperature, viscosities or ligand binding. The results shown in Fig. 1 and Table I set the basis for using this method in the investigation of protein folding/unfolding transitions.

The five derivatives (1-n)BPTI (n=15,26,41,46) and MNA-BPTI were dissolved in 6M guanidinium hydrochloride in 50% glycerol solution in 0.1M bicin buffer pH 7.5 and their fluorescence decay measured. No significant increase in average distances was observed. When ß mercaptoethanol was added, a dramatic change in

Table I. Intramolecular Distances of (1-n) BPTI

Derivative	r_m^a (A)	W_i^b (A)	r_x^c (A)
1 - 15	30.5 ± 1	7 ± 2	31.7
1 - 26	18.7 ± 1.5	9 ± 2	16.8
1 - 41	18.7 ± 1.5	10 ± 2	19.1
1 - 46	20.5 ± 1.5	9 ± 2	21.7

(a) Average distance, r2 ; (b) Width at half height of the interchromophoric distance distributions shown in Fig. 1.; (c) Distance between the labeled amino groups determined by x-ray crystallography[7].

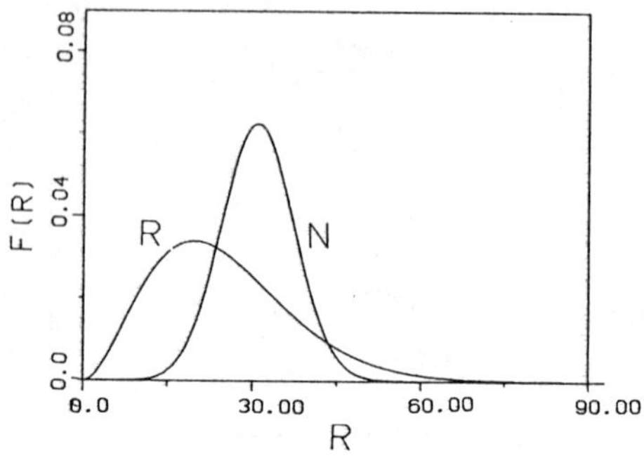

Fig. 2. Distribution of distances between residues 1 and 15 in BPTI in viscous solution (50% glycerol, -32°C) (N) and with added 6M guanidinium hydrochloride and β mercaptoethanol (R).

transfer efficiency and fluorescence decay was observed. Fig. 2 shows the denaturation of (1-15)BPTI as monitored by the excitation energy transfer. The average distance is decreased, showing that in the native state this segment is quite stretched compared to a random coil conformation. The width of the distribution is increased from 14 Å to 27 Å, a direct measurement of the transition to a random coil conformation. On the contrary, the distance between residues 1 and 26 or 1 and 46 increases upon reduction.

The high sensitivity, accuracy of distance determinations over the range of 10 Å to 80 Å, the high temporal resolution and the ability to determine well-localized distances within the polypeptide chain, make the method described above instrumental in the study of flexible biopolymers and the dynamics of their conformational transitions.

References

1. Förster, Th. (1948) Ann. Phys. (Leipzig) 2, 55-75.
2. Haas, E., Wilchek, M., Katchalski-Katzir, E. & Steinberg, I.Z. (1975) Proc. Natl. Acad. Sci. 72, 1807-1811.
3. Haas, E., Katchalski-Katzir, E., Steinberg I.Z. (1978) Biopolymers 17, 11-31
4. Amir, D. and Haas, E. (1985) Biopolymers (in press).
5. Amir, D., Levy, D.P., Levin, Y. & Haas, E. (1985) submitted.
6. Amir, D. & Haas, E. (1985) Int. J. Pept. Prot. Res. (in press).
7. Huber, R., Kulka, D., Bode, W., Schurger, P., Bartels, K., Deisenhofer, J. & Steigemann, W. (1974) J. Mol. Biol. 89, 73-101.

SYNTHESIS AND TRANSANNULAR REACTION OF CYCLOTRIPEPTIDES CONTAINING SECONDARY AMIDE GROUPS. THE ANSWER TO A PROBLEM UNSOLVED FOR MORE THAN 30 YEARS

Manfred Rothe, Manfred Fähnle, Werner Mästle, and Klaus Feige
Lehrstuhl Organische Chemie II, University of Ulm, 79 Ulm, FRG

The first attempts to synthesize cyclic peptides containing more than two amino acids were made in 1952.[1] At that time, tripeptides with CONH groups were cyclized and it was supposed, but not proved, that cyclotripeptides must be formed as reaction products. However, it was soon found out that the isolated crystalline compounds actually had the structure of cyclohexapeptides formed by dimerization. Not even a trace of cyclotripeptide could be detected in any case.

First hints to the reason of those failures resulted from the smooth formation of stable, well crystallized cyclotripeptides containing 3 N-substituted amino acids. Cyclotriprolyl was obtained by our group[2] as first example in yields exceeding 90%. This evidently shows the importance of the peptide hydrogen for the ring-closure reaction.

As 9-membered rings, cyclotripeptides are sterically characterized by a backbone conformation of the peptide with cis-amide bonds and above all with considerably restricted flexibility, though conformations with peptide bonds located close to each other are possible. According to our ideas this should lead to transannular reactions which might account for the preceding failures in addition to the well-known low formation tendency of medium-sized rings.

The key step in the interaction of two peptide bonds across the ring is the addition of a NH group to an opposite carbonyl group forming a bicyclic tetrahedral intermediate (II), a so-

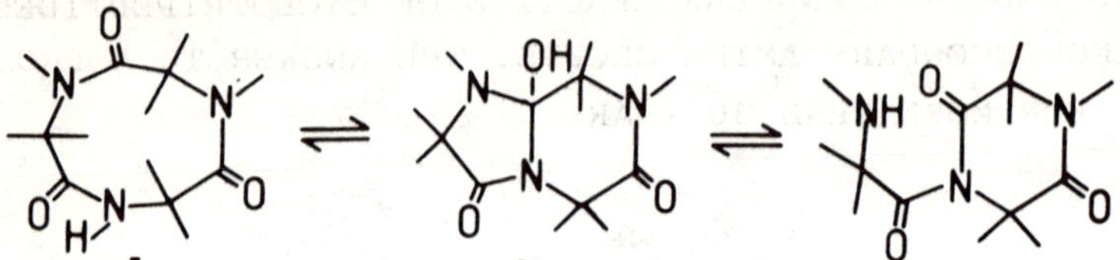

I II III

called cyclol. Azacyclols have the structure of ortho acid diamides with one acidic hydroxy group and are stable only in a few cases. Their fate determines the overall course of the reaction. Their breakdown can occur in different ways leading to a carbonyl group, either back to the cyclotripeptide (I) or by expelling an amino group to form an aminoacyl diketopiperazine (III). The tautomeric equilibrium may be shifted in favor of the cyclol or the cyclotripeptide depending on structural features of the amino acids, thus opening a synthetic access to cyclotripeptides containing CONH groups.

Two procedures were developed to synthesize aminoacyl diketopiperazines. The best route to obtain N-protected derivatives is the acylation of the ether-soluble trimethylsilyl diketopiperazines with N-protected amino acid chlorides with removal of trimethyl chlorosilane. In the special case of N-protected tripeptides with C-terminal proline, diketopiperazine ring formation can be performed under the influence of diphenylphosphoryl azide/2-mercaptopyridine[3] or EEDQ as well as by cyclization of the corresponding active esters with tert. amine. Removal of the protecting group leads to free aminoacyl diketopiperazines which are stable in certain cases, but may spontaneously react in case of a stronger nucleophilic amino group to form cyclols or cyclotripeptides. The two steps may also be combined by cyclizing unprotected tripeptides with C-terminal proline in the presence of the above-mentioned coupling reagents. In addition, the usual active ester ring closure is then proceeding via aminoacyl diketopiperazines as intermediates to yield cyclols or cyclotripeptides. This is also supported by the fact that the reaction takes place without high dilution. In this way we synthesized several all-L-cyclotripeptides of the sequence c-(Pro-X-Pro) containing Ala, Val, and Leu. If Gly or Phe were incorporated, however, the cyclotripeptides were not stable and were converted to the cyclols which were isolated in good yields. From the equilibrium mixture very small amounts of the Gly containing cyclotripeptide could be separated by silica gel chromatography.

Isomeric tripeptides of the sequence X-Pro-Pro cannot cyclize via aminoacyl diketopiperazines and cyclols, but only directly by end-to-end cyclization. In spite of these shortcomings the synthesis of c-(D-X-Pro-Pro) with Ala, Leu, and Phe was successful under mild conditions, while on the other hand, the N-terminal incorporation of L-Phe led again to the cyclol which in this case must have been formed from a less stable cyclotripeptide by transannular reaction.

It results from several cyclizations that the formation of cyclotripeptides decisively depends on sequence and configuration of the amino acids as well as on spatial requirements and flexibility of their side chains. An important factor is the incorporation of 1 or 2 proline residues favoring the formation of cis-amide bonds. A terminal D-amino acid supports a bent chain conformation favoring ring closure in contrast to peptides with an internal D-amino acid which cyclize via aminoacyl diketopiperazines, but only in small yields.

As has been mentioned, incorporation of a Gly residue shifts the equilibrium in favor of the cyclol. On the other hand, bulky side chains stabilize the 9-membered ring structure. Till now, c-(Pro-Val-Pro) is the most stable cyclotripeptide in crystalline state. In polar solvents an equilibrium with the cyclols is established again, as can be detected by ^{13}C-NMR spectroscopy. Moreover, cis-trans isomerization at the CONH bond takes place which is the prerequisite for interaction of two peptide groups across the ring and thus for cyclol formation. Accordingly, if c-(Pro-Val-Pro) is crystallized from solvents of different polarity, two stereoisomers can be obtained differing in their IR spectra. From more polar solvents, a cis-trans-cis conformer crystallizes which can be identified by an amide II band at 1515/cm. This band is missing in the all-cis conformer obtained from less polar solvents. In solution, an additional conformational equilibrium exists, as can be shown by ^{13}C-NMR studies. In this way, for the first time and entirely unexpected, a 9-membered cyclotripeptide with a trans-amide bond could be obtained, the presence of which was confirmed by X-ray analysis (Dr. T. Debaerdemaeker). Consequently, the cis-trans-cis form must correspond to a new twist conformation which is present in addition to the known[4] all-cis crown conformation.

In conclusion, we can give now an answer to a problem of peptide chemistry that has been unsolved for more than 30 years: the preceding failures in the synthesis of cyclic tripeptides are due to their low stability which further decreases with an increasing number of CONH groups. It results from the transannular formation of cyclols and their decomposition products that can be only avoided by extremely mild reaction, isolation, and separation conditions.

We gratefully acknowledge financial support by the Fonds der Chemischen Industrie.

References

1. Boissonnas, R.A., Schumann, I. (1952) Helv. Chim. Acta 35, 2229-2235.

2. Rothe, M., Steffen, K., and Rothe, I. (1965) Angew. Chem. 77, 347-348.

3. Yamada, S., Yokoyama, Y., and Shioiri, T. (1976) Experientia 32, 398-399.

4. Kessler, H., (1979) in Stereodynamics of Molecular Systems, Sarma, R.H., Ed., Pergamon Press, New York pp. 187-196.

SYNTHESIS AND CONFORMATIONAL ANALYSIS OF THREE DEHYDROPHENYLALANINE-CONTAINING CYCLIC PENTAPEPTIDES

Alvin C. Bach, II, Lila M. Gierasch, and Arnold L. Rheingold
Department of Chemistry, University of Delaware,
Newark, DE 19716

To explore the effect of trans α-β dehydrophenylalanine (ΔPhe) on reverse turns, three ΔPhe-containing model cyclic pentapeptides have been synthesized, each testing whether ΔPhe can be accommodated in a different position within a β or γ turn. For comparison, an analogous set of three cyclic pentapeptides in which D-Phe replaces the ΔPhe residue have also been prepared. The ΔPhe residue was incorporated into linear precursors using the modified Bergmann method;[1] the saturated peptides were prepared by standard solution methods.[2] All six peptides were cyclized as p-nitrophenol esters at high dilution in pyridine.

In pair 1, [cyclo(Gly-Pro-ΔPhe-D-Ala-Pro) (*I* in Scheme 1) and cyclo(Gly-Pro-D-Phe-D-Ala-Pro) (*II*)] both molecules adopt a γ turn in the D-Ala-Pro-Gly portion of the molecule and a Type II β turn in the Gly-Pro-D(Δ)-Phe-D-Ala portion. NMR data supporting this conformation are low Δδ/ΔT of the Gly and D-Ala amide protons, characteristic ^{13}C δ's of the C^{β} and C^{γ} and ^{1}H δ's for H^{α} of the Pro residues,[2] and the presence of an NOE between the Phe NH and the Pro H^{α}. This conformation places the ΔPhe or D-Phe residue into the i+2 position of the Type II β turn in each molecule.[3]

In pair 2, [cyclo(ΔPhe-Pro-Gly-D-Ala-Pro) (*III*) and cyclo(D-Phe-Pro-Gly-D-Ala-Pro) (*IV*)] both the ΔPhe- and D-Phe-containing molecules adopt an analogous β,γ-turn conformation to that found in pair 1. The γ turn is located in the

D-Ala-Pro-D(Δ)-Phe portion of the molecule and the Type II β turn in the D(Δ)-Phe-Pro-Gly-D-Ala section. NMR data supporting this conformation are principally low Δδ/ΔT of the D(Δ)-Phe and D-Ala amide protons and characteristic ^{13}C δ's of the C^{β} and C^{γ} and ^{1}H for H^{α} of the Pro residues.[2]

Molecules in the third pair [<u>cyclo</u>(Gly-Gly-ΔPhe-D-Ala-Pro) and <u>cyclo</u>(Gly-Gly-D-Phe-D-Ala-Pro)] have only one proline, and are more flexible than the other molecules studied. NMR data suggest an equilibrium between two conformations in pair 3: One conformation places the ΔPhe (*V*) [or D-Phe (*VI*)] in the i+2 position of a β turn, and the second conformation places the ΔPhe (*VII*) [or D-Phe (*VIII*)] in a position which is close to a γ turn. A structure similar to the second conformation of the ΔPhe-containing peptide was found in the solid state by x-ray crystallography (Figure 1). Other evidence for the interconverting solution conformations comes from ^{1}H and ^{13}C δ's for the Pro residue which are intermediate to the characteristic δ's observed for Pro in a comparable site in a β or γ turn.[2]

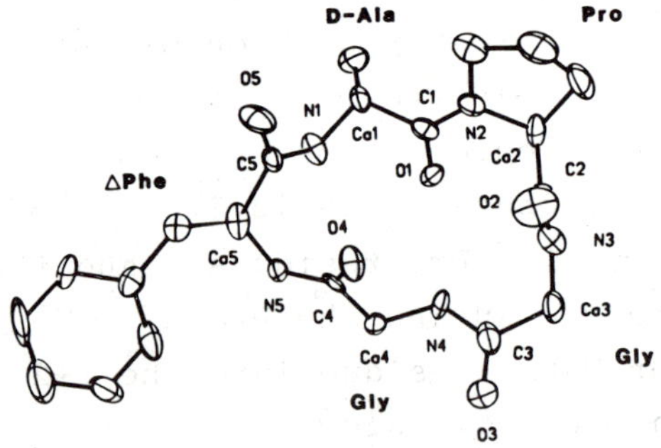

Fig. 1. X-ray crystal structure of *VII* as crystallized from acetonitrile/water (50:50). Note the Type II β turn in the D-Ala-Pro-Gly-Gly portion of the molecule.

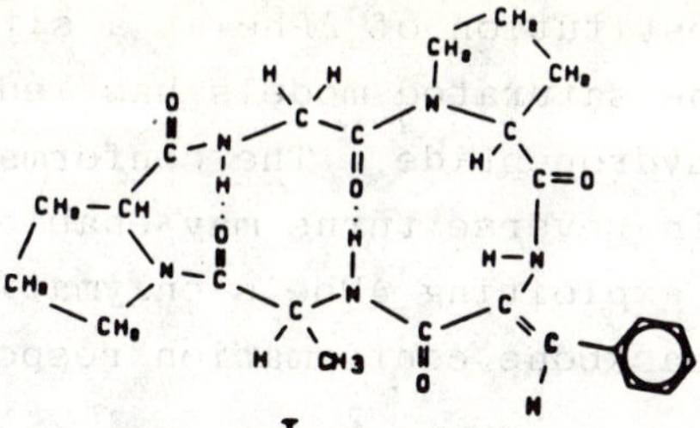

I

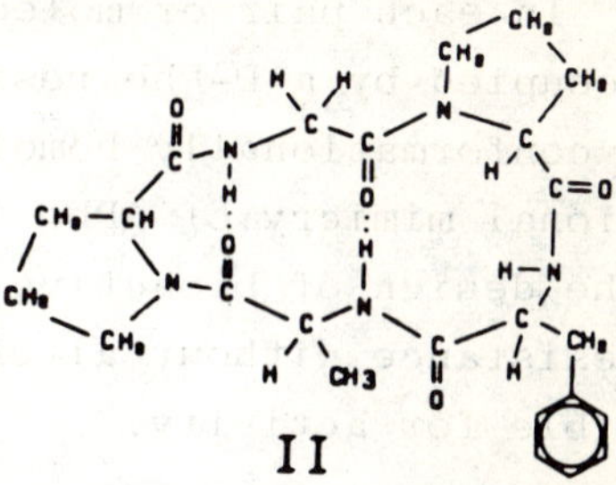

II

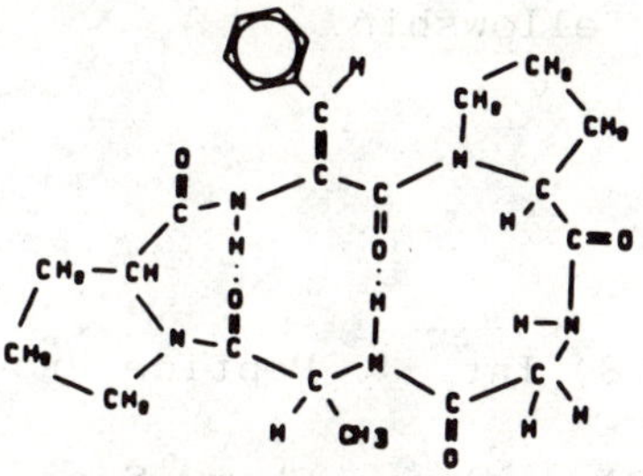

III

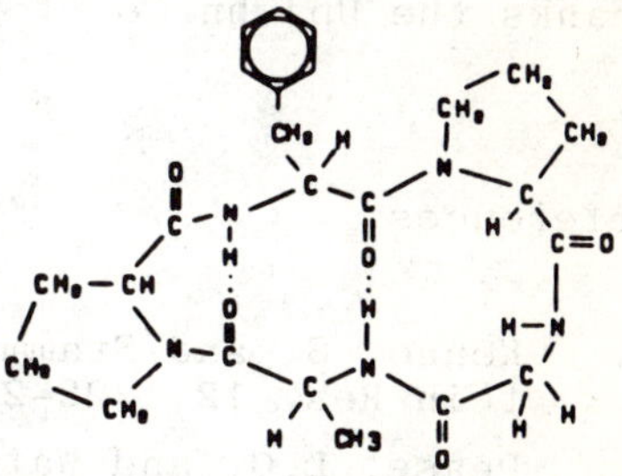

IV

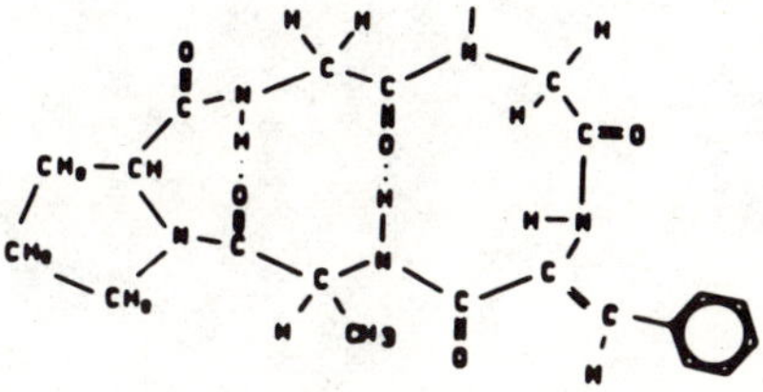

V

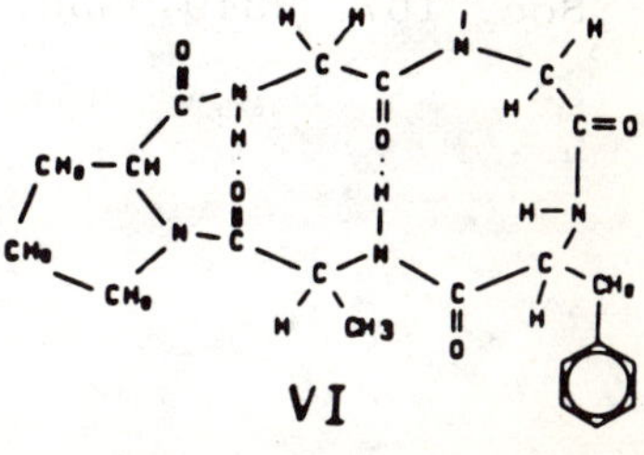

VI

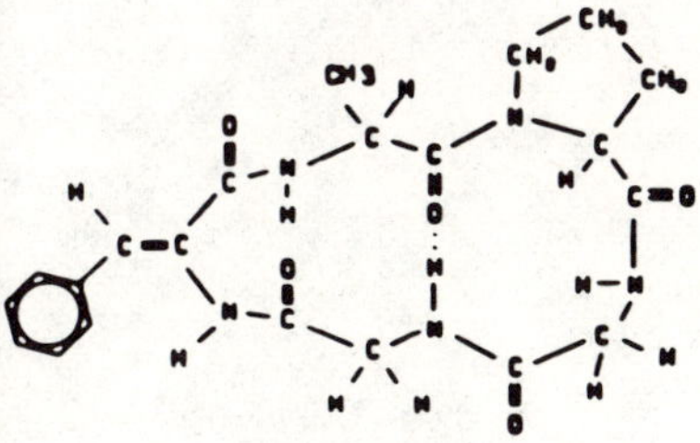

VII

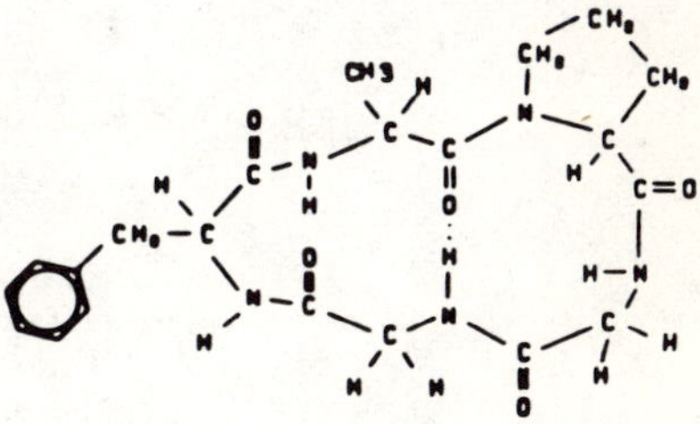

VIII

Scheme 1. Structures of cyclic pentapeptides.

In each pair of molecules, substitution of ΔPhe at a site occupied by a D-Phe residue in the saturated models has led to a conformationally homologous dehydropeptide. The conformational mimicry of ΔPhe for Phe in reverse turns may enable the design of bioactive peptides exploiting ΔPhe's enzymatic resistance without altering the backbone conformation responsible for activity.

This work was supported by N.I.H. grant GM27616. A.C.B. thanks the Upjohn Co. for a graduate fellowship.

References

1. Konno, S. and Stammer, C.H. (1978) Int. J. Peptide Protein Res. 12, 222-231.

2. Pease, L.G. and Watson, C. (1978) J. Amer. Chem. Soc. 100, 1279-1286.

3. Bach, II, A.C. and Gierasch, L.M. (1985) J. Amer. Chem. Soc. 107, 3349-3350.

BICYCLIC NONAPEPTIDES: SYNTHESIS, CONFORMATION, AND ION BINDING

Giancarlo Zanotti, Bruce Campbell, Mark A. Staples, and Elkan R. Blout
Department of Biological Chemistry, Harvard Medical School, Boston, Massachusetts 02115

Eric T. Fossel
Department of Radiology, Beth Israel Hospital, Boston, Massachusetts 02115

We report the synthesis of two homodetic bicyclic nonapeptides: cyclo-(Glu^1-X^2-Pro^3-Gly^4-Lys^5-X^6-Pro^7-Gly^8-)-cyclo-($1\gamma \rightarrow 5\varepsilon$)$Gly^9$,where X = Ala or Leu for I or II, respectively (Figure 1). These peptides extend the series of bicyclic peptides synthesized in our laboratory.[1]

The bridgeheads of these peptides impose conformational restrictions in addition to those created by the monocyclic structure, creating greater potential selectivity of interactions with metal cations. Metal complexes of these peptides may serve as models for investigating protein-ion interactions.

Bicyclic peptides I and II exhibit distinctly different conformations in solution when measured by CD, as shown in Figure 2. Both peptides show conformational sensitivity to the presence of metal cations in acetonitrile solution. Each peptide attains different conformations depending on which metal cation is added. Complexation data indicate that most of the peptide-metal complexes are sandwich complexes, with a peptide:metal stoichiometry of 2:1. The

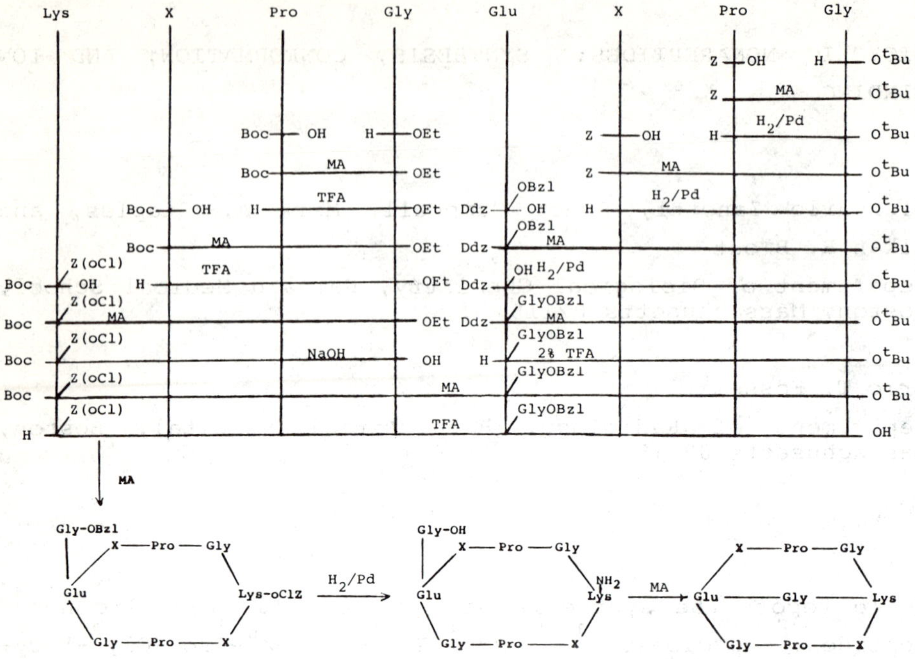

X = Ala for I (BCP-2)

X = Leu for II (BCP-3)

Fig. 1. Scheme for bicyclic nonapeptide synthesis.

apparent binding constants presented in Table I were obtained by titrating peptides I and II with metal cations, using the experimental conditions described in Figure 2.

Although a sequence-specific assignment is not yet available for the ^{1}H NMR spectrum of peptide I, the temperature dependence data (Table II) of the amide proton resonances in acetonitrile reveal that the Lys^{5}-αNH and the GlyA-αNH resonances are relatively temperature independent. These data indicate the sequestration of these protons from intermolecular interactions. Temperature-dependent shifts of the amide resonances (Δδ/ΔT) are not concentration-dependent in the range of 10 mM to 80 mM, thus indicating no significant change in aggregation state.

Table I. Binding Constants for Peptide-Metal Complexes

Peptide/cation[d]	Peptide I		Peptide II	
	1/1	2/1	1/1	2/1
	$k_1, M^{-1} \times 10^4$	$k_1k_2, M^{-2} \times 10^8$	$k_1, M^{-1} \times 10^4$	$k_1k_2, M^{-2} \times 10^8$
Metal Cation				
Li (I)	3.06	1.50	----	----
Mg (II)	complex equilibria		a	124.00
Ca (II)	----	2.52	b	b
Sr (II)	----	0.46	3.85	92.50
Ba (II)	----	7.34	----	111.00
Mn (II)	1.70	1.22	b	b
Cr (III)	2.62	0.84	c	c

a--insufficient data to calculate constants

b--wavelength-dependent complex formation

c--conformational change of the complex was not saturable with metal cation

d--stoichiometric assumption: molar ratios

Table II. Temperature Dependence of Amide Proton Resonances for Peptide II

Resonance	$\Delta\delta/\Delta T$ (Δppm x 10^{-3}) Concentration 10mM	40mM	80mM
εNH Lys (minor)	-1.50	-1.80	-1.69
εNH Lys (major)	-3.20	-4.20	-3.95
Leu A	-7.40	-6.00	-7.10
Gly A	-0.53	-0.20	-0.58
Gly B	-4.50	-4.40	-5.25
Gly C	-5.99	-5.70	-6.90
Lys	-0.22	-0.42	-0.10
Leu B	-2.32	-2.50	-2.50
Glu	-1.30	-1.35	-1.40

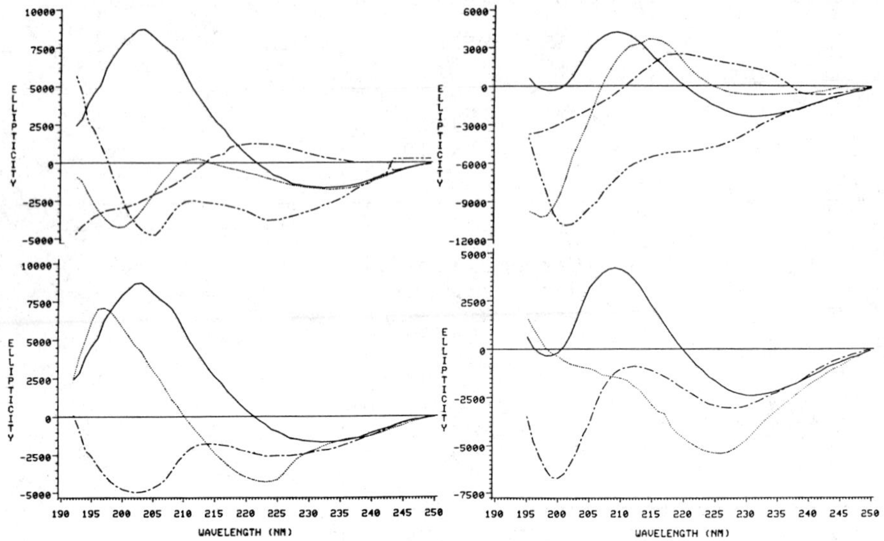

Fig. 2. CD spectra of peptide-metal complexes. Metal perchlorates were added to 50 μM bicyclic peptide in acetonitrile. Values in parentheses refer to the molar ratio of peptide:metal ion. Panels A and B show free peptide I spectra (——), and complexes with: A, Mg (II) (1.00, ...), Ca (II) (1.00, — - —), Ba (II) (1.00, — - - —); B, Mn (II) (1.25, ...) and Cr (III) (2.25, — - —). Panels C and D show free peptide II spectra (——), and complexes with: C, Mg (II) (1.00, ...), Ca (II) (1.00, — - —), Ba (II) (1.00, — - - —); D, Mn (II) (1.75, ...) and Cr (III) (1.25, — - —).

Acknowledgments

Support for this research from U.S. Public Health Service grants AM07300 and AM10794 is gratefully acknowledged.

References

1. Tolle, J.C., Staples, M.A., and Blout, E.R. (1982) J. Amer. Chem. Soc. 104, 6883-6884.

SYNTHESIS AND PROPERTIES OF SYNDIOTACTIC CYCLIC PEPTIDES CONTAINING ALPHA-AMINOFATTY ACIDS

Norikazu Nishino, Munetaka Ishiyama, and Tsutomu Fujimoto
Department of Industrial Chemistry, Kyushu Institute of Technology, Kitakyushu 804, Japan

Introduction

We are interested in the use of artificial hydrophobic amino acids in ionophore model peptides. As the fatty acid is an important element of bilayer membranes, the introduction of the nature of long chain alkyl groups into the membrane active peptide is of great interest. An example is the employment of N-decylglycine[1] for the increase of the solubility of designed ionophore in non-polar solvents. Though optically active α-aminofatty acids are known, they are not commonly incorporated in any reasonable peptide sequences. Since the natural-occurring cyclic depsipeptides, enniatins possess syndiotactic sense, and the L, D-alternating sequence in gramicidin A gives β-helix conformation whose spiral spin consists of about 6 amino acid residues,[2] we attempted to incorporate α-aminofatty acid in the syndiotactic and cyclic structure in expectation of ion binding and membrane activity.

Results and Discussion

The procedures for the preparation of optically active α-aminofatty acids are summarized in scheme 1. Alpha-aminolauric acid and α-aminomyristic acid are synthesized and resolved with A. g. acylase according to the literature.[3] The latter was obtained in much better yield and optical activity. However, this method was inapplicable for higher fatty acids due to the inactivation of acylase by the surfactants. Alpha-aminostearic acid was prepared by stereospecific hydrogenation of α, β-dehydroaminoacyl residue in cyclic dipeptide according to Kanmera et al.[4]

The incorporation of unusual amino acids with long chain alkyl groups (abbreviated as ALa, AMy, and ASt in scheme 1) into cyclic peptides was carried out in the conventional manner

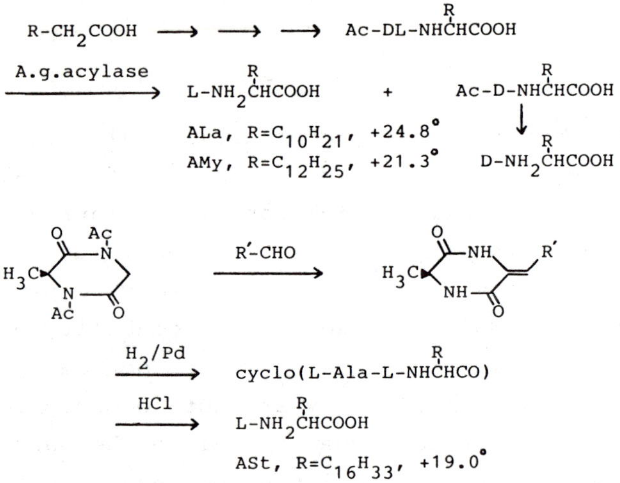

Scheme 1. Preparation of optically active α-aminofatty acids.

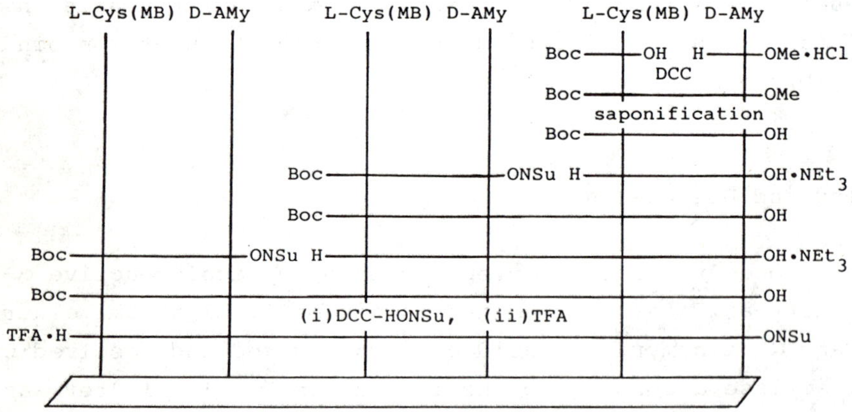

Scheme 2. Synthesis of cyclo(L-Cys(MB)-D-AMy)$_3$.

as shown in scheme 2 for cyclo(L-Cys(MB)-D-AMy)$_3$. The appropriate combination of α-aminofatty acids and usual amino acids gave other cyclic peptides, cyclo(L-Ser(Bzl)-D-ALa)$_3$, cyclo-(Gly-L-ALa)$_3$, cyclo(Gly-L-AMy)$_3$, and cyclo(Gly-ASt)$_3$. The synthesis of cyclic peptides was confirmed by field desorption mass spectrometry (JEOL DX300). All five peptides gave their M^+ ions and $(M + Na)^+$ cluster ions when excess amount of NaI was added.

These syndiotactic cyclic peptides are fairly soluble in chloroform and tetrahydrofuran (THF, excluding the Gly containing peptides), but scarcely in other organic solvents including

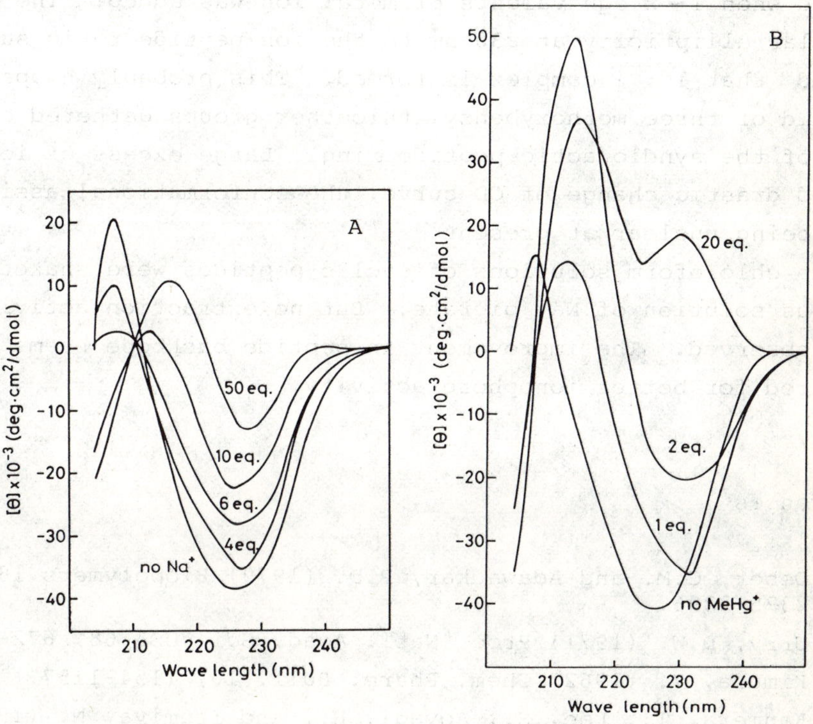

Figure. Changes in CD spectra of cyclo(L-Cys(MB)-D-AMy)$_3$ in the presence of $NaClO_4$ (A) and methylmercuric hydroxide (B). Peptide concentration, 5×10^{-4} M. Solvent; THF (A), 99% THF (B).

trifluoroethanol. Therefore CD spectra were measured in THF for cyclo(L-Cys(MB)-D-AMy)$_3$ and cyclo(L-Ser(Bzl)-D-ALa)$_3$. The former peptide changed the shape of CD curve by the addition of $NaClO_4$ in excess suggesting the weak binding with Na^+ ion (see Figure. A). In contrast with this, the CD spectrum of cyclo-(L-Ser(Bzl)-D-ALa)$_3$ nearly unchanged in the presence of large excess of Na^+ ion in spite of the similarity in chemical structure.

Since cyclo(L-Cys(MB)-D-AMy)$_3$ is furnished with possible ligand for heavy metal ion as thioether, the effects of the addition of methylmercuric hydroxide in CD spectrum was also examined. As seen in Figure. B a significant change was observed when 1~2 equivalents of metal ion was added. The plot of molar ellipticity at 230 nm to the ion/peptide ratio suggested that 1 : 1 complex is formed. This probably happens by the aid of three methoxybenzyl thioether groups gathered on one side of the syndiotactic peptide ring. Large excess of ion caused drastic change of CD curve, the conformational assignment being unclear at present.

The chloroform solutions of cyclic peptides were shaked with aqueous solution of Na^+ picrate. But no extraction activities were observed. The improvement in peptide backbone seems to be required for better ionophore activity.

References

1. Deber, C.M. and Adawadkar, P.D. (1979) Biopolymers 18, 2375-2396.
2. Urry, D.W. (1971) Proc. Natl. Acad. Sci. USA 68, 672-676.
3. Kimura, Y. (1962) Chem. Pharm. Bull. 10, 1154-1157.
4. Kanmera, T., Lee, S., Aoyagi, H., and Izumiya, N. (1980) Int. J. Peptide Protein Res. 16, 280-290.

FOLDED, HELICAL AND EXTENDED STRUCTURES OF PEPTIDES FROM $C^{\alpha,\alpha}$-DIALKYLATED α-AMINO ACIDS. SOLID - STATE AND SOLUTION CONFORMATION OF HOMO-PEPTIDES FROM $C^{\alpha,\alpha}$-DIETHYLGLYCINE.

Alfonso Bavoso, Ettore Benedetti, Benedetto Di Blasio, Vincenzo Pavone, and Carlo Pedone

Department of Chemistry, University of Napoli, 80134 Napoli, Italy

Claudio Toniolo and Gian Maria Bonora

Biopolymer Research Center, C.N.R., Department of Organic Chemistry, University of Padova, 35131 Padova, Italy

Miroslaw T. Leplawy, Adam Redlinski, and Krzysztof Kaczmarek

Institute of Organic Chemistry, Technical University, 90-924 Lodz, Poland

Introduction

Our interest in the stereochemistry of Deg ($C^{\alpha,\alpha}$-diethylglycine) peptides was stimulated by the potential usefulness of this and other $C^{\alpha,\alpha}$-dialkylated glycine residues as a new type of conformational constraint in the synthesis of agonists and antagonists of bioactive peptides[1].

We report here our results on Deg homo-peptides[2] and specifically: (i) conformational energy calculations of Ac-Deg-NHMe as a function of the relevant N-C^{α}-C' bond angle; (ii) diffraction analyses of Tfa$(Deg)_n$O*t*Bu (n=3-5); and (iii) IR absorption and ^{1}H NMR investigations of Tfa$(Deg)_n$O*t*Bu (n=2-5).

Results and Discussion

Our theoretical analysis indicates that Ac-Deg-NHMe is conformationally restricted and that the minimum energy conformation corresponds to the fully extended C_5 structure, when the N-C^{α}-C' bond angle is smaller than 108°, as experimentally observed (see below).

The results of the theoretical conformational analysis are in excellent agreement with the solid-state structural propensity of Tfa(Deg)$_n$O<u>t</u>Bu (n=2-5), as determined by X-ray diffraction. Figure 1 illustrates the molecular structure of the tetrapeptide. In the Deg residues of all homo-peptides the ethyl side chains are extended. The N-C^{α}-C' bond angles are close to 102°. This set of compounds allowed us to improve the characterization of the intramolecularly H-bonded C_5 conformation[3].

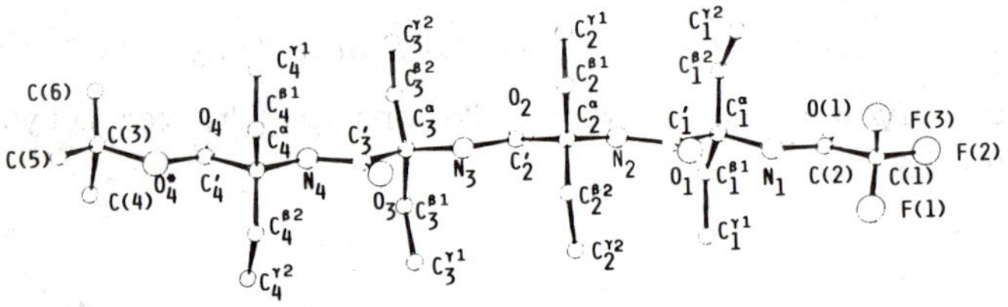

Fig. 1 Molecular structure of Tfa(Deg)$_4$ O<u>t</u>Bu

The solution conformational preferences of the Tfa(Deg)$_n$-O<u>t</u>Bu (n=2-5) homo-peptides have been determined in $CDCl_3$ by IR absorption and ^1H NMR as a function of concentration, temperature, and solvent perturbation. Intramolecular H-bonding was found to be the dominating factor. The absence of a conformational transition with increasing main-chain length and the remarkable stability to dilution, heating, and addition of a polar solvent (DMSO) is an additional relevant finding of this study. Figure 2 illustrates the IR absorption spectra (NH

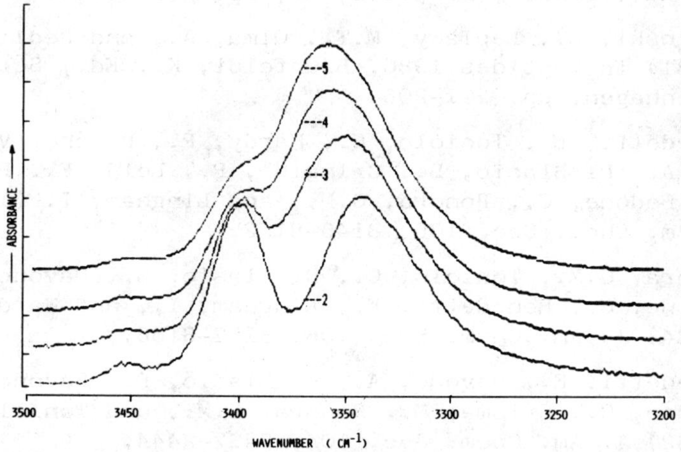

Fig. 2 IR absorption spectra in the N-H stretching region of Tfa(Deg)$_n$O*t*Bu (n=2-5) homo-peptides in $CDCl_3$ solution.

stretching region) of the four Deg homo-peptides in $CDCl_3$ (the band related to the intramolecularly H-bonded N-H....O=C group is that at lower energy). On the basis of the data on the preferred conformation of Deg homo-peptides in vacuo and in the solid-state, it seems reasonable to interpret the results in $CDCl_3$ solution as arising from multiple C_5 conformations. These findings are in agreement with those of the homo-peptides from the higher homolog Dpg (Dpg, $C^{\alpha,\alpha}$-di-n-propylglycine)[3,4], but contrast dramatically to those of the homo-peptides from the lower homolog Aib (Aib, α-aminoisobutyric acid or $C^{\alpha,\alpha}$-dimethylglycine), which have been shown to prefer the 3_{10}-helical conformation[5-7].

References

1. Spatola, A.F. (1983) in Chemistry and Biochemistry of Amino Acids, Peptides and Proteins, Vol. 7, Weinstein, B., Ed., Dekker, New York, NY, pp. 267-357.

2. Kaminski, Z., Leplawy, M.T., Olma, A., and Redlinski, A. (1981) in Peptides 1980, Brunfeldt, K., Ed., Scriptor, Copenhagen, pp. 201-206.

3. Benedetti, E., Toniolo, C., Hardy, P., Barone, V., Bavoso, A., Di Blasio, B., Grimaldi, P., Lelj, F., Pavone, V., Pedone, C., Bonora, G.M., and Lingham, I. (1984) J. Am. Chem. Soc. 106, 8146-8152.

4. Bonora, G.M., Toniolo, C., Di Blasio, B., Pavone, V., Pedone, C., Benedetti, E., Lingham, I., and Hardy, P. (1984) J. Am. Chem. Soc. 106, 8152-8156.

5. Benedetti, E., Bavoso, A., Di Blasio, B., Pavone, V., Pedone, C., Crisma, M., Bonora, G.M., and Toniolo, C. (1982) J. Am. Chem. Soc. 104, 2437-2444.

6. Toniolo, C., Bonora, G.M., Barone, V., Bavoso, A., Benedetti, E., Di Blasio, B., Grimaldi, P., Lelj, F., Pavone, V., and Pedone, C. (1985) Macromolecules, in press.

7. Bavoso, A., Benedetti, E., Di Blasio, B., Pavone, V., Pedone, C., Toniolo, C., and Bonora, G.M. (1985) Proc. Natl. Acad. Sci. USA, submitted.

SEQUENTIAL POLYPEPTIDES AS HISTONE MODELS AND THEIR COMPLEXES WITH DNA: SYNTHESIS AND CONFORMATIONAL STUDIES.

E. Panou-Pomonis, C. Sakarellos, and M. Sakarellos-Daitsiotis
Chemistry Department, University of Ioannina, 45110 Ioannina, Greece.

Introduction

The detailed knowledge of the nucleohistone molecular structure is essential in understanding their biological function since the DNA-protein interactions play a crucial role in several genetic processes. Simplified model DNA - polypeptide systems, simulating some of the properties of the lysine containing histones, have already been reported[1].Our approach to the design of the sequential polypeptides $(L\text{-Arg-X-Gly})_n$ with X= Ala (I), L-Val (II), L-Leu (III), i.e. with a hydrocarbon side-chain of various lengths and bulkiness, as models for arginine rich histones, arises from their amino-acid composition, sequence and conformational properties in solution[2]. The conformational characteristics of the polymers and their DNA complexes were investigated by circular dichroism spectroscopy.

Results and Discussion

The appropriate monomer tripeptides were synthesized by the mixed anhydride step-by-step procedure according to the Scheme 1. Pentachlorophenol esters were used for carboxyl activation and the polymerization was carried out in dimethylformamide solution in the presence of triethylamine.The toluene-4-sulfonyl group was eliminated after the polymerization by the treatment with liquid hydrogen fluoride. A satisfactory molecular

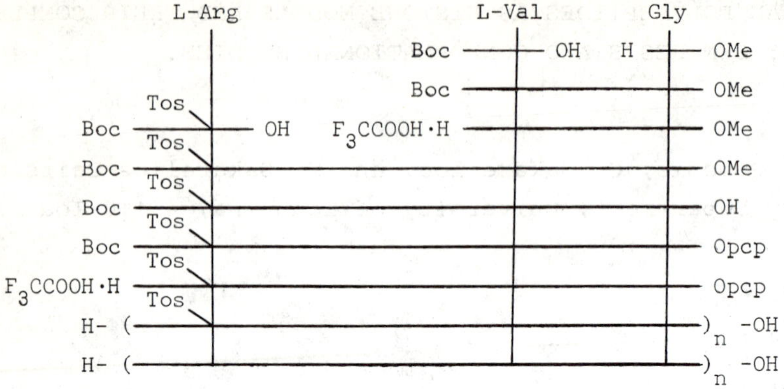

Scheme 1. Monomer and polymer synthesis

weight range of 10,000-20,000, comparable to that of arginine rich histones, was achieved. Polypeptide-DNA complexes were prepared in the desired ratio, r, Arg/nucleotide using a continuous-flow salt gradient linear dialysis.

The C.D. spectra showed that in aqueous solution, at neutral and alkaline pH, the prepared polymers I, II and III behaved as random coil. The random coil → α helix transition that basic polymers or copolymers usually undergo was not observed in our case. In trifluoroethanol or hexafluoroisopropyl alcohol-water mixtures it was indicated that the degree of helical conformation increased with the increasing bulk and the steric requirements of the side-chain of the X residue, in the order of Ala→ → Val → Leu (Fig.1A). However, even in 90% hexafluoroisopropyl alcohol or trifluoroethanol the prepared polypeptides are only partially helical. Although Ala, Val and Leu are known to form stable α helices, the fact that the prepared polytripeptides do not predominately assume any regular, repeating conformation must be mainly attributed to the periodicity of the glycine residue, which is considered as a strong α helix destabilizer. However, the inclusion of one Gly residue in a repeating trimer does not make the helical conformation thermodynamically impossible[3]. The probability then for helical conformation depends on the helical stabilization effects of the other residues. Arginine is assigned as α helix indifferent, and has a marginal effect.

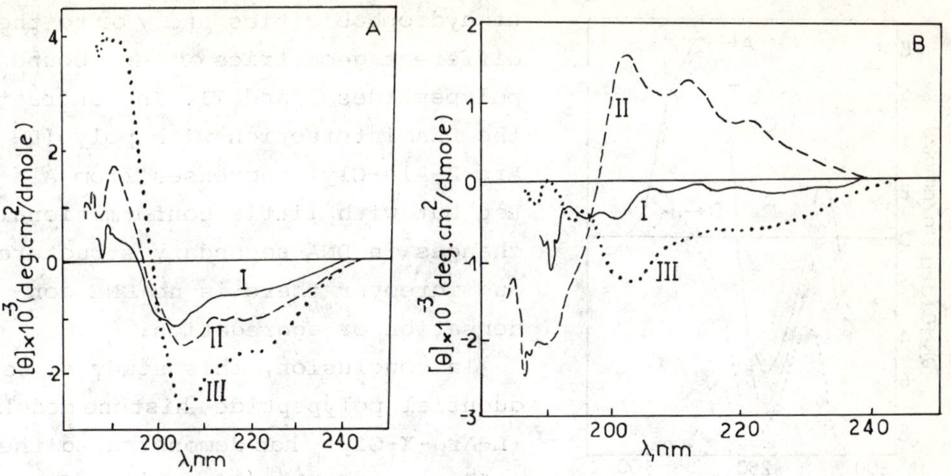

Fig.1. CD spectra of polymers I,II and III in hexafluoroisopropyl alcohol-H_2O [80:20, v/v (A) and 100:0, v/v (B)]. Concentration of polymers 3.95-3.77 mM (mean residue).

In contrast, the ability of β-structure increased with the decreasing bulk and length of the side-chain of the X residue in the order of Leu → Val → Ala. Although Leu and Val are described as more efficient β-formers compared to Ala, the periodic inclusion of the bulk side-chain of X residue, destabilizes the probability of β-structure proportionally to its bulkiness (Fig. 1A). Gly and Ala are known as β-structure indifferent, therefor have modest contribution to the β-structure formation. Poly (L-Arg-L-Val-Gly) exerts a significant preference to β-turn structure[4], compared to polymer I and II probably due to the β-branched side-chain (isopropyl-) of the valine residue (Fig.1B). Characteristics of β-turn are completely excluded from poly(L-Arg-L-Leu-Gly), which adopted the most pronounced helical conformation, under the given conditions and has to be attributed to an inverse relation between α-helix and β-turn formers.

From the CD studies of the prepared polytripeptide-DNA complexes it was found that poly (L-Arg-L-Val-Gly) exhibits pronounced structural changes (conformational transition from B to the more compact and asymmetric C form) as a function of ionic strength and polypeptide-DNA ratio, probably due to the differe-

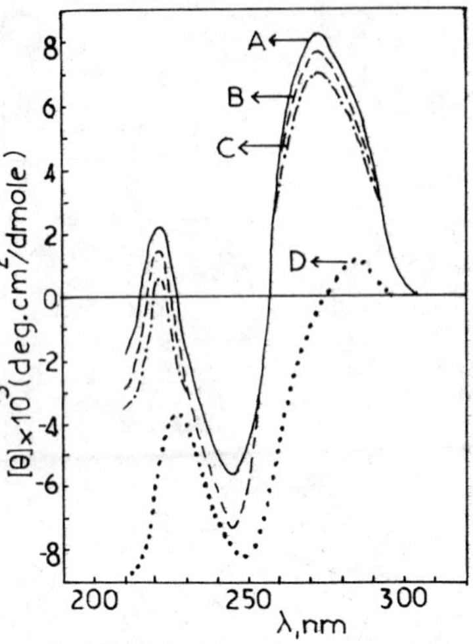

Fig.2. CD spectra of polymer-DNA complexes at r=1. A:DNA, B:I-DNA, C:III-DNA, D:II-DNA.

nt hydrophobicities and / or to the different geometries of the bound polypeptides I and II. In contrast the DNA interaction with poly (L-Arg-L-Ala-Gly) increases from Ala→ Leu but with little conformational changes in DNA secondary structure, and moreover there is no DNA condensation or aggregation.

In conclusion, this study of sequential polypeptide histone models $(L\text{-}Arg\text{-}X\text{-}Gly)_n$ has demonstrated the influence of the side-chain of the X residue (bulkiness and length, β-branching-hydrophobicity) and the repeating sequence upon the conformational properties of the polymers and their binding with DNA. It was also confirmed the significance of the hydrophobic forces, besides the arginine-phosphate charge interaction, which modulate the nature of the polypeptide-DNA complexes and their condensation into higher ordered structures as is found in chromatin.

References

1. Schwartz, A.M. and Fasman, G.D. (1977) Biochemistry **16**, 2287-2299.

2. Adler, A.J., Moran, E.C., and Fasman, G.D. (1975) Biochemistry **14**, 4179-4185.

3. Cernosek, S.F., Malin, M., Jr., Wells, M., and Fasman, G.D. (1974) Biochemistry **13**, 1252-1264.

4. Chang, C.T., Wu, C.C., and Yang, J.T. (1978) Anal. Biochemistry **91**, 13-31.

HELIX STABILIZATION IN ANALOGS OF RIBONUCLEASE A(1-13)

Eunice J. York and John M. Stewart

Department of Biochemistry,
University of Colorado School of Medicine,
Denver, Colorado 80262

Robert L. Baldwin and Kevin R. Shoemaker

Department of Biochemistry,
Stanford University Medical Center,
Stanford, California 94305

Introduction

The C-peptide of Ribonuclease A (see Table I for peptide structures) forms 30% α-helix in water near 0°C, pH 5[1,2], contrary to the Zimm-Bragg and "host-guest" predictions. This unexpected stability was thought to be due to a Glu-9→His-12 salt bridge. However, replacement of Glu-9 by Leu enhanced helix formation[3]. The α-helix has a strong dipole moment, effectively placing +0.5 charge at the N-terminus and -0.5 charge at the C-terminus. Negatively charged groups at the N-terminus and positively charged groups at the C-terminus stabilize the dipole and should enhance helix formation. Replacement of His-12 or of Glu-2 by Ala, (cf RN-12) completely destroyed the helix; substitution of Lys-1 by acetyl-Ala enhanced helicity[3]. Since RN-12 showed only 12% helix, several analogs have been synthesized to test further the helix dipole model and to investigate the role of a proposed Glu-2→Arg-10 salt bridge[4,5].

Table I. Mean Residue Ellipticity of Ribonuclease Peptides

PEPTIDE	STRUCTURE	$[\Theta]_{222}$ [a]
C-Peptide[b]	K-E-T-A-A-A-K-F-E-R-Q-H-Hse-lactone	7,000[c]
RN-12[d]	K-E-T-A-A-A-K-F-E-R-A-H-A-NH_2	2,800[e]
RN-16[d]	K-E-T-A-A-A-K-F-L-R-A-H-A-NH_2	5,700[c]
RN-21	Ac-A-E-T-A-A-A-K-F-L-R-A-H-A-NH_2	13,150[c]
RN-22	A-E-T-A-A-A-K-F-L-R-A-H-A-NH_2	8,900[c]
RN-23	Ac-A-A-T-A-A-A-K-F-L-R-A-H-A-NH_2	8,400[f]
RN-24	Suc-A-E-T-A-A-A-K-F-L-R-A-H-A-NH_2	15,150[c]
RN-25	Ac-A-R-T-A-A-A-K-F-L-E-A-H-A-NH_2	6,200[f]
RN-26	Ac-A-D-T-A-A-A-K-F-L-R-A-H-A-NH_2	12,200[c]
RN-28	Ac-A-E-T-A-A-A-K-F-L-A-A-H-A-NH_2	13,150[f]
RN-29	Ac-A-E-T-A-A-A-K-F-L-R-A-A-A-NH_2	9,500[g]

[a]Maximum value; deg cm^2/dmol. [b]See reference 2. [c]at pH 5-5.5. [d]See reference 3. [e]at pH 4.5. [f]at pH 1.0-1.2. [g]at pH 7.0; the value remains high up to pH 10.

Results and Discussion

All peptides were synthesized by the solid phase method and purified by CCD and HPLC. Amino acid analysis confirmed the composition. Helical content was determined by the magnitude of the CD band at 222nm. The C-peptide structure was modified by replacing the C-terminal Hse-lactone (from RNAse + CNBr) with C-terminal Ala-NH_2, since Hse-carboxylate destroys the helix and Met oxidizes easily. Gln-11 was replaced by Ala to avoid spontaneous hydrolysis.

The helix dipole model predicts that appropriately placed charged groups should stabilize the helix. Substitution of Lys-1 by acetyl-Ala removed two destabilizing positive charges. Further substitution of Glu-9 by Leu gave RN-21, which had >50% helix. Succinylation (RN-24) increased helix

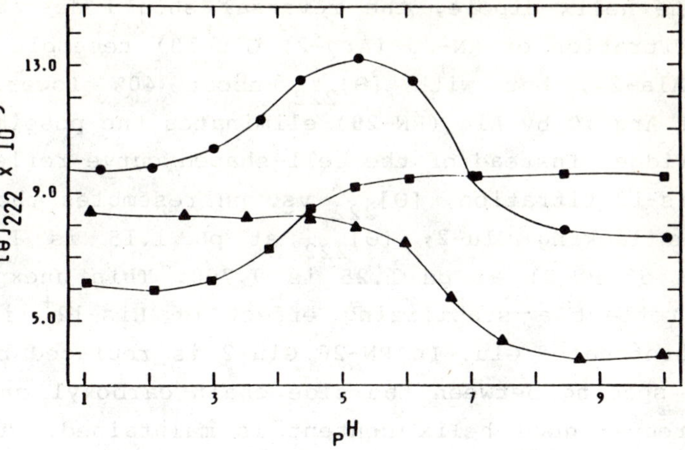

Fig. 1. pH Dependence of the mean residue ellipticity, $[\Theta]_{222} \times 10^{-3}$ deg cm^2/dmol, for RN-21 (●), RN-23 (▲) and RN-29 (■) at 3°C, 0.1M NaCl.

to about 65%. There is a steady increase in helicity as the charge on the N-terminal residue goes from +2 to -1.

Figure 1 shows the effect of pH on helicity for RN-21, -23 and -29. The bell-shaped curve of RN-21 implicates Glu-2 and His-12 as the groups ionizing between pH 2 and 7. When Glu-2 is protonated at low pH, helicity (RN-21) is similar to that of RN-23, where Glu-2 is replaced by Ala. Deprotonation of His-12 destabilizes the helix. RN-29, where His-12 is replaced by Ala, lacks the high helicity induced by a protonated His-12 but retains helicity at all pH values above the pK of Glu-2. Helix stability is clearly influenced by charged groups at or near the termini of the helix dipole.

A Glu-2→Arg-10 salt bridge has been proposed as a helix-stabilizing interaction in S-peptide(1-19)[4,5]. If this interaction is a major stabilizer of the helix, transposition of these two residues should still maintain the interaction. If the stabilization is due to electrostatic effects of charged groups on the helix dipole, the reversal should destabilize the helix. Titration of RN-25 (Arg-2, Glu-10) resembles that of RN-23 (Ala-2), but with $[\Theta]_{222}$ about 40% lower. Replacement of Arg-10 by Ala (RN-28) eliminates the possibility of a salt bridge. Instead of the bell-shaped curve reflecting Glu-2 and His-12 titration, $[\Theta]_{222}$ vs. pH resembles those of the peptides lacking Glu-2; $[\Theta]_{222}$ at pH 1.15 is 13,150, whereas that of RN-21 at pH 1.25 is 9,700. This unexpected result may reflect a stabilizing effect of His-12$^+$ in the absence of protonated Glu. In RN-26 Glu-2 is replaced by Asp to test the spacing between the side chain carboxyl and the guanidine groups; good helix content is maintained. Further peptides will be designed to test this point.

Charged group effects play a major role in stabilization of helix in short peptides in water. Modification of the C-peptide structure has led to analogs of high helix content that support the helix dipole model. (We thank Virginia Sweeney for amino acid analyses. Supported by NIH grant RR-00711 and NSF grant GP-23633 to RLB.)

References

1. Brown, J.E. and Klee, W.A. (1971) Biochemistry 10, 470-476.

2. Bierzynski, A., Kim, P.S. and Baldwin, R.L. (1982) Proc. Nat. Acad. Sci. USA 79, 2470-2474.

3. Shoemaker, K.R., Kim, P.S., Brems, D.N., Marqusee, S., York, E.J., Chaiken, I.M., Stewart, J.M. and Baldwin, R.L. (1985) Proc. Nat. Acad. Sci. USA 82, 2349-2353.

4. Rico, M., Nieto, J.L., Santoro, J., Bermejo, F.J., Herranz, J. and Gallego, E. (1983) FEBS Lett. 162, 314-319.

5. Rico, M., Gallego, E., Santoro, J., Bermejo, F.J., Nieto, J.L. and Herranz, J. (1984) Biochem. Biophys. Res. Commun. 123, 757-763.

A NICKEL BINDING PEPTIDE FROM HUMAN KIDNEY[1]

Douglas M. Templeton and B. Sarkar

Research Institute, The Hospital For Sick Children, 555 University Avenue, Toronto, M5G 1X8; and Department of Biochemistry, University of Toronto, Toronto, M5S 1A8, Canada

Introduction

Occupational exposure of human populations to nickel compounds has provided the opportunity to study the health effects of this toxic metal.[2,3] Increased incidences of respiratory and renal cancers are associated with exposure to Ni. Animal studies have demonstrated that ionic Ni(II) complexes undergo rapid renal excretion,[4] and consequently exhibit low acute renal toxicity. However, the initial accumulation of small amounts of nickel following administration of Ni salts occurs primarily in kidney,[4] and represents a carcinogenic risk in this organ. We report here the purification and characterization of a 4 kDa peptide which binds a significant portion of the ionic Ni(II) in renal cytosol.

Methods

Human kidney samples from autopsy were homogenized and the low molecular weight Ni-binding fraction obtained as described earlier.[5] A peptide binding component identified at this stage was purified by recycling on a Waters I-60 HPLC column and subsequent elution from a Waters C-18 reversed phase HPLC column with a gradient of 0-10% acetonitrile in water. Amino acid analyses were performed as previously described,[5] as were N-terminal determinations by the method of Woods and Wang.[6] Trypsinization was carried out by addition of 1 mol% of TPCK-treated trypsin (Sigma) at 0 and 12 hours, to purified peptide dissolved in NaH_2CO_3 (0.1 M, pH 8.5, 37°C) for 18 hours. Tryptic peptides were eluted from the C-18 column with tetrabutylammonium bromide as a counter ion. For Ni binding studies, $^{63}NiCl_2$ was added to aliquots of 0.1-0.2 nmol of peptide in N-ethylmorpholine (10 mM, pH 7.5) and the solution let stand for 48 hours. Then the mixture was eluted with the same buffer from an I-60 HPLC column. Peptide was determined by amino acid analysis and Ni by counting radioactivity. In some experiments, histidine was added as a competing ligand, in amounts equimolar to peptide.

Results and Discussion

We have found[5] similar chromatographic distributions of Ni on high and low pressure gel filtration of cytosol from kidneys of rats exposed to $^{63}NiCl_2$ *in vivo* and *in vitro*, and from bovine and human kidneys exposed *in vitro*. In all cases, 60-70% of the ^{63}Ni binds to low molecular weight materials. Most of this low molecular weight fraction consists of Ni bound to a heterogeneous mixture of anionic oligosaccharides. However, 30-35% of the Ni is bound to a single peptide, which has now been purified to homogeneity from human kidney, by reversed phase HPLC following addition of $^{63}NiCl_2$ to renal cytosol (Fig. 1). Alternatively, the peptide has been isolated in a similar manner without prior exposure to Ni, for use in metal binding studies. Its amino acid composition (Table I) yields a minimum molecular weight of 4 kDa. It is devoid of aromatic amino acids, and its hydrophilicity is supported by its weak interaction with the hydrocarbon-bonded reversed phase column. The content of Asx + Glx is > 25 Mol%. Since the peptide is highly anionic on high voltage paper electrophoresis at pH 6.5 and 8.9, most or all of these residues are in the acidic form. In the absence of sulfur-containing amino acids, this composition may indicate polycarboxylate chelation of Ni by Asp and Glu.

Attempts to determine the N-terminal of the peptide have been unsuccessful, suggesting blockage of the N-terminal residue. Consistent with a single Lys residue, and no Arg,

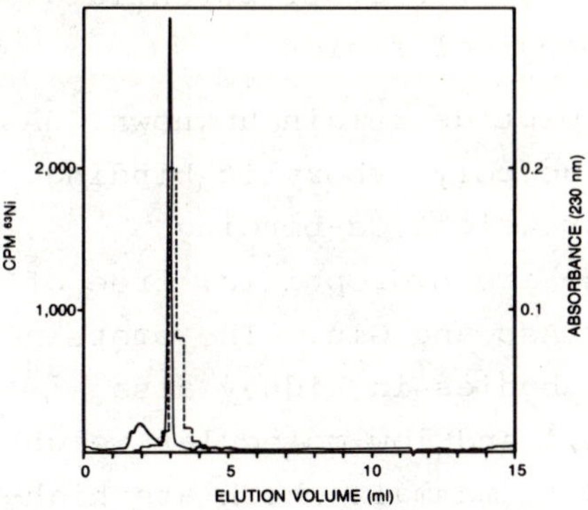

Fig. 1. Purification of Ni-binding peptide by reversed phase HPLC, showing A_{230} (——) and Ni radioactivity (----) (Ref. 5).

Table I. Amino Acid Composition of Ni-binding Peptide

Residue	Mol%	Residue	Mol%
Asx	9.3±0.2	Ala	9.5±0.2
Thr	4.0±0.2	Val	5.3±0.4
Ser	12.3±0.6	Ile	3.2±0.3
Glx	15.8±0.5	Leu	5.4±0.4
Pro	4.8±0.6	His	2.1±0.3
Gly	21.4±0.8	Lys	2.9±0.2

trypsinization yields two hydrophilic fragments. The binding of Ni by the trypsinized mixture remains intact. The best yield of the purified peptide yet achieved is ca. 10 nmol/ 50 g of human kidney. This quantity is too low for reliable equilibrium dialysis studies, which are further complicated by the low molecular weight and risk of binding of this anionic material to dialysis tubing. Therefore in order to study the Ni binding quantitatively, we have developed an HPLC technique which allows binding measurements to be performed on quantities of less than 100 pmoles. Scatchard analysis of the Ni binding in the absence of a competing ligand demonstrates 0.78 binding sites per molecule, with an apparent K_D of 1.1×10^{-5} M. In addition, there is evidence of weak binding (K_D > 0.1 M) of several more Ni atoms, presumably due to territorial binding to this polyelectrolyte. When histidine is added as a competing ligand in the binding assay, and the concentration of unbound Ni calculated from the known speciation and stability constants of the Ni(II)-histidine system,[7] the apparent K_D for binding of Ni to the peptide is found to be 10^{-9} to 10^{-11} M, demonstrating the ability of the peptide to compete successfully for Ni with other potential low molecular weight ligands in biological fluids.

The origin and function of this peptide remain unknown. A number of Ca-binding proteins provide polycarboxylic binding sites, and a family of low molecular weight Ca-binding peptides has been discussed.[8] These are hydrophilic, free of sulfur, and contain up to a quarter Asp and Glu. The protein precipitates of Pb and Bi inclusion bodies in kidney also contain high amounts of Asp and Glu,[9] and low molecular weight Cr-binding peptides have been found in mammals which are highly ionic and have isoelectric points as low as 3.5.[10] Together with the present results, these observations point to the involvement of polycarboxylic peptides in the metabolism of both essential and toxic metals.

References

1. Research supported by NSERC of Canada.
2. Sunderman, F.W., Jr. (1981) Environ. Hlth. Perspect. 40, 131-141.
3. Shannon, H.S., Julian, J.A. and Roberts, R.S. (1984), J. Nat. Cancer Inst. 73, 1251-1258.
4. Sarkar, B. (1980) in Nickel Toxicology, Brown, S.S. and Sunderman, F.W., Jr., Eds., Academic Press, London, pp. 81-84.
5. Templeton, D.M. and Sarkar, B. (1985) Biochem. J., In Press.
6. Woods, K.R., and Wang, K.-T. (1967) Biochim. Biophys. Acta 133, 369-370.
7. Glennon, J. and Sarkar, B. (1982) Biochem. J. 203, 15-23.
8. Lau, S.-J. and Sarkar, B. (1971) J. Biol. Chem. 246, 5938-5943.
9. Moore, J.F., Goyer, R.A. and Wilson, M. (1973) Lab. Invest. 29, 488-494.
10. Yamamoto, A., Wada, O. and Ono, T. (1981) Toxicol. Appl. Pharmacol. 59, 515-523.

ISOLATION AND 2D ^1H NMR OF THE NH_2 - TERMINAL 24 RESIDUES PEPTIDE FRAGMENT FROM DOG SERUM ALBUMIN AND ITS COMPLEXATION WITH COPPER AND NICKEL

B. DECOCK-LE REVEREND

Université des Sciences et Techniques de Lille, Laboratoire de Chimie Macromoléculaire, 59655 Villeneuve d'Ascq Cedex - France

B. PERLY

Centre d'Etudes Nucléaires de Saclay, I.R.D.I./D.P.G. - 91191 Gif sur Yvette Cedex - France

B. SARKAR

Research Institute, The Hospital For Sick Children, Toronto, M5G 1X8, and the Department of Biochemistry, University of Toronto, Toronto, Ontario, M5S 1A8, Canada

Introduction

The transport protein for Cu(II) in serum is albumin. Unlike human serum albumin (HSA), dog serum albumin (DSA) does not possess the characteristics of the specific first binding site for Cu(II) (1). However, the NH_2-terminal site of DSA still seems to be the preferred site where the important histidine residue in the third position in HSA is replaced by a tyrosine residue in DSA (2).

In order to get an insight of the structure of the NH_2-terminal site of DSA, we have isolated the 24 residues peptide fragment (P_{24}) from the NH_2-terminal of DSA, we have performed NMR studies and have undertaken complexation experiments with Cu(II) and Ni(II).

Methods

Dog serum albumin (fraction V) was cleaved by controlled peptic digestion. The P_{24} fragment was isolated and purified by gel chromatography on Sephadex G-25 and Cellex D ion exchange chromatography. The purity of the sample was checked by gel electrophoresis and amino acid analysis.

NMR experiments were performed on unbuffered solutions of P_{24}. The dry sample (1.5 mg) was dissolved in 1 ml $^{2}H_2O$, freeze-dried and redissolved in 0.5 ml 99.95% $^{2}H_2O$ (CEA France). For titration experiments, pH was further adjusted using dilute $NaO^{2}H$ or ^{2}HCl. Chemical shifts (downfield positive) are reported relative to internal TSP (Sodium 3-trimethylsilylpropionate-2,2,3,3,-$^{2}H_4$). Spectra were obtained at 295 K using a WM500 Brucker spectrometer operating at 500.13 MHz. High resolution spectra were obtained using 64 K data tables. An average of 100-150 transients were collected. Resolution was further enhanced by Lorentzian-Gaussian multiplication. Two dimensional shift correlation was obtained using the COSY pulse sequence (45° mixing pulse). Both time domains were processed using a squared sine bell window and the final 1 K x 1 K real matrix was presented in the magnitude mode.

Circular Dichroism spectra were recorded on a Jobin-Yvon dichrograph R.J. Mark III. The concentration of the solutions was 10^{-3}M in H_2O. For complexation studies $NiCl_2$ and $Cu(ClO_4)_2$ solutions were used.

The NMR spectra of P_{24}-Ni(II) (1:1) solution (pH = 10.7) were recorded at 360.06 MHz on a Nicolet NT-360 spectrometer at a probe temperature of $24^{o}C$.

Results and Discussion

Non-exchangeable protons were assigned using the 2D

TABLE I. ^{1}H NMR Parameters of P_{24} Fragment of DSA (in D_2O pH 4.75 from TSP=0 ppm)

	H_α	H_β	H_γ	H_δ	H_ε	Others
Glu 1	3.96	2.06	2.33			
Ala 2	4.34	1.35				
Tyr 3	4.51	3.00				H_δ 7.11
						H_ε 6.80
Lys 4	4.25	1.83	1.56	1.64	2.97	
Ser 5	4.37	3.87				
Glu 6	4.25	2	2.2			
Ile 7	4.1	1.88	0.87	0.81		
			1.58			
Ala 8	4.26	1.33				
His 9	4.56	3.1				C_2(H) 8.58
						C_5(H) 7.14
Arg 10	4.28	1.8	1.65	3.1		
Tyr 11	4.56	3.06				H_δ 7.11
		2.92				H_ε 6.80
Asn 12	4.64	2.73				
		2.79				
Asp 13	4.56	2.66				
		2.75				
Leu 14	4.25	1.6	1.55	0.8		
Gly 15	3.92					
Glu 16	4.25	2	2.2			
Glu 17	4.25	2	2.2			
His 18	4.56	3.1				C_2(H) 8.58
						C_5(H) 7.12
Phe 19	4.56	3.15				7.24-7.37
Arg 20	4.28	1.8	1.65	3.1		
Gly 21	3.87					
Leu 22	4.25	1.6	1.55	0.8		
Val 23	4.07	2	0.92			
Leu 24	4.15	1.88	1.55	0.81		

correlation matrix. Further useful informations were obtained from the spin coupling experiments and titration curves analyses. The results are summarized in Table I. Because of the overlapping of a few signals, some values are approximated. The first five residues were assigned precisely because of their importance for complexation studies.

The NMR spectra of peptide-Ni(II) (1:1) solution (pH 10.7) show clearly that the Ni(II) binds to the NH_2 terminal and the first three deprotonated amide nitrogens. The CD spectrum is also consistent with a Ni(II) (4N) square planar complex.

The CD spectra of P_{24}-Cu(II) (2:1) solutions show that Cu(II) binds to the peptide from the terminal NH_2 and then stepwise to the deprotonated peptide nitrogens as the pH is raised causing an increase in d-d transition energy. Furthermore there is no band around 400 nm where the charge transfer transition ϕO-Cu(II) is generally observed, indicating that the phenolic group is not involved in the complexation.

References

1. Appleton, D.W. and Sarkar, B. (1971) J. Biol. Chem. 246, 5040.
2. Dixon, J.W. and Sarkar, B. (1974) J. Biol. Chem. 249, 5872.

PEPTIDE SEGMENT COUPLING BY USE OF ACYL DISULFIDES

Donald Yamashiro

Laboratory of Molecular Endocrinology
University of California, San Francisco, California 94143

Introduction

An "aqueous strategy" for coupling peptide segments has been designed to allow purification of segments and their coupling in aqueous media.[1] The key element in this strategy is the specific activation of a thiocarboxyl group in the presence of most other functional groups. Peptide segments are synthesized by the solid-phase method,[2] deblocked, purified in aqueous media, blocked at non-participating amino groups with the citraconyl group (Cit)[3], and then coupled by the $AgNO_3$/N-hydroxysuccinimide (HOSu) method and deblocked under mild conditions in aqueous media. In order to increase the solubility of the these segments in aqueous media, the reaction of the thiocarboxyl group with disulfides in quanidine hydrochloride (GnHCl) was studied.

Results and Discussion

Boc-Thioalanine(Boc-Ala-SH)[4] reacts with either iodine or 2-nitrophenylsulfenyl chloride to give, respectively, $(Boc\text{-}Ala\text{-}S\text{-})_2$ (I) (62%, m.p. 152-153°) or Boc-Ala-S-S-$C_6H_4NO_2$-o (II) (63%, m.p. 106-108°). Both I and II react with H-Leu-Ⓟ in DMF at rates comparable to Boc-Ala-OSu to give, after HF cleavage, H-Ala-Leu-OH. Boc-Ala-SH reacts with 2,2'-dinitrodiphenyl disulfide in DMF in the presence of one eq. diisopropylethylamine (DIEA) to give II whereas in aqueous media (90% DMF, 1 eq. $KHCO_3$)I is formed. The unsymmetrical disulfide II is stable in DMF (20 hr, 24°) while in aqueous media (75% DMF) it undergoes partial disulfide rearrangement to I. The symmetrical disulfide I slowly hydrolyzes in basic media (80% DMF, 1 eq. $KHCO_3$) to yield Boc-Ala-SH and Boc-Ala-OH.

The % H-Leu-Ⓟ coupled by Boc-Ala-SH (1.5 eq.) in the presence of various disulfides (1.5 eq.) and DIEA (1 eq.) in DMF in 15 min were: 5,5'-dithiobis(2-nitrobenzoic acid), DTNB, 28; 2,2'-dinitrodiphenyl disulfide, 48; 2,2'-dipyridyl disulfide, 80; 2,2'dithiobis(5-nitropyridine), 97. The Izumiya racemization test[5] was applied to the coupling of Boc-Gly-Ala-SH[4] (2 eq.) to H-Leu-Ⓟ (1 eq.) by activation with 2,2'-dithiobis(5-nitropyridine) (3 eq.) in DMF (1 eq. DIEA). After HF cleavage the yield of tripeptide was 92% and racemization 0.2%.

Racemization in aqueous media is a more serious problem than in non-aqueous media with temperature, type of buffer, and pH being significant factors. Coupling in GnHCl under conditions likely to be used in an actual synthesis was therefore tested. Boc-Gly-Ala-SH (20 mM), leucine (20 mM), and DTNB (40 mM) were reacted in 5 M GnHCl at pH 6.9 ($NaHCO_3$ buffer) for 19 hr at 4° (mM, millimolar). After treatment with trifluoroacetic acid, the yield of H-Gly-Ala-Leu-OH was 38% and racemization $\lesssim$1%.

The problem of racemization can often be circumvented by solid-phase synthesis of segments large enough that Gly can be found to serve as the C-terminal residue. To further test coupling in GnHCl inhibin-like peptide, ILP-(1-31),[6] was chosen:

```
                    5                   10
H-His-Asn-Lys-Gln-Glu-Gly-Arg-Asp-His-Asp-
                   15                   20
   Lys-Ser-Lys-Gly-His-Phe-His-Arg-Val-Val-
                   25                         31
   Ile-His-His-Lys-Gly-Gly-Lys-Ala-His-Arg-Gly-OH
```

Two segments were synthesized: H-ILP-(1-25)-SH by the method of Blake[1] and H-[Lys(Msc)[27]]-ILP-(26-31)-OH (B), where Msc is methylsulfonylethyloxycarbonyl[7]. Both were purified to HPLC homogeniety, and the amino groups in the 25-peptide were citraconylated to give A. Coupling in GnHCl followed by deblocking of all amino groups were done by a one-pot procedure. Segments A (25 mM) and B (75 mM), 2,2'-dipyridyl disulfide (65 mM), and HOSu (100 mM) were reacted in 5 M GnHCl-0.2 M K_2HPO_4 at pH 7.0 for 1 hr at 24°. The Msc group was removed (0.1 N NaOH, 1 min, 24°) followed by decitraconylation (0.1 N HCl, 37°, 24 hr). Prolonged treatment in strong acid was required to remove the Cit group from the N-terminal His residue.

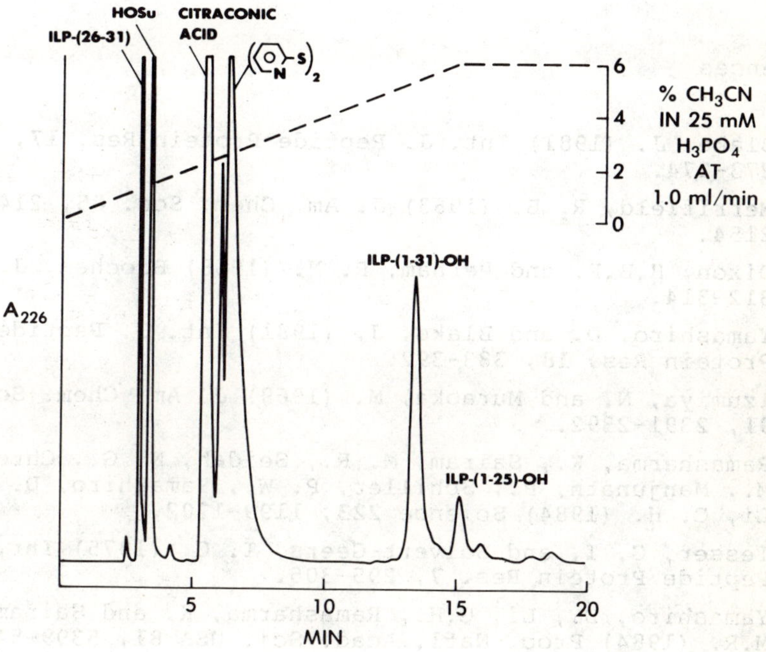

Fig. 1. HPLC of crude ILP-(1-31) after segment coupling.

The crude product analyzed by HPLC (Figure 1) showed a yield of 74% of ILP-(1-31) based on segment A. Since the thiocarboxyl group is converted to COOH in the acid deblocking, the peak corresponding to H-ILP-(1-25)-OH represents uncoupled segment A (17%). The yield of isolated product after an additional HPLC was about 50%.

The synthetic product was compared to a sample of ILP-(1-31) synthesized by the all-stepwise method.[8] They were identical in amino acid analysis, paper electrophoresis, HPLC, and HPLC of tryptic digests.

This work was supported in part by grants from the National Institute of Health (GM-30072 and GM-2907). The author thanks Prof. C. H. Li for his encouragement and support.

References

1. Blake, J. (1981) Int. J. Peptide Protein Res. **17**, 273-274.

2. Merrifield, R. B. (1963) J. Am. Chem. Soc. **85**, 2149-2154.

3. Dixon, H.B.F. and Perham, R. N. (1968) Biochem. J. **109**, 312-314.

4. Yamashiro, D. and Blake, J. (1981) Int. J. Peptide Protein Res. **18**, 383-392.

5. Izumiya, N. and Muraoka, M. (1969) J. Am. Chem. Soc. **91**, 2391-2392.

6. Ramasharma, K., Sairam, M. R., Seidah, N. G., Chretien, M., Manjunath, P., Schiller, P. W., Yamashiro, D. and Li, C. H. (1984) Science **223**, 1199-1202.

7. Tesser, G. I. and Balvert-Geers, I. C. (1975) Int. J. Peptide Protein Res. **7**, 295-305.

8. Yamashiro, D., Li, C.H., Ramasharma, K. and Sairam, M.R. (1984) Proc. Natl. Acad. Sci. USA **81**, 5399-5402.

APPLICATIONS OF THE COUPLING REAGENT BIS-(2-OXO-3-OXAZOLIDINYL) PHOSPHINIC CHLORIDE: SYNTHESIS OF Fmoc-D-ALA-MeLEU-MeLEU-MeVAL-O^{t}BU.

Roger D. Tung and Daniel H. Rich
School of Pharmacy, University of Wisconsin-Madison
Madison, Wisconsin 53706

Introduction

We have recently reported that bis-(2-oxo-3-oxazolidinyl) phosphinic chloride (BOP-Cl; 1)[1] is an efficient and convenient reagent for coupling hindered, N-alkyl amino acids,[2] and that N-protection of the carboxylate species by the Fmoc group[3] leads to faster reactions and higher yields than protection with the bulkier Boc group. To evaluate the utility of this reagent for a simplified synthesis of the cyclosporins and their derivatives,[4] we have carried out the synthesis of the tetrapeptide Fmoc-D-Ala-MeLeu-MeLeu-MeVal-O^{t}Bu. This compound is a close analog of a key fragment in the Sandoz synthesis of the immunosuppressant cyclosporin-A[5], and its ready availability by the BOP-Cl/Fmoc coupling methodology holds great promise for a simple and versatile entry to a large number of analogs of that drug.

Results and Discussion

Condensation of Fmoc-MeLeu[6] with MeVal-O^{t}Bu was carried out using 1 eq. each of the protected imino acids, with 1.1 eq. of BOP-Cl and 2.2 eq. of DIEA in CH_2Cl_2 (0°, 2 hrs) to give, after acid-base workup and chromatography, an 86% yield of dipeptide (2). A study of deprotection conditions indicated that 50% $HNEt_2/CH_3CN$ combined the desirable qualities of fast (< 1 hr) removal of the Fmoc group with good volatility of the base and solvent. Furthermore, Carpino and Han have reported that diethylamine, unlike piperidine, does not add to the exo-9-fluorenylmethylene Fmoc byproduct, thus preventing the possibility of excess base being carried into the next reaction.

Accordingly, deprotection of 2 was carried out with 50% $HNEt_2/CH_3CN$ (30 min., room temp.), followed by removal of the volatiles by rotary evaporation. The residue was taken up in CH_2Cl_2 and coupled as above with Fmoc-MeLeu (overnight, 4-6°) to afford an 82% yield of tripeptide (3) after acid-base workup and silica gel chromatography (7.5% acetone/Skellysolve B eluant). Repetition of this deprotection-coupling strategy using Fmoc-D-Ala as the carboxylate component gave an 85% yield of the fully protected tetrapeptide (4). Thus, synthesis of 4 was carried out in a 26-hr. reaction time period with an overall yield of 60% from the constituent protected amino acids.

Critical to the overall efficiency of this synthesis, however, was the availability of MeVal-O^{t}Bu in high yields. Attempts at making this compound by standard techniques (e.g., DCC/DMAP/^{t}BuOH; isobutylene/H_2SO_4 or p-TsOH) gave disappointingly low yields (< 20%). After a considerable amount of effort, it was found that the stable acid chloride of Fmoc-MeVal could be esterified with ^{t}BuOH by silver cyanide-assisted displacement[7] to yield up to 87% of Fmoc-MeVal-O^{t}Bu (5). The sequential deprotection-coupling strategy described above gave an 82% yield of 2, establishing the usefulness of 5 in the synthesis of 4.

Outlook

The synthesis of several analogs of cyclosporin A has

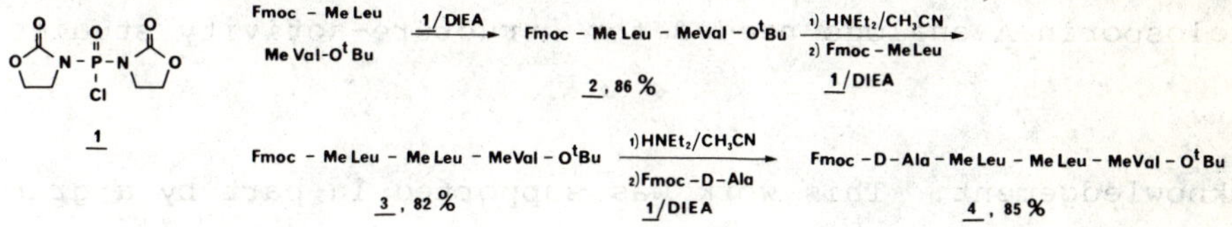
Fmoc - Me Leu
Me Val-O^{t}Bu
1/DIEA
Fmoc - Me Leu - MeVal - O^{t}Bu
2, 86%
1) HNEt$_2$/CH$_3$CN
2) Fmoc - MeLeu
1/DIEA
1
Fmoc - Me Leu - Me Leu - MeVal - O^{t}Bu
3, 82%
1) HNEt$_2$/CH$_3$CN
2) Fmoc - D - Ala
1/DIEA
Fmoc - D - Ala - Me Leu - Me Leu - MeVal - O^{t}Bu
4, 85%

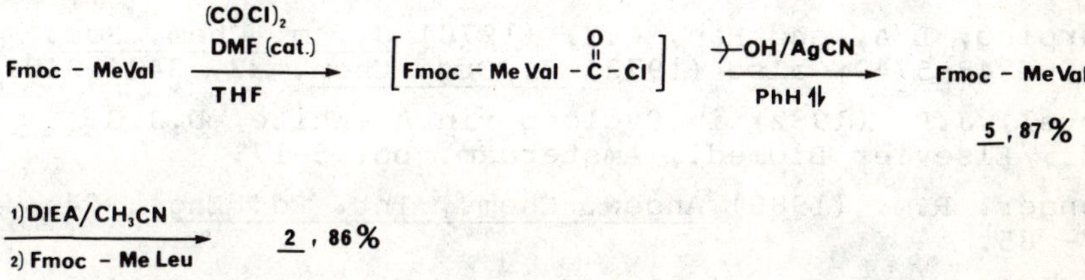
Fmoc - MeVal
(COCl)$_2$
DMF (cat.)
THF
Fmoc - Me Val - C - Cl
OH/AgCN
PhH
Fmoc - Me Val - O^{t}Bu
5, 87%
1) DIEA/CH$_3$CN
2) Fmoc - Me Leu
2, 86%

been carried out in our laboratory by Dr. M. Dhaon using the methodology developed by Wenger at Sandoz.[6,8] Biological testing of these analogs, performed by Dr. B. Dunlap, has yielded some extremely interesting information; notably, removal of 3 carbons from a single amino acid side chain in [MeLeu(3-OH)^1]-CsA from a peptide of ~ 1200 MW produces a > 100-fold loss of activity in a mixed lymphocyte culture assay. It is anticipated that the synthetic strategy described here will facilitate the synthesis of novel cyclosporin-A analogs needed for structure-activity studies.

Acknowledgement. This work was supported in part by a grant from the National Institues of Health (AM 32007).

References

1. Diago-Meseguer, J., Palomo-Coll, A.L., Fernandez-Lizarbe, J.R., and Zugaza-Bilbao, A. (1980) Synthesis 547-551.

2. Tung, R.D. and Rich, D.H. (1985) J. Am. Chem. Soc., in press.

3. Carpino, L.A. and Han, C.Y. (1970) J. Am. Chem. Soc. **92**, 5748-5749; also (1972) J. Org. Chem. **37**, 3404-3409.

4. Borel, J.F. (1982) in Cyclosporin A, White, D.J.G., Ed., Elsevier Biomed., Amsterdam, pp. 5-17.

5. Wenger, R.M. (1985) Angew. Chem., Int. Ed. Engl. **24**, 77- 85.

6. Freidinger, R.M., Hinkle, J.S., Perlow, D.S., and Arison, B.H. (1983) J. Org. Chem. **48**, 77-81.

7. Takimoto, S., Inanaga, J., Katsuki, T., and Yamaguchi, M. (1976) Bull. Chem. Soc. Japan. 2335-2336.

8. Rich, D.H., Dhaon, M.K., Dunlap, B., and Miller, S. (1985) J. Med. Chem., submitted for publication.

PENTAFLUOROPHENOL DERIVATIVES IN PEPTIDE SYNTHESIS

Lajos Kisfaludy, László Ötvös Jr., István Schőn, Miklós Lőw
Chemical Works of Gedeon Richter Ltd. H-1475 Budapest 1o,
P.O.B. 27, Hungary

Introduction

Fortunate combination of the electronic, steric, solvation, etc. properties of pentafluorophenol /PfpOH/ derivatives offers a new possibility to enhance the selectivity of some essential reactions in peptide chemistry. These are: /1/ acylation with protected /Z-, Boc-, Fmoc-/amino acid Pfp-esters, /2/formylation with For-OPfp, /3/ selective N-acetylation with Ac-OPfp, /4/ acylation with Fmoc-OPfp, /5/ amidation with Pfp-OH-amine salts, /6/fragment condensation with DCCI-PfpOH.

Results and Discussion

/1/ The usefulness of Pfp-esters in peptide synthesis was well-demonstrated in both solution[1] and solid phase techniques[2]. Generally, we can say that fast peptide bond formation obtained via Pfp-esters eliminates or minimizes the known side reactions.

2/ The most frequently used ForOH-Ac_2O reagent can result in undesirable side reactions in the presence of acid-sensitive groups. By means of For-OPfp /prepared from ForOH, PfpOH and DCCI; oil, spectroscopically characterized/ fast /max. 2o min/ and selective formylation can be effected in case of N-nucleophiles, whereas no formylation was observed with O-andS-nucleophiles, or sterically hindered N-nucleophiles,even in the presence of a tertiary base.Examples are shown in Table I.

Table I. Preparation of N-Formyl Derivatives of Amino Acids and Peptides[a]

Starting compound	Solvent	Yield,[b] %	M.p., °C	R_f	
H-Pro-NH_2	dioxane	91	193-194	0.50[e],	0.15[f]
H-Met-OMe	$CHCl_3$	90	oil	0.35[c],	0.60[d]
H-Ile-OMe	$CHCl_3$	93	oil	0.35[c],	0.55[d]
H-Thr-OBzl	$CHCl_3$	92	89-90	0.70[d],	0.70[e]
H-Val-OBu^t	$CHCl_3$	93	68-71	0.70[c],	0.85[d]
H-Pro-Leu-Gly-OEt	THF	76	oil	0.10[g],	0.50[e]
H-Met-Leu-Phe-OMe	$CHCl_3$	93	135-136	0.40[c],	0.50[f]

[a]For-OPfp: amino component = 2:1; [b]calculated for isolated For-compounds; [c]EtOAc; [d]$CHCl_3$:MeOH = 9:1; [e]$CHCl_3$:MeOH = 3:1; [f]benzene:MeOH:AcOH= 7:21; [g]EtOAc:/pyridine:AcOH:H_2O = 20:6: 11/ = 4:1

/3/ Outstanding difference in reactivity of Ac-OPfp toward amines and alcoholic hydroxyl groups in the presence and absence of a tertiary base renders the reagent a highly advantageous acetylating agent for both N,O-diacetylation and selective N-acetylation.[3] Example:

H-Thr-OBzl + AcOPfp ⟶ Ac-Thr-OBzl 82%

H-Thr-OBzl + AcOPfp ⟶(TEA) Ac-Thr/Ac/-OBzl 79%

/4/ Fmoc-OPfp /prepared from Fmoc-Cl and Pfp-OH;stable crystals,m.p. = 84-86°C;spectroscopically characterized/ is an efficient reagent for facile preparation of Fmoc-amino acids with remarkable suppression of the formation of Fmoc-dipeptides. PfpOH liberated during the acylation reaction allows the preparation of the corresponding Pfp-esters[4] without isolation of Fmoc-amino acids, in the presence of DCCI. Representative examples are given in Table II.

Table II. Preparation of Fmoc-Amino Acids and Their Pentafluorophenyl Esters

Amino acid /derivative/	Fmoc-amino acid		Fmoc-amino acid pentafluorophenyl ester	
	Yield,[a] %	M.p., °C	Yield,[a] %	m.p., °C
Ala	93	141-142	74	173-175
Val	9o	143-144	77	121-123
Trp	89	163-165	85	184-185
Cys/Bzl/	99	126-128	59	132-134
Z-Lys	89	86-88	9o	127-128
Asp/OBu^t/	78	147-148	63	97-1oo
Glu/OBu^t/	91	47-51	95	119-12o

[a]calculated for Fmoc-OPfp

/5/ N-protected amino acids and peptides can directly be converted into amides using NH_4OPfp /or substituted amine salts[5]/ after activation by DCCI or mixed anhydrides. The advantage of this amidation procedure is the easy handling of the reagent which eliminates the excess of ammonia /or amine/ and thus all base-catalysed side reactions.

Table III. Amidation of N-Protected Amino Acids and Peptides

Starting compound	Method	Purification	Yield,%	M.p. °C	/α/$_D$ c=1; DMF
Z-Gly	A	C	88	125-126	-
Z-Ala	A	D	81	126-129	-21.7[b]
Z-Val	A	C	91	196-197	+24.2
Boc-Val	A	D	77	149-15o	+13.6
Boc-Leu	A	C	76	139-14o	-11.9
Boc-Phe	A	C	96	138-139	+14.4[a]
Boc-Trp	B	C	75	152-153	-7.o[a]
Boc-Phe-Gly	A	C	59	88-91	+7.o[a]

Activation: DCCI/A/, m.a./B/; Purification by cryst./C/, by silica gel column chromatography /D/; a=EtOH, b=$CHCl_3$

/6/ The use of crystalline DCCI-PfpOH complex[6] in fragment condensation makes possible /a/ to eliminate the formation of undesired N-acyl urea derivatives and /b/ to use the starting components in 1:1 molar ration and the reagent in slight /1o-5o%/ excess. The following selected examples are given.

Z-Asn-Gly-OH + H-Ala-OBu^t 95.5%

Z-Asn-Gly-OH + H-Ala-Glu/OBu^t/-Asp/OBu^t/-Glu/OBu^t/-Ser/Bu^t/-OBu^t 8o.7%

Z-Lys/Boc/-Lys/Boc/-OH + H-Lys/Boc/-Lys/Boc/-NH_2 81%

Boc-Ser-Tyr-Ser-Met-Glu/OBu^t/-His-Phe-Arg/H/-Trp-Gly-OH + H-Lys/Boc/-Pro-Val-Gly-Lys/Boc/-Arg/H/-Arg/H/-Pro-NH_2 93%

References

1. Kisfaludy, L., /198o/ in The Peptides, Gross, E. and Meienhofer, J., Eds.,Academic Press, Vol.2. 417-44o.

2. /a/ Atherton, E., and Sheppard, R.C., /1985/ J. Chem. Soc. Chem. Commun. 165-166 /b/ Kovacs, K., and Penke, B. /1973/ in Peptides 1972. Hanson, H. and Jakubke, H. D. Eds. North Holland Publ.,Amsterdan, pp.187-188.

3. Kisfaludy, L.,Mohácsi, T., Lőw M. and Drexler, F. /1979/ J. Org.Chem. 44, 654-656.

4. Kisfaludy, L. and Schőn, I. /1983/ Synthesis 325-327.

5. Lőw, M., Rill, A. and Kisfaludy, L. in Peptides, 1984. Ragnarsson, U., Ed.,Almquist and Wiksell Int., Stockholm, pp. 267-27o.

6. /a/ Kovacs, J., Kisfaludy, L. and Ceprini, M.O. /1967/ J.Amer.Chem. Soc. 89, 183-184. /b/ Kisfaludy, L., /1981/ in Perspectives in Peptide Chemistry, Eberle, A., Geiger, R. and Wieland, T., Eds., S.Karger, Basel, pp. 58-66.

THE P-METHYLSULFINYLBENZYL GROUP, A SELECTIVELY CLEAVABLE CARBOXYL PROTECTING GROUP

James M. Samanen and Ester Brandeis
Peptide Chemistry Department, SmithKline Beckman, Inc.
1050 Page Mill Road, Palo Alto, California 94304

There is still a need to develop amine and carboxyl protecting groups that are easy to prepare, are stable to the conditions of peptide synthesis (especially DCC, Et_3N, and TFA), but are rapidly removed in high yield under mild conditions, especially while the peptide is still attached to a synthesis resin. Such protecting groups can be used for selective removal from a fully protected peptide for side chain derivatization, peptide chain elongation through the side chain, and formation of a cyclic peptide amide or ester. This paper describes the development of the p-methylsulfinyl-benzyl or B(SO) group shown below as a carboxylic acid ester. Sulfoxide reduction gives a p-methylthiobenzyl or B(S) group which is cleavable in anhydrous TFA. This report describes the synthesis of B(S) and B(SO) amino acid and peptide esters, the methods of reduction to the B(S) group, and the lability of B(S) and B(SO) groups toward various acids.

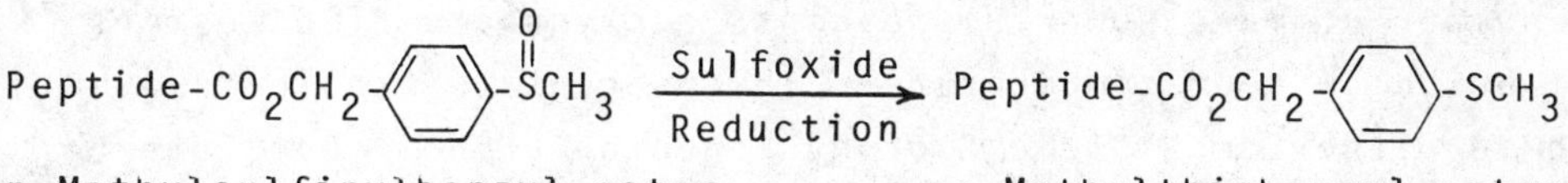

p-Methylsulfinylbenzyl ester — p-Methylthiobenzyl ester

B(SO): TFA Stable — B(S): TFA Labile

B(SO) esters may be prepared from p-methylsulfinylbenzyl alcohol <u>3</u> or p-methylsulfinylbenzyl chloride <u>5</u> by standard esterification procedures, Figure 1. Z-Phe-OB(S) and Z-Phe-OB(SO) were prepared as model B(S) and B(SO) esters for stability studies. B(SO) esters may also be obtained from B(S) esters by oxidation with metachloroperbenzoic acid.

The p-methylthiobenzyl ester of Z-Phe 7 is quite stable to glacial acetic acid (Reaction 1, Table I). In anhydrous TFA 7 solvolyzed completely in thirty minutes (Reaction 2). In preparative reactions the addition of Me_2S is recommended to trap the benzyl carbonium ion as the stable dimethylsulfonium salt. Z-Phe-OB(S), 7, was stable to anhydrous HCl in dioxane for at least 29 hours (Reaction 3).

The p-methylsulfinylbenzyl ester of phenylalanine 9 is exceptionally stable to trifluoroacetic acid. Only a trace of cleavage product phenylalanine (3.3%) was detected after two months (Reaction 4). This stability corresponds to the exposure of a B(SO) ester to at least 2880 half hour TFA treatments during peptide synthesis. As with other sulfoxides, the p-methylsulfinylbenzyl ester suffers reduction on exposure to hydrogen chloride in anhydrous dioxane (Reaction 3). Reduction was complete after four hours. Only a trace of cleavage product Z-Phe was detected after 29 hours. Partial B(SO) reduction and cleavage was observed in HBr/TFA (~ 80%) and anhydrous HF/anisole (< 10%).

The p-methylsulfinylbenzyl ester is relatively stable to reduction. Mercaptoacetic acid did not reduce Z-Phe-OB(SO). A number of more powerful reagents deoxygenate B(SO) esters at or below room temperature. The reagents of choice proved to be either Me_3SiCl/Me_2S or hydrogen chloride. Complete reduction occurred in 2 hours in the presence of excesses of both Me_3SiCl and Me_2S (Reaction 5). As mentioned above, deoxygenation by hydrogen chloride in dioxane was complete in four hours (Reaction 3). An added benefit from employment of the B(SO) group is that any oxidized methionine or S-alkylcysteine in a peptide are reduced simultaneously during

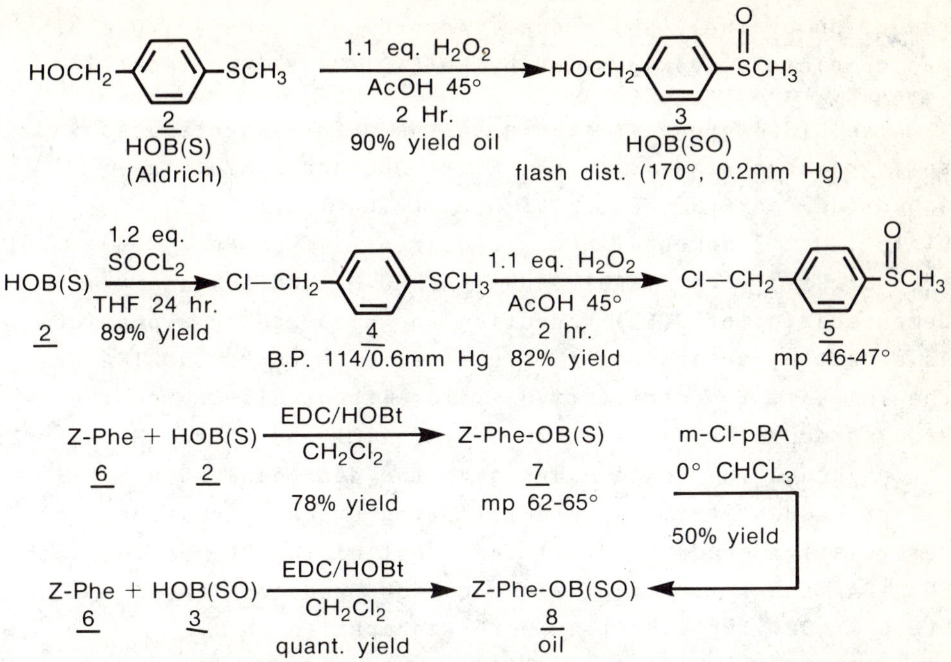

(Product structures in this paper were determined by nmr, elemental analysis and mass spectrometery.)

Fig. 1. Reagent Synthesis.

Rxn. No.	Starting Ester	Reaction Conditions	Analysis Time	(TLC yield) Products
1	7	glacial AcOH	3 d.	6, 0%
2	7	anhydrous TFA	30 min.	6, 100%
3	8	anhyd. HCl/dioxane	$<$ 4 hr.	7, 100%; 6, 0% for 29 hr.
4	9	anhydrous TFA	2 mo.	10, 3.3% (by AAA)
5	8	20 Me_3SiCl, 40 Me_2S, THF	1 hr.	7, 100%; 6, 0%
6	11	10 Me_3SiCl, 20 Me_2S, THF	4 hr.	12, 100%

Z-Phe = 6, Z-Phe-OB(S) = 7, Z-Phe-OB(SO) = 8, Phe-OB(SO) = 9, Phe = 10, Z-Met(O)OMe = 11, Z-Met-OMe = 12

Table I. Stability of B(S) and B(SO) Esters

reduction of the B(SO) group. Accordingly Z-Met(O)-OMe[2] was completely deoxygenated by Me_3SiCl/Me_2S (Reaction 6).

The B(SO) group is stable to the main reagents used in solid phase synthesis: TFA, Et_3N, DCC and symmetrical anhydride. A B(SO) ester of Boc-[(2,6-Cl_2Bzl)Tyr^1, D-Ala^2, Met^5]-enkephalin was prepared[3] from Boc-(2,6-Cl_2Bzl)Tyr-D-Ala-Gly and TFA-Phe-Met-OB(SO) as a demonstration of B(SO) reduction and cleavage in a peptide. As expected, treatment of this ester with anhydrous TFA gave the TFA salt in quantitative yield without disturbing the B(SO) group. Exposure to 20 eq. Me_3SiCl and 40 eq. Me_2S in anhydrous THF for 4 hours gave the deoxygenated B(S) ester in quantitative yield following ethanol treatment to remove silyl groups. Final treatment of the B(S) ester with 50% TFA/Me_2S followed by water quench produced TFA.[(2,6-Cl_2Bzl)Tyr^1, D-Ala^2, Met^5]-enkephalin in 95% yield. There are no indications that one's ability to assemble peptide sequences would be restricted by the presence of B(SO) groups. We envision the B(SO) ester to be particularly useful as a semipermanent carboxyl protecting group for peptide syntheses by fragment condensation, and for Asp or Glu side chain derivatization or protection. The B(S) ester constitutes a convenient alternative to the t-butyl or p-methoxybenzyl ester owing to its selective stability in anhydrous HCl/dioxane.

References

1. Drabowicz, J., Numata T. and Oae, S. (1977) Org. Prep. Proc. Int. 9, 64-83.
2. Iselin, B.M. (1961) Helv. Chim. Acta 44, 61-78.
3. Via Moon, M.W. (1982) U.S. Patent 4,251,438.

BENZYLOXYCARBONYLARGINYL- AND p-NITROBENZYLOXYCARBONYLARGINYL- GROUPS AS AMINE PROTECTION AND CHROMATOGRAPHIC HANDLES DURING PEPTIDE SYNTHESIS

John D. Glass, Angeliki Buku, Uma Roy, Diana Gazis, and Brian Hayes

Center for Polypeptide and Membrane Research and the Department of Physiology and Biophysics, Mount Sinai School of Medicine, 5th Avenue at 100th Street, New York, New York 10029.

Introduction

Before 1974, the use of enzyme-labile blocking groups in peptide synthesis was confined to a few examples of enzymatic deprotection based upon the use of chymotrypsin and thermolysin.[1,2] The broad substrate specificities of these endopeptidases and the fact that their substrate-recognition sites are not subject to reversible modification severely limited the prospects for broader application of the blocking groups. With the introduction of trypsin-labile protecting groups the situation was changed.[3-5]

The primary determinants for trypsin substrates, unlike those of chymotrypsin and thermolysin substrates, are chemically reactive entities which can be reversibly modified and which frequently are protected, in any case, during peptide synthesis. The prospects for a broadly applicable protection scheme, which follow from these facts, have stimulated the development of various chemical[6,7] and enzymatic[8] strategies for the introduction of N^{α}-protected arginyl residues into synthetic intermediates as protection of the α-amino group.

We now report a segment condensation synthesis of the neurohypophyseal peptide, oxypressin, which demonstrates some of the advantages of the Z-Arg- group as a chromophore, chromatographic handle, solubility modulator, and trypsin-labile protecting group. NO_2-Z-Arg- groups either were introduced into preformed peptide segments or during their preparation by solid phase synthesis. The NO_2-Z-Arg- group has certain chemical and spectral properties which should be of particular advantage during the semisynthesis of polypeptides using the segment condensation approach.

Results and Discussion

Z-Arg-pressinoic acid was prepared by treating a concentrated DMF solution of the diisopropylethylamine salt of pressinoic acid[9] with Z-Arg-ONp nitrate salt[6] in the presence of 1-HOBt. The reaction mixture was diluted with water containing a drop of acetic acid and the turbid aqueous solution was extracted with ether. A stream of nitrogen was used to remove traces of ether from the aqueous phase, then the mixture was heated gently (about 50-60°) to obtain a clear solution. The product separated from the aqueous solution as a white solid upon standing and cooling. About 70 mg of product was obtained from 81 mg of pressinoic acid. The product was significantly more strongly absorbed to C-18 reversed phase columns than were nitrophenol, 1-HOBt, or pressinoic acid. This could be exploited to recover a few mg of additional product from the mother liquor or to isolate the product directly from the crude reaction mixture when a synthesis was carried out on a few milligrams of material. Acid hydrolysates of the product yielded amino acids in the following ratios: Cys, 1.7; Tyr, 0.93; Phe, 1.0; Glu, 1.0; Asp, 1.0; Arg, 1.0.

Z-Arg-pressinoic acid was coupled with an excess of H-Pro-Leu-GlyNH_2 through the action of DCCI in the presence of 1-HOBt. The resultant Z-Arg-oxypressin was isolated by taking advantage of the fact that it is significantly more strongly adsorbed to C-18 reversed phase columns than are 1-HOBt, H-Pro-Leu-GlyNH_2 and other major water soluble components of the reaction mixture. Acid hydrolysates of the chromatographically isolated peptide derivative showed amino acids in the following ratios: Arg, 0.93; Cys, 1.9; Tyr, 0.91; Phe, 0.93; Glu, 1.0; Asp, 1.0; Pro, 1.1; Leu, 1.1; Gly, 0.92. Treatment of the protected peptide intermediate with trypsin rapidly removed the Z-Arg group and yielded free oxypressin which was characterized by amino acid analysis, by chromatographic comparison with authentic oxypressin, and by bioassay in the rat uterus contraction assay. About 4 mg of the final product was obtained from a reaction sequence starting with about 6 mg of Z-Arg-pressinoic acid.

The NO_2-Z-Arg- group has been applied to the N-terminus of peptide derivatives through the use of two reagents whose synthesis will be described elsewhere. NO_2-Z-Arg-ONp nitrate salt acylated the amino group of an N^{ϵ}-protected peptide fragment in the presence of 1-HOBt. NO_2-Z-Arg(Tos)-OH was used to add the NO_2-Z-Arg(Tos)- group as a final cycle in a solid phase peptide synthesis. Treatment of the resultant NO_2-Z-Arg(Tos) peptide-resin derivative with HF simultaneously cleaved the attachment to the resin and removed the tosyl group from the side chain of the arginine residue. The NO_2-Z- group was resistant to this treatment and remained on the peptide fragment. NO_2-Z-Arg-peptides were "deprotected" in two different ways, both under mild aqueous conditions. Treatment with trypsin cleaved the entire NO_2-Z-Arg- group from the peptide (albeit slowly compared to comparable deprotections of Z-Arg-peptides) whereas treatment with sodium dithionite in neutral aqueous solution[10] removed the NO_2-Z- group, leaving the Arg residue N-terminal. This is a point of synthetic flexibility which might be very useful in the course of semisyntheses using tryptic fragments of acylated proteins.

Acknowledgments

This work was supported by NIH grants GM 18752 and AM 10080.

1. Holley, R.W. (1955) J. Amer. Chem. Soc. 77, 2552-2553.
2. Ohno, M. and Anfinsen, C.B. (1970) J. Amer. Chem. Soc. 92, 4098-4102.
3. Meyers, C.A. and Glass, J.D. (1975) Proc. Nat. Acad. Sci. USA 72, 2193-2196.
4. Meyers, C.A. and Glass, J.D. (1975) in Peptides: Chemistry, Structure and Biology, Walter, R. and Meienhofer, J., Eds., Ann Arbor Science Pub., Ann Arbor, MI, 325-331.
5. Glass, J.D. (1981) Enz. Microb. Technol. 3, 1-7.

6. Glass, J.D. and Pelzig, M. (1978) Int. J. Pept. Prot. Res. 12, 75-80.
7. Blake, J. and Li, C.H. (1983) Proc. Nat. Acad. Sci. USA 80, 1556-1559.
8. Johansen, J.T., personal communication.
9. Ferger, M.F., Jones, W.C., Dyckes, D.F., and DuVigneaud, V. (1972) J. Amer. Chem. Soc. 94, 982-984.
10. Guibe-Jampel, E. and Wakselman, M. (1979) in Peptides 1978, Siemion, I.Z. and Kupryszewski, G., Eds., Wroclaw University Press, Wroclaw, Poland, 89-91.

FACILITATED FORMATION OF UNSYMMETRICAL DISULFIDE BONDS VIA THE S-(3-NITRO-2-PYRIDINESULFENYL) DERIVATIVE OF CYSTEINE

Michael S. Bernatowicz and Gary R. Matsueda
Cellular and Molecular Research Laboratory,
Massachusetts General Hospital, Boston, Massachusetts 02114

Rei Matsueda
Chemical Research Laboratories, Sankyo Co., Ltd.,
Tokyo, Japan

Specific and reversible reactions of sulfhydryl groups in proteins with S-(3-nitro-2-pyridinesulfenyl), Npys, derivatives of cysteine have suggested that this group could be useful for the chemically controlled synthesis of peptides containing an unsymmetrical disulfide bond.[1,2] The results of the following study verify this hypothesis.

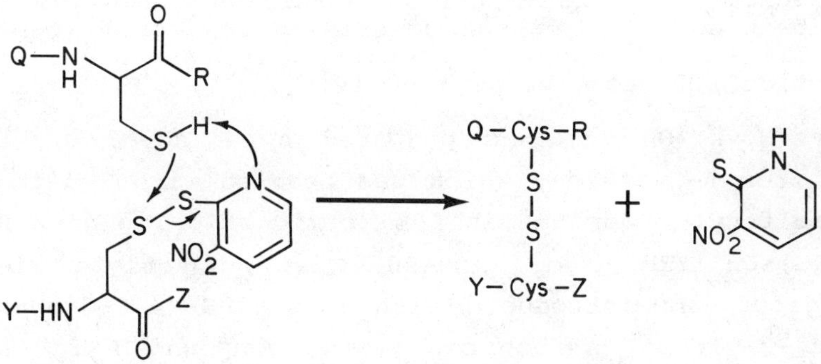

Fig. 1. Schematic representation of unsymmetrical disulfide formation. A Cys(Npys)-peptide is reacted with a second peptide containing a free thiol. Q, R, Y, and Z represent hypothetical amino acids.

In order to investigate the reaction depicted schematically in Figure 1, Boc-Cys(Npys)-OH, compound I,[2-4] was prepared and used in a conventional solid phase synthesis of a Cys(Npys)-containing 14-peptide (II). Amino acid derivative I could be used in solid phase peptide synthesis because the S-Npys group is stable in both trifluoroacetic acid at 25° and HF at 0° in the absence of thiols.[1,2,5,6] The 14-peptide (II) was found to react with a cysteine-containing 13-peptide (III), as shown in Figure 2, to give mixed disulfide (IV) at room temperature in aqueous buffers, at various pH.

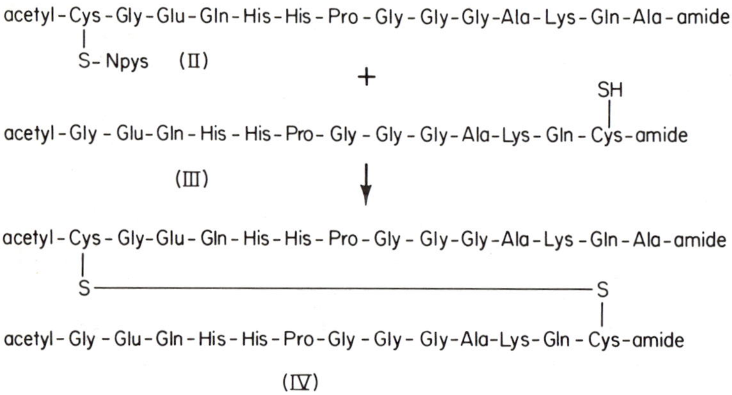

Fig. 2. Scheme for unsymmetrical disulfide bond formation between a 14-peptide containing Cys(Npys) (peptide II) and a 13-peptide containing a Cys residue (peptide III) to give peptide IV.

Boc-Cys(Npys)-OH (I) was prepared by a novel strategy starting from L-Cysteine, which was reacted with 3-nitro-2-pyridinesulfenyl chloride[3] in 90% formic acid, to give HCl·H-Cys(Npys)-OH·1/2H_2O, mp. 188-90° (dec), in 81.2% yield. The Boc-group was introduced with t-butyl-S-(4,6-dimethylpyrimidin)-2-ylthiol carbonate, giving compound I in 68.8% yield. Details of this preparation and those of peptides II and III will be published elsewhere.[7]

For the preparation of mixed disulfide IV, 10 micromoles each of peptides II and III were dissolved in 2.0 ml 1M HOAc and allowed to react at room temperature. Analytical HPLC using a C-18 column, as shown in Figure 3, indicated that the reaction gave a single product and was greater than 90% complete after 21 hours. The desired mixed disulfide IV was obtained after purification of the reaction mixture on a C-18 preparative HPLC column. Analytical HPLC experiments indicated that dithiothreitol reduction of mixed disulfide IV produced the two thiol-peptides consistent with an unsymmetrical disulfide structure. Amino acid analysis was consistent with the composition of IV.

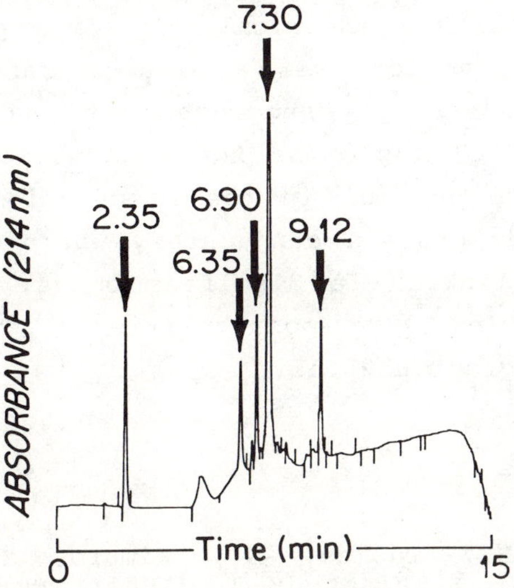

Fig. 3. Analytical HPLC of the reaction between peptides I and II after 21 hours in 1M HOAc. The peptides were eluted at a flow rate of 2 ml/min using a gradient increased linearly over 10 min from 0 to 50% acetonitrile in water in the presence of 0.1% TFA. Identity of peaks with corresponding retention times (min) are: HOAc (2.35), 3-nitro-2-thiopyridone (6.35), peptide III (6.90), the desired product IV (7.30), and peptide II (9.12).

The effects of buffer and pH on the mixed disulfide forming reaction were also studied using analytical HPLC. The rate of reaction was found to increase with increasing pH over the range studied (4.45-9.35). At pH 4.45 (0.1M KH_2PO_4), the reaction was judged to give a single product and to be greater than 90% complete after 2.5 hours. At pH 9.35 (0.1M K_2HPO_4), the starting peptides were completely consumed after 30 minutes. However, an additional undesired product, which is even more apparent in 0.1M NH_4HCO_3 at pH 7.9, is observed. The results suggest that optimum conditions exist in a phosphate buffer system between pH 4.45 and pH 9.35.

The results demonstrate that a peptide containing a cysteine residue derivatized with the Npys group can react selectively with a thiol of another cysteine-containing peptide to provide an unsymmetrical disulfide peptide. The utility of this reaction over a wide pH range in aqueous phosphate buffers strongly suggests that the chemistry of the S-Npys group holds promise for the development of reagents that will be useful in the preparation of peptide-protein or protein-protein conjugates, where the molecules of interest are linked via disulfide bonds. We are currently investigating such possibilities. This work was supported by NIH grant HL28015.

References

1. Bramson, N.H., Thomas, N., Kimura, T., Roczniak, Matsueda, R., & Kaiser, E.T. (1981) PEPTIDES: Synthesis-Structure-Function, Rich, D.H. & Gross, E. Eds., Pierce Chemical Company, Rockford IL pp. 411-415.
2. Matsueda, R., Kimura T., Kaiser, E.T., & Matsueda G.R. (1981) Chem. Lett., 737-740.
3. Matsueda, R. & Aiba, K. (1978) Chem. Lett., 951-952.
4. Boc-Cys(Npys)-OH is commercially available from Kokusan Chemical Works, 3-9 Nihonbashi-Honcho, Chuo-ku, Tokyo.
5. Ridge, R.J., Matsueda, G.R., Haber, E., & Matsueda, G.R. (1982) Int. J. Peptide Protein Res. 19, 490-498.
6. Matsueda, R., Higashida, S., Ridge, R.J., & Matsueda, G.R. (1982) Peptide Chemistry 1981, Shiori, T. Ed., Protein Research Foundation, Osaka, Japan, pp.31-36.
7. Bernatowicz, M.S., Matsueda, R., & Matsueda, G.R. (1985) Int. J. Peptide Protein Res., in press.

SOLID-PHASE SYNTHESIS OF PHOSPHOPEPTIDES: SYNTHESIS OF PHOSPHOPEPTIDES FROM THE CARBOXYL-TERMINUS OF RHODOPSIN

Anatol Arendt and Paul A. Hargrave

Departments of Ophthalmology and Biochemistry, University of Florida, Gainesville, Florida 32610

Introduction

Rhodopsin, the photoreceptor protein of rod cells in the vertebrate retina, is enzymatically phosphorylated in a light-dependent reaction. The seven serines and threonines in rhodopsin's carboxyl-terminal region may all become phosphorylated.[1]

333
Ala-Ser-Thr-Thr-Val-Ser-Lys-Thr-Glu-Thr-Ser-Gln-Val-Ala-Pro-Ala 348

We wished to synthesize various peptides from this sequence, containing PSer and PThr, in order to test the peptides as substrates for protein kinases and phosphatases. Existing methods have employed solution syntheses and employed amino acid phosphate derivatives of poor acid stability. Such procedures lead to incompleteness of reactions and poor yields of peptides.[2] We now report the synthesis of protected serine and threonine phosphates with improved stability and their use in solid phase peptide synthesis to yield phosphopeptides in high yield and purity. Three fully phosphorylated peptides, PSerGly, ValPSerLys, GluPThrPSerGlnVal and one specifically phosphorylated peptide ValSerLysThrGluPThrPSerGlnVal were obtained.

Methods and Results

We have synthesized the stable and easy to prepare N-(t-butyloxycarbonyl)-0-(diphenylphosphono)-L-serine (4a) and N-(t-butyloxycarbonyl)-0-(diphenylphosphono)-L-threonine (4b) as shown in Scheme 1. The fully protected O-phosphoserine (3a,3b) or O-phosphothreonine (3c,3d) were synthesized in excellent yield from the commercially available, inexpensive diphenylchlorophosphate (Aldrich) and esters of BocSer or BocThr. Hydrogenolysis of (3) with 5% Pd/C catalyst yielded substrates suitable for solid phase peptide synthesis. O-diphenylphosphono-serine and -threonine show excellent stability: 40% TFA in CH_2Cl_2 (10h, room T), 10% diisopropylethylamine in CH_2Cl_2 (4h, room T), and liquid HF (60 min, 0^oC).

BocNHCH(CH(R)OH)COOH —[$BrCH_2C_6H_4X$, NEt_3, X=H or NO_2]→ BocNHCH(CH(R)OH)$COOCH_2C_6H_4X$ —[$ClPO(OC_6H_5)_2$, pyridine]→

(1a) R=H (Bachem)
(1b) R=CH_3 (Bachem)

(2a) R=H X=H (Bachem)
(2b) R=H X=NO_2 (Alewood, et al[3])
(2c) R=CH_3 X=H (90%, oil)
(2d) R=CH_3 X=NO_2 (92%, oil)

→ BocNHCH($CH(R)OPO(C_6H_5)_2$)$COOCH_2C_6H_4X$ —[H_2, Pd/C]→ BocNHCH($CH(R)OPO(OC_6H_5)_2$)COOH

(3a) R=H, X=H (87%, 109-109.5^oC)
(3b) R=H, X=NO_2 (89%, 118-118.5^oC)
(3c) R=CH_3, X=H (93%, oil)
(3d) R=CH_3, X=NO_2 (95%, oil)

(4a) R=H (95%, 74-75^oC)
(4b) R=CH_3 (88%, 182-184^oC, DCHA salt)

Scheme 1. Synthesis of protected phosphoserine and phosphothreonine. Yield and melting point of compounds.

The phosphotriesters may be deprotected by treatment with tetrabutylammonium fluoride (TBAF, Aldrich) in THF-H_2O, an adaptation of the method of Ogilivie et al.[4] used in nucleotide synthesis. The reaction is slow when applied to peptides containing free amino and carboxyl groups due to their poor solubility in the organic solvent. We found it convenient to carry out the reaction with the peptide still attached to the resin. To effect removal of the phosphate protecting groups we evaluated several different concentrations of TBAF in THF-H_2O solutions in reaction with the model dipeptide BocPSer$(OPh)_2$Gly-R (R=Merrifield resin). Velocity of the reaction increased with concentration of TBAF and decreased with water concentration. The reaction was essentially complete in 90 min with 0.05 M TBAF in 95% THF (at a 1:1 molar ratio of reagent to reactant) although it was more rapid with 1M TBAF in 100% THF (and 20:1 molar ratio). Kinetics of the reaction of TBAF with phosphopeptide were determined by analysis of the phenol byproduct by HPLC (Whatman ODS-3 column, 5 x 250mm, with 0-30% B over 30 min; A= 0.1% H_3PO_4 in H_2O, B= 0.1% H_3PO_4 in CH_3CN).

Synthesis of phosphopeptides was performed manually by the solid-phase procedure of Merrifield. The side-chain functional groups of Ser, Thr and Glu were protected by benzyl groups, PSer and PThr by phenyldiesters, and the ε-amino group of Lys was protected by the 2-chlorobenzyloxycarbonyl group. Synthesis of ValPSerLys was initiated with 0.3 mmol of Boc-Lys-Merrifield resin (0.3 meq Lys/g resin) and ValSerLysThrGluPThrPSerGlnVal with 0.4 mmol Boc-Val-PAM resin (0.53 meq Val/g resin). All couplings were performed by the DCC/HOBt method with double coupling at every cycle. Each cycle of the synthesis consisted of the following operations: (1) CH_2Cl_2 wash (6 x 1.5 min); (2) deprotection with 40% CF_3COOH containing 0.1% indole (1 x 1.5 min, 1 x 30 min); (3) CH_2Cl_2 wash (6 x 1.5 min); (4) neutralization with 10% diisopropyethylamine in CH_2Cl_2 (2 x 1.5 min); (5) CH_2Cl_2 wash (6 x 1.5 min); (6) equilibration with Boc amino acid and HOBt (3 equiv. each in CH_2Cl_2-DMF, 1:1, 1 min); (7) without filtration, addition of DCC (2 equiv.); (8) coupling, 2h; (9) repeat steps 3-8. Unreacted amino groups were blocked by acetylation when necessary as judged by the ninhydrin test (1 mL acetic anhydride-pyridine mixture 1:1 v/v, 30 min). Upon completion of phosphopeptide synthesis the resin was washed with THF (6 x 1.5 min) and the phenyl protecting groups removed from PSer and PThr by treatment with TBAF. 3.0 equiv. TBAF (Aldrich 1.0M TBAF in THF) were used for each PSer or PThr in 95% THF (40 mL/mmol peptide on resin, 2h). Peptide-resin was treated with HF containing 10% anisole (0^oC, 45 min).

Crude peptides were desalted by chromatography on Bio-Gel P2 (2.5 x 100 cm) in 50% acetic acid. Peptides were purified by preparative HPLC (Whatman ODS-3 Magnum 20 column, 0.1% TFA in H_2O vs. isopropanol, 10 mL/min, monitored at 220 nm). Peptide purity was confirmed by analytical HPLC on a Whatman ODS-3 column (5 x 250 mm) using 0.1% H_3PO_4 in H_2O vs. 0.1% H_3PO_4 in CH_3CN. Amino acid compositions were as expected (by amino acid analysis and elementary analysis for nitrogen and phosphorus).

References

1. Hargrave, P. (1982) Progress in Retinal Research 1, 1-51.
2. Folsch, G. (1967) Svensk Kemisk Tidskrift 79, 38-66. b) Alewood, P.F., Johus, R.B., and Perich, J.W. (1981) PEPTIDES: Synthesis-Structure-Function, Rich, D. and Gross, E., Eds., Pierce Chemical Company, Rockford, IL, pp. 65-67. c) Alewood, P.F., Perich, J.W., and Johus, R.B. (1984) Aust. J. Chem. 37, 429-433. d) Alewood, P.F., Perich, J.W., and Johus, R.B. (1984) Tetrahedron Lett. 11, 987-990.
3. Alewood, P.F., Perich, J.W., and Johus, R.B. (1982) Synthetic Com. 12, 821-828.
4. Ogilivie, K.K., Beaucage, S.L., and Entwistle, D.W. (1976) Tetrahedron Lett. 1255-1256.

SYNTHESIS OF SULFATED CHOLECYSTOKININ OCTAPEPTIDE WHEREIN ALL SYNTHETIC STEPS ARE PERFORMED ON A SOLID SUPPORT

James D. Rosamond, Jeanne Comstock, James C. Blosser, and Susan J. Augello-Vaisey

Pennwalt Corporation, Pharmaceutical Division, Departments of Medicinal Chemistry and Pharmacology, 755 Jefferson Road, Rochester, NY 14623

Introduction

Reported solution phase syntheses of cholecystokinin octapeptide sulfate ester (CCK-8-SE) are laborious and low in overall yield[1,2] (less than 7% from phenylalanine amide). Reported solid phase methods give either very little experimental detail[3] or employ unusual protecting groups.[4] In all these cases the sulfation step is carried out with excess sulfating agent in solution phase which can lead to isolation and purification difficulties.[5] The purpose of this investigation was to develop a high yield, simplified synthetic route to CCK-8-SE in which all synthetic steps, including sulfation, are carried out on a solid support. An additional aim was to develop a simplified purification procedure.

Synthesis

The choice of protecting groups useful in a total synthesis of CCK-8-SE on a solid support is dictated by the stability of the sulfate ester group to the reagents necessary for protecting group removal. Since it is known that the sulfate ester group is stable to treatment with base and TFA (briefly) but not HF,[6] the TFA labile OtBu side chain protecting group was chosen for the aspartic acid residues.

This choice, consequently, allows the use of the TFA labile Boc protecting group for only the terminal amino acids. Thus, the remaining amino acids were protected with the base labile Fmoc group. The tyrosine hydroxyl group was not protected during coupling to allow for its sulfation prior to final deprotection.

Resins suitable for the synthesis of CCK-8-SE are likewise dictated by the stability of the sulfate ester group to the resin cleavage conditions. Because CCK-8-SE is a C-terminal amide, ammonolysis is the method of choice for cleavage of the sulfated peptide from the resin particularly since the sulfate ester group is unstable to HF treatment. Pam-resin was found to be well suited for this purpose.

The strategy adopted for the total solid phase synthesis of CCK-8-SE is outlined in Scheme 1. Aminomethyl-polystyrene (1% crosslinked, 0.7 meq. N) was coupled with Boc-Phe-4-(oxymethyl)phenylacetic acid (1.5 eq., "preactivated" with HOBt/DCC) to give Boc-Phe-OCH_2-Pam-resin. Following Boc removal, Fmoc-Tyr-Met-Gly-Trp-Met-Asp(OtBu)-Phe-OCH_2-Pam resin was secured in a similar manner by coupling the respective "preactivated" Fmoc amino acids to the resin. "Preactivated" Boc-Asp(OtBu)-OH was then coupled to the resin to generate Boc-Asp(OtBu)-Tyr-Met-Gly-Trp-Met-Asp(OtBu)-Phe-OCH_2-Pam-resin following Fmoc removal.

Sulfation of the tyrosine hydroxyl group was the next step to be accomplished. The most efficient sulfating agent was determined to be sulfur trioxide pyridine complex (40 eq. in 1:2 pyridine/DMF, 21 hr., room temperature). Sulfonation of tryptophan and tyrosine was not a serious side reaction under these conditions. Following Boc and OtBu removal (50% TFA in DCM, 30 min.), the peptide was smoothly ammonolyzed (ammonia saturated methanol, 3 days) to give crude CCK-8-SE. RP-HPLC showed this material to be composed primarily of CCK-8-SE with some contaminating CCK-8-NS (non-sulfated form). The product was readily purified by ion

Scheme 1. Solid phase synthesis of CCK-8-SE.

Boc-Phe-OCH_2-Pam-resin

↓ 1) 50% TFA/DCM
2) 5% DIEA/DCM
3) Fmoc-Asp(OtBu)-OH, HOBt, DCC

Fmoc-Asp(OtBu)-Phe-OCH_2-Pam-resin

↓ 1) 20% Piperidine/DMF
2) Fmoc-AA-OH, HOBt, DCC
3) 4 more cycles

Fmoc-Tyr-Met-Gly-Trp-Met-Asp(OtBu)-Phe-OCH_2-Pam-resin

↓ 1) 20% Piperidine/DMF
2) Boc-Asp(OtBu)-OH, HOBt, DCC
3) SO_3-Pyridine

Boc-Asp(OtBu)-Tyr(SO_3H)-Met-Gly-Trp-Met-Asp(OtBu)-Phe-OCH_2-Pam-resin

↓ 1) 50% TFA/DCM
2) 5% DIEA/DCM
3) Ammonolysis
4) Ion Exchange Chromatography

H-Asp-Tyr(SO_3H)-Met-Gly-Trp-Met-Asp-Phe-NH_2 (CCK-8-SE)

---

exchange chromatography (DEAE Trisacryl M, LKB Instruments, Inc., ammonium carbonate gradient). Pure fractions were pooled and lyophilized to give the analytically pure ammonium salt of CCK-8-SE in 23% overall yield based on stating aminomethyl-polystyrene.

Synthetic CCK-8-SE demonstrated predictable biological activity, eliciting gall bladder contraction _in vitro_ (EC_{50} 3.4 ± 0.5 nM) and exhibiting anorectic properties _in vivo_. In the latter study, synthetic CCK-8-SE inhibited food intake ($p < 0.05$) in dose related fashion during the first half hour when administered intrapertioneally in rats trained to feed during a three hour period daily (Table 1).

Table 1. Feeding inhibition of CCK-8-SE in 21 hr fasted rat.

CCK-8-SE Dose (ip)	Percent Inhibition (30 min. post dose)
3 μg/kg	25
30 μg/kg	70
300 μg/kg	95

Conclusions

A highly efficient synthesis of CCK-8-SE has been developed in which all synthetic steps including sulfation are carried out on a solid support. In contrast with solution phase sulfation methods, removal of excess sulfating agent may be easily accomplished by simple filtration due to the presence of the solid support. The product, easily purified by ion exchange chromatography, potently inhibited feeding in the fasted rat.

References

1. Ondetti, M.A., Pluscec, J., Sabo, E.F., Sheehan, J.T., and Williams, N., (1970) J. Am. Chem. Soc. 92, 195-199.
2. Pluscec, J., Sheehan, J.T., Sabo, E.F., Williams, N., Kocy, O., and Ondetti, M.A., (1970) J. Med. Chem. 13, 349-352.
3. Sipos, F., (1973) US Patent 3,714,140.
4. Gaudreau, P., Morell, J.L., St-Pierre, S., Quirion, R., and Pert, C.B., (1983) in Peptides: Structure and Function, Hruby, V.J., and Rich, D.H., Eds., Pierce Chemical Co., Rockford, IL, pp 441-444.
5. Yanaihara, N., and Sugiura, N., (1982) US Patent 4,330,466.
6. Yajama, H., Mori, Y., Koyama, K., Tobe, T., Setoyama, M., Adachi, H., Kanno, T., and Saito, A., (1976) Chem. Pharm. Bull. 24, 2794-2802.

SOLID PHASE PEPTIDE SYNTHESIS OF THE 1-9 PROTHROMBIN PRECURSOR SEQUENCE. VARIATION OF THE REACTION CONDITIONS.

Günter Benz, Richard Pottorf, and Daniel H. Rich
School of Pharmacy, University of Wisconsin-Madison
Madison, Wisconsin 53706

Introduction

The circulating factor, prothrombin, is activated in vivo by the enzyme, vitamin K-dependent carboxylase, which converts 10 N-terminal glutamic acids into γ-carboxyglutamic acids (Gla).[1] As part of our attempts to determine substrate specificity of this enzyme,[2] we have synthesized the 1-9 prothrombin precursor sequence Ala-Asn-Lys-Gly-Phe-Leu-Glu-Glu-Val by 7 different solid phase synthetic[3] procedures.[4,5,6]

Results and Discussion

Details of the different synthetic strategies are given in Scheme I. All coupling steps were monitored using the qualitative Kaiser test[7] and, in the case of the Fmoc/^{t}Bu strategy, additional monitoring by HPLC after deprotection and cleavage was carried out. All products were purified by gel filtration (Sephadex G-15) and semipreparative HPLC (Bondapak C_{18}, 10 μM) under identical conditions. Synthesis 1 utilized the conventional Merrifield technique[4] on a hydroxymethyl polystyrene resin. Boc was used as the labile protecting group and benzyl derived blocking groups were used for permanent protection. The coupling reactions were carried out using DCC, HOBT and no recouplings were necessary. The same procedures were used in synthesis 2 except that the more acid stable Pam-anchor was used.[8] Again, no recouplings were necessary, and a very pure (~ 90%) crude peptide was obtained, which after desalting and HPLC separation gave the purest (> 99%) peptide in this series.

Syntheses 3 and 4 used Fmoc and ^{t}butyl groups for protection of the peptide synthesized on the p-alkoxybenzyl polystyrene resin.[5] In synthesis 3 Gly-4 and Asn-2 had to be recoupled with DCC, HOBt. The product was obtained only in 8% yield. Synthesis 4A used symmetrical anhydrides for coupling. Pure heptapeptide 3-9 was isolated but crude peptide PT 1-9 contained a non-racemized, non-nitrile impurity (30%) poorly resolvable from product by HPLC. When Fmoc-Asn-ONp/HOBT was used (Synthesis 4B), an impurity containing two residues of Asn was obtained in about 50% yield.

Syntheses 5-7 utilized polyamide supports.[6,9] Pepsyn A (p-Hydroxymethylphenoxyacetic acid as anchor group) was used in synthesis 5 in combination with Fmoc and ^{t}butyl protection, and preformed symmetrical anhydride coupling. After two recouplings (Glu-7, Glu-6) and one acetylation (Glu-7) a 70% pure crude material was obtained, which could be further purified in the usual manner. Synthesis 6 utilized a polyamide/kieselguhr resin in a continuous flow system.[9] Preformed symmetrical anhydrides were used. Two recouplings were necessary at Leu-6 and Phe-5. The product contained a small amount (< 5%) of an unidentified second component. Synthesis 7 repeated synthesis 6 except that BOP-Cl[10] was used as coupling reagent to avoid the use of excess Fmoc-amino acids. Although BOP-Cl is poorly soluble in DMF, it worked well when preactivation was used. The purity and yield of product was as good as in synthesis 6.

All syntheses were carried out once and were not optimized for yield or purification method. In our hands, the technique using the Pam-anchor gave the purest crude PT 1-9 that required the fewest purification steps. However, the difficulties associated with syntheses 3-7 appeared to be caused by

Variations in the synthesis of the 1–9 Prothrombin Precursor

1	2	3	4	5	6	7	8	9
Ala	Asn	Lys	Gly	Phe	Leu	Glu	Glu	Val

entry	1	2	3	4	5	6	7
solid support	PS	PS	PS	PS	PA Pepsyn A	PA/SiO_2 Pepsyn KA	PA/SiO_2 Pepsyn KA
anchor	hydroxy-methyl	PAM	p-Alkoxy-benzyl alcohol	p-Alkoxy-benzyl alcohol	p-Hydroxy-methyl-phenoxy-acetic acid	p-Hydroxy-methyl-phenoxy-acetic acid	p-Hydroxy-methyl-phenoxy-acetic acid
protection scheme	Boc, benzyl	Boc, benzyl	Fmoc, ^{t}BU	Fmoc, ^{t}BU	Fmoc, ^{t}Bu	Fmoc, ^{t}Bu	Fmoc, ^{t}Bu
coupling reaction	DCC, HOBT CH_2Cl_2, DMF	DCC, HOBT CH_2Cl_2, DMF	DCC, HOBT CH_2Cl_2, DMF	pref. sym. anhydride CH_2Cl_2	pref. sym. anhydride DMF	pref. sym. anhydride DMF	BOP-Cl DMF
automation	Beckman 990 B	Beckman 990 B	Beckman 990 B	Beckman 990 B	Beckman 990 B	cont. flow	cont. flow
final cleavage	90% HF, anisole	94% HF, anisole	55% TFA, CH_2Cl_2	90% TFA, H_2O	90% TFA, H_2O	90% TFA, H_2O	90% TFA, H_2O
recoupling, acetylation	–	–	4-Gly, 2-Asn	7-Glu, 5-Phe 4-Gly 7-Glu (Ac)	7-Glu, 6-Glu 7-Glu (Ac)	6-Leu, 5-Phe	8-Glu, 7-Glu
yield crude (after cleavage)	49%	55%	90%	A 81% B 85%	41%	~ 100%	67%
purity crude (HPLC)	~ 85%	~ 90%	~ 50%	A DP 80% B DP 85%	~ 70%	~ 85%	~ 80%
yield pure	21%	35,2%	8,0%	A 19%	13,5%	42,7%	41,2%

DB: Double peak

Fmoc-Asn since homogenous PT 3-9 was easily synthesized by these methods. Up to this point, all methods were essentially equivalent.

Acknowledgements

We gratefully acknowledge financial support from Bayer AG, West Germany and a postdoctoral training grant No. DE-07031 from the National Institutes of Health.

References

1. Suttie, J.W. in The Enzymology of Post-translational Modifications of Proteins (1980), Freedman, R.B. and Hawkins, H.C., eds., Academic Press, London.

2. Dhaon, M.K., Lehrman, S.R., Rich, D.H., Engelke, J.A. and Suttie, J.W. (1984), J. Med. Chem. 27, 1196-1201 and references cited therein.

3. Merrifield, R.B. (1963), J. Am. Chem. Soc. 85, 2149-2154.

4. Merrifield, R.B. (1983) in Peptides: Structure and Function. Proceedings of the Seventh American Peptide Symposium (1983), Hruby, V.J. and Rich, D.H. eds., Pierce Chemical Company, Rockford, IL, pp. 33-44.

5. Chang, C.-D., and Meienhofer, J. (1978) Int. J. Peptide Protein Res. 11, 246-249.

6. Atherton, E., Logan, C.J., and Sheppard, R.C. (1981), J. Chem. Soc. Perkin 1 538-546.

7. Kaiser, E., Colescott, R.L., Bossinger, C.D., and Cook, P.I., (1970), Anal. Biochem. 34, 595-598.

8. Mitchell, A.R., Erickson, B.W., Ryabtsev, M.N., Hodges, R.S., and Merrifield, R.B., (1976), J. Amer. Chem. Soc. 98, 7357-7362.

9. Atherton, E., Brown, E., and Sheppard, R.C. (1981), J. Chem. Soc. Chem. Commun. 1151-1152.

10. Cabré, J., and Palomo, A.L. (1984), Synthesis 413-417; R.D. Tung and D.H. Rich, J. Amer. Chem. Soc. (1985), in press; ibid these proceedings pp

FMOC AMINO-ACID PENTAFLUOROPHENYL ESTERS - VERSATILE NEW REAGENTS IN SOLID PHASE PEPTIDE SYNTHESIS

Eric Atherton, Alison Dryland, Peter Goddard, Linda Cameron, John D. Richards, and Robert C. Sheppard
MRC Laboratory of Molecular Biology, Hills Road, Cambridge CB2 2QH, England

Preformed symmetric anhydrides of Boc and Fmoc-amino-acids are efficient acylating agents in solid phase synthesis, and have been widely applied. Our search for alternative reagents which would combine the high reactivity and general freedom from side reactions of these anhydrides with the crystallinity, ease of preparation, stability to storage and ease of manipulation more characteristic of activated esters was prompted largely by developments in continuous flow technology.[1,2] This has led to development of particularly simple and efficient methods and instrumentation[2,3] which we believe offer significant advantage over more conventional techniques. For full automation, this simplicity is best maintained if complex pre-activation procedures can be avoided. Our results have now shown that Fmoc-amino-acid pentafluorophenyl esters[4,5] may be nearly ideal in this respect.

We recently reported[5] that the difficult test decapeptide (I) is easily obtained using Fmoc-amino-acid pentafluorophenyl esters for all the peptide bond-forming steps. The technique is in routine use in our laboratory, and we now describe two additional examples. The first, the dodecapeptide (II) is a difficult tryptophan and histidine-containing sequence; the second (III) is a straightforward decapeptide used to test the efficiency of synthesis in a totally automatic assembly. Both syntheses were carried out using continuous flow techniques.

(I) H-Val-Gln-Ala-Ala-Ile-Asp-Tyr-Ile-Asn-Gly-OH

(II) H-Leu-Ala-Glu-Leu-Gly-Ala-Ser-Leu-Leu-Lys-His-Trp-OH

(III) H-Ala-Gly-Val-Ile-Phe-Pro-Leu-Ala-Val-Gly-OH

Preparation of the pentafluorophenyl esters is essentially straightforward.[4] We have not yet succeeded in preparing crystalline derivatives of Fmoc-Ser(Bu^t) or Fmoc-Thr(Bu^t), nor of Fmoc-Arg(Mtr). Activated derivatives of the latter[6] appear to be particularly unstable. On the other hand, Fmoc-His(Boc) has provided a nicely crystalline activated ester which couples without racemisation and which appears to retain its side chain protection during further chain extension.

The pentafluorophenyl esters are less reactive than analogous symmetric anhydrides, though this is much enhanced in the presence of hydroxybenzotriazole catalyst. Thus in an exploratory assembly of (III), the later stages were carried out both in the presence and absence of catalyst. Acylation times for complete reaction with only 2 equivs. of activated derivatives (ninhydrin and trinitrobenzenesulphonic acid colour tests) were 5-30 min in the presence of HOBt (including the sterically hindered Val-Ile coupling); in the absence of HOBt this last was incomplete after 75 min, brought to completion in a further 30 min by added catalyst. This level of reactivity appears to be adequate for most circumstances. In the presence of dimethylaminopyridine, pentafluorophenyl esters are effective in esterification of the first residue to hydroxymethyl resin. The usual precautions[7] regarding minimisation of racemisation by dimethylaminopyridine need to be observed.

The assembly of dodecapeptide (II) was carried out on kieselguhr-supported polydimethylacrylamide resin.[1] Pentafluorophenyl esters were used at every stage, including attachment of the hydroxymethylphenoxyacetyl linkage agent and esterification of the terminal tryptophan residue to this. Incorporation of tryptophan was 75% after 1 h and 92% overnight. A resin sample removed at this stage showed about 1% dipeptide content.[8] Reaction of this resin with Boc-Leu-ONP and acid cleavage gave a very small peak coincident with H-L-Leu-D-Trp-OH which could not be quantified. The latter is associated with use of dimethylaminopyridine catalyst[7] and applies only to addition of the C-terminal residue. The chain was extended using 3 equiv of Fmoc-amino-acid pentafluorophenyl esters, side chain protected as Boc or t-butyl derivatives as appropriate. Up to residue 8, acylations were complete after 20 min and were terminated after 40 min. At this stage reactions started to slow, and HOBt catalyst was added thereafter. Coupling of the last residue was repeated. Amino-acid analyses of the final resin and of the crude dodecapeptide detached with trifluoroacetic acid in the presence of ethane dithiol were excellent. The hplc profiles of (a) the crude product and (b) after a single ion-exchange purification on carboxymethylcellulose (72% of the total applied was recovered in the main peak) are shown in Figure 1. This is clearly a successful synthesis of a tryptophan and histidine containing peptide.

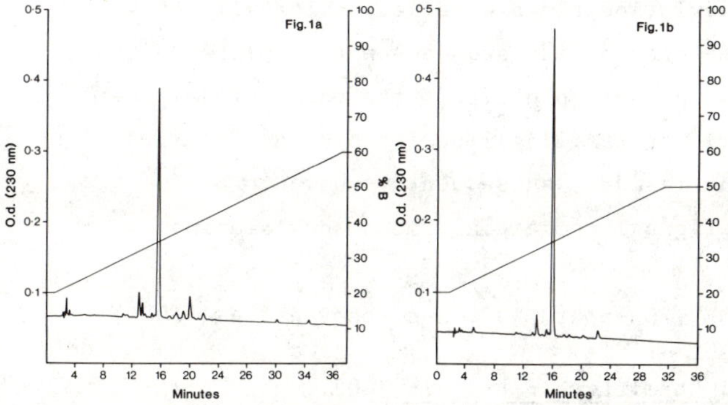

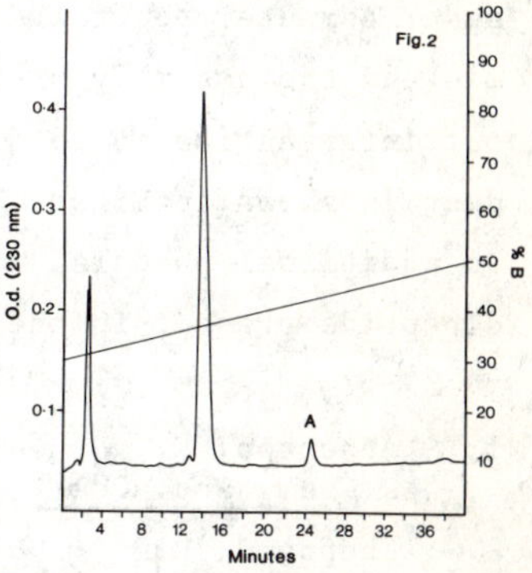

Fig. 1. Hplc of II on analytical Aquapore RP300; pump A 0.1% TFA, pump B 90% acetonitrile/10% A, flow 1.5 ml min^{-1} (a) crude product, (b) After a single ion exchange column.

Fig. 2. Hplc of crude III on Aquapore RP300; pump A 0.1% TFA, pump B 90% ethanol/10% A, flow 1.5 ml min^{-1}. The injection peak at 2 min did not contain any peptide material.

Two general points emerged from this synthesis. Cleavage of the p-alkoxybenzyl ester bond linking the tryptophan residue to the resin was exceptionally inefficient. Only 42% was detached using trifluoroacetic acid/anisole/ethane dithiol mixture. Previous experience suggests strongly that this is not due to incomplete cleavage of the ester bond but to readdition of the benzyl cation to the indole ring. This would be particularly favorable with C-terminal tryptophan, and may ocur intramolecularly. During chain extension, the effluent stream was continuously monitored spectrometrically.[3] No increase in the amount of fluorenyl derivative liberated at successive deprotection steps was observed, strongly suggesting that the histidine side chain protecting group was remaining intact. This result contrasts with earlier indications using symmetric anhydride extension[9] and is being further investigated.

Synthesis of (III) was carried out using a continuous flow synthesiser similar to that already described[2,3] but with addition of an automatic sample addition system. This permitted dissolution and introduction of solid pentafluorophenyl ester derivatives (2 equiv.) and hydroxybenzotriazole (2 equiv.) directly into the flowing stream under computer control. Acylation times were set arbitrarily at 40 min and the assembly of the decapeptide sequence was completed without intervention in 15 h. The hplc profile of the total crude product is shown in Figure 2. The trailing impurity peak A contained an additional phenylalanine residue, subsequently shown to be due to dipeptide present in the original commercial Fmoc-phenylalanine.[8]

1. Atherton, E., Brown, E., Sheppard R.C. and Rosevear, A. (1981) J.Chem. Soc. Chem. Comm, 1151.
2. Sheppard, R.C. (1983) Chemistry In Britain, 402.
3. Dryland, A. and Sheppard, R.C. (1985) J. Chem. Soc. Perkin I, in press.
4. Kisfaludy, L. and Schön, I. (1983) Synthesis, 325
5. Atherton, E. and Sheppard, R.C. (1985) J. Chem. Soc., Chem. Comm., 165.
6. Atherton, E., Sheppard, R.C. and Wade, J.D. (1983) J. Chem. Soc. Chem. Comm, 1060.
7. Atherton, E., Benoiton, N.L., Brown, E., Sheppard, R.C. and Williams, B.J. (1981) J. Chem. Soc., Chem. Comm, 336.
8. Fuller, W.D., Wigler, G.F., Chaturvedi, N.C. and Verlander, M.S. (1983) Proc. 8th Amer. Pept. Symp., (V.J. Hruby and D.H. Rich, eds.) Tucson, pp. 79-82.
9. Atherton, E., Cammish, L.E., Goddard, P., Richards, J.D. and Sheppard, R.C. In 'Peptides 1984', Proc. 18th European Peptide Symposium, Djuronasett, Sweden, 1984, p. 153.

AS PEPTIDE LENGTH INCREASES DURING SOLID PHASE SYNTHESIS THE AMINO TERMINUS pKa INCREASES. FREE IN AQUEOUS SOLUTION THE pKa DECREASES.

Charles A. Hooper, Jeffrey J. Bresch and Robert H. Reid
Department of Gastroenterology, Walter Reed Army Institute of Research, Washington, DC 20307.

Introduction

Peptides are usually synthesized with the amino group at the growing end of the chain. During the synthesis, the amino terminus (AT) is involved in two reactions, AT deprotection and condensation with the next amino acid. These usually proceed by nucleophilic acyl substitution where the AT acts as a nucleophile.[1 2] Reaction rates are directly proportional to AT nucleophilicity. Electron (e^-) release from the peptide to the AT determines the AT nucleophilicity, and the AT basicity. A high pKa value indicates strong nucleophilicity. Measuring the AT pKa indicates both e^- density, and reactivity.

Experimental

Peptides were synthesized on a solid phase support of polystyrene resin.[3] Proceeding C to N, the pKa was measured after each AT deprotection.

Several routes can determine pKa. By acid-base titration, following the pH with an H^+ ion selective electrode is difficult. Water does not wet polystyrene so the solvent changes to an organic solvent such as CH_2Cl_2. Ions, however, dissociate much less in CH_2Cl_2 than in water. Low H^+

concentrations and increased electrical impedances cause increased variability in the electrode pH measurements. Low H^+ concentrations give much higher pH values. These need to be related back to water to arrive at a meaningful pKa value.

Instead of determining the pKa from a titration curve, the pKa was calculated from the equilibrium point of a pKa governed salt reaction. After preparing the HCl salt of the AT, triethylamine (TEA) displaces a proportion of the HCl to form TEA:HCl. Determining the pKa follows the same types of pKa driven salt reactions as are followed during chloride determination monitoring of the number of free AT groups.[4 5]

Resin bed AT (pKa near 9) reaction for 10 minutes with a 5 molar pyridine:HCl (Pyr:HCl) (pKa=5.25) excess, then 2.5 fold excess for 20 minutes, prepared the AT:HCl salt. To prepare the Pyr:HCl reagent, Pyr and 37% HCl were cautiously added, 1 ml at a time maintaining an equimolar ratio, to a flask in an ice bath. Washing the resin bed removes Pyr and Pyr:HCl.

A dried 1 umole AT:HCl resin sample was analytically weighed out in a disposable Cl^- determination sample vial. Adding 200 ul CH_2Cl_2 and an accurate 2 ul of TEA (pKa=10.75) displaces a proportion of HCl from the AT:HCl (pKa near 9). A Buchler Chloridometer from Fisher Scientific measured the TEA:HCl that formed after adding 4 ml of Chloridometer 10% acetic acid in 0.64% nitric acid diluent, that extracted the TEA:HCl into aqueous acid where the Chloridometer can function, and prevented reequilibration as AgCl precipitates.

Initial AT:HCl formation was quantitated by measuring total Cl^- in 1 umole of AT. 30 ul of 30% NH_4OH displaced the HCl. Although the TEA (pKa=10.75) pKa is greater, NH_4OH (pKa=9.25) precipitates NH_4Cl, <0.0015% soluble in CH_2Cl_2, driving the reaction to >99% completion when the AT pKa is <13.8.

Results

In determining the extent of initial AT:HCl formation, NH_3 completely displaced the HCl from the AT:HCl. For proline (pKa=10.64) resin, TEA displaced almost no HCl while NH_3 displaced an equimolar amount. AT:HCl forms quantitatively.

Solid phase peptide synthesis AT pKa values were calculated from a derived equilibrium equation.

$$Ka = [H+]\ [A-]\ /\ [HA]$$

$$Ka\ AT = [HCl]\ [AT:]\ /\ [AT:HCl]$$

$$10^{(pKa\ TEA\ -\ pKa\ AT)} = [AT:]\ [TEA:HCl]\ /\ [AT:HCl]\ [TEA:]$$

Each AT in the lattice encounters a ratio of TEA:HCl to TEA:. This ratio determines the final equilibrium. Volumes divide out, leaving the equation in quantities of moles. Due to the sparsity of ions, the activity coefficients are close to 1. Let f = the fraction of HCl displaced from the AT, TAT = the total umoles of AT:HCl + AT:, and 2 ul TEA = 14.35 umoles.

$$[AT:]\ /\ [AT:HCl] = (f \times TAT)\ /\ ((1-f) \times TAT)$$

$$[TEA:HCl]\ /\ [TEA:] = (f \times TAT)\ /\ (14.35 - (f \times TAT))$$

$$10^{(pKa\ TEA\ -\ pKa\ AT)} = (f)^2\ (TAT)\ /\ (1-f)\ (14.35 - (f \times TAT))$$

Table I. Amino Terminus pKa Values Comparing

A. Solid Phase Synthesis Values in CH_2Cl_2 (Gly-Asn-Thr-Ile-Val-Ala-Val-Glu-benzyl-P)

B. To the Same Lengths of Polyglycine Free in H_2O

C. To the Ethyl or Methyl Polyglycyl Esters in H_2O

Amino Acid		Solid Phase pKa	Glycyl pKa[6]	Ester pKa[6]	Ester[6]
1.	Glu	< 9.90	9.80	7.83	Et
2.	Val	10.92	8.23	7.92	Et
3.	Ala	11.31	8.11	7.91	Et
4.	Val	11.37	8.06	7.81	Et
5.	Ile	11.44	8.02	--	--
6.	Thr	11.41	7.68	7.86	Me
7.	Asn	11.64	7.70		

Summary and Conclusions

As peptides increased from 1 to 4 amino acids in length: pKa went from <9.90 to 11.37 in CH_2Cl_2,[6] a >1.47 pKa increase, pKa went from 9.80 to 8.06 in H_2O,[6] a 1.74 pKa decrease. Changing the anion to a neutral ester makes glycine similar to the benzyl ester; the pKa in water remains nearly constant at 7.87.[6] Having greater e^- releasing potential than the ethyl ester, the benzyl ester should have a higher pKa than 7.87.

Solid phase synthesis pKa changes showed nucleophilicity increases. By the 1.47 pKa increase, acylation rates increase by a factor of 30. When amino acids couple incompletely, the AT pKa might be lower. Cl^- could give the pKa. More e^- withdrawal in the activated incoming amino acid, e^- release to the AT, and a less polar solvent could increase reactivity.

High pKa values in CH_2Cl_2 show a high e^- density around the AT nitrogen. e^- release to the AT by the neighboring amide bond makes the amide available for nucleophilic attack. Solid phase synthesis could alter the backbone to prepare analogues.

pKa differences show that solvent affects pKa; water alters the pKa of peptides. If e^- density shifts toward the AT in water as in CH_2Cl_2, water must distribute the charge density, indicating that more layers of more oriented water dipole solvation shells surround peptides than single amino acids.

References

1. Brady, S.F., Hirschmann, R., and Veber, D.F. (1977) J. Org. Chem. 42, 143-146.
2. Sheehan, J.C., and Hess, G.P. (1955) J. Am. Chem. Soc. 77, 1067-1068.
3. Merrifield, R.B. (1963) J. Am. Chem. Soc. 85, 2149-2154.
4. Dorman, L.C. (1969) Tetrahedron Letters, pp. 2319-2321.
5. Hooper, C.A., Reid, R.H., Marnane, W.G., and Seid, R.C. (1983) Eighth American Peptide Symposium, Abstract 1-24.
6. Glasstone, S., and Hammel, E.F. (1941) J. Am. Chem. Soc. 63, 243-248.

PREPARATION AND USE OF AN AMINOETHYL POLYETHYLENE GLYCOL-CROSSLINKED POLYSTYRENE GRAFT RESIN SUPPORT FOR SOLID-PHASE PEPTIDE SYNTHESIS[1]

Shmuel Zalipsky, Fernando Albericio[2], and George Barany[3]
Department of Chemistry, University of Minnesota,
207 Pleasant Street S.E., Minneapolis, MN 55455

Current reports[4] from this laboratory provide optimal chemistries for orthogonal solid-phase peptide synthesis, by use of *N*$^{\alpha}$-dithiasuccinoyl (Dts) amino acids (**1**) together with handles incorporating *p*-alkoxybenzyl (**2**) or *o*-nitrobenzyl (**3**) ester linkages. The examples for which the methodology has

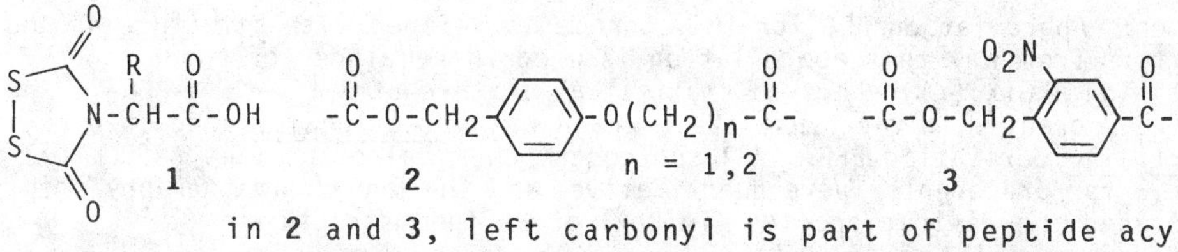

1 **2** **3**

in **2** and **3**, left carbonyl is part of peptide acyl; right carbonyl attaches to amino-support

been demonstrated were all elaborated on standard aminomethyl-copoly(styrene-1%-divinylbenzene)-resins. We attribute the very low levels of impurities that arose to extraneous functional groups in the starting resins, although inadequate solvation/swelling of the peptide-resin might be a factor also. With the goal of improving overall synthetic efficacy, we initiated studies on a new support that uniformly incorporates a derivatized polyethylene glycol (PEG) spacer, with average molecular weight ($\bar{M}$) 2000, between functional groups on the polystyrene (PS) backbone and the point of attachment of the syntheses. The present communication reports subtle but reproducible advantages of the graft (PEG-PS) polymeric supports over PS, with regard to physico-chemical properties and results of parallel model peptide syntheses.

Scheme 1. Synthesis of a Bifunctional ω-Amino Acid Derivative of PEG-2000

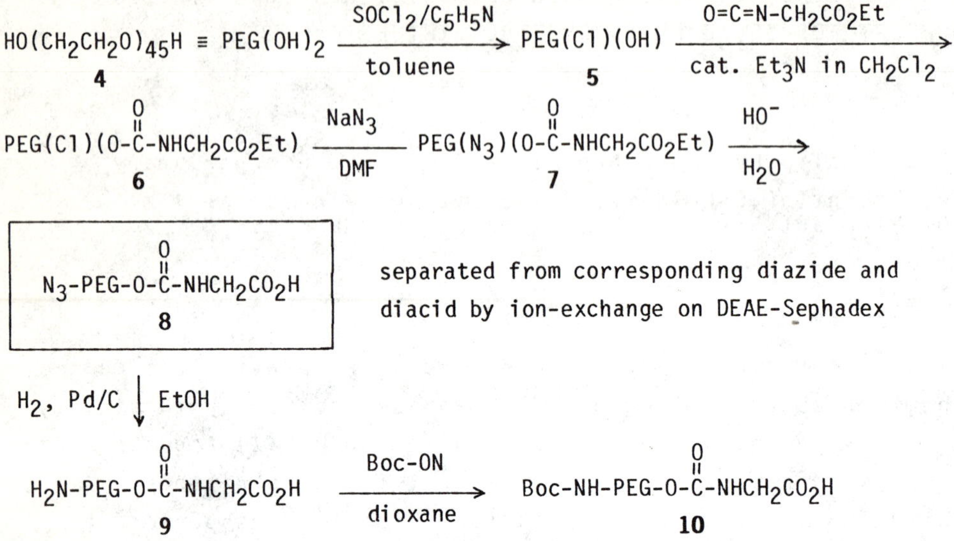

Notes: Abbreviation PEG for this Scheme is defined with structure **4**, and differs from way this abbreviation is used in remainder of paper. Notation PEG(X)(Y) refers to statistical mixture of X-PEG-X, X-PEG-Y, and Y-PEG-Y created in the first step, which was intentionally conducted to partially derivatize PEG. All subsequent chemical steps (see refs. 4, 7 and 8 for precedents) were quantitative, and the key chromatography step isolated the desired species (in box) of defined structure.

Recently, polyethylene glycol-polystyrene graft resins have been prepared and evaluated for peptide synthesis in the laboratories of Mutter[5] and Bayer[6]. In the former case, large excesses of symmetrical bifunctional soluble PEG's were reacted with functionalized PS, and difficulty was noted in grafting PEG with $\bar{M}$ > 400. In the latter case, ethylene oxide was polymerized directly onto initiator sites on PS to allow growth of chains of $\bar{M}$ 5500, but this method may be expected to give a heterogeneous distribution of grafted sites. Our method (Scheme 1) begins with bifunctional PEG (**4**), which was converted in six steps to an unsymmetrical ω-amino acid derivative (**10**). This was quantitatively attached onto aminomethyl-PS (0.6 mmol NH_2 groups/gm) by oxidation-reduction coupling[9] (Bu_3P + 2,2'-dipyridyl disulfide) to provide a graft support which is about half by weight of polyethylene glycol.

Table I. Ratio of Volume Swollen/Dry Volume of Same Bead[10]

	CH_2Cl_2	DMF	EtOAc	CH_3CN	EtOH
Boc-Gly-PS	5.8	3.6	3.7	1.4	1.2
Boc-PEG-PS	7.0	4.8	3.7	3.5	2.3

This Boc-PEG-PS was compared with a Boc-Gly-PS prepared from the same batch of starting resin. Dry beads of Boc-PEG-PS had an average diameter of 131 ± 16 μ, compared to 109 ± 14 μ for Boc-Gly-PS, for a dry volume ratio of 1.7. Swelling studies (Table I) showed slightly better results for the graft resin.

The title support[11] was obtained by TFA—CH_2Cl_2 (3:7) removal of the Boc group originally on **10**. Our usual methods gave Dts-amino acyl graft resins including the appropriate anchoring linkages (**2**, **3**) and "internal reference" amino acids for precise determinations of cleavage yields. These were either tested at that point, or extended further with Dts-amino acids to yield peptide-resins. Simultaneously, control experiments were carried out with the same reagents and identical treatments, using PS omitting the PEG spacer.

The model tetrapeptide Leu-Ala-Gly-Val assembled on PS (anchor **2**, n = 2) was 98.8% pure with 0.3-0.5% of each of the deletion peptides Ala-Gly-Val, Leu-Ala-Val, and Leu-Gly-Val. The same synthesis on PEG-PS gave product of 99.6% purity. Acidolytic cleavage of the anchor with TFA—CH_2Cl_2 (1:4), 15 min at 25 °C proceeded in 54% yield for PS and 74% for PEG-PS; both cleavage yields were >95% with TFA—CH_2Cl_2 (1:1), 1 hr. Parallel syntheses of methionine-enkephalin[12] both gave excellent results (>90% HPLC purity of crude products), with slightly better acidolytic cleavage yields (86% vs. 93%) for PEG-PS. A different kind of relevant comparison was in the yields of photolytic cleavage of Dts-Phe (anchor **3**). Using 350 nm light, TFE—CH_2Cl_2 (1:4) as solvent, 24 hr irradiation, these yields were 80% from PS and 90% from PEG-PS. On the other hand, with 6 hr irradiation, yields for release of fully protected leucine-enkephalin[4] were _ca_. 65% with both supports.

References and Notes

1. Supported by grants to G.B. from National Institutes of Health (GM 28934) and the Chicago Community Trust.

2. Recipient, postdoctoral fellowships, Generalitat of Catalunya, Spain, 1983-1984. Current address: Department of Organic Chemistry, University of Barcelona, Spain.

3. Searle Scholar, 1982; National Institutes of Health Research Career Development Award AM 01099, 1982-1987. Address inquiries to this author.

4. Preliminary summary: Albericio, F., Słomczyńska, U., Mullen, D.G., Zalipsky, S., and Barany, G. (1984) in Peptides 1984, Proceedings of the Eighteenth European Peptide Symposium, Ragnarsson, U., Ed., Almqvist & Wiksell, Stockholm, Sweden, pp. 181-184. Full papers based on this work by Barany and collaborators are in press or are being readied for publication.

5. Becker, H., Lucas, H.W., Maul, J., Pillai, V.N., and Mutter, M. (1983) Makromol. Chem. **184**, 2603-2617.

6. Bayer, E., Dengler, M., and Hemmasi, B. (1985) Int. J. Peptide Protein Res. **25**, 178-186.

7. Zalipsky, S., Gilon, C., and Zilkha, A. (1983) Eur. Polymer J. **19**, 1177-1183.

8. Barany, G. and Zalipsky, S. (1983) in Peptides: Structure and Function, Hruby, V.J. and Rich, D.H., Eds., Pierce Chemical Company, Rockford, IL, pp. 159-162.

9. Ueki, M. and Ikeda, S. (1974) Chem. Lett., pp. 25-28.

10. Modification of technique described by Sarin, V.K., Kent, S.B., and Merrifield, R.B. (1980) J. Am. Chem. Soc. **102**, 5463-5470. A Nikon video enhanced differential interference contrast microscope with a T.V. monitor was used, by courtesy of Professor D.F. Evans. Values reported are the average of ten measurements.

11. Whereas ordinary PS beads with free amino groups clump severely when applied to solid-phase synthetic protocols, this was not the case with PEG-PS at corresponding stages. In general, in every synthesis to date, swollen peptide-PEG-PS-resins showed visually superior morphologies. They were more transparent; dispersed better as fresh CH_2Cl_2 smoothly percolated through the washed resin mass; and did not adhere to silanized glass surfaces.

12. The first glycine from the C-terminal was radioactive, so to conserve material, only 1 equiv. was used for the first coupling. Whereas in the synthesis starting from PS, the ninhydrin test was slightly positive at this point, the corresponding step in the synthesis starting with PEG-PS gave a negative reaction.

AN IMPROVED SYNTHESIS OF BENZHYDRYLAMINE RESIN

William M. Bryan

Department of Peptide Chemistry, Smith Kline and French Laboratories, 1500 Spring Garden Street, Philadelphia, Pennsylvania 19101

Introduction

Most commercial benzhydrylamine resin 2 prepared via the recommended route[1] is contaminated with substantial amounts of phenylketone resin 1, due to the fact that the Leuckart reaction is difficult to force to completion, typically proceeding in 45-55% yield[2]. In addition both the Friedel-Crafts acylation and the reductive amination are sensitive to the reaction conditions making it difficult

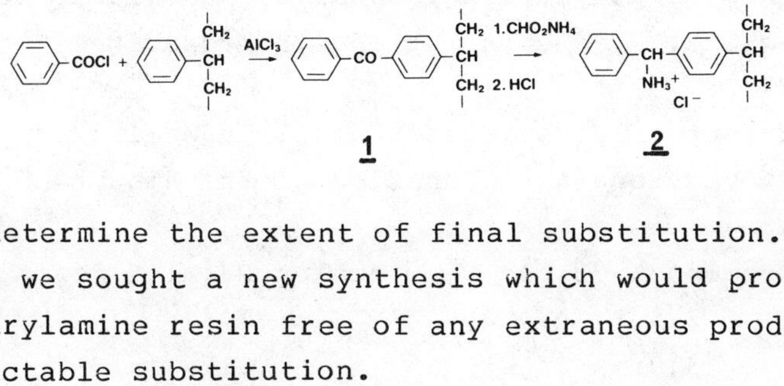

1 **2**

to predetermine the extent of final substitution. For these reasons we sought a new synthesis which would provide benzhydrylamine resin free of any extraneous products and at a predictable substitution.

Results and Discussion

Because the extent of final substitution of a resin prepared via successive reactions is dependent on the control of the first functionalization reaction we felt it necessary that this initial reaction be easily reproducible and capable of providing a predictable degree of substitution. Furthermore, the initially formed resin must contain easily replaceable functionality to ensure a homogenous final product. We find that direct amidoalkylation of 1% cross-linked polystyrene with N-(α-chlorobenzyl)phthalimide[3] 3 in the presence of stannic chloride yields benzylphthalimido resin 4. This

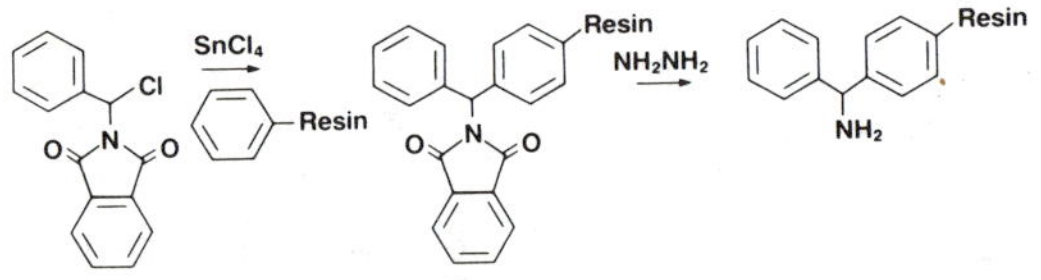

3 4 2

reaction can be easily controlled to provide substitutions of 0.4 to 1.0 mmole N/gm resin by varying the amount of 3 (Table I). We have conveniently performed this reaction on 100-200gm batches of polystyrene and have found it to be highly reproducible. Refluxing ethanolic hydrazine

Table I. Extent of Substitution of 1% Cross-linked Polystyrene as a Function of the Amount of 3

mmoles 3	mmoles $SnCl_4$	mmoles N/gm of resin 4
1.0	2.0	0.4
1.5	3.0	0.7
2.0	4.0	1.0

readily converts benzylphthalimido resin 4 to benzhydrylamine resin 2. This reaction can be easily monitored by the disappearance of the phthalimide carbonyl band in the infrared at 1710 cm^{-1} (Figure 1).

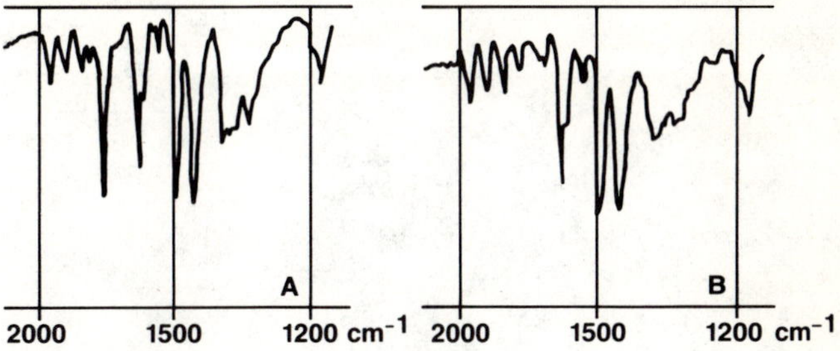

Fig. 1. Partial Infrared spectra of a KBr disk of benzylphthalimido resin 4 (A) and benzhydrylamine resin 2 (B).

Conclusion

This synthetic route provides benzhydrylamine resin free of any extraneous functionality and at a predetermined substitution. Furthermore, both the amidoalkylation and deblocking are amenable to scale-up for the preparation of 100-200gm of resin.

Acknowledgement

The author wishes to thank Edith A. Reich for the elemental nitrogen analysis.

References

1. Orlowski, R.C.; Walter, R.; Winkler, D. (1976) J. Org. Chem. 41, 3701-3705.
2. Matsueda, G.R.; Stewart, J.M. (1981) Peptide 2, 45-50.
3. Worley, J.W. (1979) J. Org. Chem., 44, 1178-1180.

SYNTHESIS OF A BASE LABILE 9-(AMINOACYLOXYMETHYL)FLUORENE-4-CARBOXAMIDOMETHYL-RESIN (I) AND ITS APPLICATION IN THE PREPARATION OF HUMAN ANGIOTENSIN II AND ATRIOPEPTIN (1-22) FRAGMENT.

Foe S. Tjoeng, Mark E. Zupec, Shad R. Eubanks and Steven P. Adams

Monsanto Co., 700 Chesterfield Village Parkway, St. Louis, MO 63198

Introduction:

Recently, the use of 9-Fluorenylmethyl (Fm) group as a base labile carboxyl protecting group (1) and as a handle in polymer-supported peptide synthesis was described (2). To study the application potential of the latter aspect of this group, we wish to report the preparation of the base labile support I and its use in the synthesis of Angiotensin II (II) and Atriopeptin (AP) 1-22 (III)(3).

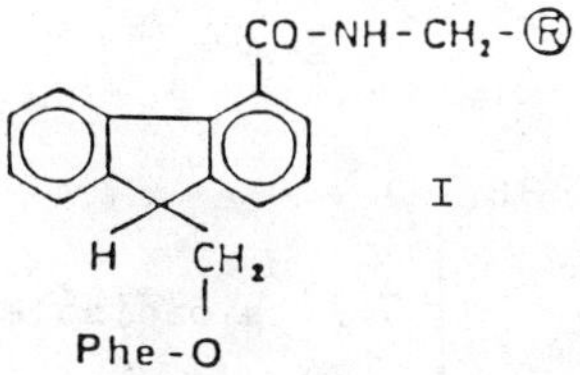

I

Asp-Arg-Val-Tyr-Ile-His-Pro-Phe II

Ser-Ser-Cys-Phe-Gly-Gly-Arg-Ile-Asp-Arg-Ile-Gly-

Ala-Gln-Ser-Gly-Leu-Gly-Cys-Asn-Ser-Phe III

(Cys residues linked by a bridge)

Scheme 1. Synthesis of 9-(Aminoacyloxymethyl)Fluorene-4-carboxamidomethyl-resin (I).

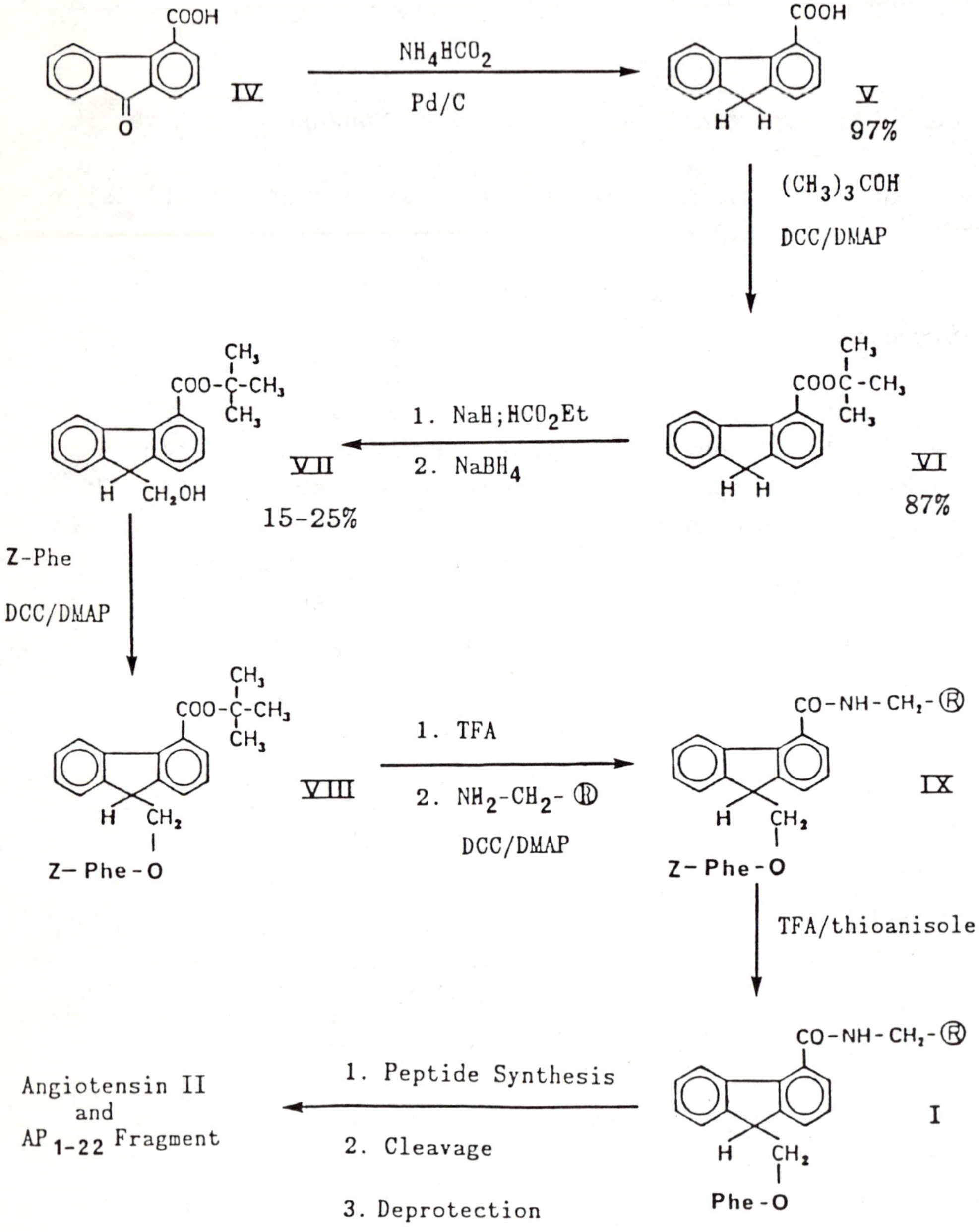

Results and Discussion:

The synthesis of Support I is outlined in the Scheme 1. 9-Fluorenone-4-carboxylic acid (IV) was converted to Fluorenyl-4-carboxylic acid (V) quantitatively by catalytic transfer hydrogenation using Pd/C and NH_4HCO_2 (16 h). Acid V was then reacted for 6 h with t-butanol and DCC in the presence of catalytic amount of DMAP to give ester VI (87%). The preparation of 9-Hydroxymethylfluorenyl-4-carboxylic acid t-butyl ester (VIII) was done according to a similar procedure described by L. Carpino (4) for the synthesis of 9-hydroxymethylfluorene. Ester VII was refluxed (24 h) with HCO_2Et/NaH in ether to form the 9-formyl derivative which was reduced with $NaBH_4$ in CH_3OH to give the 9-hydroxymethylated ester VII. The crude product was purified on a chromatotron (Silica Gel PF 254 4 mm-Disk) using $CHCl_3$ as eluent. The yields were ranging between 15-25% (3 preparations). The coupling of Z-Phe with VII for 1 h was mediated by DCC in the presence of DMAP to give ester VIII (95%). After removal of t-butyl group, the resulting acid was reacted with NH_2-CH_2-Ⓡ and DCC to afford resin IX. The support I was obtained after removal of Z-group with TFA/thioanisole (2:1; 18 h). Substitution of Phe was 0.26 mmoles per g resin. The peptides were assembled stepwise using DCC/HOBt coupling. The progress of the coupling was monitored by the qualitative ninhydrin test. Boc-amino acids (3 equiv.) were used throughout the synthesis with following side-chain protecting groups: Benzyl for Asp, Ser, Tyr; Bom for His, diZ for Arg and Acm for Cys. Protected Angiotensin II was released from resin with 15% piperidine/DMF (½h) in 75% yield*. The protecting groups were removed by catalytic transfer hydrogenation with NH_4HCOO/CH_3OH and Pd/C (4 h). The

---

* based on the amount of peptide remaining on the support.

cleavage of protected AP_{1-22} proved to be more difficult than Angiotensin II.

Treatment with 15% piperidine/DMF (½ h) gave very little release of peptide (5%). Increasing the amount of piperidine to 50% resulted only in 35% cleavage of peptide in 2 h. However, after removal of all benzyl protecting groups with HF, cyclization of the resulting [Cys(Acm)^{3,19}]-AP_{1-22} with 0.1 M I_2 in 85% HOAc afforded no desired peptide. This may be partly due to formation of aspartyl-piperidinimide. Thus, the protected peptide resin was first treated with HF/anisole (9:1) at 0°C for 1 h and cleaved with base and cyclized to give AP_{1-22}. Both peptides were purified by HPLC and sequenced and shown to be the correct material. Amino acid analyses of both peptides were in good agreement with the theoretical value.

Our preliminary study showed that this base labile support can be useful in the synthesis of natural peptide in the size of about 10 residues. With peptides larger than 20 residues, the cleavage from resin is less efficient. We believe that more work needs to be done to demonstrate its merit in the synthesis of complex peptides.

References:

1. Bednarek, M. A. and Bodansky, M. (1984) Int. J. Peptide Protein Res. **21**, 196; Kessler, H. and Siegmeier, R. (1983) Tetrahedron Lett. **24**, 281.

2. Mutter, M. and Bellof, D. (1984) Helv. Chim. Acta **67**, 2009.

3. Currie, M.G., Geller, D.M., Cole, B.R., Siegel, N.R., Fok, K.F., Adams, S.P., Eubanks, S.R., Galluppi, G.R., and Needleman, P. (1984) Science **223**, 67-69.

4. Carpino, L. A. (1980) J. Org. Chem. **45**, 4250.

A MONITORING METHOD FOR SOLID PHASE PEPTIDE SYNTHESIS USING N-α-TERTIARY-BUTOXYCARBONYL (BOC) PROTECTION FOR AMINO ACIDS

S. M. Kalbag*, E. S. Snyder, P. J. Voelker, and J. L. Hughes

Peptide Chemistry Department, SmithKline Beckman Corporation
1050 Page Mill Road, Palo Alto, CA 94304

A synthetic chemist using the solid phase peptide scheme has always sought a quantitative monitoring method which would allow him to know the course of the synthesis and make corrections to optimize the synthesis. Results of earlier work had indicated that gas chromatography (GC) could be used to detect the Boc pyrolysis product, isobutene, quantitatively.[1] By using this technique and developing a novel solid support containing a pyrolyzable internal standard, we have developed a rapid, quantitative, reproducible method of monitoring solid phase peptide synthesis.

Chemistry and Discussion of Method

An internal standard on a solid support must fulfill the following criteria. First, it has to be easily introduced into the resin synthetically. Second, both the pyrolyzable moiety and the bonds to the resin backbone have to be stable to peptide synthesis conditions. The pyrolysis product from it must be produced nearly quantitatively. Third, the olefin product from the pyrolysis has to have a GC retention time that is quantitatively distinguishable (baseline separation) from isobutene (the Boc pyrolysis product). Fourth, the retention times for both the internal standard olefin and isobutene have to be short enough in order to keep the assay time reasonable (5 to 10 minutes). Of the various groups considered, an ester of isopropyl alcohol was chosen since it met the criteria above. It has six β-hydrogens and is similar to the Boc group in pyrolytic efficiency. In addition it yields only one product, propene, which is easily separable from isobutene in GC and has a short retention time.

The isopropyl group was introduced on the resin backbone (either amino methyl or BHA resin) as α-Boc-β-(isopropyl)-asparatic acid . After removing the Boc group the derivatized phenylacetic acid group[2] was attached. Such a resin with the first amino acid residue of a desired sequence attached would have the following structure:

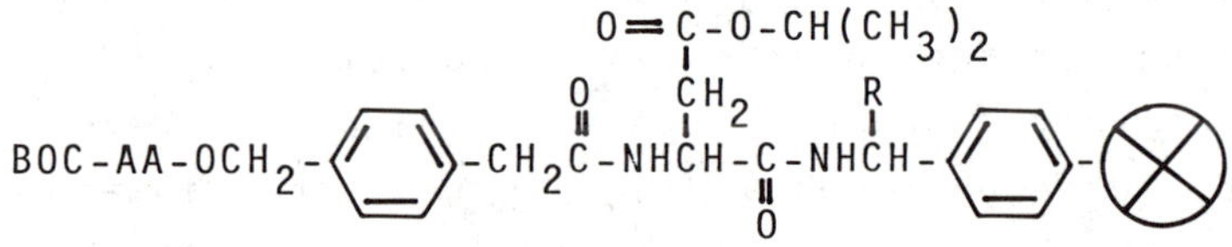

where: R is dependent on the amino resin selected.

 is the polymer support.

For the sake of brevity we have adopted a shortened designation for these resins. N-(4-oxymethylphenylacetyl)-β-isopropyl aspartyl position on BHA is designated as PAMPRO-BHA. With an attached Boc-valine residue it would be Boc-Val-PAMPRO-BHA resin.

Optimal decomposition of Boc and isopropyl ester occurred when α-Boc-(β-isopropyl)-Asp-BHA was pyrolyzed at 700°C for 20 seconds. However, pyrolysis of the same resin at 500°C for 80 seconds resulted in the optimal decomposition of Boc and isopropyl groups. At this temperature and duration no decomposition product of the branched sidechains of Boc-Val-BHA and Boc-Leu-BHA was noticed.

Application of Method to a Resin Peptide Synthesis

The model resin peptide Boc-Leu-Ala-Gly-Val-PAMPRO-BHA was synthesized to determine the applicability of this method. 4-(Boc-valyl-oxymethyl)phenylacetic acid[2] was attached to β-isopropyl-Asp-BHA . The resulting product N-[(Boc-valyl-4-oxymethyl)phenylacetyl]-(β-isopropyl) aspartyl-BHA resin was then monitored by the pyrolysis method. Monitoring of each coupling and deprotection reaction step was carried out in the following manner. A few milligrams of the washed and drained resin sample was withdrawn with a 2 mm bore flexible capillary tube, placed in a 10 mL test tube, and dried under vacuum for 5 minutes. A few beads of the resin sample was withdrawn with a melting point capillary. After spreading the beads up to ca. 1 cm from the closed end, the capillary was cleanly broken at 2 cm from the closed end. This was then placed within a platinum coil of the "Pyroprobe 150" of CDS. The pyroprobe was then placed in an interface box, which was maintained at 100°C. The interface was connected to the injector of a Beckman GC-45 through a Valco^R four port valve. Initially the valve was left open to the column (2 m x 1/8" o.d.ss., carbopack with 0.19 percent picric acid at 70 ± 5°C and 40 ± 3 ml/mm N_2 flow rate). The sample was then heated to remove residual solvent for a total of 40 seconds at 100°C by applying two consecutive 20 second electric pulses to the platinum coil. The design of CDS pyroprobe allows the platinum coil to be heated at any desired temperature and heating rate and then to be held up to a maximum 20-second heating pulse time. The valve was then closed, isolating the pyroprobe and the interface from the injector, and the sample was pyrolyzed at 500°C for 80 sec. by applying 4 consecutive 20 sec. electric pulses to the coil. The resultant gases formed in the interface were then introduced onto the column by switching the valve. Chromatograms (pyrograms) were recorded and peak areas directly integrated on a

Hewlett-Packard integrator (3390A). The pyrograms after coupling showed isobutene and propene at 2.97 and 1.06 min., respectively, in proper proportions. As a second monitoring method, the Kaiser ninhydrin test was performed at each stage of the synthesis. In addition, amino acid analysis was performed on the resin peptide VI. Hydrolysis with propionic acid: 6N HCl (1:1) for 4 hours. Molar ratios were Leu: 1.04; Ala: 1.01; Gly: 1.03; Val: 1.00.

The results for the GC assay and ninhydrin tests are summarized in Table I. The ratio of areas of isobutene and propene was determined after each coupling. The starting ratio was assumed to be 100 percent for the amount of Boc, and percent coupling for each amino acid was then calculated by comparison with the ratio from the previous coupling.

Table I. Pyrolysis Results of
Boc-Leu-Ala-Gly-Val-PAMPRO-BHA Resin

Peptide Resin	Isobut. (Area) X	Propene (Area) Y	Ratio x/y	Coupling	Deprotection	Kaiser Test (Color)
BocVal	58.41	41.59	1.4	100.0		Yellow (-)
H-Val	00.00	100.00			100.0	Purple (+)
BocGly Val	57.95	42.05	1.38	98.51		Yellow (-)
H-GlyVal	00.00	100.00			100.0	Purple (+)
BocAla GlyVal	57.31	42.69	1.34	97.1		Yellow (-)
H-AlaGly Val	00.00	100.00			100.0	Purple (+)
BocLeu AlaGlyVal	57.11	42.89	1.33	99.25		Yellow (-)

References

1. a) Kalbag, S. M., unpublished work.
   b) Hughes, J. L., unpublished work.

2. Tam, J. P., Kent, S. B. H., Wong, T. W., and Merrifield, R. B. (1979) Synthesis 12, 955-957.

GEL PHASE ^{13}C n.m.r. SPECTROSCOPY AS A METHOD OF ANALYTICAL CONTROL IN ULTRA-HIGH LOAD SOLID (GEL) PHASE PEPTIDE SYNTHESIS WITH SPECIAL REFERENCE TO LH-RH

Francis G.W. Butwell, Roger Epton, John V. McLaren, Philip W. Small and Donald A. Wellings

School of Applied Sciences, Wolverhampton Polytechnic, Wolverhampton, England, WV1 1LY, UK

Introduction

We have previously reported upon the use of gel phase ^{13}C n.m.r. spectroscopy to obtain directly hitherto inaccessible information on the removal of Boc groups and on the elaboration of peptide chains during solid (gel) phase peptide synthesis[1]. In these studies, the loading of peptide on the matrix was 0.5 mmol/g. In the case of all but the simplest peptides, long accumulation times and relatively large amounts of sample were needed to obtain useful spectra.

Fortuitously, our recent work has focussed on the development of a scheme for solid (gel) phase peptide synthesis in which a bead-form core polymer, cross-linked poly[N-[2-(4-hydroxyphenyl)ethyl]-acrylamide, is used as support matrix at a loading approaching 5.0 mmol/g, the maximum possible[2,3]. Synthesis proceeds via a series of quasi-homogeneous peptide gel networks, which are more amenable to ^{13}C n.m.r. investigation than conventional solid (gel) phase peptide assemblies.

We report here the use of ^{13}C n.m.r. spectroscopy to solve problems relevant to an ultra-high load synthesis of LH-RH. In this, Boc and Fmoc amino acids were used, together with a base-labile phenyl ester linkage at the peptide C-terminal.

Results and Discussion

For ^{13}C n.m.r. spectroscopy, 50-100 mg samples of solid (gel) phase assembly were suspended swollen in 2 cm^3 $(CD_3)_2SO$ and run on a Bruker WP80SY spectrometer operating at 20.12 MHz accumulating 50,000-200,000 pulses at 0.5 s intervals.

Examination of typical ^{13}C n.m.r. spectra (Figure 1) confirmed that reaction with a 10% $HCONMe_2$ solution of Et_2NH at 25°C effected cleavage of Fmoc within 2 h. In planning the LH-RH synthesis (Scheme 1), it was necessary to investigate if

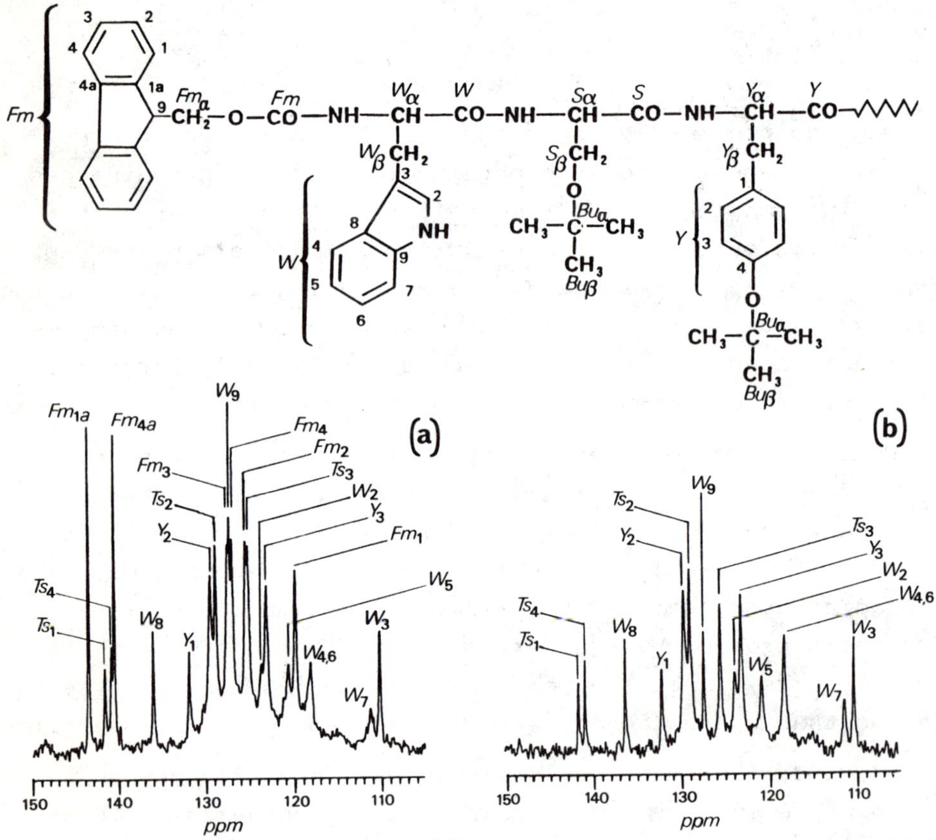

Fig. 1. Relevant region of ^{13}C n.m.r. spectrum of ultra-high load LH-RH (3-9) assembly (a) before and (b) after Fmoc removal (200,000 scans; 50 mg assembly in 2 cm^3 $(CD_3)_2SO$). Ts_1-Ts_4 relate to aromatic tosyl carbons.

[HO-C_6H_4]$_n$—Core

1. Boc-Pro-OH/DIC/DMAP ↓

[Boc-Pro-O-C_6H_4]$_n$—Core

2a. BF_3/$C_6H_5CH_2OH$ 2b. Boc-Arg(Tos)-OBt/HOBt/NMM
3a. BF_3/$C_6H_5CH_2OH$ 3b. Boc-Leu-OBt/HOBt/NMM
4a. BF_3/$C_6H_5CH_2OH$ 4b. Boc-Gly-OBt/HOBt/NMM ↓

[Boc-Gly-Leu-Arg(Tos)-Pro-O-C_6H_4]$_n$—Core

5a. BF_3/$C_6H_5CH_2OH$ 5b. Fmoc-Tyr(tBu)-OBt/HOBt/NMM
6a. Et_2NH/$HCONMe_2$ 6b. Fmoc-Ser(tBu)-OBt-HOBt
7a. Et_2NH/$HCONMe_2$ 7b. Fmoc-Trp-OBt/HOBt ↓

[Fmoc-Trp-Ser(tBu)-Tyr(tBu)-Gly-Leu-Arg(Tos)-Pro-O-C_6H_4]$_n$—Core

8a. Et_2NH/$HCONMe_2$ 8b. Boc-His(Tos)-OH/DIC ↓

[Boc-His(Tos)-Trp-Ser(tBu)-Tyr(tBu)-Gly-Leu-Arg(Tos)-Pro-O-C_6H_4]$_n$—Core

9a. BF_3.<u>m</u>$CH_3C_6H_4OH$ ↓

[BF_3.His(Tos)-Trp-Ser-Tyr-Gly-Leu-Arg(Tos)-Pro-O-C_6H_4]$_n$—Core

9b. Glp-OBt/HOBt ↓

[Glp-His(Tos)-Trp-Ser-Tyr-Gly-Leu-Arg(Tos)-Pro-O-C_6H_4]$_n$—Core

10. $NH_2CH_2CONH_2$ ↓

Glp-His-Trp-Ser-Tyr-Gly-Leu-Arg(Tos)-Pro-Gly-NH_2 [HO-C_6H_4]$_n$—Core

11a. 30%HF/Me_2S 11b. 90%HF/Me_2S ↓

Glp-His-Trp-Ser-Tyr-Gly-Leu-Arg-Pro-Gly-NH_2

Scheme 1. Ultra-high load solid (gel) phase synthesis of LH-RH using both Boc and Fmoc protecting groups and peptide detachment by 'backing-off', with $NH_2CH_2CONH_2$. Synthesis was progressed from 0.25 g core, at a loading of 4.2 mmol peptide/g, to yield, after deprotection, 1.5 g LH-RH, which was purified subsequently by standard methods concurrent nucleophilic scission of C-terminal phenyl ester linkages attaching the peptide chains to the core polymer would also occur under these conditions. Such deloading may be estimated from the diminution of a ^{13}C n.m.r. peak at 121 p.p.m. (downfield from Me_4Si), due to core phenyl group aromatic carbons ortho to the point of peptide C-terminal attachment, and the growth of a peak at 115 p.p.m., of similar ultimate intensity, due to the same carbon atoms ortho to a free phenolic hydroxyl group.

Samples of a model assembly, $[Boc\text{-}Pro\text{-}O\text{-}C_6H_4]_n$—Core, were treated with 10% $HCONMe_2$ solutions of Et_2NH, piperidine and $NH_2CH_2CONH_2$ at 25°C for 48 h. ^{13}C n.m.r. estimation of deloading was 0%, <5% and 100% respectively. Treatment of $[Boc\text{-}Ala\text{-}O\text{-}C_6H_4]_n$—Core with Et_2NH and piperidine under these reacting conditions, gave 15% and 85% deloading respectively. It follows that Fmoc removal with Et_2NH will cause peptide deloading, if any, to become significant only after many deprotection cycles. The ability of $NH_2CH_2CONH_2$ to effect scission of a phenyl ester linkage at a proline C-terminal has been exploited in a new synthesis of LH-RH (Scheme 1).

The use of N-Fmoc-O-tBu substituted hydroxyamino acids demonstrates the possibility of amino acid activation and coupling with O-tBu protected side-chains, deprotection of the latter during a later N-Boc deprotection step, and progression of the synthesis without OH side-chain protection.

References

1. Epton, R., Goddard, P., and Ivin, K.J. (1980) Polymer, 21, 1367-1371

2. Epton, R., and Williams, A. (1980) Int. J. Biol. Macromol. 3, 334-336

3. Epton, R., Marr, G., McGinn, B.J., Small, P.W. Wellings, D.A. and Williams, A. (1985) Int. J. Biol. Macromol. in press

NON-DESTRUCTIVE MONITORING OF SOLID PHASE PEPTIDE SYNTHESIS BY MASS SPECTROMETRY

Jean-Louis Aubagnac, Monique Calmes, Jacques Daunis, Btissam El Amrani and Robert Jacquier
Université des Sciences et Techniques du Languedoc, Place E. Bataillon, 34060 Montpellier cédex (France)

Introduction

Applications of polymeric supports have been considerably developed in the last few years especially in peptide synthesis. Yet analytic control of supported reactions remains a difficult problem when insoluble cross-linked polymers are used. Apart from volumetric or spectrophotometric analysis of free amino groups, only three non-destructive structural studies have been published until now using ^{19}F (1) and ^{13}C NMR (2,3).

In a previous publication (4), we have shown for the first time that the FAB ionization method and mass tandem spectrometry can be used to get mass spectra of aminoacids anchored on a polyacrylic support and to control the deprotection of supported Boc-aminoacids. This method has now been applied to the study of successive steps of a solid phase peptide synthesis.

Results and Discussion

A previously described polyacrylic resin (5) has been used in order to achieve the synthesis of the following supported peptides :

Boc-Ala-Phe-Gly-Phe-O-CH_2-CO-NH-Ⓟ 1
Ala-Phe-Gly-Phe-O-CH_2-CO-NH-Ⓟ 2
Boc-Ala-Ala-Phe-Gly-Phe-O-CH_2-CO-NH-Ⓟ 3
Ala-Ala-Phe-Gly-Phe-O-CH_2-CO-NH-Ⓟ 4

Ⓟ= polymer + arm

FAB spectra are recorded in positive and negative modes using homogeneous suspensions in glycerol.

Positive-ion mass spectra. Two different fragmentations of the ester bond between peptide and support are observed (Figure 1) :

- After the non-localized protonation of the peptide part, *alkyl-oxygen cleavage* of the ester bond occuring with an intramolecular proton transfer leaves the polyacrylic support as a neutral entity and liberates the $M+2H]^{+}$ ion : m/z 541, 441, 612 and 512 respectively for products 1-4.

- On the other hand fragmentation of the protonated species occurs at the level of the *acyl-oxygen bond*, giving rise to an acylium ion wich is always associated with one molecule of glycerol G : m/z 615, 515, 686 and 586 respectively for products 1-4.

CAD spectra allow the identification of these ions using the scheme proposed by Schlunegger (6). Table 1 summarizes for example the structural informations furnished in the case of the characteristic ion m/z 515 observed in the FAB spectrum of 2. Moreover, two other CAD spectra have been registered to complete Table 1 : they correspond to the ions m/z 219 and m/z 297 observed in the FAB spectrum of 2.

Starting with the values of Table 1 it is possible to reconstitute the peptide sequence. Ions m/z 444, 297, 347, 219 and 205 thus give the sequence Ala-Ala-Phe-Gly-Phe. Elimination of water characterizes the glycerol molecule, the presence of which is moreover established by a 92 daltons loss.

Sequential fragmentations of the primary ions have also been observed; they occur with the same mechanisms as already described in solution (7,8). However FAB spectra of non-protected peptides 2 and 4 supply more analytical informations than those of the Boc-peptides 1 and 3. This agrees with the fact that this ionization method is more conveniently applied to polar compounds.

Negative-ion mass spectra. Two cleavage mechanisms of the ester bond between peptide and support are likewise observed (Figure 1) :

- After deprotonation of the peptide part, cleavage of the *alkyl-oxygen bond* with an intramolecular proton transfer leaves the support as a neutral entity, and it then becomes possible to observe the M^{-} anions : m/z 539 (1), 439 (2), 610 (3) and 510 (4).

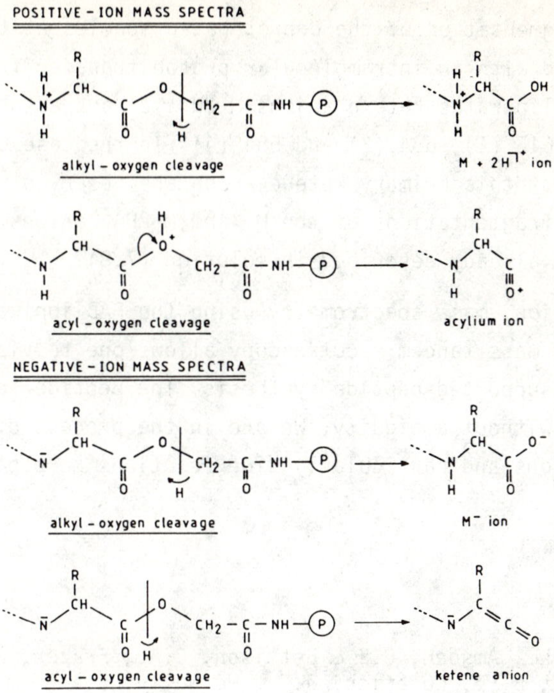

Figure 1

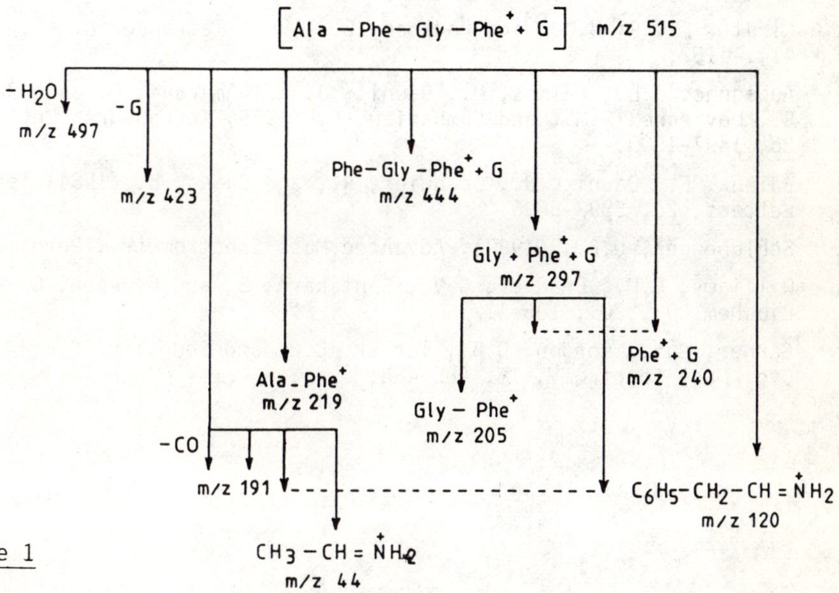

Table 1

- Fragmentation of the deprotonated species at the *acyl-oxygen bond* accompanied with an intramolecular proton transfer from the peptide to the support liberates ketene anions M-2H$^{\rceil -}$ always associated with glycerol : m/z 613 (1), 684 (3) and 584 (4). In the case of product 2, only fragments of this primary ketene anion are observed.

Sequential fragmentations of the M$^-$ and M-2H$^{\rceil -}$ anions likewise occur with mechanisms already described in solution (7,8).

In conclusion, mass spectrometry using the FAB ionization method associated with mass tandem spectroscopy allows one to visualize directly each step of a supported peptide synthesis. The peptide sequence can also be established without ambiguity. We are in the process of studying other applications and particularly side reactions in peptide synthesis.

References

1. Manatt, S.L., Amsden, C.F., Bettison, C.A., Frazer, W.T., Gudman, G.T., Lenk, B.E., Lubetich, J.F., Mc Nelly, E.A., Smith, S.C., Templeton, D.J. and Pinnell, R.P. (1980) Tetrahedron Letters, 21, 1397-1400.
2. Epton, R., Goddard, P and Ivin, K.J. (1980) Polymer, 21, 1367-1371.
3. Giralt, E., Rizo, J. and Pedroso, E. (1984) Tetrahedron, 40, 4141-4152.
4. Aubagnac, J.L., Calmes, M., Daunis, J., El Amarani, B., and Jacquier, R., Devienne, F.M., and Combarieu, R. (1985) Tetrahedron Letters, 26, 1497-1500.
5. Baleux, F., Daunis, J., Jacquier, R., and Calas, B. (1984) Tetrahedron Letters, 25, 5893-5896.
6. Schlunegger, U.P., (1980), Advanced Mass Spectrometry, Pergamon Press.
7. Williams, D.H., Bradley, C.V., Santikarn, S. and Bojesen, G. (1982) Biochem. J., 201, 105-117.
8. Garner, G.V., Gordon, D.B., Tetler, L.W. and Sedgwick, R.D. (1983) Org. Mass Spectrom., 18, 486-488.

EFFICIENCY OF SOLID PHASE SYNTHESIS USING AN AUTOMATED FLOW REACTOR: A COMPARATIVE STUDY

Yin-Zeng Liu
Shanghai Institute of Organic Chemistry, Academia Sinica, Shanghai, China

Johannes Meienhofer
Roche Research Center, Hoffmann-La Roche Inc., Nutley, New Jersey 07110

Introduction

At the Eighth American Peptide Symposium we introduced a comparatively simple computer controlled flow reactor for low pressure operation.[1] In this study we investigated whether our flow reactor operation was more efficient than manual synthesis in a rocking shaker in terms of purified product yield. We also wanted to compare the use of t-butyloxycarbonyl (t-Boc) main chain protection versus 9-fluorenylmethyloxycarbonyl[2] (Fmoc) protection in this flow reactor, as well as the use of symmetrical versus mixed anhydride coupling in Fmoc solid phase synthesis.

Results and Discussion

All experiments are listed in the Table. The peptides were purified to apparent homogeneity by reverse phase preparative HPLC on a μBondapak C_{18} column (2.3 x 30 cm). The

SUMMARY OF THE PEPTIDES SYNTHESIZED

Peptide	Starting Resin Substitution Determined by A.A. Analysis	Synthesis Mode, Strategy Coupling Method	Yield %	Amino Acid Analysis Acid Hydrolysis, 6N HCl, 110°C, 16h with phenol
Met-enkephalin H-Tyr-Gly-Gly-Phe-Met-OH	1g of Fmoc-Met-Ⓡwith 0.6 mM/g resin	**1** Flow reactor,Fmoc/tBu, symmetrical anhydride, 6 eq. AA, 13 eq. DCC	80	Gly,2.00(2); Met,0.93(1); Phe,0.98(1); Tyr,0.95(1).
Met-enkephalin	1g of Fmoc-Met-Ⓡwith 0.6 mM/g resin	**2** Manual shaker,Fmoc/tBu symmetrical anhydride	77	Gly,2.00(2); Met,0.90(1); Tyr,0.95(1); Phe,0.99(1).
Met-enkephalin	0.6g of Fmoc-Met-Ⓡwith 0.6g mM/g resin	**3** Flow reactor,Fmoc/tBu, Mixed anhydride, 3 eq. (isobutyl-chloroformate), N-methyl morpholine-15°C, 5 min.	68	Gly,2.00(2); Met,0.93(1); Tyr,0.91(1); Phe,0.98(1).
Met-enkephalin	0.6g of Fmoc-Met-Ⓡwith 0.6 mM/g resin	**4** Manual shaker,Fmoc/tBu mixed anhydride	71	Gly,2.00(2); Met,0.90(1); Tyr,0.93(1); Phe,0.97(1).
[(D)Ala^{2}]Arg^{6}-Met-enkephalin	0.8g of Fmoc-Arg(Mtr)-Ⓡ with 0.24 mM/g resin	**5** Flow reactor,Fmoc/tBu symmetrical anhydride	84	Gly,0.94(1); Ala,1.00(1); Leu,0.97(1); Tyr,0.86(1); Phe,0.97(1); Arg,0.98(1).
[(D)Ala^{2}]Arg^{6}-Met-enkephalin	0.8g of Fmoc-Arg(Mtr)-Ⓡ with 0.24 mM/g resin	**6** Manual shaker,Fmoc/tBu symmetrical anhydride	74	Gly,1.00(1); Ala,0.98(1); Met,0.89(1); Tyr,0.89(1); Phe,0.99(1); Arg,1.00(1).
[(D)Ala^{2}]Arg^{6}-Leu-enkephalin	0.5g of Fmoc-Arg(Mtr)-Ⓡ with 0.24 mM/g resin	**7** Flow reactor,Fmoc/tBu symmetrical anhydride	88	Gly,0.95(1); Ala,1.00(1); Leu,0.97(1); Tyr,0.86(1); Phe,0.97(1); Arg,0.98(1).
Dihydrosomatostatin	0.8g of Boc-Cys(Bzl)-Ⓡ with 0.38 mM/g resin	**8** Flow reactor,Boc/Bzl symmetrical anhydride	16	Asp,1.01(1); Thr,2.00(2); Gly,0.90(1); Ala,0.92(1); Phe,3.00(3); Lys,1.95(2); Ser,0.80(1); Cys,n.d.; Trp,n.d.
Dihydrosomatostatin	0.8g of Fmoc-Cys(tBu)-Ⓡ with 0.41 mM/g resin	**9** Flow reactor,Fmoc/tBu symmetrical anhydride	24	Asp,1.08(1); Thr,2.30(2); Ser,0.86(1); Gly,1.00(1); Ala,1.00(1); Phe,3.10(3); Lys,2.12(2); Cys,n.d.; Trp,n.d.

NORMAL FLOW (ML/MIN) 20
RECYCLING FLOW (ML/MIN) 20
VOLUME OF ACTIVATED AMINO ACID PER COUPLING (ML) 3

No.	Time/Min	Step	Flow ML/Min
1	2.0	DMF	20
2	3.0	CH2Cl2	20
3	2.0	TFA	20
4	6.0	TFA	5
5	2.0	CH2Cl2	20
6	2.0	TFA	20
7	6.0	TFA	5
8	2.0	CH2Cl2	20
9	2.0	NR3	20
10	2.0	CH2Cl2	20
11	2.0	NR3	20
12	2.0	PROH	20
13	3.0	DMF	20
14	.2	WASH	0
15	.18	DRAW	20
16	15.0	RECYCLE	20
17	.2	WASH	0
18	.3	WASH	20
19	2.0	DMF	20
20	2.0	NR3	20
21	2.0	PROH	20
22	3.0	DMF	20
23	.18	DRAW	20
24	15.0	RECYCLE	20
25	.2	WASH	0
26	.3	FLUSH	0
27	.5	WASH	20
28	2.0	DMF	20
29	3.0	CH2Cl2	20
30	2.0	PROH	20

PROTOCOL FOR THE BOC-SYNTHESIS

NORMAL FLOW (ML/MIN) 20
RECYCLING FLOW (ML/MIN) 20
VOLUME OF ACTIVATED AMINO ACID PER COUPLING (ML) 3

No.	Time/Min	Step
1	2.0	DMF
2	1.5	PIP
3	1.5	DMF
4	1.5	PIP
5	2.0	PROH
6	3.0	DMF
7	.3	WASH
8	.18	DRAW
9	15.0	RECYCLE
10	.3	WASH
11	.5	WASH
12	2.0	NR3
13	3.0	DMF
14	.13	DRAW
15	15.0	RECYCLE
16	.3	WASH
17	.4	FLUSH
18	.5	WASH
19	2.0	NR3
20	2.0	DMF
21	2.0	PROH

PROTOCOL FOR THE FMOC-SYNTHESIS

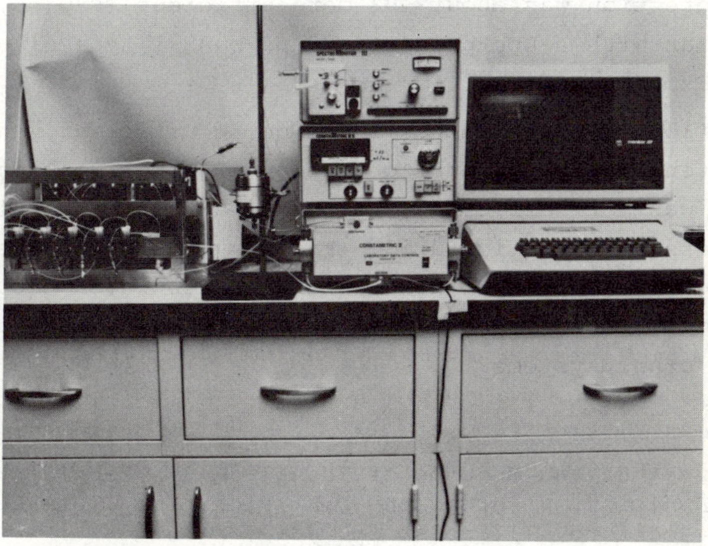

AUTOMATED FLOW REACTOR

peptides were eluted with a linear gradient of 5 to 65% acetonitrile in 0.022% trifluoroacetic acid over 180 min at 8 mL/min; detection at 220 nm. The purity of the peptides were ascertained by analytical HPLC, amino acid analysis and thin layer chromatography on precoated silica gel G-60 plates. As shown in the Table the flow reactor (Exp. 1) was more efficient 80% yield, than a rocking shaker (Exp. 2;77% yield) in Met-enkephalin syntheses, using Fmoc/tBu, symmetrical anhydride coupling. Both were somewhat less efficient in mixed anhydride coupling; (Exp. 3, 68% yield; Exp. 4, 71% yield). Comparison of symmetrical (Exp. 1, 6 eq.) versus mixed anhydride (Exp. 3, 3 eq.) coupling in the flow reactor indicated 12% more efficiency with symmetrical anhydrides (80% versus 68%). In the synthesis of Met-enkephalin analogs, Fmoc-Arg(Mtr), the flow reactor was clearly superior (Exps. 5 and 7, 84% and 88% yields) to the rocking shaker operation (Exp. 6, 74% yield). The Mtr group cleavage required two consecutive TFA treatments. The Fmoc synthesis of Dihydrosomatostatin with t-butyl side chain protection was about 10% more efficient (Exp. 9, 24% yield) than with t-Boc protection and benzyl type side chain groups (Exp. 8, 16% yield).

The flow reactor synthesis was quite efficient, and less labor intensive than manual operation. Symmetrical anhydride coupling was more efficient, but mixed anhydride coupling required less excess amino acid. Fmoc synthesis appeared more promising for scale-up synthesis than t-Boc which needs longer cycle time. The data are limited to this study and not applicable to other systems.

References

1. Jonczyk, A. and Meienhofer, J. (1983) in Peptides:Structure and Function, V. Hruby and D.H. Rich, Eds., Pierce Chemical Co., Rockford, Illinois, pp. 73-77.

2. Meienhofer, J., Waki, M., Heimer, E.P., Lambros, T.J., Makofske, R.C. and Chang, C.-D. (1979), Int. J. Peptide Protein Res. 13, 35-42.

BBr_3-CF_3COOH AS PEPTIDE RESIN DEBLOCKING AGENT-OPTIMAL CONDITION STUDIES

K. T. Wang and Y. S. Tseng
Institute of Biological Chemistry, Academia Sinica, Taipei, Taiwan, R.O.C.

Introduction

In the synthesis of alanine oligopeptides by the solid phase, we need a more simple method to cleave the peptide from the resin because we have difficulty to handle liquid HF. Pless and Bauer reported the use of boron tris(trifluoro-acetate) (BTFA) for cleavage of peptide-resin (1) and employed this reagent in the synthesis of luteinizing hormone releasing hormone (2). Felix (3) reported the use of boron tribromide in the peptide deprotection but no comment was made on the peptide-resin cleavage. In our laboratory, BTFA in TFA cleaved only 20% of penta-alanine from the peptide-resin at 0°.

We noticed that the equation will generate 3 moles of HBr when 1 mole of BBr_3 is treated with 3 moles of TFA.

$$BBr_3 \quad + \quad 3\ CF_3COOH \longrightarrow B(OOCCF_3)_3 \quad + \quad 3\ HBr$$

If this reaction occurred in the sealed vessel with excess of TFA, HBr will dissolved in the TFA and HBr in TFA is first used by Merrifield to cleave Leu-Ala-Gly-Val from the resin (4). So, we tried BBr_3 in excess TFA as the cleavage agent. In the preliminary experiment, we suspended 1 gm Boc-Ala_5-resin in TFA (10 ml) and introduced BBr_3 (0.3 ml) at 0°, then the reaction flask was capped, stirred for 3 hrs at 0°. We obtained 37.4% recovery of penta-alanine. We raised the temperature to 25°, reacted 2 hrs and obtained penta-alanine in 80% yield (crude). Then, we designed a experiment using

Boc-Ala_5-resin to study optimal condition of this BBr_3-TFA (excess) cleavage. First, we developed a HPLC separation method to monitor the cleavage reaction product that may contain smaller alanine oligopeptides (Fig. 1). Secondly, the boric acid after the cleavage should be removed by easy treatment. One hundred mg of boric acid evaporated with 5 ml methanol twice leave no residue. So, the cleavage reaction mixture is evaporated with methanol twice. This crude mixture was analyzed by HPLC and filtered through TSK gel HW40(s) column (100 cm, 2.5 cm) to collect penta-alanine fractions. The fractions were pooled and lyophylized to calculate the yield. On the other hand, the authentic penta-alanine synthesized by the solution method was treated by 5 equivalent BBr_3, 1.0 ml TFA 3 hrs at 25°C and they show only a slight decomposition (Fig. 2). The recovery of penta-alanine was 90% after gel filtration.

Results and Discussion

The cleavage was done under 10°C and 25°C with various equivalent of BBr_3 in excess TFA (Fig. 3 and Table 1, 2). The room temperature condition was considered because the solid

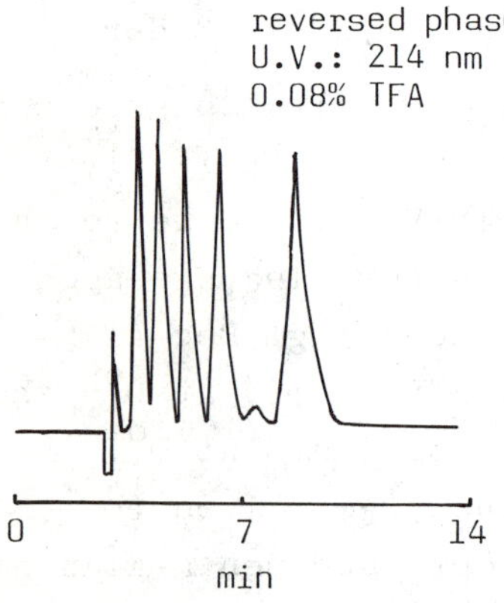

Fig. 1. HPLC of alanine oligomer

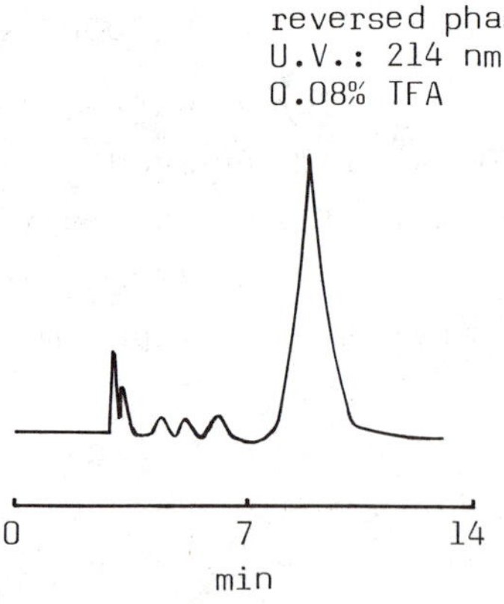

Fig. 2. Treatment of authentic Ala_5

Table 1. The cleavage of Boc-Ala_5-resin (50 mg) at 10°C.

Equivalent BBr_3	TFA	Times	Residual AA on resin (%)	Yield* (%)
5 X	1.0 ml	1 hr	26.4	57.1
5 X	1.0 ml	2 hr	32.5	65.0
5 X	1.0 ml	3 hr	30.0	54.4
10 X	2.0 ml	1 hr	21.8	50.5
10 X	2.0 ml	2 hr	10.2	65.8
10 X	2.0 ml	3 hr	18.9	66.9
20 X	3.0 ml	1 hr	11.8	42.3
20 X	3.0 ml	2 hr	21.7	40.6
20 X	3.0 ml	3 hr	6.1	46.1

Table 2. The cleavage of Boc-Ala_5-resin (50 mg) at 25°C.

Equivalent BBr_3	TFA	Times	Residual AA on resin (%)	Yield* (%)
5 X	1.0 ml	1 hr	12.5	60.4
5 X	1.0 ml	2 hr	7.1	79.3
5 X	1.0 ml	3 hr	6.9	80.5
10 X	2.0 ml	1 hr	5.5	53.6
10 X	2.0 ml	2 hr	2.8	69.8
10 X	2.0 ml	3 hr	1.8	52.7
20 X	3.0 ml	1 hr	5.8	30.8
20 X	3.0 ml	2 hr	2.6	39.3
20 X	3.0 ml	3 hr	2.7	37.1

*: After gel filtration.

phase synthesis vessel could be used without transfer of peptide-resin after the coupling of last amino acid. The best condition is 5 equivalent of BBr_3 in 1 ml TFA and reacted with 2-3 hrs. To demonstrate the usefulness of cleavage method, Leu-Ala-Gly-Val was synthesis according to Merrifield (4) and in each step, small amount of peptide-resin (20 mg) was cleaved with BBr_3-TFA (excess). The crude peptides gave excellent HPLC separation. After the synthesis, the peptide-resin was treated with BBr_3-TFA (excess) (5X, 25°C, 3 hrs) in the synthetic reaction vessel and the peptide was recovered as

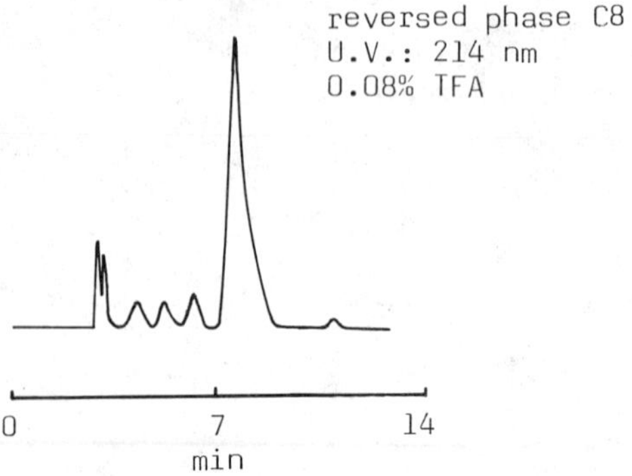

Fig. 3. HPLC of typical cleavage

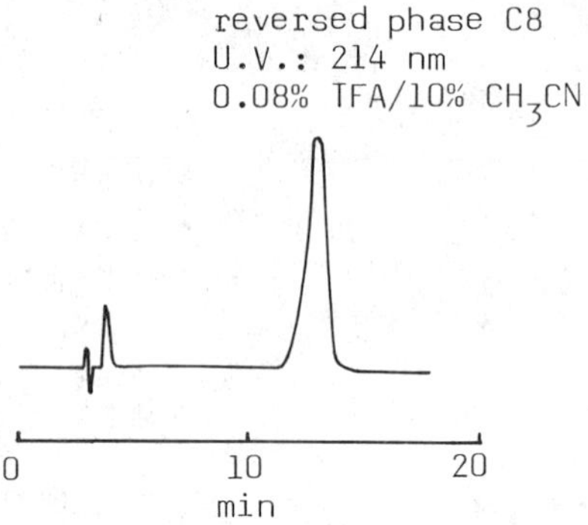

Fig. 4. HPLC of Leu-Ala-Gly-Val

described above. The yield after gel filtration was 78.6%. The residual amino acid in the resin is 5%.

Conclusion

Since many protected amino acids can be deprotected by BBr_3 or BTFA (1, 2), BBr_3-TFA (excess) at 25° could be used in the final step cleavage of the solid phase synthesis. This method may not be superior than liquid HF method but may be more convenient in many laboratories. Further studies on the side reaction are necessary.

References

1. Pless, J., and Bauer, W. (1973) Angew. Chem. Internal. Edit., 12, 147-148.
2. Bauer, W., and Pless, J. (1975) in "Peptides"; Chemistry, Structure and Biology, Walter, R. and Meienhofer, J., Eds., Ann Arbor Sci., Ann Arbor, MI, pp. 341-345.
3. Felix, A. (1974) J. Org. Chem. 39, 1427-1428.
4. Merrifield, R. B. (1963) J. Am. Chem. Soc. 85, 2149-2154.

PEPTIDE CYCLIZATIONS ON POLYSTYRENE SUPPORTS. A STUDY OF INTRASITE AND INTERSITE REACTIONS.

Alexander R. Mitchell
Lawrence Livermore National Laboratory, Livermore, CA 94550

R. B. Merrifield
The Rockefeller University, New York, NY 10021

Introduction

Initially, peptide cyclizations from substituted phenol resins were used to demonstrate effective site isolation[1,2] or "infinite dilution at finite concentration"[3] favoring intramolecular reactions. Later, Rothe *et al*[4] were unable to cyclize the pentapeptide semi-gramicidin S from a dual function support,[2] noting the formation of cyclic decapeptide (gramicidin S) and larger ring oligomers as evidence for extensive intersite reactions.

It is now recognized that considerable polymer-chain flexibility exists in low-cross-linked, solvent-swollen polystyrene beads and the balance between site isolation and interaction depends on the selection of reaction conditions as well as the specific chemistry to be undertaken on the polymeric support.[5] We investigated intersite reactions as part of a program to develop polymer supports suitable for the synthesis of cyclic peptides.[6]

Results and Discussion

Our initial investigation of intersite reactions utilized a polynitrophenol resin prepared by the acylation of aminomethylpolystyrene[7] with 3-hydroxy-4-nitrobenzoic acid. Acylation of this product with Z-Lys(Boc)-

Gly yielded starting material 1 used in Scheme 1. Removal of the Boc group and subsequent reaction in 0.5% Et_3N/DMF gave an 11% yield of cyclic dimer 3 (20 atom ring) with no detectable cyclic monomer 2 (10 atom ring). The analogous conversions in solution starting with Z-Lys (Boc)-Gly-ONp also afforded solely 3 (10%) which was characterized by elemental analysis, TLC, and MS.

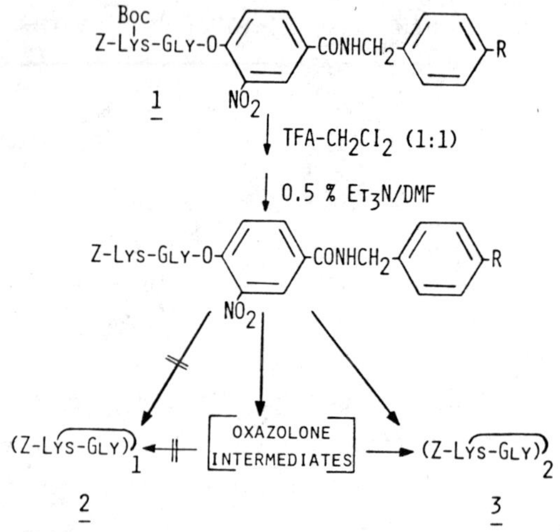

Scheme 1

The exclusive conversion of 1 to 3 seemingly confirms the observations of Rothe et al[4] demonstrating extensive intersite reactions. However, since tertiary base facilitates the formation of peptide oxazolones from peptide active esters,[8] the production of 3 from 1 can result from intersite reactions and/or cyclodimerization of peptide oxazolones released into solution from the resin.

In order to resolve the ambiguity inherent in the use of polymeric active esters, a test system was developed in which the starting material, intermediates, and products remained anchored to the resin throughout the cyclization reaction. The preparation of starting material 4 and subsequent reactions are outlined in Scheme 2. Cyclic monomer 5 (7 atom ring) results from an intrasite reaction, while linear dimer 6 and cyclic dimer 7 (14 atom ring) result from intersite reactions.

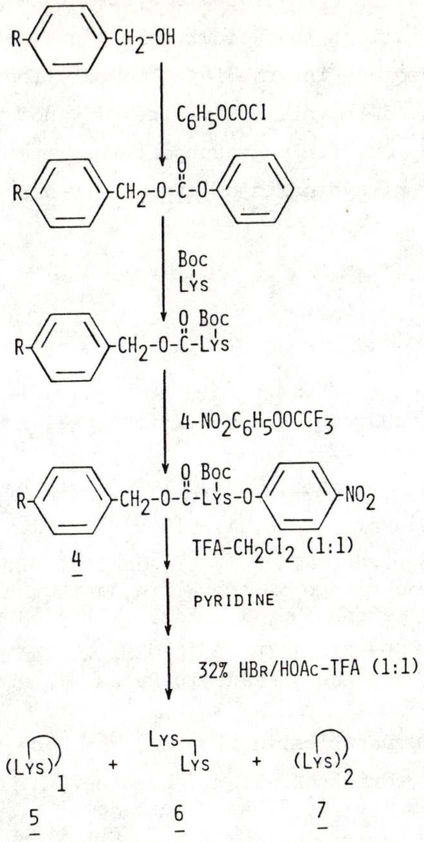

Scheme 2

Cleavage of resin-bound 5, 6, and 7 with 32% HBr/HOAc-TFA(1:1) gives the corresponding amine hydrobromides which are resolved by ion-exchange chromatography on a Beckman 120B amino acid analyzer employing a column of Dowex 50W-X4 sulfonated polystyrene eluted with pyridine acetate buffer (0.8M, pH 5.0).[9] Authentic samples of 5, 6, and 7 were obtained by conventional solution methods employing 4-nitrophenyl esters and characterization by elemental analysis, TLC, and MS. Deprotection of 4 (0.09-0.28 mmol Lys/g) in TFA-CH_2Cl_2(1:1), reaction of the product in pyridine (12 hr) and subsequent cleavage of the resin products gave 5 (80-84%), 6 (12-

19%), and 7 (1-4%).

In conclusion, we have developed a system to evaluate site isolation during peptide cyclizations that occur solely on the support and not through possible oxazolone intermediates in solution. Although intersite reactions were observed in this study, they did not proceed to the extent reported elsewhere.[4] Our findings support the report by Isied et al[10] describing an efficient synthesis of cyclo-(Gly-His)$_3$ using a bidirectional support.

References

1. Fridkin, M., Patchornik, A., and Katchalski, E. (1965) J. Am. Chem. Soc. 87, 4646-4848.

2. Flanigan, E. and Marshall, G. R. (1970) Tetrahedron Lett., 2403-2406.

3. Kraus, M. A. and Patchornik, A. (1978) Isr. J. Chem. 17, 298-303.

4. Rothe, M., Sander, A., Fischer, W., Mastle, W., and Nelson, B. (1977) in Peptides: Proceedings of the Fifth American Peptide Symposium, Goodman, M. and Meienhofer, J., Eds., Wiley, NY, pp. 506-509.

5. Barany, G. and Merrifield, R. B. (1980) in The Peptides, Gross, E. and Meienhofer, J., Eds., Academic Press, NY, Vol. 2, Part A, pp. 1-284.

6. Mitchell, A. R., unpublished results, The Rockefeller University.

7. Mitchell, A. R., Erickson, B. W., Ryabtsev, M. N., Hodges, R. S., and Merrifield, R. B. (1976) J. Am. Chem. Soc. 98, 7357-7362.

8. Bodanszky, M. (1979) in The Peptides, Gross, E. and Meienhofer, J., Eds., Academic Press, NY, Vol. 1, pp. 105-196.

9. Mitchell, A. R. and Merrifield, R. B. (1976) J. Org. Chem. 41, 2015-2019.

10. Isied, S. S., Kuehn, C. G., Lyon, J. M., and Merrifield, R. B. (1982) J. Am. Chem. Soc. 104, 2632-2634.

ΓYROSINE SPECIFIC PROTEIN KINASE INHIBITOR: SYNTHESIS OF PEPTIDE SUBSTRATES

James P. Tam, Noritake Asato and Nicholas F. Pileggi

The Rockefeller University, 1230 York Ave., New York, N.Y. 10021

Introduction

Tyrosine specific protein kinase has been shown to be associated with the transforming gene products of sarcoma-inducing retroviruses and cellular receptors for growth factors[1,2]. However, there is little knowledge of how these enzymes are involved in the control of cell transformation and metabolism, and particularly, of the *in vivo* protein substrates and specificities of these enzymes. Recently, *in vitro* peptide substrate containing tyrosine, and those containing multiple acidic amino acid residues immediately at the amino terminus of the tyrosine residue, have been used to assay a variety of tyrosyl protein kinases.

Since these acidic peptidyl substrates are prone to acid catalyzed side reactions using the conventional strong acid deptrotection method, we have developed a mild deprotection approach, gradative deprotection strategy, in their syntheses. Here, we report the gradative deprotection strategy and its application for the synthesis of peptide substrates as inhibitors for the tyrosyl protein kinase from Rous sarcoma virus.

Gradative Deprotection Strategy

The gradative deprotection strategy utilized a multidetachable benzhydrylamine resin[3,4], p-acyloxybenzhydrylamine resin, and a mild S_N2 deprotection method[5] for the removal of benzyl protecting groups. The multidetachable resin was designed to contain dual properties. The weakly electron withdrawing p-acyloxy substituent on the benzhydrylamine linkage to the resin provided the required acid stability for the repetitive CF_3CO_3H treatments during synthesis, and the subsequent S_N2 deprotection for the removal of benzyl protecting groups after the completion of the synthesis. Also concommitantly, under the S_N2 treatment, methionine sulfoxide is reduced to methionine and N^i-formyl tryptophan is deprotected to tryptophan. Furthermore, the deprotected peptide remains attaching on the resin support. This unusual design provides a simple and convenient way to remove any extraneous by-products of the deprotection mixture through washing by various solvents. Liberation of the peptide from the resin was carried out in two stages. Treatment by a nucleophile released the peptide containing a p-hydroxybenzhydrylamine handle from the resin and also converted the p-acyloxy moiety to a strongly electron donating p-hydroxy substituent on the benzhydrylamine. The handle now became susceptible to and was smoothly removed by mild acidic solvolytic treatment to give the peptide carboxamide. Thus, the gradative deprotection approach consisted of multisteps and deprotected peptides from the resin support in discrete and controlled conditions to minimize strong-acid catalyzed side reactions. A schematic gradative deprotection strategy for the 17-residue of gastrin I is shown in Fig. 1.

Peptide substrate inhibitors for tyrosyl protein kinase

Using the new deprotection strategy, we prepared several peptides, including gastrin I, its analog and caerulein, as substrates for tyrosyl protein kinase purified from Rous sarcoma virus (Table 1). All are characterized by multiple acidic amino acid residues immediately to the N-terminus of the tyrosine and are also found to be sulfated at tyrosine in physiological conditions. As shown in Table 1, these peptides served as superior substrates for tyrosine phosphorylating kinase with k_m values that are much lower than the presently used synthetic peptide substrates, such as src-peptide (residue 411-421 of Rous sarcoma virus protein kinase) and [Val^5]-angiotensin II[6]. More importantly, these peptide substrates were found be inhibitors to the autophosphorylation of the protein kinase and to the phosphorylation to the immunocomplex of the IgG heavy chain at 0.05 to 0.2 mmolar concentrations. These peptide substrate inhibitors may be useful to resolve some of

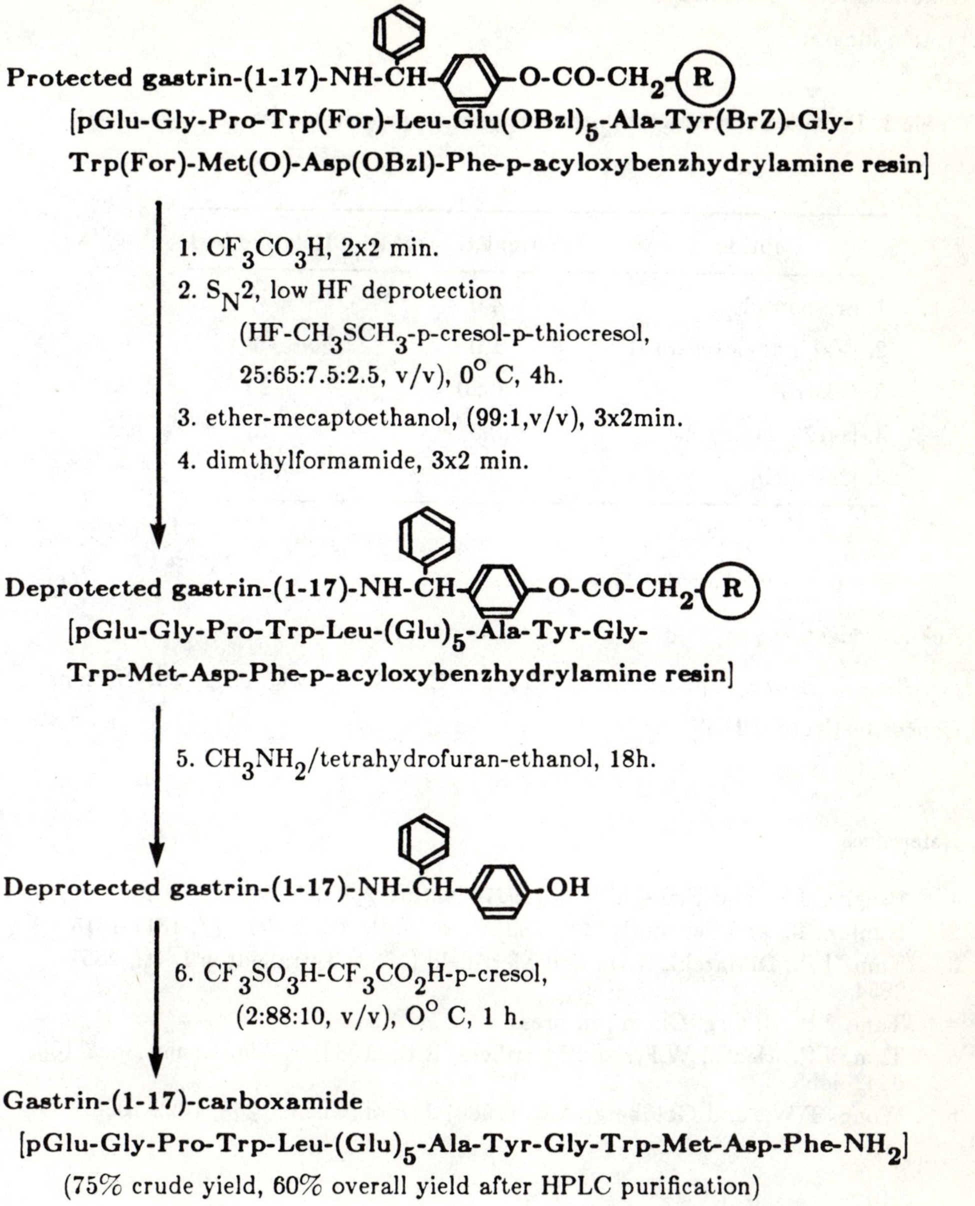

Fig. 1. Deprotection of gastrin-I by the gradative deprotection strategy.

the unanswered questions concerning the mechanism of action of tyrsoine specific protein kinase.

Table 1. Peptide substrate inhibitor for tyrosine protein kinase

Peptide	Apparent K_m(mM)	Relative to 1
1. Src-peptide	4.0	1
2. [Val^5]-angiotensin II	1.0	3
3. Gastrin I	0.20	20
4. [1-12]-Gastrin I	0.3	13
5. Caerulein	0.12	33

Acknowledgement

This work was supported by PHS grant CA 34746, awarded by the National Cancer Institute, DHHS.

References

1. Brugge, J.S., and Erikson, R.L. (1977) Nature 269, 346-348.
2. Hunter, T., and Sefton, B.M. (1980) Proc. Natl. Acad. Sci. 77, 1311-1315.
3. Tam, J.P., DiMarchi, R.D., and Merrifield (1983) Tetrahedron Lett., 2851-2854.
4. Tam, J.P., J. Org. Chem., in press.
5. Tam, J.P., Heath, W.F., and Merrifield, R.B. (1983) J. Am. Chem. Soc. 105, 6442-6455.
6. Wong, T.W., and Goldberg, A.R. (1983) J. Biol. Chem. 258, 1022-1025.

LARGE SCALE SYNTHESIS OF hGRF (1-44)NH_2 (SOMATOCRININ)

Joseph Diaz, Rémy Guégan, Pierre Cabrera, Michel Mellet, Yvette Muneaux, Pierre Perreaut, Michel Vedel and Roméo Roncucci.

Centre de Recherches CLIN-MIDY-SANOFI
34082 MONTPELLIER, FRANCE

Introduction

hGRF isolated in 1982 by R. Guillemin et al[1] from a human pancreatic tumor has the following primary structure :

H-Tyr-Ala-Asp-Ala-Ile-Phe-Thr-Asn-Ser-Tyr-Arg-Lys-Val-Leu-Gly-Gln-Leu-Ser-Ala-Arg-Lys-Leu-Leu-Gln-Asp-Ile-Met-Ser-Arg-Gln-Gln-Gly-Glu-Ser-Asn-Gln-Glu-Arg-Gly-Ala-Arg-Ala-Arg-Leu-NH_2

More recently the same authors[2] demonstrated that its structure was identical to that of hypothalamic hGRF. Clinical evaluation of this product requires large amounts of higly purified material ; for this purpose liquid phase synthesis (LPS) is more suitable. LPS of small quantities of GRF has been described by Fujii et al[3], Chino et al[4], Wakimatsu et al[5]. We have developed a procedure so to produce sufficient material for clinical trials

Synthesis

This procedure is characterized by the following features :

- Stepwise liquid synthesis of 9 subfragments [1-4], [5-11], [12-15], [16-19], [20-24], [25-27], [28-32], [33-39], [40-44]NH_2 with HPLC purity of $\geqslant$95 %, each of these fragments being protected as follows : Boc for N terminal amine, Z for lateral chain of Lys and OBzl ester group for Asp and Glu. Lateral chains of Ser, Thr and Tyr are not protected and the N^G of Arg was protonated with a strong acid.
- These fragments are then coupled in the sequence shown on Figure 1 using either Benzotriazolyloxytrisdiméthylaminophosphonium hexafluorophosphate

(BOP) or azid as carboxyl activators. Activation with BOP gave cleaner and faster coupling steps than DCC. With this process we have synthetised 300 g batches of raw protected peptide which was purified by CCD on a pilot plant apparatus, then deprotected with TFMSA/TFA as described by Fujii et al[3].

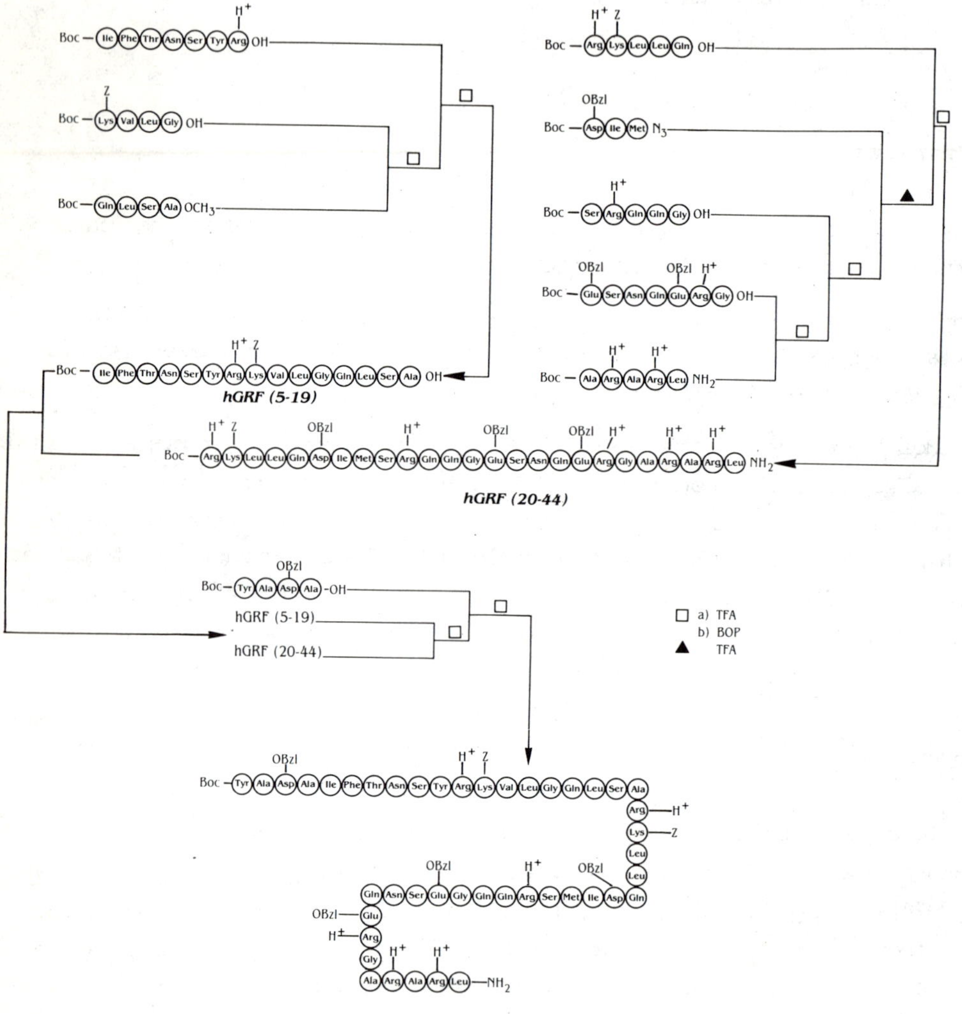

Fig. 1 : SYNTHETIC ROUTE TO PROTECTED hGRF (1-44) NH_2

Purification

Purification of the fully deprotected peptide was achieved using the following techniques : gel filtration on Sephadex column (1000 x 215 mm), ion exchange chromatography on CM Trisacryl and partition chromatography on Sephadex. Purity was determined at each step by HPLC (Table I and figure 2) and was improved significantly throughout these steps.

Table I : Improvement of hGRF (1-44)NH_2 purity

Step	pp (%)*	hGRF content (‰)**
A. after gel filtration	66.0	360
B. after ion exchange chromatography	93.5	621
C. after partition chromatography	97.0	800

*Polypeptidic purity was determined by the ratio of the main peak area to all peak areas recorded by HPLC at 195 nm.

**Represents the weight (mcg) of anhydrous base per mg of freeze-dried material.

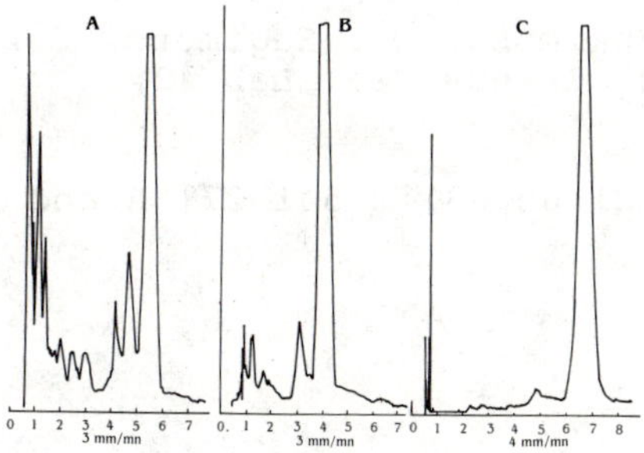

Fig. 2 : HPLC chromatogram of hGRF (1-44)NH_2 at each purification step.
Column : RP 300 (25 x 0.39 cm, 10 µm particle size) ; eluent : CH_3CN (26 %)/TEAP (pH : 3) flow rate : 2 ml/min ; detection : 195 nm.

Identification

To ensure the identity of our synthetic GRF in comparison to native hGRF further investigations were conducted.

Amino acid analysis was carried out after chemical hydrolysis ; Sanofi GRF showed amino acid ratios that corresponded to the theoretical hGRF (1-44)NH_2 values.

HPLC co-injection with a reference sample from the Salk Institute showed that the two products comigrate.

A bioassay on rat pituitary monolayer culture was performed in Pr Guillemin's laboratory with our product in comparison to the Salk reference sample. Our peptide is equipotent (within the experimental error) to the Salk reference on the secretion of GH.

References

1. Gullemin, R., Brazeau, P., Bohlen, P., Esch, F., Ling, N. and Wehrenberg, W. B. (1982) Science, 218 No, 585-7.

2. Ling, N., Esch, F., Bohlen, P., Brazeau, P., Wehrenberg, W.B. and Gulliemin, R. (1984) Proc. Natl. Acad. Sci. USA 81, 4302-4306.

3. Fujii, N., Shimokura, M., Akaji, K., Kiyama, S., and Yajima, H. (1984) Chem. Pharm. Bull. 32, 510-519.

4. Chino, N., Kumagaye, S., Masui, Y., and Sakakibara, S. (1983) in Peptide Chemistry: Munekata, E., Ed. Protein Research Foundation, Osaka (1984) pp. 285-290.

5. Wakimasu, M., Kitada, C., Yukimosa, H., Sugimoto, T., and Kobayashi, S., (1983) J. Takeda Res. Lab 42, 209-226.

6. SANOFI SA, French Patents (21.02.83) 2 541 271 A and 84 24774 A.

SYNTHESIS AND PROPERTIES OF A DISULPHIDE-BRIDGED INSULIN DIMER

Wolfgang H. Fischer, Derek J. Saunders, and Helmut Zahn
Deutsches Wollforschungsinstitut, D-5100 Aachen

Axel Wollmer
Abteilung Physiologische Chemie der RWTH, D-5100 Aachen

Introduction

The surface of insulin responsible for interaction with receptors is still subject to discussion. It has been suggested that part of the C-terminus of the B-chain is very important.[1] This very region is engaged in a thight anti-parallel contact with the corresponding region of a second molecule in the insulin dimer.

We have constructed a covalent dimer which is virtually identical to that in the crystal. In this permanent dimer the B-chain C-terminus should not be available for any other interactions. The biological activity of the derivative in vitro and the conformation as reflected in the circular dichroism were examined.

Results and Discussion

Computer graphic investigation of the monomer contacts within the insulin dimer suggested that substitution of phenylalanine B25 by cysteine followed by formation of a disulphide bridge should give a permanent dimer in which both the original relative orientation and internal conformation of the monomers are virtually unaltered.

The synthesis of the [CysB25,BysB25']insulin dimer was carried out as follows: The modified B23-30 octapeptide H-Gly-Phe-Cys(Acm)-Tyr-Thr-Pro-Lys(Tfa)-Ala-OH was synthesized by the polyamide solid phase technique.[2] The Fmoc protecting group was used as N^{α}-protection. For tyrosine and threonine side-chain protection tert.-butyl ethers were employed. The cleavage from the resin and deprotection of Tyr and Thr side chains was achieved by treatment with 90% trifluoracetic acid. The octapeptide was >95% pure by HPLC analysis after chromatography on Sephadex G-15. It was therefore used without further purification in the next step. A sevenfold excess of the peptide was reacted with bis (tert.-butyloxycarbonyl)-des(B23-30)-insulin in dimethyl-

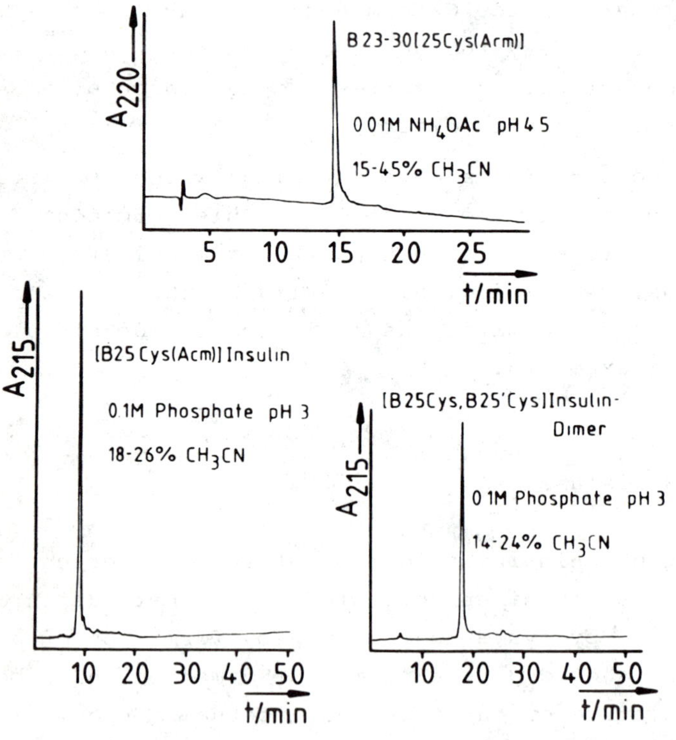

Fig. 1 RP-HPLC profiles of the synthetic derivates

formamide/glycerol(1:1) containing 20% water and 1mM $CaCl_2$ at pH 6.5 using trypsin as a catalyst. After gel filtration on Sephadex G-50, deprotection, and ion exchange chromatography on DEAE- cellulose at pH 8, pure [Cys(Acm)B25]insulin was obtained in 30% yield. Treatment with iodine in 80% aqueous acetic acid gave the disulphide-bridged dimer in 50% yield after gel filtration on Sephadex G-50.

The ability of monomeric [Cys(Acm)B25]insulin and [CysB25, CysB25']insulin dimer to stimulate lipogenesis in isolated rat adipocytes was determined. The [Cys(Acm)B25]insulin showed an activity of 3-5% (crystalline pig insulin = 100%), which is typical for B25-substituted insulins.[3,4] The activity of the dimer was only about 0.02% (Diaconescu and Brandenburg, unpublished). This confirms that insulin is active as the monomer. At least part of the dimer-forming region must be involved directly in an "informative" interaction with the receptor, or it is required in an initial interaction prior to a subsequent "informative" interaction.

As reported earlier for [LeuB25]insulin [3] substitution of phenylalanine B25 by cysteine(Acm) largely reduces biological activity without too much affecting the CD spectral properties of the native hormone (not depicted). The effect of concentration on the near ultraviolet is however reduced and seems to be due to impaired dimerization rather than to the absence of the PheB25 chromophore. Zinc ions fail to increase the spectrum further indicating that hexamers may no longer form.

In Figure 2 the strong 275nm band is an expression of the (mainly TyrB16/TyrB26 aromatic) interactions across the monomer/monomer contact being fixed by the B25-B25' cystine bridge. The far ultraviolet indicates that the accommodation of the latter may require limited rearrangements which, however (as deduced from the effect of zinc ions), appear compatible with hexamer formation.

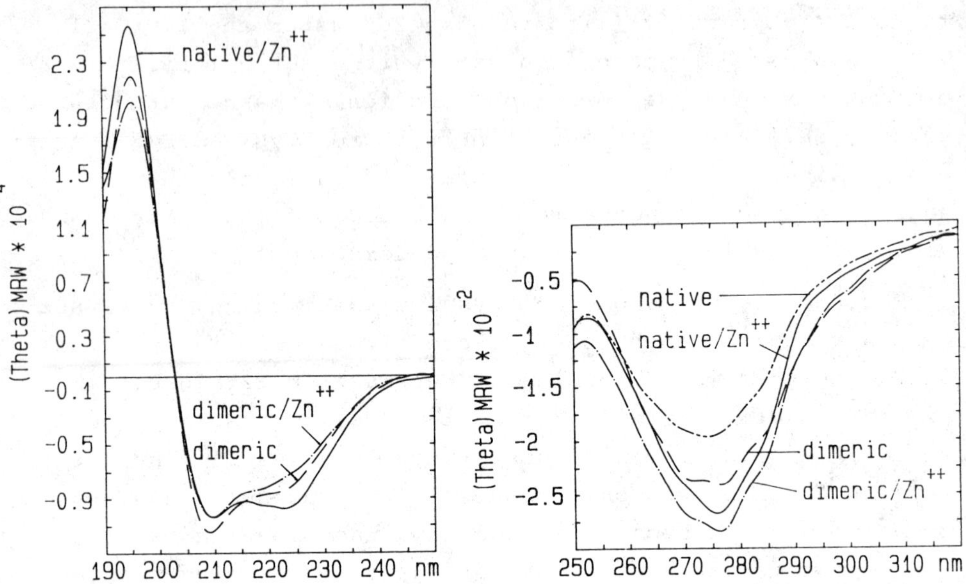

Fig. 2 CD spectra of dimeric [CysB25,CysB25']insulin and native porcine insulin in the presence and absence of $0.33Zn^{++}$/monomer. Insulin concentration = 1.6 mg/ml. Solvent: 0.025M Tris buffer pH 7.8.

Acknowledgements: Supported by DFG (SFB 113 and Wo 152/8-4) and Fonds der Chemischen Industrie.

References

1. Pullen, R.A., Lindsay, D.A., Wood, S.P., Tickle, I.J., Blundell, T.L., Wollmer, A., Krail, A., Brandenburg, D., Zahn, H., Gliemann, J., and Gammeltoft, S. (1976) Nature 259, 369-373.

2. Atherton, E., Gait, M.J., Sheppard, R.C., and Williams, B.J. (1979) Bioorganic Chemistry 8, 351-370.

3. Wollmer, A., Straßburger, W., Glatter, U., Dodson, G.G., Gattner, H.-G., Danho,W., Brandenburg, D., and Rittel, W. (1981) Hoppe-Seyler's Z. Physiol. Chem. 362, 581-591.

4. Inouye, K., Watanabe, K., and Yoshihiro, T. (1981) Biopolymers, 20, 1845-1858.

TOTAL SYNTHESIS OF BIOLOGICALLY-ACTIVE HUMAN TRANSFORMING GROWTH FACTOR TYPE ALPHA

Maqsood A. Sheikh and James P. Tam

The Rockefeller University, 1230 York Avenue, New York, N.Y. 10021

Introduction

Transforming growth factors (TGFs) represent a relatively new addition to the set of hormone-like modulators of cell growth and cell phenotype *in vitro* and were first isolated from conditioned media of retrovirally transformed cell lines[1-2]. More recent studies suggest that TGF activity might not be restricted to transformed cells and neoplastic tissues, but may play an important role in non-neoplastic tissues of embryonic and adult origin.

Human TGFα is a strong mitogenic peptide that confers phenotypic transformation to non-neoplastic indicator cells. The amino acid sequence of human TGFα, predicted from the cDNA sequence[3] of a 160 amino-acid pre-proTGFα, is a single chain protein with 50 amino-acid residues containing three disulfide linkages (Fig.1). Human TGFα differs in four residues from rat TGFα which has been synthesized recently in our laboratory[4].

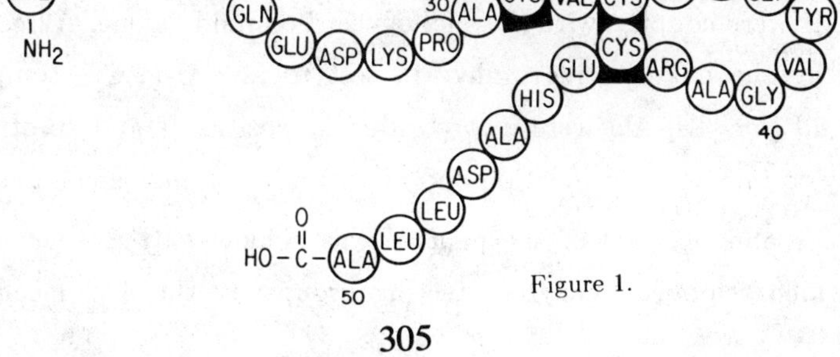

Figure 1.

Structurally and biochemically, both TGFαs are similar to epidermal growth factor (EGF), they bind and activate EGF receptor-protein kinase, and stimulate DNA synthesis and cell growth.

Since human TGFα is obtained only with difficulty from the natural source, we have undertaken the synthesis of human TGFα with the objective of providing sufficient material and performing more extensive studies on their mode of action. In this paper, we report a high yield synthesis of biologically active human TGFα.

Results and Discussion

The stepwise solid-phase synthesis, the refolding, and oxidation of human TGFα was performed essentially similar to the approach of the earlier synthesis of rat TGFα[4]. We adopted the differential acid-labile protecting group strategy using the conventional combination of t-butyloxycarbonyl group for the N^{α}-amino terminus and benzyl alcohol derivatives for the side chains. An improved, more acid-stable benzyl ester, 4-oxymethyl-phenyl-acetamidomethyl linkage, that anchored protected amino acids to the polymeric support was used to minimize loss of peptides during the repetitive acid treatments[5]. Side chain protecting groups were: Arg(Tos), Asp(OcHex), Cys(4-MeBzl), Glu(Bzl), His(Dnp), Lys(2-ClZ), Ser(Bzl), Thr(Bzl), and Tyr(BrZ). Each synthetic cycle consisted of (i) a 20 min deprotection with 50% trifluoroacetic acid/CH_2Cl_2 (ii) neutralization with 5% diisopropylethylamine/CH_2Cl_2 and (iii) double coupling with preformed symmetric anhydrides. Couplings of Boc-Asn and Boc-Gln were mediated by the preformed hydroxybenzotriazole active ester in dimethylformamide. Boc-Gly and Boc-Arg(Tos)-OH were coupled with dicyclohexylcarbodiimide alone. All couplings were monitored by the quantitative ninhydrin test to give >99.8% completion. Boc-[^{14}C]-Val and Boc-[^{3}H]-Ala were incorporated at position 1 and 46 of the sequence respectively to facilitate purification and quantitation in biological assays.

Cleavage and removal of the peptide was achieved by the new low-high HF method[6] which removed benzyl protecting groups by the S_N2 mechanism in an equi-molar solution of HF and dimethyl sulfide, and minimized the serious side reactions due to carbocations generated in the conventional S_N1 deprotection method. It also reduced the cysteinyl side reactions that might hamper the proper refolding of the multiple disulfide linkages.

After HF treatment and before any purification, the crude and reduced synthetic human TGFα was oxidized in a combination of reduced and oxidized glutathione. This avoided the formation of any polymeric materials during the purification. Crude synthetic human TGFα was purified to homogeneity in two steps: (1) gel filtration on a Bio-Gel P-10 column, (2) preparative HPLC on a C_{18} reverse-phase column (Fig. 2). Overall yield, based on starting loading of Ala to resin was 20%.

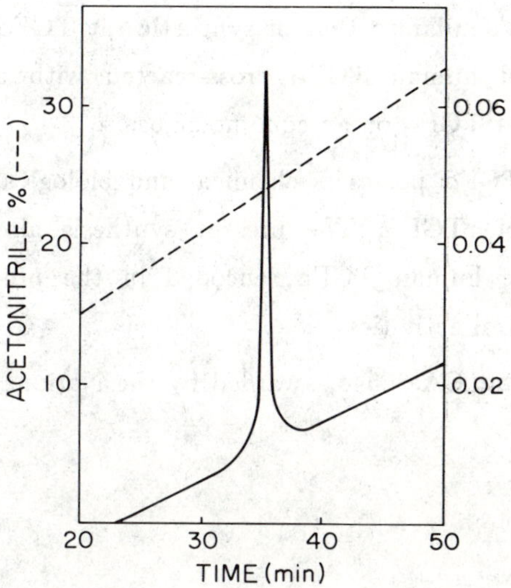

Fig. 2. HPLC of purified human TGFα. Conditions: Vydak C_{18} column (4.6x250 mm, 5 micron particle size), at a flow rate of 1.5 ml/min, eluting with a gradient of acetonitrile in 0.05% trifluoroacetic acid.

Physical characteristics of the purified human TGFα showed that it was highly homogeneous. Amino acid analysis (6N HCl hydrolysis) gave the expected theoretical molar ratios of the proposed sequence. No free thiol was detected in the synthetic human TGFα by Ellman's method of sulfhydryl determination but thiolytic reduction produced the expected six cysteinyl residues. These findings support the conclusion that synthetic human TGFα is a single polypeptide containing six cysteinyl residues in disulfide linkages and agrees with the expected chemical properties of natural human TGFα. More importantly, synthetic human TGFα eluted as a single symmetrical peak in HPLC and further confirmed the homogeneity of the material.

The biological activity of the synthetic human TGFα was investigated using three different assays. Synthetic human TGFα in the presence of TGFβ caused morphologic and phenotypic alterations of normal rat kidney fibroblasts (NRK) in monolayer culture. Furthermore, the synthetic human TGFα also caused NRK cells to lose anchorage dependence and formed colonies in soft agar. The latter property was quantified and showed to be as active as purified rat TGFα. The synthetic human TGFα also competed with $[^{125}I]$-EGF for the EGF-receptor binding on membranes of A431 human carcinoma cells. The concentration required for 50% inhibition of $[^{125}I]$-EGF binding was similar to that of synthetic rat TGFα and mouse EGF. Furthermore, iodinated human TGFα cross-reacted with a specific antibody raised against synthetic rat TGFα on an equi-molar basis.

In conclusion, the synthetic human TGFα possesses chemical and biological properties indistinguishable from those of rTGFα. The present synthesis also confirms that the 50 amino acid residue human TGFα, encoded in the prepro TGFα, possesses all the putative biological activities.

This work was supported by PHS Grant CA 36544, awarded by the National Cancer Institute, DHHS.

References

1 DeLarco, J. E. and Todaro, G. E., (1978) Proc. Natl. Acad. Sci. USA. 75, 4001-4005.

2 Marquardt H. and Todaro, G.E. (1981) J. Biol. Chem. 257, 5220-5225.

3 Derynck, R., Roberts, A. B., Winkler, M.E., Chen, E.Y. and Goeddel, D.V. (1984) Cell 38, 287-294

4 Tam, J.P., Marquardt, H., Roseberger, D.F., Wong, T.W, and Todaro, G.J., (1984) Nature (London) 309, 376-378.

5 Mitchell, A. R., Erickson, B,W., Ryabtsev, M.N.,Hodges, R.S., and Merrifield, R.B. (1976) J. Amer. Chem. Soc. 98, 7357-7362.

6 Tam, J.P., Heath, W.F., and Merrifield, R.B. (1983) J.Am. Chem. Soc. 105, 6442-6454.

TOTAL SYNTHESIS OF S-CARBAMOYLMETHYL BOVINE APOCYTOCHROME C

James Blake

Laboratory of Molecular Endocrinology,
University of California, San Francisco, California 94143

Introduction

We have previously shown that: (1) fully deprotected peptides that contain a thiocarboxyl function at their C-terminus can be synthesized by the solid-phase method;[1] (2) thiocarboxyl peptides can be purified by the standard procedures (ion-exchange, HPLC, partition chromatography) employed for other peptides; (3) the amino groups of thiocarboxyl peptides can be reversibly blocked by reaction with citraconic anhydride;[2] and (4) thiocarboxyl peptides can be specifically coupled at their C-terminus to other amino peptides by reaction with silver nitrate/N-hydroxysuccinimide. Subsequent reaction with aqueous acetic acid removes the citraconyl groups to give the desired peptide.

The above strategy has been used in a "two-segment" synthesis of a model peptide,[3] an analog of β-endorphin,[4] and human β-lipotropin.[5] We now report the synthesis of an analog of bovine apocytochrome c by a "three-segment" strategy.

Results and Discussion

The peptides [Cys(Cam)^{14,17},GlyS^{23}]-apocytochrome c-(1-23) (I), CF_3CO-[GlyS^{60}]-apocytochrome c-(24-60) (II) and CF_3CO-apocytochrome c-(61-104) (III) were synthesized by the solid-phase method. Each of the peptides was purified by chromatography on CM-cellulose, partition chromatography,[6] and/or HPLC. Peptides I and II were reacted with citraconic anhydride to block the lysine amino groups and give the corresponding peptides Ia and IIa. In the case of peptide III, reaction with citraconic anhydride was followed by reaction with 10% hydrazine/water to remove the terminal trifluoroacetyl group and give IIIa.

Peptides IIa and IIIa were coupled together by reaction with silver nitrate/N-hydroxysuccinimide in 50% DMF/water. After subsequent reaction with 10% N_2H_4/water to remove the trifluoroacetyl group from Gly-24, the peptide mixture was chromatographed on Sephadex G-50 in 30 mM ammonium bicarbonate. Uncoupled peptide IIa could be separated from an unresolved mixture of peptide IIIa and the desired 24-104 peptide. The yield of desired product was deduced to be 10-20% by diagnostic amino acid analysis of the unresolved mixture.

Three equivalents of peptide Ia were coupled to the above peptide mixture. After gel filtration and decitraconylation in aqueous acetic acid, the complex mixture was resolved by chromatography on CM-cellulose. Peptides that could be isolated and identified corresponded to the sequences 1-23, 61-104, 24-104, and (1-23)-(61-104) of apocytochrome c. In addition, a peptide that eluted in the same position as [Cys(Cam)^{14,17}]-apoCyt c derived from native cytochrome c (by consecutive reactions with silver nitrate, β-mercaptoethanol, and iodoacetamide) was shown to have the amino acid composition of the 1-104 protein. Rechromatography on CM-cellulose and HPLC gave highly purified [Cys(Cam)^{14,17}]-apocytochrome c in 0.6% yield.

The synthetic protein was shown to be identical to [Cys(Cam)^{14,17}]-apocytochrome c derived from native bovine cytochrome c by electrophoresis on paper or polyacrylamide gel, HPLC (Figure 1), and chymotryptic or tryptic map.

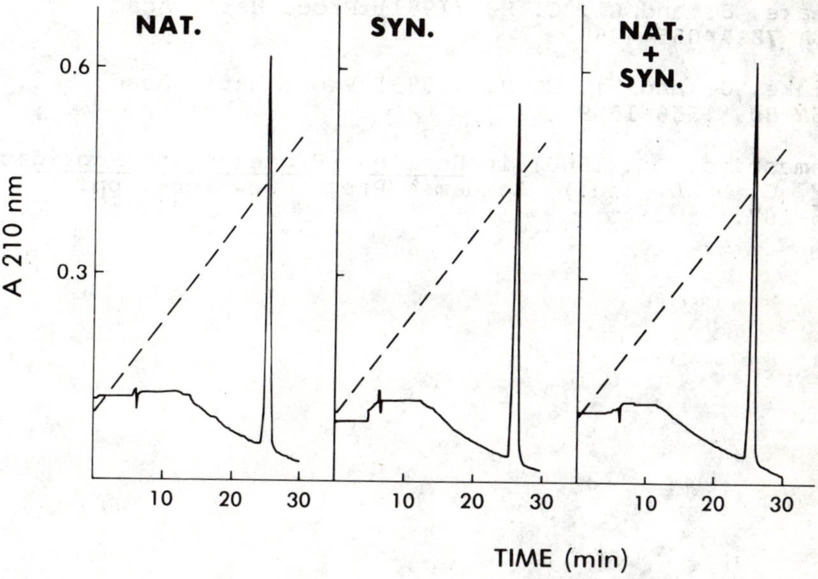

Fig. 1. HPLC of 20 μg samples of [Cys(Cam)^{14,17}]-apoCyt c derived from native cytochrome c (NAT.), total synthesis (SYN.) and a 1:1 mixture on a Vydac 201TP104 column in 2-propanol/0.1% trifluoroacetic acid (10-50%/30 min.).

Acknowledgment

This work was supported in part by National Institute of Health Grants GM-30072 and GM-2907.

References

1. Merrifield, R. B. (1963) J. Am. Chem. Soc. **85**, 2149-2154.

2. Dixon, H.B.F. and Perham, R. N. (1968) Biochem. J. **109**, 312-314.

3. Blake, J. (1981) Int. J. Peptide Protein Res. **17**, 273-274.

4. Blake, J. and Li, C. H. (1981) Proc. Natl. Acad. Sci. USA **78**, 4055-4058.

5. Blake, J. and Li, C. H. (1983) Proc. Natl. Acad. Sci. USA **80**, 1556-1559.

6. Yamashiro, D. (1980) in Hormonal Proteins and Peptides IX, C. H. Li (ed.), Academic Press, New York, pp. 25-107.

STRUCTURE DETERMINATION AND SYNTHESIS OF A NEW TRIDECAPEPTIDE FROM *PHYLLOMEDUSA ROHDEI*.

Roberto de Castiglione, Mauro Gigli, Luigia Gozzini, PierCarlo Montecucchi and Giuseppe Perseo

Farmitalia Carlo Erba SpA, Chemical Research & Development, via dei Gracchi, 35 - 20146 Milan, Italy.

Vittorio Erspamer

Institute of Medical Pharmacology, University of Rome, Città Universitaria - 00100 Rome, Italy.

Introduction

We have previously reported the occurrence of new tetra-, penta- and heptapeptides in methanol extracts of the skin of the South American frog *Phyllomedusa rohdei*. These peptides, collectively called tryptophyllins (TPH's) owing to their origin and the characterizing presence of a tryptophan residue in their sequence, were isolated from the 95% and 70% ethanol eluates from an alumina column, and synthesized by solution methods[1-4].

The present paper deals with the isolation, structure determination and synthesis of a new tridecapeptide (TPH-13) of formula

<Glu-Glu-Lys-Pro-Tyr-Trp-Pro-Pro-Pro-Ile-Tyr-Pro-Met-OH

present in the 50% ethanol eluate from the same alumina column.

Results and Discussion

An aliquot of the 50% ethanol eluate from the alumina column (corresponding to 2 g dried tissue= 10 mg lyophilized material) was chromatographed on a CM-Sephadex C-25 column (cm 1 x 20; eluent: 0.02 M $AcONH_4$ pH 4; flow rate 3 ml/h). The main pool positive to the Erlich's reagent was further purified on a DEAE Sephacel column (cm 1 x 40; eluent: 0.05 M $(NH_4)HCO_3$ pH 7.7; flow rate 2 ml/30 min). The product so obtained proved to be homogeneous by amino acid analysis, RP-HPLC (μ-Bondapak CN; 60% CH_3CN in 0.1 M HCOOH), isoelectric focusing, TLC and high voltage paper electrophoresis at pH 1.2 ($E_{1.2}$) and pH 6.5 ($E_{6.5}$).

Sequence analysis was carried out by a combination of chemical and enzymatic methods, as outlined in Figure 1.

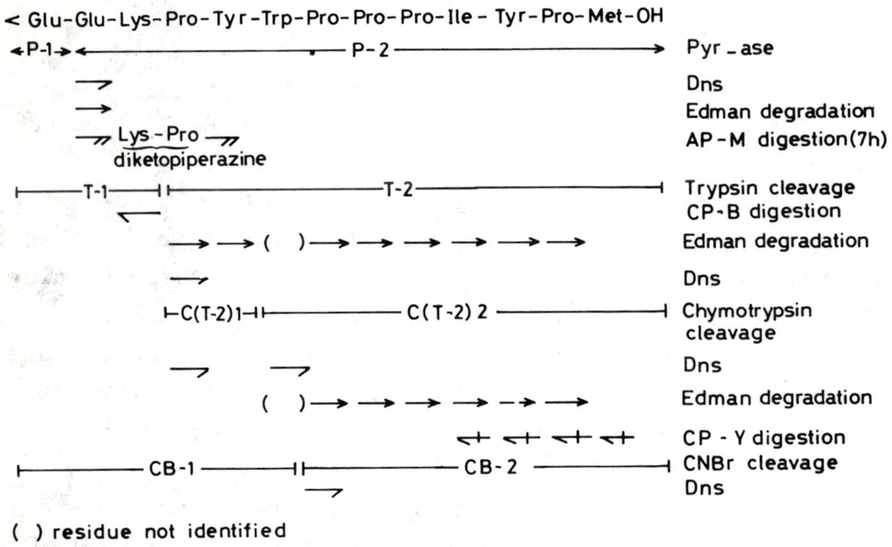

Fig. 1. Summary of the strategies employed in primary structure elucidation of TPH-13.

CP-Y digestion released methionine sulfoxide, instead of methionine: this was a possible artifact due to the isolation and purification procedures, as reported for other methionine-containing peptides[5].

The synthetic replicate was obtained by segment coupling in solution as show in Figure 2. The (1-5) and (6-13) fragments were prepared essentially by mixed anhydride and activated ester

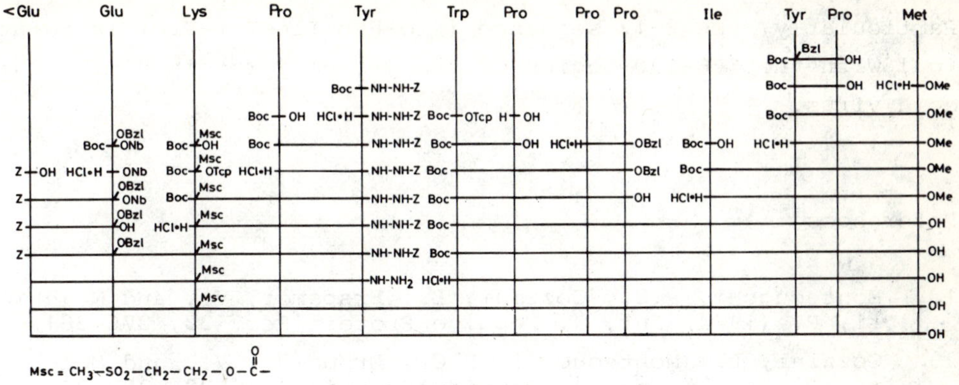

Fig. 2. Scheme of synthesis of TPH-13.

procedures, and condensed to protected TPH-13 by the azide method. After alkaline removal of the Msc protecting group, the target compound was purified by counter-current distribution in the solvent system: *n*-BuOH/EtOH/AcOH/H_2O = 5/1/1/8. By precipitation from *i*-PrOH/*i*-Pr_2O, a homogeneous compound was obtained: m.p. ca. 200°C (dec.); $[\alpha]_D^{26}$ - 78.1° (c 1, DMF); $E_{1.2}$ 0.38; $E_{5.8}$ 0.13; R_f 0.23 (*n*-BuOH/AcOH/H_2O = 4/1/1).

The synthetic peptide, after oxidation with peracetic acid, displayed the same chromatographic and degradation pattern as the natural sample. This, in turn, behaved as the synthetic peptide when reduced with mercaptoethanol. The HPLC method, however, had to be modified (gradient from 16 to 64% CH_3CN in 0.1 M HCOOH) in order to resolve the two oxidation forms[6].

Similarly to the other tryptophyllins, both the natural and the synthetic tridecapeptides were practically devoid of any activity on the usual smooth muscle preparations both *in vitro* and *in vivo*. Further pharmacological studies with the synthetic peptide are in progress and seem to indicate the presence of endocrine and behavioural effects after i.c.v. administration (G. Nisticò and P. Melchiorri, personal communications). Using a search computer programme (Intelli-Genetics) some structural homologies have been found between TPH-13 and different proteins. Particularly, the 8-12 sequence (Pro-Pro-Ile-Tyr-Pro) is identical with the 126-130 region of the *gag* polyprotein simian sarcoma virus.

References

1. Montecucchi, P.C., Gozzini, L., Erspamer, V., and Melchiorri, P. (1984) Int. J. Peptide Protein Res. 23, 276-281.

2. Gozzini, L., Montecucchi, P.C., Erspamer, V., and Melchiorri, P. (1985) Int. J. Peptide Protein Res. 25, 323-329.

3. Perseo, G., and de Castiglione, R. (1984) Int. J. Peptide Protein Res. 24, 155-160.

4. Perseo, G., Gigli, M. and de Castiglione, R. (1985) Int. J. Peptide Protein Res. 25, 316-322.

5. Houghten, R.A. and Li, C.H. (1983) Methods Enzymol. 91, 549-559.

6. Rusconi, L., Perseo, G. & Montecucchi, P.C. (1985) A.J.P. Martin Honorary Symposium, Urbino, Italy, May 27-31; Abstract C7, pp. 85-86.

SYNTHESIS OF MURINE EPIDERMAL GROWTH FACTOR

William F. Heath and R.B. Merrifield
The Rockefeller University, New York, N.Y. 10021

Introduction

Murine epidermal growth factor (mEGF) is a 53-amino acid polypeptide which is mitogenic for a number of cell types.[1] While some studies of its structure-function relationships have been done, a great deal is not known about its mode of action. The mEGF molecule possesses a complex structure containing 3 disulfide bridges and a large number of trifunctional amino acids, presenting a considerable challenge to the peptide chemist. We chose to synthesize this factor as a first step in a study of the structure-activity relationships of the molecule.

Results and Discussion

The protected peptide was assembled on the PAM resin support using the Boc group for N^{α} protection and benzyl-type protecting groups for most side chain functional groups. In addition Met(O), Trp(For), His(DNP) and Asp(OcHex) were also used. (^{3}H)Leu was introduced at position 52 to facilitate monitoring of the final product. Each residue was coupled twice as a preformed symmetrical anhydride, first in CH_2Cl_2 and then in DMF, except for Asn and Gln which were coupled using DCC/HOBt. The progress of the synthesis was followed using the quantitative ninhydrin test[2] and amino acid analysis. 4.2 g of starting aminomethyl resin (0.4 mMol/g) yielded 21 g of protected peptide resin at the end of the synthesis. The final peptide resin was obtained in 99% yield based on weight gain, ninhydrin and amino acid analysis tests.

After thiolytic removal of the DNP group and acidolytic cleavage of the amino terminal Boc group, a portion of the synthetic protected peptide was deprotected and cleaved from the resin support in 95% yield using the low/high HF procedure.[3] The low HF consisted of HF:Me_2S:p-cresol:p-thiocresol (25:65:8:2, v/v, 0° C, 2 h. The high HF was performed using HF:p-cresol:p-thiocresol (90:8:2, v/v), -10° C, 1 h. After dialysis to remove contaminating salts, the crude peptide was oxidized in the presence of air. Attempts to form the disulfide bridges using the oxidized/reduced glutathione method[4] resulted in the isolation of an EGF fraction which contained covalently bound glutathione. Analysis showed the air oxidized product to consist of a monomeric fraction and a polymeric fraction. The monomeric fraction contained one major peak (>70%) on HPLC (Fig 1), which was purified to homogeneity using low pressure liquid chromatography. The synthetic EGF coeluted with the natural material on analytical HPLC (Fig 1). The pure material gave the following amino acid analysis: Asp 7.0(7), Thr 1.6(2), Ser 5.2(6), Glu 3.1(3), Pro 2.0(2), Gly 6.2(6), Cys 6.0(6),

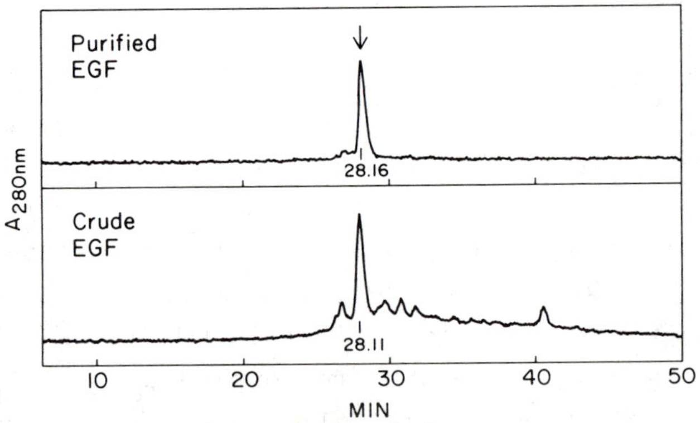

Fig. 1. Vydac C-18 HPLC of EGF. 45 min linear gradient of 26% to 38% CH_3CN into H_2O, 0.1% $CF_3CF_2CF_2CO_2H$, 1.5 ml/min. Arrow denotes elution position of natural EGF.

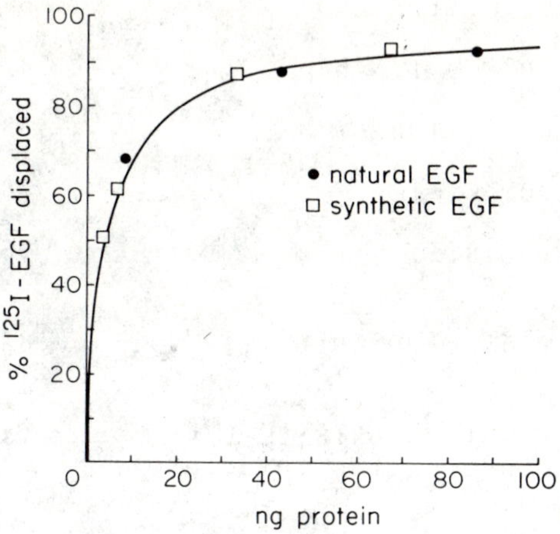

Fig 2

Radioreceptor assay on A-431 cells

Val 1.8(2), Met 0.9(1), Ile 1.9(2), Leu 4.3(4), Tyr 5.1(5), His 1.1(1), Arg 4.1(4), Trp 2.0(2).

Synthetic EGF was also identical to natural EGF in the radioreceptor assay[5] (Fig 2) and in the stimulation of growth of Sertoli cells.[6] Based on the chemical and biological data, synthetic EGF is identical to the natural material.

The nature of the polymeric fraction was investigated. The polymer consisted largely of dimers and trimers on gel permeation chromatography and had an amino acid composition comparible to that of the monomer fraction. Reduction and alkylation of the polymer produced a material which was similar to reduced and alkylated monomer on HPLC. Various refolding and oxidation procedures were used but polymer still formed. Since loss of cysteine in the HF could be a possible cause of polymer formation, the recovery of EGF cysteine after various HF treatments was studied. The results in Table I are in agreement with our earlier work,[3] and indicate that loss of cysteine in the HF does not appear to be the cause of polymer formation. Further work is necessary to answer this question.

Table I. Recovery of EGF cysteine after HF deprotection

Condition	% Recovery[1]
HF:anisole(80:20, v/v), 0° C, 1.5 h	68
HF:p-cresol(80:20, v/v), 0° C, 1.5 h	87
HF:anisole(90:10, v/v), 0° C, 1 h	58
HF:p-cresol(90:10, v/v), 0° C, 1 h	78
HF:p-cresol:p-thiocresol(90:8:2, v/v), 0° C, 1 h	95
HF:p-cresol:2-mercaptopyridine(90:10:0.1, v/v), 0° C, 1 h	82
HF:Me_2S:p-cresol:p-thiocresol(25:65:8:2, v/v), 0° C, 2 h, followed by HF:p-cresol:p-thiocresol (90:8:2, v/v), -10° C, 1 h	91

[1]Determined after performic acid oxidation and subsequent hydrolysis, and were normalized to leucine.

Acknowledgement

This work was supported by Grant AM01260 from the U.S. Public Health Service.

References

1. Carpenter, G. and Cohen, S. (1979) Ann. Rev. Biochem. 48, 193-216.
2. Sarin, V.K., Kent, S.B.H., Tam, J.P. and Merrifield, R.B. (1981) Anal. Biochem. 117, 147-157.
3. Tam, J.P., Heath, W.F. and Merrifield, R.B. (1983) J. Amer. Chem. Soc. 105, 6442-6455.
4. Ahmed, A.K., Schaffer, S.W. and Wetlaufer, D.B. (1975) J. Biol. Chem. 250, 8477-8482.
5. Delarco, J.E. and Todaro, G.J. (1978) Proc. Natl. Acad. Sci. USA 75, 4001-4005.
6. Sertoli cell assay performed by Dr. Patricia Morris, Population Council, New York, N.Y.

SYNTHESIS OF AN ARGININE-RICH RNA-BINDING FRAGMENT OF A VIRAL PROTEIN

Paul B.W. Ten Kortenaar and Godefridus I. Tesser
Department of Organic Chemistry, University of Nijmegen, Nijmegen, The Netherlands

Jaap Krüse and Marcus A. Hemminga
Department of Molecular Physics, Agricultural University, Wageningen, The Netherlands

Introduction

CCMV is an icosahedral plant virus, consisting of RNA and 180 protein subunits. It can be dissociated into protein dimers and RNA at neutral pH and high ionic strength. An arm of 25 amino acids at the N-terminus of each subunit is involved in the RNA-binding. Without RNA this part is flexible and seen by NMR, but in the presence of RNA the arm is immobilized and the resonances in NMR are broadened beyond detection (1,2). A detailed NMR-study of the RNA-protein interaction requires the availability of the N-terminal 25-peptide.

Results and Discussion

The 25-peptide, shown below, was synthesized in solution on a 500 mg scale. Arginines were introduced *without* side-chain protection to improve the solubility of the intermediates (cf. 3). Products were purified by counter-current distribution in n-BuOH-0.1 M NaCl. Protonation of arginine side-chains in carbodiimide couplings could be achieved with HOBt (4).

Ac-Ser-Thr-Val-Gly-Thr-Gly-Lys-Leu-Thr-Arg-Ala-Gln-Arg-Arg-Ala-Ala-Ala-Arg-Lys-Asn-Lys-Arg-Asn-Thr-Arg-NHMe

Synthesis of fragment (1-10) (Fig. 1) commenced with the acylation of the free base of Arg with a dipeptide azide in 20%

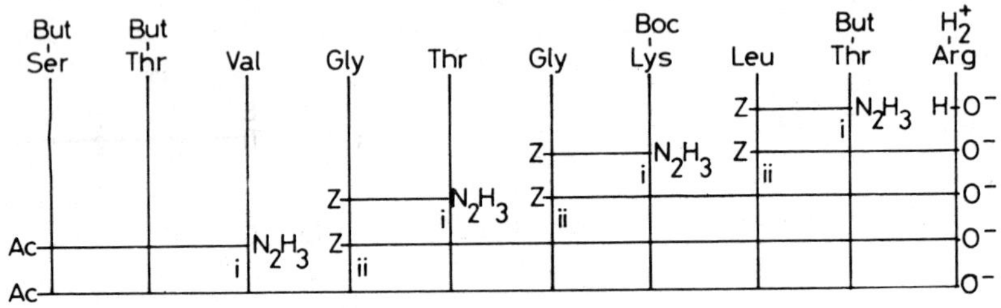

Fig. 1 Synthesis of fragment (1-10); i = H^+, NO^+; ii = H_2, Pd

aqueous DMF. The product was purified by extraction with n-BuOH-water and precipitation in ether. Coupling of the other fragments making up the decapeptide and purification of the products proceeded likewise. Yields increased from 72% for fragment (8-10) to 98% for the decapeptide due to diminishing water-solubility.

In the synthesis of derivative (11-17) (Fig. 2), complete acylation of H-(Ala)$_3$-N_2H_2Boc required the use of an excess of Z-(Arg)$_2$-OH and DCC due to lactam formation. The pentapeptide was purified by countercurrent distribution in n-BuOH-0.1 M NaCl. Salt was removed by dissolution of the product in

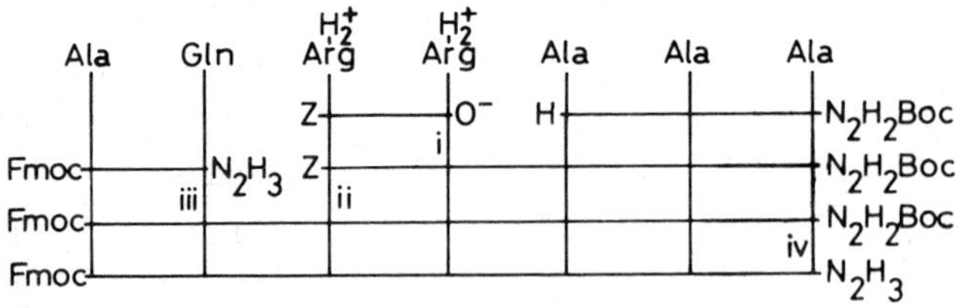

Fig. 2 Synthesis of fragment (11-17); i = DCC, HOBt; ii = H_2, Pd; iii = H^+, NO^+; iv = 3N HCl, MeOH

DMF and filtration. In the azide coupling with dipeptide (11-12) tri-n-butylamine was used as the base, since the solubility of its HCl-salt in EtOAc enables isolation of the heptapeptide (11-17) by precipitation in this solvent.

Fragment (18-25) was prepared in a stepwise manner (Fig. 3, ref. 5). Complete acylations with arginines again required the use of excesses of Z-Arg-OH and DCC. The insoluble hexapeptide

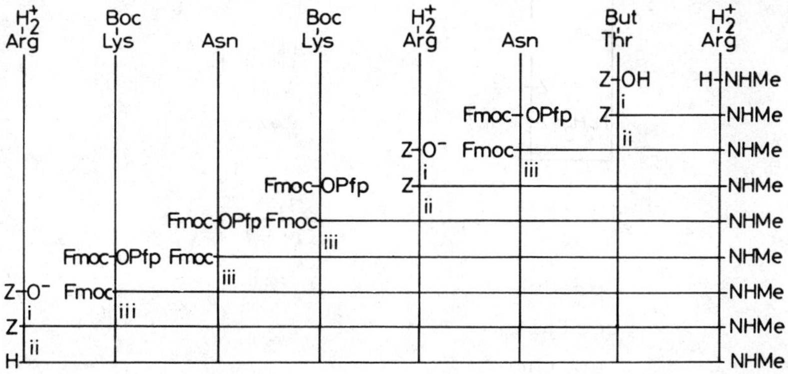

Fig. 3 Synthesis of fragment (18-25); i = DCC, HOBt; ii = H_2, Pd; iii = 10% $(CH_3)_2NH$ in DMF

(20-25) and the heptapeptide (19-25) were purified by extraction of a very dilute solution of the product in n-BuOH with water and precipitation in ether.

Assemblage of the three fragments started with the azide coupling of fragments (11-17) and (18-25). The resulting 15-peptide (11-25), isolated in 50% yield after countercurrent distribution in n-BuOH-0.1 M NaCl and desalting on Sephadex LH-20 in 25% aqueous MeOH, was coupled with the N-terminal decapeptide using DCC-HOBt (4). The proton required for the condensation was furnished by one extra equivalent of HOBt. Excess decapeptide and DCC-HOBt compensated for lactam formation. The product was obtained in 70% yield after countercurrent distribution in n-BuOH-HOAc-water (4:1:5). Removal of protective groups with TFA and chromatography on Sephadex LH-

20 in 25% aqueous MeOH provided the desired 25-peptide-amide in nearly quantitative yield.

The NMR spectrum of the synthetic peptide is in good agreement with the difference spectrum of the intact coat-protein dimers and dimers lacking the N-terminal arm (Fig. 4, the signal at 2.75 ppm in B is caused by the C-terminal methylamide).

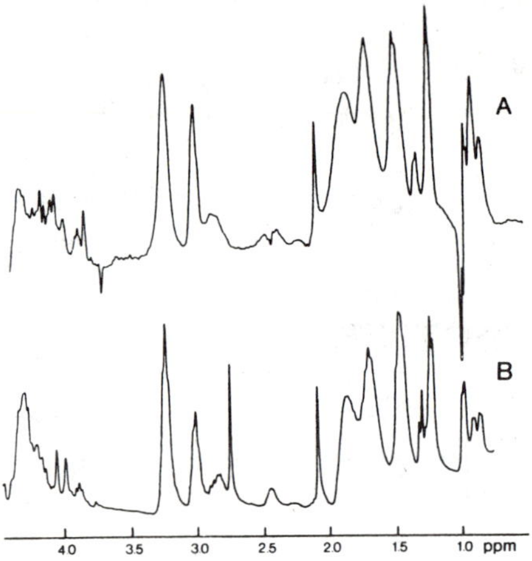

Fig. 4 ^{1}H-NMR spectra of the synthetic 25-peptide (B) and the difference spectrum of intact protein dimers and dimers lacking the N-terminal arm (A)

References

1. G. Vriend, M.A. Hemminga, B.J.M. Verduin, J.L. de Wit and T.J. Schaafsma, FEBS Lett. 134, 167-171 (1981).

2. G. Vriend, Thesis, Agricultural University, Wageningen (1983).

3. M. Bodanszky, Int. J. Pept. Prot. Res. 25, 449-474 (1985).

4. W. König and R. Geiger, Chem. Ber. 103, 788-798 (1970).

5. L. Kisfaludy and I. Schön, Synthesis 1983, 325-327.

SEMISYNTHETIC PEPTIDES AND PROTEINS

Robin E. Offord

Département de Biochimie Médicale, C.M.U., Université de Genève
CH-1211 Genève 4

Introduction

When used in relation to peptides and proteins, "semisynthetic" (or the alternative term "partially synthetic") means the use of a polypeptide of natural origin as a ready-made intermediate in a synthesis. Semisynthesis has a valid, but not exclusive, role in the production of peptides and proteins of novel structure. The contribution of total synthesis and recombinant DNA techniques is extremely well known, and the purpose of this review is to enumerate the various forms of the semi-synthetic approach, and place them in the perspective of the two other techniques.

Classically, syntheses were undertaken to confirm the result of a structure determination, to prepare a natural molecule in larger quantities than those otherwise available, or to produce an analogue for practical use or academic study. Semisynthesis and recombinant methods are largely inappropriate to the first of these aims, but have proven their value with regard to the latter two.

Types of Semisynthesis.

Table I. shows in broad outline the various types of semisynthetic approach.

Table I.

The Principal Kinds of Protein Semisynthesis

Type:	Description:	Applied to:*
1. Non-covalent	The formation of functioning complexes of protein fragments. One or more of the fragments can be replaced by a synthetic or modified version.	Pancreatic ribonuclease [1]; *S. areus* nuclease [2]; Cytochrome *c*[3]; Thioredoxin[4]; Somatotropin[5]; Phospholipase A_2[6]; Myoglobin[7]; IgA fragment F_V[8].
2. Covalent		
a) via disulphide bridges	The reoxidation of reduced chains of a multichain protein, one of the chains having been replaced by a synthetic or modified version.	Insulin[9]; Antibodies [10];
b) via the peptide bond	The chemical coupling, via alpha amino and carboxyl groups, of fragments of natural sequence with those of altered sequence. The altered sequences can be obtained by total synthesis or by semisynthesis, or any other method.	See Tables II. & III

* Not an exhaustive list. See also the other reviews cited throughout this text.

Covalent semisynthesis

Because of very strict limitations of space, the rest of this review will concentrate on semisynthesis via the formation of peptide bonds.

The main experimental requirements that confront us are: 1) a means of efficient cleavage at a few specific sites; 2) a means for separating the fragments; 3) a means of ensuring that the fragments, when coupled back together, will combine only through their alpha carboxyl and amino groups; 4) a set of reagents and procedures that enable all steps to be carried out under conditions mild enough to avoid permanently compromising the chemical and three-dimensional structure of the product. The first two of these requirements have been met simply by drawing on the techniques developed by workers in the field of protein sequencing. The third requirement (to ensure coupling through the correct groups) can be met by careful selection from the available techniques of the classical methodology of protein synthesis to bring about differential protection of side chains. Alternatively, correct coupling can now be ensured with little or no protection by using the specificity of proteolytic enzymes acting in reverse [11 12 13]. This development, which has been impatiently awaited for many years, is not restricted to small substrates: it has even been applied with success to human growth hormone [14]. The fourth requirement (mild reagents and solvents) was met by trial and error. We now know that many chemical substances that would have been expected to destroy proteins activity will not do so, whilst certain others, apparently much milder, are fatal in their effects.

That the origins of the method lie in primary structure determination becomes obvious when we look a little closer. Thus, fragment-condensation semisynthesis combines Sanger´s original strategy of sequencing (with the additional option of CNBr cleavage) with that of classical solution synthesis. Stepwise semisynthesis combines the strategy of Edman (or that of sequencing by carboxypeptidase digestion) with that of conventional stepwise synthesis.

Semisynthesis, total synthesis, or genetic engineering?

The experimental[15] and theoretical[15 16 17 18] aspects of protein semisynthesis have been extensively reviewed. The purpose of the present article, at the request of the organizers of this Syposium, is to comment on the present usefulness of the technique, particularly in the light of progress in recombinant DNA technology. Until very recently, the answer to this question was "of total synthesis, semisynthesis, and recombinant technology, only semisynthesis has produced any analogues at all that are over 100 residues in length in sufficient quantity or quality for them to be of any practical or experimental use". To this answer (which always seemed a most negative and unsatisfactory one) one was obliged to add "below the size of insulin, it is scarcely worth attempting a semisynthesis; between the size of insulin and 100 residues, semisynthesis becomes progressively more important". Now, happily, it is possible to use recombinant methods to bring about simple substitutions (i.e. changes involving amino-acid residues for which there is a codon). Many laboratories, including our own, have made usable analogues in this way. This enables one to concentrate on using semisynthesis for the modifications involving non-coded residues, and for modifications involving coded residues that happen to be particularly easily carried out by semisynthetic means.

That recombinant methods do not have it their own way even for changes of coded residues is shown by the history of the production of human insulin by semisynthetic and recombinant means. It is by no means apparent from this instance that recombinant methods are the quickest, simplest, or most efficient in terms of yield. However, rather than make an unsafe generalization from a particularly favourable case, let us look in detail at the contribution of semisynthetic proteins to improvemnts in knowledge over the last few years.

Table II. gives a number of examples of proteins that have been the object of fragment-condensation semisyntheses. Table III. cites a number of the examples of stepwise syntheses that have led to results of biological interest. It will be seen that, in many proteins, residues thought to be important for biological activity are easily accessible to the semisynthetic method. For example, we owe to semisynthetic analogues our realization of the crucial role in activity of Gly^{A1} of insulin, example, as well as for much of the concrete evidence confirming earlier suspicions as to the importance of the latter part of the B-chain. Many other contributions to our knowledge of protein structure-function relationships are schematically indicated in Tables II. and III.

If most of the analogues referred to in these Tables involve coded

Table II.

Selected Examples of Fragment-Condensation Semisyntheses

Protein	Principal operations carried out*
Insulin [15 17 19]	Enzymic and chemical re-coupling of sequences from the last eight residues of the B-chain (residues important in receptor-binding, residues implicated in activity by studies on mutant human insulins).
Proinsulin [20]	Modified open-chain form by combining natural chains with a short synthetic bridge.
Ferredoxin [15 18 17]	Preparation of analogue free of aromatic residues for functional and physical studies of Iron-sulphur clusters.
Nucleases[18]	Production of analogues of intact protein by chemical or enzymic re-forming of nicked peptide bond in non-covalent complexes.
Cytochrome _c_[15 17 21]	Recombination of CNBr fragments. Several analogues made for activity, immunological, and physical studies (e.g. Tyr^{67} changed to _p_-fluorophenylalanine). Side-chain specific reactions on separated fragments before reconstructing the protein (gives proteins modified on only some of the side-chains susceptible to the reagent in question).
Myoglobin [22]	Coupling of two synthetic and one natural fragment to give the whole sequence.
Protease inhibitors (refs [15 18])	Insertion of new residues at the protease-sensitive site. Includes the production by design of new biological specificities.
Phospholipase A_2 (refs [15 17 23])	Coupling of short peptides to a large natural fragment. Many analogues for activity and physical studies including substitution by nor-leucine.

*Not an exhaustive list. Many references are to reviews, not to primary authors amino acids, this is because it is only very recently that recombinant methods began to be reliable sources of such material. There are sufficent instances of the specific insertion at a single point of an isotopic label, or of an unatural amino acid, to give a reasonable guarantee that other such analogues will be forthcoming, and it is in this direction that attention is turning. It is easy to imagine ways in which even recombinant methods could be made to produce such materials: it is as difficult to imagine that such a move would often be worthwhile.

Table III.

Selected Examples of Stepwise Semisyntheses

Protein	Principal operations carried out*
Insulin [15 17]	Very many analogues with normal amino acids. In addition, introduction of ^{13}C, of D-amino acids in sites critical for activity, selective tritiation, selective iodination. Conversion of porcine insulin into human insulin.
Ferredoxin [15 17]	Numerous analogues for activity and physical studies of the iron-sulphur clusters.
Phospholipase A_2 (refs. [16 23])	Many substitutions for physical and activity studies. Introduction of beta-alanine, and ^{13}C amino acids (both D and L).
Glucagon [24]	Substiutions for physical and activity studies.

*Not an exhaustive list.

Conclusion

In our own work, we have often wished to insert a single isotopic change at one or other specific residue in insulin [25 26 27]. Often, the experimental system in which it was intended to use the analogue in question permitted us to insert the change at a site that was particularly easily reached. In one instance, we saw no easy semisynthetic route (the substitution of Asp for Tyr^{B16} of insulin, a proposed early site of physiological insulin degradation). This change, on the other hand, lends itself to a recombinant approach, and this is what we have chosen to do [28]. We see this as an example of the rule that applies to most fields of science, and certainly to the one in question: no single method is universally applicable.

Acknowledgements

I thank the Swiss National Science Foundation and The Stanley Thomas Johnson Foundation for financial support.

References

1. Richards, F.M. (1958) Proc. Natl. Acad. Sci. USA 44, 162-166.

2. Taniuchi, H., Anfinsen, C.B., Sodja, A. (1967) Proc. Natl. Acad. Sci. USA 58, 1235-1242.

3. Harris, D.E., Offord, R.E. (1977) Bichem. J. 161, 21-25.

4. Slaby, I., Holmgren, A. (1975) J. Biol. Chem. 250, 1340-1347.

5. Li, C.H., Blaker, J., Hayashida, T. (1978) Biochem. Biophys. Res. Comm. 82, 217-222.

6. Kihara, H., Ishimaru, K., Ohno, M., (1981) J. Biochem. (Tokyo) 90, 363-370

7. Hagenmaier, H., Ohms, J.P., Jahns, J., Anfinsen, C.B. (1978) in Semisynthetic Peptides and Proteins, Offord, R.E. and Di Bello, C., Eds. Academic Press, London and New York, pp. 23-35.

8. Gavish, M., Zakut, R., Wilchek, M., Givol, D. (1978) Biochemistry 17, 1345-1351.

9. Zahn, H., Naithani, V.K., Gattner, H.G., Büllesbach, E.E., Thamm, P.M. (1981) Naturwissenschaften 68, 56-62.

10. Rosenberg, S., Terry, W.D. (1977) Adv. Cancer. Res. 25, 323-388.

11. Laskowski, M. Jr., (1978) ref. 16, pp 255-262.

12. Inouye, K., Watanabe, K. Morihara, K., Tochino, T., Kanaya, T., Emura, J., Sakakibara, S. (1979) J. Amer. Chem. Soc. 101, 751-752.

13. Markussen, J., and Schaumburk, K., (1983) in Peptides 1982, Blaha, K., and Malon, P., Eds., W. de Gruyter, Berlin, pp. 387-394

14. Graf, L. and Li., C.H. (1981) Proc. Natl. Acad. Sci. USA 6135-6141.

15. Offord, R.E. (1980) Semisynthetic Peptides and Proteins, 235 pp., John Wiley and Sons Ltd., Chichester and New York.

16. Offord, R.E. and Di Bello, C., Eds. Semisynthetic Peptides and Proteins, 399 pp., Academic Press, London and New York

17. Sheppard, R.C. (1979) in The peptides, vol 2, , Gross, e., Maienhofer, J., Eds., Academic Press, London and New York, pp 441-484.

18. Chaiken, I.M. (1981) CRC Crit. Rev. Biochem. 11, 255-301

19. Riemen, M.W., Pon, L.A., Carpenter, F.H. (1983) 22, 1507-1513.

20. Büllesbach,E., (1982) Tetrahedron Lett., 23, 1877-1882

21. Wallace, C.J.A. , Rose, K., (1983) Biochem. J. 215, 651-658.

22. Simmerman, H.K.B., Wang, C.C., Horowitz, E.M., Berzofsky, J.A., Gurd, F.R.N. (1982) Proc. Natl. Acad. Sci. USA 79, 7739-7743.

23. van Scharrenburg, G.J.M., Puijk, W.C., de Haas, G.H., Slotboom, A.J.(1983) Eur. J. Biochem. 133, 83-89.

24. Flanders, K.C., Horowitz, E.M., Gurd, R.S., (1984) J. Biol. Chem. 259, 7031-7037.

25. Halban, P.A., Offord, R.E., (1975) Biochem. J. 151, 219-225.

26. Davies, J.G., Offord, R.E. (1985) Submitted for publication.

27. Rose, K., Gladstone, J., and Offord, R.E. (1984) Biochem. J. 220, 189-196.

28. Varley, J., Davies, J.G., Offord, R.E. Timmis, K. Paper in preparation.

SEMISYNTHESIS OF α-CHAIN OF HEMOGLOBIN S: V8-PROTEASE CATALYZED CONVERSION OF THE NON-COVALENT FRAGMENT SYSTEM TO COVALENT FORM

A. Seetharama Acharya, Shabbir A. Khan, and Ramnath Seetharam
The Rockefeller University, 1230 York Avenue, NY, NY 10021

During the course of our chemical studies on the polymerization of deoxy hemoglobin S (HbS), we wished to prepare hybrid species of HbS containing two or more mutations of the amino acid residues implicated to be part of the intermolecular contact regions. Semisynthesis appeared to be a method of choice, especially since this approach will provide the flexibility to introduce unnatural amino acids or ^{15}N or ^{13}C labelled amino acids at predetermined intermolecular contact regions.

In recent years protease catalyzed peptide bond formation has gained considerable interest in synthetic peptide chemistry. The stereoselectivity of the proteases prevents the formation of undesired products during condensation of amino acids and/or peptides even when the side chain functions are not protected. The potential of proteases to hydrolyse the pre-existing susceptible peptide bonds during fragment condensation limits the general applicability. However, an exception to this is the enzyme catalyzed reformation of peptide bonds in fragment complementing systems.[1] Native-like structure of the fragment complementing systems appears to provide the regioselectivity for the reformation of the peptide bond.

We have recently identified the region of the α-subunit of HbS corresponding to the junction of the translation

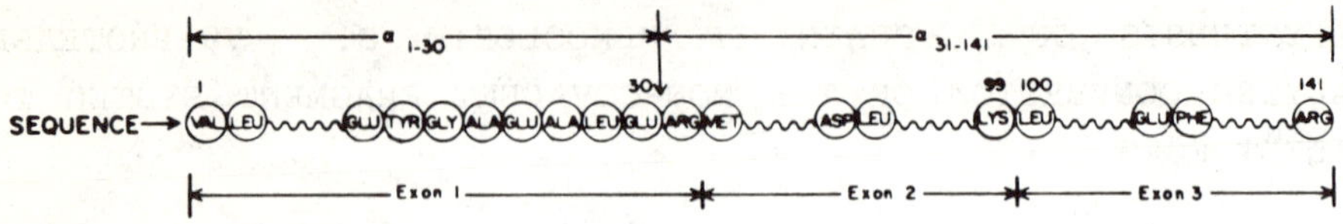

Fig. 1. Schematic representation of exon, intron junctions, and the V8-protease susceptible bond of permissible discontinuity region of α-chain.

products of Exon-1 and Exon-2 of α-globin gene as the 'permissible discontinuity region' of the polypeptide chain within its tertiary interactions (Figure 1).[2] Staphylococcus aureus V8-protease introduces a 'nick' at the peptide bond Glu(30)-Arg(31) of this region. These results prompted us to investigate the feasibility of converting the non-covalent fragment system of the α-chain to the covalent form through the V8-protease catalyzed reformation of the peptide bond Glu(30)-Arg(31).

Peptide α_{1-30} was synthesized by the solid-phase methods.[3] Starting with Boc-Glu(OBzl)-OCH_2-PAM-RESIN (0.38 mmol Glu/g resin), the peptide chain was assembled by stepwise addition of the appropriate N^{α}-Boc-amino acid derivatives. Benzyl type protecting group for the side-chain functions of Asp, Glu, Ser, and Thr, 2-ClZ for Lys, p-toluenesulfonyl for His, and formyl for Trp were used. At each step, the coupling reaction was mediated by N,N'-dicyclohexylcarbodiimide and the Boc group protection was removed with TFA-CH_2Cl_2 (1:1). ^{14}C-Leu was introduced at position 2 of the peptide to follow the semisynthesis. The fully protected peptide resin was treated with anhydrous liqiud HF/anisole (9:1, v/v) and the peptide released was purified by RPHPLC. Removal of the N^{in}-formyl group from the resulting peptide with 1 M piperidine followed by RPHPLC furnished the desired α_{1-30}.

The synthetic potential of V8-protease to generate α_{1-141} was investigated by incubating synthetic [^{14}C]-α_{1-30}

with globin derived α_{31-141} at pH 6.0 and 4 ^{o}C in the presence of n-propanol as the organic co-solvent. Semisynthesis of α-globin was followed by RPHPLC. The α_{31-141} and α_{1-141} coelute in the TFA-acetonitrile gradient conditions employed. Semisynthesis of α_{1-141} is reflected in the incorporation of the radioactivity of α_{1-30} to the protein peak. Good yields were obtained in 35% n-propanol. A time-dependent semisynthesis was observed up to about 24 h (Figure 2B), after that the yield of α_{1-141} remained nearly constant up to 72 h. In the absence of V8-protease incorporation of radioactivity into protein does not occur (Figure 2A). Tryptic peptide mapping of the semisynthetic material established the V8-protease catalyzed reformation of Glu(30)-Arg(31).

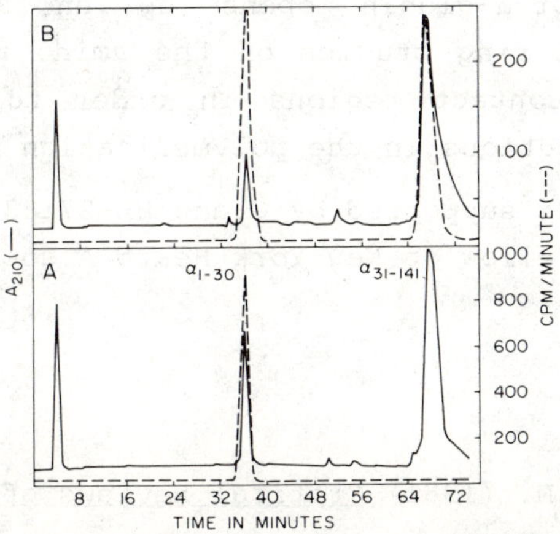

Fig. 2. Semisynthesis of α-globin. HPLC of a mixture of (^{14}C) α_{1-30} and α_{31-141} incubated without (A) and with (B) V8-protease.

On incubation of shorter amino terminal fragments α_{1-23} or α_{1-27}, both of which contain Glu as the carboxy terminal residue, with α_{31-141} under conditions used for the semisynthesis of α_{1-141}, V8-protease did not catalyse the formation of either the peptide bond Glu(23)-Arg(31) or

Glu(27)-Arg(31). This clearly reflects the high regio-selectivity in the V8-protease catalyzed reformation of peptide bonds. In order to establish whether this selectivity is a reflection of a recognition by V8-protease of a unique sequence preceeding the Glu(30), the tripeptide α_{28-30} (Ala-Leu-Glu) was incubated with α_{31-141}. The reformation of Glu(30)-Arg(31) in this mixture also could not be demonstrated. The high selectivity in the condensation of α_{1-30} with α_{31-141} suggests that native-like non-covalent interaction between the fragment α_{1-30} and α_{31-141} exists which brings out appropriate stereochemical orientation of the carboxy and amino termini of the nicked region of the globin chain, thus facilitating the V8-protease catalyzed semisynthesis of α-globin. This demonstration of semisynthesis of α-globin opens up an avenue for the molecular engineering studies of the amino acid residues of intermolecular contact regions in order to decipher their relative contributions in the polymerization process.

This work is supported by Grant HL-27183 to ASA. ASA is an Established Fellow of New York Heart Association.

References

1. Chaiken, I.M. (1981) Critical Reviews of Biochem. **11**, 255-301.

2. Seetharam, R., Dean, A., and Acharya, A.S. (1984) Fed. Proc. **43**, 1840.

3. Barany, G., and Merrifield, R.B. (1981) in The Peptides: Analysis, Synthesis, Biology. Gross, E., and Meienhofer, J. Eds., Academic Press, New York, pp. 1-284.

PEPSIN CATALYSED TRANSPEPTIDATION OF LEU-LEU-NH_2

Mary K. Lutek, Charles M. Deber and Theo Hofmann

Research Institute, Hospital for Sick Children, Toronto, M5G 1X8, and Department of Biochemistry, University of Toronto, Toronto, M5S 1A8, Ontario, Canada

Introduction

The detailed mechanism of action of acid proteases remains a subject of active investigation. Within this family of proteinases, which includes pepsin, chymosin, renin, cathepsin D and some fungal enzymes, those which have been studied have two active aspartyl residues, molecular weights of about 35,000; a specificity for cleaving peptide bonds between large hydrophobic side chains; and a common set of inhibitors. In addition, the three-dimensional structures of four of these enzymes have been demonstrated to be very similar.[1]

Porcine pepsin is known to catalyse three types of reactions. Along with hydrolysis of polypeptides, it can facilitate the transpeptidation of both N- and C-terminal amino acids of certain substrates.[2,3] Studies on the mechanism of the transpeptidation reactions have been carried out with the additional aim of extrapolating the findings to the hydrolysis reaction. ^{13}C-NMR investigations of the binding of a pepstatin analogue to pepsin[4] suggested that similar experiments with ^{13}C-enriched peptide substrates might be a fruitful approach to the problem. Studies in our laboratory using the pepsin substrate Leu-Tyr-NH_2, enriched 99% with ^{13}C at the carbonyl carbon of leucine, have enabled the detection of some intermediate products (vide infra) assumed to be common to all acyl transpeptidation reactions.[5] To demonstrate the generality of these observations, we report here parallel experiments using ^{13}C-enriched Leu-Leu-NH_2.

Results and Discussion

Leu-Leu-NH_2, labelled with ^{13}C (99%) at the carbonyl carbon of the N-terminal leucine, was synthesized by mixed anhydride coupling of ^{13}C-t-BOC-leucine and leucine amide, followed by removal of the BOC-protecting group by HCl/ethyl acetate treatment.

A ^{13}C NMR spectrum of the intermediate products of the reaction of porcine pepsin with Leu-Leu-NH_2 is shown in Figure 1. The chemical shifts of the non-substrate peaks were identical to those found when Leu-Tyr-NH_2 was incubated with pepsin under identical conditions.[5] The appearance of carbonyl resonances from leucine monomer (176.4 ppm vs. external dioxane), dimer (179.7, 170.7 ppm), and trimer (173.7 ppm, middle leucine; others presumably homosteric with Leu-Leu) is also consistent with results from other studies.[2b] Free Leu-NH_2 apparently diffuses away from the enzyme and does not reenter the reaction scheme. Resonance assignments were made by comparison with ^{13}C-enriched Leu-Leu (both Leu's labelled) and Leu-Leu-Leu (middle Leu labelled) which we synthesized for this purpose. As when Leu-Tyr-NH_2 was the substrate, *Leu-*Leu was formed in greater quantities and earlier in the reaction than free leucine, a result confirming that transpeptidation is initially a predominant reaction. Peak "X" (174.4 ppm) is the first non-substrate resonance to appear; its integrated intensity is about 10% of the total intensity of total ^{13}C-Leu resonances, a value qualitatively consistent with full occupation of available active sites under the initial reaction conditions of ca. 1/10 pepsin/substrate. It's observation is followed by the appearance of the resonance for the internal carbonyl of Leu-Leu-Leu (L-*L-L in Fig. 1). Peak "X" is then the first to disappear,

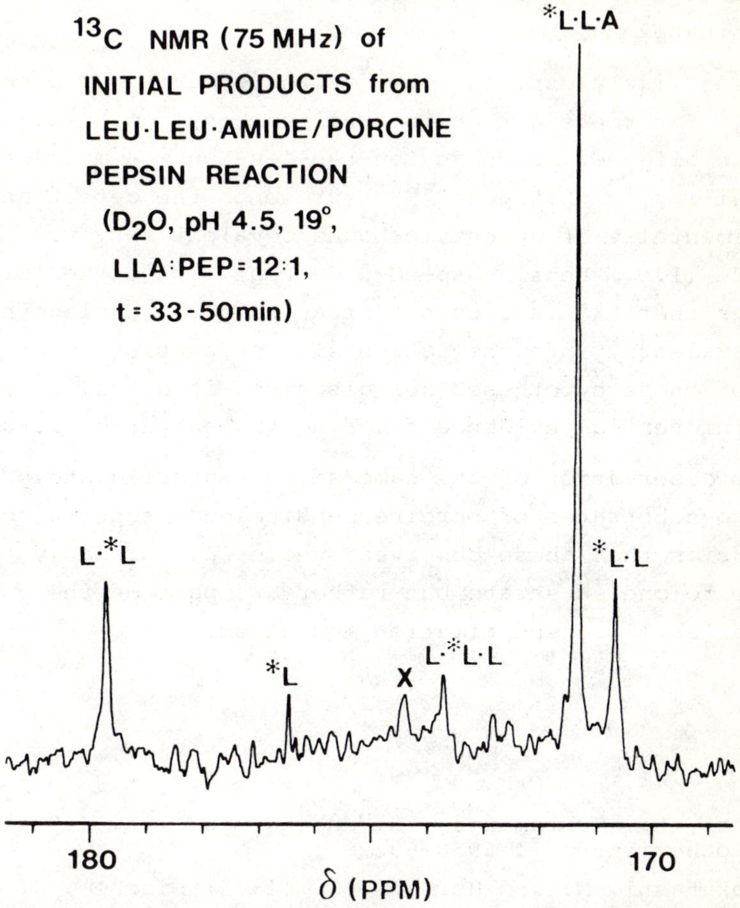

Fig. 1. ^{13}C NMR (75 MHz) spectrum of the intermediate products formed during incubation of porcine pepsin with Leu-Leu-NH_2. In the experiment shown, 10 mg (41 µmoles) Leu-Leu-NH_2 (enriched 99% with C-13 in the N-terminal leucine carbonyl carbon) is incubated with 100 mg (3.3 µmoles) porcine pepsin in 1.0 ml 99.8% D_2O at pH 4.5 and 19°C. Asterisks indicate to which C-13 labelled carbonyl carbon the peak is assigned (i.e., *L-L is the N-terminal carbonyl carbon peak of Leu-Leu) (see text). The spectrum shown is one of a series of spectra, each consisting of 2000 scans taken over a period of 17 minutes, and was recorded between t=33 and t=50 minutes after the addition of pepsin.

shortly followed by the loss of the Leu-Leu-Leu peak. Eventually the products Leu-Leu, and finally Leu monomer predominate.

Among likely explanations for the identity of peak "X" is that it may represent an internal carbonyl of a larger leucine oligomer, such as Leu-Leu-Leu-Leu. A more attractive possibility is that peak "X" represents the detection experimentally of an enzyme-bound covalent anhydride intermediate (i.e., Pepsin-Asp-CO-O-CO-Leu). A further alternative is that "X" is leucine (or an oligomer of leucine) bound non-covalently into the active site of pepsin. The merits of each of these hypotheses are discussed in detail elsewhere,[5] with inferential evidence favoring the anhydride structure.

The observation of the same set of intermediate products for two substrates of porcine pepsin lends support to the conclusion that these observed species are probably not unique to one substrate but rather are part of the enzyme's general acyl transpeptidation mechanism.

References

1. Bott, R., Subramanian, E. and Davies, D.R. (1982) Biochemistry 21, 6956-6962.

2.a. Takahashi, M. and Hofmann, T. (1975) Biochem. J. 147, 549-563.

b. Takahashi, M., Wang, T.T. and Hofmann, T. (1974) Biochem. Biophys. Res. Commun. 57, 39-46.

3.a. Neumann, H., Levin, Y., Berger, A. and Katchalski, E. (1959) Biochem. J. 73, 33-41.

b. Fruton, J.S., Fuji, S. and Knappensberger, M.H. (1961) Proc. Natl. Acad. Sci. U.S.A. (47, 759-761.

c. Silver, M.S. and Stoddard, M. (1972) Biochemistry 11, 191-200.

4. Rich, D.H., Bernatowicz, M.S. and Schmidt, P.G. (1982) J. Am. Chem. Soc. 104, 3535-3536.

5. Deber, C.M., Lutek, M.K. and Hofmann, T. (1985) submitted to J. Biol. Chem.

ON THE INDISPENSABILITY OF ARG^{38} IN COMPLEXES OF COMPLEMENTARY CYTOCHROME C FRAGMENTS

Hans A. Boots and Godefridus I. Tesser
Catholic University of Nijmegen, Department of Organic Chemistry, Toernooiveld, 6525 ED NIJMEGEN, The Netherlands

Introduction

Horse cytochrome c contains a heme residue and consists of 104 amino acids. Protection of the ε-amino functions of the 19 lysyl residues using the methylsulphonylethyloxycarbonyl (1) group followed by tryptic digestion produces only the fragments (1-38) and (39-104) (2). The obtained fragments were enzymatically and chemically processed (Scheme 1) and subsequently deprotected by a short treatment with base and then used for complexation. Each of the relevant combinations of the obtained fragments gave a tight complex, which could be purified by gelfiltration (pH 6.9).

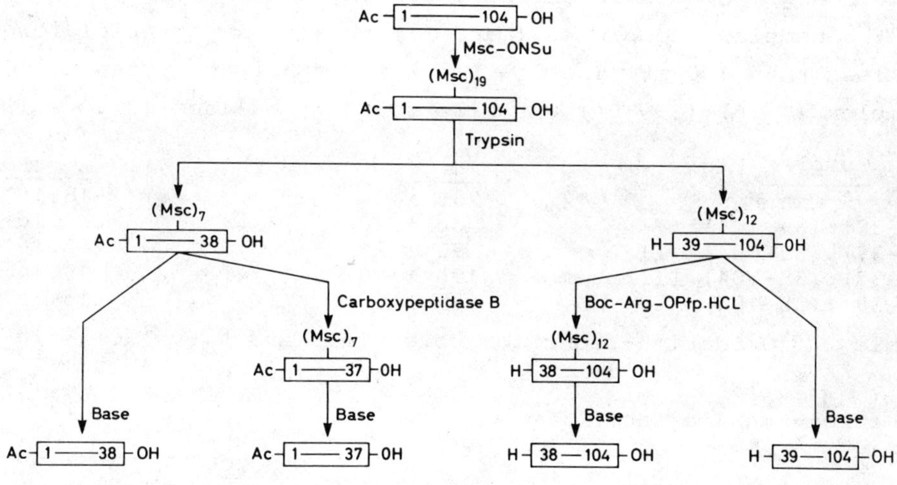

Scheme 1 Strategy for the preparation of the complexes

Results and discussion

The obtained fragments were (1-37), (1-38), (39-104) and (38-104). The former two compounds contain the heme residue, which causes the complexation formation. Spectral properties of the complexes are compared in Table 1.

		ϵ 695 nm (millimolar)	$\frac{A415.5_{red}}{A409_{ox}}$	$\frac{A550_{red}}{A520_{red}}$	$\frac{A550_{red}}{A528_{ox}}$	$\frac{A409_{ox}}{A530_{ox}}$
CYTOCHROME C		0.77	1.22	1.74	2.47	9.47
(1-38):(39-104)	I	0.62	1.21	1.62	2.45	10.99
(1-37):(39-104)	II	0.61	1.24	1.64	2.56	11.20
(1-37):(38-104)	III	0.66	1.29	1.68	2.56	10.69
(1-38):(38-104)	IV	0.52	1.22	1.64	2.48	10.66

Table 1 Comparison of spectral properties of cytochrome c and the four complexes (pH 6.9, 21 °C)

Redox potentials

If Arg^{38} occurs in the complexes including (38-104) (Complexes III and IV) the redox potential is about 190 mV (Table 2). If Arg^{38} occurs in the heme bearing fragment or is absent in the complex (Complexes I and II) the redox potential lowers to less than 100 mV. The low value for the redox potential of complex (1-38):(39-104) was also found by Wallace (3).

Complex		E_m (mV)	K_m
cytochrome c		265 ± 10	18.10^{-8}
(1-38):(39-104)	I	80 ± 20	37.10^{-8}
(1-37):(39-104)	II	90 ± 20	63.10^{-8}
(1-37):(38-104)	III	190 ± 10	18.10^{-8}
(1-38):(38-104)	IV	185 ± 10	13.10^{-8}

Table 2 Oxidation-reduction potentials and K_m-values

Electron-transfer activity

The four complexes mediated the transfer of electrons from ascorbic acid to cytochrome c oxidase and to oxygen, using tetramethylparaphenylenediamine as an intermediate electron carrier (Fig. 1). The K_m-values (Table 2) differ but not widely, reflecting the order of their redox potentials. The activity of complex II, lacking Arg^{38} is conspicuous and allows the conclusion that Arg^{38} is not necessary for electron transfer to

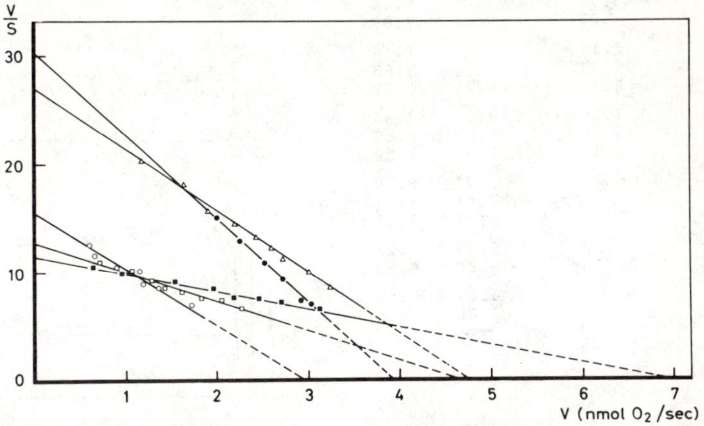

Fig. 1 Eady-Hofstee plots for the reaction of cytochrome c oxidase with cytochrome c (Δ), complex I (□), complex II (■), complex III (o) and complex IV (●)

cytochrome c oxidase. The negligible activity of cytochrome c bearing a modified Arg^{38} residue found by Wallace and Rose (4) should thus be understood as depending on their test criterion, where also the activity of the reductase is implied.

High resolution 500 MHz proton NMR-spectroscopy

From the comparison of the chemical shifts of some aromatic residues of horse cytochrome c and complexes I and II (Table 3) it is shown that the two complexes have a comparable surrounding of the heme group that differs to some extent of the native protein. The absence of the signal of Ile^{57} (Me^{δ}) at -0.4 ppm (Fig. 2) indicates the formerly present loop, comprising the fragment 38-58, to be broken.

		cytochrome c	I	II
Trp[59]	C_2	6.98	6.99	6.98
	C_4	7.59	7.60	7.66
	C_5	6.68	6.64	6.70
	C_6	5.73	6.06	6.14
	C_7	7.07	7.14	7.09
Phe[10]	o	7.11	7.13	7.14
	m	6.71	6.71	6.68
	p	6.27	6.21	6.23
Phe[36]	o	7.39	7.36	7.36
	m	6.89	6.88	6.85
	p	7.11	7.16	7.20
Phe[82]	o	6.71	6.71	6.70
	m	7.40	7.41	7.41
	p	7.23	7.24	7.27
Tyr[74]		7.22	7.19	7.23
		6.65	6.80	6.82
Tyr[67]		-	6.64	6.55
		-	5.06	5.05

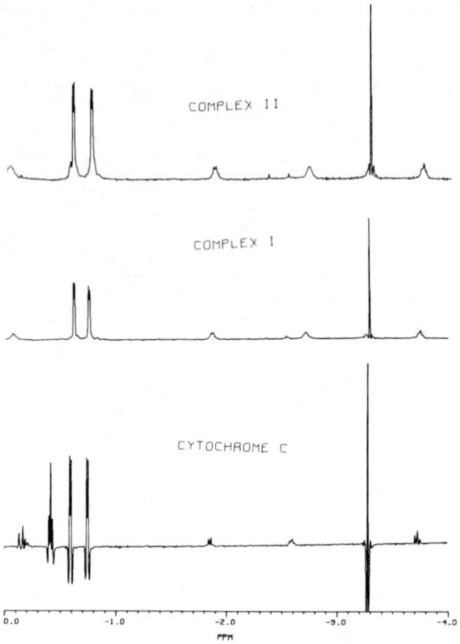

Table 3 Comparison of NMR-spectra of cytochrome c and the
Fig. 2 complexes I and II (at pH 6.9 and at 40 °C)

Acknowledgement

This investigation has been carried out under the auspices of the Netherlands Foundation for Chemical Research (S.O.N.) and with financial aid from the Netherlands Organization for the Advancement of Pure Research (Z.W.O.).

References

1. Boon, P.J. and Tesser, G.I. (1985) Int. J. Peptide Protein Res. 25, 510-516.
2. Westerhuis, L.W., Tesser, G.I. and Nivard, R.J.F. (1979) Recl. Trav. Chim. Pays-Bas 98, 109-112.
3. Wallace, C.J.A. (1984) Biochem. J. 217, 595-599.
4. Wallace, C.J.A. and Rose, K. (1983) Biochem. J. 215, 651-658.

STUDY OF THE LIPID-BINDING DOMAIN OF PHOSPHOLIPASE A_2 USING SEMISYNTHESIS AND CHEMICAL MODIFICATION

A.J. Slotboom, G.J.M. van Scharrenburg, F. van der Wiele, W. Atsma and G.H. de Haas
State University of Utrecht, Laboratory of Biochemistry
Padualaan 8, 3584 CH Utrecht, The Netherlands

Introduction

Pancreatic phospholipase A_2[1] (M_r 14,000; PLA) has its highest catalytic activity when present in high-molecular weight enzyme-lipid complexes ($M_r \approx 75,000$). The amino acid residues directly involved in lipid binding are located in the N-terminal region, the regions around residues 20, 31 and 69, and the C-terminal region (Figure 1). These residues form a hydrophobic edge, the lipid-binding domain, around the active site cavity. This paper describes the effects of chemical modification and amino acid substitutions in the lipid-binding domain

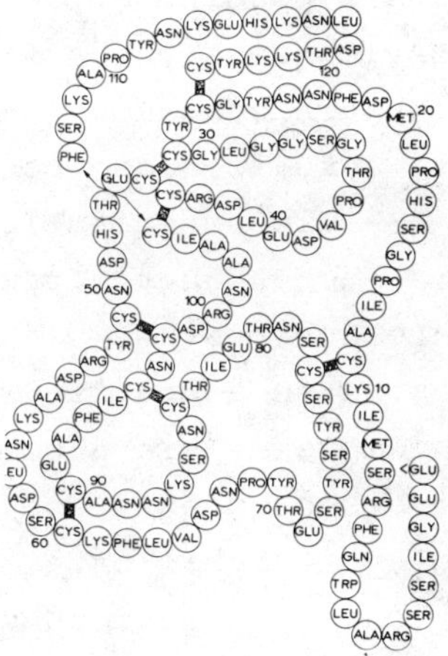

Fig. 1. Primary structure of porcine pancreatic prophospholipase A_2. Limited proteolysis liberates the active PLA by cleavage of the N-terminal heptapeptide.

on PLA-lipid complex formation and catalytic activity.

Results and Discussion

(i) Preparation and properties of phospholipase A_2 mutants

In porcine PLA Arg^6 was substituted for Glu and in the bovine PLA Asn^6 by Arg. The required ε-amidinated porcine des(Ala^1-Ser^7)- and bovine des(Ala^1-Gly^7)-PLA fragments were prepared as described previously[2]. For the porcine PLA analogs the peptides t-Boc.A.L.W(For).Q.F.R.A, and t-Boc.A.L.W.-Q.F.E.A, and for the bovine PLA analog the peptide t-Boc.A.L.-W(For).Q.F.R.G were prepared by SPPS[2]. After purification by reversed-phase HPLC, the peptides were covalently coupled to the respective PLA fragments using the mixed carbonic anhydride procedure. Subsequent removal of protecting groups and purification gave the desired mutants.

Using neutral micellar L-dioctanoyllecithin as substrate an almost 3-fold increase of the V_{max} value was observed upon substitution of Asn^6 by Arg in the bovine PLA. In contrast, a 2-fold decrease of the V_{max} value was found for the substitution of Arg^6 by Glu in the porcine PLA. The $Asn^6 \longrightarrow$ Arg substitution in bovine PLA improves the affinity for binding of the neutral micellar substrate analog (n-octadecenylphosphocholine) 13-fold, while a 10-fold decrease in affinity was observed for the $Arg^6 \longrightarrow$ Glu substitution in porcine PLA.

Pancreatic PLA has a much higher affinity for acidic than for neutral zwitterionic substrates. In contrast to neutral substrates, acidic substrates aggregate with pancreatic PLA to high molecular weight complexes via a comicellization pathway already below the cmc[3]. Due to this complex formation the enzyme is able to display its full catalytic activity also at submicellar concentrations of acidic substrate. The concentrations of the acidic substrate analog 1,2-diheptylcarbamoylglycero-3-sulfate (cmc 1.6 mM) required for aggregation to high molecular weight complexes of native porcine, Glu-6 porcine, native bovine and Arg^6 bovine PLAs are 130, 550, 300 and 840 μM, respectively. Due to these differences, the separation of a mixture of native and Glu^6 porcine PLAs on a Sephadex G-100 column equilibrated with 130 μM of the substrate analog, is possible. Similarly, a mixture of native and Arg^6 bovine PLAs can be separated at 300 μM substrate analog. Native porcine and Arg^6 bovine PLA elute at a position in accordance with the high molecular weight complex of 75,000, while Glu^6 porcine and native bovine elute at a position corresponding to their monomeric molecular weights of 14,000. It can thus be concluded that the presence of a positively charged residue at position 6 is favourable for the formation of high molecular weight protein-lipid complexes with optimal catalytic activity.

(ii) Preparation and properties of Lys-116-ε-N-palmitoyl phospholipase A_2

Reaction of methylacetimidate with porcine pancreatic prophospholipase A_2 (Prec) bound to micelles of the negatively charged substrate 1,2-dioctanoylglycero-3-sulfate results in amidination of 8 out of the 9 Lys residues. In the resulting protein (Amprec-8) the Lys residue protected was identified to be predominantly Lys-116. Reaction of Amprec-8 with N-hydroxysuccinimide ester of palmitic acid gave Lys-116-ε-N-palmitoyl Amprec-8 (Pal-Amprec-8), which upon tryptic activation gave Pal-Ampa-8. This enzyme has a V_{max} value of 1450 μequiv min^{-1} mg^{-1} using L-dioctanoyllecithin micelles as compared to 1400 and 2170 for Ampa-9 (fully ε-amidinated PLA) and Ampa-8, respectively. Phospholipase A_2 preferentially hydrolyzes substrates present at organized lipid-H_2O interfaces. In general, the susceptibility of substrates to the action of PLA is highly dependent on the packing density of the substrate molecules. Therefore, the use of the monomolecular surface layer technique is required, allowing the kinetic analysis of the hydrolysis of substrate spread at the air-H_2O interface at different packing densities. Porcine pancreatic PLA can easily penetrate monolayer of L-didecanoyllecithin up to ≈10 dynes/

cm (Figure 2). The covalent introduction of the palmitoyl moiety in Ampa-8 was found to have a dramatic effect on the penetration power: Pal-Ampa-8 can penetrate monolayers up to ≈ 32 dynes/cm (Figure 2). As a matter of fact, it was found that even long chain lecithin present in the outer membrane of the intact human erythrocyte cell is hydrolyzed by Pal-Ampa-8 at a much higher rate and with even greater efficiency than by the best penetrating snake venom PLA (*Naja naja*). Concomitantly with this degradation all cells had turned echinocytic. It has to be mentioned that under the same conditions Ampa-8 or native pancreatic PLA have no such effects at all. Clearly the introduction of the palmitoyl moiety tremendously improves the penetrating capacity of the pancreatic PLA and transforms it into a "membrane bound"-like enzyme.

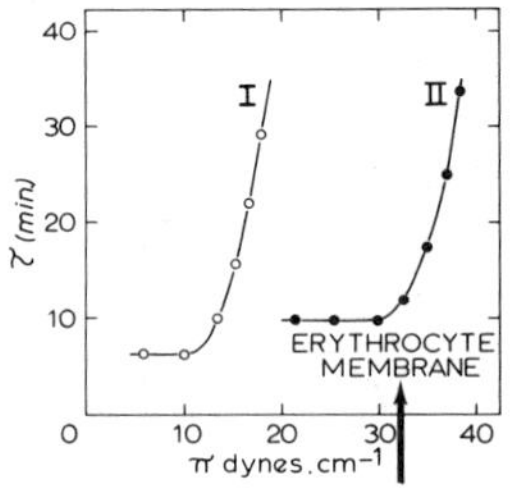

Fig. 2. Influence of surface pressure of a didecanoyllecithin monolayer on the induction time (τ) of the kinetics of hydrolysis by Ampa-9 (I) and Pal-Ampa-8 (II) at pH 8.0.

References

1. Slotboom, A.J., Verheij, H.M., and de Haas, G.H. (1982) in Phospholipids: New Biochemistry, Hawthorne, J.N. and Ansell, G.B., Eds., Elsevier, Amsterdam, Vol. 4, pp. 359-434.

2. van Scharrenburg, G.J.M., Puijk, W.C., Seeger, P.R., de Haas, G.H., and Slotboom, A.J. (1984) Biochemistry 23, 1256-1263.

3. Hille, J.D.R., Egmond, M.R., Dijkman, R., van Oort, M.G., Jirgensons, B., and de Haas, G.H. (1983) Biochemistry 22, 5347-5353.

REVERSIBILITY OF TRYPTIC HYDROLYSIS IN A SERIES OF DISULFIDE LOOP SUBSTRATES

John D. Glass, Irving L. Schwartz, Susan Fairman, Suseann Forbes, Chandra S. Pande, and Angeliki Buku

Center for Polypeptide and Membrane Research and Department of Physiology and Biophysics, Mount Sinai School of Medicine, 5th Avenue at 100th Street New York, New York 10029

Introduction

The activation of human plasminogen(s) occurs when a single Arg-Val bond joining residues 560 and 561 (Glu-plasminogen numbering) is cleaved by the specific actions of urokinase, tissue plasminogen activators, or certain other specific proteases[1]. This peptide bond lies within a small disulfide loop joining the side chains of Cys residues in positions 557 and 565 of the plasminogen sequence, so there is no activation fragment released upon hydrolysis of the bond. Reasoning that conformational constraints in the disulfide loop might explain the specificity of activating enzymes for this one arginyl bond among the many arginyl bonds in plasminogen, Ganu and Shaw [2] synthesized a nonapeptide amide disulfide loop corresponding to the sequence of the human plasminogen activation loop (Fig.1, Peptide I).

CH_2-CHX-Pro-Gly-Arg-Y-Gly-Gly-Cys-NH_2
|
S ——————————————————┘

Peptide I X = NH_2, Y = Val-Val

Peptide II X = H , Y = Val-Val

Peptide III X = H , Y = Val

Peptide IV X = H , Y = Val-Val-Val

Figure 1. Plasminogen Activation Loop Analogs.

The peptide proved to be a substrate of neither urokinase nor tissue plasminogen activator, but it is an excellent substrate of trypsin. We have shown earlier [3] that tryptic hydrolysis of the Arg-Val bond in this peptide is readily reversible in the presence of high concentrations of 1,4-butanediol so long as the disulfide bond in the peptide is intact. Reduction of the disulfide bond renders tryptic hydrolysis of the Arg-Val bond irreversible as far as determined by the chromatographic assay being used. These results were expected in a qualitative sense, but the speed and efficiency of the intramolecular condensation were notable. These results raised questions with ramifications ranging from the control of fibrinolysis to the prospects for a broadly applicable system of protease-catalyzed peptide fragment coupling.

Results and Discussion

We have used solid phase methods to prepare larger quantities of the human plasminogen activation loop peptide as well as a series of its analogs in which the terminal amino group is formally replaced by a hydrogen atom and in which the native Val-Val sequence is either retained, replaced by a single Val residue, or replaced by a Val-Val-Val sequence. Simple isocratic HPLC systems were worked out for rapid separations of each cyclic peptide and corresponding acyclic hydrolysis product. These chromatographic systems were calibrated with the hydrolyzed and unhydrolyzed peptides so that reaction mixtures could be quickly analyzed and the relative concentrations of the hydrolyzed and unhydrolyzed peptides could be measured. For example, buffered solutions of trypsin containing varying percentages of 1,4-butanediol were added to small aliquots of lyophilized peptide and small samples of the reaction mixtures were then analyzed on calibrated C-18 reversed phase HPLC columns developed isocratically with appropriate mixtures of CH_3CN / 0.1% aqueous trifluoroacetic acid. In this way it was possible to measure the equilibrium ratios of each cyclic peptide and its tryptic hydrolysis product in different concentrations of 1,4-butanediol, as a function of pH, and as a function of substrate structure. In most instances these measurements have been made by following the tryptic reactions to equilibrium in both the hydrolytic and condensation directions. The results are recorded in the following tables.

Table I. Equilibrium Percentage of Peptide in the Condensed (Unhydrolyzed) Form at pH 7 in the Presence of Trypsin and Varying Concentrations of 1,4-Butanediol.[a]

	Butanediol (v/v) %			
	60	70	80	90
Peptide I	15(11)	21(17)	35(34)	60(64)
Peptide II	20(15)	34(29)	49(46)	74(67)
Peptide III	39	55	67	92
Peptide IV	10(16)	19(22)	36(44)	64(68)

[a]The numbers outside the parentheses were measured in the hydrolytic direction (ie approaching equilibrium from unhydrolyzed peptide) and the numbers within the parentheses were measured approaching equilibrium from the hydrolyzed peptide.

Table II. Equilibrium Percentages of Peptide I in the Condensed Form in the Presence of 85% (v/v) Butanediol and Buffers of Varied pH[a]

pH 6	pH 7	pH 8
40	42	21

[a]Measurements were made in the hydrolytic direction, starting with unhydrolyzed peptide.

The reversible enzymatic hydrolysis of the Arg-Val bond in these plasminogen activation loop analogs may have important implications for two rather different areas of research. On the one hand, it is of some interest to know whether or not the activation of plasminogen itself is reversible. If the zymogen activation is reversible it will be of some importance to determine whether or not reversible activation of the zymogen occurs to a significant extent under physiological conditions, and perhaps as part of a physiological control mechanism.

On the other hand, the plasminogen activation loop peptides are probably the simplest models of intramolecular enzymatic peptide bond formation which have so far been reported. A systematic structure-activity study in this family of peptides is technically feasible and should provide insight into the factors which control the kinetics and thermodynamics of intramolecular and intracomplex fragment coupling reactions. Such information will be invaluable in designing broadly applicable methods of semisynthesis based upon the enzymatic coupling of peptide fragments.

Acknowledgments

This work was supported by NIH grants GM 18752 and AM 10080.

References

1. Sottrup-Jensen, L., Zajdel, M., Claeys, H., Petersen, T.E., and Magnusson, S.(1975) Proc. Nat. Acad. Sci. USA 72, 2577-2581.
2. Ganu, V.S. and Shaw, E. (1982) Int. J. Pept. Prot. Res. 20, 421-428.
3. Ganu, V.S. and Glass, J. D. (1983) in Peptides 1982, Blaha, K. and Malon, P., Eds. Walter de Gruyter, Berlin, 403-406.

ENZYMATIC SYNTHESIS OF ARGININE PROLINE PEPTIDE BONDS USING CLOSTRIPAIN AS A CATALYST

Anders J. Andersen
Carlsberg Biotechnology Ltd., Tagensvej 16, DK-2200 Copenhagen N, Denmark

Introduction

Since the application of using proteases as catalysts in synthesis of peptide bonds for the first time was demonstrated in 1937 (1) several reports have been published on the subject. Specially within the last decade an increasing interest has been paid to the field and a great number of different bond formations between different amino acids have been demonstrated to take place (2, 3, 4, 5). Thus, protases are now used to prepare peptides in multigram amounts both in stepwise synthesis as well as in fragment couplings.

However, until now protease catalyzed formation of peptide bonds in which proline donates its imino group have not been demonstrated in spite of several attempts (6, 7, 8, 9, 10).

We have recently investigated the application of the arginine specific protease Clostripain from Clostridium Histolyticum for its potential in peptide synthesis. During this work we discovered that this enzyme is an effective catalyst for the formation of arginyl-proline peptide bonds. Thus, using the substrate Z-arg-OMe and pro-R^1 as peptide nucleophiles high yields of Z-arg-pro-R^1 were obtained. With substrates at a concentration of 0,05-0,9 M and proline derivatives or proline dipeptides at 1 M and clostripain concentrations of 10^{-6} M, yields of 50-90% have been obtained.

These results demonstrate for the first time that proteases can catalyze the synthesis of amino acid-proline peptide bonds. The present studies suggest that clostripain can be used both in stepwise synthesis as well as in fragment condensations. Furthermore, the studies suggest that other, not yet investigated proteolytic enzymes might also catalyze similar peptide bond formations.

Materials and Methods

Z-arg-OH was purchased for Bachem, Switzerland and esterified by the HCl-MeOH method. Clostripain was obtained from Sigma Chemical Company, St. Louis, USA. All other amino acid- and peptide derivatives have been prepared by standard procedures. Synthesis reactions were performed in a pH-stat and followed by HPLC, using reverse-phase C_{18} columns and TEAP/CH_3CN mixtures as eluant. Yields are determined by HPLC using the Z-group as chromophore at 254 nm.

Results and Discussion

The data in Table I show that coupling between Z-arg-OMe and different proline dipeptides proceeds in remarkable high yields. Between the five different nucleophiles tested, it is only the one with leucine in the C-terminal P_2 position that differs substantially with respect to the yield. Whether this is due to the hydrofobic nature of leucine or the ethylester group - or both - have not yet been investegated. The lower concentration of pro-leu-OEt under the reaction does not account for the lower yield because lowering the concentrations of e.g. pro-gly-NH_2 to 0.5 M does not effect the yield. It should be noted from Table I that no effect on coupling yield is seen between using respectively pro-gly-NH_2 and pro-gly-OH. This result is in contrast to our experience so far. Normally the effect of replacing an amino acid amide in the C-terminal position of a nucleophile results in a lower yield.

TABLE I: CLOSTRIPAIN CATALYZED PEPTIDE SYNTEHSIS USING Z-arg-OMe AS SUBSTRATE AND PRO-R[1] AS NUCLEOPHILES.

Amine components	product	yields %
pro-NH_2	Z-arg-pro-NH_2	86
pro-gly-OH	Z-arg-pro-gly-OH	85
pro-gly-NH_2	Z-arg-pro-gly-NH_2	89
pro-leu-OEt*	Z-arg-pro-leu-OEt	62
pro-NH-Et	Z-arg-pro-NH-Et	84

Substrate: 100 mM, Nucleophile: 1 M, Clostripain: 0,5-3 μM, pH = 8.5, 0-10% MeOH, *pro-leu-OEt: 0,5 M

From Table II it can be seen that the substrate concentration can be raised to near the same level as the nucleophile without dramatic effect on the yield. This might be of importance in special cases of fragment condensations, where the nucleophile is the more expensive and therefore an optimal utilisation is desired.

TABLE II: CLOSTRIPAIN CATALYZED SYNTHESIS OF Z-arg-pro-gly-NH_2 USING DIFFERENT CONCENTRATIONS OF Z-arg-OMe AND pro-gly-NH_2 AS NUCLEOPHILE

Concentrations of Z-arg-OMe	Concentrations of pro-gly-NH_2	Ratio nucleophile/substrate	Yield %
0.05 M	1.0 M	20	89
0.1 M	1.0 M	10	89
0.2 M	1.0 M	5	88
0.5 M	1.0 M	2	79
0.38 M	0.5 M	1.3	72

Clostripain: 0,5-3 μM, pH = 8.5

Semilar couplings, as shown in Table I were studied using trypsin and the arginine specific enzyme trombin. In all reactions no synthesis of peptide bonds were observed.
At the moment systematic investigations of the amino acid ester hydrolysis specificity of clostripain is performed in our laboratory to find out whether this enzyme, like e.g. trypsin (11), can be used in coupling procedures - step wise as well as in fragment condensations - using substrates with C-terminal amino acids other than arginine.

Acknowledgement

The excellent technical assistent of Anette Møller is gratefully acknowledged.

References

1. Bergman, M: Traenkel-Conrat, H.: J.Biol.Chem., 124, 1-6 (1938)
2. Fruton, S.: Adv. Enzymol. (ed. A. Meister), vol. 53, p. 239. Wiley, New York (1982)
3. Morihara, K.: Prot.Nucl.Acid, Enz. 26, 1975 (1981)
4. Widmer, F., Breddam, K. and Johansen, J.T., Carlsberg Res. Commun. 46, 96 (1981)
5. Breddam, K., Widmer, F. and Johansen, J.T., Carlsberg Res. Commun. 48, 231 (1983)
6. Widmer, F., Breddam, K. and Johansen, J.T., Carlsberg Res. Commun. 45, 453 (1980)
7. Oka, T. and Morihara K., J.Biochem, 88, 807 (1980)
8. Oka, T. and Morihara K., J.Biochem, 82, 1055 (1977)
9. Oka, T. and Morihara K., J.Biochem, 84, 1277 (1978)
10. Morihara K. and Oka, T., Biochem,J., 163, 531 (1977)
11. Chaiken, I.M., Komorya, A., Ohno, M. and Widmer, F., F.Appl.Biochem.Biotechnol., 7, 385 (1982)

ENZYMATIC SYNTHESIS OF D-AMINO ACID CONTAINING PEPTIDES

Pia Thorbek and Fred Widmer
Carlsberg Biotechnology Ltd., Tagensvej 16, DK-2200 Copenhagen N, Denmark

Introduction

While the high substrate stereospecificity of proteases in hydrolytic reactions is well established, relatively little is known about the nucleophile stereospecificity of these enzymes in peptide synthesis, i.e. about their ability to couple D-amino acid derivatives and D-amino acid containing peptides. Earlier reports indicated that some enzymes, e.g. CPD-Y, did not couple D-nucleophiles (1,2), while others, e.g. trypsin and chymotrypsin, did so, although at an unpractically low yield (3,4). Recently, however, it was reported that chymotrypsin catalyzed the coupling of D-$LeuNH_2$ ($leuNH_2$*) to ZTyrOMe in over 90% yield (5) using high concentration of reactants. The preparative, enzymatic synthesis of D-amino acid containing peptides appeared thus to be practically feasable for at least some cases. We therefore decided to establish the generality of this observation by testing - on several enzymes - a series of D-amino acid amides and esters as well as some di- and tripeptide amides with D-residues located in different positions.

Materials and Methods

The optical purity of the nucleophiles, being absolutely critical in these experiments, was $\geq$ 98%, as determined (6) by RP-HPLC on Vydac columns (RP-18;5µm;4,5x250mm) using TEAP/CH_3CN buffer. Amino acid amides and esters were derivatized into diastereomeric compounds before analysis. Reactions were

* Abbreviations: Amino acids abbreviated in upper case refer to the L-enantiomer and those in lower case refer to the D-enantiomer. T: trypsin; CT: chymotrypsin; P: papain; PPSE: postproline specific endoprotease; E: elastase; V8: S.aur.V8 protease.

followed and yields determined by AAA and HPLC using the "common-chromophore" method (7).

Results and Discussion

The data in Table I on the chymotryptic coupling yields with L- and D-amino acid amides and esters show that this enzyme can be used to prepare L/D peptides as had been shown earlier (3-5). In all instances, the D-isomer coupled in a lower yield than the corresponding L-isomer at equal concentration. The differences are strongly side chain dependent: 11 and 77% for 0.1 M $serNH_2/SerNH_2$, or 6 and 72% for 0.1 M argOEt/ArgOEt, versus 75 and 89% for 0.1 M $pheNH_2/PheNH_2$. Increasing the concentration of the D-isomers resulted in an increased yield, occasionally reaching the level obtained with the L-isomer.

Results in Table II on the coupling yields of L- and D-$LeuNH_2$ with various enzymes and substrates demonstrate that chymotrypsin is no exception among the serine- and thiol endoproteases in its ability to couple D-$LeuNH_2$. The serine carboxypeptidases CPD-Y and CPD-W, however, do not couple D-$LeuNH_2$. This confirms our earlier observations on the high nucleophile stereospecificity of CPD-Y (1,2).

Using diastereomers of di- and tripeptides as nucleophiles (Table III and ref. 6), we found that the yields with the endoproteases depend strongly on the position of the D-residue in the peptide. For peptides having the D-amino acid in the N-terminal P_1' position, no synthesis - even at higher concentrations than shown in Table III - has been seen so far. For di- and tripeptides, having the D-residue in the C-terminal P_2' or P_3' positions, the yields were acceptable (6) and often equal to those for the all L-isomer. For the tripeptide Leu-$pheLeuNH_2$ which has the D-residue in the middle P_2' position the coupling yields were very low. It will be interesting to see how longer peptides with D-residues in the middle positions will couple.

In summary, our experiments show that the lack of total nucleophile stereospecificity reported for CT and T (3-5) is a

TABLE I: SIDE-CHAIN AND CONCENTRATION DEPENDENCE OF YIELDS WITH L- AND D-NUCLEOPHILES IN CHYMOTRYPTIC COUPLING USING AcTyrOEt AS SUBSTRATE

NUCLEO-PHILE	PRODUCT	% YIELD AT THREE NUCL.-CONCENTR. 0.1 M	0.25 M	1.0 M
SerNH$_2$	AcTyrSerNH$_2$	77	-	-
serNH$_2$	AcTyrserNH$_2$	11	29	51
AlaNH$_2$	AcTyrAlaNH$_2$	79	-	-
alaNH$_2$	AcTyralaNH$_2$	13	22	38
PheNH$_2$	AcTyrPheNH$_2$	89	-	-
pheNH$_2$	AcTyrpheNH$_2$	75	88	-
LeuNH$_2$	AcTyrLeuNH$_2$	88	-	-
leuNH$_2$	AcTyrleuNH$_2$	40	54	70
LeuOMe	AcTyrLeuOMe	21	42	69
leuOMe	AcTyrleuOMe	7	18	41
ArgOEt	AcTyrArgOEt	72	78	-
argOEt	AcTyrargOEt	6	10	17

Substrate:10mM;CT:0.5μM;pH8.5;20%EtOH

TABLE II: COMPARISON OF COUPLING YIELDS OF L- AND D-LeuNH$_2$ WITH VARIOUS ENZYMES AND SUBSTRATES

ENZYME	SUBSTRATE	% YIELD LeuNH$_2$	leuNH$_2$
Carboxypeptidase Y	BzTyrSerOEt	86	0
Carboxypeptidase W	AcTyrOEt	91	0
Chymotrypsin	AcTyrOEt	88	40
Elastase	BzGlyCys(Acm)MetOMe	78	12
Papain	BzTyrSerOEt	97	4
Postpro.Spec.Endopr.	AcTyrProOBzl	50	20
S.Aur.V8 Protease	BzPheLysSerLeuAspOEt	83	34
Trypsin	BzArgOEt	74	11

Substr.:10mM;LeuNH$_2$/leuNH$_2$:100mM;Enz.:0.5-5μM;pH9.0;DMF:5-20%

TABLE III: COMPARISON OF COUPLING YIELDS OF DIASTEREOMERIC PEPTIDES WITH VARIOUS SUBSTRATES AND ENZYMES

EN-ZYME	SUBSTRATE	NUCLEO-PHILE	PRODUCT	YIELD (%)
T	BzArgOEt	Cys(Acm)TyrOEt	BzArgCys(Acm)TyrOEt	74
		cys(Acm)TyrOEt	BzArgcys(Acm)TyrOEt	0
		Cys(Acm)tyrOEt	BzArgCys(Acm)tyrOEt	84
PPSE	AcTyrProOBzl	LeuAlaNH$_2$	AcTyrProLeuAlaNH$_2$	24
		leuAlaNH$_2$	AcTyrProleuAlaNH$_2$	0
		LeualaNH$_2$	AcTyrProLeualaNH$_2$	34
P	BzTyrSerOEt	LeuPheLeuNH$_2$	BzTyrSerLeuPheLeuNH$_2$	65
		leuPheLeuNH$_2$	BzTyrSerleuPheLeuNH$_2$	0
		LeupheLeuNH$_2$	BzTyrSerLeupheLeuNH$_2$	3
		LeuPheleuNH$_2$	BzTyrSerLeuPheleuNH$_2$	60
CT	AcTyrOEt	LeuPheLeuNH$_2$	AcTyrLeuPheLeuNH$_2$	76
		leuPheLeuNH$_2$	AcTyrleuPheLeuNH$_2$	0
		LeupheLeuNH$_2$	AcTyrLeupheLeuNH$_2$	5
		LeuPheleuNH$_2$	AcTyrLeuPheleuNH$_2$	79
		{LeuPheLeuNH$_2$ (9)	AcTyrLeuPheLeuNH$_2$	73
		{LeupheLeuNH$_2$ (1)	AcTyrLeupheLeuNH$_2$	0

Substr.:10mM;Nucl.:100mM;Enzymes:0.5-5.0μM;pH9.0;DMF:10-40%

general property of serine- and thiol endoproteases. These enzymes can therefore be used in coupling reactions with D-amino acid derivatives and di- and tripeptides with D-residues at the C-terminal. For enzymatic synthesis of D-amino acid containing peptides it is important to use optically highly pure D-isomers since the L-isomers will generally couple better than the D-isomers.

The isomeric peptides should in principle be hydrolyzed by the enzymes used for their formation (we found that $ZTyrleuNH_2$ is slowly cleaved by CT). It should thus be possible to predict, from the available literature (e.g. 8 and 9) on the hydrolysis of D-amino acid containing peptides, the ability of enzymes to couple D-components.

The results shown here further indicate that for syntheses where the all-L peptides are desired, enzymic coupling of chemically made fragments (10) will result in a stereochemically improved product, because nucleophiles with a D-residue in P_1', or in P_2' of tripeptides, do not couple at all, or at much lower yields than the all-L isomers. Such D-isomers will thus - particularly if present as impurities at low concentrations - be partially or fully eliminated.

References

1. Widmer, F., Breddam, K. and Johansen, J. T. (1981) in: Peptides 1980, Brunfeldt, K., Ed., Copenhagen, pp. 46-55.
2. Widmer, F., Breddam, K. and Johansen, J. T. (1981), Carlsberg Res. Commun. 46, 97-106.
3. Oka, T. and Morihara, K. (1977) J. Biochem. 82, 1055-1062.
4. Morihara, K. and Oka, T. (1977) Biochem. J. 163, 531-542.
5. Petkov, D. D. and Stoineva, I. B. (1984) in: Peptides 1984, Ragnarsson, U., Ed., Almquist, Stockholm, pp. 205-208.
6. Thorbek, P. and Widmer, F., in preparation.
7. Widmer, F. and Johansen, J. T. (1979) Carlsberg Res. Commun. 44, 37-46.
8. Morihara, K., Oka, T., and Tsuzuki H. (1969) Biochem. Biophys. Res. Commun. 35, 210-214.
9. Bizzozeco, S. A., Baumann, W. K. and Dutler, H. (1982) Eur. J. Biochem. 122, 251-258.
10. Widmer, F., Bayne, S., Houen, G., Moss, B. A., Rigby, R. D, Whittaker, R. G. and Johansen, J. T. (1984) in: Peptides 1984, Ragnarsson, U., Ed., Stockholm, pp.193-200.

STRUCTURAL FEATURES OF SYNTHETIC PEPTIDE SUBSTRATES FOR TRANSGLUTAMINASES

J.J. Gorman
CSIRO, ANAHL, P.O. Bag 24, Geelong, Vic. 3220. AUSTRALIA

J.E. Folk
Enzyme Chemistry Section, NIDR, NIH. Bethesda M.D. 20205, USA.

Introduction

Transglutaminase enzymes catalyse the aminolysis of peptide bound glutamine residues (equation 1).

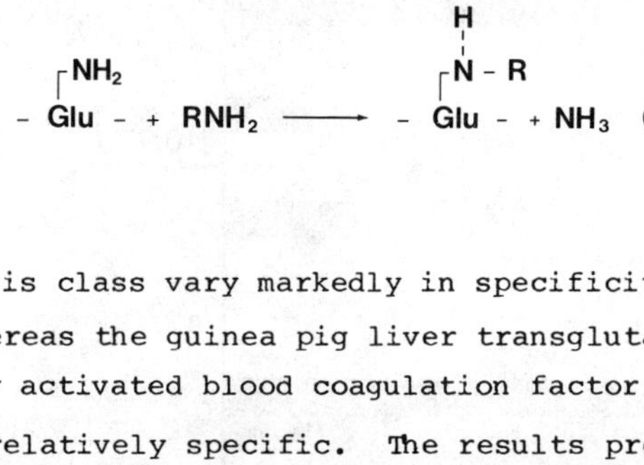

(1)

Enzymes of this class vary markedly in specificity for glutamine substrates.[1] Whereas the guinea pig liver transglutaminase has a fairly broad specificity activated blood coagulation factor XIII (plasma transglutaminase) is relatively specific. The results presented here eminate from studies designed to define the structural features which influence this specificity.[2,3]

Results and Discussion

Glutamine 167 of β-casein is particularly susceptible to factor XIIIa. Structural features adjacent to glutamine 167 which may promote this specificity have been examined. Analogues of the 15-peptide homologous to the sequence of 161 to 175 of β-casein,

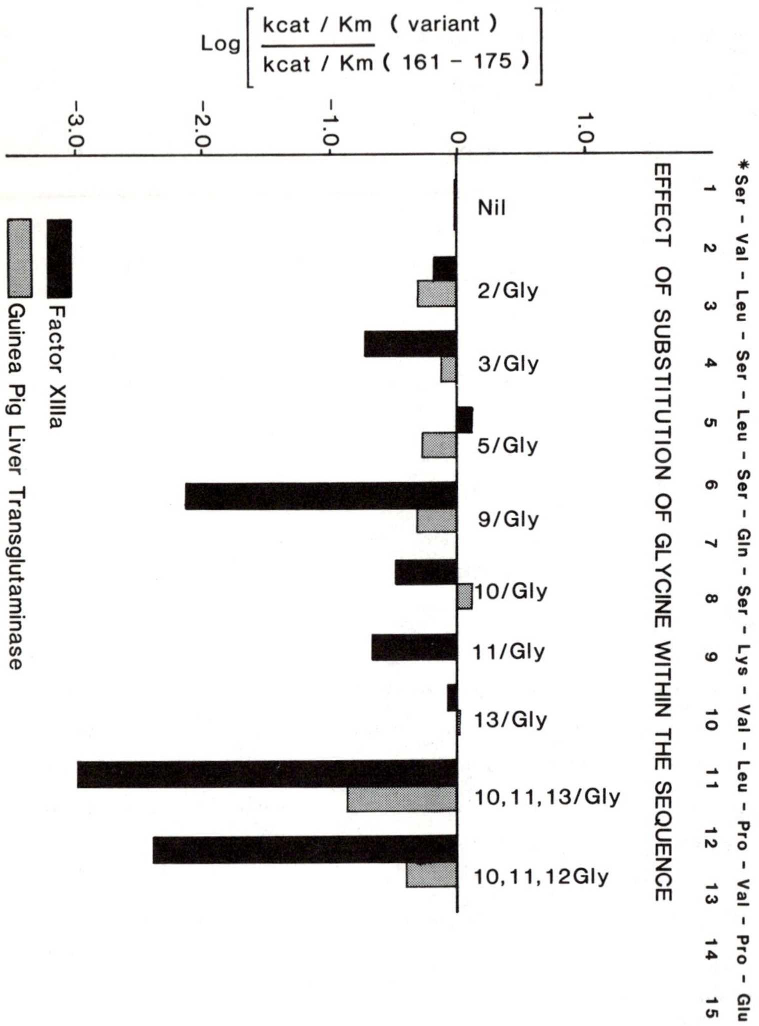

* This peptide corresponds to residues 161–175 of β–Casein

Fig 1

1	2	3	4	5	6	7	8	9	10	11	12	13	14	15
Ser	Val	Leu	Ser	Leu	Ser	Gln	Ser	Lys	Val	Leu	Pro	Val	Pro	Glu

were synthesized using glycine replacements (Figure 1). Replacements of Lys (9), Val (10) or Leu (11) substantially reduced the catalytic efficiency of the sequence for the plasma enzyme. Although the native 15-peptide was equivalent as a substrate for the guinea pig liver transglutaminase and plasma enzyme, replacements of residues 9, 10 or 11 were relatively inconsequential to the liver enzyme. Other replacements were of marginal influence on either enzyme.

The next objective was to test whether the variations in efficiency observed above were due specifically to side chain removals or because of greater conformational freedom allowed by glycine. This was achieved using variants of the decapeptide amide (Figure 2).

1	2	3	4	5	6	7	8	9	10
Leu	Gly	Leu	Gly	Gln	Gly	Lys	Val	Leu	Gly NH_2

This data indicates the importance of one or more of the serine residues to the efficiency of factor XIIIa. The effect of replacement of lysine by glycine was confirmed, however, a variety of the side chains at position 7 (e.g. alanine, histidine, arginine or leucine) were acceptable to Factor XIIIa. Conformational rigidity and not side chain specificity is apparently important at position 7. However, variation of positions 8 and 9 to alanine or by reversal of the Val Leu sequence significantly decreases the efficiency of Factor XIIIa and to a lesser extent the liver enzyme. This apparently reflects a conformational determinant of substrate specificity formed by specific hydrophobic amino acid side chains held in a particular conformation.

This information may aid design of peptide substrates and inactivators capable of differentiation between enzymes of the transglutaminase class.

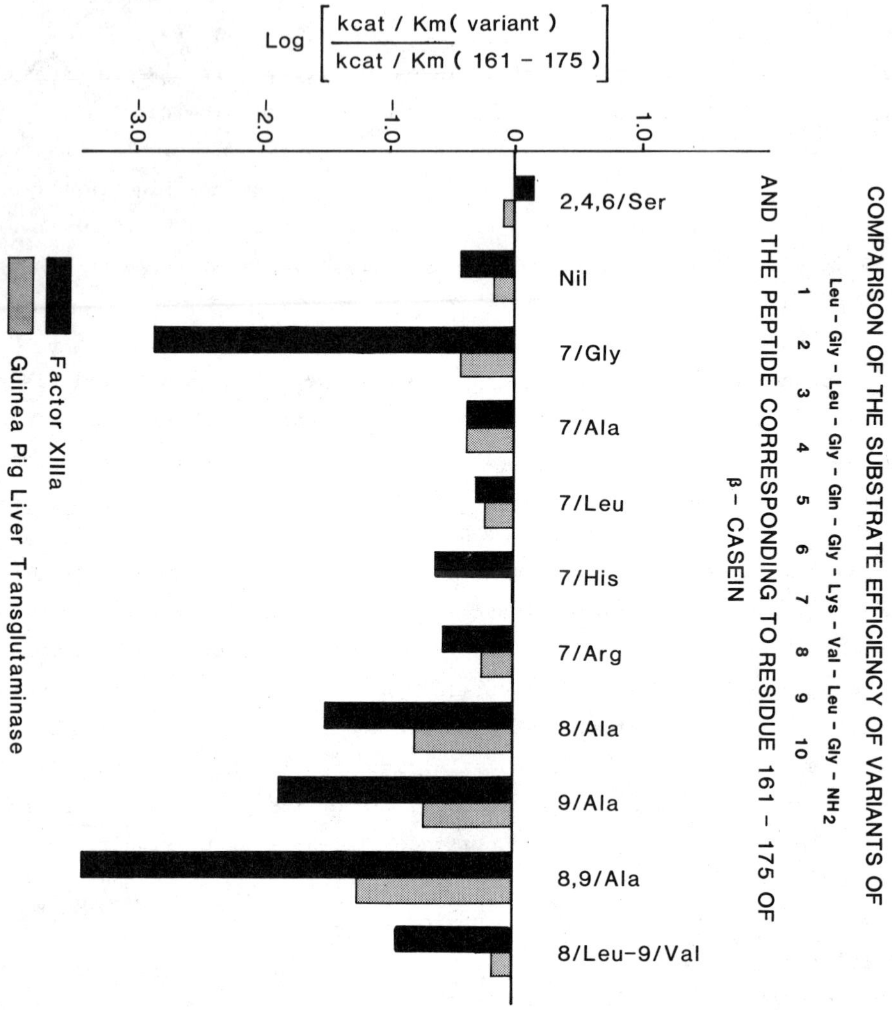

Fig 2

References

1. Folk, J.E., (1983) Advances in Enzymology and Related Areas in Molecular Biology, Edit. Meister, A., John Wiley and Sons Inc. pp. 1-55.

2. Gorman, J.J., and Folk, J.E. (1981) J.Biol.Chem. 256, 2712-2715.

3. Gorman, J.J., and Folk, J.E. (1984) J.Biol.Chem. 259, 9007-9010.

CHEMICAL CLEAVAGE OF PEPTIDES BEFORE GLUTAMIC ACID RESIDUES

Shabbir A. Khan and Bruce W. Erickson
The Rockefeller University, New York, New York 10021.

Structural characterization of a protein usually involves selective cleavage of its polypeptide chain(s) into fragments, which are analyzed by two-dimensional mapping or are separated and subjected to Edman degradation. Tryptic cleavage after lysine and arginine and cyanogen bromide cleavage after methionine are the most widely used fragmentation procedures in protein sequence studies. Staphylococcus aureus V8 protease cleaves rather selectively after glutamic acid (Glu) residues, although cleavage after aspartic acid is occasionally seen.

We have recently devised a two-step procedure for efficient chemical cleavage of model peptides before glutamic acid residues.[1] First, an internal glutamic acid residue is cyclized into an internal pyroglutamic acid (Glp) residue through activation of the side-chain carboxyl group. Second, under basic hydrolytic conditions, the peptide bond preceding the internal pyroglutamic acid residue is cleaved selectively (Scheme 1).

Internal glutamic acid residues can be converted into internal pyroglutamic acid residues by reaction with N,N-carbonyldiimidazole (CDI) in dimethylformamide[1] (Scheme 1). Polypeptides and proteins, however, would require use of aqueous solvents. Attempts to cyclize internal glutamic acid residues into internal pyroglutamic acid residues in model peptides using CDI or a water-soluble carbodiimide in water gave very low yields. Cyclization of an internal glutamic

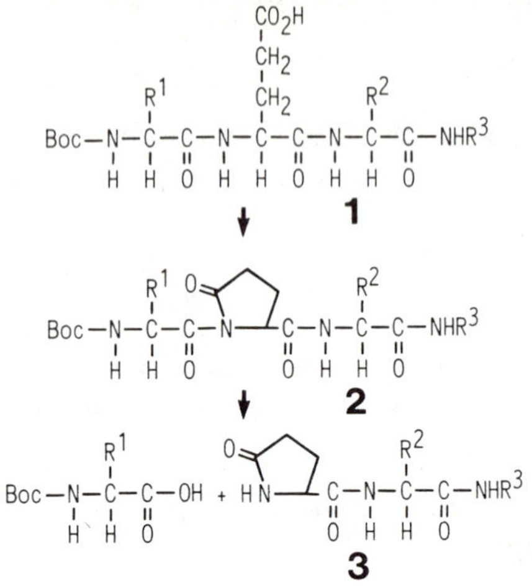

Scheme 1. Synthesis (1 → 2) and regioselective hydrolysis (2 → 3) of tripeptides containing an internal pyroglutamic acid residue.

acid residue into an internal pyroglutamic acid was also observed during liquid HF treatment of a benzyl ester[2] and a sugar ester[3] of internal glutamic acid residues. The active intermediate in these cyclizations is probably an acylium ion (C-C≡O$^+$), which could also be formed directly from a carboxyl group.

We found that reaction of Gly-Glu-Val-NH_2 or Ala-Glu-Val-NH_2 with 9:1 (v/v) trifluoromethanesulfonic acid/trifluoroacetic acid (TFMSA/TFA) gave Gly-Glp-Val-NH_2 or Ala-Glp-Val-NH_2, respectively. The apparent first-order rate constants k_1 were 11.6×10^{-5} s^{-1} and 11.8×10^{-5} s^{-1}, respectively. As anticipated, cyclization of the glutamic acid residue to the desired pyroglutamic acid residue proceeded smoothly with no other products being detected. The half-time for cyclization of both tripeptides was about 100 min at 24 °C (Figure 1). After 24 h at 24 °C, the analytical yields were >99%. TFMSA has been used previously as an alternative to HF for removal of protecting groups from synthetic peptides.[4] TFA was present because it is an

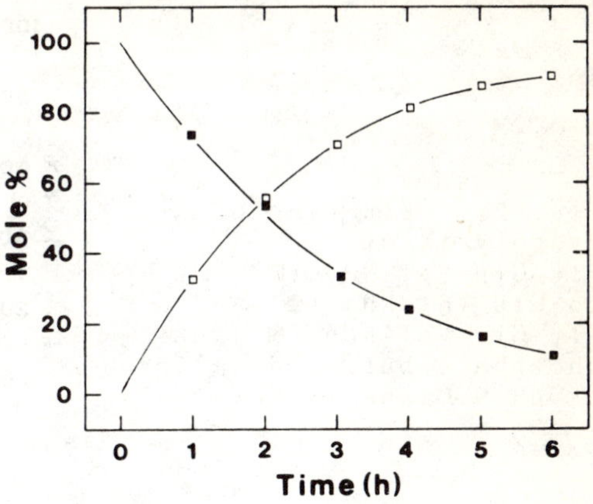

Fig. 1. Kinetics of TFMSA/TFA catalyzed conversion of Gly-Glu-Val-NH_2 (■) into Gly-Glp-Val-NH_2 (□) at 24 °C.

excellent solvent for peptides and proteins. When TFMSA (acidity H_o = -13) was replaced by methanesulfonic acid, a weaker acid (H_o = -6.5), the half-time for cyclization of Gly-Glu-Val-NH_2 was about 15.5 h (k_1 = 1.24 x 10^{-5} s^{-1}) and the reaction was only about 60% complete after 24 h. Thus treatment with TFMSA/TFA at room temperature is an efficient method for converting internal glutamic acid residues into pyroglutamic acid residues. This method could be conveniently used as the first-step of the two-step cleavage procedure with proteins.

The second step involves cleavage of the Xxx-Glp peptide bond under basic conditions. We found that about 88% of Gly-Glp-Val-NH_2 fragmented to Gly and Glp-Val-NH_2 with a half-time of about 3.7 h at pH 9.0 and 24 °C (Figure 2). The rate contant k_1 for alpha cleavage with fragmentation was 5.20 x 10^{-5} s^{-1} and gamma cleavage with ring opening to Gly-Glu-Val-NH_2 was 0.71 x 10^{-5} s^{-1}. The alpha/gamma ratio was 7.3:1. In general, gamma cleavage would produce overlapping fragments.

Fig. 2. Kinetics of hydrolysis of Gly-Glp-Val-NH_2 (□) to Glp-Val-NH_2 (●) and Gly-Glu-Val-NH_2 (■) in phosphate-buffered saline at pH 9.0 and 24 °C.

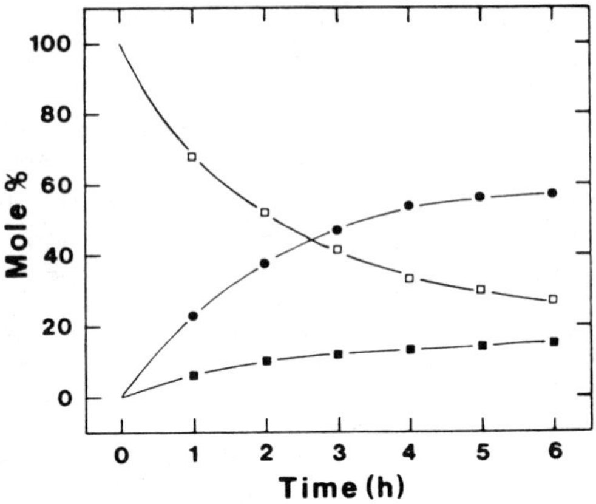

Thus, acid-catalyzed cyclization of internal glutamic acid residues into internal pyroglutamic acid residues followed by base-catalyzed cleavage of the peptide bonds in proteins could result in fragmentation of the peptide chain. Then the mixture of peptide fragments could be mapped or treated with pyroglutamyl aminopeptidase to remove the amino-terminal pyroglutamic acid residues. In the latter case, the unblocked peptide fragments could be separated and subjected to automated Edman degradation.

References

1. Khan, S. A., and Erickson, B. W. (1984) J. Am. Chem. Soc. 106, 798-799.

2. Feinberg, R. S., and Merrifield, R. B. (1975) J. Am. Chem. Soc. 97, 3485-3496.

3. Khan, S. A., and Erickson, B. W. (1982) J. Biol. Chem. 257, 11864-11867.

4. Yajima, H., Fujii, N., Ogawa, H., and Kawatani, H. (1974) J. Chem. Soc. Commun., 107-108.

MODIFICATION AND CLEAVAGE OF PEPTIDES AT ASPARAGINE AND GLUTAMINE RESIDUES

James K. Blodgett and G. Marc Loudon
Department of Medicinal Chemistry and Pharmacognosy
School of Pharmacy, Purdue University
West Lafayette, Indiana 47907

Introduction

Previous studies in our laboratory[1,2] have demonstrated that the reagent I,I-(bis[trifluoroacetato-O]phenyl)iodine, which we abbreviate "PIFA," is very useful for the conversion of aliphatic amides into amines.

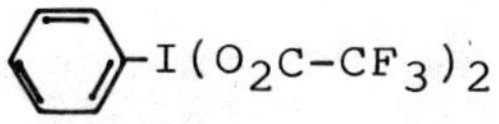
PIFA

In this work we demonstrate that this reaction is readily extended to the amide side-chains of Gln and Asn residues in peptides, and that the resulting peptides with modified side-chains can be specifically cleaved at the modified residues.

Results and Discussion

The reagent PIFA readily effects the oxidative rearrangement of carboxamide side-chains in peptides. Thus Gln residues are converted into DABA (2,4-diaminobutanoic acid) residues and Asn residues into DAPA (2,3-diaminopropanoic acid) residues.

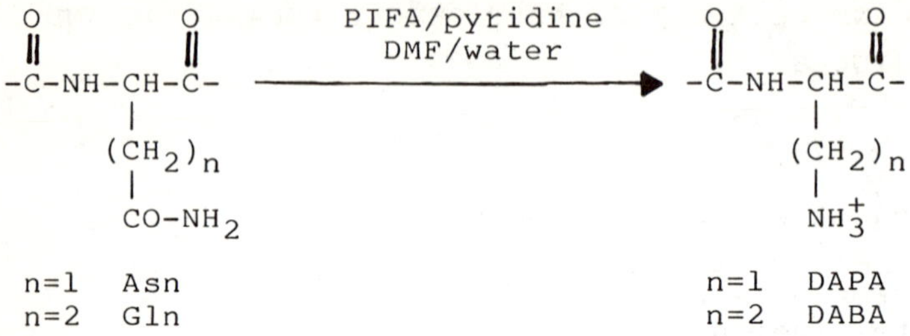

(1)

This reaction is carried out in 1:1 DMF-water with sufficient pyridine to bring the pH to an observed reading of 3.36. (Higher pH values precipitate a μ-oxo dimer of PIFA[2].) The reaction proceeds over 5-7 hr in virtually quantitative yield. Holt and Milligan[3] have reported that iodobenzene diacetate (IBDA) effects a more rapid rearrangement. In fact this is true if the reagents are allowed to establish their own pH values (1.65 for PIFA, 4.77 for IBDA). However, at a common pH (3.36), PIFA reacts twelve times faster, as we would expect from its better leaving group. Moreover in more concentrated peptide solutions (0.25 M) the higher pH associated with IBDA could lead to side reactions, such as urea formation, although at the peptide concentrations used by us ($\leq$0.14 M) this has not been a problem. Thus the apparent superiority of IBDA is a pH effect, a point that is clear from our mechanistic work[2].

It had been shown earlier in the literature that peptides containing a DABA residue could undergo both cleavage[4] and transpeptidation[4d] under alkaline conditions.

Cleavage:

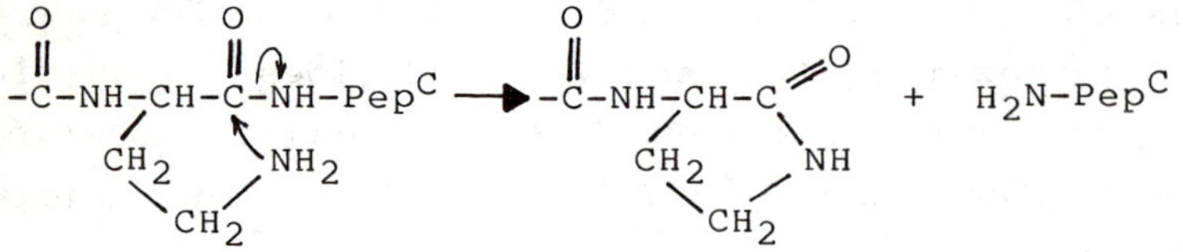

(2)

Transpeptidation:

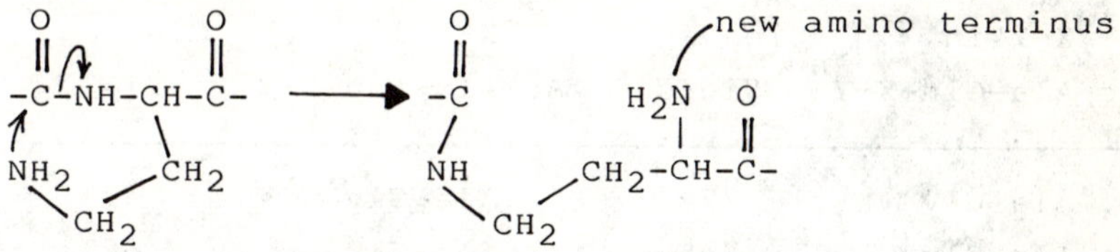

(3)

In most previously examined cases, however, the peptides under investigation were designed so that they could react in only one of the two ways; no peptide in which the two modes of reaction were pitted against each other has been systematically examined. Accordingly, we examined transpeptidation of a number of DABA-containing peptides of the form Ac-Ala-DABA-X-Y, where X and Y are amino acid residues, and found that transpeptidation is the preferred mode of reaction. The transpeptidation equilibrium requires buffers known to be bifunctional catalysts[5], and, for DABA peptides, lies about 4:1 in favor of the transpeptidized product, a result in quantitative agreement with the known effect of ammonium ion pK_a on aminolysis equilibria[6]. The optimum conditions for transpeptidation are pH 9.7, 0.25 M bicarbonate buffer; under these conditions transpeptidation is established in about 24 hr for DABA peptides. (The details of DAPA transpeptidation are under investigation.) Competing with transpeptidation is about 10% of the cleavage pathway.

Once transpeptidation has occurred, a new α-amino terminus is formed (Eq 3). Thus the transpeptidized peptide can be cleaved with the Edman reagent. The net result of modification, transpeptidation, and cleavage, is a peptide cleavage at Gln and Asn residues (Table I). Control experiments with authentic transpeptidized peptides have shown that the yields in this Table reflect largely the transpeptidation yields.

Table I. Cleavage of Peptides at Gln and Asn Residues

X-Y in Ac-Ala-Gln-X-Y	Cleavage Yield	X-Y in Ac-Ala-Asn-X-Y	Cleavage Yield
Ala-Gly	77%	Ala-Gly	62%
Ile-Gly	71%	Ile-Gly	66%
Gly-Gly	62%	Gly-Gly	46%
Phe-Gly	69%	Phe-Gly	52%
Ala-Phe	74%	Ala-Phe	63%

References

1. Loudon, G. M., Radhakrishna, A. S., Almond, M. R., Blodgett, J. K., and Boutin, R. H. (1984) J. Org. Chem. 49, 4272-4276

2. Boutin, R. H., and Loudon, G. M. (1984) J. Org. Chem. 49, 4277-4284

3. Holt, L. A., and Miligan, L. A. (1981) Aust. J. Biol. Sci. 34, 395-400

4. (a) Barrass, B. C., and Elmore, D. T. (1957) J. Chem. Soc. 4830-4834 (b) Lipson, M. A., and Sondheimer, E. (1964) J. Org. Chem. 29, 2392-2394 (c) LeQuesne, W. J., and Young, G. T. (1952) J. Chem. Soc. 594-597 (d) Poduska, K., Katrukha, G. S., Silaev, A. D., and Rudinger, J. (1965) Coll. Czech. Chem. Comm. 30, 2410-2433.

5. (a) Satterthwait, A. C., and Jencks, W. P. (1974) J. Am. Chem. Soc. 96, 7031-7044 (b) Smith, V. F., Jr., and Schmir, G. L. (1975) J. Am. Chem. Soc. 97, 3171-3177 (c) Cox, M. M., and Jencks, W. P. (1981) J. Am. Chem. Soc. 103, 580-587 (d) Lee, Y.-N, and Schmir, G. L. (1979) J. Am. Chem. Soc. 101, 3026-3055, 6277-6282 (e) Cunningham, B. A., and Schmir, G. L. (1966) J. Am. Chem. Soc. 88, 551-558; (1967) ibid. 89, 917-922

6. Fersht, A. R., and Requena, Y. (1971) J. Am. Chem. Soc., 93, 3499-3504

PROPERTIES OF INTERFERON-α_2 ANALOGUES PRODUCED FROM SYNTHETIC GENES

Roger Camble, Michael D.Edge, and Valerie E.Moore

ICI Pharmaceuticals Division, Mereside, Alderley Park, Macclesfield, Cheshire, United Kingdom. SK10 4TG

Introduction

As new strategies and methods for the chemical synthesis of peptides have been devised many of them have been tested in syntheses of proteins over 100 amino-acid residues long. However, successful synthesis of a parent protein has rarely been exploited further to produce a series of analogues for structure-function studies. In the last 10 years recombinant DNA technology has opened a new biosynthetic approach to the synthesis of native proteins. Recently, techniques such as hybrid gene formation, site-directed mutagenesis and total gene synthesis have been developed to construct new genes coding for novel analogues of native proteins[1]. We have shown that genes long enough to encode interferon-α_1 (IFN-α_1), and IFN-α_2 (166 and 165 amino-acid residues respectively[2,3]) can be synthesised chemically and expressed in Escherichia coli[4,5,6]. In principle synthetic genes can be used to prepare analogues with any genetically coded amino-acid substituted at any number of desired positions throughout the length of the polypeptide chain. Using IFN-α_2 as a model system we are exploring the potential of this approach to generate protein analogues for structure-activity studies.

Objectives

Can analogues be produced at a reasonable rate ?

There is much excitement at present about protein engineering and the ability, using computer modelling techniques, to predict the effects of structural changes in proteins on their functions. However, experience with small-molecule medicinal chemistry suggests a substantial number of analogues covering a range of structural variants is still likely to be required before any therapeutically significant change in biological profile can be anticipated.

Do the products give coherent structure-function profiles?

When planning a high throughput of compounds, time, manpower and material requirements preclude rigorous purification and detailed structural analysis of every analogue protein. Knowing the coding gene sequence for the target analogue to be correct, would protein extracts provide consistent target sequence-activity patterns when purified only to a stage where bacterial contaminants do not interfere in the bioassays? Analogues with profiles of interest can then be characterised in full.

Can the different biological activities of IFN-α_2 be separated?

IFN-α_2 has well-defined antiviral, antiproliferative and immunomodulatory activities. Increased antiviral selectivity provided a target which could be assessed by comparing ratios of titres in different assays (possible with extracts) rather than by having to determine absolute potencies. Increased selectivity of action in an analogue might alleviate some of the undesirable side-effects observed clinically with purified natural IFN and recombinant native IFN-α_2.

Synthetic Route

Synthetic genes were produced from up to 68 overlapping oligonucleotides by a succession of enzymatic ligation steps [4,6]. Each gene was joined directly to the expression plasmid vector pSTP1[7] downstream from the *trp* promoter and introduced into *E.coli*. Alernatively, a short gene fragment covering the analogue modifications was used with a plasmid vector already containing the rest of the IFN-α_2 coding sequence. Plasmid DNA was isolated from clones containing gene inserts and subjected to restriction enzyme analysis to eliminate anomalous constructs. Extracts from 100ml cultures of three selected clones were screened for IFN-like activity in antiviral and NK2-monoclonal antibody[8,9] immunoradiometric (NK2-IRMA) assays. In the absence of either biological or immunological activity the presence of any new protein was detected by labelling of plasmid-encoded proteins in E.coli mini-cells[6]. From these data a single clone was selected for each analogue and the structure of the analogue gene confirmed by DNA sequencing. Extracts from larger scale cultures (1-4 litres) of the analogue clone were prepared and the protein isolated and bioprofiled.

By using common double-stranded segments for those regions of genes not being changed, up to 30 different analogue genes can be prepared at a time. The methodology is still being refined but the equivalent of 1-2 analogues a week have been produced using this general protocol; in total over 100 structural variants of IFN-α_2 have been examined.

Protein Products - isolation and characterisation

In most cases about 0.1% of the protein produced by analogue clones was IFN-related product. Each analogue was isolated either by simple fractionation on a size exclusion column (removes over 80% of unrelated bacterial protein but gives a product still less than 1% pure) or by purification on NK2 or YOK monoclonal antibody affinity supports[9] (gives just IFN-related proteins and allows direct assessment of target analogue product by PAGE-SDS). With few exceptions, for a given analogue, the degree of purity of the sample did not affect significantly the ratios between titres in the human antiviral, antiproliferative and natural-killer cell-stimulating bioassays and the NK2-IRMA assay used for profiling. Thus the presence of E.coli contaminants was not critical for initial structure-activity interpretations.

However, a correct analogue gene sequence does not guarantee a monocomponent analogue protein product. In purified samples of different analogues we have detected heterogeneities from retention of a variable amount of N-terminal initiator methionine residue, from monomer species with incomplete or incorrect disulphide bridge formation, from oligomer species linked in most cases through intermolecular disulphides, or from truncated species probably produced by bacterial protease activity. Thus changes programmed into the gene may indicate only the minimal structural difference of the protein product from the parent IFN-α_2. In spite of these complications, comparison of activity profiles between different analogues provided remarkably coherent target sequence-activity patterns.

Structure-Activity Results

IFN-α_2 is a single-chain protein of 165 amino-acids[3] cross-linked by two disulphides[10]. We prepared analogues exploring several different themes (Table 1), designed both to assess the "synthetic methodology" and to explore structure-activity relationships. In the absence of reliable 3-dimensional structure data, ideas were based primarily on considerations of the sequence of IFN-α_2 and on sequence comparisons between different human IFN species.

Table I. Types of Analogues Examined

1. Truncations / deletions
2. Replacement of residues of a specific amino – acid
3. Movement of cysteine residues
4. IFN – α_2/α_1 hybrids
5. IFN – α_2/γ hybrids
6. Clusters of conserved amino – acids

Residues not important for activity

Up to 10 residues can be removed from either end of the polypeptide chain with retention of all four activities. With analogues further truncated at either end, correlation of structure with activity is complicated by degradation problems but it does seem that the NK2 monoclonal antibody can still recognise the product from the gene for IFN-α_2(29-165). Replacement of residues 10-15 by six alanines reduces

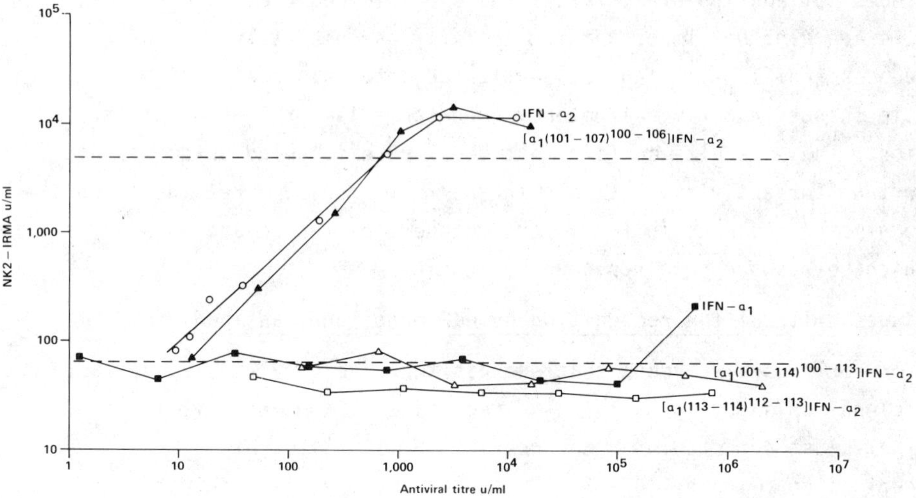

Fig 1. Cross – Reaction of IFN – α_2/α_1 Hybrids With NK2.

Table II Specific Activities of Purified IFNs and Hybrid Analogues

Target Protein	Specific Activity (U / mg)	
	Antiviral	NK_2 – IRMA
IFN – α_2	2×10^8	2.5×10^9
IFN – α_1	8×10^6	4×10^3
$[\alpha_1(101-114)^{100-113}]$ IFN – α_2	6×10^8	$<1.4 \times 10^4$ a.
$[\alpha_1(101-107)^{100-106}]$ IFN – α_2	2×10^8	1×10^9
$[\alpha_1(113-114)^{112-113}]$ IFN – α_2	2×10^8	$<1.3 \times 10^4$ a.
$[\gamma(98-114)^{98-114}]$ IFN – α_2	2×10^8	$<4.3 \times 10^4$ a.
$[\gamma(98-107)^{98-107}]$ IFN – α_2	1×10^8	1×10^9
$[\gamma(95-101)^{101-107}]$ IFN – α_2	5×10^8	2×10^9

a. No activity detected at the highest concentrations used.

but does not abolish bioactivity. The specific residues in positions 98-114 are also not important for activity as completely unrelated sequences from IFN-γ[11] can be substituted into this region with little change in antiviral potency (see Table II) or in bioactivity ratios. These results confirm that the 1-98 disulphide bridge is not important for activity. Activity in the four assays is retained when all five glycine residues in IFN-α_2 are changed to alanine and when all five methionine residues are replaced by leucine.

Residues critical for recognition by NK2 monoclonal antibody

In early work we found that whereas NK2 (the only widely available monoclonal antibody against IFN-α) recognised IFN-α_2 easily, IFN-α_1 was only just detectable in the NK2-IRMA at 10^5 times higher concentration (Figure 1, Table II) and was not retained on NK2-Sepharose columns. These two IFN-α species differ at 27 positions and IFN-α_1 has an additional residue inserted at position 44. Seven of these changes occur between positions 100-113 of IFN-α_2. The hybrid analogue [Met^{100},Glu^{102},Glu^{103},Arg^{104},Gly^{106},Asn^{112},Ala^{113}]IFN-α_2 (or [α_1(101-114)$^{100-113}$]IFN-α_2) showed biological activity indistinguishable from IFN-α_2 but was not recognised by NK2. The amino-acids critical for antibody recognition were further pinpointed with two more analogues. [α_1(101-107)$^{100-106}$]IFN-α_2 behaved like IFN-α_2 in every respect but [Asn^{112},Ala^{113}]IFN-α_2 was again not recognised by the antibody (Figure 1). It seems that the presence of this Asn-Ala sequence in IFN-α_1 is sufficient to account for its lack of reaction with NK2, and that some part of the corresponding Lys-Glu sequence in IFN-α_2 probably makes an important contribution to the epitope for this antibody. The behaviour of the IFN-α_2/γ hybrids with NK2(Table II) is consistent with this general pattern.

Importance of conserved residues for activity

IFN-α and IFN-β show similar biological profiles and appear to compete for the same cell-surface receptors. If structural elements critical for activity do exist in IFN-α_2, preferred places to look for them might be among the 35 amino-acid residues strictly conserved among

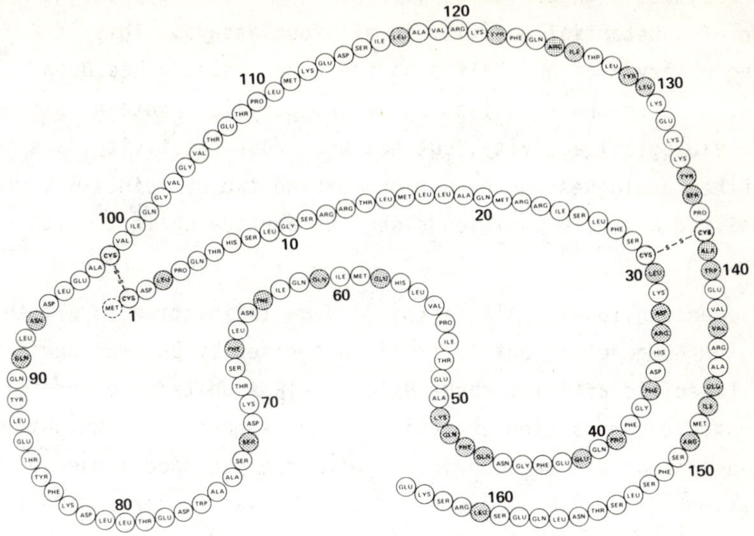

Fig 2. The Primary Sequence of HuIFN − α_2 Indicating the Conserved Residues

all reported species of IFN-α and of IFN-β (Figure 2). There is no reason that every conserved residue should be essential for function or for maintaining the active conformation of the molecule (indeed residues 3 and 161 were already eliminated by our results with truncated analogues) so we set out to change clusters of 3 to 7 residues at a time. Then only the residues in analogues showing a loss or separation of activity need be studied in more detail. Clusters were selected in the simplest possible manner, sequentially along the chain, but were chosen to fit into the structural elements of a speculative ab initio conformational model[12]. Except for the two conserved cysteines and the one conserved alanine, residues were replaced by alanine, thus in effect, attempting to identify essential side-chains by deleting them. Both oligonucleotide purification and gene assembly proved difficult for some of these planned analogues. This may reflect secondary structure formation provoked by self-complementary GC dimers in proximal alanine codons.

Two analogues of particular interest involve conserved residues around the 29-138 disulphide bridge. The integrity of this bridge was reported to be important for activity[13]. However, replacement of both cysteine residues with serine to give [$Ser^{29,138}$]IFN-α_2 leads to retention of substantial activity in all four assays. This is a surprising observation and, although the gene sequence has been confirmed, further characterisation of the purified protein is in progress. Biological activity, but not NK2 cross-reactivity, was lost in several other analogues where just one of the two cysteines was replaced or was shifted a single position along the peptide chain in either direction.

The second analogue, [$Ala^{30,32,33}$]IFN-α_2 is inactive in all three bioassays on human cells but is still recognised by and was purified using NK2(specific activity about $9x10^8$NK2-IRMA units/mg of protein). Both amino-acid composition analysis and the sequence of the N-terminal 33 amino-acid residues are consistent with the proposed sequence. In collaboration with us, F.Marcucci and E.De Maeyer in Paris have shown that this analogue is also devoid of antiviral and cell-growth inhibitory activity on bovine MDBK cells. In receptor binding assays measuring displacement of radiolabelled IFN-α_2 the analogue did not bind to the receptor on human WISH cells but did bind to the receptor on bovine MDBK cells with an affinity similar to IFN-α_2. Finally they have demonstrated that [$Ala^{30,32,33}$]IFN-α_2 is a competitive antagonist of the antiviral and anticellular activity of IFN-α_2 on MDBK cells but has no effect on these activities in human cells.

We have just completed a series of nine single residue replacement analogues providing three different substitutions at each of the three positions 30,32, and 33. These should help to define further the contribution of each of the three conserved residues to the loss of recognition by human cells and to the absence of biological responses after binding to receptors on bovine cells. It will also be important to see if activity can be restored selectively in the different biological assays. We can say already that both antiviral and antiproliferative effects on human cells are especially sensitive to changes at the arginine residue in position 33. Purified samples of [Ala^{33}]IFN-α_2 and even [Lys^{33}]IFN-α_2 show a 500-fold decrease in potency relative to IFN-α_2 in these two bioassays but give similar specific activities in the NK2-IRMA.

Towards analogues with non-coded structural features

Once an amino-acid residue of interest has been identified, the 20 genetically-coded L-amino-acids provide only a limited selection of the structural variations traditionally open to the peptide chemist. To extend protein structure-activity studies to analogues with other structural features will require application to the biosynthetic product of either protein modification or protein-semi-synthesis techniques. Both approaches might be facilitated by appropriate tailoring of the parent protein produced in bacteria. For example, to assist subsequent semi-synthesis, inconvenient cleavage sites present in the parent protein can be eliminated by substitution with other amino-acids which are still compatible with retention of biological activity. Thus, as mentioned above, we have prepared the analogue in which all five methionine residues in IFN-α_2 are replaced with leucines. In a second phase, methionine residues could be re-introduced into this analogue at any desired positions along the chain. Cyanogen bromide cleavage of the resulting bacterial product should then generate fragments specifically designed for convenient semi-synthetic re-assembly around a chemically synthesised segment containing non-coded structural elements.

Conclusions

We are convinced that the synthetic gene approach can produce target analogues with the throughput required to make useful progress in a protein structure-activity programme. The approach has the advantage that any number of changes can be introduced simultaneously anywhere in the sequence. In the case of IFN-α_2, the majority of analogues of diverse types have proved sufficiently robust when prepared in E.coli that coherent structure-activity patterns are obtained without recourse to multi-step purification protocols or complex protein folding procedures. This will probably not be true for some other proteins of interest. Antibody reagents provide a simple one-step method of purification needed to monitor heterogeneity in target analogue-related proteins.

References

1 For a review of such work with IFN proteins see Edge,M.D. and Camble,R.(1984) Biotechnology and Genetic Engineering Reviews 2, 215-252.

2 Mantei,N., Schwarzstein,M., Streuli,M., Panem,S., Nagata,S. and Weissmann,C.(1980) Gene 10,1-10.

3 Streuli,M., Nagata,S. and Weissmann,C.(1980) Science 209, 1343-1347

4 Edge,M.D., Greene,A.R., Heathcliffe,G.R., Meacock,P.A., Schuch,W., Scanlon,D.B., Atkinson,T.C., Newton,C.R. and Markham,A.F.(1981) Nature 292, 756-762.

5 De Maeyer,E., Skup,D., Prasad,K.S.N., De Maeyer-Guignard,J., Williams,B., Meacock,P., Sharpe,G., Pioli,D., Hennam,J., Schuch,W. and Atherton,K. (1982) Proc. Natl. Acad. Sci. USA. 79, 4256-4259.

6 Edge,M.D., Greene,A.R., Heathcliffe,G.R., Moore,V.E., Faulkner,N.J., Camble,R., Petter,N.N., Trueman,P., Schuch,W., Hennam,J., Atkinson,T.C., Newton,C.R. and Markham,A.F.(1983) Nucleic Acids Research 11, 6419-6435.

7 Windass,J., Newton,C.R., De Maeyer-Guignard,J., Moore,V.E., Markham,A.F. and Edge,M.D.(1982) Nucleic Acids Research 10, 6639-6657.

8 Secher,D.S. and Burke,D.C.(1980) Nature 285, 446-450.

9 Available from Celltech Ltd., Slough, England.

10 Wetzel,R. (1981) Nature 289, 606-607.

11 Gray,P.W. and Goeddel,D.V. (1982) Nature 298, 859-863.

12 Sternberg,M.J.E. and Cohen,F.E. (1982) Internat. J.Biol. Macromol.4, 137-144.

13 Wetzel,R., Levine,H.L., Estell,D.A., Shive,S., Finer-Moore,J., Stroud,R.M. and Bewley,T.A. (1982) in Chemistry and Biology of Interferons: Relationship to Therapeutics,UCLA Symp. Mol.Cell.Biol. 25, Merigan,T. and Friedman,R., Eds., Academic Press, New York pp.365-376.

14 Marcucci,F. and De Maeyer,E., paper in preparation

ISOLATION AND CHARACTERIZATION OF "NATIVE" AND REC. EGLIN C FROM E. COLI, SELECTIVE PROTEINASE INHIBITORS FOR HUMAN LEUCOCYTE ELASTASE, CATHEPSIN G AND CHYMOTRYPSIN.

W. Märki, H. Rink, P. Sieber, H.P. Schnebli, and M. Liersch
Pharmaceutical Division, CIBA-GEIGY Ltd., CH-4002 Basel

F. Raschdorf, and W. Richter
Central Function Research, CIBA-GEIGY Ltd., CH-4002 Basel

Introduction

In the search for a potent and specific inhibitor of human leucocyte elastase for possible therapeutic use we became interested in eglin c, the elastase/cathepsin G inhibitor from the leech Hirudo medicinalis which was discovered in 1977[1]. The structure of eglin c consists of a single polypeptide chain of 70 amino acids. Striking features of this molecule are the lack of sulfur containing amino acids and its outstanding resistance against denaturation by heat and acid.

The total synthesis of a 232 base-pair coding sequence of eglin c from six fragments synthesized by the phosphotriester method and enzymatic assembly was performed. The gene was cloned and expressed in high yield in E.coli under the transcriptional control of the trp promoter[2].

E G L I N C (Hirudo medicinalis)

```
                                     10                                      20
H-Thr-Glu-Phe-Gly-Ser-Glu-Leu-Lys-Ser-Phe-Pro-Glu-Val-Val-Gly-Lys-Thr-Val-Asp-Gln-
                                     30                                      40
Ala-Arg-Glu-Tyr-Phe-Thr-Leu-His-Tyr-Pro-Gln-Tyr-Asp-Val-Tyr-Phe-Leu-Pro-Glu-Gly-
                                     50                                      60
Ser-Pro-Val-Thr-Leu-Asp-Leu-Arg-Tyr-Asn-Arg-Val-Arg-Val-Phe-Tyr-Asn-Pro-Gly-Thr-
                                     70
Asn-Val-Val-Asn-His-Val-Pro-His-Val-Gly-OH
```

Results and Discussion

Cultures of transformed E. coli pML 163 (host W 3110△102) were grown and the cells harvested and disrupted either with lysozyme or mechanically. Biologically active products were obtained, isolated and purified to homogeneity using conventional chromatography. The purification procedure included acid extraction, ionexchange-chromatography (CM and DEAE 53), gelfiltration and RP-HPLC on a Vydac 218TP54 (330 Å pore size) column (for identification only).

The recombinant eglin c and authentic eglin c comigrated on SDS-PAGE with an apparent molecular weight of 6000. However biosynthetic eglin c, when analyzed by RP-HPLC showed a higher retention time than the natural protein. Since the expression product had an amino acid composition indistinguishable from eglin c (data not shown) and showed an pI of 5.5 versus 6.5 the possibility that eglin c had an additional methionine residue (N-terminus) could be excluded.

The strategy for polypeptide structural analysis established chemical molecular weights for the intact molecules and tryptic peptides by FAB-MS (figure 1) [3]. The FAB measurements with accurate mass determination (RbI/CsI external mass calibration)

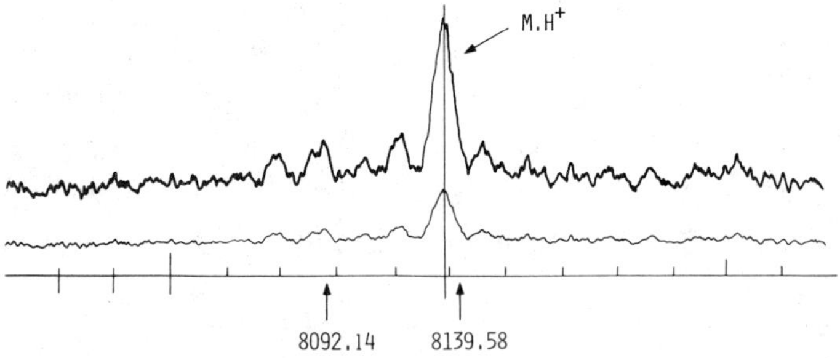

$C_{373}H_{550}N_{96}O_{107}$ ($M.H^+$ calc. = 8092.04, $M.H^+$ found = 8091.79 ± 0.28)

Recombinant EGLIN C ($M.H^+$ found = 8133.74 ± 0.40, Δ = 42 dalton)

Fig.1. Conditions: thioglycerol matrix, ZAB-HF mass spectrometer (VG-Analytical Ltd.), Xenon bombardment, 3 keV, scan linear mode.

defined the molecular weight for recombinant eglin c as 8133.74 (△42 dalton compared with native eglin c). Thus a detailed analysis had to be directed towards uncovering the nature and site of the structural deviation. Consequently, both eglin c molecules were enzymatically digested with trypsin providing identical fragments based on RP-HPLC and FAB-MS of the crude hydrolysates, except for the N-terminal fragment T_1.The nominal mol. weights were 951 and 909 respectively. The mass difference corresponds either to a propyl or acetyl residue.

The identification of the substituent X was tackled in three independent ways: i) N-peracetylation with Ac_2O/MeOH of tryptic peptide T_1 for both eglin c and subsequent FAB/MS analysis of resultant diacetates[3]; ii) detection of free acetic acid after partial hydrolysis of the recombinant peptide with 0.03 M HCl by gas chromatography and iii) by ^{1}H-NMR analysis of the isolated fragment T_1 (acetyl group, δ 2.15 ppm, singlett, 3 H).

The analytical data provide strong evidence that the expressed eglin c in E. coli is modified to N^{α}-acetyl eglin c in a posttranslational event. This blockage of the N-terminal amino acid threonine accounts very well for unsuccessful microsequencing. The assigned structure and molecular mass of the N-terminal fragment T_1 (mol.wt.= 951) has finally been confirmed by

Reversed Phase HPLC of Tryptic Digests

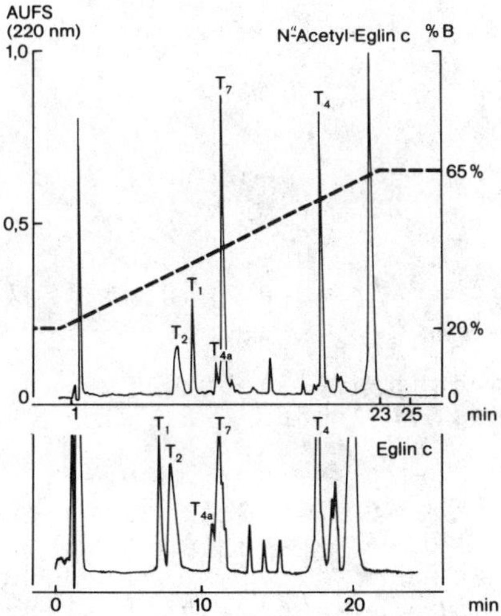

Conditions:
Stat. Phase: Nucleosil (MN) C 18 5 μm (4,6 × 120 mm)
Eluent: 0.1% Tfa (A); Acetonitrile/H_2O 8:2 + .07% Tfa (B)
Flow: 1,2 ml/min

comparison with its synthetic replicate, based on HPLC and MS.

This type of modification of a recombinant DNA derived product has not been observed in E. coli before. However of the characterized N-terminal blocking groups of natural proteins in eukaryotes acetyl is known to be the most common one.

Table. Association Rate Constants (k_{ass}) and Inhibition Constants (Ki) for Inhibitor Interactions with leucocyte elastase

Inhibitor	k_{ass} (M^{-1} sec^{-1})	Ki (M)
native Eglin c	$1.4 \pm 1.2 \times 10^{7}$	8.3×10^{-11}
rec N-Acetyl Eglin c	1.8×10^{7}	7.8×10^{-11}
alpha 1-PI	$2.4 \pm 1.9 \times 10^{7}$	

k_{ass} and Ki were determined as described elsewhere[4].

From this data it can be concluded that both eglin c (native and modified) from E. coli have the same biological properties as the authentic protein, isolated from the leech. Thus eglin c is an excellent inhibitor for human leucocyte elastase and cathepsin G, reacting as rapidly as alpha 1-PI with these enzymes and is of therapeutic potential for emphysema, septicaemia, shock and other inflammatory disorders.

References

1. Seemüller, U., Meier, M., Ohlsson, K., Müller, H.-P., and Fritz, H. (1977) Hoppe Seyler's Z. Physiolog. Chemie 358, 1105-1117 .

2. Rink, H., Liersch, M., Sieber, P., and Meyer, F. (1984) Nucl. Acids Res. 12, 6369-6387 .

3. Richter, W., Raschdorf, F.,and Märki, W. (1984),Int.Symp. on MS in Health and Life Sci., San Francisco, Sept.9.-13.

4. Virca, G.D., Bodmer, J.L., Metz, G., Bieth, J.G., Maschler, R., and Schnebli, H.-P. (1985) B.B.A. (subm.).

COUPLING OF LARGE PROTECTED PEPTIDE FRAGMENTS IN TRIFLUOROETHANOL: SYNTHESIS OF THYMOSIN α_1

Arthur M. Felix, Ching-Tso Wang and Theodore J. Lambros

Peptide Research Department, Roche Research Center, Hoffmann-La Roche Inc., Nutley, New Jersey 07110

Introduction

The coupling of protected peptide fragments by the dicyclohexylcarbodiimide(DCC)-1-hydroxybenzotriazole (HOBt) method[1,2] is known to suppress N-acylurea formation and minimize the danger of racemization. Repetitive stepwise synthesis,[3] as well as the coupling of large N^{α}-acyl peptides,[4] have been carried out successfully by this method. The efficiency of the DCC-HOBt coupling results from the rapid reaction of HOBt with intermediate O-acylisoureas. This forms the HOBt-peptide esters which rapidly acylate free amino-terminal peptides. Other side reactions, including dehydration of the ω-carbonamide groups of asparagine or glutamine, can be avoided by the DCC-HOBt procedure.[2,5] Low levels of racemization have been observed using various solvents,[6] including: dimethylformamide (DMF), dimethylsulfoxide (DMSO), tetrahydrofuran (THF), methylene chloride (CH_2Cl_2), acetonitrile (CH_3CN) and hexamethylphosphoramide (HMPA).

Condensation of large peptide fragments frequently fails to produce more than trace amounts of products.[7] The use of excess acylating segment or prolonged periods of coupling have been frequently used to overcome this problem.[8] Low rates of reaction have also been attributed to poor solubility in organic solvents and/or low molar concentrations of the fragments. This results in irreversible intramolecular reactions or significant amounts of side reactions of the activated carboxyl component with solvent.[9] Alternative solvents with enhanced solvating power such as phenol, cresol and hexafluoroisopropanol have been used successfully for coupling reactions in several cases.[10,11] It was recently reported that trifluoroethanol (TFE) improved coupling efficiency in solid phase peptide synthesis and enhanced the determination of free amino groups in peptide-resin.[12] These observations were attributed to improved solvation of peptide, but the magnitude of the effect was projected to be sequence dependent. Trifluoroethanol has also been used as solvent for selective acidolytic cleavage of N^{α}-trityl- and Boc-protecting groups, for the selective oxidation of cysteine residues[13,14] and for the deprotection of peptide phenyl esters[15] by the peroxide oxidation method.[16]

Synthesis of Thymosin α_1

Thymosin α_1, an N^{α}-acetyl octaeicosapeptide, isolated from calf thymus,[17] and reported to enhance parameters of immune function, has been synthesized both by classical procedures in solution[18,19] and by solid-phase methods.[20-22] Recently, an alternate solution synthesis using fragments with tert-butyl side chain protecting groups was devised for the purpose of scaling-up the process.[15] Assemblage of the intermediate fragments led to the preparation of the two key intermediates (N^{α}-acetyldecapeptide, 1, and octadecapeptide, 2,)[23] shown in Figure 1. DCC-HOBt coupling of these intermediates was studied in various solvents under a variety of reaction conditions.

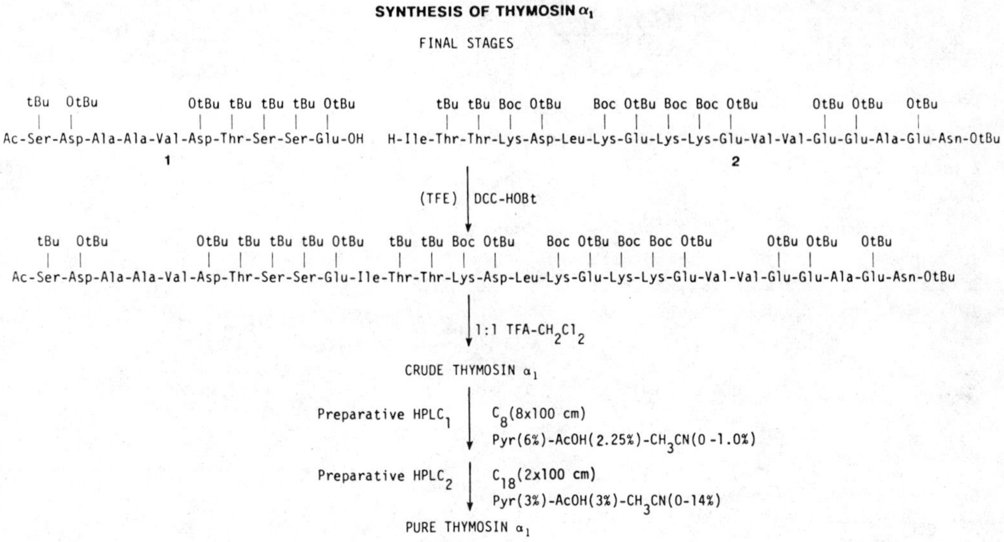

Fig. 1 Final stages for the synthesis of thymosin α_1 by fragment condensation

Results and Discussion

The N^{α}-acetyldecapeptide, 1, was insoluble in $CHCl_3$, slightly soluble in 1-methyl-2-pyrrolidinone (MPD), DMF, DMSO and soluble in hexafluoroisopropanol (HFIP), HMPA and TFE. The solubility of the protected octadecapeptide, 2, was generally greater than that of 1 and DCC-HOBt coupling reactions were carried out in each of the above solvents under standardized conditions outlined in Table I. Following deprotection (1:1 TFA-CH_2Cl_2), yields of thymosin α_1(1-28) and thymosin α_1 (11-28) [from unreacted 2] were quantitatively determined for each reaction mixture by analytical hplc.

Although both starting materials are insoluble in $CHCl_3$, the coupling reaction proceeded in this solvent[24] and thymosin α_1 was obtained in 49.3% yield. Coupling in TFE

Solvent	Solubility Start	Solubility Final	Product (% Yield) Thymosin α_1(11-28)	Product (% Yield) Thymosin α_1(1-28)
Hexafluoroisopropanol	Soluble	Homogeneous	> 60	2.1
Hexamethylphosphoramide	Soluble	Gel	< 2	4.1
1-Methyl-2-pyrrolidinone	Sl.sol.	Gel	12.5	22.2
DMF-DMSO (1:1)	Sl.sol.	Gel	23.1	30 8
Trifluoroethanol	Soluble	Homogeneous	17.2	43.8
Chloroform	Insoluble	Suspension	25.9	49.3

(a) Standard conditions: Ac-(1-10)-OH (2.3 eq); DCC(5.5 eq); NMM(16 eq); 20°C; 24h.

Table I. Solvent Study: Coupling of Ac-(10)-OH and H-(11-28)-OtBu in Various Solvents by the DCC-HOBt Procedure[(a)]

proceeded in slightly lower yield (43.8%) but was the solvent of choice since the crude thymosin α_1 contained fewer contaminants than that obtained from the $CHCl_3$-mediated reaction (Figure 2). Lower yields of thymosin α_1 (20-30%) were obtained when the coupling was carried out in DMF-DMSO or MPD. Solubility of protected fragments is not the major factor in the coupling reaction. Although the starting materials were soluble in HMPA and HFIP, only trace amounts (<5%) of thymosin α_1 were observed after DCC-HOBt coupling. The very low yield in HMPA may be attributed to a side reaction with protected octadecapeptide, 2, since it is almost completely consumed. In contrast, the low yield of thymosin α_1 in HFIP is related to nonreactivity of 2 in this solvent since >60% of thymosin α_1(11-28) is recovered as the major product.

Coupling of the two fragments was evaluated in TFE in which variations were systematically made in equivalencies of protected decapeptide, DCC, HOBt and N-methylmorpholine (NMM) and in temperature and coupling time (Table II). Excess of acylating fragment, 1, improved the coupling yield from 26% (1 equivalent) to 43.8% (2.3 equivalents). Larger excesses of 1

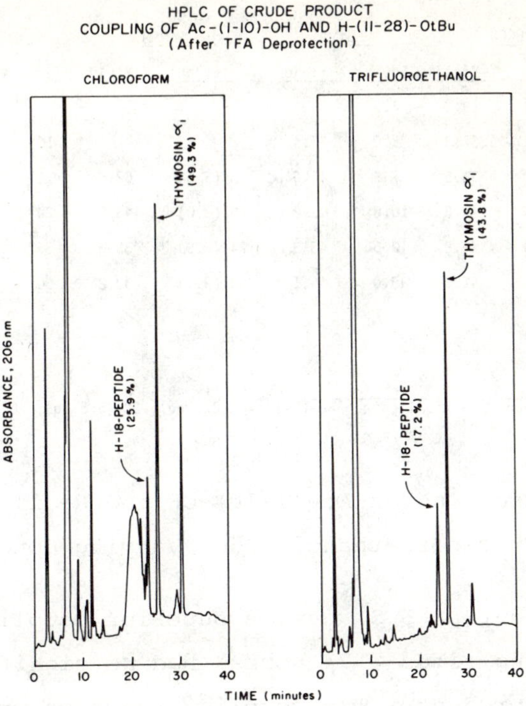

Fig. 2. Analytical hplc of crude product (after TFA deprotection) from the coupling of protected decapeptide and protected octadecapeptide in left (chloroform) and right (trifluoroethanol). Coupling conditions: Ac-(1-10)-OH (2.3 eq); DCC (5.5 eq); HOBt (10 eq); NMM (16 eq); 20°C; 24h. HPLC conditions: Column: Lichrosorb RP-8 (0.4x25 cm) [5μ]; Eluant: (A) 0.1M $HClO_4$ (pH 2.5) (B) Acetonitrile; Gradient 10%(B) to 30%(B) in 30 min; Detection: 206 nm; Flow rate: 1 mL/min.

gave only minimal improvement in yield. Similar observations were made with DCC and HOBt for which optimum yields were obtained, with excesses of 5.5 equivalents and 10.0 equivalents, respectively. On the other hand, use of large excesses of NMM suppressed the DCC-HOBt coupling in TFE significantly. Although 16 equivalents of NMM (which gives an apparent pH 8.0) was used for the comparison studies, the use of 8 equivalents

Ac-(1-10)-OH		DCC		HOBt		NMM					
Equiv	% Yield	Equiv	% Yield	Equiv	% Yield	Equiv("pH)	% Yield	Temp	% Yield	Time(h)	% Yield
1.0	26.0	1.0	19.8	2.5	36.1						
1.3	24.1	2.5	37.3	5.0	35.4			-10°	39.0	6	37.9
1.8	39.1	4.0	38.5	7.5	42.6	8 (5.5)	62.5	2°	40.4	12	40.4
2.3	**43.8**	**5.5**	**43.8**	**10.0**	**43.8**	**16 (8.0)**	**43.8**	**20°**	**43.8**	**24**	**43.8**
2.5	49.1	7.0	44.8	12.5	43.3	24 (8.9)	39.3	35°	51.1	34	43.8
3.0	49.3	8.5	42.9	15.0	49.1	48 (9.1)	30.2	50°	55.4	47	46.4
										57	44.1
										72	44.8

(a) Standard conditions are highlighted [Ac-(1-10)-OH(2.3 eq); DCC(5.5 eq); HOBt(10 eq); NMM (16 eq); 20°C; 24h]

Table II. Coupling of Ac-(1-10)-OH and H-(11-28)-OtBu in Trifluoroethanol under Varying Conditions[(a)]

of NMM (apparent pH 5.5) gave a superior coupling yield (62.5%). Longer coupling times (>24 hours) had no significant effect on the yield of product. The coupling reaction proceeded in better yield at higher temperatures; at 50° the yield of thymosin α_1 (55.4%) was significantly greater than that observed at 20° (43.8%). The thymosin α_1 was evaluated by tryptic digestion followed by hplc and by glass capillary chromatography and was determined to be free from racemization.[23] Further confirmation of structure was provided by amino acid analysis, optical rotation, ^{1}H-nmr and ^{13}C-nmr spectroscopy and FAB mass spectroscopy.

The optimum conditions for the DCC-HOBt coupling of 1 and 2 in TFE include (1) excess acylpeptide, 1 (>2.3 eq); (2) excess DCC (>5.5 eq); (3) excess HOBt (>10.0 eq); (4) coupling time >24 hours; and (5) excess NNM (8.0 eq). Yields are significantly better at higher temperatures but studies have not been evaluated at temperatures >50°C. These optimum conditions, achieved for the synthesis of thymosin α_1, may serve as a guideline for the DCC-HOBt coupling of other large protected fragments with low solubility in conventional solvents.

References

1. König, W. and Geiger, R. (1971) in Peptides: 1969, Scoffone, E., Ed., North Holland, pp. 17-22.

2. König, W. and Geiger, R. (1970) Chem. Ber. 103, 788-798.

3. Fletcher, G.A. and Young, G.T. (1972) J. Chem. Soc., Perkin Trans. 1, pp. 1867-1874.

4. Kamber, B., Riniker, B., Sieber, P., and Rittel, W. (1976) Helv. Chim. Acta 59, 2830-2840.

5. Jäger, G., König, W., Wissmann, H., and Geiger, R. (1974) Chem. Ber. 107, 215-231.

6. Kemp, D.S., Trangle, M. and Trangle, K. (1974) Tet. Lett. 2695-2696.

7. Meienhofer, J. (1979) in Peptides, Analysis, Synthesis, Biology, Vol. 1, Gross, E. and Meienhofer, J., Eds.; Academic Press, New York, pp. 197-239.

8. Bodanszky, M. (1971) in Prebiotic and Biochemical Evolution, Kimball, A. and Oro, J., Eds., North Holland Amsterdam, pp. 217-222.

9. Meienhofer, J. (1981) Biopolymers 20, 1761-1784.

10. Geiger, R., Jäger, G., König, W., and Volk, A. (1969) Z. Naturforsch, Part B, 24, 999-1004.

11. Tzougraki, C., Makofske, R., Gabriel, T., Michalewsky, J., Meienhofer, J., and Li, C.H. (1980) Int. J. Pept. Protein Res. 15, 377-398.

12. Yamashiro, D., Blake, J., and Li, C.-H. (1976) Tet. Lett. 1469-1472.

13. Kamber, B., Hartmann, A., Jöhl, A., Märki, F., Riniker, B., Rittel, W., and Sieber, P. (1975) in Peptides, Chemistry, Structure and Biology, Walter, R. and Meienhofer, J., Eds., Ann Arbor Science, Ann Arbor, Michigan, pp. 477-485.

14. Riniker B., Kamber, B., and Sieber, P. (1975) Helv. Chim. Acta 58, 1086-1094.

15. Felix, A., Heimer, E., Wang, C.-T., Lambros, T., Swistok, J., Ahmad, M., Roszkowski, M., Confalone, D., and Meienhofer, J. (1983) in Peptides: Structure and Function, Hruby, V.J. and Rich, D.H., Eds., Pierce Chem. Co., Rockford, IL, pp. 889-893.

16. Kenner, G.W. and Seeley, J.H. (1972) J. Am. Chem. Soc. 19, 3259-3260.

17. Goldstein, A., Low, T., McAdoo, M., McClure, J., Thurman, J.B., Rossion, J., Lai, C.-Y., Chang, D., Wang, S.S., Harvey, C., Ramel, A., and Meienhofer, J. (1977) Proc. Natl. Acad. Sci. USA 74, 725-729.

18. Wang, S.-S., Kulesha, I.D., and Winter, D.P. (1979) J. Am. Chem. Soc. 101, 253-254.

19. Birr, C. and Stollenwerk, U. (1979) Angew. Chem. Int. Ed. 18, 394-395.

20. Wong, T.W. and Merrifield, R.B. (1980) Biochem. 19, 3233-3238.

21. Wang, S.-S., Makofske, R., Bach, A., and Merrifield, R.B. (1980) Int. J. Pept. Protein Res. 15, 1-4.

22. Colombo, R. (1981) J. Chem. Soc. Chem. Commun. 1012-1013.

23. Felix, A., Heimer, E., Wang, C.-T., Lambros, T., Swistok, J., Roszkowski, M., Ahmad, M., Confalone, D., Scott, J., Parker, D., Meienhofer, J., Trzeciak, A., and Gillessen, D. (in press) Int. J. Pept. Protein Res.

24. A. Trzeciak and D. Gillessen (private communication).

THE IMPACT OF SECONDARY STRUCTURE FORMATION IN PEPTIDE SYNTHESIS

Manfred Mutter, Karl-Heinz Altmann, Dieter Bellof, Andreas Flörsheimer, Jürgen Herbert, Martin Huber, Birgit Klein, Ludovik Strauch and Thomas Vorherr

Institute of Organic Chemistry, University of Basel, CH-4056 Basel, Switzerland

Hans-Ulrich Gremlich

Medical Chemistry Research Dept., E. Merck, Postfach 4419, D-6100 Darmstadt, Germany

Introduction

Considerable progress has been achieved in the chemistry and methodology of peptide synthesis; however, the various strategies for building up a peptide chain still suffer from intrinsic limitations, e.g. low solubility or lack of quantitative conversions. Unfavourable 'polymer effects' due to interactions between the crosslinked support and attached peptide chain have been considered as the source of chain-length dependent coupling problems in stepwise solid-phase synthesis. However, it has been shown that conformational transitions of the growing peptide chain may cause a dramatic decrease in the reactivity[1]. Thus, conformationally induced changes in the physico-chemical properties of a growing peptide chain were revealed to be a general source of trouble. Further evidence for these conformational effects is presented in this work with special attention on their impact in polymer-supported synthesis.

Results and Discussion

The close interrelation between conformation and physicochemical properties of a peptide was brought to light in the synthesis and conformational investigation of homo- and cooligopeptides[2,3]. Generally, but especially in the case of hydrophobic peptides exhibiting low solubility, the use of polyethylene glycol (PEG) as solubilizing C-terminal protecting group (liquid-phase-method[4]) proved to be a valuable tool in the conformational analysis of peptides[3,5]. The solubility of a growing non-attached peptide chain strongly depends on the preferred conformation, as shown previously[1]: in particular, the transition random-coil (r.c.) ⟶ β-structure is paralleled by a drastic decrease of the solubility. With PEG-peptides, the same observable transition is paralleled by a decrease of the reactivity and the β-structure aggregation is directly detectable by gel permeation chromatography[6]. Such aggregation must inevitably interfere with stepwise synthesis. For example, the β-structure forming peptide oligo-(L-Val)-PEG (Figure 1,B) clearly reflects its aggregated state in the coupling kinetics: addition of excess carboxylic component does not result in a significant increase of the coupling yields (Figure 1,A): a repeated precipitation and redissolution provoking redistribution of hidden terminal amino groups appears to be a condition for quantitative reaction. Obviously, the analytical controls for complete reaction suffer from the same conformational restraints as the coupling reaction itself[7]. Indeed, a drastic decrease in the sensitivity of the ninhydrin reaction is observed with increasing chain length, as shown in Figure 2. In view of such a drop of the reactivity of terminal amino groups even in homogenous solution, the reliability of quantitative coupling tests during stepwise synthesis of long peptide chains must be re-evaluated.

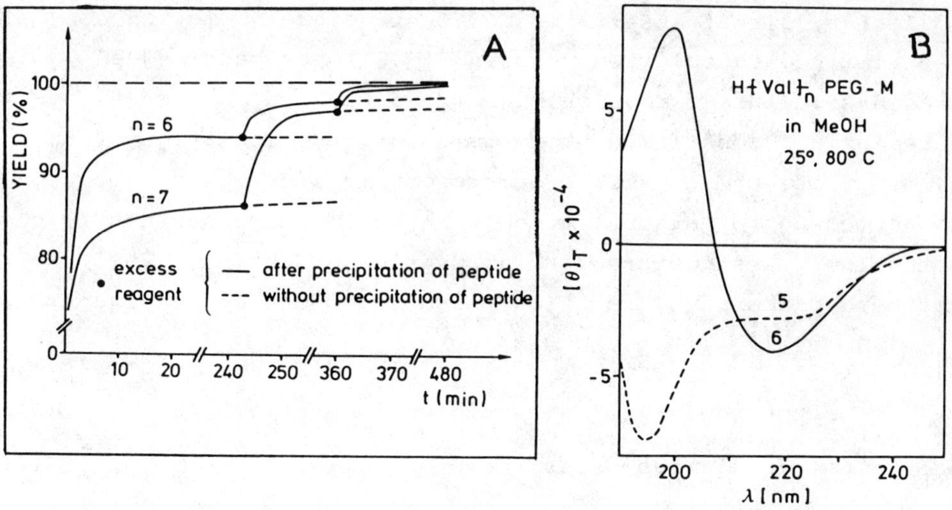

Fig. 1 Coupling kinetics of Fmoc-Val ⟶ $H\{Val\}_n$ PEG-M, in CH_2Cl_2/DMF (A); CD-spectra of $H\{Val\}_n$ PEG-M in MeOH at 25° and 80° C (B).

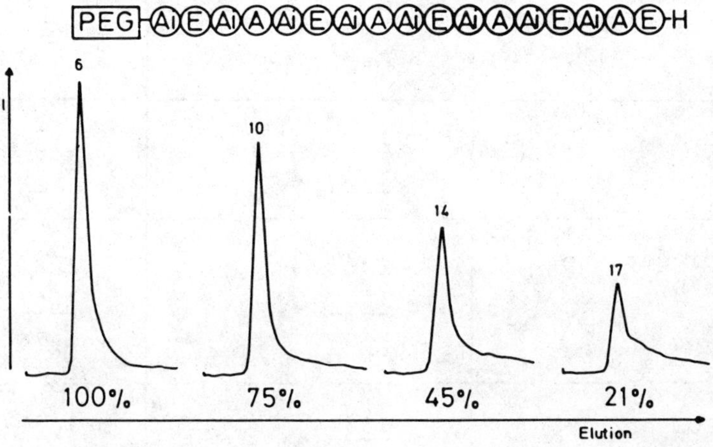

Fig. 2 Decrease in the sensitivity of the Stein-Moore-reaction of PEG-bound 17-peptide (Ai = aminoisobutyric acid) and precursors. Aliquots of 1 μM PEG-peptide for the indicated stages were directly subjected to an automatic amino acid analyser.

So far, direct observation of β-structure formation has been limited to solution coupling[1,3]. It has been argued[8], that β-structure formation might be decreased or even eliminated using crosslinked supports due to the isolation of the sites of attachment . In order to obtain information about the impact of conformational effects in solid-phase synthesis, the potentially β-structure forming peptide H-(Thr(Bu^t)-Val]$_4$-R has been prepared by solution - and solid-phase-techniques:

R_I = PEG-M (polyethylene glycolmonomethylether, $\bar{M}_r$~5000)[4] (I)

R_{II} = polystyrene - 1 % divinylbenzol[7] (II)

R_{III} = polyacrylamide-crosslinked[9] (III)

The influence of the mode of attachment (R) upon the conformational properties of the growing peptide chain has been investigated by CD and FT-ATR-IR[10] studies. The results are summarized in Table I.

H-[Thr(Bu^t)-Val]$_4$-R	SOLID STATE KBr, (cm^{-1})		swollen in CH_2Cl_2	
R_I = PEG-M	1625 1690	β	1625 1655	β a) r.c. a)
R_{II} = Polystyrene-1% DVB	1639 1688 1655	β r.c.	1633 1687 1676	β
R_{III} = Polyacrylamide, crosslinked[9]	1637 1653	β r.c.	1633 1675	β

Table I: FT-ATR-IR[10]-absorption bands (amide I-region[11]) of H-[Thr(Bu^t)-Val]$_4$-R ; a) conventional IR of solution in CH_2Cl_2;—— strong, ---- weak absorption band

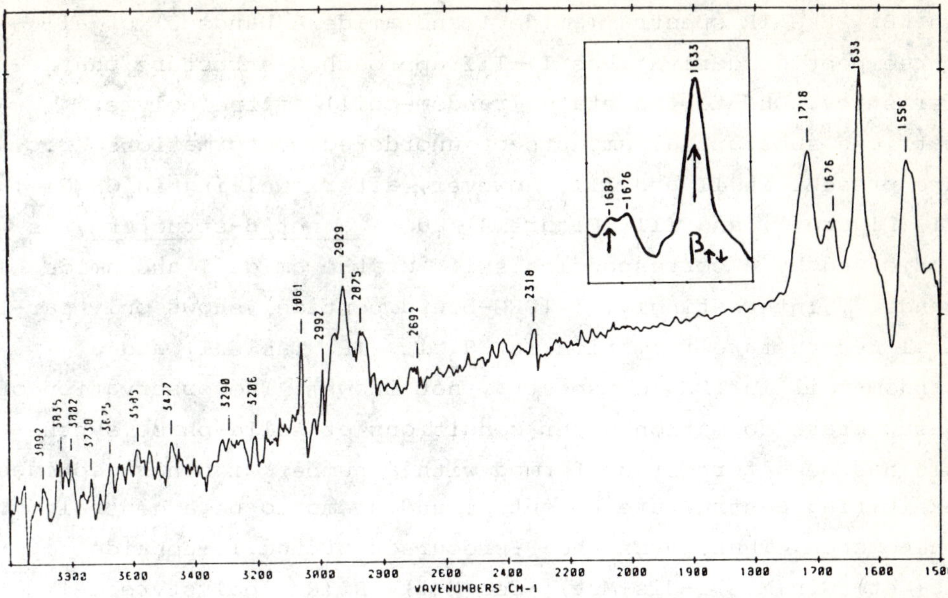

Fig. 3 FT-ATR-IR-spectra of H[Thr(Bu^t)-Val]$_4$ PS; (PS = polystyrene-1% DVB).

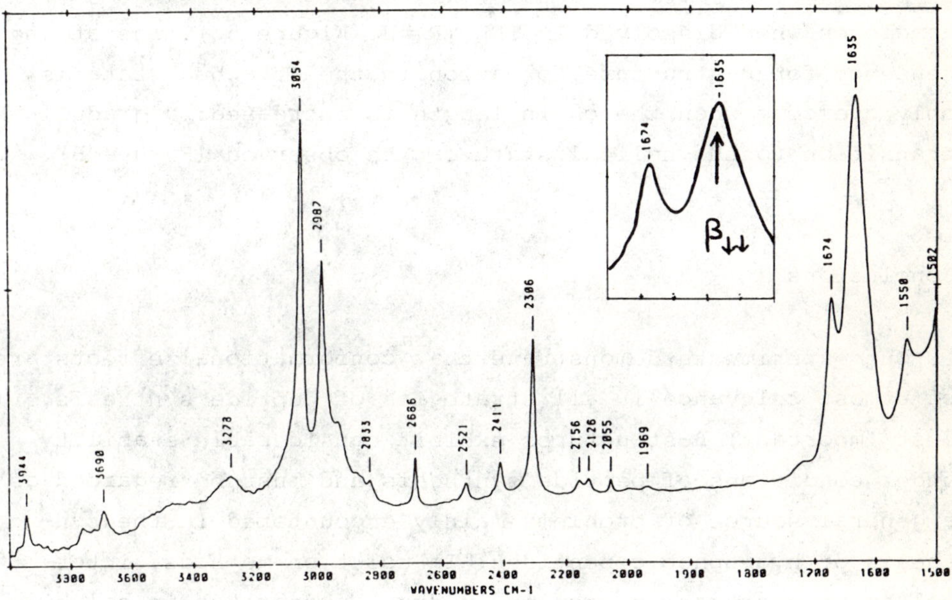

Fig. 4 FT-ATT-IR-spectra of H[Thr(Bu^t)-Val]$_4$ PA; (PA = polydimethylacrylamide, crosslinked[9]).

In KBr, the IR spectra (amide I and amide A bands[11]) of all three peptide derivatives I - III approach β-structure characteristics. Due to the state (random-coil) of the polymer matrices substantial amounts of unordered conformations (r.c.) are present in II and III. However, after swelling in CH_2Cl_2, II (Figure 3) and III (Figure 4) adopt a net β-structure, as indicated by a corresponding shift of the amide I and amide A bands[11]. Interestingly, I (PEG-bound peptide) shows only partial β-structure formation in CH_2Cl_2 (IR, Table I) and a random-coil in TFE (CD spectra, not shown). The observation of β-structure formation under conditions of solid-phase synthesis has been further confirmed with a number of other peptides exhibiting β-structure potential and seems to be a general phenomenon. Thus, even the 9-precursor of the 18-peptide Ac$\leftarrow$Met$\rightarrow_4$Glu$\leftarrow$Met$)_2$-His$\leftarrow$Met$)_3$-Ser$\leftarrow$Met$\rightarrow_6$ R (R = polystyrene-1% DVB) adopts a β-structure when attached to a swollen polymer support (Figure 5A), although it is helical like its higher homologes when dissolved in TFE (R = H, Figure 5C). The strong tendency for β-structure formation in the attached state is only overcome when the chain length is increased: a gradual transition to the helical state can be observed (Figure B).

Conclusions

Our experiments demonstrate that conformational effects are of utmost relevance in all strategies of peptide synthesis. Most important, β-structures exhibit considerable stability under conditions of peptide synthesis and must be regarded as a general source of problems mainly encountered in the synthesis of hydrophobic peptides. In solution methods, intermolecular aggregation results in drastic decrease in solubility. In solid-phase synthesis, conformational transitions might be the origin of both incomplete coupling reactions and

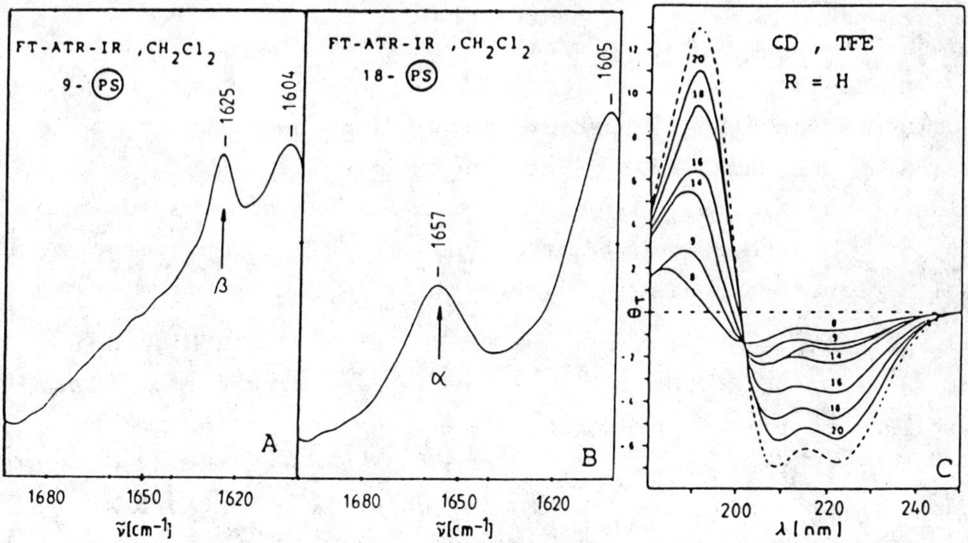

Fig. 5 FT-ATR-IR spectra of AC$[\text{Met}]_2$ Ser$[\text{Met}]_6$ PS (PS = polystyrene-1% DVB, A) and of Ac$[\text{Met}]_n$ Glu$[\text{Met}]_2$ His-$[\text{Met}]_3$ Ser$[\text{Met}]_6$ PS(n = 4,B), swollen in CH_2Cl_2; CD-spectra of the nonattached homologous peptide (n = 6, R = H) and its precursors in TFE.

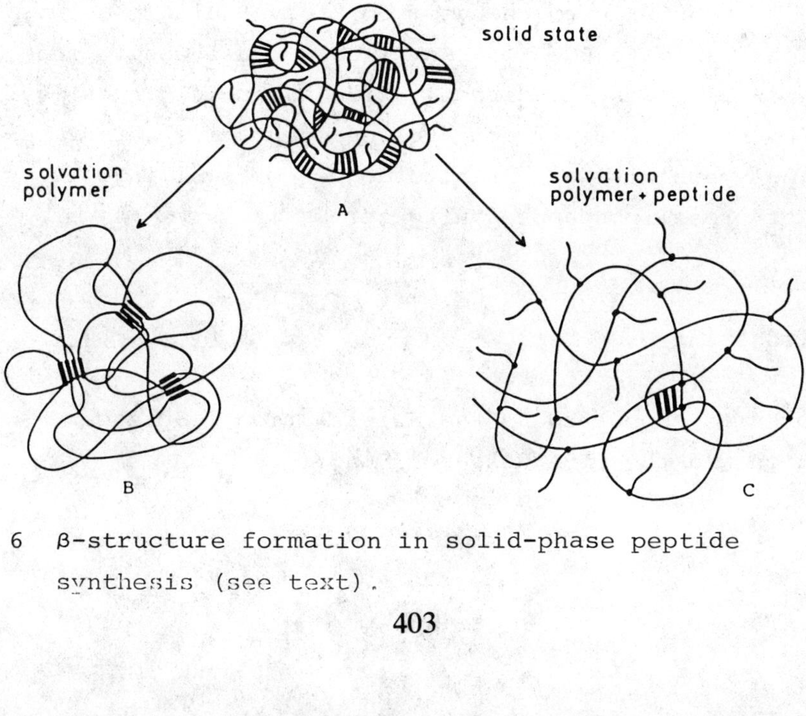

Fig. 6 β-structure formation in solid-phase peptide synthesis (see text).

deficiency of analytical controls.

In the light of these results, the polymer-support used in solid-phase synthesis must exhibit maximum swelling in solvents which at the same time garantee optimum solvation of the peptide chain (Figure 6 C). Under conformational aspects, polymer-supports of the polyacrylamide-type[9] are clearly preferable in solid-phase synthesis.

For example, hydrophobic polymers which show a high degree of swelling in apolar solvents favour ß-structure formation, viz. aggregation of hydrophobic peptides due to the freely interpenetrating coils (Figure 6 B). On the other hand, polar supports show high swelling in solvents which destabilize or even disrupt ß-structures[12] (Figure 6 C). However, the existence of ß-structure promoting sites within the statistically functionalized resin (Figure 6 C) appears to be an intrinsic problem, which can only be solved by new types of polymeric supports[13,14]. Consequently, a major step forward in overcoming unfavourable conformational effects in solid-phase peptide synthesis is to change from truly "solid-phase-conditions" (Figure 6 A) towards "liquid-phase-conditions". This conclusion results directly from a careful investigation of PEG-bound peptides, which are representative for cross-linked polymers with an infinite degree of swelling, i.e. the number of crosslinks approaching zero.

Acknowledgement

This work was supported by the Swiss National Science Foundation and the CIBA Research Fund.

References

1. Mutter, M., Pillai, V.N.R., Anzinger, H., Bayer, E. and Toniolo, C. (1980) PEPTIDES: Synthesis-Structure-Function, Proceedings of the Sixteenth European Peptide Symposium, Brunfeldt, K. Ed., Scriptor, Copenhagen, pp. 660-665.

2. Maser, F., Klein, B., Mutter, M., Toniolo, C. and Bonora G.M. (1983) Biopolymers 22, 223-240.

3. Pillai, V.N.R. and Mutter, M. (1981) Acc. Chem. Res. 14, 122-130.

4. Mutter, M. and Bayer, E. (1979) THE PEPTIDES: Analysis Synthesis-Biology, Vol. 2, Meienhofer, J. and Gross, E. Eds. Willey, New York, pp. 285-333.

5. Mutter, M. and Altmann, K.-H. (1985) Int. J. Peptide Prot. Res., in press.

6. Schmitt, J. and Mutter M. (1983) Biopolymers 22, 1849-1852.

7. Birr, Ch. (1978) ASPECTS OF THE MERRIFIELD PEPTIDE SYNTHESIS, Springer-Verlag, New York 1978.

8. Merrifield, R.B., personal communication.

9. Arshady, R., Atherton, E., Clive, D.L.J. and Sheppard, R.C. (1981) J. Chem. Soc., Perkin Trans. I, 529-537.

10. Harrick, N.J. (1967) INTERNAL REFLECTION SPECTROSCOPY, Interscience, New York.

11. Krimm, S. (1983) Biopolymers 22, 217-225.

12. Toniolo, C., Bonora, G.M., Maser F. and Mutter, M. (1983) J. Chem. Soc., Chem. Comm., 1298-1299.

13. Becker, H., Lucas, H.-W., Maul, J., Pillai, V.N.R., Anzinger, H. and Mutter, M. (1982) Makromol. Chemie 183, 217-223.

14. Bayer, E., Dengler, M. and Hemmasi, B. (1985) Int. J. Peptide Prot. Res. 25, 178-186.

DIFFICULT SEQUENCES IN STEPWISE PEPTIDE SYNTHESIS: COMMON MOLECULAR ORIGINS IN SOLUTION AND SOLID PHASE?

Stephen B. H. Kent

Division of Biology 147-75, California Institute of Technology
Pasadena, California 91125

In both stepwise and segment solution synthesis, the most common difficulty is poor solubility of protected peptide intermediates. Systematic studies of this problem have been carried out in several laboratories, particularly using IR spectroscopy to characterize the state of peptide chains.[1] From this work it has been concluded that the phenomena of slow, incomplete couplings and insolubility of protected intermediates are related at a molecular level by the tendency of some protected peptides to form intermolecular beta-sheet aggregates which both hinder couplings and tend to make the peptides precipitate. Although it is known that this tendency to form beta-sheet aggregates is sequence dependent, the rules governing such dependence are poorly understood and the occurrence of the problem is not yet predictable.[2]

The use of soluble polymers to increase the solubility of covalently attached peptide chains in organic solvents has only limited efficacy in that such polymers do not strongly interfere with intermolecular beta-sheet formation. In fact, suitable peptide sequences can pull out of solution the normally-soluble polymers to which they are attached.[3] Thus, poor solubility of protected peptide intermediates is a continuing problem in the solution synthesis of maximally-protected peptides.

In stepwise solid phase synthesis, it has frequently been stated that "solubility problems do not exist." Yet, at the same time the phenomenon of "difficult" sequences is well known and has been reported in the literature for more than a decade. The nature of the "solid phase" is now well understood and, thanks to recent progress in perfecting the chemistry of chain assembly, the characteristics of these difficult syntheses are becoming well-defined. Based on these data, it can now be seen that the phenomena of sequence-dependent synthetic difficulties have a common mechanistic origin in solution and on the solid phase. Because of the favorable properties of the supports commonly used for solid phase peptide synthesis, the problems are minimized. With suitably optimized chemistry, virtually any sequence can now be synthesized by stepwise solid phase procedures.

Nature of the Resin Support in Solid Phase Peptide Synthesis

In the solid phase method, the growing protected peptide chain is covalently attached to a solid support for convenience of intermediate purification. This "solid" support is a loosely crosslinked interpenetrating polymer network, either styrene-divinylbenzene or crosslinked polyacrylamide, which imbibes organic solvents to become highly swollen. Synthesis of the covalently attached peptide chain occurs throughout the interior of the swollen beads.

A large body of evidence indicates that the interior of the peptide-resin is a highly mobile environment comparable at a molecular level to free solution, except that large scale translational motion (diffusion) is restricted. Inhomogeneities within the swollen peptide-resin are averaged out rapidly on the nmr time scale (i.e., within microseconds).[4] The statistical mechanics of interpenetrating polymer networks have been well worked out by Flory,[5] and it can be rigorously shown that the swollen polymer and the covalently attached peptide chains mutually enhance one another's solvation. This has been confirmed by experimental observation of the huge swelling increases that occur in the course of assembly of long protected peptide chains on crosslinked resin supports.[6] Thus, interference by the resin support is not the explanation for the difficulty in assembling some peptide chains on solid supports.

Chemistry of Solid Phase Peptide Synthesis

Over the past few years, many of the problems of stepwise solid phase synthesis attributed to putative physical shortcomings of the resin have been shown to be due to chemical side reactions. Thus, trifluoroacetylation was due to extraneous reactive functionalities on the resin not to physical entrapment of TFA,[7] and the occurrence of free peptides shorter than the target peptide was due to impurities in protected amino acids from some sources rather than to steric hinderance within the resin.[8] Chronic occurrence of deletion peptides was due to reversible Schiff's base formation with aldehyde moieties within the resin rather than to temporary steric occlusion of the growing peptide chain.[9] These chemical side reactions have been minimized by the use of chemically clean, stable resins[10] (Table I).

Table I. Levels of Chronic Chemical Side Reactions in Stepwise Solid Phase Peptide Synthesis

STANDARD RESIN		CLEAN, STABLE RESIN
1.0%	CHAIN LOSS	0.01%
1.8%	TRIFLUOROACETYLATION	0.02%
0.6%	DELETION	0.05%
0.6%	TERMINATION	0.02%
4.0% per residue		0.1%

This has resulted in a dramatic improvement in synthetic yields to greater than 99.8% per amino acid residue. Using this improved chemistry, many peptides can be synthesized in excellent yield and high purity.

Difficult Sequences

With the background of chronic chemical problems removed, it becomes more obvious that there are in fact substantial difficulties in the synthesis of some peptides by stepwise solid phase methods. Thus, several difficult syntheses have been recognized[11] and survey of a large number of syntheses in our laboratory showed that about 20% of them contained regions that gave poor coupling yields under standard synthetic conditions in dichloromethane (2-20% incomplete).[12] We have proposed that these difficult sequences occurred because the peptides had a strong tendency to form intermolecular beta-sheet aggregates (i.e., precipitate) rather than interact with the solvent (remain in solution).

Observed characteristics consistent with this explanation include: the problem is sequence rather than residue specific. This is most dramatically illustrated by

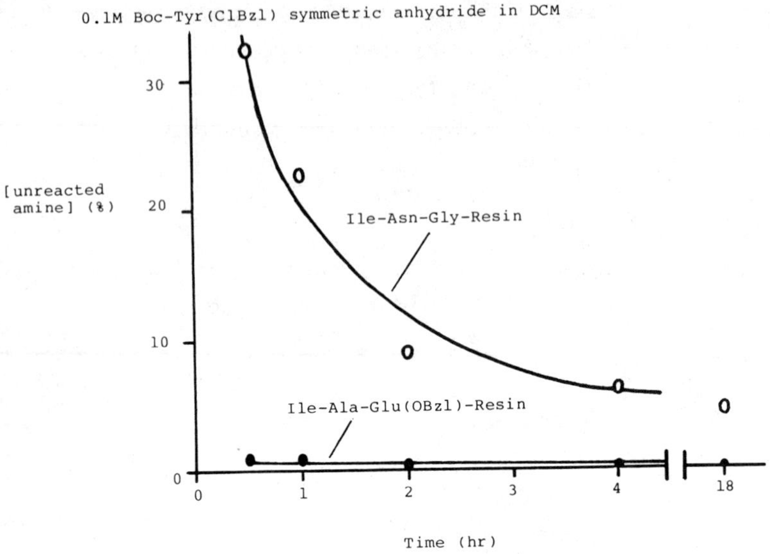

Fig. 1. Reaction of Boc-Tyr(ClBzl) Symmetric Anhydride with Ile-Asn-Gly-Resin and Ile-Ala-Glu(OBzl)-Resin.

the experiment shown in Figure 1, where both reactions are between activated Boc-Tyr in solution and tripeptides with N-terminal isoleucine residues. Every variable of the two experiments was held constant (even the same batch of resin was used) except for the internal sequence of the two peptides. The dramatic difference in coupling rates in dichloromethane was paralleled by differences in molecular mobilities reflected in the nmr spectra of the two peptide-resins. The slower coupling peptide-resin showed evidence of intermolecular aggregation resulting in broadened nmr spectral lines and restricted molecular tumbling rates.[4] Other incoming activated amino acids (e.g., Boc-Ala) also reacted slowly with this poorly solvated peptide.

Observed sequence-dependent coupling difficulties occurred for several consecutive residues with maximal frequency at chain lengths of about 15 residues and were rare beyond 20 residues.[12] This is consistent with the known maximal tendency of amino acid copolymers to form beta-sheet aggregates at chain lengths of 4-18 residues, and for the same copolymers to adopt helix/random coil conformations at chain lengths of 20 or more residues. Difficult sequences in solid phase synthesis show a strong quantitative dependence on the number of peptide chains per gram of polystyrene resin: higher loadings of peptide on the resin exaggerate the problem, while lower loadings minimize it.[13]

Thus, it can be clearly seen that the phenomenon of difficult sequences in stepwise solid phase peptide synthesis is due to the properties of the peptide, not to the properties of the resin support used. Any tendency to aggregate and become poorly solvated is dramatically reduced for peptides attached to swollen interpenetrating polymeric network supports,[6] but can still affect the course of a synthesis. This ability of these crosslinked supports to assist the solvation of attached peptide chains is in contrast to the inability of soluble linear polymers to interfere with intermolecular aggregation of attached peptide chains, and is dramatically illustrated by the successful stepwise assembly of poly-isoleucine on a Gly-(styrene-divinylbenzene) resin, using dichloromethane as solvent. Although some reduction in reaction rate and yield was observed at intermediate chain lengths (Figure 2), even in these cases it was possible to obtain close to quantitative yields on recoupling in the same solvent. In analogous experiments with uncrosslinked polymer supports, it was not possible to keep the poly-isoleucine in solution even with increasing amounts of polar solvents such as DMF and DMSO.[3]

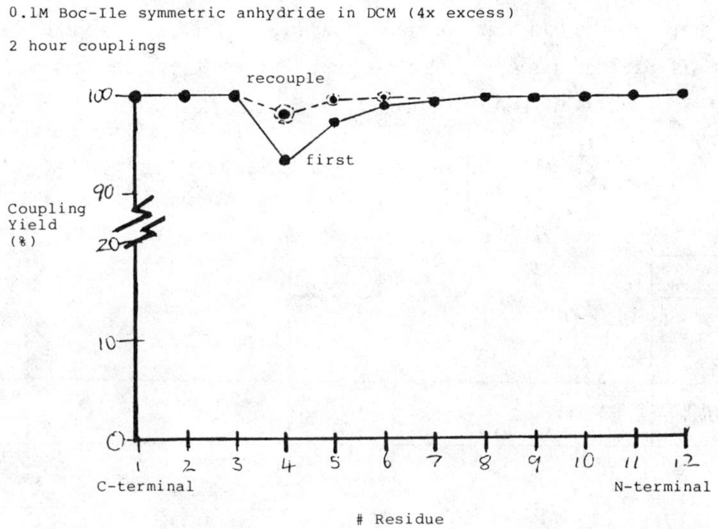

Fig. 2. Assembly of Ile_{12}-Gly-OCH_2-Pam-(S-DVB) in DCM. Coupling Yields from Quantitative Ninhydrin Monitoring, Confirmed by Quantitative Edman Degradation.

Even with crosslinked resin supports, couplings can be solvent dependent. For the difficult reaction shown in Figure 1, the use of DMF, a much better solvent for protected peptides, completely eliminates the differences in reaction rates and nmr behavior for the two peptide resins. The loading dependence of the difficult coupling is also eliminated by the use of DMF.[13] Thus, the use of better solvents for protected peptides helps minimize the occurrence of sequence-dependent difficult couplings. It should be noted that there are no reliable reports of incomplete removal of the N^{α}-Boc group in stepwise solid phase synthesis, because the 60% TFA in DCM used provides an excellent medium for the maximal solvation of peptide-resins.

Practical Consequences

This understanding of the origin of sequence-dependent coupling problems has contributed to a practical stepwise solid phase chemistry of considerable generality for the assembly in high yields of protected peptide chains. We have developed a peptide synthesizer, the design of which was based on the above rationale, that effects such synthetic chemistry.[14] The activated amino acid derivative (preformed symmetric anhydride) is formed separately by DCC activation in dichloromethane, and then exchanged into DMF solution, after which it is added to the peptide-resin neutralized in DMF (Figure 3). This rational approach to a general

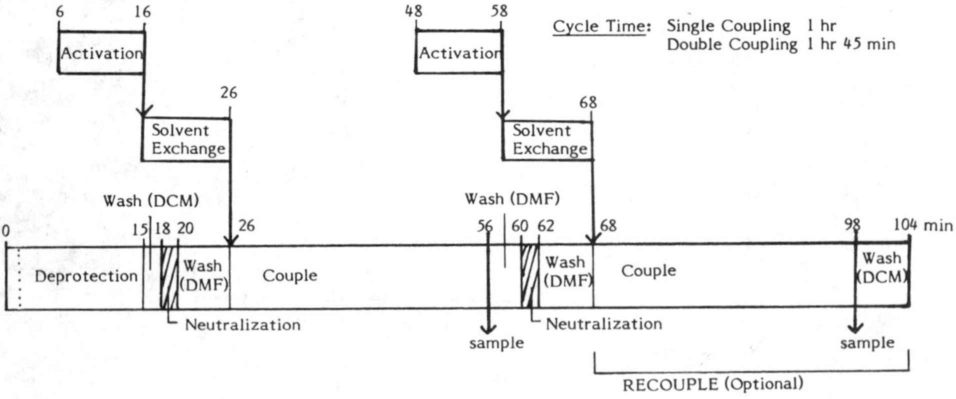

Fig. 3. Parallel Activation and Solvent Exchange to Give Boc-Amino Acid Symmetric Anhydride in DMF.

stepwise solid phase synthesis, in conjunction with the use of clean, stable resins and high purity solvents and reagents, has proved to be very effective for a wide range of target peptides. Occasional sequence-dependent difficulties in chain assembly are still observed, but with considerably reduced frequency and seriousness. Quantitative data on one of the most difficult sequences so far encountered is shown in Table II.

Table II. Sequence-dependent Coupling Difficulties in the Assembly of the Protected Peptide Chain Human Growth Hormone Releasing Factor (1-27)

1				5					10
Tyr-	Ala-	Asp-	Ala-	Ile-	Phe-	Thr-	Asn-	Ser-	Tyr-
99.02	99.23	99.13	99.14	99.10	99.11	99.22	98.28	98.14	96.38
99.00	99.21	99.23	99.21	99.03	99.20	98.99	99.03	98.45	96.81
								- - -	- - -
11				15					20
Arg-	Lys-	Val-	Leu-	Gly-	Gln-	Leu-	Ser-	Ala-	Arg-
92.69	98.06	97.66	98.99	99.02	91.29	97.04	99.07	99.38	99.00
96.99	98.95	98.63	99.42	99.52	95.31	98.73	99.03	99.64	98.94
- - -	- - -	- - -			- - -	- - -			
21				25		27			
Lys-	Leu-	Leu-	Gln-	Asp-	Ile-	Met-(Me) BHA-Resin			
99.35	99.67	99.70	99.76	99.67	99.57	99.69			
99.65	99.72	99.81	99.84	99.92	99.91	99.92			

Chain Assembly Yields
(% reaction based on quantitative ninhydrins):

FIRST COUPLE	98.88% per residue
SECOND COUPLE	99.19% per residue
Increment	+0.31% per residue

Summary

Difficult sequences in stepwise solid phase synthesis originate in the properties of the peptide *not* the resin support. The mechanism is the same as for sequence-dependent synthetic difficulty in solution synthesis: formation of intermolecular beta-sheet aggregates. Swollen interpenetrating polymer network supports to which the peptide chain is attached in solid phase synthesis help minimize this tendency to intermolecular aggregation; this can be further minimized by the exclusive use of better solvents for protected peptide chains in the course of chain assembly.

References

1. Baron, M.H., DeLoze, C., Toniolo, C., and Fasman, G.D. (1978) *Biopolymers* **17**, 2225-2239.
2. Pillai, V., and Mutter, M. (1981) *Acc. Chem. Res.* **14**, 122.
3. Abd El Rahman, S., Anzinger, H., and Mutter, M. (1980) *Biopolymers* **19**, 173.
4. Live, D.H., and Kent, S.B.H. (1983) *Peptides: Structure and Function*, Hruby, V.J., and Rich, D.H., Eds., Pierce Chem. Co., Rockford, pp. 65-68.
5. Flory, P. J. (1979) *Macromolecules* **12**, 119.
6. Sarin, V.K., Kent, S.B.H., and Merrifield, R. B. (1980) *J. Am. Chem. Soc.* **102**, 5463-5470.
7. Kent, S.B.H., Mitchell, A.R., Engelhard, M., and Merrifield, R. B. (1979) *Proc. Natl. Acad. Sci. USA* **76**, 2180-2184.
8. Kent, S.B.H., and Merrifield, R. B. (1983) *Int. J. Peptide Protein Res.* **22**, 57-65.
9. Kent, S.B.H. (1983) *Peptides: Structure and Function*, Hruby, V.J., and Rich, D.H., Eds., Pierce Chem. Co., Rockford, pp. 99-102.
10. Kent, S.B.H. (1980) *Biomedical Polymers*, Goldberg, E.P., and Nakajima, A., Eds., Academic Press, New York, pp. 213-242.
11. (a) Westall, F.C., and Robinson, A.B. (1970) *J. Org. Chem.* **35**, 2842-2844.

    (b) Hancock, W.S., Prescott, D.J., Marshall, G.R., and Vagelos, R.P. (1972) *J. Biol. Chem.* **247**, 6224-6233.

    (c) Atherton, E., Woolley, V., and Sheppard, R.C. (1980) *J.C.S. Chem. Comm.*, 970-971.
12. Meister, S.M., and Kent, S.B.H. (1983) *Peptides: Structure and Function*, Hruby, V.J., and Rich, D.H., Eds., Pierce Chem. Co., Rockford, pp. 103-106.
13. Kent, S.B.H., and Merrifield, R.B. (1981) *Peptides 1980*, Proc. 16th European Peptide Symp., Brunfeldt, K., Ed., Scriptor, Copenhagen, pp. 328-333.
14. Kent, S.B.H., Hood, L.E., Bielan, H., Meister, S., and Geiser, T. (1984) *Peptides 1984*, Proc. 18th European Peptide Symp., Ragnarsson, U., Ed., Almqvist & Wiksell, Stockholm, pp. 185-188.

DETECTION OF PROBLEM SEQUENCES IN SOLID PHASE SYNTHESIS

Eric Atherton and Robert C. Sheppard

NRC Laboratory of Molecular Biology, Hills Road, Cambridge CB2 2QH, England.

Incomplete acylation is relatively easy to detect by qualitative or quantitative colour tests though it is difficult to separate effects of individual residues (e.g. steric hindrance of Val, Ile; decomposition of activated arginine before acylation is complete, etc.) from more fundamental sequence effects. Using polydimethylacrylamide resins and Fmoc-amino acids, we have observed marked slowing of acylation in a number of solid phase syntheses which appear to have their origin in structural transitions within the resin matrix. Frequently this has been associated with visible changes in resin structure, the highly expanded gel shrinking and becoming more granular in nature. Only in one case (see below) has a substantial change in tactics been necessary to complete the synthesis.

The recent development of continuous flow methods[1,2] has given some new insight into the problem. Spectrometric monitoring of the deprotection as well as acylation steps is now possible. There are substantial problems in quantitative interpretation of the spectrometric data to the accuracy required by solid phase synthesis,[2] but the general shape of the deprotection profiles has proved to be a sensitive indicator. The figure shows three deprotection profiles from a recent synthesis of a sequence (6) related to the adenovirus tail fibre. Profile (a) is normal and is typical of the first 12 residues. The following deprotection reaction gives a lower broader peak (b), further broadened in the next (c). Subsequent profiles were similar to this last. It should be emphasised that the spectrometric profile reflects both the rate of cleavage of the Fmoc group and the rate of diffusion of the ultra violet absorbing product out of the resin matrix. In either event, it most likely indicates some form of structural transition within the resin-peptide complex. A number of other examples are collected in the Table.

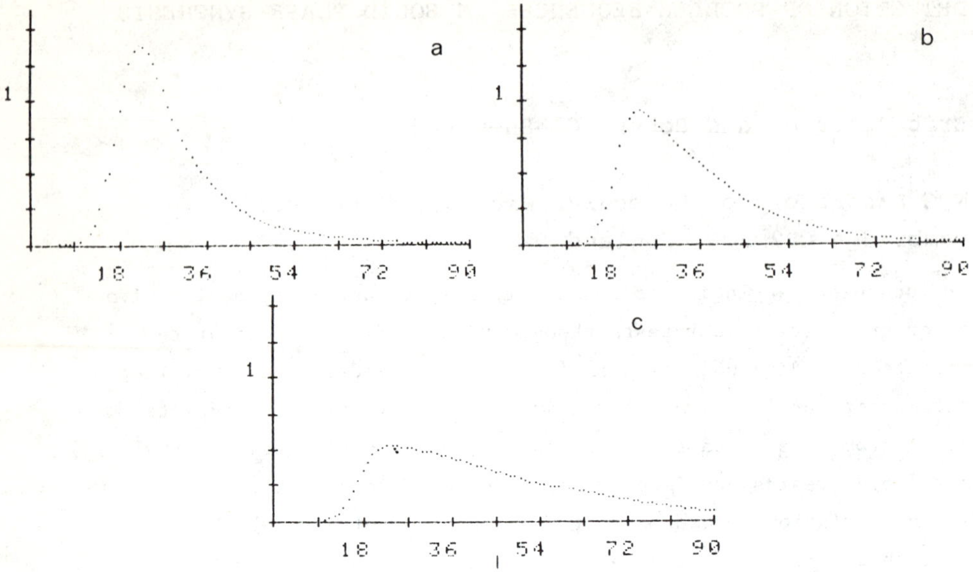

15 14 13 12 11 10 9 8 7 6 5 4 3 2 1

(1) 28←Arg-Lys-Gln-Met-Ala-Val-Lys-Lys-Tyr-Leu-Asn-Ser-Val-Leu-Thr

(2) H-Ser-Gln-Glu-Gly-Asn-Thr-Met-Lys-Thr-Asn-Asp-Thr-Tyr

(3 Val-Gln-Ala-Ala-Ile-Asp-Tyr-Ile-Asn-Gly

(4) Ser-Val-Tyr-Ala-Glu-Glu-Val-Lys-Pro-Phe-Pro

(5) Leu-Asp-Gly-Trp-Gln-Val-Ile-Ile-Thr-Asp-Asp-Tyr

(6) 33←Lys-Leu-Ser-Val-Ala-Thr-Lys-Gly-Pro-Leu-Thr-Val-Ser-Asp-Gly

(7) Leu-Ala-Glu-Leu-Gly-Ala-Ser-Leu-Leu-Lys-His-Trp

(8) Ser-Asp-Ser-Ala-Gln-Gly-Ser-Asp-Val-Ser-Leu-Tyr-Ala

(9) 39←Glu-Arg-Glu-Asp-Leu-Ile-Ala-Tyr-Leu-Lys-Lys-Ala-Thr-Asn-Glu

(10) 26←Leu-Val-Lys-Trp-Ile-Ile-Asp-Thr-Val-Asn-Lys-Phe-Thr-Lys-Lys

All the continuous flow examples (1-7) were brought to completion, requiring only extended deprotection (as indicated by the spectrometric record) or extended or repeated acylation (as indicated by colour tests). The three last examples were earlier and used more conventional discontinuous techniques. The cytochrome c part sequence (9) showed severe resin shrinkage when the chain was extended to the region indicated. It was brought to completion by increasing the polarity of the side chain protecting groups (Lys(Boc) changed to Lys(Tfa); Met to Met(O)), and reducing the loading on the resin.[3]

The haemolysin sequence (10) did not show marked resin collapse. Amino-acid incorporation was apparently incomplete in the region indicated (there are analytical problems because of the Trp-Ile-Ile sequence), but remarkably it behaved differently when the assembly was repeated under continuous flow conditions using the different kieselguhr-supported polydimethylacrylamide resin. The assembly was seemingly completed satisfactorily as judged by amino-acid analysis, though no biologically active product was obtained.

Interpretation of the data in the slide in terms of individual sequences is difficult. In general however, we would favour explanations based on the concepts regarding peptide-resin structure which we discussed in 1971.[4] Then we considered that the three states (a-c) might contribute to the peptide resin complex depending on the nature of the resin, the permeating, solvating fluid, and of course the peptide sequence and its protecting groups. Later we added a fourth possible state (d) in which the poor solvation and collapse of the peptide chain (intramolecular association) depicted in (b) is replaced by intermolecular association (d). This last state would most likely be characterised by resin shrinkage (there is effectively increased cross-linking of the resin) whereas (b) would not. State (c) with poor solvation of the polymer backbone could also result in resin shrinkage or poor initial swelling. This is unlikely to contribute substantially in the polydimethylacrylamide series as the resin is very highly solvated indeed by dimethylformamide. Our early thoughts were that this state might intervene when a polar support

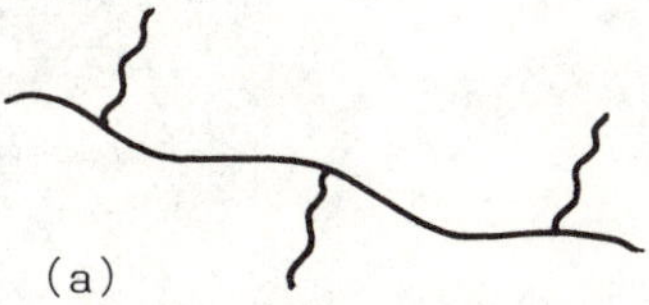
(a)

(b)

(c)

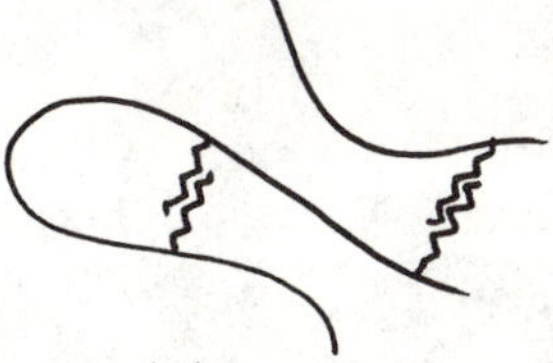
(d)

(eg. polydimethylacrylamide) was permeated by a non-polar solvent (eg dichloromethane), or vice versa with polystyrene and dimethylformamide. It is of great interest that this is apparently not the case, and that improved coupling has been obtained when dimethylformamide is used (presumably admixed with residual dichloromethane) as the permeating liquid in polystyrene-based syntheses.[5] Probably the beneficial effect of added dimethylformamide reflects dominantly the improved solvation of the protected peptide chain and the soluble component, and the kinetically improved reaction medium that it provides.

1. Atherton, E., Brown, E., Sheppard, R.C. and Rosevear, A.(1981), J. Chem. Soc. Chem. Comm. 1151; Sheppard, R.C. (1983) Chem. Brit. 402.
2. Dryland, A. Sheppard, R.C. J. Chem. Soc. Perkin I, in press.
3. Atherton, E., Wooley, V. and Sheppard, R.C. (1980) J. Chem. Soc. Chem. Comm., 970.
4. Sheppard, R.C. in 'Peptides 1971' (1973) Proc. 11th Europ. Peptide Symp., Vienna, 1971; North Holland Pub. Co., Ansterdam, p.111.
5. Kent, S.B.H. and Merrifield, R.B. in 'Peptides 1980', Proc. 16th Europ. Peptide Symp., Helsingor, Denmark, 1980. Scriptor, Copenhagen 1981, p.328.

SELF-ASSOCIATION AND SOLUBILITY OF PEPTIDES. A SOLVENT-TITRATION STUDY OF PEPTIDES RELATED TO THE C-TERMINAL DECAPEPTIDE SEQUENCE OF PORCINE SECRETIN

C. Toniolo, G.M. Bonora, and V. Moretto

Biopolymer Research Center, C.N.R., Department of Organic Chemistry, University of Padova, 35131 Padova, Italy

M. Bodanszky

Department of Chemistry, Case Western Reserve University, Cleveland, Ohio 44106

Introduction

IR absorption is a useful technique to quantitatively titrate the extent of self-association in peptides. The structural transition from unordered, solvated species to partially or fully ordered, intermolecularly H-bonded species, eventually forming a regular ß-structure, is paralleled by a marked decrease in solubility.[1] These phenomena can have an adverse effect on rates of aminolysis in peptide synthesis. Therefore, the knowledge of the relationship between structure and solubility is of paramount importance in planning a correct strategy of synthesis. In this communication we describe an extension of the application of the IR absorption technique to examine the equilibria between unassociated and self-associated species of the C-terminal decapeptide segment of porcine secretin -Arg-Leu-Gln-Arg-Leu-Leu-Gln-Gly-Leu-Val-NH_2 and related short sequences. Both N^{α}-protected and N^{α}-deprotected peptides were examined.

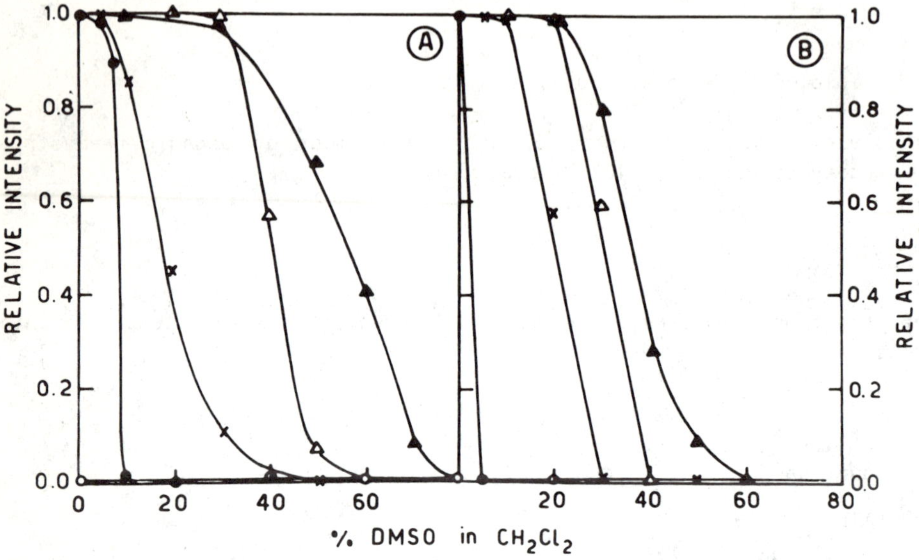

Fig. 1. Relative intensity of the amide-I C=O stretching band related to strongly self-associated molecules in the IR absorption spectra of: (A) Z-protected C-terminal dipeptide (--O--), tripeptide (--●--), tetrapeptide (--x--), pentapeptide (--△--), and hexapeptide (--▲--); and (B) Boc-protected analogs, in CH_2Cl_2-DMSO mixtures as a function of increasing percentages of DMSO. Conc. 2×10^{-2} M.

Results and Discussion

The present study was performed in CH_2Cl_2, a commonly used solvent in peptide synthesis, by adding increasing amounts of either DMSO or HMPA, known to form effective H-bonds with the NH groups of the peptide moieties, thereby disrupting (peptide) N-H ... O=C (peptide) H-bonds. We

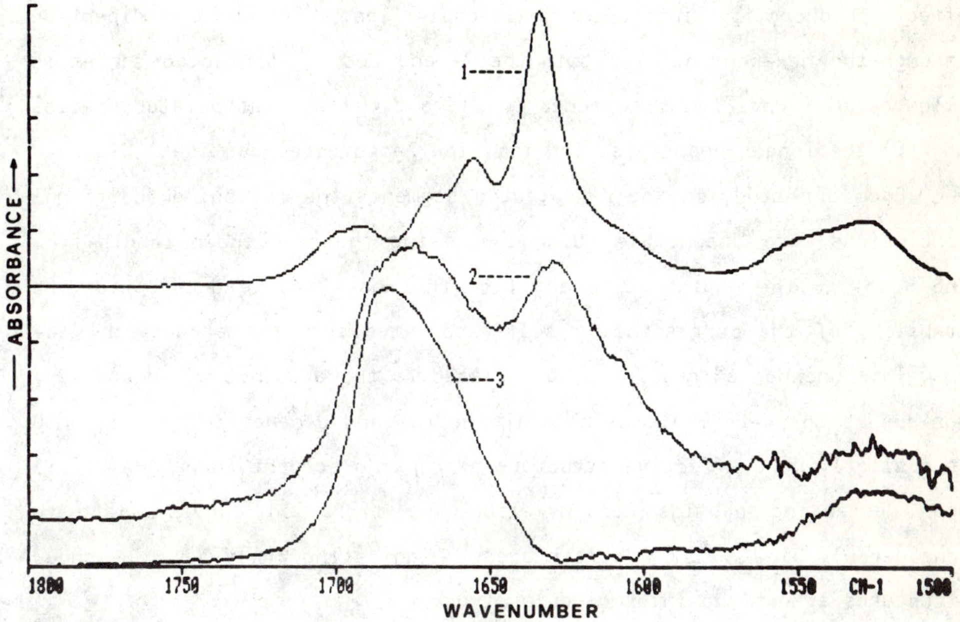

Fig. 2. Original, computer-drawn IR absorption spectra in the 1800-1500 cm^{-1} region of $^{+}H_2$-Leu-Gln-Gly-Leu-Val-NH_2 trifluoroacetate salt in CH_2Cl_2 (curve 1), CH_2Cl_2-DMSO 90:10 (v/v) (curve 2), and CH_2Cl_2-DMSO 90:10 (v/v) in the presence of 1% NMM (curve 3). Conc. 2×10^{-3} M.

have followed the disappearance of the amide-I C=O streching band of strongly intermolecularly H-bonded peptide molecules (near 1630 cm^{-1}).[2] For each peptide the relative IR absorption intensity was calculated from the area of the 1630 cm^{-1} band, taking as 1.0 relative intensity the value observed in CH_2Cl_2 or in the CH_2Cl_2-X (X, structure-disrupting solvent) mixture containing the lowest percentage of X required to dissolve the peptide, if it is not completely soluble in CH_2Cl_2 at the concentration to be examined.

Typical solvent-titration curves are shown in Fig. 1, where the effect produced by increasing main-chain length (from the dipeptide through the hexapeptide) for both the Z- and Boc N^{α}-protected series is illustrated. More DMSO is needed to disrupt the self-associated species of: (i) the longest peptides, and (ii) the Z-protected peptides.

From our study on the Z-protected segments the following additional conclusions were drawn: (i) Upon increasing the main-chain length from the hexa- to the heptapeptide and from the nona- to the decapeptide the stability of the corresponding self-associated species markedly decreases. This unexpected result may be related to the presence of an Arg(NO_2) residue at the N-terminus of both the hepta- and decapeptide. (ii) DMSO is a sligtly more effective structure-disrupting solvent than HMPA. (iii) Upon decreasing peptide concentration the titration curves shift to considerably lower DMSO percentages, indicating that the H-bonded structures are of the intermolecular type.

Fig. 2 shows the beneficial influence of added base (NMM, N-methyl morpholine) on the spectrum of the N^{α}-deprotected pentapeptide salt. The band characteristic of strongly intermolecularly H-bonded molecules completely disappears upon addition of 1% NMM (to a CH_2Cl_2-DMSO 90:10, v/v, solvent mixture). Comparable results were found upon addition of triethylamine.

References

1. Mutter, M., Pillai, V.N.R., Anzinger, H., Bayer, E., and Toniolo, C. (1981) in Peptides 1980, Brunfeldt, K., Scriptor, Copenhagen, pp. 660-665.

2. Miyazawa, T. and Blout, E.R. (1961) J. Am. Chem. Soc. 83, 712-719.

ENHANCEMENT OF COUPLING EFFICIENCY IN SOLID PHASE PEPTIDE SYNTHESIS BY ELEVATED TEMPERATURE

James P. Tam

The Rockefeller University, 1230 York Ave., New York, N.Y. 10021

Introduction

A poorly understood problem during the synthesis of long and complex peptide by the solid phase method is the inability ocasionally to bring a particular coupling reaction to completion. During the synthesis of big gastrin, transforming growth factor and other peptides in our laboratory by the solid phase method[1], such difficulties in some coupling reactions were observed. In all cases, 1.5 to 2.0% of unreacted amino groups of the peptide chains were detectable after the completion of a double coupling cycle by the quantitiative ninhydrin monitoring method[2]. Moreover, these unreactive amino groups were refractory and could not be significantly overcome by alternative activating method, prolonged coupling time (>18 h), or polar aprotic solvents (e.g. dimethylformamide) in the coupling reaction.

Because these unreactive amino groups are reactive to and detected by the reagents present in the ninhydrin method at 100° C, it is likely that, at elevated temperature, the refractory unreactivity of the coupling reaction can be overcome. Based on this rationale, the development of a new coupling protocol using elevated temperature to overcome the refractory, low level of unreactive amino group in the coupling step in solid phase peptide synthesis is described.

Results and Discussion

Peptides shown in Table I were synthesized by a double coupling protocol using preformed symmetrical anhydride, in CH_2Cl_2 and dimethylformamide as the first and second coupling solvents respectively. The loading of the peptide on the resin was usually about 0.4 mmol/g resin. Coupling efficiency, prior to the difficult coupling reaction, as monitored by the quantitative ninhydrin method, averaged 99.52% and a background of 0.34%, giving an average coupling efficiency of 99.86%. In the examples shown in Table I, the couple efficiency averaged only 98% and could not be increased further by prolonged coupling. However, coupling by preformed symmetrical anhydride (prepared in CH_2Cl_2) at 50° C in 1-methyl-2-pyrrolidinone for 1 h, improved the coupling yield to give >99.7% completion and essentially reduced the level of imcomplete coupling from 1.7-2.2% to the background level of 0.3%. 1-Methyl-2-pyrrolidinone was preferred over dimethylformamide for its stability at elevated temperature.

Table 1. Coupling efficiency after coupling at elevated temperature

Sequence‡	2nd	Coupling yield (%)† coupling after 50°C	bkgrd.
....Tyr(BrZ)--ValGlyValArg(Tos)..	98.0	99.6	0.4
...Glu(OBzl)--Glu(OBzl)Lys(ClZ)..	98.3	99.7	0.3
...Asp(OBzl)--PheGlyPheProGln..	98.1	99.7	0.3
Asp(OcHex)--GlyAsnGlnPheLys..	97.8	99.8	0.2

† by quantiative ninhydrin method; ‡ the coupling reaction is between the two N-terminal residues.

The examples represented in Table I did not conform to any predictable pattern of chain length. These peptides occurred with chain length of 12 to 60 residues. Furthermore, the N-terminal amino acid residues are to limited to β-branched amino acids which are known to give slower coupling rates.

Difficulties of coupling in the solid phase peptide synthesis have been attributed to chemical and physical reasons. Schiff base formation[3] and physical aggregation due to hydrogen bonding[4,5] might account for these coupling difficulties. The enhancement of coupling efficiency at elevated temperature by symmetrical anhydride in polar aprotic solvent provides indirect evidence for the physical nature of the coupling difficulties and an useful solution to overcome this refractory unreactive problem in the coupling reaction.

Acknowledgement

This work was supported by PHS grant CA34746, awarded by the National Cancer Institute, DHHS.

References

1. Sarin, V.K., Kent, S.B.H., Tam, J.P. and Merrifield, R.B. (1981) Anal. Biochem. 237, 927-936.
2. Kent, S.B.H. and Merrifield, R.B. in "Peptides 1980", K. Brunfeldt, ed., Scriptor, Copenhagen, 1981, pp. 328-333.
3. Pillai, V.N. and Mutter, M. (1981) Acct. Chem. Res. 14, 122-130.
4. Kent, S.B.H. and Merrield, R.B. (1983) Int. J. Peptide Prot. Res. 22, 57-56.
5. Meister, S.M. and Kent, B.H. in "Peptides, Structure and Function, Proceedings of the Eighth American Peptide Symposium", V.J. Hruby and D.H. Rich, ed., Pierce Chemical Co., Rockford, Ill., 1983, pp.103-106.

ASYMMETRIC SYNTHESES OF NON-PROTEINOGENIC AMINO ACIDS

Ulrich Schöllkopf

Institut für Organische Chemie der Universität Göttingen
Tammannstraße 2, D-3400 Göttingen FRG

I Introduction

There is a demand for opticallv active - if possible optically pure-uncommon amino acids both for pure and applied organic or bioorganic chemistry. Since asymmetric synthesis is - at least in principle - the shortest and most economic way to optically compounds, it is a challenge for the synthetic organic chemist, to develop asymmetric syntheses of amino acids.

Strategy

Our approach is based on the following concept. 1. From a racemic lower amino acid and a chiral auxiliary an heterocycle is built up, that is CH-acidic adjacent to the potential amino group and that contains two sites susceptible to hydrolysis. 2. An electrophile is introduced diastereoselectively via the anion of the heterocycle. 3. Subsequntly the heterocycle is cleaved by hydrolysis to liberate the heterocycle. 3. Subsequently the heterocycle is cleaved by hydrolysis to liberate the chiral auxiliary and the new optically active amino acid. This lecture deals with the use of metalated bis lactim ethers 3 of 2,5-diketopiperazines 1.[1] - The bis-lactim ether 2 gives with butyllithium or LDA (THF, -70°C) the lithium compounds 3. Electrophiles react with 3 to give the adducts 4, whereby chirality is transfered from C-6 to C-3. $E^{\oplus}$ enters at C-3 trans to R^1 at C-6. The degree of asymmetric induction (=de=$(D_1 - D_2)/D_1 + D_2)$ $\cdot$ 100) exeeds in many cases 95 % and reaches up to 99 %. The products 4 can be hydrolyzed liberating the optically active target molecules 6 and the esters 5 (the chiral auxiliaries). The esters 5 and 6 are separable either by fractional distillation or - eventually after further hydrolysis to the amino acids - by chromatography.

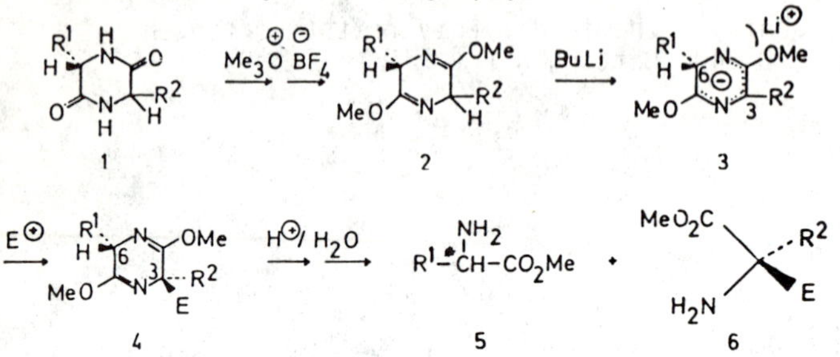

1, 2 3	a	b	c	d	e
R^1	iPr	iPr	tBu	iC_4H_9	CH_3
R^2	Me	H	H	H	H

3 (R)-α-Methyl Amino Acid Esters 9 from 3a and Alkyl Halides

2a Yields with butyllithium (or LDA) regiospecifically 3a. This reacts with alkyl halides with virtually complete asymmetric inductions to give the products 7. In the ^{1}H NMR only (6S,3R)-diastereomers are detectable. Acid hydrolysis of 7 liberates (besides methyl L-valinate 8) the α-methyl esters 9 which are enantiomerically pure by ^{1}H NMR-standard.[2]

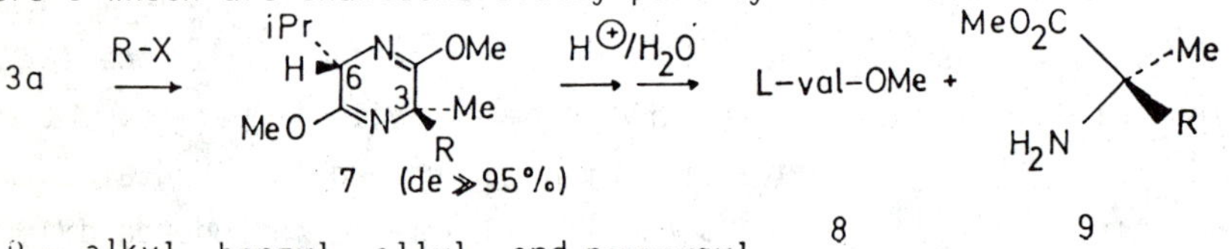

R = alkyl-,benzyl-,allyl- and propargyl

We assume that the ion pairs 3 contain a planar anion with the lithium cation situated near N-1. Furthermore, we postulate a mobile equilibrium between two diastereomeric in pairs 10 and 11, which lies far on the left side because of steric reasons. Due to attractive complexation between $Li^{\oplus}$ and X-R, 10a reacts via 12a‡ to (3R,6S)-7 and 11a via 13a‡ to (3S,6S)-7. 10a reacts faster than 11a, since 12a‡ is relatively strain free, wheras 13a‡ is strained due to steric congestion "at the bottom side".

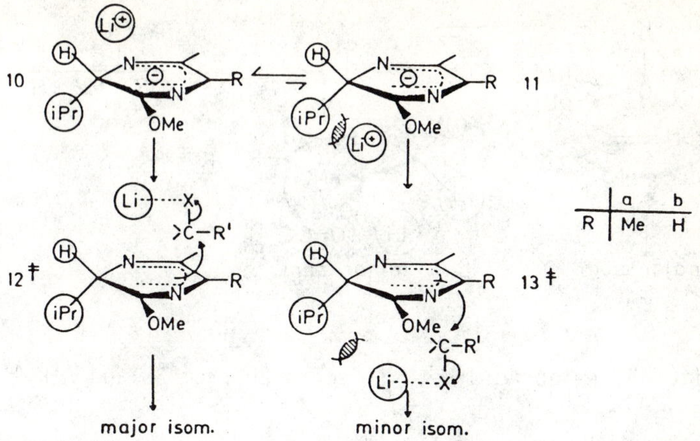

4 (R)-α-Alkenyl Alanine Methyl Esters 16 from 3a and Ketones

Ketones add to 3a with high induction (d.e. > 95 %) to give the adducts 14. After dehydration 14 → 15, hydrolysis of 15 yields the (R)-α-alkenyl alanine esters 16 (enantiomerically pure by ^{1}H NMR-standard).[3]

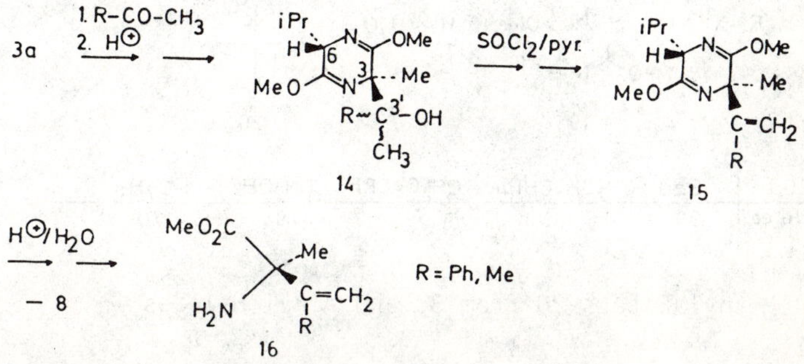

The diastereofacial bias of 3a toward carbonyl compounds can be explained by a model analogous to the one put forward in 3.2. TS 17a ‡, leading to the major isomer, is of lower energy than TS 18a ‡ which is strained due to steric hindrance at "the bottom side".

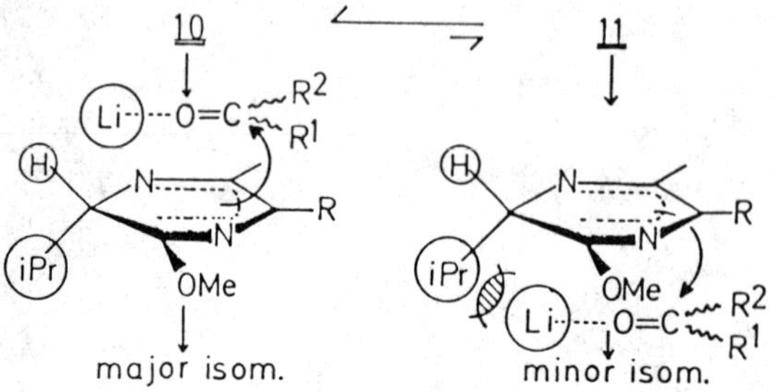

17‡ 18‡

R	a	b
	Me	H

5 (α-Unsubstituted) Amino Acid Methyl Esters 20 from 3b and Alkyl Halides

The bis lactim ether 2b of cyclo(L-val-gly) 1b is lithiated by butyllithium regiospecifically in the glycine part to give 3b. This reacts with alkyl halides to afford the (3R)-products 19 with de-values from 70- > 95 %.[4] On hydrolysis,the products 19 are cleaved to methyl L-valinate 8 and (R)-amino acid methyl esters 20.[4] The results are rationalized on the basis of the TSs 12b‡ and 13b‡ .

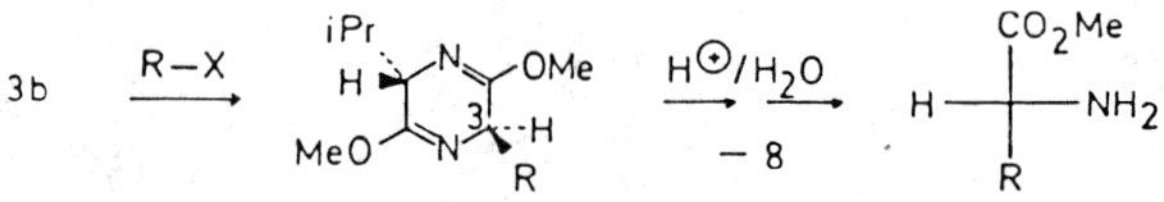

19 20

19,20, R	Bz	$CH_2CH{=}CHPh$	$CH_2C{\equiv}CPh$	CH_2OBz	n-C_7H_{15}
% de.ee	92	>95	>95	94	70

6 Amino Acid Methyl Esters 20 from 3c and Alkyl Halides

Bulkier than iso-propyl is tert-butyl. Hence it is not surprising, that 3c reacts with all alkyl halides, tried so far - apart from methyl iodide - with de >95 %.[5]

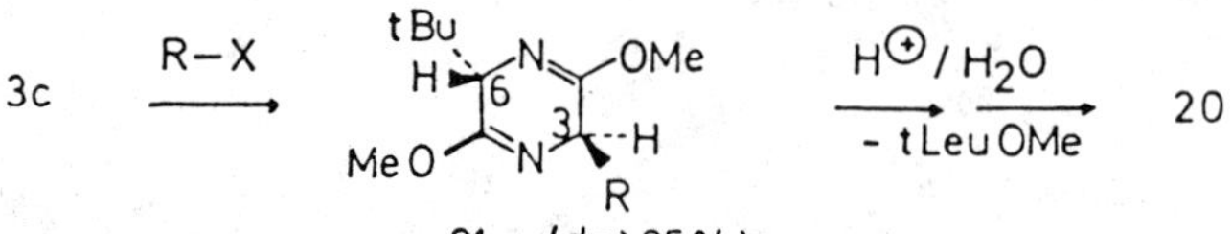

21 (de>95%)

7 (R)-β-Hydroxy Valine 23 and (R)-β-Methylene Phenylalanine Ester 25 from 3b and Acetone resp. Acetophenone

Acetone and acetophenone afford with 3b the (3R)-adducts 22 with de > 95%. From 22a practically optically pure (R)-β-hydroxy valine 23 is obtainable[6], from 22b (R)-β-methylene phenylalanine methyl ester 25 (via 24).[7]

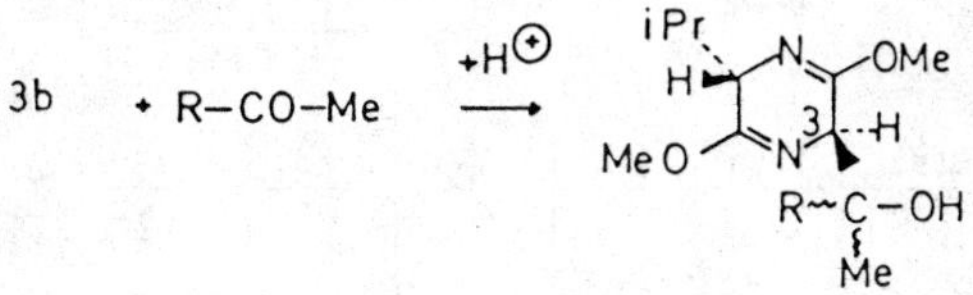

22	a	b
R	Me	Ph

22

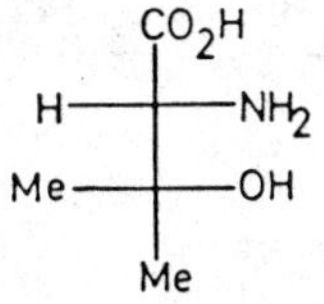

23

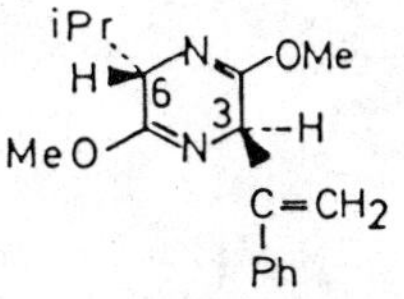

24

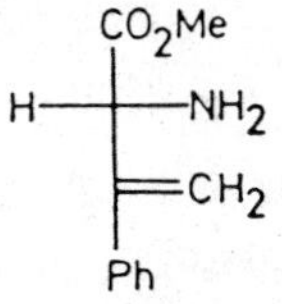

25

These results can be rationalized on the basis of the model concept depicted above. TS $17b^{\ddagger}$ is of considerably lower energy than $18b^{\ddagger}$.

8 (2R)-3-Substituted Serines 27

8.1 Addition of Aldehydes to the Lithium Compound 3b

Compared with ketones (cf.7), aldehydes react with the lithium compound 3b with somewhat lower diastereroselectivity.[8] The induction at C-3 (de at C-3) are listed in the table as well as the (3R,3'S) : (3R,3'R)-ratios.

The diastereoface selection with regard to the anion of 3b is best explained on the basis of the TS $17b^{\ddagger}$ and $18b^{\ddagger}$. The enantioface selection at the carbonyl group can be rationalized on the basis of the chair like TSs $29a^{\ddagger}$ and $30a^{\ddagger}$.[8] The (3R,3'S)-epimers are formed predominantly, because the 1,3-diaxial R OMe - and the R Li-repulsion in $30a^{\ddagger}$ outweighs the 1,2 R H-repulsion in $29a^{\ddagger}$. - As described in ref.[8], (2R)-3-substituted serine methyl esters 27a - or serines 27b - can be obtained from the compounds 26.

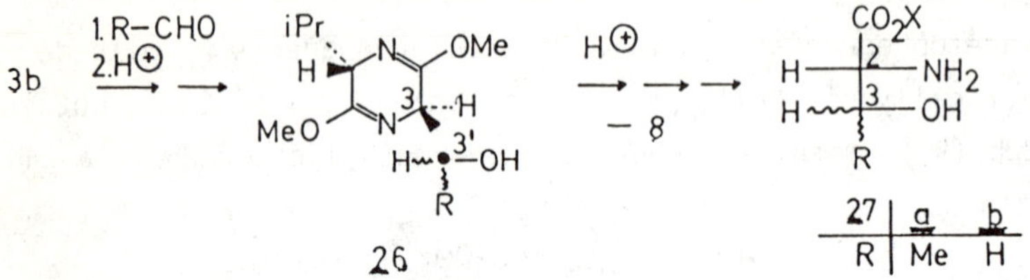

26	R	(3R,3'S)	(3R,3'R)	(3S,3'S)	(3S,3'R)	% de C-3	u : l a)
a	Ph	16	13	1	1	86	1,2:1
b	iPr	55	10	1	1	94	5,5:1
c	tBu	35	18	2	1	85	2 : 1
d	Me	19,5	4,7	1	1	85	4 : 1

a) (3R,3'S):(3R,3'R)

8.2 Addition of Aldehydes to a Titanium Derivative of 2b

All factors, that render the TSs 17b‡, 18b†, 29† and 30‡ more compact should enhance both the diastereoface selection with respect to the anion and the enantioface selection with regard to the carbonyl group. Consequently,exchange of lithium for metals with shorter matel-oxygen- and metal-nitrogen-bonds should lead to an higher degree of de at C-3 and to an higher u : l -ratio in 26. Exchange of lithium for tris(dialkyl-amino)titanium has a dramatic effect. The titanium compound 28 yields with aldehydes essentially diastereomerically pure (3R,3'S)-adducts 26.[9]

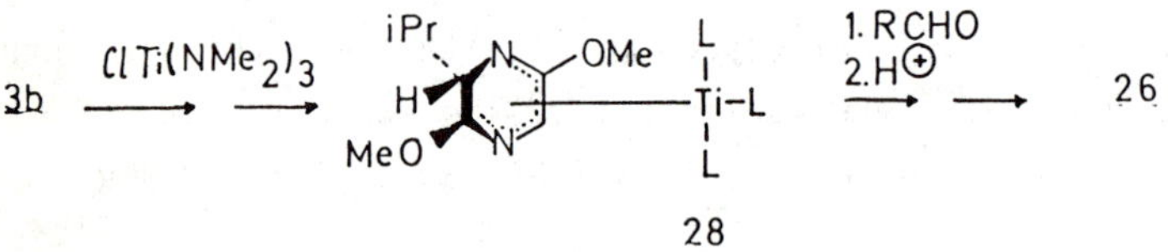

26	R	(3R, 3'S)	(3R,3'R)	(3S, 3'S)	(3S, 3'R)	% de C-3	u:l b)
a	Ph	32	1	— a)	— a)	>99	32:1
b	iPr	151	2,3	1	1	97	65:1
c	Me	88	0,6	1	1	95,6	146:1

a) Not detectable any more with capillary GLC. b) (3R,3'S):(3R,3'R)

As for the transition states $29^{\ddagger}$ and $30^{\ddagger}$: With M = $Ti(NR_2)_3$, the TSs are more compact than with M = Li. Hence, $29b^{\ddagger}$ and $30b^{\ddagger}$ differ more in energy than $29a^{\ddagger}$ and $30a^{\ddagger}$.

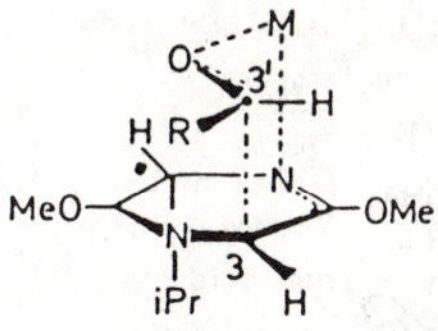

$29^{\ddagger}$

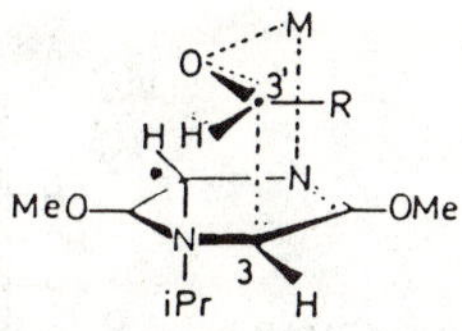

$30^{\ddagger}$

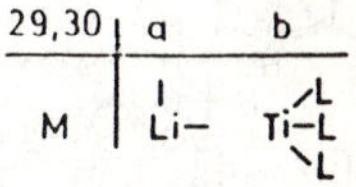

L = NMe_2

So far achiral aldehydes were discussed. With chiral aldehydes double stereodifferentiation comes into play. Results are described obtained with 28 and (R)- and (S)-glyceraldehyde. As expected, based on the TS-models $29b^{\ddagger}$ and $30^{\ddagger}$ - whereby $29b^{\ddagger}$ is the overall dominating low energy TS - and on the Felkin-Anh-model of carbonyl addition. (S)-glyceraldehyde react with exceedingly high diastereoselectivity in any respect to form 32, but (R)-glyceraldehyde reacts with high de at C-3 (> 96 %) to form 31, but with a lower threo/erythro ratio (3R,1'R):(3R,1'S) of ca. 19, compared with ca. 128 for (S)-glyceraldehyde. With (S)-glyceraldehyde, TS $29b^{\ddagger}$ is a totally matched pair - arrangement of the glyceraldehyde moiety with resepect to the heterocyclic anion is favorable and carbonyl attack is Felkin - whereas TS $30b^{\ddagger}$ is a totally mismatched pair - arrangement of the aldehyde moiety with respect to the heterocyclic anion is poor and carbonyl attack is anti-Felkin. Hence, the energy difference of the two transition states is relatively large.

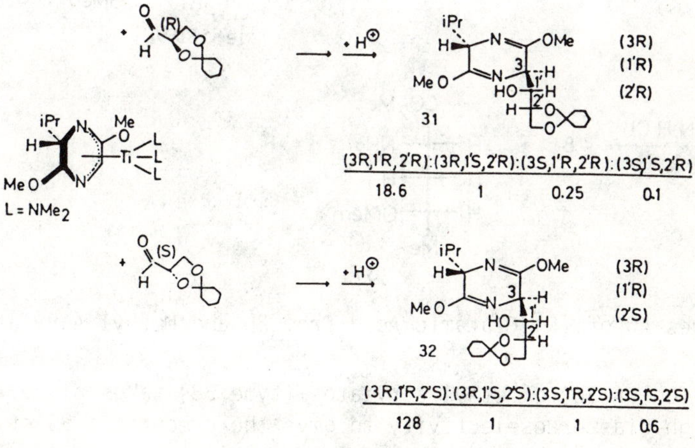

On hydrolysis (subsequent to O-acetylation[10] of 31 and 32 the derivatives 33 of D-2-amino-2-deoxy-xylonic acid and 34 of L-2-amino-2-deoxy-arabinoic acid are obtained.

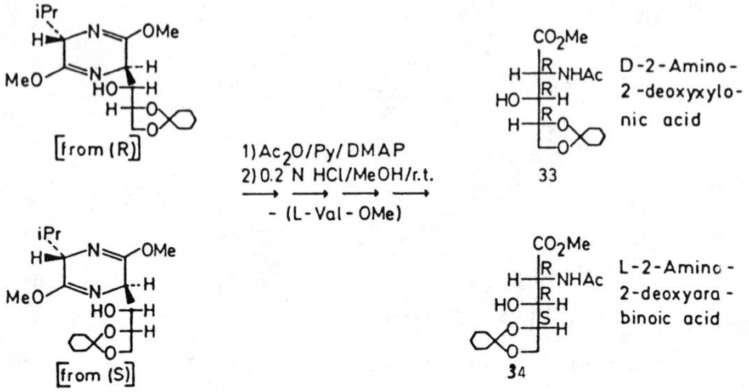

9 (R)-Homoserines from 3b and Epoxides

Epoxides (type 35) react with 3b after addition of BF_3-etherate to give the addition products of type 36. The de-values are ca. 50 - 97 %.[11] The hydroxy group of the adducts has to be Mem-protected.[11] Hydrolysis proceeds smoothly to give the O-Mem-protected (R)-homoserine methyl esters 37.

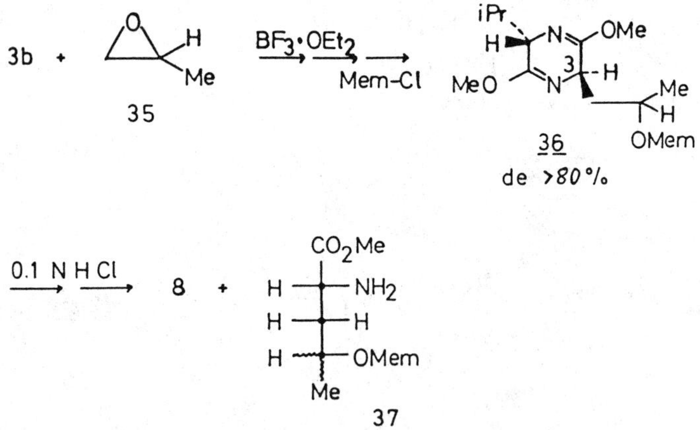

10 Derivatives 40 of (R)-Glutamic Acid from 3b and Methyl Acrylates

Michael-addition of 3b to methyl acrylates (type 38) takes place with exceedingy high diastereoselectivity to give the precursors 39 of

(D)-glutamic acids. This is an intriguing results since, for instance, methyl acrylate is a relatively small electrophile. Probably, in the transition state of the Michael-addition the π-system of the heterocyclic anion an the π-system of the α,β-unsaturated ester are arranged parallel to each other forming a kind of π-complex in which the ester moiety is turned inside (cf.41 and 42). In this arrangement the electrophile comes close to the chiral inducing center, rendering the energy difference of the relatively strainfree "topside" TS and of the strained "bottomside" TS relatively large.

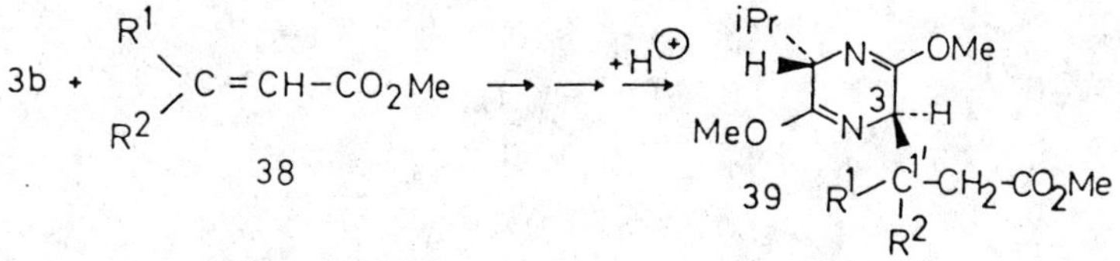

39

R^1	R^2	(3R):(3S)	(1'S):(1'R)
H	H	200:1	
Me	Me	140:1	
H	Ph	225:1	19:1

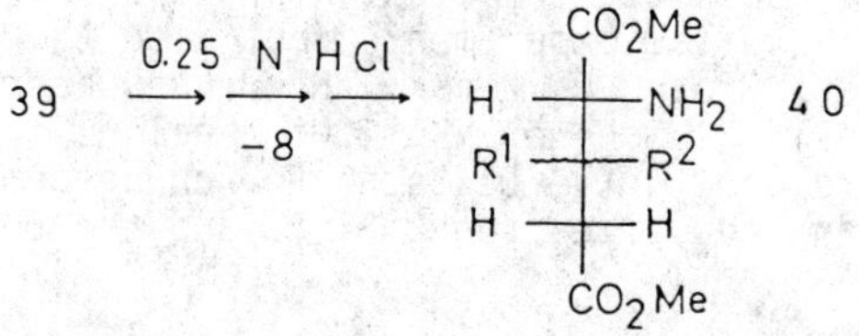

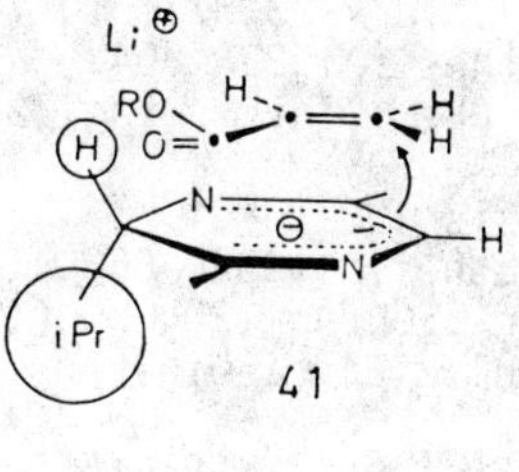

41

Strainfree TS‡

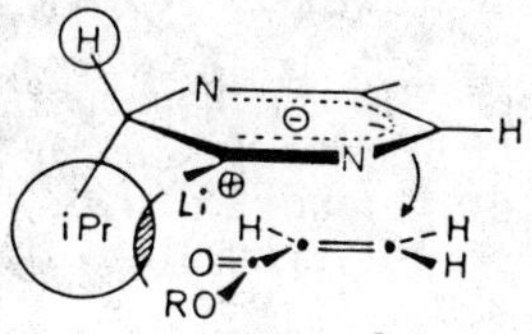

42

Highly strained TS‡

11 (R)-Cyclohexenyl Glycine 46 from 3b and Cyclohexanethione

Thioketones (type 43) also react with 3b with virtually complete asymmetric induction. For instance, thiocyclohexanone 43 gives the addition product 44 (after S-methylation) with ca. 97 % asymmetric induction. Surprisingly , when treated with Raney-Ni, 44 undergoes a regioselective elimination of methanethiol forming the Hofmann-olefin 45 exclusively. From 45, (R)-(-)-cyclohexenyl glycine 46 can be obtained, which is enantiomerically pure by NMR-standard.[12]

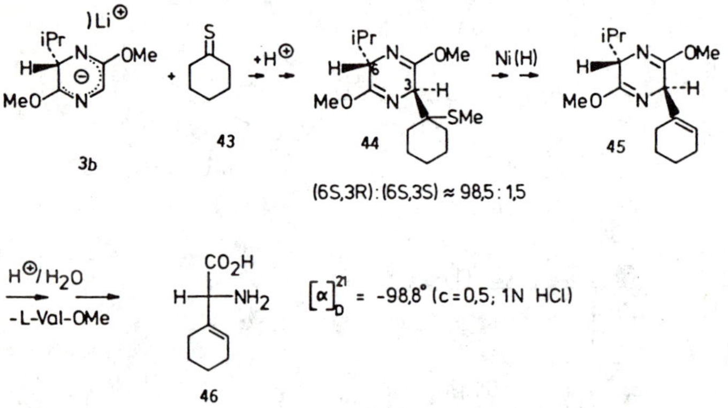

Acknowledgement - Thanks for enthusiastic cooperation are due to W.Hartwig, who carried out the inital cruzial experiments, but also to U.Busse, Y. Chiang, C.Deng, M. Grauert, U. Groth, R. Lonsky, H.-J. Neubauer, J.Nozulak, D.Pettig, K.-H. Pospischil, R. Scheuer, K.-O.Westphalen. Thanks for supporting these studies are due to the Deutsche Forschungsgemeinschaft, the Fonds der Chemischen Industrie, the BASF AG and the Degussa AG.

References

1. Review on previous results: Schöllkopf, U., Pure & Appl.Chem. 55, 1799 (1983).Compounds 2 are available from Merck-Schuchardt.
2. Schöllkopf, U. et al, Synthesis 1981,696; ibid 1983,37;ibid 1983,675.
3. Schöllkopf, U.,Groth,U. and Chiang,Y.,Synthesis 1982, 864.
4. Schöllkopf, U.,Groth,U. and Deng,C.,Angew.Chem.,Int.Ed.Engl.20,798 (1981).
5. Schöllkopf, U. and Neubauer,H.-J., Synthesis 1982, 861.
6. Schöllkopf,U.,Nozulak,J. and Groth,U., Synthesis 1982, 868.
7. Schöllkopf, U.and Groth,U., Angew.Chem.,Int.Ed.Engl.20, 977 (1981).
8. Schöllkopf, U.,Groth,U.,Gull,M.-R.and Nozulak,J.,Liebigs Ann.Chem. 1983, 1133.
9. Schöllkopf, U.,Nozulak,J. and Grauert,M.,Synthesis 1985, 55.
10. Grauert,M.and Schöllkopf,U.,Liebigs Ann.Chem.1985, in press.
11. Schöllkopf,U. and Gull,M.-R., Synthesis 1985, in press.
12. Schöllkopf,U.,Nozulak,J.and Groth,U.,Tetrahedron 40, 1409 (1984).

SYNTHESIS OF (R)- AND (S)-1-AMINO-[2,2-$^{2}H_2$]CYCLOPROPANE-1-CARBOXYLIC ACIDS: A DOUBLE VIOLATION OF SCHÖLLKOPF'S RULE

Pullachipatti K. Subramanian and Ronald W. Woodard*

Department of Medicinal Chemistry, College of Pharmacy University of Michigan, Ann Arbor, Michigan 48109-1065

Introduction

The achiral natural product 1-aminocyclopropane-1-carboxylic acid (ACC) is the key intermediate in the bioconversion of methionine to the fruit-ripening hormone ethylene.[1] Stereospecifically labeled ACCs have been reported as 1:1 mixtures of enantiomers and utilized to probe the mechanism of ethylene biosynthesis.[2] Recently two syntheses of (R)- and (S)-1-amino-[2,2-$^{2}H_2$]cyclopropane-1-carboxylic acids have appeared,[3] however, they are multistep procedures and require an independent method of assessing the % of enantiomeric excess (% ee) of the isomers. Herein, we describe a simple, two-step synthesis of (R)- and (S)-[2,2-$^{2}H_2$]ACC based on the Schöllkopf[4] approach to asymmetric synthesis.

Results and Discussion

We chose as our chiral agent R-(+)-2-methyl-3-phenylalanine (1), which is prepared from phenylacetone[5] in a convenient four-step sequence in 49 % overall yield by the asymmetric Strecker synthesis first described by Weinges and co-workers.[6] This amino acid is chosen because it lacks an α-hydrogen and therefore the bis-lactim ether derived from it can only be alkylated at the C-6 position. The phenylalanine derivative is esterified with methanol-HCl[7] and the resulting methyl ester 2 is condensed[8] with N-t-Boc-glycine to give the dipeptide methyl ester 3. Following the procedure of Nitecki[9], the dipeptide methyl ester 3 is cyclized, after treating with formic acid, by boiling in a mixture of sec-butanol and toluene (2:1)[10] to yield the 2,5-diketopiperazine 4. The piperazine 4 is then converted into the bis-lactim ether 5 by treatment with trimethyloxonium tetrafluoroborate.[11]

The bis-lactim ether 5 upon treatment with butyllithium in tetrahydrofuran (THF) at -78°C gives the lithio derivative which reacts with regiospecifically deuterium labeled 2-haloethyl triflates[12-14] to displace selectively the triflate group to afford a mixture of diasteriomers 6a and 6b (~4 : 1) in high yields. Surprisingly, the 2-haloethyl group enters at C-6 cis to the bulky benzyl group at C-3 (in violation of Schollkopf's rule) to induce the R-configuration at C-6.

The configuration at C-6 of these alkylated products is assigned based on their ^{1}H-nmr spectra. The ^{1}H-nmr spectrum of the parent compound 5 shows signals at δ 2.84, 3.66 (AB, d, J = 20.53 Hz) due to the C-6 hydrogens. Based on the ^{1}H-nmr studies of Kopple[15] on the diketopiperazine cyclo[phe-gly], the ^{1}H-nmr signal at δ 2.84 of 5 is assigned to the hydrogen atom trans to the benzyl group at position C-3 and the other signal at δ 3.66 cis to the benzyl group. In the case of the major diastereomer 6a, the C-6 hydrogen signal appears at δ 3.87 (S) indicating a cis addition of the 2-bromo-1,1-dideuteroethyl group has occurred and that the major diastereomer has the (6R)-configuration. The (6S)-configuration of the minor isomer is also confirmed by ^{1}H-nmr.

This violation of Schöllkopf's rules cannot be explained by a unimolecular Sn1 type ionization of the 2-haloethyltriflate to a 2-haloethyl cation, which would act as the alkylating species, since no apparent scrambling of the deuteriums is observed which one would expect if an ethyl bromonium cation is formed. It is assumed that the α - anion attacks the 2-haloethyl triflate in an Sn2 process to displace the triflate ion as a leaving group. In order to further investigate this analogous reversal in stereochemistry, the lithio derivative of 5 is treated at -78°C with methyl triflate and the C-6 stereochemistry is analyzed by ^{1}H-nmr. These results are shown in Table 1 along with the results from the reaction of the lithio derivative of 5 at -78°C with DCl. As can be seen from the Table only the triflates give analogous results.

The synthesis of the title compounds is completed via an intramolecular Schöllkopf addition performed on the mixture of diastereomers 6a and 6b in THF at -78°C by treating with butylithium. An unusual ring closure occurs to give predominantly 7a (7a:7b, 3:1) in which the addition of the alkyl group is cis to the bulky benzyl group (anti-Schöllkopf's addition).

Table I. Diastereomeric Excess of Various Alkylation Reactions

Alkylating Agent	First-Alkylation Product (%)		Second-Alkylation Product (%)	
	Trans	Cis	Trans	Cis
$BrCH_2CD_2OTf$	22	78	27	73
$ClCH_2CD_2OTf$	20	80	27	73
$BrCD_2CH_2OT$	20	80	28	72
CH_3OTf	50	50		
DCl / D_2O	90	10		

The mode of addition in this intramolecular cyclization, however, may be rationalized by considering the probable transition state. The standard trans approach in the cyclization involves greater steric interaction between the methylene group attached to C-6 and the C-3 benzyl group than the cis approach which would be favored in this case thus leading to the predominant formation of 7a from 6a and 6b. The (6S)-configuration of 7a is derived from high field ^{1}H-nmr data. The methylene protons at δ 0.13, 0.5 (AB, d, J=4.12 Hz) which are located within the shielding cone of the aromatic ring (benzyl group) suffers an upfield shift of Δδ ~ 0.6 ppm as compared to those at δ 0.69, 1.13 (AB, d, J = 4.10 Hz) in the minor isomer 7b.

Upon hydrolysis (0.25 N HCl, room temperature, 48 h) the adduct 7a and 7b is cleaved to (R)-2-methyl-3-phenylalanine methyl ester and (S)-1-amino-[2,2-$^2H_2$]cyclopropane-1-carboxylic acid methyl ester. These esters were further hydrolyzed (6N HCl, reflux, 6 h) to give the corresponding amino acids which could be separated by chromatography. Similarly, the (R)-isomer of [2,2-$^2H_2$]ACC is obtained from the reaction of the lithio derivative of 5 and 2-bromo-2,2-dideuteroethyl triflate.

Although the Schöllkopf rules of addition are unpredictive in this study, the use of the bis-lactim to synthesize the target compounds is successful since they are obtained in reasonable % ee. The use of 2-methyl-3-phenylalanine as the chiral agent also allows one to use the ^{1}H-nmr of the product bis-lactim to determine the % ee directly.

Acknowledgment

This work is supported by NIH GM 30097.

References

1. Adams, D.O. and Yang, S.F. (1979) Proc. Nat. Acad. Sci. (U.S.A.) 76, 170-174.

2. Adlington, R.M., Baldwin, J.E., and Rawling, B.J. (1983) J. Chem. Soc. Chem. Commun. 290-293.

3. Hill, R.K., Prakash, S.R., Wiesendanger, R., Angst, W., Martinoni, B., Arigoni, D., Liu, H.-W., and Walsh, C.T. (1984) J. Am. Chem. Soc. 106, 795-796.

4. Schöllkopf, U, Nozulak, J., and Groth, U. (1984) Tetrahedron 40, 1409-1417.

5. Hass, H.B., Susie, A.G., and Heider, R.L. (1950) J. Org. Chem. 15, 8-14.

6. Weinges, K., Gries, K., Stemmle, B.,and Schrank, W. (1977) Chem. Ber. 110, 2098-2105.

7. Stein, G.A., Bronner, H.A.,and Pfister III, K. (1955) J. Am. Chem. Soc. 77, 700-703.

8. Wang, S.S., Tam, J.P., Wang, B.S.H., and Merrifield, R.B. (1981) Int. J. Peptide Protein Res. 18, 459-467.

9. Nitecki, D.E., Halpern, B.,and Westley, J.W. (1968) J. Org. Chem. 33, 864-866.

10. One equivalent of morpholine is added, otherwise cyclization is difficult under Nitecki's conditions.

11. Schöllkopf, U, Hartwig, W., Pospischil, and K.H., Kehne, H. (1981) Synthesis, 966-969.

12. Hogg, J.L. and Schowen, R.L. (1974) J. Pharm. Sci. 63, 1620-1623.

13. Ott, D.G. (1981) Synthesis with Stable Isotopes, John Wiley and Sons, New York pp. 33-35.

14. Beard, C.D., Baum, K., and Grakauskas, V. (1973) J. Org. Chem. 38, 3673-3677.

15. Kopple, K.D.,and Marr, D.H. (1967) J. Am. Chem. Soc. 89, 6193-6200.

NOVEL CONFORMATIONALLY CONSTRAINED AMINO ACIDS AS LYSINE-9 SUBSTITUTIONS IN SOMATOSTATIN ANALOGS

Ruth F. Nutt,[+] Paul E. Curley,[+] Steven M. Pitzenberger,[+] Roger M. Freidinger,[+] Richard Saperstein* and Daniel F. Veber[+]

Merck Sharp & Dohme Research Laboratories, [+]West Point, Pennsylvania 19486 and *Rahway, New Jersey 07065

The importance of a primary aliphatic amine functionality with a critical distance from the peptide backbone has been established for the position-9 amino acid in somatostatin and small ring analogs such as cyclo(Pro-Phe-D-Trp-Lys^9-Thr-Phe) (I).[1] A hydrophobic interaction between the hydrocarbon chain of lysine and the side chain of Trp has been demonstrated in solution by NMR and may also be an important feature for high biological activity.[2] In order to further probe the steric and hydrophobic requirements for the position-9 side chain, the following cyclic dibasic amino acids were synthesized and evaluated as constrained lysine analogs in I: cis and trans 4-aminocyclohexylglycine (AChxGly), cis and trans 4-aminocyclohexylalanine (AChxAla), and para-aminomethylphenylalanine (p-AmPhe).

The selectively protected dibasic amino acids trans and cis DL-AChxGly, trans and cis L-AChxAla, DL-p-AmPhe were synthesized as described in Scheme 1. Boc-4-aminophenylglycine (**1**) was prepared by reaction of 4-aminophenylglycine with one equivalent of Boc-ON. Separation of the trans and cis isomers **3-6** was carried out by silica gel chromatography and structure assignments were made by NMR. Derivative **11** was prepared by base catalyzed alkylation of diethylacetamidomalonate with α-bromo-p-tolunitrile. The cyclic hexapeptides **14-18** (Table I) were prepared by methods described

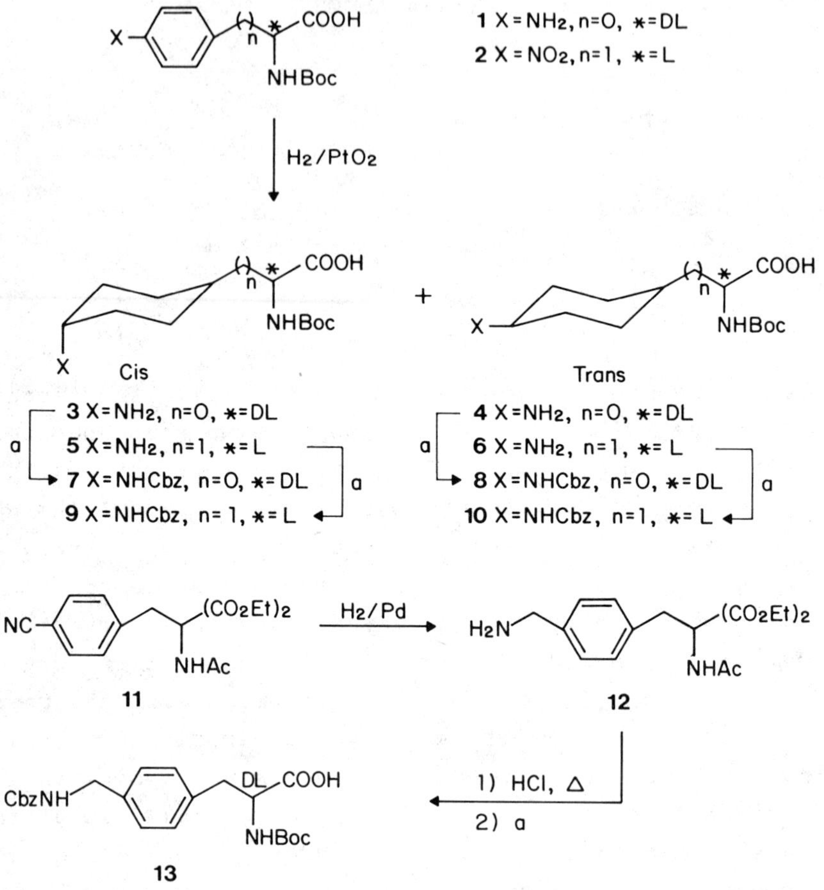

[a]Steps carried out similar to procedures described in ref. 1.

Scheme 1. Synthetic routes to dibasic amino acids.

previously[1] using **7, 8, 9, 10** and **13** in the solid phase synthesis of the peptide-resins. The optically pure **14, 15,** and **18** were obtained by silica gel chromatography of the diastereomeric cyclic hexapeptide mixtures, and their structures were assigned after CD and NMR comparisons with I.

The five somatostatin analogs **14-18** were examined for bioactivity as inhibitors of insulin, glucagon, and growth hormone release (Table I). Substitution of lysine in I with cis and trans AChxGly (**14** and **15**) and the aromatic side chain containing amino acid p-AmPhe (**18**) results in reduction of potencies to $\leq$10% of that for I. Substitution with AChxAla (**16** and **17**), however, results in high potencies with the trans isomer **17** being 10 times more active than I as an inhibitor of insulin and glucagon release. The high *in vivo* potency of **17** may be partially attributed to enhanced metabolic stability, since total resistance to digestion by trypsin was observed. Analog **17** also exhibited increased hydrophobicity ($p \frac{oct}{H_2O} = 11.4$) compared to I ($p \frac{oct}{H_2O} = 2.9$), which may reduce clearance rate.

Structural and conformational properties of the analogs were examined in attempts to correlate them with bioactivity. Side chain length of the dibasic amino acid seems to be optimal with a $C_\alpha \rightarrow NH_2$ distance of 5.9-6.8 A^o (Table I).

Table I. Hormone Release Inhibition[a] by cyclo(D-Trp-Y-Thr-Phe-Pro-Phe)

Compd	Y	Insulin	Glucagon	GH (*in vitro*)	Distance $C_\alpha \rightarrow NH_2$ (A^o)
14	cis AChxGly	<0.07	<0.07	0.007	4.6
	Orn[1]	0.49	0.16	0.16	4.85
15	trans AChxGly	<0.07	<0.07	0.02	5.81
16	cis AChxAla	1.4	2.2	0.5	5.96
I	Lys	5.2	8.0	1.7	6.19
17	trans AChxAla	57	56	1.3	6.76
18	p-AmPhe	0.5	0.7	0.15	7.66

[a]Bioassay protocol described in Veber, D.F. *et al.* (1978) *Proc. Natl. Acad. Sci. USA* **75**, 2636-2640; activities relative to somatostatin=1.

Side chain length alone, however, does not explain the low bioactivity of **15**, which has a basic side chain more favorable in length than the ornithine containing analog. Solution conformational studies by CD and NMR show no differences between the analogs, all having similar backbone and side chain conformations. The characteristic proximity of the indole ring to the basic side chain as evidenced by the presence of highly shielded γ-protons, is retained in all analogs. Distinguishing features, however, between the active analogs and the inactive **15** are the lack of rotational constraint and steric bulk at the β-carbon of the basic side chain. One or both features seem to be important for attaining the bioactive side chain conformation or an efficient interaction with the receptor.

In conclusion, several conformationally constrained dibasic amino acids have been synthesized and represent useful tools for elucidating important structural and conformational features of key basic residues in bioactive peptides. In somatostatin analogs it was shown that steric and/or conformational constraint at the β-carbon of the position-9 amino acid greatly reduces biological activity. Trans aminocyclohexylalanine as lysine replacement in cyclo-(Pro-Phe-D-Trp-Lys-Thr-Phe) has resulted in increased potency, metabolic stability, and hydrophobicity.

References

1. Nutt, R.F., Veber, D.F., Curley, P.E., Saperstein, R., and Hirschmann, R. (1983) Int. J. Peptide Protein Res. **21**, 66-73.

2. Arison, B.H., Hirschmann, R., and Veber, D.F. (1978) Bioorg. Chem. **7**, 447-451.

THE DIFFERENTIATION OF π- AND τ-DERIVATISED HISTIDINES

John H. Jones, Andrew E. Derome, Daniel L. Rathbone, and David W. Thomas

The Dyson Perrins Laboratory, University of Oxford, South Parks Road, Oxford OX1 3QY, UK

Roberto Colombo, Fabio Colombo, and Francesca Fontana

Department of Biomedical Sciences and Technology, CNR and Scientific Institute San Raffaele, via Olgettina 60, 20132 Milan, Italy

Introduction

The differentiation of π- and τ-derivatised histidine residues (1,2) is a long-standing problem[1,2]. Various experimental

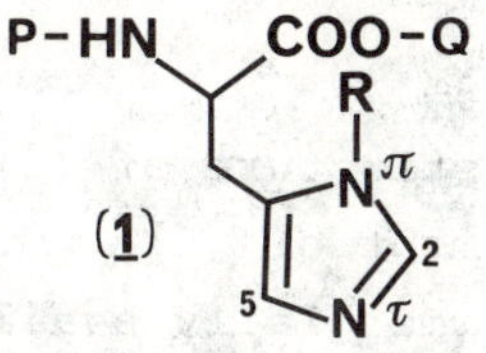

(1)

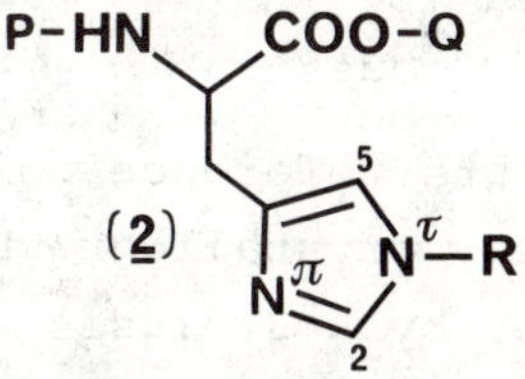

(2)

	P	Q	R
(3)	Boc	H	$BzlOCH_2$
(5)	H	H	Me
(6)	Z	Me	Bu^tOCH_2
(8)	Z	Me	$MeOCH_2CH_2OCH_2$

	P	Q	R
(4)	H	H	Me
(7)	Z	Me	Bu^tOCH_2
(9)	Z	Me	$MeOCH_2CH_2OCH_2$
(10)	Boc	H	Bzl

approaches to the question can be found in the literature, but no general unequivocal methods are available. In practice the commonest way of making π - τ assignments has been simply to assume that in the reactions of histidine side-chains with equimolecular amounts of electrophilic reagents the main product will be that arising from attack at the least hindered, i.e. τ, position. In our experience this is a valid assumption. However, if an excess of alkylating agent is employed, under some circumstances the mixture of monoalkyl isomers and disubstituted product can contain comparable amounts of each of the monoalkyl isomers. In the extreme case of the unhindered highly reactive methylating agent methyl methanesulphonate, it has been reported that the second order rate constant for first reaction at the π-nitrogen of N(α)-acetylhistidine methylamide is significantly greater than that for first reaction at the τ-nitrogen, and the generalisation that τ-substituted products are invariably predominant when histidine side-chains react with electrophiles has been contradicted[3]. For many years the response to the problem in the field of peptide synthesis was to ignore it. Since location at the π-position is a fundamental *desideratum* for histidine side-chain protection[4], we have sought to disentangle some of the confusions over the question. To establish a more secure and general basis for π-τ assignments, we have devised and tested two simple methods which we now briefly report.

Results and Discussion

Our absolute reference point is the structure of (3), which rests on X-ray crystallography[5]. Treatment of (3) with excess CH_3I followed by hydrolysis gave the commercially available (Sigma) amino-acid described as "3-methyl-L-histidine" which is thus confirmed as N(τ)-methyl-L-histidine (4). It follows that the isomer "1-methyl-L-histidine" (Sigma) is N(π)-methyl-L-histidine (5). Compounds (4) and (5) are thus subsidiary reference structures, which can be cleanly separated and easily identified by chromatography. Most of our work has so far

Table I. The Differentiation of π- and τ-Derivatised Histidines by Conversion to im-Methylhistidines (4) and (5).

Structure assigned	Proportions of (4) and (5) produced (4),%	(5),%
(6)	97	3
(7)	0	100
(8)	100	0
(9)	0	100

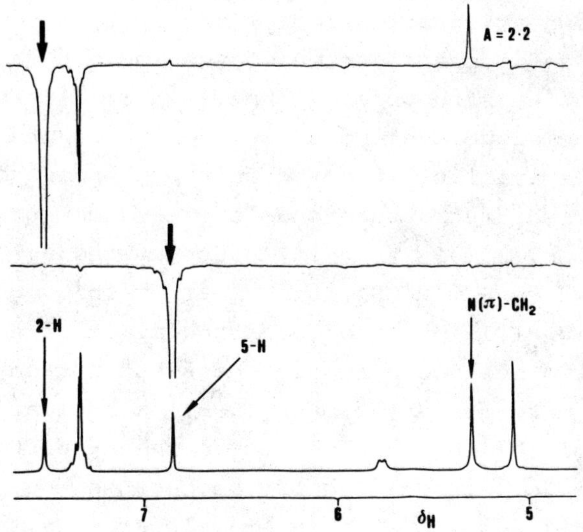

Figure 1. Nuclear Overhauser effects in the n.m.r. spectrum of compound (8).

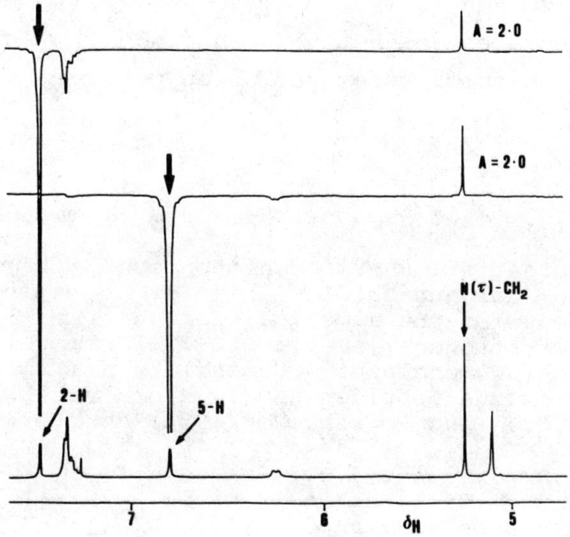

Figure 2. Nuclear Overhauser effects in the n.m.r. spectrum of compound (9).

been done on an amino-acid analyzer[6], but h.p.l.c. and g.l.c. procedures gave equally good results. Conversion of a derivative of unknown orientation to (4) or (5) by methylation and deprotection and identification of the im-methylhistidine produced defines the structure. We have applied this simple distinction in numerous instances[6], two of which are shown in Table I. The availability of a fair number of compounds of firmly established structure enabled the demonstration of another criterion, namely that in π-substituted structures of type (1) in which R = R'CH_2, a nuclear Overhauser enhancement of the CH_2 signal is observed on irradiation of the proton (2-H) between the two heterocyclic nitrogens but not on irradiation of the other more distant proton (5-H); a nuclear Overhauser enhancement of the CH_2 is observed in the τ-substituted isomer on irradiation of either ring proton because they are equidistant (Figures 1 and 2). As an example of a well-known compound of hitherto incompletely defined structure, we have examined N(α)-t-butoxycarbonyl-N(im)-benzyl-L-histidine of commercial origin (Sigma). It proved to be a pure single isomer: the n.O.e. criterion indicated it to have the τ-benzyl structure (10), consistent with the known tendency of this intermediate to lead to gross racemization on coupling.

References

1. Matthews,H.R. and Rapoport,H. (1973) J.Amer.Chem.Soc. 95, 2297-2303.
2. The differentiation has been much confused by the currency in biochemical and chemical circles of completely contrary conventions for the numerical designation of the two heterocyclic nitrogens; see: IUPAC-IUB Joint Commission on Biochemical Nomenclature, Nomenclature and Symbolism for Amino Acids and Peptides, Eur.J.Biochem. (1984) 138, 9-37.
3. Calleman,C.J. and Poirier,V. (1983) Acta Chem. Scand. B 37, 809-815.
4. Fletcher,A.R., Jones,J.H., Ramage,W.I. and Stachulski,A.V. (1979) J.Chem.Soc. Perkin Trans. I, 2261-2267.
5. Brown,T., Jones,J.H. and Wallis,J.D. (1982) J.Chem.Soc. Perkin Trans. I, 3045-3048.
6. Colombo,R., Colombo,F., Derome,A.E., Jones,J.H., Rathbone, D.L. and Thomas,D.W. (1985) J.Chem.Soc. Perkin Trans. I, in the press.

SYNTHESIS AND BIOLOGICAL PROPERTIES OF NOVEL TRYPTOPHAN-MODIFIED PEPTIDES: PREPARATION OF 2-SUBSTITUTED TRYPTOPHAN DERIVATIVES

Stephen F. Brady,* Richard Saperstein,[+] and Roger M. Freidinger*

Merck Sharp & Dohme Research Laboratories, West Point,* PA 19486 and Rahway,[+] NJ 07065

The importance of tryptophan to biological function in many peptides is well recognized, and therefore, substitution within the indole portion of this amino acid is of considerable interest. Incorporation of ring-substituted tryptophan residues into peptides has typically been accomplished during assembly of the chain.[1] However, peptides with 2-substituted tryptophan have been relatively inaccessible, largely due to the lack of availability of suitable precursors.[2] A notable exception is the 2-oxindolylalanyl residue, which is readily obtained from tryptophan by a simple oxidative conversion.[3] Herein we wish to report the application of this methodology to the preparation of 2-chlorotryptophan in peptides.

At the outset our efforts were focused on obtaining the 2-oxindolyl derivative **II** from the highly potent somatostatin analog **I**, in order to assess the effect of loss of indole on biological properties. We demonstrated that, under the conditions detailed in Scheme 1, reaction was specific for tryptophan, and that, in addition to **II**, a significant amount of analog **III** could be isolated. Chlorination of indole in DMSO-HCl represents a heretofore unreported finding.

The reaction was conveniently monitored by the drop in Trp indole absorption (278 nm), concurrent with the appearance of oxindole (252 nm). Starting material was gone within

Scheme 1. Oxidative Chlorination

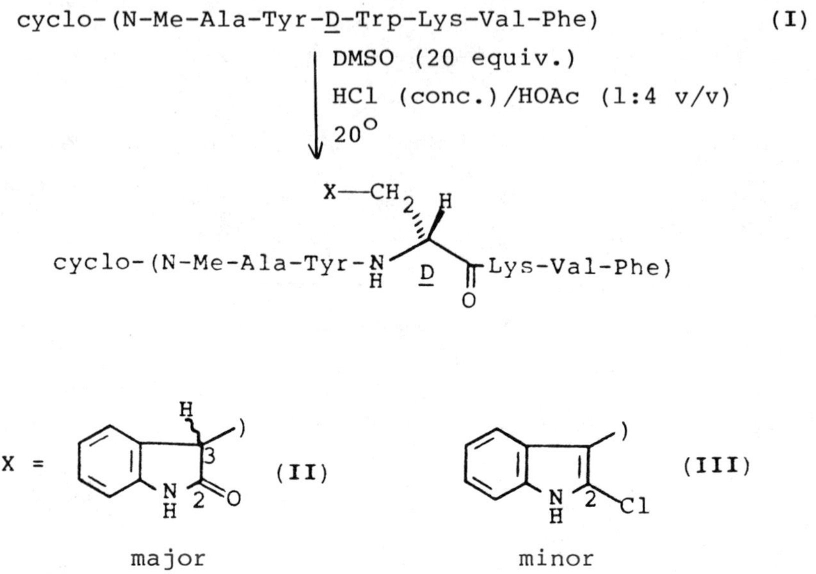

15 min, replaced by two products (*ca*. 4:1 ratio by TLC), which were easily separated by chromatography on silica gel.

The major product (45% yield) was characterized as follows: u.v. λ_{max} (2*N* HOAc) 252 nm (ε 7200); FD-MS m/e 825 (calc. M+H). On HPLC this material showed two nearly equal peaks (*ca*. 94% of total) very close together. The 360 MHz PMR spectrum (CD_3OD) showed considerable differences from that of the parent structure **I**, notably the lack of indole-shielded Lys $^{\gamma}CH_2$ (0.55 ppm in **I** *vs*. 1.40 ppm in **II**) and the presence of a number of doubled signals (N-Me-Ala $^{\alpha}CH_3$ (0.25 ppm, overlapping d); Tyr 2,6 (*ortho*)-CH (6.95/6.97 ppm, d/d 6.50/6.56 ppm, d/d); N-Me-Ala $^{N}CH_3$ (2.33 ppm, overlapping s)). Each of the two compounds, after separation by preparative reverse phase HPLC, reverted to the original mixture within 24 h. These findings support structural assignment of the two components to 3-position epimeric oxindoles (see **II**) and demonstrate the lability of the 3-position with respect to epimerization, and also the existence of the oxindole moiety predominantly as the keto tautomers.

The minor product (11% yield) was identified as the 2-chlorotryptophan derivative **III**: u.v. λ_{max} (2N HOAc) 275 nm (ε 8700); FD-MS m/e 843 (calc. M+H); Cl found 4.09% (none ionic), calc. 3.92% (acetate salt); HPLC 98% one peak. The PMR spectrum exhibited the shielded Lys $^{\gamma}CH_2$ (0.45 ppm), evidence for intact indole ring, and the 2,6 (ortho)-CH pattern of Tyr (7.02, 6.72 ppm, d,d). Significant by its absence was the characteristic Trp C^2H singlet (6.99 ppm in I), the sole difference in the aromatic region.

Further study showed that both **II** and **III** are stable under the conditions of oxidation (thus, chloroindole does not arise from oxindolyl). Various attempts to increase the yield of **III** failed; the product ratio remained unchanged with different reagent proportions, lower temperature, or added co-solvents. Thus, we conclude that the process of chlorination, which we presume to occur *via* direct attack of a $Cl^{\oplus}$ species[4] on the Trp indole, must be limited by a very low concentration of reactive species. We feel, nonetheless, that our procedure represents a practical means of obtaining 2-chlorotryptophan derivatives. The cleanness of the process recommends its utility for simple, one-step modification of peptidyl trypotophan.*

The biological properties of analogs **I-III** are summarized in Table I. Not unexpectedly, the oxindolyl analog **II** is considerably less potent than **I**. Interestingly, compound **III** also shows substantially reduced potency, in contrast to our findings with other somatostatin analogs having halo-substituted tryptophan.[1]

---

*Oxidation of a simple model structure, H-Gly-Trp-Gly-OH, afforded, in parallel to the results with **I**, two products in a *ca.* 5:1 ratio (TLC), lending support for the generality of the process.

Table I. Potencies for Hormone Release Inhibition[a]

Compd	Insulin	Glucagon	GH *in vitro*
I (Trp)	107	53.2	52.3
	(49.8, 249)	(20.6, 186)	(35, 75)
II (2-oxindolyl)	0.89	0.94	1.4
	(0.4, 1.85)	(0.2, 3.5)	(0.9, 2.3)
III (2-chloroindolyl)	11.1	4.7	2.6
	(5.4, 24.0)	(1.3, 18.9)	(1.5, 4.0)

[a]Bioassay protocols were as cited in reference 1; potencies are relative to somatostatin=1 (95% confidence limits in parentheses).

Acknowledgements

We wish to thank Mr. Y. C. Lee for elemental analyses, Mr. J. Smith for mass spectra, Drs. D. W. Cochran and S. M. Pitzenberger for PMR determinations, and Mr. C. Homnick and Mrs. S. Fitzpatrick for amino acid and HPLC analyses. We also thank Dr. D. Veber for his support.

References

1. Nutt, R.F., Saperstein, R., and Veber, D.F. (1983) PEPTIDES: Structure and Function, Proceedings of the Eighth American Peptide Symposium, Hruby, V. and Rich, D., Eds., Pierce Chemical Company, Rockford, IL pp. 345-348.

2. Phillips, R.S. and Cohen, L.A. (1983) Tetrahedron Lett. **24** (50), 5555-5558.

3. Savige, W.E. and Fontana, A. (1976) J. Chem. Soc. Chem. Commun. 599-600; *ibid* (1980) Int. J. Pept. Protein Res. **15**, 285-297.

4. *Cf*. Rynbrandt, R.H. (1971) Tetrahedron Lett., 3553-3555.

SYNTHESIS OF NOVEL SYMMETRIC DIAMINO ACIDS

P. Anantha Reddy and Bruce W. Erickson
The Rockefeller University, New York, New York 10021

Introduction

Symmetrically branched polypeptides may be useful new biomaterials in many analytical, biological and synthetic applications. As part of our protein engineering project[1,2] to synthesize novel proteins with selective binding affinity or catalytic activity, we have prepared several symmetrically branched polypeptides using novel symmetric diamino acids as covalent linkers. We outline here the chemical synthesis of three such novel symmetric diamino acids:

3-(Aminomethyl)-4-aminobutyric acid (Aab, 4)

3,5-Bis(aminomethyl)benzoic acid (Baa, 8)

3,5-Bis(2-aminoethyl)benzoic acid (Bab, 11)

The aliphatic linker Aab, an isomer of ornithine, and the aromatic linkers Baa and Bab each have a plane of symmetry (dotted line). The two amino groups are not distinguishable when they bear identical substituents. Since the tertiary carbon of Aab is prochiral, derivatives of Aab will be chiral when the substituents on both amino groups are not identical.

Results and Discussion

Aab (4) was synthesized by two similar routes (Scheme 1).

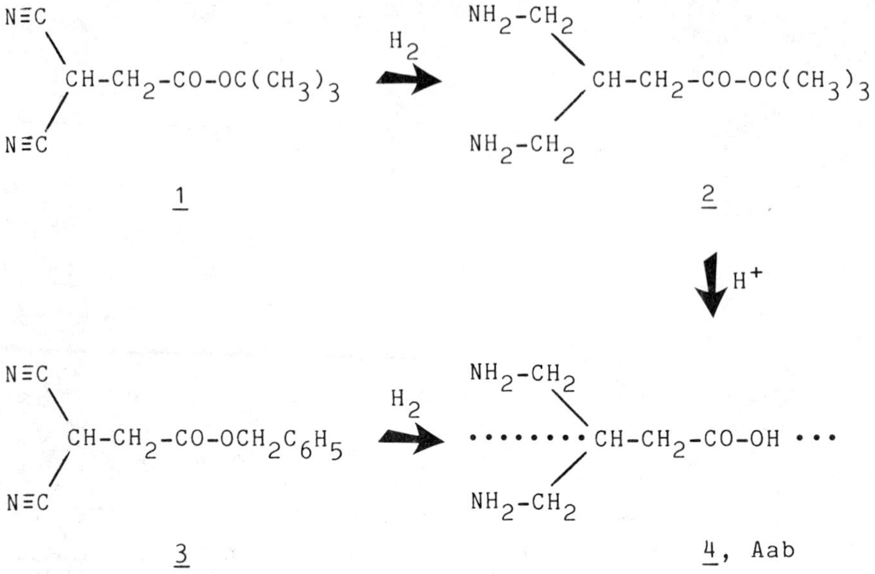

Scheme 1. Synthesis of 3-(aminomethyl)-4-aminobutyric acid (Aab, 4).

First, malononitrile ($CH_2(CN)_2$) was condensed with tert-butyl chloroacetate in the presence of sodium hydride in tetrahydrofuran to give tert-butyl 3,3-dicyanopropionate (1) in 75% yield. Hydrogenation of 1 over platinum oxide in glacial acetic acid followed by acidolysis of diamino ester 2 afforded Aab in 80% yield. Alternatively, malononitrile was condensed with benzyl 2-bromoacetate in the presence of sodium hydride in tetrahydrofuran to furnish benzyl 3,3-dicyanopropionate (3) in 80% yield. Then simultaneous hydrogenation and hydrogenolysis of 3 over platinum oxide in glacial acetic acid provided Aab in quantitative yield.

Baa (8) was synthesized (Scheme 2) from methyl 3,5-bis(bromomethyl)benzoate (5), which was prepared[3] from 3,5-dimethylbenzoic acid in two steps. Treatment of the dibromide 5 with potassium phthalimide in dimethylformamide[4,5]

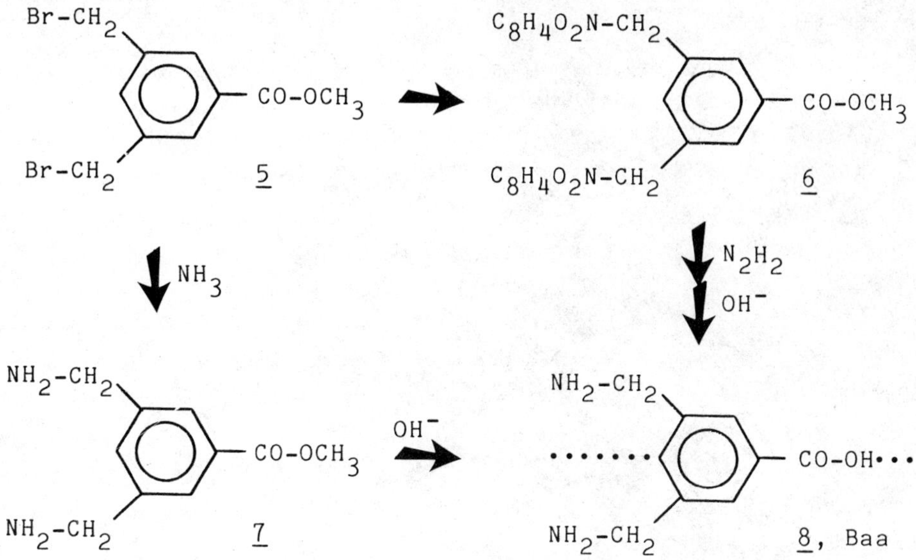

Scheme 2. Synthesis of 3,5-bis(aminomethyl)benzoic acid (Baa, 8).

yielded the bis(phthalimido) derivative 6, which upon hydrazinolysis followed by alkaline hydrolysis afforded Baa in 45% over-all yield. Alternatively, the key intermediate 7 was obtained in 50% yield by treatment of the dibromide 5 with methanolic ammonia.

Bab (11) was also prepared (Scheme 3) from methyl 3,5-bis(bromomethyl)benzoate (5). Reaction of this dibromide with potassium cyanide and 18-crown-6 in dry actonitrile[6] gave methyl 3,5-bis(cyanomethyl)benzoate (9) in 90% yield. Hydrogenation of 9 over platinum oxide in glacial acetic acid produced the bis(aminoethyl) derivative 10, which upon alkaline hydrolysis afforded Bab in 80% over-all yield.

The N,N'-bis(*tert*-butoxycarbonyl) derivative of Aab couples efficiently during solid-phase peptide synthesis. The corresponding derivative of Bab is being used in the synthesis of betabellin,[1,2] a chemically engineered protein.

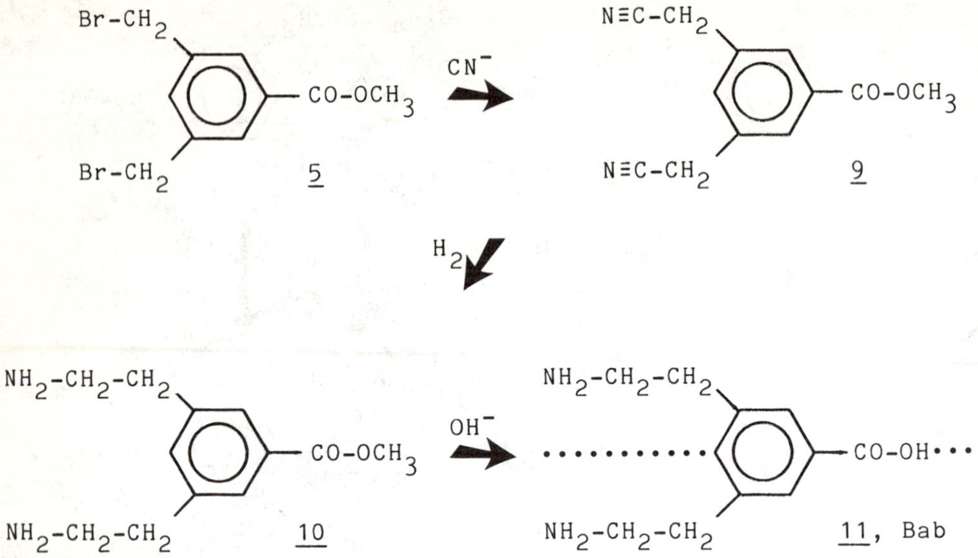

Scheme 3. Synthesis of 3,5-bis(2-aminoethyl)benzoic acid (Bab, 11).

This work was supported by NIH research grants AI 18362 and CA 33168 and by research contract N000014-83-K-0600 from the Office of Naval Research and the Army Research Office.

References

1. Unson, C.G., Erickson, B.W., Richardson, D.C., and Richardson, J.S. (1984) Fed. Proc. 43, 1837.
2. Daniels, S.B., Unson, C.G., Richardson, J.S., Richardson, D.C., and Erickson, B.W. (1985) Fed. Proc. 44, 694.
3. Markovac, A., and Lamontagne, M.P. (1980) J. Med. Chem. 23, 1198-1201.
4. Ing, H.R., and Manske, R.H.F. (1926) J. Chem. Soc., 2348-2351.
5. Benoiton, L. (1968) Can. J. Chem. 46, 1549-1552.
6. Cook, F.L., Bowers, C.W., and Liotta, C.L. (1974) J. Org. Chem. 39, 3416-3418.

METHODS OF SYNTHESIS AND SOLID PHASE COUPLING OF KETOMETHYLENE DIPEPTIDE ISOSTERES

Cris M. Olsen, Amrit K. Judd, and Ronald G. Almquist

Life Sciences Division, SRI International, 333 Ravenswood Avenue, Menlo Park, California 94025, USA

Introduction

Peptidase stability of biologically active peptides may be achieved by replacement of the amide link of a dipeptide by a dipeptide isostere containing a ketomethylene (Km) group. We have found it desirable to investigate various methods of preparing ketomethylene dipeptides, and the requirements for their incorporation into larger peptides by solid phase peptide synthesis.

Results and Discussion

The synthesis of the ketomethylene analogs of L-Pro-Gly, 3 and 6, are outlined in Scheme 1. Conversion of 1.3 equiv. of 4-bromo-1-butene to the Grignard reagent followed by dropwise addition to the thiopyridyl ester[1] of Boc-L-Pro (THF, 0°C) afforded 2 in 67% yield. A solution of 2 in acetone at 0°C was treated with an aqueous solution of RuO_2 (0.05 equiv.) and $NaIO_4$ (5 equiv.) to give Boc-L-Pro-KmGly, 3, in 63% yield. The Fmoc protected Pro-KamGly (Kam means ketalmethylene) was prepared from 2 in three steps. Removal of the Boc group (50% TFA/CH_2Cl_2) followed by dropwise addition of this trifluoroacetate salt to a solution of Na_2CO_3 and excess Fmoc-Cl gave 4 in 80% yield. The ketone was protected as its ethylene ketal (ethylene glycol, toluene, pTsOH, reflux, 73% yield) to give 5 and the olefin oxidized as for 2 to give the product 6 in 56% yield. The ketal was removed by acid treatment, yielding the ketone, after the entire desired decapeptide analog had been assembled.

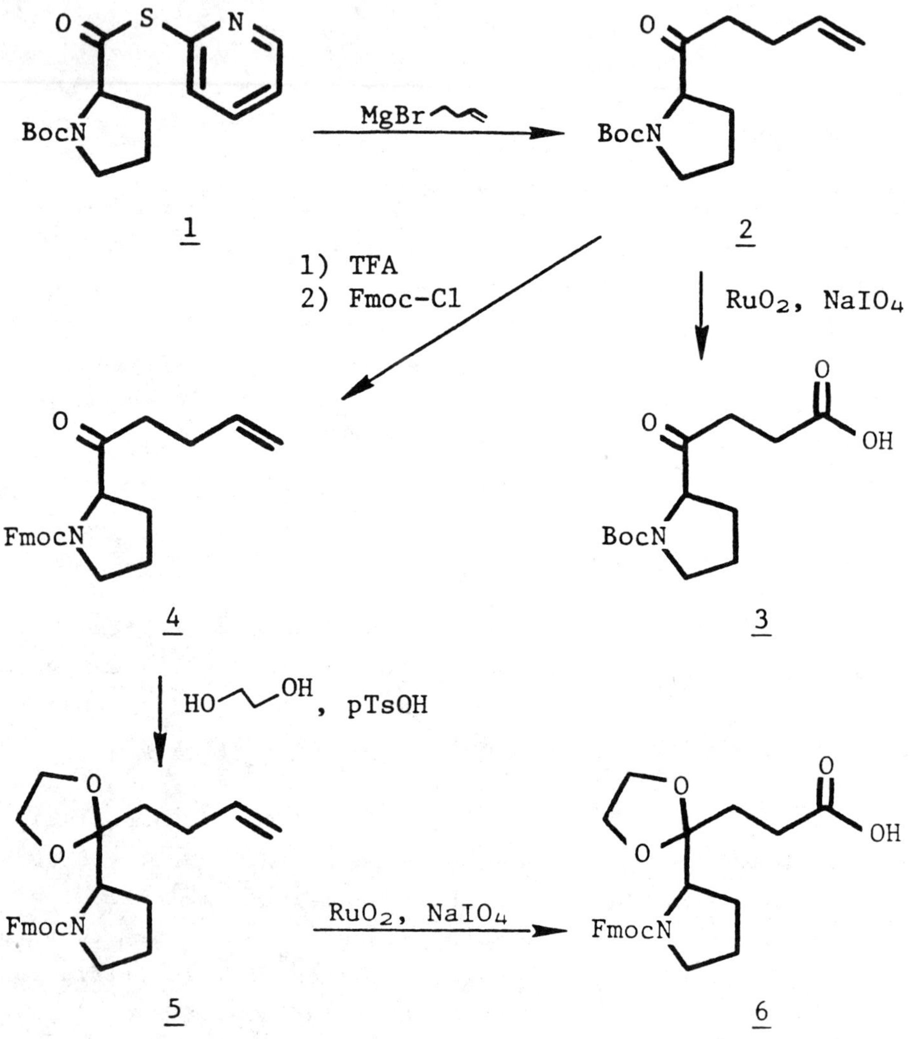

Scheme 1.

Incorporation of Pro-KmGly into a decapeptide analog was completed using two synthetic approaches. Fmoc-Pro-KamGly, 6, was attached to benzhydrylamine resin as its preformed hydroxybenzotriazole (HOBt) active ester. The remainder of the synthesis was completed using Fmoc amino acids and the procedure described by Chang et al.[2] The completed peptide was cleaved from the resin and all blocking groups, including the ketal, were removed with anhydrous HF. The resulting crude peptide was purified by gel filtration followed by preparative reversed-phase HPLC.

The synthesis was also completed using Boc-ProKm-Gly, 3, as the ketomethylene containing dipeptide. Since the ketone is unprotected, precautions needed to be taken to prevent its interaction with free amino groups. After Boc-Pro-KmGly had been coupled to benzhydrylamine resin as its HOBt ester, the next two amino acids were coupled as a dipeptide, using HOBt and DCC as coupling reagents, at a pH less than 7.0. Coupling at higher pH's lead to poor incorporation of the dipeptide. Coupling to the Pro-KmGly-resin complex with a single, activated Boc amino acid was easily accomplished but subsequent TFA deprotection caused termination of the synthesis. Model studies indicated the formation of the cyclic Schiff's base as the cause of termination. The formation of this cyclic product was prevented by coupling the two amino acids adjacent to the ketomethylene containing dipeptide as their Boc protected dipeptide. Following synthesis of this tetrapeptide resin complex the subsequent Boc amino acid couplings were completed at pH 7 as their preformed HOBt active esters. The final peptide was cleaved, deblocked, and purified as in the Fmoc approach.

Fmoc-Pro-KamGly, 6, was also incorporated onto *p*-alkoxybenzyl alcohol resin using DCC and dimethylaminopyridine. Subsequent deprotection of the Fmoc-Pro-KamGly-resin complex with piperidine caused rapid loss of the dipeptide from the resin due to the formation of a compound analogous to a diketopiperazine.

Conclusions

Ketomethylene dipeptides may be prepared by Grignard reactions on thiopyridyl esters of Grignard stable N protected amino acids. The ultimate carboxyl function of a Km-dipeptide may be masked by an olefin synthon as for compounds 3 and 6. Solid phase peptide synthesis with ketomethylene dipeptides required protection of the ketone as the ketal when Fmoc synthesis protocols are used. Utilization of Boc protection requires coupling of a dipeptide to the amino terminus of the Km-dipeptide, and that this coupling be done below pH 7 and subsequent couplings be done at pH 7.

References

1. Jennings-White, C., and Almquist, R.G. (1982) *Tetrahedron Lett*. **23**, 2533-2534.

2. Chang, C.D., Felix, A.M., Jimenez, M.H., and Meienhofer, J. (1980) *Int. J. Pept. Protein Res*. **15**, 485-494.

SYNTHESIS AND INCORPORATION OF FLUOROAMINO ACIDS INTO PEPTIDES[1]

Simon J. Hocart and David H. Coy

Peptide Research Laboratories, Department of Medicine,
Tulane University Medical School, New Orleans, LA 70112

Introduction

3-Fluoro- and 3,3-difluoro amino acids are known to be potent enzyme inhibitors and metabolic antagonists to the naturally occurring amino acids[2] and consequently are of interest for incorporation into peptides. However, the reported methods for the preparation of fluoroamino acids, including fluorodehydroxylation by SF_4 in HF,[3] and azirine[4,5] and aziridine[6,7] ring opening in HF-pyridine, involved hazardous reagents and had low reported yields which made the methods impractical for the preparation of sufficient quantities for use in peptide synthesis. The aim of this work was to develop a method for the preparation of 3-fluoro- and 3,3-difluoroamino acids from the readily accessible 3-hydroxyl precursors.

Results and Discussion

Most 3-fluoroamino acids exist as four epimers, namely the threo and erythro, D and L isomers, whereas the 3,3-difluoro amino acids exist as D and L isomers only. Consequently, we chose to study the synthesis of 3-(4-chlorophenyl)-3,3-difluoroalanine since this compound was of value for incorporation into luteinizing hormone releasing hormone (LH-RH) antagonists, the most potent of which contain 4-chlorophenyl-D-alanine in position 2, which are of interest as potential contraceptive agents.[8]

Our strategy was to prepare the 3-hydroxyamino acid, which after suitable protection was to be oxidised to the 3-oxo intermediate, which in turn could be fluorinated by, hopefully, facile reaction with diethylaminosulphur trifluoride (DAST).[9] The starting amino acid, 3-hydroxy-3-(4-chlorophenyl)alanine, was prepared by the reaction of 4-chlorobenzaldehyde with glycine.[10] Initially, the functional groups were protected by acetylation and esterification to give the acetyl methyl ester. This compound could be oxidised to give methyl N-acetyl-3-oxo-3-(4-chlorophenyl)alaninate, in high yield, using chromium trioxide in glacial acetic acid. However, this compound proved to be unstable to DAST, giving multiple ninhydrin positive spots on analysis by TLC.

In an analogous preparation of 4-fluoroproline and 4,4-difluoroproline, the amino acid precursors were protected as their benzyloxycarbonyl benzyl esters, to give the fluoro acids in good yield.[11] However, this protection scheme proved to be inadequate with our amino acid, and again, no useful reaction was obtained with the fluorinating reagent. It should be noted that proline, a secondary amino acid, does not have an amido hydrogen present after amino protection. Consequently, we prepared benzyl N,N-phthaloyl-3-oxo-3-(4-chlorophenyl)alaninate. This compound proved to be remarkably stable to DAST, as judged by the lack of any ninhydrin reaction by TLC, but unfortunately, the compound reacted very slowly to give the difluoro product. The conversion could not be enhanced by using large excesses of reagent, elevated temperatures, or catalysis with boron trifluoride. Hydrolysis of the reaction mixture afforded a small sample of 3-(4-chlorophenyl)-3,3-difluoroalanine which eluted as a single symmetrical peak on an LKB amino acid analyser using citrate buffers (t_R(F_2Cpa), 79.7 min; t_R(D-Cpa), 70.2 min; t_R(Arg), 82.9 min).

Table I. Antiovulatory Activity of Several Position 9 LH-RH Analogues

	Analogue	Antiovulatory* activity
I	Pro^9	77% @ 1 (13)
II	$FPro^9$	46% @ 1 (11)
III	F_2Pro^9	0% @ 1 (9)
IV	$NMeAla^9$	13% @ 3 (7)
V	$HPro^9$	86% @ 3 (7)
		0% @ 1 (6)

*expressed as the percentage of (n) rats prevented from ovulating at a dose of x micrograms.

The fluoroproline analogues and several other proline analogues were incorporated into the LH-RH antagonist [N-Ac-D-Nal^1,D-Cpa^2,D-Trp^3,D-Arg^6,Phe^7,(Pro^9),D-Ala^{10}]-LH-RH (I),[8] as their N-t-butyloxycarbonyl derivatives by standard solid phase peptide synthesis.[12] The peptides were dissolved in a 40% propan-1,2-diol - saline vehicle and assayed for antiovulatory activity in a standard rat assay.[13] The results are given in Table I.

The monofluoro analogue (II) and the difluoro analogue (III) exhibited a decrease in activity as compared to the parent peptide (I) which correlated with the increasing fluorine content, and hence hydrophobicity, of the residues. The NMeAla analogue (IV) was much less active despite its steric similarity to proline. The HPro analogue (V) was also much less active then the parent peptide, which may be related to the increased hydrophilicity of the molecule.

Acknowledgement

This work was supported by NICHD contract HD-4-2832.

References

1. Abbreviations used in this paper for amino acids, protecting groups and peptides follow the recommendations of the IUPAC-IUB Commission on Biochemical Nomenclature and Symbols as described in Eur. J. Biochem. (1972) 27, 201-207, and J. Biol. Chem. (1975) 250, 3215-3216: D-Nal, 3-(2-naphthyl)-D-alanine; D-Cpa, 4-chlorophenyl-D-alanine; NMeAla, N-methylalanine; F_2Pro, 4,4-difluoroproline; FPro, 4-fluoroproline; HPro, 4-hydroxyproline; F_2Cpa, 3-(4-chlorophenyl)-3,3-difluoroalanine

2. Kollonitsch, J. (1978) Israel J. Chem. 17, 53-59.

3. Kollonitsch, J., Marburg, S., and Perkins, L.M. (1979) J. Org. Chem. 44, 771-777.

4. Wade, T.N., and Kheribet, R. (1980) J. Org. Chem. 45, 5333-5335.

5. Wade, T.N., and Guedj, R. (1979) Tetrahedron Letts. 41, 3953-3954.

6. Wade, T.N. (1980) J. Org. Chem. 45, 5328-5333.

7. Alvernhe, G.M., Ennakoua, C.M., Lacombe, S.M., and Laurent, A.J. (1981) J. Org. Chem. 46, 4938-4948.

8. Hocart, S.J., Nekola, M.V., and Coy, D.H. (1985) J. Med. Chem. in press.

9. Middleton, W.J. (1975) J. Org. Chem. 40, 574-578.

10. Greenstein, J.P., and Winitz, M. (1961) in Chemistry of the Amino Acids, J. Wiley & Sons, New York, pp 2509-2592.

11. Sufrin, J.R., Balasubramanian, T.M., Vora, C.M., and Marshall, G.R. (1982) Int. J. Pept. Protein Res. 20, 438-442.

12. Coy, D.H., Coy, E.J., Schally, A.V., Vilchez-Martinez, J.A., Delseljuk, L., Carter, W.A., and Arimura, A. (1974) Biochemistry 13, 323-326.

13. Vilchez-Martinez, J.A., Coy, D.H., Coy, E.J., Arimura A., and Schally, A.V. (1976) Endocrine Res. Commun., 3, 231-241.

SYNTHESIS AND PROPERTIES OF SOME PEPTIDES CONTAINING A BICYCLIC DIPEPTIDE UNIT WITH SEMI-RIGID β-TURN CONFORMATION

Ukon Nagai and Kazuki Sato

Mitsubishi-Kasei Institute of Life Sciences
Machida-shi, Tokyo 194, JAPAN

Introduction

Use of the analogs with restricted conformational freedom is a promising approach to study bioactive conformation of peptides as exemplified by high activity of some cyclic analogs of bioactive peptides such as somatostatin[1], enkephalin[2], α-MSH[3] and some others. Here, we would like to report a new type of conformational restriction. Recently, we have reported the synthesis of compound 1[4]. The significant feature of 1 consists in the backbone conformation which is semi-rigid and similar to that of type II' β-turn (see A). The skeleton of 1 will be called BTD hereinafter; e.g. compound 1 is expressed as Pht-BTD-OH using the nomenclature. BTD can be incorporated anywhere in a peptide chain since it is a sort of dipeptide, and the peptide analog containing BTD would be forced to take on β-turn conformation at the point of substitution. If such an analog of a bioactive peptide exhibit any biological activity, valuable informations would

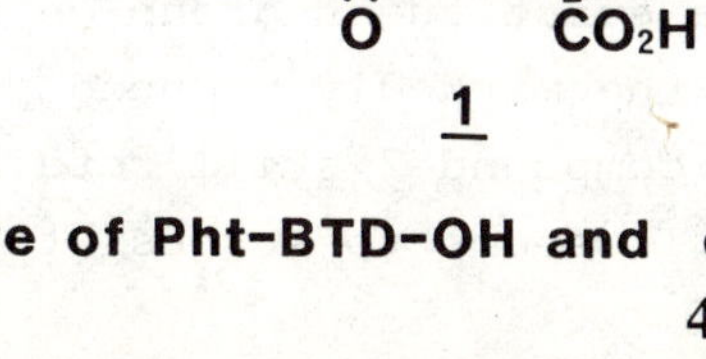

1

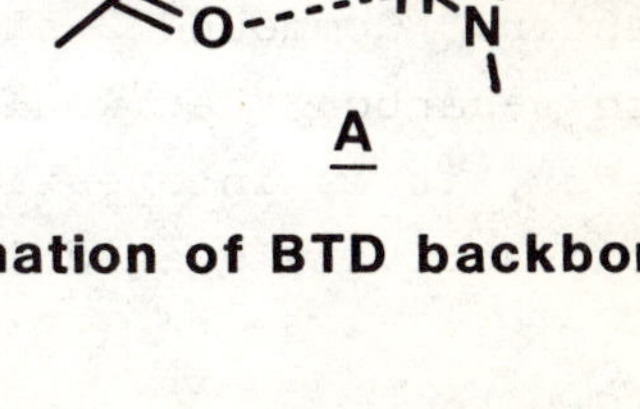

A

Fig.1

Structure of Pht-BTD-OH and conformation of BTD backbone

be obtained about the bioactive conformation from the results. The synthesis and properties of some peptides containing the BTD unit will be described briefly.

Results and Discussion

Enkephalin Analog

First, BTD was incorporated into enkephalin; i.e. [BTD^{2-3}, Leu^5]-enkephalinamide (2) was synthesized. Starting from Leu-NH_2, the peptide chain was elongated toward N-terminal by stepwise coupling of N-protected amino acids or BTD. The final product was purified by passing through a column of Sephadex LH-20.

The structure of 2 was confirmed by amino acid analysis (Phe:Leu:Tyr = 1.00 : 1.13 : 0.93) and FAB-MS spectrum, which gave the molecular ion peak at m/z 639 (MH^+) and fragment ion peaks at m/z 481, 346, 318, 226, 199, and 136. The peaks can be reaily interpreted from the structure, H-Tyr-BTD-Phe-Leu-NH_2•HCl.

Biological activity of 2 was tested by the ability to inhibit the electrically stimulated contraction of the isolated Guinea pig ileum. The inhibition was 20% at the dose of $2x10^{-5}$ M and reversible with naloxone (10^{-5} M). The potency is ca. 1/500 of [Leu^5]-enkephalin. Inhibition of [^3H]-DALAMID binding to the rat membrane receptor was also observed at the same order of concentration (IC_{50} $3x10^{-5}$ M). The weak potency may be explained by the fact the replacement of Gly^3-residue of enkephalin with any other amino acid residue causes marked or complete loss of the opioid activity[5]. This suggest that the presence of any substituent on the third α-carbon would obstruct interaction with the receptor. Compound 2 also has a substituent group on the third α-carbon that would cause the markedly reduced potency. However, it is interesting that compound 2 still retains the opioid activity even a little. It seems to suggest that the active conformation involves β-turn at the 2-3 position.

Gramicidin S analog

Second, two BTDs were incorporated into gramicidin S(GS), that is, $[BTD^{4-5,4'-5'}]$-GS was synthesized. Boc-L-Val-L-Orn(Z)-L-Leu-N_2H_3 was coupled with H-BTD-OH via the azide to afford the pentapeptide equivalent, Boc-L-Val-L-Orn(Z)-L-Leu-BTD-OH (3). A half portion of 3 was converted to the N-hydroxysuccinimide ester (4), and the other half was deprotected by acid treatment to yield the free form (5). Compounds 4 and 5 were coupled to give the decapeptide equivalent (6), Boc(L-Val-L-Orn(Z)-L-Leu-BTD-$)_2$OH. Compound 6 was then subjected to esterification with N-hydroxysuccinimide, N-terminal deprotection, cyclization, and removal of the side-chain benzyloxycarbonyl groups to result in the formation of the desired product (7), cyclo(L-Val-L-Orn(Z)-L-Leu-BTD-$)_2$· 2HBr. The structure of 7 was confirmed by elemental analysis and mass spectrum (MH^+ 1049).

CD spectrum of 7 was closely similar to that of GS indicating that they have similar conformation in solution. Antimicrobial activity of 7 was also compared with that of GS, and they showed the same activity (3.13 µg/ml against both B. subtilis and S. aureus as expressed by minimal inhibitory concentration). The results demonstrated that the bioactive conformation of GS must involve two β-turns of type II' since 7 cannot take on the other conformation at least at the corner positions.

CD Spectrum of Ac-BTD-NHMe

Third, N-Ac-BTD-NHMe (8) was synthesized and its CD spectrum was measured in order to test the possibility that some derivatives of BTD may be useful as a standard substance which gives the CD spectrum of type II' β-turn. It seems interesting that 8 showed the CD spectrum almost antipodal to that of type II β-turn reported by Woody[6] based on theoretical calculation.

References

1. Veber, D.F., Holly, F.W., Nutt, R.F., Bergstrand, S.J., Brady, S.F., Hirschman, R., Glitzer, M.S., and Saperstein, R. (1979) Nature 280, 512-514

2. Mosberg, H.I., Hurst, R., Hruby, V.J., Galligan, J.J., Burks, T.F., Gee, K., and Yamamura, H.I. (1982) Biochem. Biophys. Res. Commun. 106, 506-512.

3. Sawyer, T.K., Hruby, V.J., Darman, P.S., and Hadley, M.E. (1982) Proc. Natl. Acad. Sci. USA 79, 1751-1755.

4. Nagai, U. and Sato, K. (1985) Tetra. Lett. 26, 647-650.

5. Morley, J.S. (1980) Ann. Rev. Pharmacol. Toxicol. 20, 81-110.

6. Woody, R.W. (1974) in Peptides, Polypeptides and Proteins, Blout, E.R., Bovery, F.A., Goodman, M., and Loton, N., Eds., John Wiley, New York, pp. 338-350.

A SURVEY OF THE SUITABLE ROUTES TO O-GLYCOPEPTIDES

André A. Pavia

Laboratoire de Chimie Bioorganique, Faculté des Sciences,
33, rue Pasteur, 84000 Avignon, France.

Introduction

Although the role of carbohydrates has not been fully elucidated it appears more and more obvious that they contribute for a large part to the properties of glycoproteins as well as to the biological functions which they perform[1,2,3]. The growing awareness that model glycopeptides may help to investigate both the primary structure of native glycoproteins and certain fundamental aspects such as recognition or binding phenomena has aroused the interest of chemist for this type of compounds. As carbohydrate chemistry has made remarquable progress[4] over the past few years it is now possible to envisage the selective chemical synthesis of well-defined carbohydrate-peptide conjugates which may be used as starting material for the preparation of both artificial antigens and immune absorbers or utilized in many other areas of biological interest.

Results and discussion

Basically, there are two types of carbohydrate peptide linkages. The N-glycosidic type involves exclusively β-N-acetylglucosamine (β-GlcNAc) and the amide side-chain of asparagine. The O-glycosidic type which exists in a large variety of forms in both α- and β-anomeric configurations involves either serine (Ser), threonine (Thr), hydroxyproline (Hyp) or hydroxylysine (Hyl) and/or N-acetylgalactosamine (D-GalNAc), galactose (D-Gal), mannose (D-Man), glucose (D-Glc), xylose (D-Xyl), L-arabinose (L-Ara) and L-fucose (L-Fuc). Representative examples of such linkages are those between α-D-GalNAc

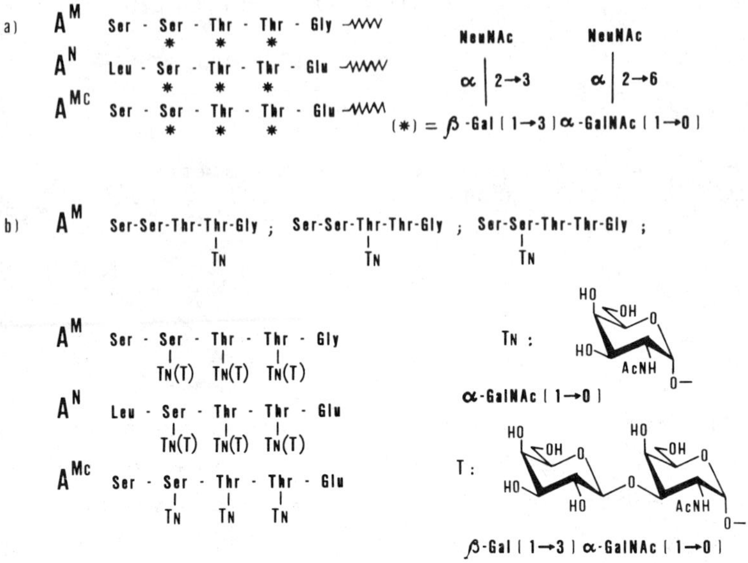

Fig. 1 *a) Amino terminal portion of human glycophorin A (M, N, M^C);*
b) T and T_N O-glycopeptides relevant to human glycophorin A

and Ser/Thr identified in a large number of animals glycoproteins or between β-Xyl and Ser found in proteoglycans.

This paper deals with the chemical methods available today to achieve the two following goals : i) the stereoselective attachment of saccharides to aminoacids or simple peptides, ii) the synthesis in reasonnable quantities (10 to 100 mg scale) of well-defined naturally occurring O-glycopeptides carrying a cluster of carbohydrate residues.

The difficulties and the potentialities of this type of chemistry will be illustrated through the synthesis of amino-terminal portions (see fig. 1) of human glycophorin A, the main erythrocyte membrane glycoprotein which is known to express the MN blood-group specificity and to serve as receptor for influenza virus, certain lectins and *Plasmodium falciparum*, the vector of malaria.

I – *METHODS OF GLYCOSYLATION* : Three fundamental procedures have proved to be suitable to achieve a glycosidic bond. Two of them are related to the so-called Koenigs-Knorr reaction in which a glycosyl halide is used to link a carbohydrate residue to an hydroxylic substrate. The third one represents a peculiar modification of the well-known acid-catalyzed Fischer reaction. In this case the glycosyl halide is replaced by a reducing sugar as a starting material.

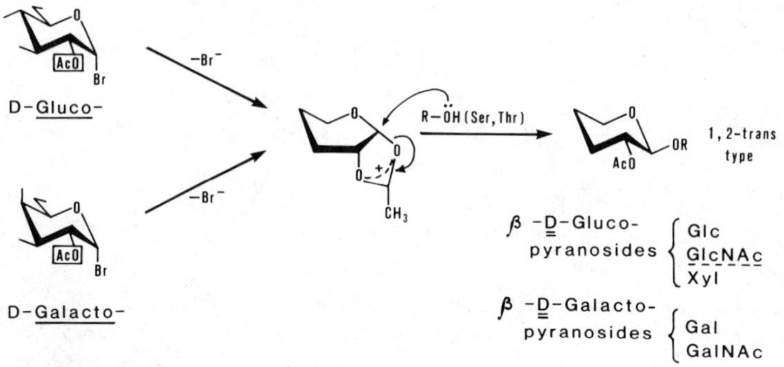

□ : Neighbouring active group , -OAc , -OBz , -NHAc .
Catalyst : $Hg(CN)_2$, $HgCl_2$, $AgClO_4$, $AgCO_3$, Ag triflate
Solvents : CH_3NO_2 , CH_3NO_2/Toluene

Fig 2 Neighbouring Group Assisted Procedure

1 The neighbouring group assisted procedure. This procedure allows the stereoselective synthesis of β-glycosidic linkages in the D-gluco seric i.e., in D-Glc, D-GlcNAc, D-Xyl and in the D-Galacto seric i.e., in D-Gal and D-GalNAc. In this procedure, the reaction proceeds through the intermediacy of a cyclic dioxocarbocation generated *in-situ* after the elimination of a bromide ion via an active catalyst[5]. In the second step the nucleophile attacks the intermediate at the anomeric carbon from the upper side leading to the 1,2 *trans* glycoside as shown below (fig. 2a). The presence of a neighbouring group active substituent at carbon-2 e.g., acetate or benzoate is responsible for the stereoselectivity of the reaction.

However in the D-manno seric where the configuration at C-2 is opposite to that of D-gluco and D-galacto type molecules the same mechanism leads to an acyloxonium intermediate of inverted configuration. In this case the corresponding *trans* opening of the intermediate following the same rule as above results in the formation of a glycoside with the α-anomeric configuration

2) The *in-situ* anomerization procedure. The preparation of an α-D-glycosidic linkage raises much more difficulties. In order to prevent the neighbouring group assistance just described, it is essential that the glycosyl halide used as starting material does not bear any acyl or acetamido group at position-2 of the pyranose ring. This approach is based on the possibility of producing an equilibrium, via suitable catalysts[6,7], between the α- and the β-halide. This equilibrium is reached through the intermediacy of different ion pairs (Fig. 3)

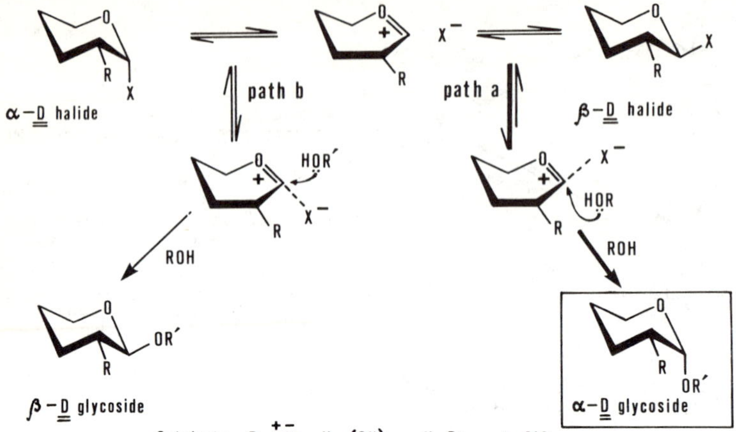

Catalysts : $R_4\overset{+}{N}\overset{-}{Br}$, Hg $(CN)_2$, Hg Br_2 , Ag ClO_4 ...
Solvents : CH_2Cl_2
R , a non participating group : OCH_2Ø or N_3 ...

Fig. 3 In-situ Anomerization Procedure

At the equilibrium, the more stable α-glycosyl halide is predominant, the reason for that lies in the existence of destabilizing interactions -the so-called anomeric effect- in the β-anomer. Considering now the kinetics of the glycosylation step it can be observed that the less stable β-halide (top right) reacts much faster than the α-anomer (top left). By taking advantage of this difference in reactivity one can control the stereoselectivity. Under certain conditions the reaction can be driven preferentially or even exclusively towards the α-D-glycopyranoside (bottom right).

3 -Acid-catalyzed procedure. Aldehydes and ketones react in anhydrous alcoholic solution saturated with hydrogen chloride with the formation of acetals. Under the same conditions, reducing sugars which are already hemiacetals establish an equilibrium in which cyclic acetals (glycosides) predominate. In the Fischer procedure the equilibrium is displaced towards the formation of the glycoside by using a large excess of alcohol. In practice the alcohol is used as solvent. For this reason the method is limited to the preparation of simple alkyl glycosides. In 1981 we proposed a new procedure[8] basically related to the earlier Fischer method. In this approach a O-benzyl derivative of a reducing sugar is reacted with an alcohol in the presence of trifluoromethanesulfonic anhydride to give in very good yield a α:β mixture of glycopyranosides in which the α-glycoside

predominates. The reaction is carried out in CH_2Cl_2 or CH_2Cl_2/CH_3CN mixture at relative low temperature and the condensation occurs within a few minutes.

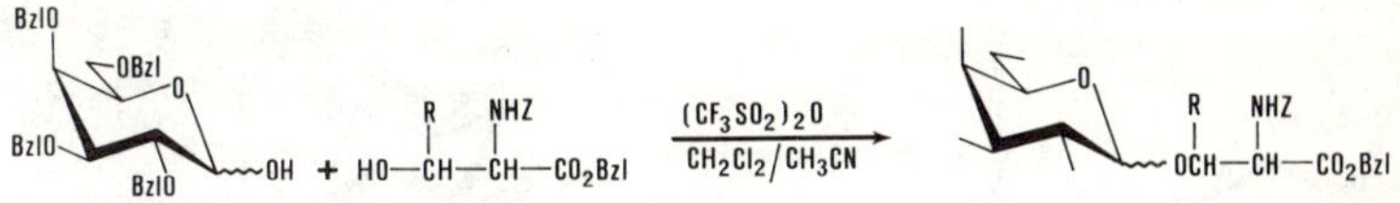

SUGAR / A.A. RATIO : 1/3
TEMP.° : 0°C
REACTION TIME : 20 TO 30 MINUTES
YIELD : 60 TO 90 % OF THE ANOMERIC MIXTURE OF GLYCOSYL-AMINOACIDS
STEREOSELECTIVITY : PREFERENTIALLY α ($\alpha/\beta \simeq 75/25$)

Fig. 4 Overall reaction and experimental conditions used for the preparation of glycosyl-aminoacids by the trifluoromethanesulfonic anhydride method.

The reaction was shown to proceed through a cyclic oxonium ion resulting from the protonation of the anomeric hydroxyl group[9]. Reaction of the nucleophile with the highly reactive intermediate leads preferentially to the formation of the more stable (anomeric effect) α-glycoside.

The effeciency of the method is based on the peculiar property of trifluoromethane sulfonic acid generated *in situ* to form with water a stable, insoluble hydronium trifluoromethanesulfonate. Therefore, in this procedure $(CF_3SO_2)_2O$ play the double role of source of proton and water scavenger.

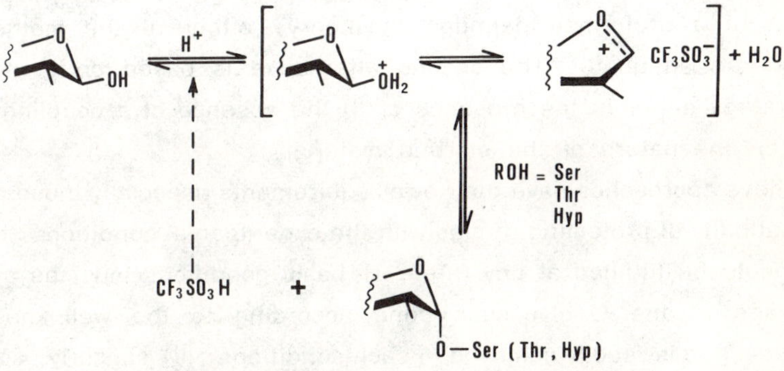

$$(CF_3SO_2)_2O + 3H_2O \longrightarrow 2\ CF_3SO_3^-\ H_3O^+$$

Fig. 5 Acid-catalyzed glycosylation (TF_2O method)

The main advantages of this procedure are the following : simplicity, rapidity, good yield, reasonnably good α-stereoselectivity and possibility to recover the excess of aminoacid.

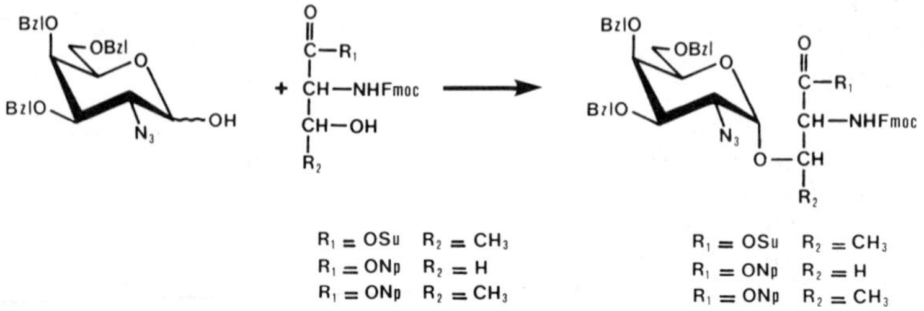

Fig. 6. Preparation of glycosyl aminoacids active esters, key compounds in the synthesis of portions of glycophorin A[10].

It is worthnoting that experimental conditions of this reaction are compatible with the presence of an active ester on the aminoacid involved in the condensation. In the last 3-4 years, all naturally occurring O-glycosidic carbohydrate peptide linkages have been prepared in this way. More importantly, the procedure was applied to the synthesis of glycosyl aminoacid active esters[10] which represent key compounds for the stepwise synthesis of more sophisticated glycopeptides, such as those reported in Fig. 1.

II - *GLYCOPEPTIDE SYNTHESIS*. Two approaches can be envisaged for the synthesis of O-glycopeptides. The block coupling strategy deals will the glycosylation of a preformed peptide. In theory, either of the methods of glycosylation are adequate. The second alternative is based on a stepwise coupling strategy either in the presence or in the absence of a coupling agent depending on the nature of the starting material.

Each of these approaches have their own requirements especially those related to the compatibility of protecting groups with the experimental conditions, but the following should be fulfilled at any rate : i) basic conditions must be avoided since cleavage of the O-glycosidic bond according to the well-known β-elimination reaction is readily achieved in such conditions ; ii) similarly, aqueous acidic conditions must be precluded since O-glycosidic linkages are readily hydrolyzed in the presence of acid ; iii) protecting and/or activating groups of the peptide moieties should be stable towards the chemical reagents and experimental conditions required for the glycosidation step.

1. Glycosylation of a preformed peptide : (block coupling strategy). To be suitable this route implies the following requirements : i) the method of attachment of the saccharide(s) residue(s) should be highly stereoselective. This is especially true whith peptides containing several points of glycosylation ; ii) because the effeciency of the glycosylation step depends on the reactivity of the hydroxyl group, this reactivity should not be significantly altered by the incorporation of Ser or Thr in the peptide chain ; iii) the peptide must be soluble in the usual solvent (CH_2Cl_2 or CH_2Cl_2/CH_3CN) in which glycosylation reaction are performed. In fact the block synthesis strategy can be only envisaged for the preparation of short glycopeptides containing no more than two hydroxyaminoacids. Longer peptide chain will cause dificulties and the step by step synthesis is preferable. Some example are shown in fig. 7.

2. Stepwise coupling strategy. A stepwise coupling strategy using either glycosylaminoacids, in the presence of a coupling agent or glycosyl aminoacids active esters as building units has more potential than the above block synthesis. The active ester peptide coupling method is very well suited for the preparation of glycopeptides especially those which include in their sequence a cluster of monosaccharides. The synthesis of the aminoterminal glycophorin A^N, and A^{Mc} (see Fig. 1), responsible of the so-called T_N specificity are reported in figure 8 as an illustration of this approach. It is clear that in this case the synthesis of

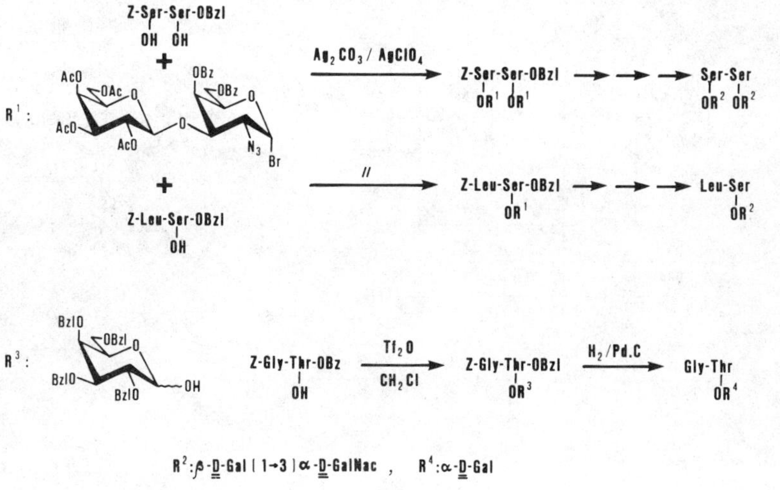

Fig. 7 Block synthesis of Leu-Ser11, Ser-Ser11 and Thr-Gly10.

serine and threonine activo esters must be achieved by the acid-catalyzed method, since none of the other methods allow the preparation of glycosylaminoacids active esters.

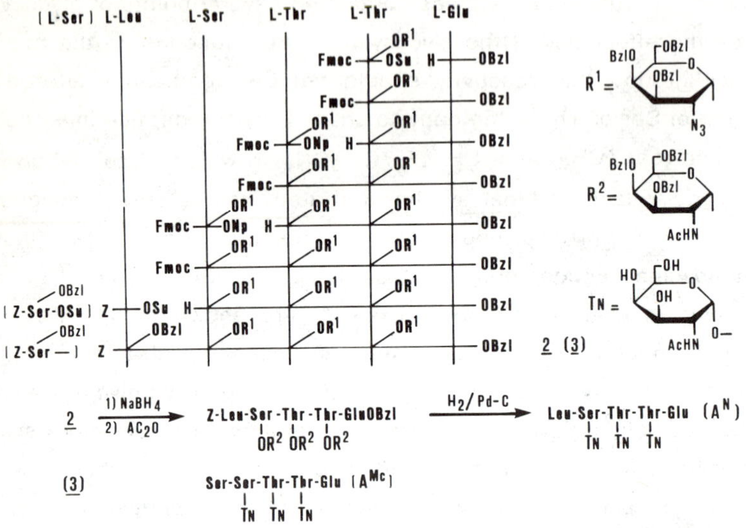

Fig. 8 Synthesis of T_N glycopeptides related to human glycophorin A^N and A^{Mc} (see ref.[13]). The corresponding A^M glycopeptide has been described elsewhere[12].

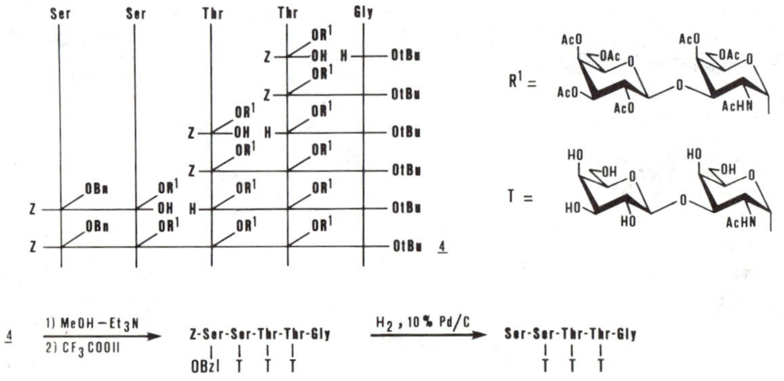

Fig 9 Synthesis[15] of the O-glycopentapeptide (T antigénicity) related to N-terminal portion of glycophorin A^M. The same strategy was followed for the preparation of the A^N analog[15].

It is worthnoting that in this approach there is no problem of solubility, benzyl derivatives of sugars being readily soluble in the usual solvents of peptide synthesis. On the contrary benzyl substituents may improve the solubility of certain peptide sequences which would otherwise cause some difficulties. The synthesis of building units consituted of an oligosaccharide linked to an aminoacid active ester presents severe difficulties. The number of steps and the variety of the experimental conditions make the blocking-deblocking strategy a serious task ,consequently, the active ester coupling method become unsuitable. For the preparation of glycopeptides with larger oligosaccharide moities a step by step condensation of non activated glycosylaminoacids in the presence of a coupling agent has proved to be more convenient. Fig 9 presents the route which has been follow to obtain the N-terminal sequence of glycophorin A^M and A^N carrying the so-called T-immunospecificity, i.e. three β-Gal(1→3)-α-D-GalNAc, disaccharides O-linked to Ser-2, Thr-3 and Thr-4. E.F.D.Q. was used as coupling agent.

Conclusion

The easiest approach to the synthesis of glycopeptides carrying a cluster or monosaccharides is the stepwise active ester coupling method beginning at the C-terminus and proceeding towards the N-terminus. It is worthwhile mentionning that once protecting and activating group have been carefully selected the synthesis of glycopeptides does not present more difficulties than that of the parent peptide. For the synthesis of more sophisticated glycopeptides carrying one or several O-linked oligosaccharides, the best approach remains the step by step synthesis ; however, it is preferable to use a peptide coupling agent such an EEDQ or DCC. The major difficulties to be faced in this approach will be related to the building up of the oligosaccharide moiety. A good knowledge of the carbohydrate chemistry is required. Finally the so-called block coupling strategy i.e., the direct linking of oligosaccharide (s) to the preformed peptide is of limited applications.

References

1. Sharon, N., (1975) Complex Carbohydrates, Their Chemistry, Biosynthesis and functions, Addison-Wesley, Reading, Massachusetts.

2. The glyconjugates (Horowitz, M.I. and Pigman, W. eds), Academic Press, New-York, vol. I, 1977, Vol. II, 1978.

3. Montreuil, J., (1980) in Adv. Carbohydr. Chem. Biochem., 37, 157-223.

4. Paulsen, H., (1982), Angew. Chem., 94, 184-201 ; Angew. Chem. Int. Ed. Engl. 21, 155-173.

5. Paulsen, H., (1983), Sel. Goal Synth. Effic., Proc. Worshop Conf. Hoechst, 14th, Bartmann, W., Trost, B., eds, Verlag Chem., Weinheim, R.F.A., 169-190

6. Lemieux, R.U., Hendricks, K.R., Stick, R.V., and James, J., (1975), J. Am. Chem. Soc., 97, 4056-4062.

7. Paulsen, H., and Lockhoff, O., (1978), Tetrahedron Lett., 4027-4030.

8. Lacombe, J.M., Pavia, A.A., and Rocheville, J.M., (1981); Can. J. Chem., 59, 473-481.

9. Pavia, A.A., and Ung-Chhun, S.N., (1981), Can. J. Chem., 59, 482-489.

10. Lacombe, J.M., and Pavia, A.A., (1983), J. Org. Chem., 48, 2556-2563.

11. Paulsen, H., Paal, M., and Schuttz, M., (1983), Tetrahedron Lett. 24, 1759-1762.

12. Ferrari, B., and Pavia, A.A., (1983), Int. J. Pept. Prot. Res., 22, 549-559.

13. Ferrari, B., and Pavia, A.A., (1985), Tetrahedron, in press.

14. Dill, K., Douglas Carter, R., Lacombe, J.M., and Pavia, A.A., (1985), Biochem. J. submitted.

15. Bencomo-Verrez V., and Sinaÿ, P., (1984) Glycoconjugate J., 1, 5-8.

CARBOHYDRATE DERIVATIVES OF PEPTIDES. SYNTHESIS OF A GLYCATED HEPTAPEPTIDE SEQUENCE OF HUMAN HEMOGLOBIN.

John Halstrøm and Viggo Kruse
Novo Research Institute, Novo Allé
DK-2880 Bagsvaerd, Denmark

Introduction

Modification of exposed amino groups in proteins by non-enzymatic glucosylation (glycation) has received much attention recently, particularly in connection with diabetes[1].

The initial product of the reaction with glucose is a Schiff base, which is known to undergo Amadori rearrangement to yield, ultimately, a 1-deoxy-D-fructosyl (FRC) amine[1]. In addition to amino termini, the ε-amino groups of some lysine residues are important sites of glycation.

During the development of an immunoassay for glycated human hemoglobin, it was decided to synthesize the N^{ε}-FRC substituted sequence 13-19 of the α-chain.

Results and discussion

In order to minimize side reactions, the synthesis was carried out by fragment condensation in solution (Figure 1).

The D-fructose residue of Lys(FRC) is reported to survive acidolytic removal of N^{α}-formyl[2], and N^{α}-benzyloxycarbonyl (Z)[3] by 2N HCl/H_2O at 100° and 4.5 N HBr/AcOH at 20°, respectively. For N^{α}-protection during the glycation, tert. butyloxycarbonyl (Boc) was selected, because it is removed by 1N HCl/AcOH at 20°. For protection of the secondary ε-amino function resulting from the Amadori rearrangement, 9-fluorenylmethoxycarbonyl (Fmoc) was chosen.

The Z group of Boc-Lys(Z)-Val-Gly-Ala-OH (I) was removed by catalytic hydrogenolysis, and the resulting Boc-Lys-Val-Gly-Ala-OH (II) was treated with a 50% excess of D-glucose in boiling methanol for 4 hours. After chromatography on silica, Boc-Lys(FRC)-Val-Gly-Ala-OH (III) was obtained in 35% yield, crystallized from methanol.

Selective blocking of the secondary ε-amino function was achieved by reaction with Fmoc-OSu, and Boc-Lys(FRC,Fmoc)-Val-Gly-Ala-OH (IV) was obtained after chromatography on silica and crystallization from 2-propanol.

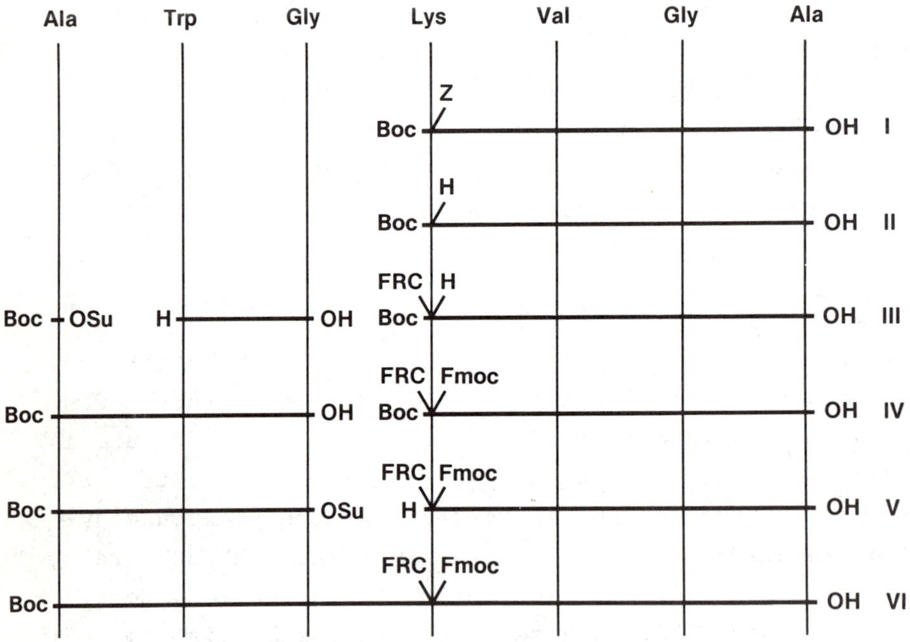

Fig. 1. Scheme of synthesis. FRC is 1-deoxy-D-fructosyl.

^{1}H NMR spectra of peptides III and IV in methanol show that the D-fructose residue is present in both.

The purity of the peptides was monitored by HPLC. Amino acid analyses are shown in the table.

Table. Amino Acid Analysis

Peptide	II	III	III^{x}	IV^{x}	A^{x}	A	B
Gly	1.01	1.01	1.00	1.00	2.02	2.00	2.01
Ala	1.00	0.99	1.00	1.00	2.03	2.00	2.00
Val	1.00	0.99	0.93	0.93	0.96	1.00	0.99
Lys	0.99	0.51	0.78	0.76	0.19	0.22	0.97
Trp					1.02	0.58	0.84
Fur		++++	++	+	++	+++	0

Hydrolysis: 24 h at 111°C in 6N HCl or 4N MSA(x).
Fur = furosine (estimated; no standard available)

For the carbohydrate-containing peptides III and IV, the recovery of lysine is low, due to formation of furosine[1] and pyridosine[1] under the conditions of acid hydrolysis.

Removal of the Boc group by 1N HCl/AcOH for 30 min at 20° resulted in a heterogeneous product. Reaction with Boc-Ala-Trp-Gly-OSu in aqueous THF at pH 8.5 yielded a mixture, from which the two major components, heptapeptides A and B, were isolated by chromatography on silica, followed by preparative HPLC.

Judged by ^{1}H and ^{13}C-NMR spectra, neither of the two heptapeptides A or B appear to contain intact carbohydrate.

Amino acid analysis, on the other hand, shows considerable furosine formation for A (see table), and a surprisingly low recovery of lysine, whereas in B furosine formation is negligible and the lysine value is close to 1.

To clarify these results, A and B were analyzed by plasma desorption MS. The molecular ion (MH^{+}) found for A is 1156.3 and for B 1049.7 with an accuracy of ±1 mass unit.

If the MH^+ for A is taken to be 1155.3 instead of 1156.3, this value corresponds to 18 mass units less than the theoretical MH^+ value of 1173.3 for the desired glycated N^{α}-Boc,N^{ε}-Fmoc heptapeptide (VI).

Complete dehydration of Lys(FRC) to furosine requires a loss of 3 mol H_2O. Although dehydration during MS cannot be entirely excluded, the loss of 1 mol H_2O indicates that a furosine precursor is formed during treatment with 1N HCl/AcOH. From this precursor furosine is formed in high yield on acid hydrolysis. Thus, not only intact Lys(FRC) gives rise to the formation of furosine[1].

The MH^+ value of 1049.7 for B is 38.5 mass units higher than the theoretical value of 1011.2 for the non-glycated N^{α}-Boc,N^{ε}-Fmoc heptapeptide. This difference of 38.5 may, within the accuracy of the method, be explained by a C_3H_4 substitution of the lysine ε-amino group.

In conclusion, Boc group cleavage from glycated peptides should be avoided, as it seems to cause dehydration and cleavage of the D-fructose residue.

Acknowledgement

We wish to thank M. Christensen, L. Permin, and H. Thøgersen of this institute for the amino acid and NMR analyses, and P. Roepstorff of the University of Odense, Odense, Denmark for the MS analysis.

References

1. Schleicher, E. and Wieland, O.H. (1981) J. Clin. Chem. Clin. Biochem. 19, 81-87.
2. Finot, P.A. and Mauron, J. (1969) Helv. Chim. Acta 52, 1488-1495.
3. Rucklidge, G.J., Bates, G.P., and Robins, S.P. (1983) Biochim. Biophys. Acta 747, 165-170.

NOVEL SIDE-CHAIN TO SIDE-CHAIN LINKED CYCLIC OPIOID PEPTIDE MONOMERS AND DIMERS: SYNTHESIS AND ACTIVITY PROFILES

Peter W. Schiller, Thi M.-D. Nguyen, Carole Lemieux and Louise A. Maziak

Laboratory of Chemical Biology and Peptide Research
Clinical Research Institute of Montreal
Montreal, Quebec, Canada H2W 1R7

Introduction

The synthesis of cyclic enkephalin analogs sofar has resulted in μ-receptor-selective[1], non-selective[2] and δ-receptor-selective compounds[3]. In an effort to further improve selectivity we designed analogs cyclized via side-chain amino and carboxyl groups. In the solid-phase synthesis of the latter analogs on the benzhydrylamine resin[4], the peptide segment to be cyclized was assembled using Fmoc-protection for the α-amino group, and Boc and <u>tert</u>. butyl protection for the side-chains of Lys (or Orn) and Glu (or Asp), respectively. Following side-chain deprotection with TFA/CH_2Cl_2 cyclization on the resin with DCC/HOBt as coupling agents yielded the desired cyclic monomers (e.g. compound <u>1</u>) as well as the side-chain-linked antiparallel cyclic dimers (e.g. compound <u>1a</u>) in

H-Tyr-D-Orn-Phe-Asp-NH_2 (<u>1</u>)

H-Tyr→D-Orn→Phe→Asp-NH_2 (<u>1a</u>)
H_2N-Asp←Phe←D-Orn←Tyr-H

various proportions. Analogs were tested in the guinea pig ileum (GPI) assay (μ-receptor-representative) and the mouse vas deferens (MVD) assay (δ-receptor-representative), and in binding assays based on displacement of [^{3}H]DAGO (μ-selective) and [^{3}H]DSLET (δ-selective) from rat brain membrane sites.

Results and Discussion

In the binding assays (Table I) the cyclic tetrapeptide analog 1 showed excellent μ-receptor selectivity due to very weak affinity for the δ-receptor, as indicated by the high ratio of the binding inhibition constants (K_i^{δ}/K_i^{μ}= 213). Compound 1 was more μ-selective than its linear correlate 1b, suggesting that its high selectivity is a consequence of the conformational restriction introduced through ring closure. In contrast to 1 and 1b the corresponding cyclic dimer 1a was non-selective because of its higher affinity for the δ-receptor resulting from the different conformational constraint present in the dimeric structure. Transposition of the Orn (Nva) and Asp residues produced peptides (2, 2a and 2b) showing qualitatively similar potency relationships, whereas reversal of the configuration in position 4 of cyclic monomers 1 and 2 led to compounds (3 and 4) with significantly lower μ-receptor selectivity. Expansion of the 13-membered ring structure present in 1 through insertion of an additional methylene group in the 2- and 4-position side-chains resulted in a more flexible cyclic analog (15-membered ring) which showed little preference for μ-receptors over δ-receptors due to a 500-fold increase in δ-receptor binding affinity. This result suggests that variation in the degree of conformational restriction can produce drastic shifts in the selectivity profile of peptides. The analogous ring expansion in 2 had a less dramatic effect (compound 6). The cyclic pentapeptide analog 7 was highly potent in both the [^{3}H]DAGO and the [^{3}H]DSLET binding assay and, therefore, is non-selective. Unexpectedly, the corresponding open chain analog 7b showed quite pronounced μ-receptor selectivity, indicating that in some cases conformational restriction can also produce a decrease in receptor selectivity. In comparison with morphiceptin, cyclic monomer 1 has a 13-times higher K_i^{δ}/K_i^{μ}-ratio, whereas it is about 5 times less μ-selective than DAGO in the binding assays.

Table I. Receptor Binding Affinities of Opioid Peptide Analogs

	Compound	$[^3H]$DAGO K_i^{μ} [nM]	$[^3H]$DSLET K_i^{δ} [nM]	K_i^{δ}/K_i^{μ}
1	H-Tyr-D-Orn-Phe-Asp-NH_2	10.4 ± 0.7	2'220 ± 65	213
1a	(H-Tyr-D-Orn-Phe-Asp-NH_2)$_2$	25.6 ± 1.9	42.2 ± 2.4	1.65
1b	H-Tyr-D-Nva-Phe-Asn-NH_2	11.7 ± 0.7	441 ± 15	37.7
2	H-Tyr-D-Asp-Phe-Orn-NH_2	9.62 ± 2.52	1'320 ± 150	137
2a	(H-Tyr-D-Asp-Phe-Orn-NH_2)$_2$	85.9 ± 15.5	129 ± 12	1.50
2b	H-Tyr-D-Asn-Phe-Nva-NH_2	42.9 ± 15.9	2'760 ± 90	64.3
3	H-Tyr-D-Orn-Phe-D-Asp-NH_2	21.7 ± 3.2	422 ± 17	19.4
4	H-Tyr-D-Asp-Phe-D-Orn-NH_2	40.5 ± 1.4	115 ± 22	2.84
5	H-Tyr-D-Lys-Phe-Glu-NH_2	1.43 ± 0.07	4.36 ± 0.46	3.05
6	H-Tyr-D-Glu-Phe-Lys-NH_2	0.994 ± 0.329	49.3 ± 3.3	49.6
7	H-Tyr-D-Lys-Gly-Phe-Glu-NH_2	1.31 ± 0.21	0.690 ± 0.025	0.527
7a	(H-Tyr-D-Lys-Gly-Phe-Glu-NH_2)$_2$	5.33 ± 0.48	23.2 ± 1.3	4.35
7b	H-Tyr-D-Nle-Gly-Phe-Gln-NH_2	0.628 ± 0.037	23.4 ± 2.8	37.3
8	Morphiceptin	22.9 ± 0.6	382 ± 87	16.7
9	DAGO	1.22 ± 0.12	1'280 ± 90	1'050
10	[Leu^5]enkephalin	9.43 ± 2.07	2.53 ± 0.35	0.268

Very similar potency relationships were observed in the GPI- and MVD-assays. However, cyclic monomers 1, 3 and 7 were much more potent in the μ-receptor-representative bioassay (GPI) than was expected on the basis of their affinities determined in the μ-receptor-selective binding assay ($[^3H]$DAGO displacement). Thus, compared to its linear correlate 7b, cyclic analog 7 was 9 times more potent in the GPI assay but less than half as potent in the $[^3H]$DAGO binding assay. This result may be explained with an enhanced "efficacy" ("intrinsic activity") of 7 as a consequence of the conformational constraint present in this cyclic analog. The concomitant affinity loss and "efficacy" enhancement observed with the cyclic peptide may be due to the fact that part of the receptor binding energy is used to induce a more productive conformational change in the receptor protein. Because of its increased "efficacy" at the μ-receptor cyclic analog 1 shows a more than three times higher IC50(MVD)/IC50(GPI)-ratio than DAGO and, therefore, is a highly selective μ-agonist.

Acknowledgements

Supported by grants from the MRCC (MT-5655) and the QHF.

References

1. DiMaio, J., Nguyen, T.M-D., Lemieux, C., and Schiller, P.W. (1982) J. Med. Chem. 25, 1432-1438.

2. Schiller, P.W., Eggimann, B., DiMaio, J., Lemieux, C., and Nguyen, T.M.-D. (1981) Biochem. Biophys. Res. Commun. 101, 337-343.

3. Mosberg, H.I., Hurst, R., Hruby, V.J., Gee, K., Yamamura, H.I., Galligan, J.J., and Burks, T.F. (1983) Proc. Natl. Acad. Sci. USA 80, 5871-5874.

4. Schiller, P.W., Nguyen, T.M.-D., and Miller, J. (1985) Int. J. Peptide and Protein Res. 25, 171-177.

CONFORMATION-BIOLOGICAL ACTIVITY RELATIONSHIPS OF CONFORMATIONALLY CONSTRAINED DELTA SPECIFIC CYCLIC ENKEPHALINS

Victor J. Hruby*, L.-F. Kao*, Lane D. Hirning+ and T.F. Burks+
Departments of Chemistry* and Pharmacology+, University of Arizona, Tucson, Arizona 85721

Introduction

We previously reported that the cyclic highly conformationally constrained 2,5-bis-penicillamine analogues of enkephalin, [D-Pen[2],D-Pen[5]]E and [D-Pen[2],L-Pen[5]]E, have highly specific and potent delta-receptor selective activities.[1] The unique biological activity profiles[1-3] of these compounds led us to examine the conformational properties responsible for these activities and to further examine conformational structure-biological activity relationships.

Experimental

The bis-penicillamine analogues of Table I were synthesized using methods previously reported.[1,4] Purification was accomplished by gel filtration on Sephadex G-10 followed by partition chromatography[4] or preparative reverse phase HPLC. Purity was assessed by TLC, HPLC, amino acid analysis, and fast atom bombardment (FAB) mass spectrometry. The guinea pig ileum (GPI) and mouse vas deferens (MVD) assays were performed as previously reported.[1] Nuclear magnetic resonance (nmr) spectra were recorded on a Bruker WM-250 nmr spectrometer with D_2O/H_2O as solvent. Chemical shifts were measured in ppm downfield from internal TSP-d_4.

Results and Discussion

Previous studies[1,4] and nmr investigations (see below) suggested that the stereostructural properties of the 4 position were important to the bioactivity of DPDPE. We therefore prepared DPDPE analogues substituted in the 4 position with residues that were further conformationally constrained (Table I) including a N-MePhe[4] analogue, tetrahydroisoquinoline carboxylate (Tic) and substituted Tic[4] analogues, and a cyclic aliphatic amino acid, α-aminocyclohexane carboxylate, analogue. We also have examined the structural requirements of the C- and N-terminal residues for high delta receptor specificity. The results for some of these analogues are given in Table I. The substitution of the Phe[4] amide proton for a methyl group causes a 10 fold loss in delta receptor specificity and a 270 fold drop in potency at the MVD receptor. A minor or no loss in potency is observed when the aromatic side chain of Phe is fixed in a gauche conformation by the use of a Tic or 7-NO_2-Tic residue in the 4 position. More importantly, these latter

Table I. Delta Opioid Receptor of Conformationally Constrained [D-Pen[2],D-Pen[5]]enkephalin Analogues

DPDPE Analogue	IC$_{50}$(nM) GPI	IC$_{50}$(nM) MVD	IC$_{50}$(GPI)/IC$_{50}$(MVD)
[D-Pen[2], D-Pen[5]]E	6,930	2.19 ± 0.30	3,160
[D-Pen[2], N-MePhe[4], D-Pen[5]]E	~180,000	596 ± 182	312
[D-Pen[2], Tic[4], D-Pen[5]]E	>3 x 10^7	1500	>2,000
[D-Pen[2], Tic(7-NO_2)[4], D-Pen[5]]E	>3 x 10^7	494 ± 134	>6,000
[D-Pen[2], Acc[4], D-Pen[5]]E	>1 x 10^7	>2 x 10^6	-
[DOPA[1], D-Pen[2], D-Pen[5]]E	91,200	307 ± 75	297
[D-Pen[2], D-Pen[5], Arg[6], Phe[7]]E	4,170	262 ± 50	18
Normorphine	91 ± 19	540 ± 113	0.17

[a]Abbreviations: E = enkephalin; Tic = Tetrahydroisoquinoline carboxylate; Acc = α-aminocyclohexane carboxylate

analogues are of similar delta receptor selectivity as DPDPE. These results and conformational analysis indicate that the biologically important conformation for the Phe^4 side chain is the gauche(-) conformation. The greatly reduced potency of the Acc^4 analogue at the MVD and GPI receptors suggests that an aromatic residue is preferred by the delta opioid receptor. The loss of specificity and potency by the $DOPA^1$ and the Arg^6, Phe^7 analogues are consistent with the need for specific stereostructural requirements of the aromatic ring in position 1 and the C-terminal carboxylate since the conformation remains unchanged in these analogues.

The unambiguous nmr assignments for all protons and carbons of DPDPE were obtained by a combination of ^1H and ^{13}C 1D nmr, by ^1H-^1H and ^{13}C-^1H COSY, ^1H-^1H delayed COSY, ^{13}C 2D-J-resolved and attached proton test experiments, and by pH and temperature dependent studies. The nmr data were consistent with a constrained conformation for DPDPE and the further constrained analogues reported here. For example, the large chemical shift differences of the Gly^3 α-protons can be accounted for by a fixed conformation of the peptide backbone

Table II. Non-Equivalence of Diastereotopic Gly^3 α-Hydrogens in [D-Pen^2, D-Pen^5]Enkephalin Analogues

DPDPE Analogue[a]	δ Gly^3 αH (ppm)	Δδ (ppm)
[D-Pen^2, D-Pen^5]E	3.54, 4.34	0.80
[D-Pen^2, N-$MePhe^4$, D-Pen^5]E	3.35, 3.38	0.03
[D-Pen^2, Tic^4, D-Pen^5]E	3.68, 5.04	1.36
[D-Pen^2, $Tic(7\text{-}NO_2)^4$, D-Pen^5]E	3.75, 5.15	1.40
[D-Pen^2, Acc^4, D-Pen^5]E	3.67, 4.51	0.84
[$DOPA^1$, D-Pen^2, D-Pen^5]E	3.48, 4.29	0.81
[D-Pen^2, D-Pen^5, Arg^6, Phe^7]E	3.54, 4.35	0.81

[a]See Table I for abbreviations

in the 14-membered ring in which the anisotropic effects of the adjacent peptide bonds cause an upfield shift for one of the diastereotopic Gly^3 Hs and a downfield shift for the other. The $\Delta\delta$ values for these protons in the Acc^4 analogue and increased $\Delta\delta$ values for the Tic^4 analogues (Table II) indicate that the side chain aromatic of Phe^4 is not contributing significantly to the non-equivalence of the Gly^3 αHs in DPDPE. Analysis of $^3J_{\alpha CH-\beta CH}$ and $^2J_{\beta CH-\beta CH}$ values indicate that only the trans and gauche(-) conformations are allowed for the Phe^4 and Tyr^1 side chain groups in DPDPE. The conformation was further evaluated by analysis of chemical shift and coupling constant data, by temperature studies, and by $^1H-^1H$ 2D-NOE (NOESY) experiments. A conformation consistent with this data has been obtained and its relationship to the high delta opioid receptor can be proposed. These proposals and the minimum energy conformation are being evaluated by conformational and dynamic calculations (Hruby, Kao, Pettitt, and Karplus, unpublished results).

This work was supported by grants from the U.S. Public Health Service and the National Science Foundation. VJH is a recipient of a Guggenheim Foundation Fellowship (1984-1985).

References

1. Mosberg, H.I., Hurst, R., Hruby, V.J., Gee, K., Yamamura, H.I., Galligan, J.J., and Burks, T.F. (1983) Proc. Natl. Acad. Sci. USA **80**, 5871-5874.

2. Galligan, J.J., Mosberg, H.I., Hurst, R., Hruby, V.J., and Burks, T.F. (1984) J. Pharmacol. Exp. Therap. **229**, 641-648.

3. Hirning, L.D., Mosberg, H.I., Hurst, R., Hruby, V.J., Burks, T.F., and Porreca, F. (1985) Neuropeptides 5, 383-386.

4. Hurst, R. (1984) M.S. Thesis, University of Arizona.

THE SYNTHESIS AND SAR OF ORALLY ACTIVE ENKEPHALIN ANALOGS WITH MODIFIED N-TERMINAL TYROSINE RESIDUES.

Donald W. Hansen Jr., Robert H. Mazur, Michael Clare
Searle Research and Development, Division of G.D. Searle & Co., 4901 Searle Parkway, Skokie, Illinois 60077, U.S.A.

Introduction

The knowledge that the tyramine portion of both the rigid opiates and the enkephalins is critical for analgesic activity prompted us to investigate structural modifications of the Tyr^1 in two of our enkephalin analog series. Initially we looked at the effect of N-alkylation on a tetrapeptide prototype characterized structurally by having a 2-adamantyl amide at the C-terminus. Biologically this series is typifyed by having activity in our writhing mouse assay but no activity in a mouse hot plate screen. Calculations predict that though alkylation of the N-terminal nitrogen alters both its steric accessability and basicity, the conformations of the peptide backbone are little affected. The introduction of alkyl groups at various positions on the Tyr^1 ring is predicted to have much more significant effects on the local backbone conformations. We chose to apply this latter modification to a second enkephalin series characterized by having a C-terminal 6-aminohexanoic acid methyl ester and excellent activity in both writhing mouse and mouse hot plate assays.

Methods and Results

The N-methylation of protected(Boc) amino acids was carried out by the method of Benoiton[1], which is not applicable to the synthesis of the corresponding N-ethyl derivatives. N-ethylation of Boc protected aminoacids without racemization was accomplished by the method described in our recent publication[2]. The compounds in Tables I and II below were synthesized from the requisite protected amino acids and peptide fragments by solution coupling methods (isobutylchloroformate) using mixed anhydride procedures.

Table I. N-Terminal Adamantyl Tetrapeptide Biology Data

Compounds[a]	Opiate[b] Binding	W. sc	Mouse[c] ig	Hot P.[d] iv
Morphine	$1.4x10^{-9}$	0.40	4.00	A
Tyr Gly Gly Phe Met	$1.1x10^{-8}$	I	I	I
Tyr Gly Gly Phe Leu	$3.9x10^{-8}$	I	I	I
Phe (D)Met Gly Phe NH2Ad	$3.5x10^{-7}$	I	I	I L
Me - Phe (D)Met Gly Phe NH2Ad	$8.6x10^{-7}$	I	I	I L
Et - Phe (D)Met Gly Phe NH2Ad	$1.3x10^{-6}$	2.22	I	I
Tyr (D)Met Gly Phe NH2Ad	$1.3x10^{-10}$	3.40	0.24	I
Me - Tyr (D)Met Gly Phe NH2Ad	$2.9x10^{-9}$	8.35	0.41	A
Et - Tyr (D)Met Gly Phe NH2Ad	$3.7x10^{-9}$	I	1.30	I
Tyr(2,6Me) (D)Met Gly Phe NH2Ad	$9.0x10^{-11}$	7.68	0.93	I

[a] All compounds are HCl salt.
[b] Displacement of 1.0nM ^{3}H-naloxone in whole rat brain less cerebellum, in the absence of sodium ion.
[c] Inhibition of PBQ (.025% W/V, 0.1ml/10g body weight, ip). Dosage levels = 10 mg/kg.
[d] Increase in response (paw - lick) latencies to 55°C. hot plate in mice. Dosage levels = 50 mg/kg. ED_{50}'s are sc.

Ring alkylated tyrosines were initially prepared through modifications of published procedures[3]. Recently we have developed alternate syntheses of both racemic and optically active modified tyrosines which involves either the symmetric or asymmetric reductive amination of the necessary pyruvic acid precursor.

Conformational analyses run on full tetrapeptide models proved unrewarding because of the large numbers of minima arising from the many rotational degrees of freedom.

However, it was noted that minima calculated for the N-terminal dipeptide fragments fell into well defined families. Molecular mechanics calculations with full geometric relaxation reveal that in ca. 75% of the lowest minima families both Tyr^1 sidechain torsional angles lie close to 90°. When these minima are mapped onto rigid opiate templates such as metazocine or morphine derived from x-ray data, there is good topographical overlap between the N-terminal moiety and the rigid template functions. That is to say, though the phenolic ring and the amino nitrogen occupy much the same space, there is no such correspondence between intervening atoms and bonds connecting them. The fit is actually better than among the opiate templates themselves. We believe that the observed enhanced binding in the 2,6-Me_2Tyr analogs arises as a result of a closer fit to this template than for the simple Tyr cases.

Table II. Tyr^1 Ring Alkylated Pentapeptide Biology Data

Compounds[a]					Opiate[b] Binding	W. Mouse[c] sc	ig	HP[d] iv
Morphine					$1.4x10^{-9}$	0.40	4.00	A
Tyr	(D)Met	Gly	Phe	NHR	$1.0x10^{-8}$	0.93	I	A
Tyr(2,6Me)	(D)Met	Gly	Phe	NHR	$7.1x10^{-13}$	0.33	I	22
Tyr	(D)Met(O)	Gly	Phe	NHR	$8.4x10^{-9}$	1.68	I	A
Tyr(2,6Me)	(D)Met(O)	Gly	Phe	NHR	$4.0x10^{-13}$	0.24	1.85	8.0
(D)Tyr(2,6Me)	(D)Met(O)	Gly	Phe	NHR	$2.4x10^{-9}$	I	1.40	120
Tyr(2,3,6Me)	(D)Met	Gly	Phe	NHR	$1.3x10^{-8}$	3.90	2.02	I
Tyr(2,3,6Me)	(D)Met(O)	Gly	Phe	NHR	$6.5x10^{-7}$	I	6.76	A
Tyr(3tBu)	(D)Met	Gly	Phe	NHR	$7.0x10^{-7}$	I	I	-
(D)Tyr(3tBu)	(D)Met(O_2)	Gly	Phe	NHR	$2.4x10^{-6}$	I	A	I
m-Tyr(2,4Me)	(D)Met(O)	Gly	Phe	NHR	$7.0x10^{-8}$	I	A	A

[a] All compounds are HCl salts. R=$(CH_2)_5CO_2Me$.
[b,c] and [d] are identical with those described in Table I.

CONCLUSION:

Adamantylamide tetrapeptides: As anticipated, tyrosine in the N-terminal position is generally required for analgesic activity, but the significant sc writhing mouse activity of the analog containing Et-Phe^1 is noteworthy. Both the in vitro and in vivo activity of Tyr^1 analogs are relatively insensitive to N-terminal alkylation, though extension from N-methyl to N-ethyl eliminates oral writhing activity.

Tyr^1 Ring Alkylated Pentapeptides: The introduction of alkyl groups onto the Tyr^1 ring produces dramatic increases in both in vitro and in vivo activities in both series of compounds. In particular the 2,6-dimethyl substitution pattern is remarkable in its effect on opiate binding. Several of these compounds possess opiate binding IC_{50}'s of 10^{-13}, approximately 10,000 the potency of morphine. These materials are, to our best knowledge, the most potent opiate binding peptides reported to date. In addition, the presence of a (D)$Met(O)^2$ gives rise to analogs with very good oral writhing activity.

Acknowledgements

Thanks are due to our Searle colleagues: D. Tyner, R. Kochman, M. Nevins, A. Staplefeld, C. Sepulveda, S. Arnolde, L. Prieto, E. Drower, D. Hammond, J. Dygos, E. Yonan, K. Mclaughlin and R. Nosal.

References

1. McDermott,J., Benoiton,N.L.(1973)Can. J.Chem.51,1915-1919; Cheung,S.T.; Benoiton,N.L.(1977)Can. J. Chem. 55,906-910.

2. Hansen Jr.,D.W., Pilipauskas,D.(1985)J. Org. Chem. 50,945-950.

3. Abrash,H.I.; Niemann,C. (1963)Biochemistry 2,947-953.

OPIOID RECEPTOR BINDING AFFINITY AND EFFICACY OF METKEPHAMID ANALOGS MODIFIED AT THE PHENYLALANINE RESIDUE

Paul D. Gesellchen, Robert T. Shuman, Robert C. A. Frederickson,[1] and Martin D. Hynes

The Lilly Research Laboratories, Eli Lilly and Company, Indianapolis, IN 46285

Metkephamid, a minimally modified analog of methionine enkephalin, exhibits potent opioid receptor binding, high efficacy in isolated tissue assays and produces analgesia in animal models with little potential for development of physical dependence.[2] These observations led to the progression of metkephamid to clinical trials where it has been shown to be an efficacious analgesic.[3,4] We now report the synthesis and pharmacological profile for a series of metkephamid analogs with modified phenylalanine residues.

Materials and Methods

Substituted phenylalanine amino acids were synthesized and resolved (carboxypeptidase A) as previously described.[5,6] Peptides were synthesized by solid-phase methodology.[7] Diastereomeric peptides were separated by reversed-phase liquid chromatography.[8] Opioid receptor binding affinity was determined in a receptor assay using $[^3H]$-naloxone and $[^3H]$-[D-Ala^2,D-Leu^5]-enkephalin (DADLE) as the μ and the δ ligands, respectively.[9] In vitro efficacy was determined by measurement of the inhibition of the electrically induced contractions in the mouse vas deferens assay.[2] Analgesia was determined in the mouse hot plate assay using an escape jump as the measure of response latency.[9]

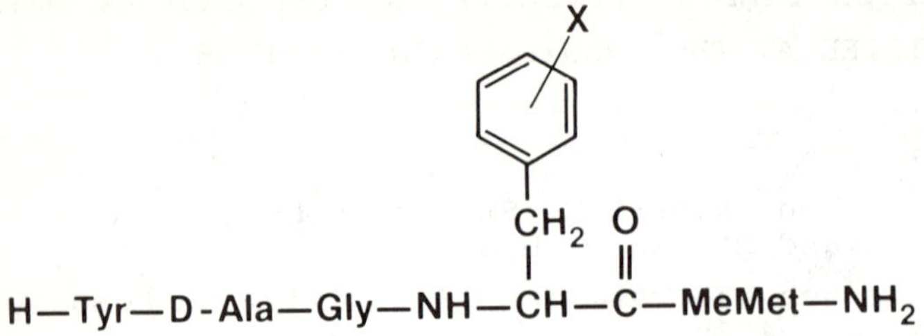

Fig. 1. Structure of Metkephamid (X = H) and Analogs.

Results and Discussion

Nineteen analogs (Figure 1) of metkephamid are reported in which the aromatic ring of the phenylalanine residue has been substituted. The peptides were tested for their ability to bind to opioid receptors (Tables I and II). The high affinity of the majority of the analogs implies that the aromatic nucleus of the phenylalanine residue in metkephamid binds to a region which is capable of accepting a range of molecular volumes and electronegativities. These analogs, like metkephamid, have equal affinity for the μ and δ receptors. Thus, aromatic substitution does not dramatically influence receptor selectivity.

Table I. Receptor Binding*; IC_{50} (nM) vs [^{3}H]-Naloxone

X	F	Cl	Br	I	NO_2	CH_3	OCH_3	OH	CF_3	Ph
Ortho	3.5	3.5	---	---	---	---	---	---	20.	---
Meta	3.0	3.0	1.7	2.0	---	6.0	7.0	7.5	16.	>100
Para	0.8	5.5	3.8	7.0	6.0	---	---	95.	2.7	---

*Metkephamid: IC_{50} = 2.5 nM

Table II. Receptor Binding*; IC_{50} (nM) vs [^{3}H]-DADLE

X	F	Cl	Br	I	NO_2	CH_3	OCH_3	OH	CF_3	Ph
Ortho	6.5	5.0	---	---	---	---	---	---	25.	---
Meta	3.2	1.7	1.5	10.	---	4.5	5.0	3.6	9.5	400
Para	1.1	6.0	4.5	14.	2.4	---	---	340	4.5	---

*Metkephamid: IC_{50} = 4.4 nM

Although most of the analogs are more potent than metkephamid in the mouse vas deferens assay (Table III), the overall range of potencies for active analogs is similar to that measured in the binding assays (approximately 20 fold). The data also suggest that substitution in the para position with small electronegative groups improves efficacy.

Table III. Mouse Vas Deferens*; IC_{50} (nM)

X	F	Cl	Br	I	NO_2	CH_3	OCH_3	OH	CF_3	Ph
Ortho	14.1	9.9	---	---	---	---	---	---	17.5	---
Meta	2.8	0.71	0.49	2.1	---	2.7	2.9	9.9	1.8	2100
Para	0.77	2.2	2.2	12.7	0.64	---	---	390	1.5	---

*Metkephamid: IC_{50} = 8.9 nM

Analgesic activity (Table IV) reveals that the active meta position analogs are essentially equipotent with metkephamid. The meta position is thus uniquely susceptible to substitution without concomitant changes in affinity or efficacy. The

Table IV. Mouse Hot Plate - Jump Assay*; ED_{50} (mg/kg-sc)

X	F	Cl	Br	I	NO_2	CH_3	OCH_3	OH	CF_3
Ortho	2.5	2.2	---	---	---	---	---	---	0.90
Meta	0.29	0.11	0.36	0.13	---	0.33	0.15	0.75	0.33
Para	0.023	0.67	1.6	>30	0.16	---	---	>10	0.42

*ED_{50}: Metkephamid = 0.36 mg/kg-s.c., Phe(3Ph) >30 mg/kg-s.c.

Phe(3Ph) analog is an exception (steric?) to this rule. The para position analogs exhibit an amplified size/activity relationship initially observed in the vas deferens assay. Exact reasons for the low affinity of the para hydroxy phenylalanine [Phe(OH)] analog are not known; however, data presented here eliminate steric and lipophilic factors from consideration. Activity differences between the Phe(F) analog and the Phe(I) analog cannot be attributed to poor pharmacokinetics as the active Phe(3I) analog is predicted to possess a similar lipophilicity to the inactive isomeric Phe(I) analog. Para fluoro phenylalanine has been incorporated into numerous other enkephalin derivatives and they have exhibited higher in vivo potencies than unsubstituted analogs. The differences in the relative potencies measured in the in vitro and in vivo assays emphasize the importance of using several models when studying structure-activity relationships.

References

1. Present address: Monsanto Co., St. Louis, MO 63167.
2. Frederickson, R. C. A., Smithwick, E. L., Shuman, R. T., and Bemis, K. G. (1981) Science **211**, 603-605.
3. Calimlim, J. F., Wardell, W. M., Sriwatanakul, K., Lasagna, L., and Cox, C. (1982) Lancet **1**, 1374-1375.
4. Bloomfield, S. S., Barden, T. P., Mitchell, J. (1983) J. Clin. Pharmacol. Ther. **34**, 240-247.
5. Shuman, R. T. and Gesellchen, P. D. (1982) U. S. Patent 4,322,340.
6. Gesellchen, P. D. and Shuman, R. T. (1981) U. S. Patent 4,265,808.
7. Gesellchen, P. D. and Shuman, R. T. (1981) U. S. Patent 4,283,329.
8. Gesellchen, P. D., Tafur, S., Shields, J. E. (1979) PEPTIDES: Chemistry, Structure and Biological Function, Gross, E. and Meienhofer, J., Eds., Pierce Chemical Company, Rockford, IL pp. 117-120.
9. Hynes, M. D. and Frederickson, R. C. A. (1982). Life Sciences **31**, 1201-1204.

PUTATIVE OPIOID ANTAGONISTS: SYNTHESIS AND BIOLOGICAL PROPERTIES OF D-ALA^2 MET-ENK-AMIDE ANALOGS WITH UNUSUAL TYR RESIDUES

Amrit K. Judd, Lawrence R. Toll, John A. Lawson, Edward T. Uyeno, Willma E. Polgar, and Gilda H. Loew
SRI International, 333 Ravenswood Avenue, Menlo Park, CA 94025, U.S.A.

Introduction

In the search for a clinically useful, nonaddicting analgesic, many potent long-lasting analogs of Leu- and Met-enkephalin have been synthesized but in general they have appreciable addiction liability. By analogy with nonpeptide opiates, enkephalin analogs with both agonist and antagonist activity could be more promising candidates for clinical use.

We report here the successful synthesis of two novel μ-selective peptide antagonists designed by incorporating two modified tyrosine residues, m-Tyr and β-CH_3-m-Tyr into D-Ala^2-Met^5-enkephalin amide. The rationale for the modification of tyrosine residues was based upon the postulated similarities of peptide opioids with a particular class of μ-selective non-peptide opioids, 3-phenylpiperidines, and the known requirement of a (m-OH)phenyl for antagonism in that class.[1]

Materials and Methods

Synthesis. The L-meta-tyrosine was prepared by the stereospecific hydrolysis of the racemic ethyl ester by chymotrypsin. The β-methyl analogs were prepared by a new synthetic procedure to be described elsewhere. Ultimate separation of the four enantiomers was effected by HPLC followed by resolution by treatment with Protease Type VIII. Tyr-D-Ala-Gly-Phe-Met-NH_2, m-Tyr-D-Ala-Gly-Phe-Met-NH_2 and β-CH_3-m-Tyr-D-Ala-Gly-Phe-Met-NH_2 peptides were synthesized in our laboratory by solution method and the purification of final products achieved by HPLC or partition chromatography. The β-CH_3-m-tyrosine peptide investigated has L-configuration on the C_α atom but has unknown configuration at the C_β atom. Opiate receptor binding assays and computer-assisted data analysis were performed as described previously.[2] The mouse tail-flick assay was used to measure agonism and antagonism.

Results and Discussion

Receptor affinities and maximum binding capacities for the best 5-site fit to the data for the four labeled ligands and the three peptides studied are given in Table 1. As shown in this table, all the compounds bind with highest affinity to the same site. Thus this site appears to be similar to the μ_1 site described by Pasternak.[3] In addition, all the peptides also have fairly high affinity for another site with high affinity for naloxone, which we are calling μ_2. In contrast to DADL and D-Ala^2-met-enk-amide which have high affinity at a third site labeled δ, the two new peptides have quite a low affinity at this site and are very μ-selective.

The results of whole animal tests for analgesic activity by i.c.v. administration of the three peptides using the mouse tail-flick assay are given in Table 2. D-ala^2-met-enk-amide is the most potent peptide agonist, with potency 1/3 that of morphine. Both modifications made in the tyrosine residue of this analog diminish its agonist potency, the m-Tyr analog has 1/50 and the β-CH_3-m-Tyr analog 1/10 its agonist activity. In general, the relative analgesic agonist potency of the three peptide analogs parallels their relative affinity at the "μ_1," but not the "μ_2" receptor, lending support to previous

Table I. Receptor Affinities and Maximum Binding Capacities of (D-Ala^2-Met^5)-Enkephalin Amide Analogs for a 5-Receptor Site Model

	K_D (nM)				
	Site 1 "μ_1"	Site 2 "μ_2"	Site 3 "δ"	Site 4 "κ"	Site 5
Naloxone	0.60	3.0	20.8	0.5	100
DADL	1.80	14.9	1.3	312	8,333
EKC	0.60	4.3	15.0	0.003	588
DHM	0.30	12.7	119	1,515	83
_p_-Tyr	0.53	5.0	0.40	1,600	476
_m_-Tyr	13	10	50	5,260	435
β-CH_3-_m_-Tyr	19	50	200	370	179
B_{max} pmol/g	4.2	20	3.7	2.3	157

evidence presented by Pasternak[3] that the μ_1 receptor is responsible at least for relief from heat-induced pain.

Antagonist activity against morphine-induced tail-flick inhibition was found for all three peptides in the order _m_-Tyr > _p_-Tyr > β-CH_3_m_Tyr, with _m_Tyr having about 1/5 the potency of nalorphine. Thus they are all "mixed" agonist antagonists with varying ratios of agonist to antagonist activity. _m_Tyr is six times more potent as an antagonist, _p_-Tyr is ten times more potent as an agonist, and β-CH_3-_m_-Tyr is about equally potent for both activities.

In common with the 3-phenylpiperidines then, changes in the phenyl ring and at the β(2)-position of tyrosine do modulate the extent of agonism and antagonism in these D-ala^2-met-enk-amide peptides. Other modifications of the 3-CH_3 3-phenyl piperidines which enhance antagonist potency are the presence of a 2-CH_3 group and an N-phenyethyl substituent.

Corresponding changes in the tyrosine residue involve an α,β diCH_3 m-OH tyrosine and an N-phenethyl substituent on the terminal amino group. These analogs have been synthesized and preliminary binding and animal testing done. Results show that in the peptides an α CH_3 group apparently greatly diminishes receptor affinity and an N-phenethyl greatly diminishes solubility leading to a less promising spectrum of activity. Therefore the most promising peptide is the β-CH_3-m-Tyr analog, with equal agonist and antagonist activities, which could be a clinically useful analgesic.

Table II. Analgesic and Narcotic Antagonist Potencies of (D-Ala[2], Met[5])-Enkephalin Amide Analogs Evaluated in the Mouse Tail-flick Test

COMPOUND	AGONISM		ANTAGONISM	
	ED_{50} μmol/kg	95% Conf. Lim. μmol/kg	Ant. AD_{50} μmol/kg	95% Conf. Lim. μmol/kg
p-Tyr[1]	0.214	0.074- 0.618	2.283	1.512-3.453
m-Tyr[1]	10.302	5.630-18.854	1.684	0.785-3.596
β-CH_3-m-Tyr[1]	2.394	1.203- 4.764	3.530	2.263-5.507
Nalorphine	19.320	7.983-46.754	0.322	0.184-0.563
Morphine	0.063	0.026- 0.151	---	--- ---

References

1. Iorio, M., and Casy, A.F. (1978). *J. Med. Chem.* **21**, 812-815.

2. Toll, L., Keys, C., Polgar, W., and Loew, G. (1984). *Neuropeptides* **5**, 205-208.

3. Wolozin, B. and Pasternak, G.W. (1981). *Proc. Natl. Acad. Sci. USA* 78, 4141-4145.

CYCLOPROPYL PHE^4-ENKEPHALINS

C. Mapelli, H. Kimura and C. H. Stammer
Department of Chemistry, University of Georgia
Athens, Georgia 30602

We have now completed the first incorporation of a cyclopropyl amino acid into a bioactive peptide hormone. The synthesis, resolution and incorporation of cyclopropyl phenylalanine (∇Phe) into the crucial fourth position of Leu^5-enkephalin has been accomplished. The biological activities of these conformationally restricted peptides show clearly that the brain binding receptors prefer only one of the isomers, the (+)-∇^{Z}Phe, to all of the others. This is also confirmed by the in vivo tail-flick assays.

The syntheses of the E- and Z-isomers of the rigid cyclopropyl phenylalanine residues required different chemical approaches. The E-compound was best prepared by the addition of phenyl diazomethane to a dehydroalanine derivative,[1] while the Z-isomer resulted from the treatment of the well known 4-benzylidene-2-phenyl oxazolone with diazomethane.[2] Both of these synthetic methods have their weak and strong points and newer methods are now under investigation; in particular, those which will lead to the desired optically active amino acids. The syntheses of the title peptides are outlined in Figures 1 and 2. There was nothing exceptional in the chemistry of blocking, coupling and deblocking of these cyclopropyl amino acids, even though the rates of some of these reactions was somewhat slower, apparently for steric reasons, than those of natural phenylalanine.

A correlation between structure, conformation and bioactivity is more difficult. Inspection of models

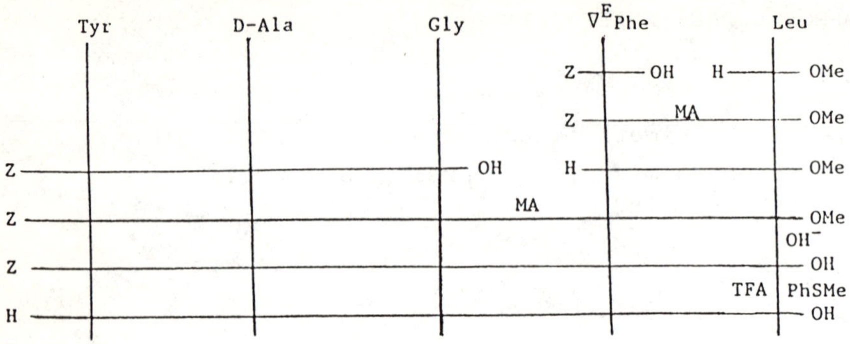

Fig. 1. Synthesis of D-Ala[2],∇EPhe,Leu[5]-Enkephalin
MA = mixed anhydride; TFA = CF_3CO_2H; PhSMe = thioanisole

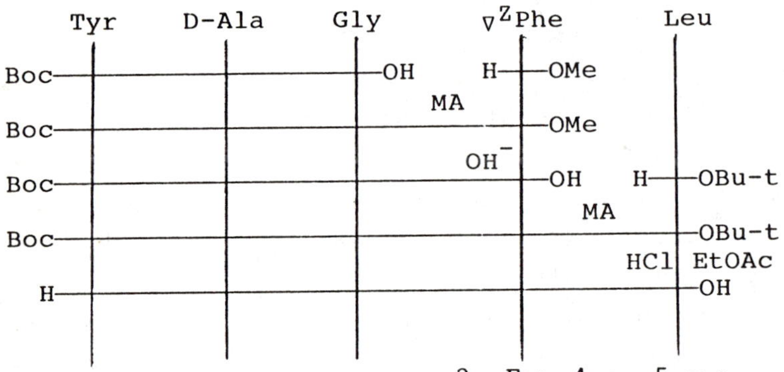

Fig. 2. Synthesis of D-Ala[2],∇EPhe[4],Leu[5]-Enkephalin.

indicates that when the β-phenyl group is cis to the amino function of the Phe residue (Z-isomer), rotation about the N-C_α bond (the phi angle) will be severely restricted. Similarly, when the large group is eclipsed with the carbonyl group (E-isomer), C_α-C' bond rotation (the psi angle) will be limited to small values. In general, these restrictions will favor the formation of turns in the peptide chain at the point of insertion of the cyclopropyl amino acid. It is well known that early NMR studies of enkephalins indicated that a β-turn was present in the enkephalin chain.[3] The insertion of a ∇-Phe residue into this chain is therefore consistent with a β-turn as a requirement for bioactivity and, thus, does not reduce activity when the "right" isomer is correctly placed.

Circular dichroism peaks and bioactivities of the four isomeric peptides are summarized in Table I. It is difficult to draw any firm conclusions from these data, since the absolute configurations of the ∇Phe residues have not been determined and the presence of aromatic sidechain absorption overshadows the peptide backbone absorption which might have allowed some conformational interpretations to be made. NMR studies with a view toward establishing the conformations of these peptides are now underway.

COMPOUND	$[\theta]_T \times 10^{-4}$ (222-230 nm)	Rat Brain Binding IC_{50}(nm)	Tail-flick Latency (MPE%) ($p<0.001$) Concentration(I.C.V.) $10^{-8}M$	$10^{-7}M$
(-)-∇^EPhe-ENK	-1.4	13,300	5	92
(+)-∇^EPhe-ENK	+9.3	8,900	0	35
(-)-∇^ZPhe-ENK	-6.3	30.7	20	62
(+)-∇^ZPhe-ENK	+9.3	1.2	42	63

Table I. CD extrema and bioactivities.

In summary, the high bioactivity of at least one of the cyclopropyl peptides and the stability to enzymolysis (chymotrypsin, thermolysin, carboxypeptidase-Y) of cyclopropyl peptides indicate that this modification of amino acids could lead to the synthesis of useful medicinal agents in the future.

References

1. Suzuki, M., Gooch, E.E. and Stammer, C. H. (1983) Tetrahedron Letters, **24**, 3839-3840.

2. King, S. W., Riordan, J. M., Holt, E. M. and Stammer, C. H. (1982) J. Org. Chem. **47**, 3270-3273.

3. Smith, G. D., and Griffin, J. F., (1978) Science, **199**, 1214-1216; Khaled, M.A., Long, M.M., Thompson, W.D., Bradley, R.J., Brown, G.B. and Urry, D.W. (1977) Biochem. Biophys. Res. Commun. **76**, 224.

SYNTHESIS AND CONFORMATIONAL STUDIES OF ENKEPHALINLIKE CYCLIC PEPTIDES AND DEPSIPEPTIDES[1]

Christian Zechel[a] and Horst Kessler

Institut für Organische Chemie, Universität Frankfurt
D-6000 Frankfurt, Germany
[a]Present address: Department of Chemistry, University of Arizona, Tucson, Arizona 85721

Rolf Geiger

Hoechst AG, D-6000 Frankfurt, Germany

Solution phase peptide synthesis afforded five conformationally restricted analogs of Leu-enkephalin. Conformational restriction was achieved by side chain - carboxyterminus cyclization[2]. For this purpose the bifunctional amino acids D-Orn and D-Ser were incorporated into position 2 of the peptide sequence. The structure of the described peptides corresponds to the scheme shown below:

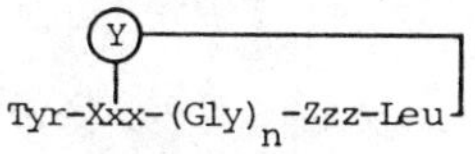

	Xxx	n	Zzz	
c1	D-Orn	1	Phe	(Y=N)
c2	D-Orn	1	MePhe	(Y=N)
c3	D-Orn	1	Tic+	(Y=N)
c4	D-Ser	1	Phe	(Y=O)
c5	D-Ser	2	Phe	(Y=O)

+Tic = 1,2,3,4-tetrahydroisoquinoline carboxylic acid

The crucial steps in the synthesis of the cyclodepsipeptide c4 are the formation of the ester linkage between D-Ser and Leu and the cyclization to a relatively strained 13-membered ring. For the formation of the ester bond at the very beginning of the synthesis (see Figure 1) the reagent combination n-propylphosphonic acid anhydride (PPA)[3]/DMAP was used[4]; the diastereomerically pure depsitripeptide 2 was obtained in quantitative yield and after selective cleavage of the Ddz-

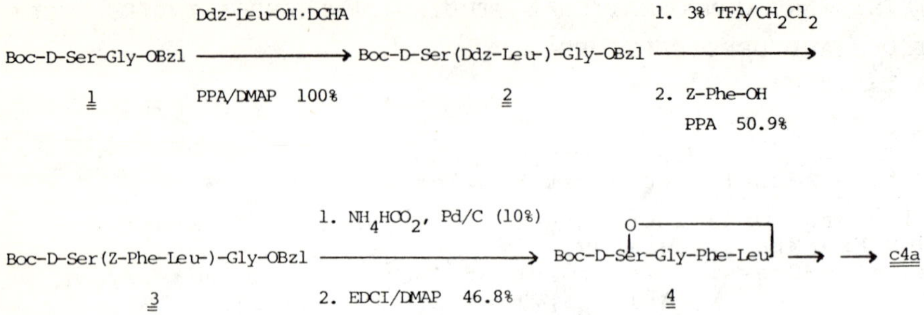

Fig. 1. Synthetic scheme of the cyclodepsipeptide c4

group converted to the tetrapeptide 3. Attempts to synthesize the key intermediate 3 directly by esterification of 1 with Z-Phe-Leu-OH were less successful because of incomplete reaction and considerable racemization of the Leu-residue. The cyclization was carried out with the aid of EDCI/DMAP[5]; these reagents provide the necessary strong activation of the C-terminal Gly-residue.

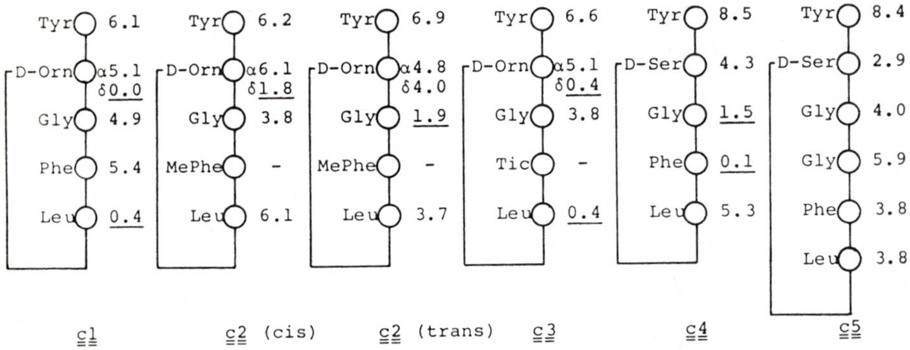

Fig. 2. NH-chemical shift temperature gradients of c1-c5 ($-\Delta\delta/\Delta T$ of c2 measured in the range of 295-320 K)

The conformational properties of these five compounds were examined by ^{1}H, ^{13}C- and ^{15}N-NMR-spectroscopy. The $-\Delta\delta/\Delta T$-values of NH-chemical shifts (Figure 2) reveal that only c1, c3 and c4 match the criteria of conformational homogeneity[6]. Remarkably the MePhe4-compound c2 (differing from the Tic4-

analog **c3** only by the cleavage of a single C-C-bond) exists in two conformations due to cis/trans-isomerism at the Gly-MePhe-bond (cis/trans 1:4.5). The NMR-data (^{1}H, ^{13}C and ^{15}N) of **c1** and **c3** are very similar, suggesting that both peptides have the same backbone conformation which is characterized by two intramolecular hydrogen bonds (Gly^3-CO←HN-Leu^5, thus forming a γ-bend, and D-Orn^2-CO←HN$^\delta$-D-Orn^2)[7]. In **c3** the fixed aromatic side chain of the Tic^4-residue causes large anisotropy effects which give rise to considerable high field shifts of all Leu-resonances in the ^{1}H-spectrum. From the NH-temperatures coefficients it follows that the smaller ring system of **c4** does not contain the γ-loop (Leu^5-NH is externally oriented). The absence of this structural feature is confirmed by heteronuclear NMR-spectra as well. The resonance of the Phe^4-amide nitrogen atom, which is adjacent to the internally bonded Gly^3-CO-group in **c1**, experiences a high field shift in analog **c4**, in which Gly^3-CO is not involved in intramolecular hydrogen bonding[8]. Similarly a comparison of the ^{13}CO-data (assignments by H,C-COLOC[9]) shows that only the δ-value of Gly^3-CO (virtually the same in **c1** and **c3**) is significantly lowered in **c4**, while the chemical shifts of the remaining carbonyls are unchanged compared to **c1** and **c3**[10].

In the GPI-assay the very flexible D-Orn^2-MePhe^4-analog exhibits the highest activity, while the rigid Phe^4-compound is less potent. The potency drops further in the most rigid Tic^4-peptide indicating that the aromatic side chain of residue 4 is fixed in a "wrong" conformation for receptor binding. As to the depsipeptides, both show little (**c4a**) or no (**c5a**) activity at least in the μ-selective GPI-assay. The lack of

		IC_{50} |nm|	activity^a
c1a	HCl·H-Tyr-cyclo (-N$^\delta$-D-Orn-Gly-Phe-Leu-)	2.5	35.2
c2a	HCl·H-Tyr-cyclo (-N$^\delta$-D-Orn-Gly-MePhe-Leu-)	0.05	1760
c3a	HCl·H-Tyr-cyclo (-N$^\delta$-D-Orn-Gly-Tic-Leu-)	15.5	5
c4a	TFA·H-Tyr-D-Ser(O-)-Gly-Phe-Leu (cyclic, O bridged to Leu)	10	8.8
c5a	TFA·H-Tyr-D-Ser(O-)-Gly-Gly-Phe-Leu (cyclic, O bridged to Leu)	1354	0.07

Table 1. Activity of c1a - c5a in the GPI assay
^{a}relative to Met-Enkephalin (=1)

activity in c5a might be caused by an unfavorable distance of the aromatic rings due to the incorporation of a second Gly-residue[12].

This work was supported by the Deutsche Forschungsgemeinschaft and the Fonds der Chemischen Industrie. C.Z. thanks the Fonds der Chemischen Industrie for a scholarship.

References

1. Peptide conformations 37.
36. Kessler, H. and Bernd, M., Liebigs Ann. Chem., in press.

2. Hruby, V.J. (1982) Life Sci. **31**, 189-199.

3. Wissmann, H. and Kleiner, H.-J. (1980) Angew. Chem. Int. Ed. **19**, 133.

4. Kessler, H., Kutscher, B., Mager, B. and Grell, E. (1983) Liebigs Ann. Chem. 1541-1550.

5. Kessler, H., Kutscher, B. (1985) Tetrah. Lett. **26**, 177.

6. Kessler, H. (1982) Angew. Chem. Int. Ed. **21**, 512-523.

7. Kessler, H., Hölzemann, G. and Zechel, C. (1985) Int. J. Prot. Pept. Res. **25**, 267-279.

8. Kricheldorf, H.R. (1981) Org. Magn. Res. **15**, 162-173.

9. Kessler, H., Griesinger, C., Zarbock, J. and Loosli, R. (1984) J. Magn. Res. **57**, 331-336.

10. This fact is remarkable since the chemical shifts of carbonyl resonances should be influenced by several parameters besides hydrogen bonding (e.g. γ-effects[11]).

11. Tonelli, A. (1984) Biopolymers **23**, 819-835.

12. Schiller, P.W. and St.-Hilaire, J. (1980) J. Med. Chem. **23**, 290-295.

NMR AND COMPUTER SIMULATIONS OF A DIASTEREOMERIC PAIR OF CYCLIC RETRO-INVERSO ENKEPHALIN ANALOGS

Nancy J. Mammi, Moises Hassan, Judd Berman and Murray Goodman
Department of Chemistry, University of California, San Diego
La Jolla, California 92039

Introduction

The conformations of two diastereomers of the cyclic retro-inverso modified enkephalin analog, H-Try-c[-D-A_2bu-Gly-gPhe-L,D-mLeu-], have been studied by proton nmr at 360 MHz, and computer simulations. They each exhibit high potency in the guinea pig ileum (GPI) assay and high μ-receptor specificity.[1] The temperature dependence of the chemical shifts of amide protons was measured for the compounds in DMSO-d_6 and in solutions of varying concentrations of H_2O in DMSO-d_6 to elucidate hydrogen bond donors. Computer simulations of minimum energy conformations were used to identify the hydrogen bond acceptors.

Materials and Methods

The diastereomeric mixture of the cyclic enkephalin analogs was prepared in our laboratory and the isomers were separated by reversed phase HPLC.[2] Spectral assignments were made based on two-dimensional shift correlation, two-dimensional relayed coherence transfer, and one-dimensional nOe spectra. 15 mM solutions were prepared in DMSO-d_6. The H_2O titration/temperature studies were carried out by adding H_2O to the DMSO-d_6 solutions and obtaining spectra at 5 temperatures over a range of 20-65°C for several solvent compositions.

The computer simulations involved minimization of the potential energies of the molecules using a steepest descent method followed by a modified Newton-Raphson method.[3] The starting conformations for the energy minimizations were obtained from molecular dynamics simulations and from systematically adjusted conformations.

Results and Discussion

The temperature coefficients of the amide protons of the diastereomers are shown in table I. These data reveal one intramolecularly hydrogen bonded proton in each diastereomer, the D-A_2bu γ-NH. During H_2O titration/temperature studies, the hydrogen bond involving this proton was not disrupted, even in solutions of 0.83 mole fraction H_2O in DMSO-d_6.

Table I. Temperature Coefficients in -ppb/K of Amide Protons in DMSO-d_6 Solutions and H_2O/DMSO-d_6 Solutions of L-mLeu Diastereomer (left) and D-mLeu Diastereomer (right).

	Mole Fraction H_2O			
Residue	X=0.0	X=0.22	X=0.42	X=0.83
D-A_2bu	4.6	4.9	5.4	a
Gly	5.3	5.5	5.9	5.4
gPhe[1]	5.9	5.1	5.1	5.1
gPhe[2]	4.6	4.9	5.2	4.1
D-A_2bu (γ)	0.3	1.4	1.4	1.8

	Mole Fraction H_2O			
Residue	X=0.0	X=0.36	X=0.52	X=0.83
D-A_2bu	4.3	4.8	5.0	5.8
Gly	4.7	5.5	5.5	5.7
gPhe[1]	10.2	8.9	8.4	7.1
gPhe[2]	3.5	4.8	4.9	5.8
D-A_2bu (γ)	2.1	2.2	2.0	2.0

[a] D-A_2bu exchange broadened extensively at high H_2O concentrations. gPhe[1] is adjacent to Gly, and gPhe[2] is adjacent to mLeu.

Computer simulations of minimum energy conformations produced 11 energy minima for the L-mLeu diastereomer and 8 energy minima for the D-mLeu diastereomer. In table II, the intramolecular hydrogen bonds and the relative potential energies of the minima are presented.

Table II. Hydrogen Bonds observed in the Minimum Energy Conformations of the L-mLeu Diastereomer (left) and the D-mLeu Diastereomer (right).

Minimum	Energy[a]	Hydrogen Bonds		
I	0.0	Gly NH	<--->	Tyr CO
		gPhe^1 NH	<--->	D-A$_2$bu CO
II	1.3	Gly NH	<--->	Tyr CO
		gPhe^1 NH	<--->	D-A$_2$bu CO
III	1.9	D-A$_2$bu γ-NH	<--->	L-mLeu^1 CO
		gPhe^1 NH	<--->	D-A$_2$bu CO
IV	4.1	gPhe^1 NH	<--->	D-A$_2$bu CO
V	4.3	Gly NH	<--->	Tyr CO
VI	5.4	Gly NH	<--->	Tyr CO
		gPhe NH^1	<--->	D-A$_2$bu CO
		D-A$_2$bu γ-NH	<--->	mLeu^1 CO
VII	6.8	Gly NH	<--->	Tyr CO
VIII	7.3	gPhe^1 NH	<--->	D-A$_2$bu CO
		D-A$_2$bu γ-NH	<--->	mLeu^1 CO
IX	8.2	gPhe^1 NH	<--->	Tyr CO
X	10.1	none		
XI	17.8	D-A$_2$bu γ-NH	<--->	D-A$_2$bu CO

Minimum	Energy[a]	Hydrogen Bonds		
I	0.0	Gly NH	<--->	Tyr CO
		gPhe^1 NH	<--->	D-A$_2$bu CO
II	1.7	Gly NH	<--->	Tyr CO
		gPhe^1 NH	<--->	D-A$_2$bu CO
		D-A$_2$bu γ-NH	<--->	mLeu^1 CO
III	5.8	Gly NH	<--->	Tyr CO
IV	5.8	gPhe^1 NH	<--->	D-A$_2$bu CO
		D-A$_2$bu γ-NH	<--->	mLeu^1 CO
V	9.0	Gly NH	<--->	Tyr CO
		D-A$_2$bu γ-NH	<--->	D-A$_2$bu CO
		gPhe^2 NH	<--->	mLeu^1 CO
VI	10.5	D-A$_2$bu γ-NH	<--->	D-A$_2$bu CO
		D-A$_2$bu γ-NH	<--->	mLeu^1 CO
		gPhe^1 NH	<--->	D-A$_2$bu CO
VII	11.7	Gly NH	<--->	Tyr CO
VIII	19.5	D-A$_2$bu γ-NH	<--->	D-A$_2$bu CO
		gPhe^2 NH	<--->	mLeu^1 CO
IX	23.7	none		

[a] Energies are relative to the absolute minimum in Kcal/mol. gPhe^1 is adjacent to Gly, gPhe^2 is adjacent to mLeu, mLeu^1 is adjacent to gPhe, and mLeu^2 is adjacent to D-A$_2$bu side chain.

From the computer simulations, we have deduced the hydrogen bond acceptors for D-A$_2$bu γ-NH, identified by nmr experiments to be intramolecularly hydrogen bonded. For both diastereomers, two hydrogen bond acceptors are possible, the carbonyl of D-A$_2$bu to form a C_7 ring and the mLeu carbonyl adjacent to Phe to form a C_6 ring. ORTEP drawings of these conformations are shown in Figure 1.

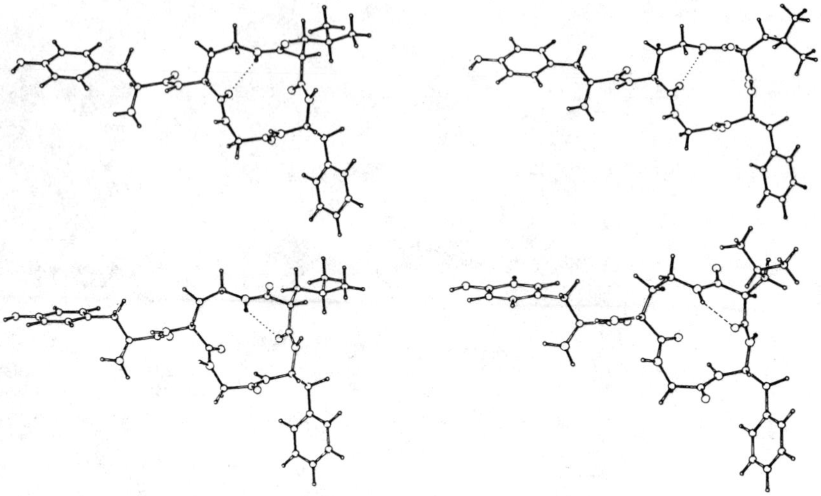

Fig. 1. ORTEP drawings of the preferred conformations of the L-mLeu diastereomer (left) and the D-mLeu diastereomer (right).

Ackowledgment

This work was supported by NIH AM 15410.

References

1. Berman, J., Goodman, M., Nguyen, T. and Schiller, P. (1983) Biochem. Biophys. Res. Comm., 115, 864-870.

2. Berman, J. and Goodman, M. (1984) Int. J. Peptide Protein Res., 23, 610-620.

3. Dauber, P., Goodman, M., Hagler, A., Osguthorpe, D. and Stern, P. (1981) ACS Symposium Series, No. 173, Supercomputers in Chemistry, Lykos, P. and Shavitt, I., Eds., pp. 161-191.

SYNTHESIS AND PHARMACOLOGY OF DIPEPTIDES RELATED TO DES[GLY[3]] ENKEPHALIN: MODIFICATION OF THE C-TERMINAL AMIDE

B.A. Morgan, J. Singh* and S.J. Ward
Sterling-Winthrop Research Institute, Rensselaer, NY 12144

Introduction

Both dermorphin (Tyr-ala-Phe-Gly-Tyr-Pro-Ser-NH_2) isolated from frog skin,[1] and morphiceptin (TyrProPheProNH_2), a synthetic tetrapeptideamide based on a sequence from β casein,[2] deviate from the typical opioid sequence in that the important tyrosine and phenylalanine residues are linked by a single amino acid residue. Recently, we have shown that the dipeptide amide (1) [Tyr-ala-NH$(CH_2)_3$Ph] is the minimal structural requirement[3] for the dermorphin series, being equipotent to normorphine in the guinea pig ileum preparation in vitro. We now report the synthesis and pharmacology of a series of analogs of the dipeptide (1) in which the C-terminal phenylpropyl moiety has been modified.

Methods

Chemistry - Chiral amines (9a), (11a), and (9b) and (11b) were prepared from optically pure L- and D-alanine derivatives, respectively [Figure 1]. N^{α}-Boc alanine methyl ester (6) was reduced with diisobutylaluminum hydride to give aldehyde[4] (7). Condensation of the aldehyde with the benzylidenetriphenylphosphonium ylid gave exclusively the trans-allylic amine derivative (8) in high yield. Careful removal of the

*Address correspondence to this author.

BOC group from compound (8) with 4N HCl in ethyl acetate gave pure chiral allylic amine hydrochloride (9). However, when dioxane was used for acidolysis of the BOC group from (8a), HCl addition to the olefin (9a) was observed in 60% yield. Hydrogenation of the olefin (8) followed by deprotection of the BOC group gave the optically pure amine hydrochloride salt (11).

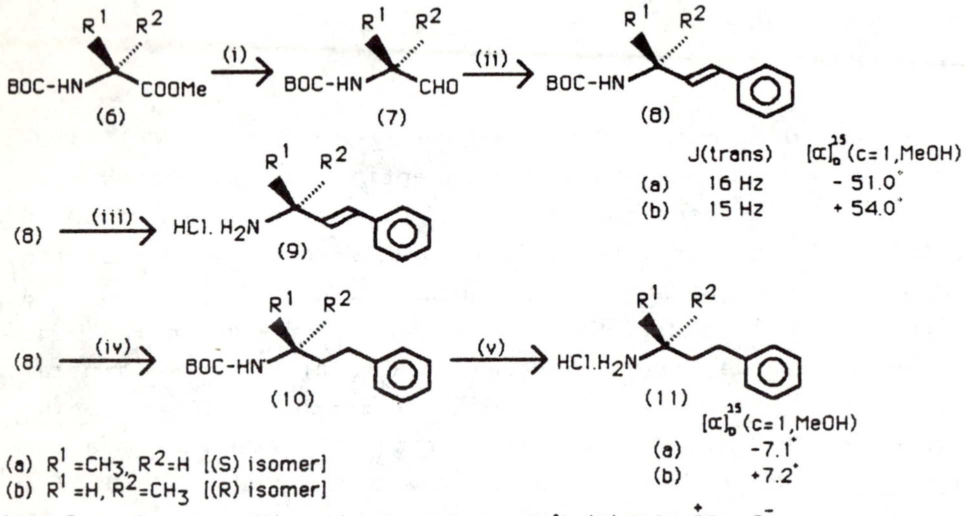

Fig. 1. Reagents and conditions: (i) DIBAL, Toluene, -70°C; (ii) $PhCH_2\overset{+}{P}Ph_3\ \bar{B}r$, n-BuLi, THF, -78°C->rt; (iii) 4N HCl/EtOAc, (iv) H_2, Pd/C, EtOAc, (v) 4N HCl/p-dioxane.

The chiral amines (9a), (9b), (11a) and (11b) were coupled with the dipeptide Boc-Tyr(tBu)-ala-OH (12) by the

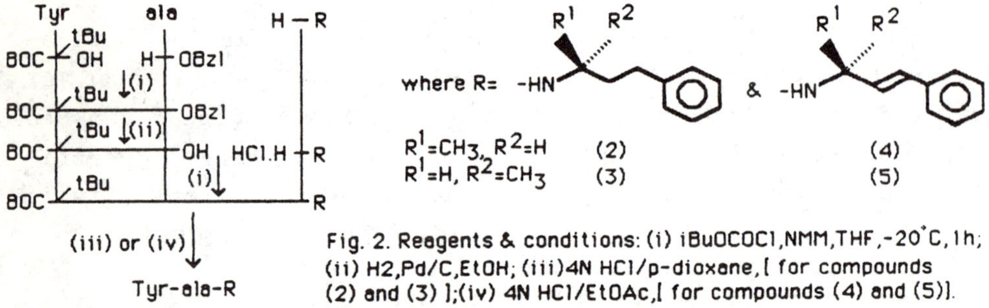

Fig. 2. Reagents & conditions: (i) iBuOCOCl, NMM, THF, -20°C, 1h; (ii) H2, Pd/C, EtOH; (iii) 4N HCl/p-dioxane, [for compounds (2) and (3)]; (iv) 4N HCl/EtOAc, [for compounds (4) and (5)].

mixed anhydride method. Subsequent removal of the protecting groups gave the peptides (2), (3), (4) and (5), respectively [Figure II]. All purified peptides were characterized by elemental analysis, ^{1}H-NMR, analytical RPLC and tlc.

Biology - The peptides were tested for agonist potency in vitro in the guinea pig ileum (GPI)[5] and mouse vas deferens (MVD)[6] preparations.

Results and Discussions

Four analogs (2-5) of the dipeptide (1) are reported in which the C-terminal phenylpropyl moiety has been modified. Agonist potencies of these analogs relative to normorphine in both GPI and MVD are shown in Table 1. Ke values[5] for the interaction of dipeptides with naloxone were also determined, and indicated that the agonists interact with μ receptors in both the GPI and MVD (data not shown).

Substitution of a methyl group α- to the nitrogen in the pro(S) position of (1) was found to have no effect on agonist activity as is shown by the equal potencies of analogs (1) and (2). However, analog (3) with a methyl group in the pro(R) position of (1), showed somewhat lower activity (3 fold, GPI; 5 fold MVD). When conformational constraints were placed on the phenylpropyl portion of the analog (2) by the introduction of an olefin group, a similar reduction (5 fold, GPI; 1.3 fold, MVD) in the activity was observed. However, a substantial reduction (57 fold, GPI; 17 fold MVD) in activity was observed when an olefin group was introduced in analog (3).

These data indicated that a methyl substituent in the pro (S) position of the phenylpropyl group does not interfere with activity at the receptor. In contrast, a substituent in the pro(R) position has some effect on activity which is reflected in a 3-5 fold reduction in the potency. Analogs containing an olefin group are more sensitive to subtle structural modifications compared to their saturated analogs (rel. potency ratios (4)/(5) = 30 vs (2)/(3) = 3; GPI). The (S)-methyl olefin (4) is 1-4 times less potent than the corresponding saturated analog (2). This would suggest that phenylpropyl group adopts an essentially extended conformation at the receptor. The analog (5), with the (R) methyl substituent and the olefin group, was found to be least potent, indicating that the combination of methylation at the pro(R) position and introduction of the olefin inhibits interaction with the opiate receptor.

Table 1

Cmpd	Structure	Guinea Pig Ileum[a]	Mouse Vas Deferens[a]
1	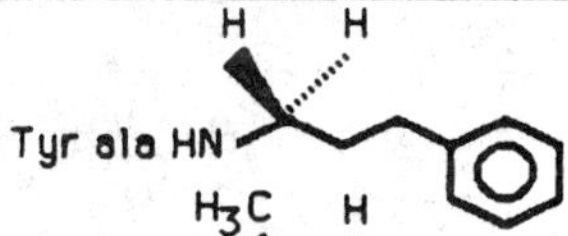	120	85
2	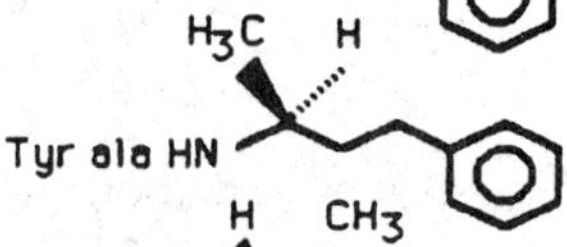	105	65
3	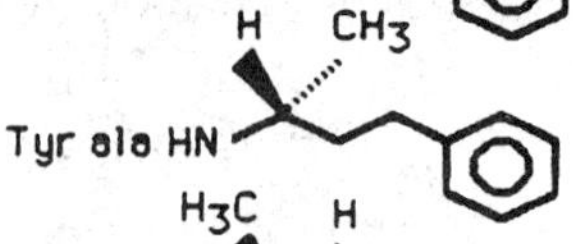	35	12
4	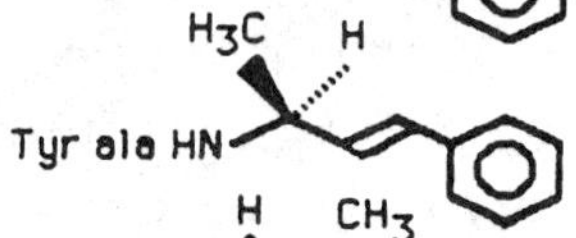	20	50
5	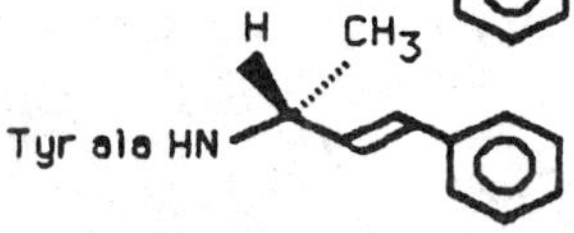	0.7	<3

(a) Potencies relative to Normorphine = 100.

Acknowledgements

We thank Ms. D.M. LoPresti, Ms. D.M. Wescoe and Mr. W. Cumiskey for carrying out GPI and MVD *in vitro* pharmacology experiments.

References

1. Erspamer, V. and Melchiorri, P. (1980) Growth Hormone and Other Biologically Active Peptides, Pecile, A. and Miller, E.E., Eds., Exerpta Medica, New York, pp. 185-200.

2. Chang, K-J., Killian, A., Hazum, E., Cuatrecasas, P., and Chang , J-K. (1981), Science 212, 76-77.

3. Casiano, F.M., Cumiskey, W.R., Gordon, T.D., Hansen, P.E., McKay, F.C., Morgan, B.A., Pierson, A.K., Rosi, D., Singh, J., Terminiello, L., Ward, S.J., and Wescoe, D.M. (1983) in Peptides: Structure and Function, Proceedings of the Eighth American Peptide Symposium, Hruby, V. and Rich, D., Eds., Pierce Chemical Company, Rockford, IL, pp. 311-314.

4. Rich, D.H., Sun, E.T. and Boparai, A.S. (1978). J. Org. Chem., 43, 3624-3626.

5. Kosterlitz, H.W., and Watt, A.J. (1968), Brit. J. Pharmacol. Chemother., 33, 266-276.

6. Hughes, J., Kosterlitz, H.W. and Leslie, F.M. (1975), Brit. J. Pharmacol., 53, 371-381.

ENKEPHALIN TRIPEPTIDE AMIDES

D. Sarantakis[*], H. Fletcher, 3rd.[*], R. Tasse[+] and Wm. Dvonch[+]

*Medicinal Chemistry Department, Wyeth Laboratories, Inc.
Research Division, P.O. Box 8299, Philadelphia, PA 19101
+Experimental Therapeutics Department, Wyeth Laboratories, Inc.

Introduction

It was of interest to us to prepare tripeptide amide analogs of enkephalin for biological evaluation as analgesics. There is evidence that it is necessary to retain the benzyl side chain of the phenylalanine residue of the enkephalin molecule for significant biological activity[1], also there is no obvious need, from conformational considerations[2,3], for the presence of the carbonyl group of phenylalanine.

We describe here a series of biologically potent enkephalin tripeptide amides designed on the basis of the above reasons.

Results and Discussion

The protected tripeptides (1) and (2), Table A, were prepared by classical methods of peptide chemistry. The piperazine containing components, 4-methyl-S-α-(phenyl-methyl)-1-piperazineethanamine (5) and N, 4-dimethyl-S-α-(phenylmethyl)-1-piperazine ethanamine (6) were prepared from the corresponding L-phenylalanyl piperazine amides (3) and (4) by reduction with 1M-BH_3/THF. A typical procedure for the synthesis of (5) is as follows: 1-L-Phenylalanine-4-methyl-piperazine (11 g) was dissolved in THF (100 ml) and treated with 200 ml of 1M-BH_3/THF. The solution was refluxed for 2 hours, the borates were destroyed by 20 ml of 2N-aqHCl, the excess THF was distilled off and the aq. solution was refluxed for 20 hours, then, left to stand for 4 days. The solid which separated was filtered off and the filtrate was evaporated to dryness. The residue was crystallized from ethanol-ether (1:3, v/v) to yield 9.8% of the hydrochloride salt. Infra red spectrum does not show any carbonyl absorption.

TABLE A

No.	Compound	TLC System, Support, R_f
1	BZL BOC-Tyr-D-Ala-Gly-OH	A (Sil. gel) 0.22
2	BZL BOC-N-Tyr-D-Ala-Gly-OH CH_3	B (Sil. gel) 0.29
3	$CH_2C_6H_5$ CBZ-NHCHCON NCH$_3$	B (Sil. gel) 0.41
4	$CH_2C_6H_5$ BOC-NCHCON NCH$_3$ CH_3	A (Sil. gel) 0.70
5	$CH_2C_6H_5$ H_2NCHCH$_2$N NCH$_3$ ·3HCl	C (Avicel) 0.29
6	$CH_2C_6H_5$ CH$_3$NHCHCH$_2$N NCH$_3$ ·3HCl	C (Avicel) 0.31
7	BZL $CH_2C_6H_5$ BOC-Tyr-D-Ala-Gly-NHCHCH$_2$N NCH$_3$	B (Alox) 0.5
8	BZL $CH_2C_6H_5$ BOC-N-Tyr-D-Ala-Gly-NCHCH$_2$N NCH$_3$ CH_3 CH_3	C (Sil. gel) 0.48

TLC Systems: A = $CHCl_3$-CH_3OH, 4:1, v/v B = $CHCl_3$-CH_3OH-CH_3CO_2H, 9:1:0.5, v/v
C = n-Butanol-H_2O-CH_3CO_2H, 4:1:1, v/v

The tripeptide carboxylic acid (1) or (2) was coupled with (5) or (6) by the DCC/HO-BTA method to afford (7) or (8). The BOC-group was removed by TFA-anisole and the O-benzyl protecting group by catalytic hydrogenation. The free tripeptide enkephalin analogs (I) and (III), Table B, were purified by Silica gel 60/$CHCl_3$-CH_3OH elution, chromatography or Sephadex LH-20 gel filtration (elution with 10% acetic acid) depending on the quantity of the preparation.

The thiomorpholine and morpholine containing congeners (V) to (X), were prepared in a similar manner.

The phenylbenzoquinone induced writhing test for analgesia[5] and opiate receptor assay[6] were employed for biological evaluation.

TABLE B

No.	Compound	Phenylbenzoquinone induced Writhing test (mouse, sc) ED_{50} (mg/kg) (95% conf. lim.) peak time	Opiate receptor binding assay Relative Displacement Potency (Morphine 1)
I	H-Tyr-D-Ala-Gly-NHCH($CH_2C_6H_5$)CH_2N(ring)NCH_3	1.05 (0.62-1.79) 15 min. 4.60 (3.1-6.9) 30 min.	3.1
II	CH_3NH-Tyr-D-Ala-Gly-NHCH($CH_2C_6H_5$)CH_2N(ring)NCH_3	6.2 (4.3-8.9) 30 min.	3.4
III	CH_3NH-Tyr-D-Ala-Gly-N(CH_3)CH($CH_2C_6H_5$)CH_2N(ring)NCH_3	0.62 (0.37-1.03) 30 min.	2.1
IV	H-Tyr-D-Ala-Gly-Phe-N(CH_3)$CH_2CH_2N(CH_3)_2$	4.86 (3.48-6.78) 15 min.	1.7
V	H-Tyr-D-Ala-Gly-Phe-N(ring)S	10 mg/kg, 40% analgesia, 15 min.	1.2
VI	H-Tyr-D-Ala-Gly-NHCH($CH_2C_6H_5$)CH_2N(ring)S	6.85 (4.11-11.45) 15 min.	6.7
VII	H-Tyr-D-Ala-Gly-Phe-N(ring)S^+O^-	1.87 (0.7201.64) 15 min.	1.8
VIII	H-Tyr-D-Ala-Gly-NHCH($CH_2C_6H_5$)CH_2N(ring)S^+O^-	2.12 (1.13-3.98) 15 min.	2.8
IX	CH_3NH-Tyr-D-Ala-Gly-N(CH_3)CH($CH_2C_6H_5$)CH_2N(ring)S	6.58 (4.16-10.39) 15 min.	2.3
X	H-Tyr-D-Ala-Gly-NHCH($CH_2C_6H_5$)CH_2N(ring)O	10 mg/kg, 40% analgesia, 15 min.	2.4

All the compounds reported in Table B show good opiate receptor binding activity. The most potent analog for analgesia is (III) which incorporates N-methyl groups on the tyrosyl and phenethylamine moieties of the compound. The only conformer containing a β-turn in (III) is that one which involves the D-Ala[2] carbonyl group and the protonated at physiological pH piperazinyl-1-nitrogen atom. There is strong evidence that a β-turn of the above type is accepted by the morphine receptors from the fact that the cyclic disulfide D-Cys[2,5]-Enkephalin amide prepared first by Sarantakis[7] and later by Schiller et al[8], possess potent analgesic and opiate receptor binding activity. This cyclic enkephalin analog can attain only a β-turn centered at the Gly[3],Phe[4] residues of the compound. Surprisingly the tetrapeptides (IV) and (VII) which cannot stabilize such a β-turn through hydrogen bonding between the C-terminal tertiary amidic nitrogen and the carbonyl group of D-Ala show potent analgesic activity. In this case, either the β-turn is induced by the receptor or a different conformation is accepted by the receptor. The alternative is a β-turn which has a hydrogen bond between the amino group of phenylalanine and the carbonyl group of tyrosine. The same situation exists for the tripeptide amide H-Tyr-D-Ala-Gly-$N(CH_3)CH_2CH_2C_6H_5$ which retains potent biological activity but it cannot stabilize any β-turn through hydrogen bonding[9]. The fact that the sulfoxide congener (VII) is more potent than the thiomorpholine analog (V) may be due to better penetration of the Blood-Brain-Barrier (BBB).

In conclusion, a new series of tripeptide phenethylamide analogs of enkephalins is presented, which are endowed with potent analgesic activity. Interaction of these analogs with the opiate receptors may involve more than one type of β-turn of the enkephalin structure or induction of the proper orientation of the side chains of the amino acids by the receptor.

References

1. Roques, B.P., Gacel, G., Fournie-Zaluski, M-C., Senault, B., and Lecomte, J-M. (1979) Eur. J. Pharmacology, 60, 109-110.

2. Roques, B.P., Garbay-Jaureguiberry, C., Oberlin, R., Anteunis, M., and Lala, A.K., (1976) Nature, 262, 778-779.

3. Jones, C.R., Gibbons, W.A., and Garsky, V., (1976) Nature, 262, 779-782.

4. Konig, W., and Geiger, R., (1970) Chem. Ber., 103, 788-791.

5. Siegmund, E., Cadmus, R. and Go, L., (1957) Proc. Soc. Exp. Biol. Med., 95, 729-731.

6. Chang, J-K, Fong, B.T.W., Pert, A. and Pert, C.B., (1976) Life Sci., 18, 1473-82.

7. Sarantakis, D., US Patent 4,148,786, (1979) CA 91 39885W.

8. Schiller, P.W., Eggimann, B., DiMaio, J., Lemieux, C., and Nguyen, T. M.-D, (1981) Biochem. Biophys. Res. Commun., 101, 337-343.

9. Morgan, B.A., Bouer, J.D., Guest, K.P., Handar, B.K., Metclaff, G., and Smith, C.F.C., (1977) in Peptides: Goodman, M., Meienhofer, J. (eds) John Wiley and Sons. pp. 111-113.

CONFORMATIONAL STUDIES OF DERMORPHIN FROM FT-IR, LASER RAMAN, CD, CONFORMATIONAL ENERGY CALCULATIONS AND MOLECULAR MODELING

Rajendra S. Bhatnagar, N. Pattabiraman, and Keith R. Sorensen
University of California, San Francisco, CA 94143

Timothy W. Collette and Lionel A. Carreira
University of Georgia, Athens, GA 30602

V. Renugopalakrishnan
Harvard Medical School, Boston, MA 02115

Rao S. Rapaka
National Institute on Drug Abuse, Rockville, MD 20857

Introduction

Dermorphin, Tyr^1-D-Ala^2-Phe^3-Gly^4-Tyr^5-Pro^6-Ser^7-NH_2 is a potent μ-receptor agonist, with some affinity for δ-receptor

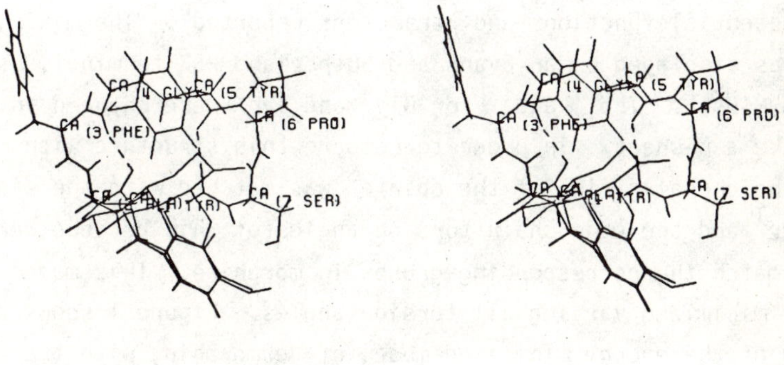

Fig 1. Stereo view of dermorphin with the structure of morphine superimposed. See text for explanation.

sites[1]. It has been suggested that dermorphin may have flexible, extended conformation[2,3]. It is likely that the opioid peptides and opiates may assume a similar pharmacophoric conformation at the receptor site[4] and, in fact, our studies on dermorphin show that it may present a molecular perspective similar to that of morphine (Figure 1).

Molecular Modeling Studies:

Since D-Ala and Pro usually appear as the second residues in ß-turns, and since ß-turns have been shown to play a role in receptor selectivity[5], we considered models for dermorphin that contained this conformation for residues Tyr^1-D-Ala^2-Phe^3-Gly^4 (turn 1) and Tyr^5-Pro^6-Ser^7-NH_2 (turn 2). In the initial studies, molecular modeling was carried out using the program AIMS (Ames Interactive Molecular Modeling program)at NASA, Ames. The ϕ and ψ of Gly^4 and Tyr^5 were fixed at 180^0. We built four types of ß-turns for 1 and 2 independently and calculated the energy using published[6] potential functions and parameters. Since the total energy of the four types of ß-turns were about the same, hydrogen bond energy was used as a criterion for assigning the type of bend. By this criterion, a type III ß-turn was found to be favored at 1, and a type II ß-turn at 2. The side chains were assumed to be in standard conformations and the model thus obtained was subjected to extensive energy minimization varying the torsion angles using the potential functions and parameters reported[6]. The minimized structure was displayed on an Evans and Sutherland PS2 terminal, using the program MIDAS[7]. The ϕ and ψ for Gly^4 and Tyr^5 were changed in order to make a ß-sheet. In order to compare this structure with morphine, the phenolic ring of the opiate[8] was matched with the side chain of Tyr^1 and the side chain torsion angles of this residue were changed to match the corresponding groups in morphine. This model was then energy minimized varying all torsion angles. Figure 1 shows a stereo view of the energy minimized model of dermorphin, with the morphine structure superimposed. The hydrogen bonds are shown as dotted lines. Turn 1 is stabilized by a hydrogen bond as well as by a favorable electrostatic interaction between the NH of Tyr^5 and the C=O of Tyr^1. The distance between the C_β atoms of Tyr^1 and Tyr^5 is approximately 8Å. As seen in Figure 1, Tyr^5 is close to the six-membered ring in the morphine molecule. We also minimized a model with an extended conformation where all torsion angles were assumed to be 180°. Interestingly, the total energy of the extended structure was less favorable than the compact structure with the two ß-turns. The folded structure makes it possible to bring the two Tyr residues closer than possible in the extended structure.

Conformational Studies of Dermorphin in Solution

Vibrational spectroscopy of dermorphin: The theoretically derived structure of dermorphin was supported by our studies using FT-IR and Raman spectroscopy. The amide I band in the IR spectrum at $1663cm^{-1}$ and in Raman spectrum at $1666cm^{-1}$ confirmed the presence of ß-turns[5,9]. The IR active shoulder at $1678cm^{-1}$ and the Raman band at 1684^{-1} also indicated ß-sheet structures[9]. The amide III band in the FT-IR spectrum manifested as a doublet, at $1261cm^{-1}$ and $1238cm^{-1}$ respectively and in the Raman spectrum, at $1262cm^{-1}$ and $1242cm^{-1}$ respectively. The doublet also indicated ß-turn and ß-sheet structures [5,9]. The two Tyr residues contributed a doublet at $827cm^{-1}$ and $850cm^{-1}$ (more intense band) respectively. The ratio of the intensity of the Tyr doublet of 1.3 is indicative of exposure of Tyr residues, confirming the structure derived from modeling studies.

CD of dermorphin: The CD spectra of dermorphin were recorded at 25° in water, methanol, and in trifluoroethanol. The spectrum in each solvent was characterized by two peaks. CD spectra of dermorphin in TFE at 45° and 60° revealed that dermorphin is capable of existing in ordered conformation in solution, and it is surprisingly stable in aqueous solutions. CD spectra of dermorphin was also obtained at several concentrations in trifluoroethanol and the results suggested that intermolecular association may contribute to the generation of order in the peptide.

We have preciously proposed that a ß-turn conformation may be recognized by δ-receptors and a ß-sheet structure by μ-receptors[5,10]. Our theoretical and experimental studies suggest that dermorphin assumes a conformation characterized by the presence of ß-turns as well as a ß-sheet. This structure is characterized by a Tyr ring oriented in a configuration superimposable on morphine. Hence, this folded "morphine-like" conformation of dermorphin explains its biological activity.

References

1. Feuerstein, G., (1985) in NIDA Research Monograph, Rapaka, R.S., and R. L. Hawks Eds, DHHS Publication, Rockville, MD, in Press
2. Salvadori, S., Tomatis, R., Gibbons, W.A., Tancredi, T., and Temusi, P.A. (1983) Peptides: Structure and Function, Proceedings of the Eighth American Peptide Symposium, Hruby, V.J., and Rich, D.H., Eds., Pierce Chemical Company, Rockford, Il, pp 785-788.
3. Arlandini, E., Ballabio, M., de Castiglione, R., Giola, B., Mainati, M.L., Perseo, G., and Rizzo, V (1985) Int. J. Peptide Protein Res., 25,33-46.
4. Schiller, P.W. (1984) in The Peptides, Udenfriend, S., and Meinhofer, J., Eds., Academic Press, San Diego, pp 219-268.
5. Renugopalakrishnan, V., Rapaka, R.S., Collette, T.W., Carreira, L.A., and Bhatnagar, R.S. (1985) Biochem. Biophys. Res. Commun. 126, 1029-1035.
6. Momany, F.A., McGuire, R.F., Burgess, A.W., and Scheraga, H.A. (1975) J. Phys. Chem., 79, 2361-2381.
7. Huang, C., Ferrin, T., Gallo, L., and Langridge, R., (1985) MIDAS User Manual, Computer Graphics Laboratory, Univ. California, San Francisco
8. Gylbert, L. (1970) Acta Cryst. B29, 1630-1635.
9. Tu, A.T. (1982) in Raman Spectroscopy in Biology: Principles and Applications, John Wiley, New York.
10. R. S. Rapaka, V. Renugopalakrishnan, and R. S. Bhatnagar (1985), American Biotechnology Laboratory (In Press).

INTERPRETATION OF THE MU ACTIVITY OF SEVERAL DERMORPHIN FRAGMENTS IN TERMS OF A NEW RECEPTOR MODEL.

Annalisa Pastore,Teodorico Tancredi[+] and Piero A.Temussi
Dipartimento di Chimica,Universita'di Napoli,via Mezzocannone 4,80134 Napoli,and + ICMIB,Arco Felice,Italy.

Severo Salvadori and Roberto Tomatis
Istituto di Chimica Farmaceutica,Universita' di Ferrara,Italy.

Introduction

Much of the work on the structure-activity relationship (SAR) of opioid peptides has been based on a comparison with the structure of morphine,in spite of the fact that some of these peptides interact with the delta rather than the mu receptor and,most of all,in spite of the fact that they are often more active than morphine itself. In order to discuss the SAR of (conformationally flexible) opioid peptides it is necessary to delineate a self-consistent topological model of the receptor site using molecular molds that are at least as active as the most active peptides and more rigid.

Results and Discussion

A systematic analysis of the structures of selected morphine-like opioids and of the potent mu agonists of the fentanyl family showed[1] that an idealized mu agonist can be depicted as in figure 1.

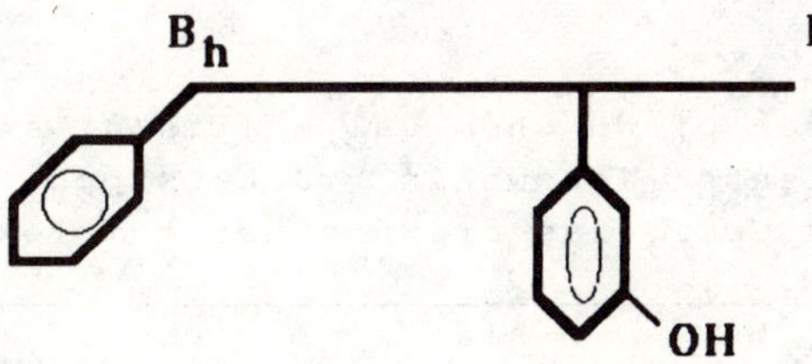

Figure 1. Idealized mu agonist showing the five main features.

In other words we propose that mu opioids can be characterized by essentially five features : 1) a rigid T-shaped backbone (common to all opium alkaloids),2) a hard base (Bh),3) a soft base (Bs),4) a hydrogen bond donor on the stem of the T (the OH group of morphine or of Tyr) and 5) an aryl ring adjacent to Bh. This second aromatic ring corresponds closely to the P subsite proposed by Portoghese[2] but its orientation with respect to the other four features can be defined more accurately since it is based on the structure of a single molecule,that is methyl-fentanyl.

The SAR of dermorphin,a natural mu opioid eptapeptide[3] and of several shorter analogs find a simple interpretation on the basis of our model[1] . We concentrated our experimental and theoretical efforts on the N-tetrapeptide fragment of dermorphin, Tyr-D-Ala-Phe-Gly-NH_2 (henceforth called D-tetra) and of its analogs since these compounds,in spite of their short sequence,can have activities higher than those of morphine and of natural enkephalins[4] . The presence of a D-residue and the location of a Phe residue in third position favor the attainment of low-energy conformations in which the two aromatic rings are placed in a relative position very similar to that of the two aromatic rings of methylfentanyl[1] . A typical representative of this family of conformations is the beta-turn previously proposed[5] for enkephalins. The actual accessibility of such a conformation for D-tetra was confirmed experimentally by an nmr study in chloroform of its complex with 18-crown-6-ether[6] . The combination of the complexation of the charged nitrogen and of the low polarity of the solvent mimics the physico-chemical environment of the receptor much better than water or DMSO. Table I shows that the NH chemical shifts are more widespread in chloroform than in DMSO and their temperature coefficients point to the existence of a hydrogen bond involving the NH of ^{4}Gly (whose coefficient is larger than 2.4 ppb,the value typical of exposed protons in chloroform).

Table I. Comparison of NH chemical shifts(δ) and temperature coefficients (shown in brackets,ppb) of D-tetra in DMSO and $CDCl_3$ (as crown ether complexes).

Solvent	Tyr	D-Ala	Phe	Gly
DMSO		8.48 (-6.5)	8.49 (-5.0)	8.38 (-4.0)
$CDCl_3$	7.29 (-1.0)	8.01 (-2.8)	7.77 (-3.0)	8.34 (-5.4)

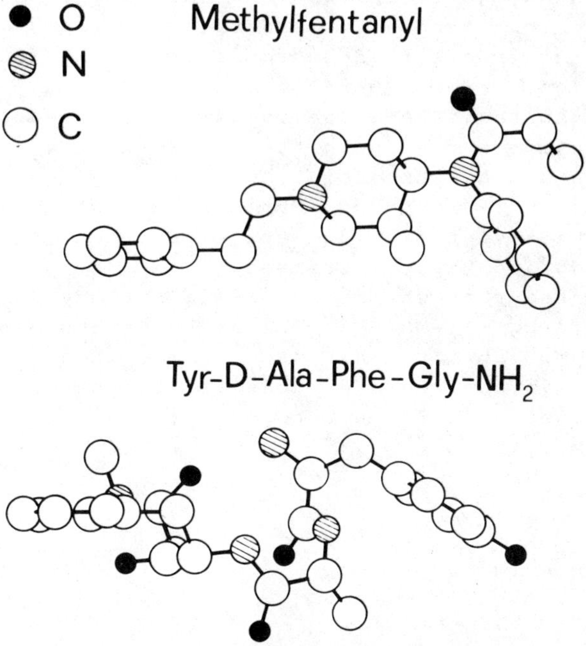

Figure 2. Molecular models of methylfentanyl and of D-tetra.

The similarity of the molecular models of D-tetra in a beta turn conformation and of methylfentanyl,as shown in figure 2 is striking and can form the basis for the interpretation of many apparently unrelated data both on analogs of D-tetra and on other opioids[1] .For instance,it was observed that even a moderate lenghtening and/or an increase of the flexibility of the second residue,leading to a destabilisation of the beta turn,is paralleled by a dramatic decrease of the activity. Thus,substitution of D-Ala with D- or L-O-Ala(i.e.2-aminopropioxy acid)leads to a 1000 fold decrease of the relative mu potency,as measured by the GPI test[7] .All modifications of the fourth residue(whose sidechain is not crucial for the stability of the the beta turn)have little or no influence on the activity. Addition of a guanidino terminus always proved advantageous;for instance the relative GPI potency jumps from 100 of D-tetra to 460 of H_2N-C(=NH)-Tyr-D-Ala-Phe-Gly-NH_2 .A simple explanation of this fact is that two extra single bonds between the positive nitrogen and the stem of the T make the structure closer to that of methylfentanyl.

On the other hand, modifications that change the orientation of the third residue generally lead to a decreased activity ; thus substituting Phe with dehydrophenylalanine lowers the relative GPI potency from 100 to 1[8].

In order to subtantiate these ideas we designed a linear peptide lacking Tyr(a residue hitherto considered essential) that ought to retain a substantial mu activity. According to our model the OH group of Tyr, although generally very important, can be absent provided the separation between the aromatic (T) ring and Bh is increased.

It was predicted that whlist Phe-D-Ala-Phe-Gly-NH_2 should be nearly inactive, H_2N-C(=NH)-Phe-D-Ala-Phe-Gly-NH should have a mu activity comparable to that of D-tetra. Indeed, when these compounds were synthesized and subjected to the GPI test, the relative potencies, with respect to D-tetra, were found to be 3 and 21 respectively.

References

1. Temussi,P.A., Esposito,G.,Lelj,F., Pastore, Salvadori,S., Tancredi,T.and Tomatis,R.(1985) J.Med.Chem.,submitted.

2. Portoghese,P.S.,Alreja,B.D.and Larson,D.L. (1982) J.Med. Chem.,24,782-787.

3. Erspamer,V.and Melchiorri,P. (1980) in Growth Hormone and other Biologically Active Peptides,Pecile,A.and Muller, E.E.,Eds.,Excerpta Medica,Amsterdam,p.185.

4. Salvadori,S.,Sarto,P.and Tomatis,R.(1981) Int.J.Peptide Protein Res.,19,536-541.

5. Bradbury,A.F.,Smith,D.G.,Snell,C.R.,Birdsall,N.J.M., and Hulm,E.C.(1976) Nature,260,624-625.

6. Pastore,A.,SalvadoriS.,Tancredi,T.Temussi,P.A.and Tomatis R.(1984)in Peptides 1984,Ragnarsson,U.,Ed.,Uppsala,p.333

7. Tomatis,R.Salvadori,S.and Sarto,P. (1981) Il Farmaco, 36,937-941.

8. Salvadori,S.Marastoni,M.Balboni,G.and Tomatis,R. (1985) Il Farmaco, in press.

SYNTHESIS AND BIOLOGICAL EVALUATION OF PEPTIDE E, [D-ALA^2] PEPTIDE E AND CARBOXY-TERMINUS FRAGMENTS

E.P. Heimer, A.M. Felix, M. Ahmad, T.J. Lambros, and S. Spector

Peptide Research Dept., Roche Research Center, Hoffmann-La Roche Inc., & Roche Inst. of Molecular Biology, Nutley, NJ 07110

D.M. Helmeste and C.H. Li

University of California, San Francisco, Laboratory of Molecular Endocrinology, San Francisco, CA 94143

Introduction

Peptide E (PE), a 25-amino acid peptide comprising residues 206-230 of bovine preproenkephalin contains Met-EK and Leu-EK segments at the amino- and carboxy-terminii, respectively.[1,2]

We recently reported on the synthesis and some biological activities of PE[3,4] and observed that it is 30 times more potent than Met-EK in the guinea pig ileum assay (GPI) and binds preferentially to the mu (μ) and kappa (κ) opiate receptor sites. PE sequences are completely conserved between the bovine, human, and rat species. Fragments containing shorter sequences have been reported[5,6] and their synthetic counterparts PE(1-12), (1-20) and (1-22) retain the activity of PE in the GPI assay and led to the postulation that the amino-terminus region of the molecule is essential for biological activity.

Intraventricular (ICV) injection (mice) with PE(1-22) produced a substantial analgesia comparable to that of morphine[7]

which is in contrast to the lack of analgesia seen after the ICV injection of PE.[4] Intracisternal administration of PE produced a profound long lasting analgesic effect.[8] We have observed that mice treated with PE (ICV) showed unusual behavioral changes manifested by hyperactivity, polydipsia and hyperphagia. The tetrapeptide sequence, PE(14-17) [Trp-Met-Asp-Tyr] is homologous to the satiety peptide cholecystokinin (30-33) which is also a potent stimulator in the release of insulin and other islet hormones in the perfused pancreas.[9] To explore the possible role of PE in its opioid, neurotransmitting and gastrointestinal properties a systematic series of carboxy-terminus analogs have been designed and synthesized.

Results and Discussion

Synthesis of PE and fragments was carried out by the solid phase procedure starting with Boc-Leu-O-CH_2-PAM resin. Aliquots of the growing peptide-resin were removed to permit the preparation of all the PE analogs which were purified by preparative hplc (Table I). The 2-stage HF cleavage[10] circumvented the additional step requiring alkaline removal of Trp(For) and resulted in a more homogeneous crude material. The peptides were purified by preparative hplc and shown to be essentially homogeneous in two analytical hplc systems. The opiate receptor binding assay (Table I and Figure 1) shows that PE is slightly less potent in competing for ^{3}H-βhEP binding compared to βhEP itself. Substitution of D-Ala^2 for Gly^2 decreases affinity slightly compared to peptide E. This substitution has also been found to reduce binding affinity of β-camel EP and confirms the importance of amino acid residues in sequences 1-5 for binding potency. In agreement with the predicted importance of residues 1-5, PE(2-25) and PE(6-25)

Synthetic Peptides	Binding Activity on Brain Opioid Receptor IC50* (nM)	Relative Potency	Guinea Pig Ileum Activity Relative Potency†
Human β-Endorphin (βhEP)	2.9 ± 0.3	100	
Peptide E(PE)	3.4 ± 0.4	87	1.0
[D-Ala^2]-PE	4.7 ± 0.4	63	0.37
PE-(14-25)	236 ± 34	1.3	NR
[Glu^{18}]-PE-(14-25)	>1μM	<0.2	NR
PE-(2-25)	>1μM	<0.2	NR
PE-(6-25)	>1μM	<0.2	NR

* Mean ± SEM, n = 3-4

† Potency in the electrically stimulated guinea pig ileum relative to Peptide E which is defined as 1.0. NR indicates no response.

Table I. Bioassay Results

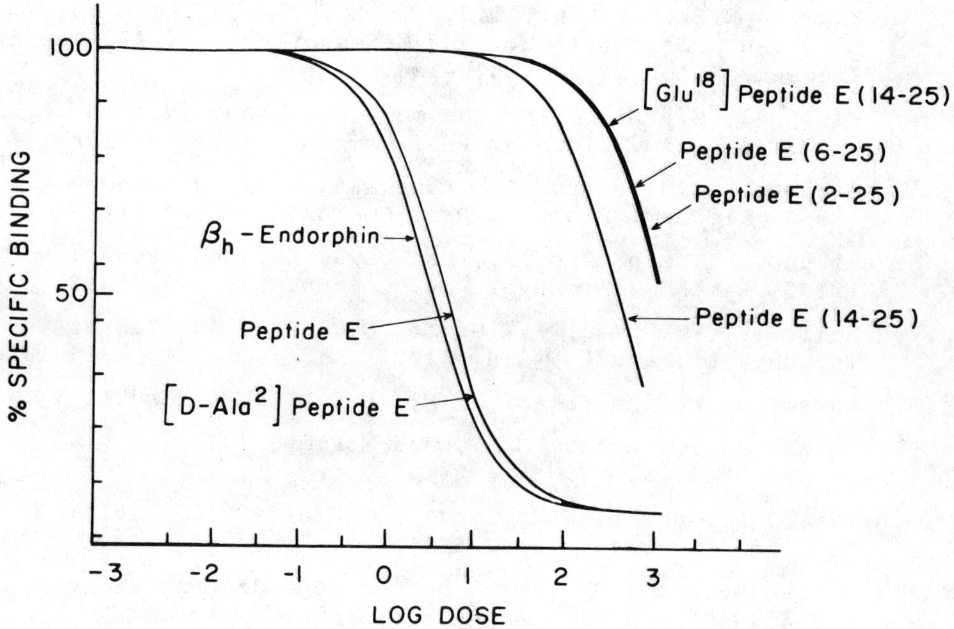

Fig. 1. Displacement of ^{3}H-βhEP by Peptide E and Analogs

have negligible binding affinity and PE(14-25) has some affinity which is lost in [Glu^{18}]-PE(14-25). Replacement of Glu by the non-charged Gln may directly affect binding or alter conformation. Although D-Ala^2 analogs of Met-EK give increased potency in the GPI assay,[11] [D-Ala^2]-PE showed substantial loss in potency compared to PE. The remaining fragments of peptide E were not active and it was concluded that Tyr^1 was essential for biological activity in these assays.

References

1. Gubler, U., Seeburg, P., Hoffmann, B.J., Gage, L.P., and Udenfriend, S. (1982) Nature (London) 295, 206-208.
2. Kilpatrick, D.L., Taniguchi, T., Jones, B.N., Stern, A.S., Shively, J.E., Hullihan, J., Kimura, S., Stein, S., and Udenfriend, S. (1981) Proc. Natl. Acad. Sci. USA 78, 3265-3268.
3. Heimer, E.P., Lambros, T.J., Felix, A.M., Fleminger, G., Li, C.-H., Westphal, M., and Meienhofer, J. (1983) Arch. Biochem. Biophys. 225, 518-524.
4. Westphal, M., Li, C.-H., Heimer, E.P., and Meienhofer, J. (1983) Biochem. Biophys. Res. Commun. 114, 1094-1088.
5. Mizuno, K., Minamiro, N., Kanagawa, K., and Matsuo, H. (1980) Biochem. Biophys. Res. Commun. 97, 1283-1290.
6. Mizuno, K., Minamiro, N., Kanagawa, K., and Matsuo, H. (1980) Biochem. Biophys. Res. Commun. 95, 1482-1488.
7. Höllt, V., Haarman, I., Grimm, C., Herz, A., Cankat-Tulunay, F., and Lo, H.H. (1982) Life Sci. 31, 1883-1886.
8. Nakamura, K. and Takamiya, M. (unpublished results).
9. Holst, J.J., Jensen, S.L., and Morley, J.S. (1980) Nature 284d, 33-38.
10. Tam, J.P., Heath, W.F., and Merrifield, R.B. (1983) J. Am. Chem. Soc. 105, 6442-6455.
11. Morgan, B.A. (1979) Amino-acids, Peptides and Proteins 10, 474-489 and references therein.

AMINO-TERMINAL FRAGMENTS OF BOVINE PRO-OPIOMELANOCORTIN (POMC): O-GLYCOSYLATION AND CYSTINE BRIDGE ORIENTATION

Monica A. Seger and Hugh P.J. Bennett

Endocrine Laboratory, Royal Victoria Hospital and Department of Medicine, McGill University, Montreal, Quebec H3A 1A1

The amino-terminal of POMC contains four residues of cysteine, namely in positions 2, 8, 20 and 24. It was previously assumed that cystine bridges linked the residues 2 and 8 in a manner analogous to calcitonin, and also linked 20 and 24. The highly-conserved amino-terminal is separated from the rest of the molecule by the acidic joining peptide which is rapidly cleaved in the bovine pituitary to give forms consisting of 77 residues (frequently referred to as 16K fragment). These forms have a potential cleavage site at $Arg_{49}Lys_{50}$, which is virtually uncleaved in the anterior lobe of the pituitary gland. In the intermediate lobe, where processing is more extensive, smaller forms corresponding to residues 1 to 49[1], Lys γ_3MSH[2] and Lys γ_1MSH[3] are found. We have purified all the major forms of the amino-terminal or 16K fragment to determine whether alternate cystine bridging or post-translational modifications might be responsible for the differences in the extent of processing of POMC in the two lobes of the bovine pituitary.

The extraction of amino-terminal fragments of POMC was facilitated by the use of reversed-phase and ion-exchange fractionation procedures as described previously.[4] Peptides were purified by reversed-phase high performance liquid chromatography (RP-HPLC). Three forms of 16K fragment were purified from extracts of the neurointermediate pituitary.

These corresponded in amino acid composition to $16K_{1-49}$ and two forms of $16K_{1-77}$ (i.e. $16K_{1-77}A$ and $16K_{1-77}B$). $16K_{1-49}$ was found to contain no amino-sugars. $16K_{1-77}A$ and B were found to be glycopeptides with the following amino-sugar contents (in moles sugar/mole peptide) - $16K_{1-77}A$: 2.6GlcNH$_2$, 1.7 GalNH$_2$; $16K_{1-77}B$: 1.9 GlcNH$_2$, 0.4 GalNH$_2$. A single form of 16K fragment was purified from extracts of the anterior pituitary. This form had a similar amino sugar content to that of $16K_{1-77}A$ i.e. 2.5 GlcNH$_2$, 1.2 GalNH$_2$. Each form of $16K_{1-77}$ was subjected to tryptic digestion. This gave rise, in each case, to a series of tryptic fragments which corresponded exactly to that predicted by the primary sequence of bovine POMC.[5] These tryptic peptides included the amino-terminal 1 to 49 fragment and the carboxyl-terminal 65 to 77 fragment. Amino sugar analysis demonstrated that $16K_{1-77}$ from the anterior lobe and $16K_{1-77}A$ from the neurointermediate lobe were glycosylated at sites contained within both these tryptic peptides. In contrast $16K_{1-77}B$ was clearly only glycosylated within the 65 to 77 fragment. The amino sugar content of these tryptic peptides was compatible with the presence of an O-glycosylation site at threonine residue 45 and a N-glycosylation site at asparagine residue 65 as predicted from the observations of Seidah et al.[6]

In order to establish the nature of the cystine bridging within the various forms of 16K, each was subjected to trypsin and V_8 protease digestion. Incubations were performed in ammonium bicarbonate buffer which had previously been saturated with argon in order to minimize cleavage of the cystine bridges during digestion.[1] This combination of proteolytic enzymes cleaves the 1 to 49 sequence between residues 4 and 5 (V_8 site), 14 and 15 (V_8 site) and 22 and 23 (tryptic site).[1] Examination of the fragments produced demonstrated that cystine bridges linked residues 2 and 24 and linked residues 8 and 20 in every case. This loop form of secondary structure does not appear to influence processing since it is common to all forms of 16K and presumably POMC found in both lobes of the pituitary.

The lack of O-glycosylation of $16K_{1-49}$ found in the neurointermediate lobe would suggest that this post-translational modification may influence the extent of biosynthetic processing of $16K_{1-77}$. Thus the presence or absence of O-glycosylation of residue 45 may determine the extent of cleavage at $Arg_{49}Lys_{50}$. This hypothesis is summarized in Figure 1.

PROPOSED SCHEME FOR PROCESSING OF 16K FRAGMENT IN THE BOVINE PITUITARY

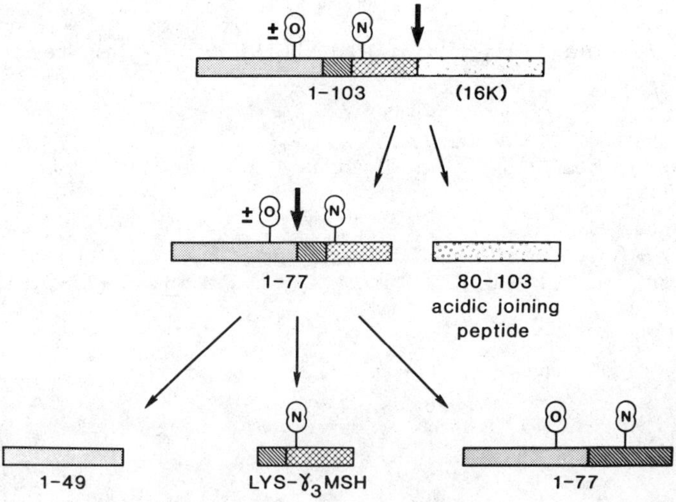

Figure 1.

Proposed biosynthetic processing of $16K_{1-77}$ of POMC in the intermediate lobe of the bovine pituitary. According to this scheme only those forms of $16K_{1-77}$ lacking O-glycosylation at threonine residue 45 are processsed to form $16K_{1-49}$ and $Lys^1\gamma_3MSH$. O-glycosylation of $16K_{1-77}$ is complete in the anterior lobe, thus no $16K_{1-49}$ or $Lys\gamma_3MSH$ is synthesized.

This work was supported by an operating grant from the Canadian MRC and a scholarship from the Fonds de la recherche en Santé du Québec.

References.

1. Bennett, H.P.J. (1984), Biochem. Biophys. Res. Commun., 125, 229-236.

2. Browne, C.A., Bennett, H.P.J. and Solomon, S. (1981), Biochem. Biophys. Res. Commun., 100, 336-343.

3. Bohlen, P., Esch, F., Shibasaki, T., Baird, A., Ling, N. and Guillemin, R. (1981) FEBS Letters, 128, 67-70.

4. James, S. and Bennett, H.P.J. (1985) J. Chromatogr. 326, 329-338.

5. Nakanishi, S., Inoue, A., Kita, T., Nakamura, M., Chang, A.C.Y., Cohen, S.N. and Numa, S. (1979) Nature, 278, 423-427.

6. Seidah, N.G. Rochemont, J., Hamelin, J., Lis, M., and Chrétien, M. (1981) J. Biol. Chem., 256, 7977-7984.

POTENT CONFORMATIONALLY CONSTRAINED ANALOGS OF GnRH

Jean Rivier, Janos Varga, John Porter, Marilyn Perrin, Yaira Haas, Anne Corrigan, Catherine Rivier, Wylie Vale.

Clayton Foundation Laboratories for Peptide Biology. The Salk Institute, 10010 N. Torrey Pines Road, La Jolla, CA 92037

Scott Struthers and Arnold Hagler

The Agouron Institute, 505 So. Coast Boulevard, La Jolla, CA 92037

Introduction

The synthesis and biological activities of several cyclic GnRH analogs have been reported [1-2]. Their observed low potency <u>in vivo</u> as antiovulatory agents in the rat left open the question of whether these analogs were binding to the physiologic receptor or had as yet unrecognized pharmacokinetic properties. Theoretical simulations, (including molecular dynamics, energy minimization, and template forcing) of GnRH and a cyclic antagonist, cyclo [-Δ^3Pro^1,4ClDPhe^2,DTrp^{3,6},NMeLeu^7, βAla^{10}-]-GnRH led us to propose a putative receptor binding conformation for the antagonist [3]. The iterative application of theoretical simulations, synthesis of cyclic analogs, and biological testing has now uncovered two classes of potent conformationally constrained GnRH peptides.

Materials and Methods

Peptide synthesis was achieved using SPPS. Cyclic structures involving disulfide bridging were obtained after air oxidation under dilute conditions; those bridges involving peptide bond formation were achieved using azide coupling of the unprotected peptide amide, [i.e. a)hydrazinolysis of the benzylester of Asp or Glu on the resin, b) HF cleavage, c) azide formation, and d) cyclization]. Purification was achieved using preparative HPLC [4]. Unnatural amino acids were synthesized, resolved and protected in our laboratory. Peptides were characterized by HPLC for purity; amino acid composition and optical rotation were obtained. Biological evaluation included an *in vitro* test, a binding assay, an antiovulatory assay (AOA) and several other *in vivo* assays [5] (See below and Table I).

Results and Discussion

Peptide structures and biological data are shown in Table I: these encompass two new classes of active cyclic GnRH analogs: a) Decapeptide analogs which were cyclized through side chains at position 4 and 10 (cpds 4,5,7) or shorter analogs (cpds 8,9) which still exhibit strong binding affinity b) end to end cyclic GnRH decapeptides (cpds 11-13) which were found to exhibit partial agonism even with histidine at position 2. While previous antagonists have employed removal of L-His^2 or its substitution by D-amino acids, as the means of selectively impairing transducing ability of the peptide, these latter analogs demonstrate that this can also be achieved through the introduction of conformational constraints (see cpds 11-13). Analogs which were cyclized through side chains at positions 4 and 9 (cpd 2) were considerably less potent.

Whereas the biological activity of cyclic analogs containing a disulfide bridge may be explained as the result of their ring opening *in vitro* or *in vivo*, this argument cannot be used for cyclic structures resulting from amide bonding. On the contrary, one may argue that those cyclic analogs may be more resistant to enzymatic degradation. Hence, their biological efficacy was tested in several *in vivo* bioassays. Duration of action was assessed in castrated rats after either sc or iv injection. Following iv injection, cpds 1(5,50 μg), 4(25,250 μg), 5(25,250 μg), and 6(250 μg) significantly inhibited LH release for 16-24 hrs. When administered sc, cpds 4 and 5, but not 6 and 7 (100 μg), lowered plasma LH levels for 11 hours while cpd 4 was the only one still active at 26 hrs. However, the possibility that cpd 7 may be marginally active at 26 hrs led us to inject it (250 μg) on diestrus I in the cycling rat. Under those conditions it was found to be fully potent, thus suggesting delayed absorption. Oral activity was evaluated in castrated male rats:

TABLE I. STRUCTURES AND BIOLOGICAL CHARACTERIZATION OF GnRH ANALOGS

	Compounds	Relative Potencies <u>In Vitro</u>*	K_D** (nM)	AOA***
1.	Ac-D2Nal-4FDPhe -DTrp -Ser-Tyr-DArg-Leu-Arg-Pro-Gly-NH_2	0.81	0.57 ± .090	1.0 (0/10)
2.	——————-Dap- ——————-Asp-Gly-NH_2		200	2,000 (10/10)
3.	———-4ClDPhe-D3Pal-Ser-Arg-D3Pal- ———-Pro-DAla-NH_2	0.91	0.24 (.17-.34)	0.5 (0/10)
4.	——————-Dap- ——————-Asp-NH_2	0.25	4.5 (3.4-5.8)	25 (0/10)
5.	——————-Cys- ——————-Cys-NH_2	0.04	0.85 ± .050	10 (0/10)
6.	—————-(Ac)Dap- ——————-Asn-NH_2	0.003	1.0 (.64-1.6)	50 (3/7)
7.	Ac-Δ3Pro-4FDPhe- DTrp -Dap-Tyr-D2Nal- ————-Asp-NH_2	0.25	0.80 (.59-1.1)	2,500 (10/10)
8.	Ac- ——————————-NH_2	0.008	5.6 (3.9-8.1)	2,000 (3/3)
9.	Ac- ————————-NH_2	0.001	100	
10.	pGlu-His-Trp-Ser-Tyr-Gly-Leu-Arg-Pro-Gly-NH_2		7.2 ± 1.60	
11.	c[-βAla- ————-D2Nal- ————-]		5.0 (2.4-8.4)	
12.	c[-Gaba- ——————————-]		9.1 (6.2-13)	
13.	c[-Eaca- ——————————-]		0.44 ± .060	

* <u>In vitro</u> potencies, [Vale, W., Grant, G., Amoss, M., Blackwell, R. and Guillemin, R. (1972) Endocrinology 91:562], for GnRH antagonists (cpds 1-9) were normalized to that of [Ac-Δ3Pro[1]-4FDPhe[2],DTrp[3,6]]-GnRH = 1.0. Cpds 11-13 were partial agonists achieving 30% to 60% maximum stimulation of LH release.

** Mean K_D ± SEM; range given in parentheses. The K_D values were calculated, as described in Perrin, M.H., Haas, Y., Rivier, J.E., and Vale, W.W. (1983) Molecular Pharmacology 23:44.

*** Under AOA is given the amount of peptide (μg) injected on noon of proestrus to cycling rats. The number of ovulating rats over the total number of animals tested is given in parentheses.

----- Abbreviations: D2Nal = β(2-Naphtyl)DAla; D3Pal = β(3-pyridyl)DAla; Dap = 2,3-diamino propionic acid; Gaba = γ amino butyric acid, Eaca = 6 amino caproic acid; 4FDPhe = 4 Fluoro-DPhe all cyclic peptides (cpds 4,5,6 at 3 mg each) were inactive when tested 5 hrs after administration with the exception of cpd 1 (1 mg).

In summary, we described cyclic analogs which for the first time had high affinity for the GnRH receptor and were very potent in an AOA. Those designed to be more hydrophillic (cpds 4,5) exhibited, in vivo, similar patterns of action as the linear analogs whether hydrophyllic (cpds 1,3) or hydrophobic [1]. Cpd 7, which was more hydrophobic, showed delayed action. Ongoing studies should allow us to further document those differences in pharmacokinetics. Finally, the first evidence that conformational restrictions per se could result in the generation of analogs with strong binding affinity but partial ability to trigger a receptor's response has been presented (cpds 11-13).

Acknowledgements

Research supported by NIH grant HD13527 and contract NO1-HD-4-2833. Research conducted in part by the Clayton Foundation for Research, California Division. (W. Vale and C. Rivier are Clayton Foundation Investigators). We wish to thank R. Kaiser, J. Dykert, L. Spaulding, G. Berg and G. Morgan for technical assistance and R. Hensley for manuscript preparation.

References

1. Rivier, J., Rivier, C., Perrin, M., Porter J., and Vale, W. (1981) In: LHRH Peptides as Female and Male Contraceptives, Zatuchni, G.I., J.D., Shelton, J.J. Sciarra (eds.), Harper and Row, Philadelphia, PA., pp. 13-23.

2. Seprodi, J., Coy, D.H., Wilchez-Martinez, J.A. et al. (1978). J. Med. Chem. 21:993-995.

3. Struthers, R.S., Rivier J. and Hagler A. (1984) Annals of NY Academy of Sciences, B. Venkataraghavan and R. J. Feldman (eds), In: Macromolecular Structure and Specificity: Computer Assisted Modeling and Applications 439:81-96, and references therein.

4. Rivier, J., McClintock, R., Galyean, R. and Anderson, H. (1984) J. of Chromatography 288:303-328.

5. Rivier, C., Vale W. and Rivier J. (1983) J. Med. Chem. 26:1545-1550 and refs therein.

LHRH-PHOTOAFFINITY LABELLING ANALOGUES: SYNTHESIS AND BIOLOGICAL APPLICATIONS OF ANALOGUES WITH DIFFERENT LABEL POSITIONS

Zbigniew Mackiewicz, Ginette Lagacé, Michelle Beauchesne, Serge Bélisle, Diego Bellabarba, Nicole Gallo-Payet, Jean-Guy Lehoux and Emanuel Escher

Department of Pharmacology and Service of Endocrinology, Medical Faculty, University of Sherbrooke, Sherbrooke, Quebec, Canada J1H 5N4

Introduction

The receptor of LHRH (also called GnRH or LRF) has been the subject of much investigation on pituitary membranes. In several studies (1,2,3) this receptor has been identified with photoaffinity labelling peptides, having the label moiety attached to the N-ε of *lys*^{6}-LHRH.* Because this position of the peptide is very unlikely to be in close contact with the receptor, due to the big tolerance to modifications, new compounds with different label positions are worthwhile to study. In angiotensin II-receptor labelling studies it was also observed that the molecular weight of the receptor depends on the agonistic or antagonistic nature of the label (4). We therefore prepared five new analogues with agonistic or antagonistic properties and containing the labelling moiety either in pos. 2, 4, 5 or 7 of LHRH.

<u>Syntheses</u>: With classical solid phase methods on an automatic peptide synthesizer Peptomat (our laboratory) were prepared the following sequences: pGl-His-Trp-Ser-Tyr-Gly-Leu-Arg-Pro-Gly-NH_2, LHRH; [*ala*^{6},(4'-NO_2)Phe^{7}]LHRH,<u>1</u>; [*phe*^{2,3,6},(4'-NO_2)Phe^{7}]LHRH,<u>2</u>; [(4'-NO_2)*phe*^{6}]LHRH,<u>3</u>; [*phe*^{2,6},(4'-NO_2)*phe*^{3}]LHRH,<u>4</u>; and [Ac-Pro^{1},(4'-NO_2)*phe*^{2},β*np*^{3,6}]LHRH,<u>5</u> (β*np* = β-D-Naphthylalanine).

* Abbreviation for D-amino acids, D-Lys = *lys*

Boc Gly, the first amino acid, was esterified to chloromethylated polystyrene (0.54 meq/g) and the TFA-BOC protection scheme was used. Side chain protection was either of the Bzl-type (Ser and Tyr) or Tosyl (His and Arg). For chain elongation, either symmetrical anhydrides or DCC-HOBT activation was used, depending on the residue to be coupled. The protected peptide was cleaved from the resin by ammonolysis (NH_3 sat at 0^oC in DMF/2-propanol 1:1) at 30^oC for 6 days and filtered over LH20 with DMF. The protecting groups were cleaved by liquid HF (0^oC, 60 min, 10% anisole and .1% AcTrp), the N-terminal glutamine cyclized for peptides 1-4 by incubation in 25% AcOH at RT overnight, lyophilized and again filtered over LH20/DMF before application on preparative HPLC (2.5 x 40 cm, C18 30 μ Macherey Nagel, Darmstadt, Germany). The peptides were eluted with a gradient from 20 to 70% acetonitrile/H_2O, 0.05% in TFA. The nitrophenylalanyl-precursors 1-5 were converted to the azidophenylalanyl peptides 6-10 by catalytic hydrogenation followed by diazotation and azide formation (5) and again purified on preparative HPLC. The final peptides were recovered by lyophilization from the HPLC fractions previously buffered with 2N NH_4OAc. All products were at least 98% pure on analytical HPLC, had exact amino acid analysis and were pure in at least two different TLC systems.

Biological assays: All compounds were tested on their binding capacities on rat pituitary membranes and on human placental membranes against iodinated Buserelin [(3'-^{125}I)Tyr^5,(O-t-But) *ser*^{6}]LHRH. The binding parameters from scatchard analysis and the half maximal displacement concentration ED_{50} are presented in table 1.

Labelling studies: For the first radiolabelling studies, peptide 6 was iodinated with the iodogen method and purified by HPLC (6). This product, normally around 30 μCi, was characterized with a Scatchard analysis against non-radioactive 6 and compared to a Scatchard analysis of iodinated Buserelin

Table 1. Binding parameters of LHRH analogues.

		Human placental membranes		Rat pituitary membranes	
		A	B	A	B
	Buserelin, [(O-*t*-But)*ser*^{6}]LHRH	6.25	6.04	9.54	9.41
1	[*ala*^{6},(4'-NO_2)Phe^7] "	6.24*	5.91	-	-
2	[phe^{2,3,6},(4'-NO_2)Phe^7] "	6.10*	6.17	-	-
3	[(4'-NO_2)*phe*^{6}] "	6.39	5.77	-	-
4	[*phe*^{2,6},(4'-NO_2)*phe*^{4}] "	4.32	4.70	-	-
5	[AcPro^1,(4'-NO_2)*phe*^{2},βnp^{3,6}]"	-	-	-	-
6	[*ala*^{6},(4'-N_3)Phe^7] "	8.62	5.41	8.53	7.40
7	[*phe*^{2,3,6},(4'-N_3)Phe^7] "	9.18	6.14	8.24	6.60
8	[(4'-N_3)*phe*^{6}] "	8.50	5.15	9.13	7.57
9	[*phe*^{2,6},(4'-N_3)*phe*^{4}] "	8.47	5.17	8.05	6.89
10	[AcPro^1,(4'-N_3)*phe*^{2},βnp^{3,6}] "	8.29*	6.19	9.19*	8.71

A: log Ka in [M^{-1}] B: -log ED_{50}

*: non-linear scatchard plot, a second, lower affinity site is present.

Fig. 1 (left) and 2 (right): SDS-PAGE of photolabelling.

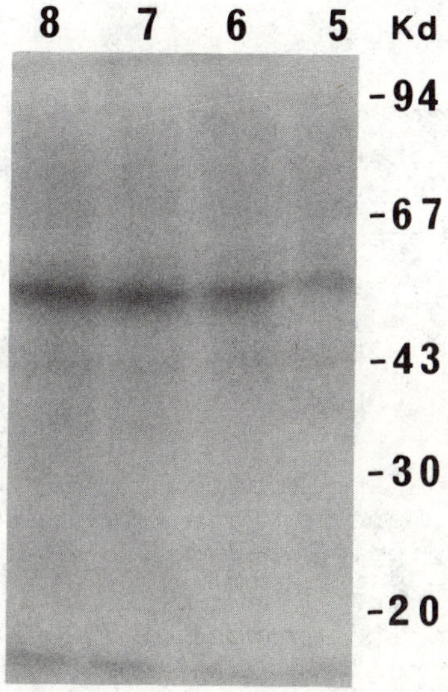

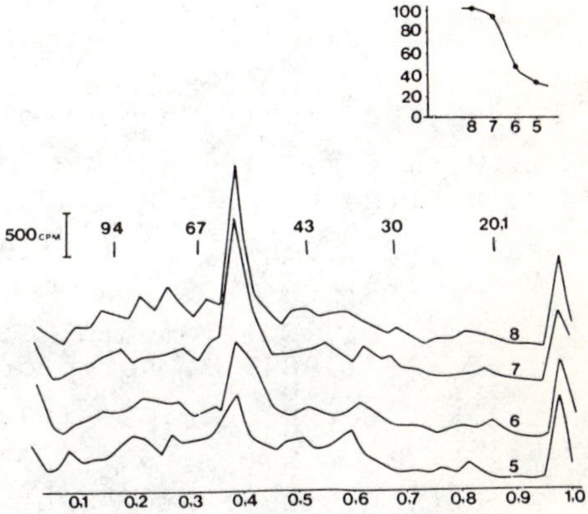

Numericals 5-8 are -log [LHRH] used to compete the labelling of ^{125}I-6 on human placental membranes. Fig. 1 is an autoradiography of a 6% SDS-PAGE, Fig.2 is the radioactivity profile of the lanes cut into strips, MW-markers are indicated in Kd, ordinate (Fig. 2) is radioactivity, abscissa is the gel Rf-value. Fig. 2 inset is the relation of the labelling in % of the 58 Kd protein (ordinate) versus the protective LHRH concentration (abscissa) in -log [LHRH].

against 6; both produced identical curves and parameters (see table 1). The other azidopeptides 7-9 were iodinated in the same manner, however peptide 10 had to be labelled in buffered 50% acetic acid because of poor solubility in water.

The photolabelling experiments with iodinated 1 on human placental membranes are presented in Fig. 1 and 2; they clearly show a competitive and specific labelling of a 58'000 dalton protein, presumably the receptor of LHRH on the placenta. Identical results were obtained with iodinated 8 but not with 9 which did not produce any specific labelling; studies with peptides 7 and 10 are currently under way.

The molecular weight observed of this denatured LHRH-receptors is in good agreement to the molecular weight found on other LHRH-bearing tissues (3): 60K on rat pituitary membranes, 60K and 54K on ovarian and testicular membranes. However also the angiotensin II receptor (4) and the ACTH-receptor (7) is in the same range if denatured and reduced.

References:

1. Hazum, E. (1981) Endocrinology 109, 1281-1283.

2. Nikolics, K., Teplan, I., Ramachandran, J. (1984) Int. J. Peptide Protein Res. 24, 430-436.

3. Hazum, E., Schwartz, I. (1984) Biochem. Biophys. Res. Comm. 125, 532-538.

4. Escher, E., Guillemette, G., Marie, J., Guillon, G., Jard, S. (1984) J. Receptor Res. 4, 267-281.

5. Laczko, E., Escher, E. (1981) Helv. Chem. Acta 64, 621-627.

6. Escher, E. (1983) in "Synthesis and Application of Isotopically Labelled Compounds"; W.P. Duncan & A.B. Susan Ed., Elsevier-North Holland, Amsterdam, pp. 231-234.

7. Sircar, S., Escher, E., Bouthillier, F., Lehoux, J.-G. (1985) Commun. on the 5th Intl. Spring Symposium, "Neural and Endocrine Peptides and Receptors '85", Washington, D.C., May 28-31.

A CYCLIC HEXAPEPTIDE LH-RH ANTAGONIST

Roger M. Freidinger, C. Dylion Colton, William C. Randall, Steven M. Pitzenberger and Daniel F. Veber
Merck Sharp & Dohme Research Laboratories, West Point, PA 19486

Richard Saperstein, Edward J. Brady and Byron H. Arison
Merck Sharp & Dohme Research Laboratories, Rahway, NJ 07065

Luteinizing hormone-releasing hormone (LH-RH), Glp-His-Trp-Ser-Tyr-Gly-Leu-Arg-Pro-Gly-NH_2, has been the subject of intensive studies as a novel approach to fertility control.[1] Despite the synthesis of hundreds of agonist and antagonist analogs, it has not proven possible to significantly reduce the size of the peptide while retaining activity. Based on conformational considerations, a cyclic hexapeptide has now been designed which, although low in potency, is the simplest antagonist of LH-RH synthesized to date.

Several of the existing analogs have suggested the presence of a Tyr-Gly-Leu-Arg Type II' beta turn in the receptor bound conformation of agonist and/or antagonist analogs.[2] Another peptide for which experimental evidence supports this turn type is somatostatin, and highly potent cyclic hexapeptide analogs have been designed.[3] In the latter cases, the key tetrapeptide unit (e.g., Phe-D-Trp-Lys-Thr) can be effectively constrained to the active beta turn conformation by a dipeptide unit composed of Phe followed by a secondary amino acid (Type VI beta turn).[4] It therefore appeared possible to stabilize the LH-RH Type II' beta turn in a cyclic hexapeptide, and the structure selected to test this possibility was cyclo-(Tyr-D-Trp-Leu-Arg-Trp-Pro) 1. D-Trp was chosen to replace Gly[6] since this substitution favors the beta turn and is known to enhance potency. Trp was chosen to precede Pro to potentially recover hydrophobic binding lost by deletion of residues 1-3 at the amino terminus of LH-RH. Phe at the comparable position in somatostatin cyclic hexapeptides is an important binding element. Recent computer studies of LH-RH and a cyclic decapeptide antagonist appear to be generally in accord with these design concepts.[5] Since deletion of His^2 in LH-RH generally leads to antagonists, the cyclic hexapeptide was expected to be an antagonist.

The linear precursor peptide was synthesized by the solid phase method with Tyr at the carboxy terminus and removed from the resin and cyclized to **1** according to methods previously reported.[4]

Compound **1** was evaluated for effects on LH secretion alone and in the presence of LH-RH in rat pituitary cells _in vitro_. At doses as high as 10^{-4}M, the analog did not affect LH release indicating a lack of agonist activity. The rates of LH production at a series of LH-RH and cyclic peptide concentrations were measured. The compound is antagonistic with a graded response as its concentration is increased. LH release did not reach 100% maximum response even in the presence of the lowest dose (10^{-8}M) of cyclic peptide. LH-RH-mediated LH secretory rates were reduced by 50% at a molar dose ratio of about 400 (antagonist: LH-RH). This level of potency is three times greater than the early decapeptide antagonist (D-Phe^2)-LH-RH.[6] This data was fitted by nonlinear least squares to competitive, uncompetitive, and noncompetitive inhibition models. The best fit was found for the competitive model (Fig.), but experimental error was great enough to preclude a definite assignment of the mechanism. The parameters giving the best fit (average for three experiments) to the competitive model were $K_d = 2 \pm 1 \times 10^{-9}$M (LH-RH) and $K_i = 8 \pm 4 \times 10^{-7}$M (**1**).

It is evident that cyclic hexapeptide **1** is able to achieve a conformation which binds to the LH-RH receptor. Studies

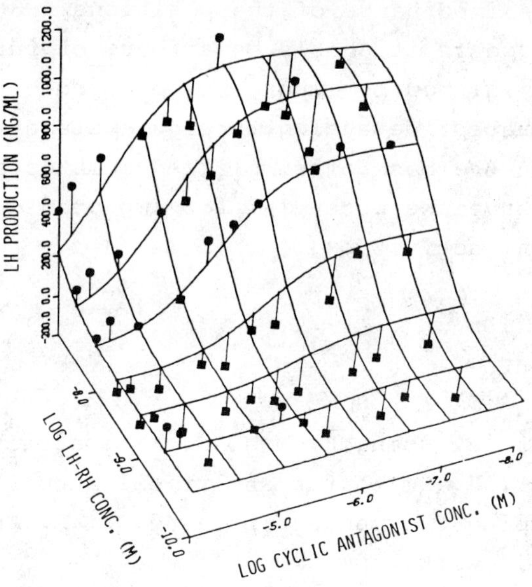

Figure. Data for inhibition of LH-RH induced LH release *in vitro* fit to a competitive inhibition model.

in solution indicate that **1** and the somatostatin analog cyclo-(Phe-D-Trp-Lys-Thr-Phe-Pro) **2** have similar backbone conformations. The circular dichroism spectra in pH 7.4 phosphate buffer in the 190-240 nm region are virtually superimposable. The NMR spectra of **1** in both D_2O and $\underline{d}_6$-DMSO are similar and consistent with a single major conformer. The spectrum in $\underline{d}_6$-DMSO has been completely assigned with the aid of 2-dimensional COSY and NOE spectra. A study of the chemical shifts of the alpha NH's from 23-48° indicated that the Tyr and Arg protons are relatively solvent shielded. The NH-$C^{alpha}H$ coupling constants and side chain upfield shifts are in accord with the predicted two turn conformation. Nuclear Overhauser enhancements show proton proximities consistent with Type II' and VI turns as was observed for **2**.

The present results show that using a conformational approach it is possible to delete 6 of the 10 amino acids of LH-RH and still retain receptor binding. This result further emphasizes the importance of the positions 5-8 beta turn in LH-RH, and suggests it should be a focus of further design studies. It will now be important to improve the potency of the cyclic hexapeptide analog through systematic modification as was done in the case of simplified somatostatin analogs. Such analogs have the potential for improved duration of action and oral activity.

Acknowledgements

We thank Mr. C. Homnick, Mrs. S. Fitzpatrick, Mrs. J. Murphy, and Mr. J. Smith for analytical support, Mrs. M. Banker for typing the manuscript, and Dr. P. Anderson for his support of this work.

References

1. LH-RH Peptides as Male & Female Contraceptives, Zatuchni, G.I., Shelton, J.D., and Sciarra, J.J., Eds., Harper & Row, Philadelphia (1981).

2. Freidinger, R.M., Veber, D.F., Perlow, D.S., Brooks, J.R., and Saperstein, R. (1980) Science 210, 656-658 and references cited therein.

3. Veber, D.F., Freidinger, R.M., Perlow, D.S., Paleveda, W.J., Jr., Holly, F.W., Strachan, R.G., Nutt, R.F., Arison, B.H., Homnick, C., Randall, W.C., Glitzer, M.S., Saperstein, R., and Hirschmann, R. (1981) Nature 292, 55-58.

4. Freidinger, R.M., Perlow, D.S., Randall, W.C., Saperstein, R., Arison, B.H., and Veber, D.F. (1984) Int. J. Peptide Protein Res. 23, 142-150.

5. Struthers, R.S., Hagler, A.T., and Rivier, J. (1984) in Conformationally Directed Drug Design, Vida, J.A. and Gordon, M., Eds., American Chemical Society, Washington, DC, pp. 239-261.

6. Rees, R., Foell, T., Chaie, S., and Grant, N. (1974) J. Med. Chem. 17, 1016-1019.

EFFECTS OF D-AMINO ACID SUBSTITUENTS ON SITES OF LHRH ANALOG CLEAVAGE BY RENAL TISSUE

George FLOURET, Frank A. CARONE, AND Eleanor FERRONYALKA,
Northwestern University Medical School, Chicago, IL 60611

Darryl R. PETERSON,
The Chicago Medical School, North Chicago, IL 60064

Introduction

In previous studies tritiated luteinizing hormone-releasing hormone (LHRH) was incubated with brush border membranes from renal tubules (rabbit), microperfused into isolated renal proximal tubules (rabbit) <u>in vitro</u>[1], and microinfused into renal proximal and distal tubules (rat) <u>in vivo</u>[2]. HPLC analysis[3] of collection fluid, bathing medium or urine, where appropriate, showed that the renal brush border of the proximal tubule degrades LHRH to pGlu-His-Trp-Ser (4), pGlu-His-Trp (3), and pGlu-His (2). We postulated that brush border peptidases act by either: a. cleavage of LHRH at Ser^4-Tyr^5, yielding tetrapeptide 4, followed by carboxypeptidase action on the latter, yielding tri and dipeptides 3 and 2, or; b. cleavage of LHRH at multiple sites simultaneously by several endopeptidases[4]. To discern between the two pathways, we synthesized tritiated [D-Ser^4]LHRH, [D-Ser^4, D-Trp^6]LHRH, [D-Trp^6]LHRH, and studied their cleavage in the above renal systems.

Materials and Methods

a. <u>Peptide Synthesis</u>. ^{3}H-[D-Ser^4]LHRH, I, ^{3}H-[D-Trp^6]LHRH, II, and ^{3}H-[D-Ser^4, D-Trp^6]LHRH, III, were prepared by coupling

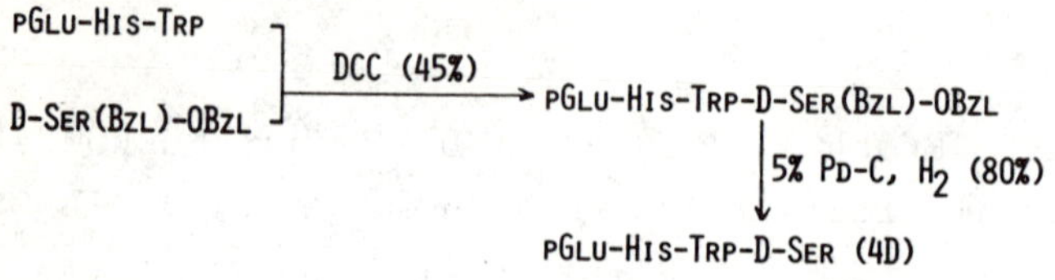

Fig. 1. Synthetic scheme for pGlu-His-Trp-D-Ser.

[3, 4-^{3}H]-pyroglutamic acid, in the presence of DCC and HOBt, with the pertinent free nonapeptides amides. The latter were assembled by a solid phase method as described previously[3]. Side chain functionalities were protected by the use of Boc-Arg(Tos), Boc-Tyr(Bzl) and Boc-Ser(Bzl) or Boc-D-Ser(Bzl). pGlu-His-Trp-D-Ser-Tyr (5D) was also prepared by the solid phase method. The synthetic schemes for pGlu-His-Trp-D-Ser (4D) and pGlu-His-Trp-D-Ser-Tyr-Gly (6D) are shown in Figures 1 and 2 respectively.

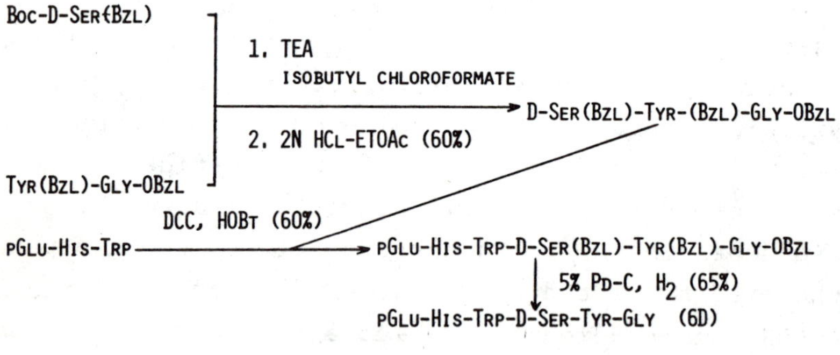

Fig. 2. Synthetic scheme for pGlu-His-Trp-D-Ser-Tyr-Gly.

b. <u>Analysis of tritiated analogs and metabolites</u>.

Experimental samples to be analyzed were injected into an HPLC system along with standards of metabolites 4D, 5D, 6D and analog I, II, III, and eluted as previously described for the analysis of LHRH and metabolites[3]. The eluent was collected in scintillation counting vials at 1-minute intervals, and the radioactivity in each vial was counted. The radioactive fractions detected co-chromatographed with metabolites or analog standards.

c. Renal microperfusion of tubules. Proximal straight tubule segments from rabbit kidneys were microperfused as previously described[1]. A final bath sample was taken. The perfusate was collected continuously in a collection micropipette and is referred to as the collection fluid.

d. In vivo studies. Tritiated LHRH analogs were microinfused into proximal and distal tubules of rats, as previously described[2]. Urine was collected during the infusion period and for the following 5 minutes. All injection sites were localized by tubular injection of latex and subsequent microdissection.

Results and discussion.

^{3}H-[D-Ser^4]LHRH was substantially degraded by rabbit brush border membranes, microperfused rabbit proximal tubular segments *in vitro*, and microinfused rat proximal tubules *in vivo*, with the major metabolite being 6D in all three instances. In addition, rabbit brush border membranes and proximal tubular segments *in vitro* yielded metabolite 3, and rat proximal tubules *in vivo* yielded metabolite 2. Thus, D-Ser blocked enzymatic cleavage at Ser^4-Tyr^5 but allowed cleavage at the new locus Gly^6-Leu^7. The degradation of LHRH by brush border membranes[1] was studied as a function of time. This revealed the formation of very small amounts of pGlu-His-Trp-Ser-Tyr-Gly(6) but always substantially less than metabolite 4, indicating that the rate of enzymatic cleavage at position 4 greatly exceeds that at position 6.

^{3}H-[D-Trp^6]LHRH was not cleaved at position 6 or 4 by isolated brush border membranes or by microinfused rat proximal tubules *in vivo*; however, metabolite 2 and small amounts of metabolite 3 were generated. ^{3}H-[D-Ser^{4}, D-Trp^{6}]LHRH was not cleaved by brush border membranes but was degraded to a small extent to metabolite 2 by microinfused proximal tubules.

These findings support the view that renal endopeptidases cleave LHRH at multiple sites. A renal brush border peptidase has been described which should yield metabolites 2, 4 and pGlu-His-Trp-Ser-Tyr-Gly[5]. Renal Angiotensin I-converting enzyme cleaves LHRH to tripeptide 3 [6]. Thus, if the latter enzyme is present in renal brush border, these enzymes could cause the cleavage pattern detected in our experiments.

Acknowledgements

Supported in part by NIH Grants HD-08669, AM 14610 and AM 27646.

References

1. Stetler-Stevenson, M.A., Flouret, G., and Peterson, D.R. (1981) Am. J. Physiol. 241, F117-F122.
2. Stetler-Stevenson, M.A., Flouret, G., Nakamura, S., Gulczynski, B., and Carone, F.A. (1983) Am. J. Physiol. 244, F628-F632.
3. Stetler-Stevenson, M.A., Chang Yang, D., Lipkowski, A., McCartney, L., Peterson, D., and Flouret, G. (1981) J. Med. Chem. 24, 688-692.
4. Flouret, G., Stetler-Stevenson, M.A., Carone, F.A., and Peterson, D.R. (1984) in "LHRH and its Analogs" (Vickery, B.H., Nestor, J.J., Jr., and Hafez, E.S.E., eds) MTP Press. pp 397-410.
5. Kerry, M.A., and Kenny, A.J. (1974) Biochem. J. 137, 477-478.
6. Skigdel, R.A., and Erdos, E.G. (1985) Proc. Nat. Acad. Sci. USA 82, 1025-1029.

POTENT LHRH AGONISTS CONTAINING $N^{G},N^{G'}$-DIALKYL-D-HOMOARGININES

John J. Nestor, Jr., Ram Tahilramani, Teresa L. Ho, Georgia I. McRae, and Brian H. Vickery

Institutes of Bio-Organic Chemistry and Biological Sciences, Syntex Research, Palo Alto, California 94304

Introduction

During the course of numerous luteinizing hormone-releasing hormone (LHRH) analog studies, it became clear that substitution of very hydrophobic D-amino acids in position 6 gave the most potent analogs *in vivo*.[1] This was attributed to a combination of factors including increased receptor affinity, protection from proteolysis and prolonged duration of action due to depoting of the analog in the body. Analogs containing very hydrophobic amino acids [e.g. Nal(2)] can bind to hydrophobic carrier proteins and be released later to the circulation.[2] For example, [D-Nal(2)^{6}]LHRH (nafarelin) is 80% bound to serum albumin in equilibrium dialysis experiments.[3] Such hydrophobic analogs also bind to cell membranes and may show a prolonged biological half-life ($t_{1/2}$) in that manner. We now report a series of LHRH analogs which exhibit high biological activity but which contain D-hydrophilic amino acids in position 6.

Analog Design

We recently reported a series of synthetic amino acids designed for phospholipid membrane interaction [$N^{G},N^{G'}$-dialkylhomoarginines; hArg(R_2)], which were prepared from Lys.[2] The rationale for this design involves an ionic interaction between the guanidine function of hArg(R_2) and the negatively charged phosphate head group of the phospholipid, which is further stabilized by hydrophobic interactions of the alkyl chains on the guanidine with the membrane. These amino acids might be expected to provide two beneficial effects when incorporated into hormone analogs: depoting of the analog on cell membranes for increased $t_{1/2}$ and increased apparent receptor affinity. The latter effect might be expected due to the hypothesis that ligand-membrane interaction is a critical step preceding ligand-receptor binding.[4] Accordingly, a series of these amino acids was incorporated into the LHRH structure.

It was previously shown that LHRH analogs containing D-basic amino acids in position 6 could be more potent than the native hormone ([D-Arg^6]LHRH, 4x LHRH; [D-Arg^6,Pro^9-NHEt]-LHRH, 17x LHRH; *in vitro*).[5] The incorporation of D-hArg(Et_2) into these structures leads to analogs which are substantially more potent *in vivo* (Table I), with **5** being in the range[2] of the most potent LHRH analogs reported. In contrast to our observations with the highly hydrophobic LHRH agonists, in which the addition of the hydrophobic NHEt modification reduces potency, for these more hydrophilic compounds it results in a substantial potency increase. These results are consistent with the early studies with relatively hydrophilic LHRH agonists.[5] When the NHEt was replaced by the more hydrophobic (and longer) NHPr (**6**), the potency was halved, probably due to conformational factors.

More hydrophobic amino acid substitutions gave reduced potency. Unsymmetrical alkyl substitutions (**4**,**8**,**9**), designed to allow continued electrostatic interaction despite the increased chain length and steric hindrance, were less potent, either with or without the NHEt modification. The exceptionally hydrophobic analogs **11** and **12** also had reduced potency. As seen previously with very hydrophobic analogs,[5] the decrease is not great and they remain very potent compounds.

Table I. Biological Activities of Analogs

#	Compound		HPLC;[a] k'	Potency[b]
1	LHRH	(Standard)		1
2	[D-hArg(Et$_2$)^6]LHRH		0.51	40
3	[D-hArg(CH$_2$CH$_2$)^6,Pro^9-NHEt]LHRH		0.55	130
4	[D-hArg(Me,Bu)^6]LHRH		0.74	40
5	[D-hArg(Et$_2$)^6,Pro^9-NHEt]LHRH		0.82	150
6	[D-hArg(Et$_2$)^6,Pro^9-NHPr]LHRH		1.05	70
7	[D-Trp^6,Pro^9-NHEt]LHRH	(Standard)	1.36	100
8	[D-hArg(Me,Bu)^6,Pro^9-NHEt]LHRH		1.37	80
9	[D-hArg(Me,hexyl)^6]LHRH		1.67	70
10	[D-Nal(2)^6]LHRH	(Standard)	1.80	200
11	[D-hArg(hexyl$_2$)^6]LHRH		15.3	50
12	[D-hArg(hexyl$_2$)^6,Pro^9-NHEt]LHRH		∿18	80

[a]k' = (ret. vol.-void vol.)/void vol.; 40% CH_3CN, 0.03 M in NH_4OAc, pH 7; Altex 5μ, C_{18}. [b]Measured by 2x daily sc injection (saline-0.1% BSA) in 2-week rat estrus suppression assay.[6]

It was previously suggested that LHRH agonistic potency was directly related to hydrophobicity[7] and that the most potent analogs would be more hydrophobic than 7. While the most potent analogs in the hydrophobic series (e.g. [D-Nal(2)^6]LHRH) were in accord with this suggestion, the most potent analogs in the present series are substantially more hydrophilic than 7. The hydrophobic analogs may achieve the necessary long $t_{1/2}$ and high receptor affinity for high *in vivo* potency by purely hydrophobic mechanisms of interaction with membranes and carrier proteins. This new class of analogs may achieve similar results through membrane affinity, caused by combined electrostatic (hydrophilic) and hydrophobic interactions, and they therefore have a very different and lower optimum hydrophobicity.

Conclusions

The incorporation of D-hArg(R_2) in position 6 of LHRH results in a series of extremely active LHRH agonists (to 150x LHRH potency). The optimum hydrophobicity for this series lies in a range substantially lower than that for the earlier, hydrophobic LHRH analogs (e.g. [D-Nal(2)^6]LHRH). The D-hArg(R_2) substitution may provide increased receptor affinity and tissue depoting through a stabilized interaction with cell membranes by a mechanism different from that operating in the hydrophobic LHRH series.

References

1. Nestor, J.J. Jr. (1984) LHRH and Its Analogs: Contraceptive and Therapeutic Applications, Vickery, B.H., Nestor, J.J. Jr., and Hafez, E.S.E., Eds., MTP Press Ltd., Lancaster pp. 3-10.
2. Nestor, J.J. Jr., Ho, T.L., Tahilramani, R., Horner, B.L., Simpson, R.A., Jones, G.H., McRae, G.I., and Vickery, B.H., ibid., pp. 23-33.
3. Chan, R.L. and Chapman, M.D. (1985) Biochem. Biophys. Res. Commun. 127, 673-679.
4. Schwyzer, R., Gremlich, H.-U., Gysin, B., and Sargent, D.F. (1983) Peptides: Structure and Function, Hruby, V. and Rich, D.H., Eds., Pierce Chemical Company, Rockford, IL pp. 657-664.
5. Rivier, J., Ling, N., Monahan, M., Rivier, C., Brown, M., and Vale, W. (1975) Peptides: Chemistry, Structure, and Biology, Walter, R. and Meienhofer, J., Eds., Ann Arbor Science Publishers, Ann Arbor, MI pp. 863-870.
6. Nestor, J.J. Jr., Ho, T.L., Simpson, R.A., Horner, B.L., Jones, G.H., McRae, G.I., and Vickery, B.H. (1982) J. Med. Chem. 25, 795-801.
7. Nadasdi, L. and Medzihradszky, K. (1981) Biochem. Biophys. Res. Commun. 99, 451-457.

SUBSTITUTION OF ARG^5 FOR TYR^5 IN GNRH ANTAGONISTS

Roger W. Roeske and Nishith C. Chaturvedi
Department of Biochemistry, Indiana University School of Medicine, Indianapolis, Indiana 46223

Jean Rivier, Wylie Vale, John Porter and Marilyn Perrin
The Salk Institute, P.O. Box 85800, San Diego, California 92138

Introduction

Analogs of GnRH, both agonists and antagonists, are interesting because of their potential role in contraception and in the treatment of prostate cancer and other gonadotropin-related disorders.

As a result of conformational energy minimization calculations[1,2] it was proposed that both agonists and antagonists of GnRH have a β-turn at residues 5, 6 and 7. This was supported strongly by the finding[3] that an analog having a γ-lactam as a conformational constraint at residues 6 and 7 had good agonist activity. In an antagonist, such as Acetyl-D-xxx^1,D-4-Cl-Phe^2, D-Trp^3,D-xxx^6,D-Ala^{10}-GnRH, a variety of residues, e.g., D-Trp^6 D-Leu^6 or D-Arg^6 is compatible with high antagonist activity.

Hypothesis

It seemed to us likely that when the antagonist binds to the receptor, the side chain of residue 6 is not in contact with a specific group, but interacts with components of the cell membrane, either the negatively-charged head groups of the lipids or the hydrocarbon chains. This would explain the

TABLE I. ANTIOVULATORY ACTIVITY OF GnRH ANALOGS

GnRH: pGlu-His-Trp-Ser-Tyr-Gly-Leu-Arg-Pro-Gly-NH_2

	Analog of GnRH	Dose (ug/rat,sc)	No. rats ovulated [a]	K_d [b]
I	Ac-D-Nal (2)^1,D-α-Me-4-ClPhe^2,D-Trp^3,D-Arg^6,D-Ala^{10}	0.5	4	8.0
II	Ac-D-Nal(2)^1,D-α-Me-4-ClPhe^2,D-Trp^3,Arg^5,D-Tyr^6,D-Ala^{10}	1.0	6	
		2.5	0	2.8
III	Ac-D-Nal(2)^1,D-α-Me-4-ClPhe^2,D-Trp^3,Arg^5,D-Arg^6,D-Ala^{10}	1.0	9	
		2.5	1	4.0
IV	Ac-D-Nal(2)^1,D-α-Me-4-ClPhe^2,D-Trp^3,Arg^5,D-Nal(2)^6,D-Ala^{10}	2.5	5	17
V	Ac-D-Nal(2)^1,D-4-ClPhe^2,D-Trp^3,D-Arg^5,D-Nal(2)^6,D-Ala^{10}	10.0	10	220
VI	Ac-D-Nal(2)^1,D-4-ClPhe^2,D-Trp^3,Lys^5,D-Nal(2)^6,D-Ala^{10}	5.0	5	
		10.0	0	6.5
VII	Ac-D-Nal(2)^1,D-4-ClPhe^2,D-Trp^3,His^5,D-Nal(2)^6,D-Ala^{10}	5.0	2	
		10.0	0	6.1
VIII	Ac-D-Nal(2)^1,D-4-ClPhe^2,D-Trp^3,Glu^5,D-Arg^6,D-Ala^{10}	10.0	9	2.7
IX	Ac-D-Nal(2)^1,D-4-FPhe^2,D-Nal^{3,6},Arg^5	7.5	10	71
X	Ac-Δ^{3}Pro^1,D-4-FPhe^2,D-Nal^{3,6},Arg^5	2.5	6	1.7
XI	Ac-D-Nal(2)^1,D-4-ClPhe^2,D-Trp^3,Arg^5,D-Tyr^6,Arg^7,Leu^8,D-Ala^{10}	5.0	10	-
XII	Ac-D-Nal(2)^1,D-4-ClPhe^2,D-Trp^3,D-Leu^6,Arg^7,D-Ala^{10}	2.5	8	-

a ten rats in each test

b dissociation constant, nanomolar, in vitro binding assay

Nal(2) is 2-naphthylalanine

high activity of antagonists having either a hydrophobic residue or a positively-charged residue in position 6 and the low activity of antagonists having D-Glu in position 6.

In order to determine whether the side chain of residue 5 was in a similar environment, we synthesized a series of antagonists having a positively-charged residue in place of Tyr^5. Although many substitutions for Tyr^5 had been made previously, none had a side chain that would be ionized at pH 7.

Synthesis and Biological Assay

All of the analogs were synthesized by the solid phase method on an MBHA resin support, 0.35 meq N/g, using a Beckman 990 automated peptide synthesizer. Coupling reactions were carried out for 2 hours with a 2.5-fold excess of Boc amino acids, HOBt and DCC in DMF: CH_2Cl_2, 1:1. The deblocking mixture was 50% TFA in $CHCl_3$ and 5% thioanisole. After removal of the peptides from the resin by HF at 0° for 45 min, they were extracted with 50% aq HOAc and lyophilized. The crude peptides were purified by the procedure of Gesellchen et al.[4] on C_{18} silica gel. The analogs were tested for antiovulatory activity as described by J. Yardley et al.[5]

Discussion

Analog II, which differs from the reference compound I only in transposition of residues 5 and 6, is about half as active as I. This combination of Arg^5 D-Tyr^6 is slightly more effective than Arg^5,D-Arg^6, which is more active than Arg^5,D-Nal^6 (III and IV). A D-Arg residue is clearly detrimental to antagonist activity, as expected. The combination of Lys^5, D-Nal^6 or His^5,D-Nal^6 is somewhat less active than Arg^5, D-Nal^6. A Glu^5 substitution reduces activity markedly (VIII).

Transposition of residues 6 and 7 (XII) is much less favorable than the 5,6 exchange, possibly because this results in two adjacent Arg residues, which might be especially susceptible to proteases. Two exchanges in the same molecule $-Arg^5$, $D-Tyr^6$, Arg^7,Leu^8 - are definitely unfavorable for activity.

In general the results support the hypothesis that the side chains of residues 5 and 6 are not interacting with specific groups in the receptor and are interchangeable.

A distinct advantage of the Arg^5,$D-Tyr^6$ combination is its low activity compared to Tyr^5,$D-Arg^6$ in releasing histamine from rat mast cells[6], thus lowering the danger of side effects when these compounds are administered in vivo.

Acknowledgements

This work was supported by the NICHD and by the Showalter Trust of I.U. Foundation. We thank R. Naqvi and M. Lindberg for the rat antiovulatory assays, which were done at the Mason Institute, under contract with the Contraceptive Development Branch of the NICHD.

References:

1. Momany, F. (1976) J. Am. Chem. Soc. 98, 2990-2996.
2. Momany, F. (1978) J. Med. Chem. 21, 63-68.
3. Freidinger, R., Veber, D., Perlow, D., Brooks, J. and Saperstein, R. (1980) Science 210, 656-658.
4. Gesellchen, P., Tafur, S., and Shields, J. (1979) in Peptides: Structure and Biological Function, Gross, E., and Meienhofer, J., Eds., Pierce Chemical Co., Rockford, Il, pp. 621-624.
5. Yardley, J., Foell, J., Beattie,C., and Grant, N. (1975) J. Med. Chem. 18, 1244-1247.
6. Hook, W., Karten, M., and Siraganian, R. (1985) Abstracts of the 69th Annual meeting of FASEB, Fed. Proc. 44, Abstract No. 5336.

MICROBIAL FERMENTATIONS AS SOURCE OF NON-PEPTIDIC PEPTIDE RECEPTOR LIGANDS

Georg Albers-Schönberg, Ray S. L. Chang, Victor J. Lotti, Tsing-bau Chen, Richard L. Monaghan, Jerome Birnbaum, Edward O. Stapley, Michael A. Goetz, Maria Lopez, Arthur A. Patchett, Jerrold M. Liesch, Otto D. Hensens and James P. Springer

Merck Sharp & Dohme Research Laboratories, Rahway, N. J. 07065

Peptides are generally not ideal drugs because they are not well absorbed and rapidly metabolized. Extensive modifications, guided by conformational analyses, of some peptides have been done[1)], but no one has achieved the equivalent of modeling morphine on the basis of the enkephalin structures. As an alternative, the screening for natural products can lead to the discovery of completely novel and unpredictable leads. To give you some idea of what such a program looks like and how one can use microbial fermentations as a source of non-peptidic peptide receptor ligands, I want to take you briefly through the history of our natural products program at MSDRL. Its sustained success has not come easily, and I will point out some of the problems, especially in the difficult area of receptor assays. Nevertheless, our recent discovery of a potent, non-peptidic cholecystokinin antagonist[2)] will illustrate that even in this area the systematic screening of fermentation broths can generate a lead.

As other such programs, ours started a long time ago in the antibiotics area but became successful only after mode of action based assays had been introduced. The desired mode of action was inhibition of the biosynthesis of the bacterial cell wall. The discoveries that resulted were phosphomycin[3)]. the cephamycins[4)], and the first carbapenem, thienamycin[5)]. Because of these successful searches—all three have led to marketed products—we have since conducted most of our screening with carefully selected mode of action based assays. The original rational for the mode of action strategy was that one can select specific biochemical targets which promise efficacy and minimize toxicity. However, such assays can detect also compounds of low potency that nevertheless act specifically and can be improved upon by chemical modification.

One exception to the mode of action strategy, which is of interest here, was the search for a new antiparasitic, anthelmintic compound for which we used an in vivo assay and which led to the discovery of the avermectin macrolides[6]. The exact mode of action of these extremely potent compounds is still not known, but they appear to act by potentiating post-synaptic conductance through the GABA mediated chloride channel[7]. In the vertebrate CNS this is a very complex receptor that contains several other binding sites including that for the benzodiazepins which affect the GABA-site and for which just recently an endogenous peptide ligand has been reported[8]. It will be very interesting to see this picture develop and to learn whether or not the avermectins actually are ligands to a peptide neurotransmittor or neuromodulator receptor.

Antibacterial cell-wall biosynthesis inhibition covers a great number of enzymatic steps and is basically still a classical antibacterial assay. The next step in the development of the mode-of-action strategy was then to screen with just one purified enzyme. This has been done successfully by Hamao Umezawa[9] in Japan who searched for inhibitors of bacterial β-lactamases. In our laboratories, A. Alberts and A. A. Patchett and their associates decided to apply fermentation broth screening to the search for inhibitors of cholesterol biosynthesis in man. Specifically, they chose as target the reduction of hydroxymethylglutaric acid to mevalonic acid by HMGCoA reductase which in a short time led to the discovery of mevinolin in a fungal culture[10]. The same compound and its desmethyl derivative compactin have also been discovered in similar screening by A. Endo[11] at Tokyo Noko University.

The search for enzyme inhibitors was also our first step in the search for natural products from fermentation broths which was not directed against infectious diseases but against malfunctions in our own biochemistry. The logical next step in this direction as well as in the further development of mechanism-of-action based assays would then be to use also receptor binding, as differentiated from enzyme inhibition, assays. This may seem a small step, but it is not, for two reasons. First: Enzymes are comparatively quite well understood today but our understanding of hormone receptors is only in its beginning. Second: The discovery of a cholesterol biosynthesis inhibitor in a microorganisms does not necessarily surprise. Triterpene derived products are not uniquely mammalian but ubiquitous in nature, and even fungi may need means for controlling their levels. But are there reasons to anticipate that microorganisms manufacture ligands to mammalian receptors? Some information relevant to this question is emerging. J. Roth[12)] in a recent paper summarizes the observations - by radioimmuno assay, chromatographic behaviour and carefully chosen bioassays - of microbial products similar to insulin, somatostatin, ACTH and endorphins, and others including cholecystokinin. Roth hypothesizes that these peptides are microbial messenger molecules which structurally and somehow functionally have been preserved throughout evolution. Other microorganisms may produce metabolites, peptidic or non-peptidic, to simulate or interfere with the function of these peptides. A somewhat more pragmatic view is that secondary microbial and plant metabolites originate, for whatever purposes, in enzymatic reactions from the fundamental, chiral biochemical starting materials. This makes them an attractive collection of structures to test for biological activity. Sometimes we will find a useful medicine, sometimes only a laboratory tool. More important are such questions as, whether one can find a small molecule agonist or antagonist to a substantially larger peptide hormone. This may simply depend on the size of the binding site. If most of the peptide sequence is needed, and not only for conformational reasons, then this hormone receptor may not be a good screening target. Nevertheless, the discovery of mevinolin created much enthusiasm among our pharmacologists for natural products screening and receptor assays are now used whenever an appropriate target has been identified. R. Chang and V. Lotti of our Pharmacology Department initiated the cholecystokinin project and developed the assay[13)] that yielded the antagonist which I will describe to you.

Not unexpectedly, receptor assays for the screening of fermentation broths present considerable problems. The uncertainties in interpreting radioligand/receptor binding data are well known[14)] and aggravated if the ligand is in as crude a state as in fermentation broth. These broths are cultures in very complex media of newly isolated, virtually unknown organisms whose growth and metabolic patterns vary widely from culture to culture. Many produce detergent-like compounds, lipids and fatty acids that interfere non-specifically with many receptors or the nearby areas of the membrane. The metal ion requirements for receptor binding and microbial growth may be incompatible. Degradation of the radioligand by microbial peptidases can be a problem. Assays using membrane preparations generally work better than whole-cell assays which are very sensitive to cytotoxic broth components. Although we have no direct evidence for its occurrence, receptor internalization is a potential problem in whole cell receptor assays. The receptors themselves also vary. Some are very sensitive to the just mentioned non-specific interferences, and assays using them seem never to detect an even remotely interesting compound. Others seem very insensitive against interferences but also, week after week, turn in negative screening data. Others, however, are not unduly sensitive to interferences and discover typical secondary metabolites that bind competitively and in a dose dependent manner to the receptor.

We generally assume that these secondary metabolites are present in broth in 10^{-4} to 10^{-6}M concentrations. Compounds that are only detectable in these concentrations are not very potent if compared to peptide hormones or neurotransmitters but may be valid leads for medicinal-chemical improvement provided they show selectivity for the receptor. Unfortunately, in this potency range, competitive and dose dependent binding does not yet seem to imply specificity. In order to include at least some specificity criterium at the level of the primary screen, it is useful to test the broths side-by-side against two receptors—although not necessarily two subtypes of the same receptor because these may behave very differently with respect to interferences. We have also seen cases in which the activities in the two assays of such a tandem could be chromatographically separated and we have seen at least one case in which non-specific binding stimulation masked a specific binding inhibition by another broth component. Such situations can be very cumbersome if, for instance, fermentation yields have to be improved for the isolation and identification of a minor constituent.

Secondary, functional assays are an absolute requirement. They again help to determine specificity and they differentiate between agonists and antagonists. They may be biochemical reactions that are coupled to the receptor, isolated organ preparations, or even in vivo experiments.

Meaningful biological characterization, however, requires a pure chemical entity. Unfortunately, after successful isolation, it all too often turns out that one is dealing with one of the many already known compounds on which generations of natural products chemists have earned their degrees or which have been discovered because of some other biological activity. We cannot avoid these disappointments but we can make the best of them. The fullest use of modern isolation and identification techniques helps us to move fast, to build our files, understand our assays, and minimize duplication. Also, the experienced medicinal chemist and pharmacologist can often guess what the merits of a structure are and guide the investment of limited resources to only the most promising leads.

In sum, there are no rules for success which go beyond a well tuned collaboration between biologists, microbiologists and chemists, and one's Company's wholehearted committment.

Turning then to the practical example, cholecystokinin is a peptide hormone of the upper intestinal tract which regulates gall bladder contraction, gut motility and the secretion of digestive enzymes, such as amylase, in the pancreas[15)]. It belongs to the growing group of intestinal peptides which have also been observed in the central nervous system where, however, it's function and detailed distribution are still unknown, although high-affinity binding sites have been demonstrated[16)]. Cholecystokinin is a 33 amino acid peptide amide. The C-terminal octa- and tetrapeptides also occur naturally. The former retains all of the activities of the full peptide while the latter has other activities[17)]. A sulfated tyrosine in position 27 appears to be necessary for full biological activity[17)]. The C-terminal pentapeptide amide is identical to the C-terminal of gastrin, the hormone which regulates gastric acid secretion.

Cbz-CCK-(27-32)NH_2 is a potent CCK antagonists[18]. Known non-peptidic CCK antagonists are proglumide, benzotript and 2,2'-N,O-dibutyryl-cGMP[19].

The pancreatic assay was established as previously described by Chang and Lotti[13]. Specific binding was defined as the difference between total and non-specific binding of 60-80 pM ^{125}I-CCK-33 in the absence and presence of 1 μM CCK-8. Active broths were tested for specificity for the pancreatic receptor in an analogous assay using brain tissue membranes and in a gastrin assay as described by Praissman[20]. Of the tested broths, one, which subsequently was identified as an Aspergillus alliaceus culture[21] (ATCC 20655/6), appeared interesting in terms of potency and specificity. It showed half-maximal inhibition of ^{125}I-CCK-33 binding to the pancreatic receptor at about 1% whole broth equivalent in assay solution. Benzotript and proglumide have IC_{50} values of 100 and 600 μM corresponding to about 30 and 180 μg/ml. If the activity of the A. alliaceus fermentation had been due to a compound of similar MW and affinity, it would have had to be present in the 10^{-2}M concentration range, 100-fold higher than the upper limit of the 10^{-4} to 10^{-6}M range which we mentioned before as typical for secondary metabolites in broths. In addition the broth showed no comparable displacement of ^{125}I-CCK-33 and ^{125}I-gastrin from their respective brain and gastric gland receptors. Proglumide and benzotript have roughly equal affinities to all three receptors and dibutyryl-c-GMP only about a 20-fold higher affinity for the pancreatic than for the brain CCK and Gastrin receptor. As pointed out earlier, cholecystokinin-like immunoreactivity has been observed in microbial cells[12] and many of the active components which we find in fermentations are indeed at least partly peptidic compounds. Thus, there was considerable risk of finding another peptidic antagonist. Nevertheless, this A. alliaceus metabolite clearly merited chemical isolation and identification.

The active compound, which we subsequently named Asperlicin to denote its origin from an Aspergillus species and its antagonistic or "lytic" activity, was purified by solvent extraction and chromatography on silica gel, LH-20 resin and ODS reverse phase[21]. Broth concentrations were determined to be $5 \cdot 10^5$M, in the expected range. The structure was determined by mass spectrometry, NMR and X-ray analysis[22]. FAB and high-resolution MS provided the elemental composition $C_{31}H_{29}N_5O_4$. Hydrolysis followed by GC/MS analysis of the hydrolysate identified leucine and anthranilic acid in 1:2 molar ratio. Allowing for several equivalents of water to be eliminated in the condensation of these substructures, the remaining atoms add up to a modified tryptophan. ^{13}C and ^{1}H NMR spectra confirmed these conclusions and added a tertiary hydroxy group and a third 1,2-disubstituted benzene ring (part of the modified tryptophan) to the list of substructures. The mass spectral fragmentation pathways were then established by B^2/E and B/E linked-scan metastable peak analysis and readily interpreted in terms of structure 1. X-ray crystallographic analysis confirmed the structure and added the absolute stereochemical information. Asperlicin thus bears some structural resemblance to the previously reported[23] fungal metabolite tryptoquivaline. This substance, however, up to 100 μM does not measurably bind to the CCK receptor, nor does Asperlicin show any of the tremorgenic properties of tryptoquivaline.

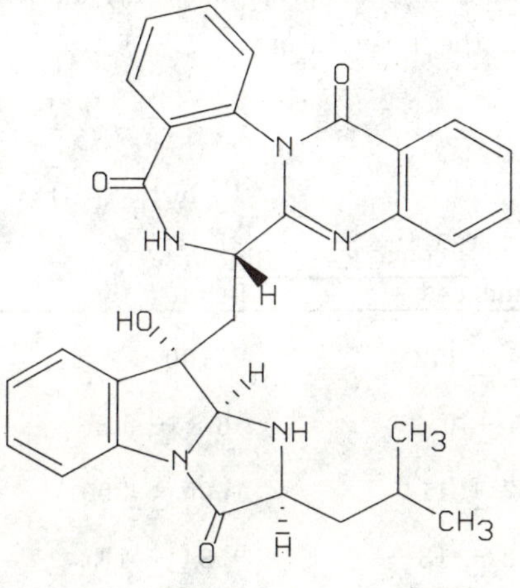

FIGURE I. ASPERLICIN

Finally, the selectivity and relatively high potency, compared to the known non-peptide antagonists, of pure asperlicin for the peripheral CCK receptor was demonstrated in in vitro binding (Table I.) as well as functional in vitro and in vivo assays. Scatchard analysis of ^{125}I-CCK-33 binding to the peripheral receptor in the presence and absence of 1.3 μM asperlicin suggests competitive binding of asperlicin with K_i 0.9 ± 0.2 μM. Asperlicin effectively antagonized CCK-8 induced contractions of the isolated guinea pig ileum and gall bladder (Schild plot analysis) in good agreement with the in vitro K_i value, but even at much higher concentrations (13 μM) did not antagonize contractions induced by acetylcholine, histamine, Substance P, pentagastrin or electrical stimulation. In the anesthetized guinea pig, 12 mg/kg i.v. asperlicin, given 1 to 4 hours in advance, effectively antagonized CCK induced gall bladder contractions while 50 mg/kg i.v. proglumide had no effect.

To summarize then, our experiences in discovering through systematic natural product screening not only novel antibiotics and potent enzyme inhibitors but now also selective receptor ligands, are encouraging. Asperlicin is now the subject of a medicinal chemistry program, and the results of this project will be published in the near future.

	IC_{50} (μM)		
	^{125}I-CCK-33 Binding		^{125}I-Gastrin
	Pancreas	Brain	Gastric Gland
Asperlicin	1.4 ± 0.2	> 100	> 100
CBZ-CCK (27-32) NH_2	3.5 ± 0.4	6.8 ± 0.8	1.0 ± 0.2
Dibutyryl-c-GMP	87 ± 11	1600 ± 300	1200 ± 400
Benzotript	102 ± 18	84 ± 13	59 ± 24
Proglumide	600 ± 58	875 ± 125	900 ± 200

Table I. Effect of CCK antagonists on ^{125}I-CCK-33 binding to pancreatic and brain membranes and of ^{125}I-gastrin binding to gastric gland membranes.

References

1. D. F. Veber, R. M. Freidinger, D. S. Perlow, W. J. Paleveda Jr., F. W. Holly, R. G. Strachan, R. F. Nutt, B. H. Arison, C. Homnick, W. C. Randall, M. S. Glitzer, R. Saperstein and R. Hirschmann, Nature 292, 55-58 (1981); D. G. Hangauer, P. Gund, J. D. Andose, B. L. Bush, E. M. Fluder, E. F. McIntyre and M. Smith, in Macromolecular Structure and Specificity, Computer Assisted Modeling and Applications. Ann. N.Y. Acad. Sci., 439, 124 (1985).

2. R. S. L. Chang, V. J. Lotti, R. L. Monaghan, J. Birnbaum, E. O. Stapley, M. A. Goetz, G. Albers-Schönberg, A. A. Patchett, J. M. Liesch, O. D. Hensens and J. P. Springer, Science, in press.

3. D. Hendlin, E. O. Stapley, M. Jackson, H. Wallick, A. K. Miller, F. J. Wolf, T. W. Miller, L. Chaiet, F. M. Kahan, E. L. Foltz and H. B. Woodruff, Science 166, 122-123 (1969); B. G. Christensen, W. J. Leanza, T. R. Beattie, A. A. Patchett, B. H. Arison, R. E. Ormond, F. A. Kuehl, Jr., G. Albers-Schönberg and O. Jardetzky, ibid, 123-124.

4. E. O. Stapley, M. Jackson, S. Hernandez, S. B. Zimmerman, S. A. Currie, S. Mochales, J. M. Mata, H. B. Woodruff and D. Hendlin, Antimicrob. Ag. Chemotherap. 2, 122-131 (1972); T. W. Miller, R. T. Goegelman, R. G. Weston, I. Putter and F. J. Wolf, ibid. 132-135; G. Albers-Schönberg, B. H. Arison and J. L. Smith, Tetrah. Lett. 1972, 2911-2914; R. Nagarajan, L. D. Boeck, M. Gorman, R. L. Hamill, C. E. Higgins, M. M. Hoehn, M. W. Stark and J. G. Whitney, J. Am. Chem. Soc. 93 2308-2310 (1971).

5. J. S. Kahan, F. M. Kahan, R. Goegelman, S. A. Currie, M. Jackson, E. O. Stapley, T. W. Miller, A. K. Miller, D. Hendlin, S. Mochales, S. Hernandez, H. B. Woodruff and J. Birnbaum, J. Antibiot. 32, 1-12 (1979); G. Albers-Schönberg, B. H. Arison, O. D. Hensens, J. Hirshfield, K. Hoogsteen, E. A. Kaczka, R. E. Rhodes, J. S. Kahan, F. M. Kahan, R. W. Ratcliffe, E. Walton, L. J. Ruswinkle, R. B. Morin and Burton G. Christensen, J. Am. Chem. Soc. 100, 6491-6499 (1978).

6. R. W. Burg, B. M. Miller, E. E. Baker, J. Birnbaum, S. A. Currie, R. Hartman, Y.-L. Kong, R. L. Monaghan, G. Olson, I. Putter, J. B. Tunac, H. Wallick, E. O. Stapley, R. Oiwa and S. Omura, Antimicrob. Ag. Chemother. 15, 361-367 (1979); T. W. Miller, L. Chaiet, D. J. Cole, L. J. Cole, J. E. Flor, R. T. Goegelman, V. P. Gullo, H. Joshua, A. J. Kempf, W. R. Krellwitz, R. L. Monaghan, R. E. Ormond, K. E. Wilson, G. Albers-Schönberg and I. Putter, ibid. 368-371; J. R. Egerton, D. A. Ostlind, L. S. Blair, C. H. Eary, D. Suhayda, S. Cifelli, R. F. Riek and W. C. Campbell, ibid. 372-378; G. Albers-Schönberg, B. H. Arison, J. C. Chabala, A. W. Douglas, P. Eskola, M. H. Fisher, A. Lusi, H. Mrozik, J. L. Smith and R. L. Tolman, J. Am. Chem. Soc. 103, 4221-4224 (1981).

7. S. S. Pong, R. Dehaven, C. C. Wang, Biochim. Biophys. Acta. 646, 143-150 (1981) S. S. Pong, C. C. Wang, J. Neurochem. 38, 375-379 (1982).

8. P. Ferrero, A. Guidotti, B. Conti-Tronconi, E. Costa, Neuropharmacol. 23, 1359-1362 (1984).

9. H. Umezawa, Enzyme Inhibitors of Microbial Origin, University of Tokyo Press, (1972).

10. A. W. Alberts, J. Chen, G. Kuron, V. Hunt, J. Huff, C. Hoffman, J. Rothrock, M. Lopez, H. Joshua, E. Harris, A. Patchett, R. Monaghan, S. Currie, E. Stapley, G. Albers-Schönberg, O. Hensens, J. Hirshfield, K. Hoogsteen, J. Liesch and J. Springer, Proc. Nat. Acad. Sci. 77, 3957-3961 (1980).

11. A. Endo, FEBS Letters 72, 323 (1976) and A. Endo, M. Kuroda and K. Tanzawa, J. Antib. 32 852 (1979).

12. D. LeRoith, J. Shiloach, M. Berelowitz, L. A. Frohman, A. S. Liotta, D. T. Krieger and Jesse Roth, Federation Proceedings 42, 2602-2607 (1983).

13. R. S. L. Chang, V. J. Lotti and T. B. Chen, Biochemical Pharmacol. 33, 2334-2335 (1984).

14. I. M. Klotz, TIPS 4, 253-255, (1983) and I. M. Klotz, Science 217, 1247-1249 (1982); E. Burgisser, TIPS 5, 142-144 (1984).

15. V. Mutt, in Gastrointestinal Hormones, G. B. J. Glass ed., Raven Press, New York, 1980, pp. 169-221.

16. R. B. Innis, S. H. Snyder, Proc. Nat. Acad. Sci. 77, 6917-6921 (1980); A. Saito, H. Sankaran, I. D. Goldfine, J. A. Williams, Science, 208, 1155-1156 (1980).

17. J. F. Rehfeld, L. I. Larsson, N. R. Goltermann, T. W. Schwartz, J. J. Holst, S. L. Jensen, J. S. Morley, Nature 284, 33-38 (1980).

18. M. Spanarkel, J. Martinez, C. Briet, R. T. Jensen and J. D. Gardner, J. Bio. Chem. 258, 6746-6749 (1983).

19. J. D. Gardner, R. T. Jensen, Am. J. Physiol. 246, G471-G476 (1984).

20. M. Praissman, M. E. Walden, C. Pellecchia, J. Receptor Res. 3, 647-665 (1983).

21. Manuscript in preparation.

22. Manuscript in preparation.

23. J. Clardy, J. P. Springer, G. Büchi, K. Matsuo and R. Wightman, J. Am. Chem. Soc. 97, 663 (1975); J. P. Springer, Tetrahedron Lett. 1978, 339-342.

INVOLVEMENT OF GLY^{29} IN CONFORMATION, STRUCTURE-ACTIVITY RELATIONSHIPS AND ENZYMATIC DEGRADATION OF CCK_{26-33}.

Christiane Durieux, Bruno Charpentier, Isabelle Menant, Didier Pelaprat, Gilles Gacel and Bernard P. Roques.

Département de chimie organique (UA498 CNRS, U266 INSERM) UER des Sciences Pharmaceutiques et Biologiques, 4 avenue de l'Observatoire 75006 Paris France.

Introduction

CCK8 (CCK_{26-33} fragment of cholecystokinin) has been shown to exist in various areas of the central nervous system (1-3). Both biological studies and multiplicity of pharmacological effects involving CCK_8 (4,1) suggest a neurotransmitter role for the peptide and a possible heterogeneity of CCK receptors.

In a previous structure-activity relationships study, it has been shown that as well in CCK_{26-33} related peptides (1) as in Boc [$diNle^{28,31}$] CCK_{27-33} analogs (3,4) (5,6) replacement of Gly^{29} by several residues was able to modify or abolish the different peripheral effects of native CCK_8 without dramatically altering the binding parameters at central level (Table I). Moreover, this crucial role of Gly^{29} was corroborated by the NMR conformational study performed on [Nle^{28}, Aib^{29}, Nle^{31}] CCK_{27-33} in which the obtained conformation is largely different from that of CCK_8 (5,7). In addition , preliminary enzymatic degradation studies have demonstrated the implication of Gly^{29} as a preponderent degradation pathway in CCK_8 metabolism (8). So in order to investigate in details the importance of Gly^{29} in the biological relevance of CCK_8, we report in the present paper : i) the degradation of [^3H] Boc[$diNle^{28,31}$] CCK_{27-33} by rat brain slices ii) the synthesis and biological activities of cyclic and linear CCK_8 analogs.

Results and discussion

Enzymatic degradation : previous studies on the enzymatic degradation of CCK_8 related peptides were performed on synaptic plasma membranes from pig brain cortex, allowing us to demonstrate the involvement of the neutral endopeptidase "enkephalinase" and of thiol proteases. To establish the putative physiological implication of these enzymes, the degradation of CCK was studied on slices of rat brain. These preparations were used to rule out the problems of contamination by cytoplasmic or lysosomal enzymes. In this study, a highly potent tritiated analog of CCK_8:[^{3}H] Boc [$Nle^{28,31}$] CCK_{27-33} was used : Boc Tyr(SO_3H)[^{3}H] Nle-Gly-Trp-[^{3}H] Nle- Asp-pheNH_2.

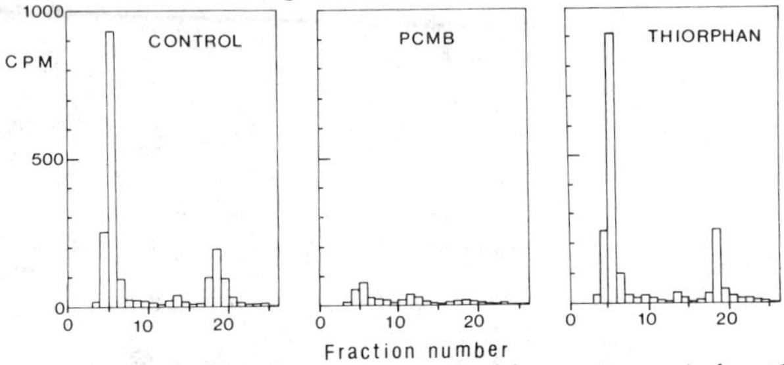

Figure 1 : HPLC characterization of metabolites formed in the degradation of [^{3}H] Boc [$Nle^{28,31}$] CCK_{27-33} (10nM) by rat brain cortical slices (1.5mg protein/ml).

The metabolites formed were isolated and characterized by HPLC by comparison with synthetic fragments. The degradation on striatal slices, containing high level of enkephalinase was compared with brain cortex slices, exhibiting a high concentration of CCK_8.

Figure 1 shows the elution pattern obtained after 15 min of incubation at 37°C in the presence or in the absence (control) of inhibitors. The major peak, eluted in fraction 17-19 corresponded to [Nle] CCK_5 and the weak peak, (fractions 14-15) to [Nle] CCK_4 in cortex as in striatal slices (not shown here). The fragments eluted in the void volume corresponded to 2 or more cleavages of Boc[Nle] CCK_{27-33}. The primary pathway for the degradation of Boc[diNle] CCK_{27-33} seems to involve the cleavage of Nle-Gly bond leading to [Nle] CCK_5, this cleavage was abolished by PCMB, a thiol protease inhibitor according to the results obtained with synaptic plasma membranes (8). Thiorphan, a potent inhibitor of enkephalinase, appeared unable to prevent the degradation process. Enkephalinase activity occurred only when PCMB-sensitive enzymes were inhibited.

Synthesis and Biological activities :firstly, in order to reinforce the folding of the N-terminal part of CCK_8 evidenced by NMR conformational study (7), a more constrained peptide (5) was synthetized by replacement of Gly^{29} by Lys^{29} and cyclization by amide bond formation between side-chains of this residue and of Asp^{26} (scheme I).

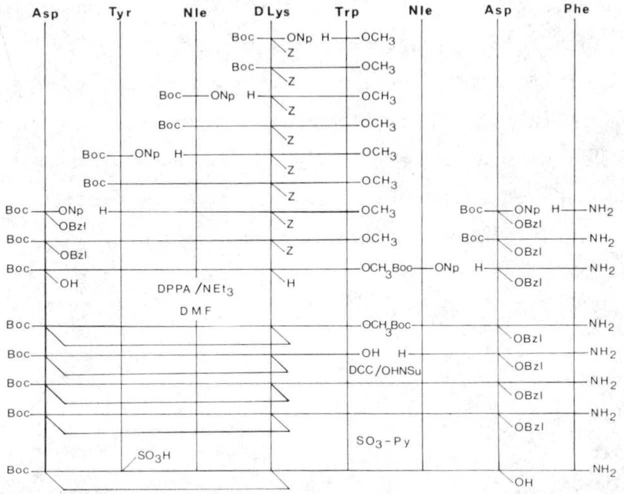

Scheme I : Synthesis of Boc-Asp-Tyr(SO_3H)-Nle-D-Lys-Trp-Nle-Asp-phe-NH_2.

The biological properties of this compound are reported on Table I

COMPOUNDS									Agonist potency - EC_{50} (M) Amylase release	GPI	Gall bladder	Binding - KI (M) mouse brain tissue	Antagonist potency - IC_{50} (M) GPI
CCK_8	Asp	Tyr(SO_3H)	Met	Gly	Trp	Met	Asp	Phe NH_2	1.0 10^{-9}	2.0 10^{-9}	6.0 10^{-9}	5.5 10^{-10}	
1	Asp			DAla					0.9 10^{-9}	3.0 10^{-7}	1.8 10^{-7}	1.5 10^{-7}	
2	Boc		Nle	Gly		Nle			1.0 10^{-9}	2.8 10^{-9}	3.2 10^{-9}	1.6 10^{-9}	
3				Aib					> 10^{-5}	> 10^{-5}	> 10^{-5}	4.6 10^{-9}	NT
4				LAla					> 10^{-5}	> 10^{-5}	> 10^{-5}	7.0 10^{-9}	NT
5	Boc Asp			DLys					1.0 10^{-7}	> 10^{-5}	> 10^{-5}	1.2 10^{-7}	∽ 10^{-5}
6									1.0 10^{-7}	> 10^{-5}	> 10^{-5}	3.0 10^{-7}	∽ 10^{-5}
7	Z Asp			D Lys(εZ)					> 10^{-6}	> 10^{-5}	> 10^{-5}	2.9 10^{-8}	∽ 10^{-6}
8								NH_2	∽ 10^{-6}	> 10^{-6}	> 10^{-5}	NT	∽ 10^{-5}

Table I : Biological activities of CCK_8 analogs .

The values are the mean five independant determinations. NT = not tested.
Binding experiments with [^3H] Boc [diNle ^{28,31}]CCK_{27-33} : 0.4nM(9).

The apparent discrepancy between the binding properties of 5 on central receptors and the lack of activities on peripheral organs prompted us to test possible antagonist activity of 5 on GPI. Indeed, as presumed, the cyclic analog was able to inhibit the GPI contractions evoked by CCK_8 (3nM) with an $IC_{50} \sim 10^{-5}$ M. This effect could be related to the cyclic structure of the N-terminal part of the peptide, including Tyr (SO_3H) or to the modification of Gly^{29}. Synthesis and biological studies of linear analogs of 5 (6 and 7) nicely evidenced that antagonist property of this series of compounds is due to the replacement of Gly by Lys. Furthermore the hydrophobic aromatic Z group is able to increase both CNS receptor recognition and antagonist effect. In contrast, removal of Phe^{33} (compound 8), a modification well known to induce antagonist properties on peripheral tests (10) has not an additional antagonist effect in this series of compounds. Compound 5 represents the first described cyclic analog of CCK_8. Such type of cyclisation as well as introduction of large residue such as Lys seem to induce antagonist properties which might be related to a decreased degree of freedom in these compounds.

References

1. Innis, R.B., Correa, F.M., Uhl, G.R., Schneider B. and Snyder, S.H. (1979) Proc. Natt. Acad. Sci. USA., 76 : 521-525.
2. Vanderhaeghen J.J., Lostra, F., D. Mey, J. and Gilles, G. (1980) Proc. Natl. Acad. Sci. USA. 77 : 1190-1194.
3. Hokfelt, T., Rehfeld, J.F., Skirboll, L., Ivemark, B., Goldstein, M. and Markey, K. (1980) Nature 285 : 476-478.
4. Morley, J.E. (1982) life Sci. 30 : 479-493.
5. Gacel, G., Durieux, C., Fellion, E., Fournié-Zaluski, M.C., Begué, D., Menant, I., Rossignol, P. and Roques B.P. (1984) In proceeding of the 18 th european peptide symposium, edited by U. Ragnarson, Stockholm : Almqvist and Wiksell international 383-385.
6. Ruiz Gayo, M., Daugé, V., Menant, I., Begué, D., Gacel, G. and Roques, B.P. (1985) Peptides 6 in press.
7. Durieux, C., Belleney, J., Lallemand, J.Y., Roques, B.P. and Fournié-Zaluski, M.C. (1983) Biochem-Biophys. Res. Commun. 114 : 705-712.
8. Mc Dermott, J.R., Dodd, P.R., Edwardson, J.A., Hardy, J. A. and Smith, A.I. (1983) Neurochem. Intern. 5 : 641-647.
9. Pelaprat, D., Zajac, J.M., Gacel, G., Durieux, C., Morgat, J.L., Sasaki, A. and Roques Life Sci. submitted.
10. Spanarkel, M., Martinez, J., Briet, C., Jensen, R. and Gardner, J.D. (1983) J. Biol. Chem. 258 : 6746-6749.

EVIDENCE FOR A CHOLECYSTOKININ-OCTAPEPTIDE RECEPTOR ON GUINEA PIG TRACHEA

S. Bunting and H.H. Holzgrefe

Lipids Research, The Upjohn Company, Kalamazoo, Michigan 49001.

A.E. deVaux, D.J. Staples and T.K. Sawyer

Biopolymer Chemistry, The Upjohn Company, Kalamazoo, Michigan 49001.

Introduction

Cholecystokinin (CCK), originally found in the gut, has also been shown to be present in other tissues including the brain and peripheral autonomic nerves. Immunoreactive material has also been detected in cat lung and nasal mucosa.[3] In many instances, the octapeptide of CCK (CCK-8) has been shown to possess equivalent biological activity to CCK and is found in significant amounts in tissue extracts. Cholecystokinin octapeptide contracts gall bladder and guinea pig ileum.[2] We have shown that CCK-8 also contracts guinea pig isolated trachea and that the structural requirements for activity parallel those described for other systems.

Methods

Male Hartley guinea pigs (160-300g) were killed by cervical dislocation. The tracheae were removed, and individual zig-zag strips were prepared.[1] Tissues were suspended under 2g of tension in a 10ml Bennett organ bath containing Krebs-Henseleit solution maintained at 36.5±0.5°C.

and gassed with 5% CO_2-95% O_2. Force changes were measured isometrically. Cumulative dose-response curves (DCR's) were obtained for CCK C-terminal peptides and histamine. Parallel single-dose control DRC's for these agonists demonstrated that tachyphylaxis did not occur. As leukotriene E_4 (LTE_4) demonstrated a profound tachyphylaxis of long duration, DRC's for this agonist were performed in parallel in six tissues, adding a single dose of LTE_4 to each. At the end of all DRC's, a single supra maximal dose of histamine ($2x10^{-5}$M) was given to each tissue. All data were calculated either as percent of individual maximum or as percent maximal histamine response. The data are presented as a mean$\pm$SEM.

Results and Discussion

Guinea pig trachea in vitro was dose dependently contracted by CCK-8 (EC_{50}, $5.6x10^{-8}$M, n=23) and caerulein (EC_{50}, $1.3x10^{-8}$M, n=7). The dose-response curves of both compounds were parallel and each achieved a similar maximum contractile force. Caerulein was approximately 4 times more active than CCK-8. Antagonists to histamine, acetylcholine, 5-hydroxytryptamine and catecholamines had no effect on responses to CCK-8. [See Figure 1 below].

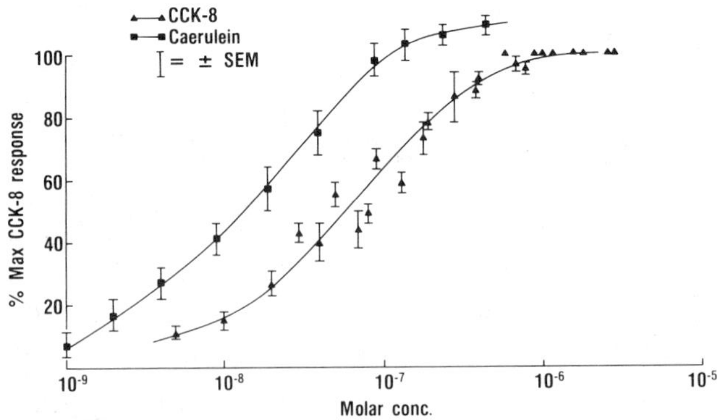

The presence of a sulphated tyrosine residue was an important determinant for contractile activity of CCK-8; its absence resulted in a marked reduction in potency (cf. EC_{50} CCK-8 sulphated = $5.6 \times 10^{-8}\underline{M}$; EC_{50} CCK-8 desulphated = 1.6×10^{-5}M). [See Figure 2 below]

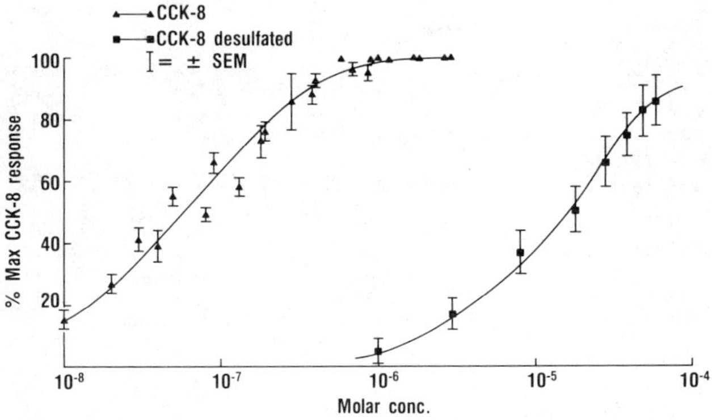

Reduction in peptide size to less than 8 amino acids, resulted in reduced potency; for example, CCK-7 either as the sulphated or desulphated form, was considerably less potent than CCK-8. [See Figure 3 below]

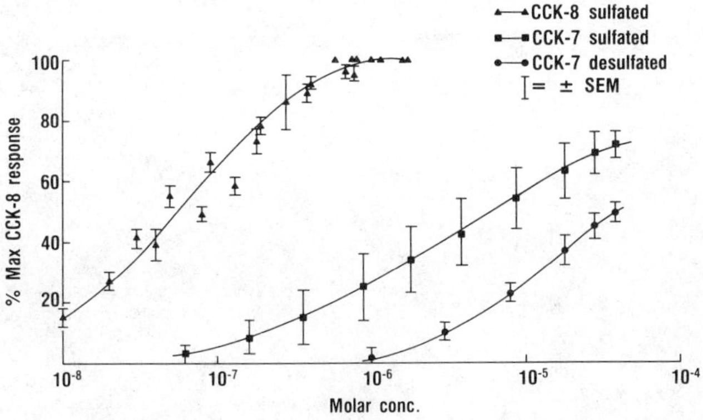

The CCK-tetrapeptide (CCK-30-33) and tripeptide (CCK-31-33) amides were both inactive at 10^{-4}M. A [$Nle^{28,31}$] substituted CCK-8 analog (U-67827E) was as potent as natural CCK-8. [See Table 1 below]

CCK C-Terminal Peptides	*Threshold Activity* GPT*
1. CCK-(31-33) Met-Asp-Phe-NH_2	Inactive @ 10^{-4}M
2. CCK-(30-33) Trp-Met-Asp-Phe-NH_2	Inactive @ 10^{-4}M
3. CCK-(27-33) Tyr-Met-Gly-Trp-Met-Asp-Phe-NH_2	10^{-5}M
4. CCK-(27-33) Tyr (SO_3H)-Met-Gly-Trp-Met-Asp-Phe-NH_2	6 X 10^{-6}M
5. CCK-(26-33) Asp-Tyr-Met-Gly-Trp-Met-Asp-Phe-NH_2	10^{-5}M
6. CCK-(26-33) Asp-Tyr (SO_3H)-Met-Gly-Try-Met-Asp-Phe-NH_2	10^{-8}M
7. U-67827E Ac-Asp-Tyr (SO_3H)-Nle-Gly-Trp-Nle-Asp-Phe-NH_2	10^{-8}M
8. U-71403 CHO-Met-Asp-Phe-NH_2	Inactive @ 10^{-4}M

* Threshold activity refers to the concentrations of [peptide] at which significant guinea pig tracheal contraction in vitro was stimulated and a ten-fold dilution of which was inactive.

Comparison of EC_{50} doses (calculated as the dose required to give 50% of the maximum contraction possible for each compound) for histamine, LTE_4 and CCK-8 showed that CCK-8 was approximately 10 times more potent than histamine and 10 times less potent than LTE_4. At a maximum dose, however, CCK-8 generated less force than either histamine or LTE_4. Like histamine CCK-8 was most active on guinea pig tracheal tissue with little or no activity on rat and rabbit trachea.

References

1. Emerson, J., and MacKay, D. (1979) J. Pharm. Pharmacol. 31:798.

2. Gardreau, P., Morell, J.L., St. Pierre, S. Quirion, R., and Pert, C. (1983) PEPTIDES: Structure and Function (V.J. Hruby, and D.H. Rich, eds.), Pierce Chemical Co., Rockford, IL, pp. 441-444.

3. Polak, J.M., and Bloom, S.R. (1982) Exper. Lung Res. 3:313-328.

CHOLECYSTOKININ ANALOGS CONTAINING NON-CODED AMINO ACIDS

Kalmán Kovács, Gábor K. Tóth, Márta Zarándi and Botond Penke
Department of Medical Chemistry

Tibor Kádár and Gyula Telegdy
Department of Pathophysiology

Ferenc Hajnal and János Lonovics
1 st Department of Medicine
Szeged University Medical School, Szeged, Hungary

Introduction

Tyrosine sulphate ester plays an important role in the biological actions of many well-known peptides, e.g. cholecystokinin (CCK), caerulein and gastrin-II. The possibility exists for substitution of the tyrosine sulphate residue with p-sulphophenylalanine (Phe(SA), such a caerulein analogue was reported earlier[1]. Similar substitution of other amino acid residues by an amino sulphonic acid is interesting e.g. norleucine-sulphonic acid in an enkephalin analogue[2] or p-sulphophenylalanine in an angiotensin[3]. The use of cysteic acid for similar substitutions of amino acids is obvious[4]. We earlier reported that the aspartyl residue is not necessary for gastrinic or CCK-like activity[5]. It can be replaced by serine sulphate ester. On the basis of this observation, we replaced the Asp residues in position 7 of CCK-8 or in position 3 of tetragastrin by Phe(SA). On the basis of literature data[6] we replaced Gly to Phe(SA) in pentagastrin expecting an increased biological potency. Similar subtitution was made in CCK-7; tyrosine sulphate ester was replaced by Phe(SA). Iminodiacetic acid (IDA), the N-carboxymethyl analogue of Gly can likewise be used to to replace coded amino acids. This imino acid can be useful for two purposes: substitution with N-alkylated amino acids/ such as sarcosine or other N-methylamino acids / generally causes an enhanced enzymic resistance. On the other hand, IDA can substitute aminodicarboxylic acids such as Asp or Glu. Earlier papers reported that positions 1 and 4 are important in the metabolic pathways of CCK-8[7.8]. On the basis of these results we replaced the Asp and Gly residues in positions 1 and 4 of CCK-8 by IDA.

Peptide synthesis

The peptides were prepared in liquid phase as described in the synthesis of CCK-8[9]. L-sulphophenylalanine was synthesized from L-Phe by direct sulphonation. During the syntheses the -amino group was protected with Boc. For the introduction of Phe(SA) into peptide molecules both the active ester and the DCC condensation methods were applied. Both methods permitted peptide synthesis without isolation difficulties. IDA was protected with Boc and reacted with DCC, to yield the cyclic anhydride, which was used for the synthesis of compounds 7 and 9. For preparation of the dimeric compound 8, Boc-iminodiacetic acid bis-pentafluorophenyl ester was applied. The analogues were purified as previously described[10]. TLC and HPLC investigations in different solvent systems revealed that all peptides were homogeneous.

Bioassay methods

Gastrinic activities were assayed on perfused rat stomach by conductometric titration of the HCl liberated in response to i.v. administration of the peptides solution[11]. CCK-like activity measurements were carried out in vitro by the method of Berry and Flower[12] as modified by Lonovics[13] on isolated rabbit gall bladder strips. Anticonvulsive effects against picrotoxin-induced seizures were measured in mice i.p. administration of the peptide solution by the method of Kádár[14] essentially as described by Zetler[15].

Results and Discussion

Tyr sulphate could be replaced by Phe(SA) in CCK-7 without substantial decrease of the biological potencies. Substitution of Asp in position 7 by Phe(SA) resulted in practically no change in CNS effect, but the cholecystokinetic activity was strongly decreased. Although Gly could be replaced by Tyr-sulphate ester[6] or by Phe(SA) giving an increase of gastrin-like activity, these replacements gave analogs almost inactive in CNS-response. Replacement of Asp in gastrin peptides (3,4) decreased all the biological potencies, without a total loss af activities. This lends support to our theory[5] that Asp could be replaced by electronically similar anion-containing non-coded amino acids such as Ser- and Thr-sulphate ester. The introduction of IDA to pentagastrin (7,8) caused the total loss of gastrin activity, however, the anticonvulsive potency was increased.

COMPOUND	PEPTIDE SEQUENCE	IN VIVO GASTRINIC act. rel.%	IN VITRO CHOLECYS-TOKINETIC act.rel.	ANTICONVULSICE ACT. RELATIVE % tonic seizure	time until death
tetragastrin	H-Trp-Met-Asp-Phe-NH_2	80	0	54*	43*
pentagastrin	Boc-β-Ala-Trp-Met-Asp-Phe-NH_2	100	0	113*	104*
CCK-8	H-Asp-Tyr(SE)-Met-Gly-Trp-Met-Asp-Phe-NH_2	0	100	100*	100*
1	Boc-Phe(SA)-Met-Gly-Trp-Met-Asp-Phe-NH_2	-	20	61*	102*
2	Boc-Phe(SA)-Trp-Met-Asp-Phe-NH_2	100	-	0	6
3	Boc-Trp-Met-Phe(SA)-Phe-NH_2	7	-	20	21
4	H-Met-Gly-Trp-Met-Phe(SA)-Phe-NH_2	5	-	29	58*
5	H-Asp-Tyr(SE)-Met-Gly-Trp-Met-Phe(SA)-Phe-NH_2	-	3	106*	96*
6	Boc-Phe(SA)-Met-Gly-Trp-Met-Phe(SA)-Phe-NH_2	-	5	69*	33*
7	Boc-IDA-Trp-Met-Asp-Phe-NH_2	0	-	120*	82*
8	Boc-IDA[-Trp-Met-Asp-Phe-NH_2]$_2$	0	-	154*	112*
9	H-IDA-Tyr(SE)-Met-Gly-Trp-Met-Asp-Phe-NH_2	-			

All of our results are in accordance with the hypothesis that the peripheral, central gastrin and CCK receptors are different.

References

1. Agosti,A., Orlandini,F. and de Castiglione,R., (1971)Experientia 27, 412-413.
2. Bajusz,S., Ronai,Z., Szekely,J.I., Turan,A., Juhasz,A., Patthy,A., Berzetei I. and Miglecz,E. (1981) in Peptides 198o K. Brunfeldt Ed. Scriptor Copenhagen pp. 4o3-4o8.
3. Escher,E. and Bernier,M. (1981) in Peptides 198o K. Brunfeldt Ed. Scriptor Copenhagen pp. 457-462.
4. Hubbuch,A., Danho,W. und Zahn,H. (1979) Liebigs Ann. Chem. 776-783.
5. Penke,B., Hajnal,F., Lonovics,J., Holzinger,G., Kadar,T., Telegdy,G. and Rivier,J. (1984) J.Med.Chem. 845-849.
6. Hirst,B.H., Reed,J.D., Shaw,B., Smeaton,L.A., Blair,E.L. and de Castiglione,R. (1979) In: "Hormonal Receptors in Digestion and Nutrition," Rosselin,G. Fromageot,P. Bonfis,S.,Eds. Elsevier/North Holland, Amsterdam 1979 pp. 413-418.
7. Bodanszky,M., Tolle,J., Gardner,J.D., Walker,M.D. and Mutt, V. (198o) Int. J. Peptide Protein Res. 4o2-411.
8. Penke,B., Rivier,J. (1983) in Peptides: Structure and Function Hruby V.J. and Rich D.H. Eds. Pierce Chemical Co. Rockford IL. pp. 119-122.
9. Tóth,G.K., Penke,B., Zarándi,M. and Kovács,K., (1985) Int. J.Peptide Protein Res. in press
1o. Tóth,G.K., Penke,B., Janáky,T., Kovács,K. and Rivier,J., (1985) Chromatography, the State of the Art Kalász,H. and Ettre, L.S. Eds. Akadémiai Kiadó, Budapest pp. 287-298.
11. Halter,F., Kohler,B., Smith,G.M. (1971) Helv.Med. Acta Suppl. 113-
12. Berry,H., Flower,J. (1971) Gastroenterology 6o. 4o9-42o
13. Lonovics,J., Guzman,G., Dewitt,P., Hejtmancik,K.E. Suddith, R.L., Rayford, P.L. Thompson,J.C. (198o)Gastroenterology, 79, 817-822.
14. Kádár,T., Pesti,A., Penke,B., Tóth,G.K., Zarándi,M. and Telegdy,G. (1983) Neuropharmacology 22, 1223-1229.
15. Zetler,G. (198o) Eur.J.Pharmac.65, 297-3oo.

CONFORMATIONAL PROPERTIES OF GASTRIN RELATED PEPTIDES

M.T.Foffani & E.Peggion

Biopolymer Research Center, Department of Organic Chemistry, University of Padua, Via Marzolo 1, 35131 Padua, Italy.

S.Mammi & M.Goodman

Department of Chemistry, University of California at San Diego, La Jolla, 92093 USA.

G.Boverman, L.Moroder, &E.Wuensch

Max-Planck Institute fur Biochemie, Abteilung Peptidchemie, 8033 Martinsried bei Munchen, Federal Republic of Germany.

Introduction

Recent studies on gastrin fragments of increasing chain length[1] have shown that elongation of the glutamic acid sequence at the N-terminus from 2 to 4 residues causes a conformational change in trifluoroethanol (TFE) solution, which involves the peptide backbone and the aromatic side-chains as well. A correlation has been observed between increase of biological potency upon chain elongation and peptide folding tendency in TFE, suggesting the possibility that the structure of the gastrin hormones in this solvent medium is of biological importance. In view of this fact we have undertaken systematic conformational studies on synthetic gastrin fragments using CD and NMR spectroscopy.

Results and Discussion

HG-4 (Trp-Nle-Asp(O-tBu)-Phe-NH_2). The 360 MHz ^1H NMR spectra of the C-terminal tetrapeptide fragment have been recorded in DMSO-d_6 and TFE-d_3 containing 2% D_2O (v/v). In DMSO all resonances are consistent with results reported in the literature[2,3]. The temperature coefficients of the amide protons do not provide evidence for the presence of intramolecular hydrogen bonds and confirm the conclusion of Feeney at al.[3] that the peptide conformation is random in this solvent medium.
The NMR spectrum in TFE-d_3 + 2% D_2O (v/v) is reported in Figure 1. With the exception of Nle C^{α} H and C^{δ} H_2, and of Trp C^4 H, all resonances move downfield with respect to those observed in DMSO. The phenyl protons appearsas a doublet, suggesting a limited degree of rotational freedom of the aromatic ring. An interaction between Phe and Nle side-chains could account for the restricted rotational freedom of the phenyl group and also for the upfield shift of the Nle resonance (shielding effect of the ring current). These data are in favor of the hypothesis, based upon CD results

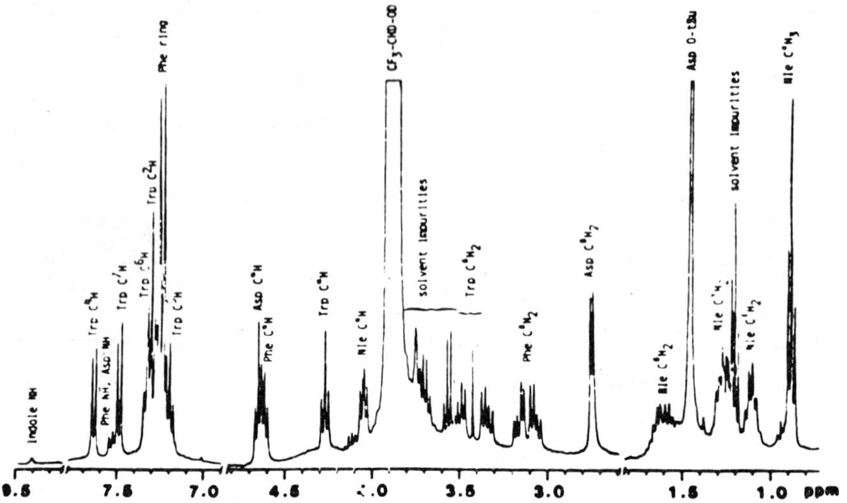

Fig.1.360 MHz ^{1}H NMR spectrum of the gastrin tetrapeptide in TFE-d_3+2%D_2O.

and conformational energy calculations[2,4] that the tetrapeptide assumes a partially folded structure in TFE.

Gastrin Octapeptide (pGlu-Ala-Tyr-Gly-Trp-Nle-Asp-Phe-NH_2). The 360 MHz ^{1}H NMR spectrum of the octapeptide in DMSO-d_6 in the aromatic and amide NH region is shown in Figure 2. With respect to the results obtained with the tetrapeptide, there is a consistent upfield shift of the aromatic Trp resonances, due to the fact that the N-terminal amino group of this residue becomes acylated in the octapeptide. The resonances of the amide protons of the C-terminal pentapeptide sequence -Gly-Trp-Nle-Asp-Phe-NH_2 agree with those recently reported by Durieux et al.[5] for the cholecystokinin fragment CCK 27-33, containing the same sequence. One exception is the Asp NH resonance, which, in our case, is shifted by $\sim$ 0.25 ppm. This is probably due to the fact that in our gastrin fragment Met is replaced by Nle, with possible effects on the resonance of the adjacent Asp NH. The temperature coefficients of all amide protons also in this case do not indicate the presence of intramolecular hydrogen bonds. The peptide conformation remains random in DMSO. The insolubility of this fragment in TFE-d_3 and the strong tendency to aggregation at concentration levels well below the sensitivity of NMR prevented us from obtaining meaningful data in this solvent medium.

Des-Trp^1,Nle^{12}-Minigastrin (Leu-(Glu)$_5$-Ala-Tyr-Gly-Trp-Nle-Asp-Phe-NH_2. The 360 MHz ^{1}H NMR spectrum of the minigastrin analog in DMSO-d_6 (not shown) indicates that there is no change of the resonance position of the aromatic protons with respect to those observed for the octapeptide. We therefore conclude that there is practically no difference in the conformational preference of these compounds in DMSO, where they assume a random structure.

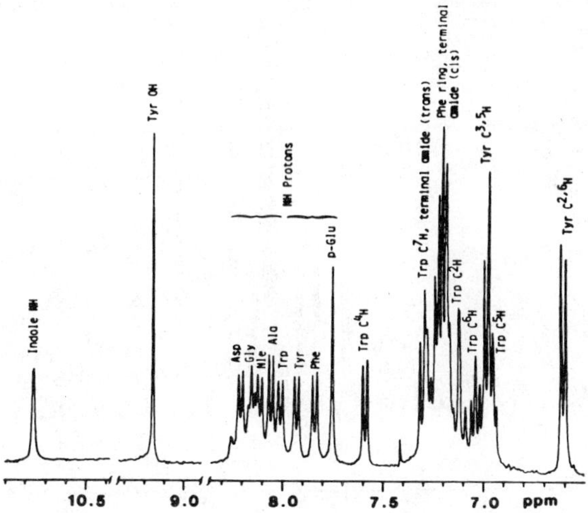

Fig.2. 360 MHz ^1H NMR spectrum of the gastrin octapeptide in DMSO-d_6.

The minigastrin analog is slightly soluble in TFE-d_3 containing 2% D_2O. By using a special procedure we have been able to prepare a solution with a sufficiently high concentration to carry on NMR experiments. The NMR spectrum in the aromatic region (not shown) exhibits the indole NH resonance at 10.47 ppm, i.e. downfield by ~1.1 ppm with respect to that observed for the tetrapeptide in the same solvent. Furthermore the indole NH does not exchange with D_2O after three weeks, while all amide resonances exchange almost completely. These observations led initially to conclude that the Trp side-chain is involved in a strong intramolecular hydrogen bond. However, due to the poor solubility of the minigastrin analog in TFE-d_3 , there was the risk that these results were affected in some way by intermolecular self-association. To clarify this point we have studied by CD and NMR spectroscopy the conformational properties of minigastrin in non-deuterated TFE-H_2O mixtures. By CD measurements we have shown that the ordered conformation of the hormone, described in our previous work[1] remains stable up to a water content of about 20% (v/v). In such a medium the solubility of the minigastrin analog is much higher than in the perdeuterated solvent system. We have been able to prepare hormone solutions with concentrations higher than 5.0×10^{-3}M (on peptide residue basis) without intermolecular association, and to run CD and NMR experiments on the same solution. By using solvent peak suppression techniques we obtained the NMR spectrum shown in Figure 3. The indole NH resonance is now located at 9.45 ppm, indicating that the hydrogen bonded structure involving this group, previously observed, was due to intermolecular association. By temperature-dependence studies we have found that in the spectrum of Figure 3 there are 5 resonances in the amide NH region with temperature coefficients $\leqslant 3.0 \times 10^{-3}$ ppm/°C.

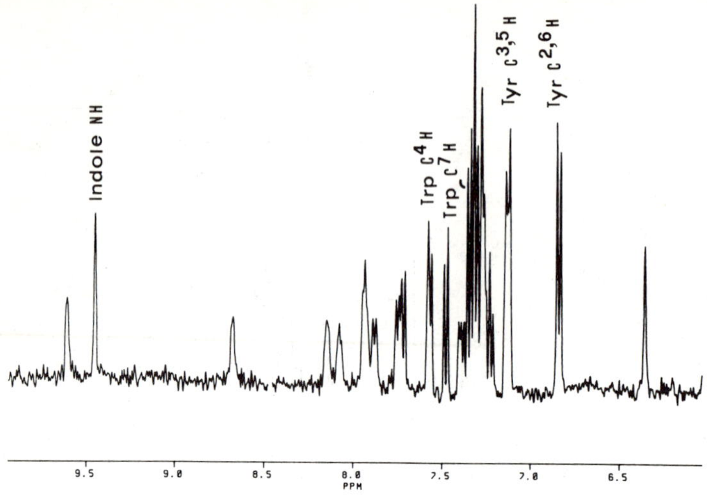

Fig.3. 400 MHz ^1H NMR spectrum of des-Trp^1,Nle^{12}-minigastrin in TFE+10%H_2O.

These resonances are located at 8.67 ppm, 7.93 ppm, 7.75 ppm, 7.70 ppm, and 7.56 ppm, and are indicative for the presence of intramolecular hydrogen bonds. Temperature-independent amide resonances have been observed also in the gastrin dodecapeptide and undecapeptide, with -(Glu)$_4$- and -(Glu)$_3$- sequences respectively at the N-terminus. Quite obviously, in order to test our hypothesis of the presence of a β-turn located in the central part of the molecule, and of a helical segment comprising the -(Glu)$_5$- sequence at the N-terminus of minigastrin all amide NH resonances have to be precisely assigned. Work is in progress in this direction and will be reported elsewhere.

References

1. Peggion,E., Foffani,M.T., Wuensch,E., Moroder,L., Goodman,M., & Mammi,S., (1985) Biopolymers, 24, 647-666.
2. Bleich,H.E., Cutnell,J.D., & Glasel,J.A., (1976), Biochemistry, 15, 2455-2466.
3. Feerey,J., Roberts,G.C.K., Brown,J.P., Burgen,A.S.V., & Gregory, H., (1972), J.Chem.Soc., Perkin Trans?, 2, 601-604.
4. Abillon,E., Phan Van Chuong,P., & Fromageot,P., (1981), Int.J.Peptide Protein Res., 17, 480-485.
5. Durieux,C., Belleny,J., Lallemand,J.Y., Roques,B.P., & Fournie-Zaluski, M.C., (1983), Biochem.Biophys.Res.Comm., 114, 705-712.

EXAMINATION OF THE CONFORMATIONAL REQUIREMENTS OF GLUCAGON AT ITS RECEPTOR

John L. Krstenansky, Beat Gysin, Dev Trivedi and Victor J. Hruby

Department of Chemistry, University of Arizona
Tucson, Arizona 85721

Introduction

While much work on the structure-activity relationships of glucagon has been done,[1,2] few analogs have been synthesized for the purpose of assessing glucagon's receptor-bound conformational requirements.[1,3-5] We have synthesized a number of glucagon analogs in order to determine the role of various regions of the glucagon molecule in glucagon's receptor binding and transduction.

Results and Discussion

The synthesis of the reported peptides was accomplished by solid-phase techniques on Merrifield or p-methylbenzhydrylamine resins. N^{α}-Boc amino acids were coupled as preformed symmetrical anhydrides. Cleavage of the protected peptides by HF was followed by purification: dialysis, cation-exchange chromatography, gel filtration, and preparative reverse-phase HPLC. Identity and purity was assessed by amino acid analysis, tlc, HPLC, and α-chymotryptic peptide mapping.

Chou-Fasman calculations for glucagon predict β-turns at the 2-5, 10-13, and 15-18 residues and a β-sheet or α-helix for the C-terminal 19-27 region.[6] The crystal structure is mostly α-helical (10-25 region) with a less regular helix for

Table I. Biological Activity of Analogs (Rat Liver Membranes)

	Compound	Adenylate Cyclase		Receptor Binding	
		EC_{50} (nM)	relative potency[a]	IC_{50} (nM)	relative potency
I	glucagon (G)	8.4	100(100)	3.6	100
II	[W^{1}]-G	19	44 (35)	17	22
III	[P^{3}]-G	500	2 (40)	250	2
IV	[F^{13}]-Gamide	10	81	3.3	110
V	[F^{10}]-Gamide	7.0	120 (85)	3.0	120
VI	[F^{10}]-G	70	12 (84)	20	18
VII	[F^{10,13}]-G	78	11 (89)	35	10
VIII	[A^{11}]-G	31	27 (40)	15	24
IX	[Aoc$^{11-13}$]-Gamide	2000	.4(100)	1300	0.3
X	[Ahx^{17,18}]-G	5000	.2 (10)	5000	0.1
XI	[K^{17,18},E^{21}]-G	1.2	700(100)	0.7	510
XII	[E^{15},K^{17,18}]-G	79	11 (80)	16	22

[a]Potency relative to glucagon = 100% with maximal activation of adenylate cyclase relative to glucagon = 100% in parenthesis.

another four residues at each end.[7] Glucagon in the presence of a lipid-water interphase is reported to adopt an α-helical structure for the 10-14 and 17-29 regions. Incorporation of a model amphiphilic α-helix in the C-terminal region of glucagon resulted in a compound with very weak affinity.[3] Efforts to extend the C-terminal α-helix by making the Arg^{17,18} residues hydrophobic leads to an analog reported to retain full potency and activity.[4] However, when we placed the hydrophobic residue α-aminohexanoic acid in the 17 and 18 positions (Compound X) both potency and activity were greatly reduced. Compound XI was designed from Chou-Fasman calculations[6] to enhance the α-helical potential of the 19-27 region ($\langle Pa \rangle = 1.230$ and $\langle Pb \rangle = 1.070$ relative to I: $\langle Pa \rangle = 1.180$ and $\langle Pb \rangle = 1.150$) and the β-turn potential at 15-18 ($\langle Pt \rangle = 1.255$ relative to I: $\langle Pt \rangle = 1.197$) while introducing minimal change. This analog was a full agonist with a 500-700% increase in potency.

Compound XII was an attempt to further extend the α-helix into the 14-18 region (⟨Pa⟩ = 1.162 relative to I: ⟨Pa⟩ = 0.990), but these seemingly small changes resulted in a partial agonist. The C-terminal amides are more potent than the corresponding C-terminal carboxylic acids (compare Compounds V and VI).

Substitution of Phe for Tyr at the 10 and 13 positions (IV to VII) tests the importance of the phenolic groups, increases the lipophilicity, and increases the α-helical potential in the 10-13 region (VII: ⟨Pa⟩ = 1.047 and ⟨Pt⟩ = 0.910 relative to I: ⟨Pa⟩ = 0.827 and ⟨Pt⟩ = 1.180). In both positions potency is decreased implying that the phenolic groups are involved with glucagon's receptor interaction. The Ser^{11} residue plays a role in receptor activation (Compound VIII) since both potency and efficacy are affected by this substitution. Replacement of the Ser^{11}, Lys^{12}, and Tyr^{13} residues with a flexible spacer of similar length, ω-aminooctanoic acid, yielded Compound IX which was a full agonist even though it has greatly reduced potency. Since the biological message of glucagon must be contained elsewhere in the molecule, partial agonism from 11-13 region changes must be due to the ability of these structural modifications to affect the interaction of the "message" portion with the receptor; modifications of the Lys^{12} residue have led to similar conclusions.[9] The N-terminal region of glucagon is the most sensitive to modification (Compounds II and III), and it is the most likely location of the biological message of the glucagon molecule.[1,9-10]

Therefore, the model that is currently suggested for glucagon-receptor interaction leading to transduction follows: the C-terminal region of glucagon is the primary site for initial receptor recognition, and involves a specific interaction with the hormone as an amphiphilic α-helix in this region. The helix is "broken" by a turn in the 15-18 region with an additional binding site involving the 10-13 region (possibly with helical structure as is present in the crystal structure and the lipid-water interface conformation utilizing a lipophilic binding area of glucagon involving the Tyr^{10}, Tyr^{13} and other residues). The nature of the binding of the 10-13 region determines the orientation of the "message" N-terminal region to the receptor thereby influencing its ability to activate the hormone receptor complex.

This work was supported by U.S. Public Health Service Grant AM 21085 (VJH) and the Swiss National Science Foundation (BG).

References

1. Hruby, V.J., Krstenansky, J.L., McKee, R., and Pelton, J.T. (1985) in Hormonal Control of Gluconeogensis, Kraus-Friedman, N., Ed., CRC Press, Cleveland, OH.

2. Bromer, W.W. (1983) in Glucagon I: Handbook of Experimental Pharmacology Vol. 66/I, Lefebvre, P.J., Ed., Springer-Verlag, Berlin, FRG, pp. 1-22.

3. Musso, G.F., Kaiser, E.T., Kezdy, F.J., and Tager, H.S. (1983) in Peptides: Structure and Function, Hruby, V.J. and Rich, D.H., Eds., Pierce Chemical Company, Rockford, IL, pp. 365-368.

4. Epand, R.M. and Liepnieks, J.J. (1983) J. Biol. Chem. **258**, 203-207.

5. Mojsov, S., Lu, G., Merrifield, R.B., and Iwanij, V. (1983) in Peptides: Structure and Function, ibid., IL, pp. 373-376.

6. Chou, P.Y. and Fasman, G.D. (1975) Biochem. **14**, 2536-2541.

7. Sasaki, K., Dockerill, S., Adamiak, D.A., Tickle, I.J., and Blundell, T. (1983) Nature (London) **257**, 751-757.

8. Braun, W., Wider, G., Lee, K.H., and Wüthrich, K. (1983) J. Mol. Biol. **169**, 921-948.

9. Bregman, M.D., Trivedi, D., and Hruby, V.J. (1980) J. Biol. Chem. **255**, 11725-11733.

10. Rodbell, M., Birnbaumer, L., Pohl, S.L., and Sunby, F. (1971) Proc. Natl. Acad. Sci. USA **68**, 909-913.

SECRETIN-GLUCAGON HYBRIDS: SYNTHESIS AND BIOLOGICAL ACTIVITY

David Andreu and Bruce Merrifield
The Rockefeller University, New York, NY 10021.

Introduction

Analogs of glucagon having some receptor binding affinity but little or no potency would be useful in assessing the relative contribution of hyperglucagonemia to diabetes. The search for such competitive inhibitors of glucagon[1] is of further interest since it may lead to compounds of potential therapeutic value as complements of insulin in diabetes treatment.

In our research on glucagon structure-activity relationships we have explored the possibility that secretin-glucagon hybrids might exhibit some of the above properties. In spite of their considerable sequence homology, secretin and glucagon have quite different and unrelated physiological roles. In particular, secretin does not bind to hepatic cells as glucagon does, nor does it elicit any glucagon-like activity. It is interesting to speculate, however, whether appropriate modifications of the secretin molecule tending to make it more glucagon-like might lead to peptides capable of binding the glucagon receptor without concurrently stimulating the adenyl cyclase system.

Previous work from our laboratory[2] has applied improved solid-phase methods to produce synthetic, fully active glucagon in good yields and high purity. We have used similar procedures to prepare four peptides (I-IV, Figure 1) representing several degrees of secretin-glucagon sequence hybridation.

Synthesis

Peptide amides I and II were assembled stepwise on a

p-methylbenzhydrylamine resin and acids III and IV on an oxymethyl resin. The protection scheme was identical to our synthesis of glucagon[2]. All peptides except I were assembled automatically in a computer-controlled version of a commercial synthesizer[3]. Final cleavage and deprotection were done by low-high HF[4], in 80-90% yields. In each case, the desired peptide accounted for ca. 70% of all crude material, as judged by HPLC. Purification was done in a single step by low-pressure liquid chromatography on C_{18}-silica. Overall synthesis-purification yields ranged from 35% to 40%. The purified peptides were homogeneous by HPLC and gave amino acid compositions consistent with theory.

	5 10 15
Secretin	His-Ser-Asp-Gly-Thr-Phe-Thr-Ser-Glu-Leu-Ser-Arg-Leu-Arg-Asp-
Peptide I	——— Asp ——————— Glu-Tyr —— Arg-Tyr-Arg ——
Peptide II	——— Asp ——————— Glu-Tyr —— Arg-Tyr-Arg ——
Peptide III	——— Asp ——————— Glu-Tyr —— Lys-Tyr-Leu ——
Peptide IV	——— Asp ——————— Glu-Tyr —— Arg-Tyr-Leu ——
Glucagon	His-Ser-Gln-Gly-Thr-Phe-Thr-Ser-Asp-Tyr-Ser-Lys-Tyr-Leu-Asp

	20 25
Secretin	Ser-Ala-Arg-Leu-Gln-Arg-Leu-Leu-Gln-Gly-Leu-Val-NH_2
Peptide I	—— Ala —— Leu —— Arg-Phe-Leu —— Trp —— Val-NH_2
Peptide II	—— Ala —— Leu —— Arg-Leu-Leu —— Gly —— Val-NH_2
Peptide III	—— Arg —— Ala —— Asp-Phe-Val —— Trp —— Met-Asn-Thr-OH
Peptide IV	—— Arg —— Ala —— Asp-Phe-Val —— Trp —— Met-Asn-Thr-OH
Glucagon	Ser-Arg-Arg-Ala-Gln-Asp-Phe-Val-Gln-Trp-Leu-Met-Asn-Thr-OH

Fig. 1. Amino acid sequences of secretin, glucagon and secretin-glucagon hybrids. Invariant residues indicated by a line.

Biological Activity

Receptor-binding assays of both native glucagon and peptides I-IV were performed on purified rat liver plasma membranes in 1% BSA, 1 mM DTT, 25 mM Tris.HCl, pH 6.95. Results are shown in Figure 2. Peptide concentrations required to displace 50% of membrane-bound [^{125}I]glucagon were obtained from the curves and are given in Table I. For native glucagon, this value is 1.6×10^{-9} M, in good agreement with data from other sources.

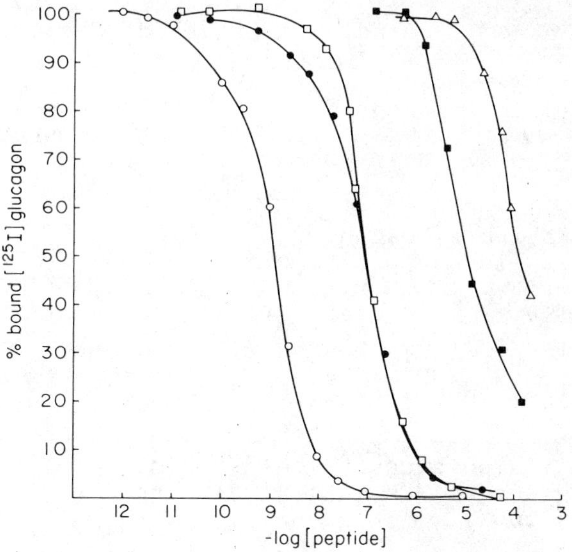

Fig. 2..Displacement of [^{125}I]glucagon from rat liver plasma membranes by natural glucagon (○) and peptides I (■), II (△), III (●) and IV (□).

Of the other peptides, only III and IV showed significant levels of receptor binding, with half-maximal displacement concentrations in the 10^{-7} M range. Peptides I and II did not completely displace labeled glucagon from the membrane even at concentrations near 10^{-3}M.

Peptides I-IV were also tested for their activation of rat liver adenyl cyclase (data not shown). All four failed to give any detectable response even up to 10^{-5} M concentrations, which indicates an activity of less than 0.001% of glucagon. In particular, III and IV did not show activation levels concomitant with their binding behavior in that concentration range.

Discussion

From a wide range of potentially interesting secretin-glucagon hybrids, we have chosen to study two resembling secretin (I and II) and two more glucagon-like. Our results show that peptides III and IV, totally homologous to glucagon from residues 13 to 29, retain certain affinity for the receptor, whereas I and II do not. Therefore, an intact C-terminus seems

Table I. Binding Affinity of Secretin-Glucagon Hybrids.

Peptide	Half-maximal Displacement Concentration (nM)	Relative Binding Affinity (%)*
Native glucagon	1.6	100.0
$[Tyr^{10,13}, Phe^{22}, Trp^{25}]$secretin (I)	11,000	0.014
$[Tyr^{10,13}]$secretin (II)	130,000	0.0012
$[Asp^{3}, Glu^{9}]$glucagon (III)	100	1.6
$[Asp^{3}, Glu^{9}, Arg^{12}]$glucagon (IV)	71	2.2

* Calculated as (half-maximal displacement concentration of glucagon)/(half-maximal displacement concentration of peptide) x 100%

to be a necessary though not sufficient requirement for binding. This conclusion is in agreement with a number of other results[5]. Interestingly, although III and IV require concentrations about 50 times higher than glucagon to displace ^{125}I-labeled glucagon from its receptor, they are nevertheless inactive against adenyl cyclase at these concentrations and can therefore be properly categorized as competitive inhibitors of glucagon.

References

1. Bregman, M.D., Trivedi, D., and Hruby, V.J. (1980) J. Biol. Chem., 255, 11725-11731.
2. Mojsov, S., and Merrifield, R.B. (1984) Eur. J. Biochem., 145, 601-605.
3. Whitney, D.B., Silverman, P., Tam, J.P., and Merrifield, R.B., manuscript in preparation.
4. Tam, J.P., Heath, W.F., and Merrifield, R.B. (1983) J. Am. Chem. Soc., 105, 6442-6455.
5. Bromer, W.W. (1983) in Handbook of Experimental Pharmacology (Lefebvre, P.J., ed.) vol. 66/I, pp. 1-22, Springer-Verlag, Berlin.

STUDIES LEADING TO ORALLY ACTIVE ANTAGONISTS OF THE ANTIDIURETIC (V_2) AND VASOPRESSOR (V_1) RESPONSES TO ARGININE VASOPRESSIN (AVP).

M. Manning, A. Misicka, S. Stoev, E. Nawrocka, W. A. Klis, A. Olma, K. Bankowski, B. Lammek and W. H. Sawyer, Department of Biochemistry, Medical College of Ohio, Toledo, OH and Department of Pharmacology, College of Physicians and Surgeons of Columbia University, New York, NY.

The most potent of our original AVP V_2/V_1 antagonists, [1-(β-mercapto-β,β-cyclopentamethylenepropionic acid), 2-O-ethyltyrosine, 4-valine] arginine vasopressin ($d(CH_2)_5Tyr(Et)VAVP$)[1], has been modified using the Merrifield method[2] to give analogs having: 1. Substitutions (single and combined) of L and D-Amino acids[3-5] (including non-amino acids in the tail). 2. Deletions in the ring and in the tail. We present here some highlights of our recent findings. Examples of some new selective AVP V_2 antagonists with substitutions at positions 2 and 4 are given in Table I.

Table I. SOME SELECTIVE - D-Leu[2] - ANTIDIURETIC ANTAGONISTS

```
                1         2   3   4   5   6   7   8   9
  (CH2)5>C-CH2-CO-D-Leu-Phe-X-Asn-Cy-Pro-Arg-Gly-NH2
         |                        |
         S------------------------S
```

Antagonist		Antiantidiuretic		Antivasopressor		Selectivity
	X^4	ED[b]	pA_2[c]	ED[b]	pA_2	Ed Ratio
1.	Abu	4.7	7.22	101	5.86	21
2.	Val[a]	1.2	7.79	26	6.45	22
3.	Thr	9.1	6.87	220	5.52	24
4.	Ile	2.3	7.50	67	6.01	29

[a] Data from ref. 3
[b] The effective dose (ED) is the dose (in nmoles/kg) of antagonist which reduces the agonistic response to AVP by 50%.
[c] Estimated in vivo pA_2 values represent the negative logarithms of the "effective dose" divided by the estimated volume of distribution (67 ml/kg)
[d] ED ratio = antivasopressor ED/antiantidiuretic ED.

Position 9 tolerates a wide variety of substitutents. Table II. lists a series of $d(CH_2)$[D-Phe[2],Ile[4],X-NH_2[9]] AVP analogs where X = Ala, Orn, Ser, Val, Arg, Tyr, Phe, Ile, Thr, D-Ala, Pro. All analogs exhibited potent V_2/V_1 antagonism.

TABLE II. SOME 9 SUBSTITUTED ANTIDIURETIC ANTAGONISTS[a]

```
          1            2   3   4   5   6   7   8   9
(CH2)5>C-CH2-CO-D-Phe-Phe-Ile-Asn-Cy-Pro-Arg-X
       S————————————————————————————S
```

POSITION 9	ANTIANTIDIURETIC ED[b]	ANTIANTIDIURETIC pA_2	ANTIVASOPRESSOR ED[b]	ANTIVASOPRESSOR pA_2
Gly-NH_2	0.46	8.24	0.99	7.86
Ala-NH_2	0.31	8.38	1.5	7.67
Orn-NH_2	0.46	8.18	1.1	7.80
Ser-NH_2	0.55	8.10	0.54	8.10
Val-NH_2	0.57	8.11	1.6	7.66
Arg-NH_2	0.67	8.05	0.55	8.11
Tyr-NH_2	0.95	7.85	3.7	7.27
Phe-NH_2	1.5	7.69	3.8	7.28
Ile-NH_2	1.9	7.58	6.3	7.04
Thr-NH_2	2.7	7.39	1.7	7.64
D-Ala-NH_2	2.9	7.38	1.9	7.55
Pro-NH_2	4.4	7.19	5.7	7.12

[a]From ref. 6 and 7 and this communication.
[b]For definition of ED and pA_2 see footnote to Table I.

Deletions in Ring Lead to Loss of Antagonistic Potencies. The potent and selective V_2 antagonist $d(CH_2)_5D$-Ile^{2},Ile^{4})AVP[5] was systematically modified by deleting amino acids from positions 2, 3, 4, and 5, individually and in different combinations. None of the resulting nine analogs exhibited any V_2 antagonism. Some exhibited only weak agonism (less than 0.015 U/mg).

Deletions in the tail lead to retention of antagonistic potencies as long as positive charge is retained. Removal of the C-terminal glycine or C-terminal glycinamide from the potent V_2/V_1 antagonist $d(CH_2)_5D$Tyr(Et)VAVP[4] resulted in compounds which are as potent as the parent compound[6, 7, 8a]. Replacement of the arginine residue in $d(CH_2)_5D$-Tyr(Et)VAVP by Gln, Leu or Cit resulted in analogs which exhibit drastic reductions in anti V_2 potency[6]. Remarkably, deletion of the Pro^{7} residue from $d(CH_2)_5D$Tyr(Et)VAVP led to full retention of both anti V_2 and anti V_1 potencies (anti V_2 ED=1.1 nmoles/kg; pA_2=7.82; anti V_2 ED=0.56; pA_2=8.09). This analog had previously been found to exhibit a 50% reduction in anti V_2 potency[8b]. Differences in the assays used may account for this discrepancy. Thus, it would appear that for $d(CH_2)_5D$-

Tyr(Et)VAVP[4] although a positive charge in the tail is required for V_2 antagonism, its position need not be fixed. These findings provide very useful clues for the design of more potent orally active antagonists than those reported below.

Orally Active Antidiuretic Antagonists. Replacement of Gly-NH_2 by ethylenediamine in d(CH_2)$_5$[D-Tyr(Et)Ile[4]]AVP and in d(CH_2)$_5$[D-Phe[2],Ile[4]]AVP[9] resulted in analogs which exhibited oral activity (i.e., more than 50% inhibition of V_2 response) in doses of 112 and 125 times the I.V. dose in 3 of 6 and 8 of 12 trials respectively (Table III).

TABLE III. ANTAGONISTIC POTENCIES OF ETHYLENEDIAMINE (Eda)[9] SUBSTITUTED ANALOGS (ADMINISTERED INTRAVENOUSLY (I.V.) AND ORALLY) AND THOSE OF RELATED Gly-NH_2[9] ANALOGS.

```
          1        2   3   4   5   6   7   8   9      Y = NH-CH2-CONH2 or
(CH2)5>C-CH2-CO-X-Phe-Ile-Asn-Cy-Pro-Arg-Y                -NH-CH2-CH2NH2
       |                      |
       S----------------------S
```

			ANTIANTIDIURETIC				ANTIVASOPRESSOR	
			ED[b]	pA_2[b]	ED[b,c,d]	ORAL ED	ED[b]	pA_2[b]
	X^2	Y^9	(I.V.)		(ORAL)	I.V. ED		
1.	D-Phe[a]	Gly-NH_2	0.46	8.24			0.99	7.86
1a.	D-Phe	Eda	0.6	8.10	~75[d]	~125	0.88	7.89
2.	D-Tyr	Gly-NH_2	1.4	7.7			0.55	8.08
2a.	D-Tyr	Eda	1.1	7.81			0.73	7.93
3.	D-Tyr(Et)	Gly-NH_2	1.2	7.75			1.2	7.78
3a.	D-Tyr(Et)	Eda	0.8	8.00	~90[e]	~112	0.32	8.33

[a]From ref. 9.
[b]For definition of ED and pA_2 see footnote to Table I.
[c]More than 50% inhibition
[d]No. of Responses = 8; No. of Trials = 12
[e]No. of Responses = 3; No. of Trials = 6

CONCLUSION. Our S/A findings to date on AVP V_2 antagonists are summarized in Table IV. Promising leads for enhancement of 1) V_2 antagonistic potency, 2) anti V_2/V_1 selectivity together with 3) leads for the design of orally active antagonists have been developed. Some of these compounds may be of value for the treatment of hyponatremia resulting from the syndrome of inappropriate secretion of antidiuretic hormone (SIADH)[10].

Table IV. SUMMARY OF S/A FINDINGS ON AVP ANTIDIURETIC ANTAGONIST $d(CH_2)_5Tyr(Et)VAVP^1$ [a]

I. SUBSTITUTIONS (L or D)

A. Non-variable residues: 1, 3, 5.

B. Variable residues: 2, 4*, (6) 7, 8, 9. *L-only

II. DELETIONS

A. Ring: none tolerated in $d(CH_2)_5[D\text{-}Ile^2,Ile^4]AVP^6$.

B. Tail: well tolerated in $d(CH_2)_5D\text{-}Tyr(Et)VAVP^4$: requirement only for positive charge.

III. ORAL ACTIVITY: ~100 times I.V. dose to date. (Table III)

[a]From ref. 9, this communication and unpublished data.

ACKNOWLEDGEMENTS. We thank Ms. Valerie Murphy for her help in manuscript preparation and Ms. N. C. Wo and Mr. W. V. Greim for performing most of the bioassays. This work was supported in part by research grants from the National Institute of General Medical Sciences (GM25280) and the National Institute of Arthritis, Diabetes and Digestive and Renal Diseases (AM-01940).

REFERENCES

1. Sawyer, W. H., Pang, P. K. T., Seto, J., McEnroe, M., Lammek, B. and Manning, M. (1981) Science 212, 49-51.
2. Merrifield, R. B. (1963) J. Am. Chem. Soc. 85, 2149-2154.
3. Manning, M., Klis, W. A. Olma, A., Seto, J., and Sawyer, W. H. (1982) J. Med. Chem., 25, 414-419.
4. Manning, M., Olma, A., Klis, W. A., Kolodziejczyk, A. M., Seto, J., and Sawyer, W. H. (1982) J. Med. Chem. 25, 45-50.
5. Manning, M., Nawrocka, E., Misicka, A., Olma, A., and Klis, W. A. (1984) J. Med. Chem. 27, 423-429.
6. Manning, M., Nawrocka, E., Misicka, A., Klis, W. A., Olma, A., Kruszynski, M., Kolodziejezyk, A. M., Bankowski, K., Seto, J., and Sawyer, W. H. (1984). Proceedings of 18th European Peptide Symposium. Ulf Ragnarsson Ed. Almqvist and Wiksell Intnl. Stockholm pp. 101-405.
7. Manning, M., Olma, A., Klis, W. A., Kolodziejczyk, A. M., Nawrocka, E., Misicka-Kesik., A., Seto, J., and Sawyer, W.H. (1984) Nature 308, 652-653.
8. (a) Huffman, W. F. and Moore M. L. (1984) United States Patent No. 4,469,679. (b) Ali, F. E. and Huffman, W. F. (1984) U. S. Patent No. 4,481,193.
9. Manning, M. and Sawyer, W. H. (1985) Vasopressin, Schrier, R. W. Ed., Raven Press, New York, pp. 131-144.
10. Schrier, R. W., (1985) New. Eng. J. Med. 312, 1121-1123.

MOLECULAR SITES OF OXYTOCIN INACTIVATION IN THE RAT UTERUS: *IN VITRO* INVESTIGATION WITH ENZYME PROBES USING OIL-IMMERSION TECHNIQUE

Vladimir Pliška, Guido Jutz and Stefan Beck

Eidgenössische Technische Hochschule, Institut für Tierproduktion
CH-8092 Zürich, Switzerland

Introduction

Inactivation of a peptide in the vicinity of its receptors (receptor compartment) plays a dominating role in the duration of its biological effect[1]. Because of the inaccessibility of this tissue compartment, this process cannot be measured directly. However, modified oil-immersion technique[2,3] enables a recording of the inactivation process by pharmacological means. We have used this technique to investigate the oxytocin inactivation in the rat uterus receptor compartment. Analogues of oxytocin structurally protected against putative inactivation enzymes were employed as enzyme probes[3].

Methods

Modification of the original method[2] was described earlier[3]. Phasic contractions of a rat uterus strip were inhibited by potassium depolarization. Stimulation was achieved by oxytocin analogues listed in Table I. Kinetic analysis of the decay phase based on a two-compartment model enables estimates of rate constants of inactivation in the receptor compartment, and of peptide transport from the medium into the receptor compartment.

In vitro inactivation of peptides using minced uterine tissue (50 mg/ml) was carried out in a phosphate buffered balanced salt solution, pH 7.4. Initial peptide concentration was 0.1 to 1 μM; its decay was assayed on rat uterus *in vitro*. Rate constants of inactivation were obtained by an exponential fit.

Index of persistence (I_p)[1,4] reflects the disappearance rate of a drug from its receptor compartment in an *in vivo* experiment. Values indicated in Table I were computed from literature data[5].

Results and Discussion

Characteristic time-response profiles of oil-immersion and wash-out relaxation curves are shown in Figure 1. Nonlinearities occur particularly in the after-stimulation phase[6] (initial paek response). The initial decay in

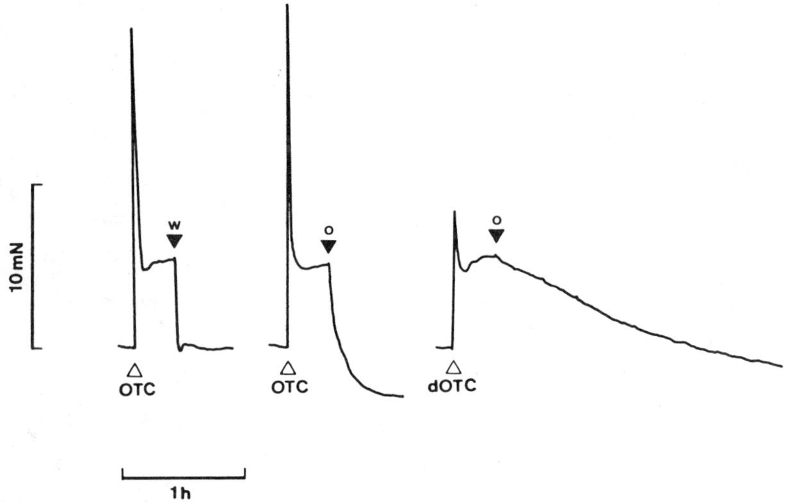

Fig. 1. Wash-out (w) and oil-immersion curves (o) for oxytocin (OTC) and deaminooxytocin (dOTC) in 10 nM concentrations. Isometric tension of the muscle (recorded with a transducer) in millinewtons (mN).

oil-immersion, on the other hand, seems to follow an exponential course. The decrease of the basal tension below the pre-stimulation value is most likely caused by small changes in internal muscle tension ofter oil-immersion; it can be ignored for purposes of the kinetic analysis. The organ, however, is not in an intimate contact with oil; it is surrounded by a tiny water layer (thickness estimated to about 0.3 mm) which supplies the residual peptide

Table I. Inactivation and Transport of Oxytocin Analogs in Rat Uterus: Rate Constant Ratios *Oxytocin/Analog*

Peptide[a]	Inactivation *in vitro* oil immersion	Inactivation *in vitro* minced tissue	I_p [b]	Transport[c] water→tissue
OTC	1	1	1	1
dOTC	4.71	1.97	1.67	2.03
[Hpm¹]-OTC	3.00	1.55		1.28
[Agl⁹]-OTC	0.54	1.54		1.18
[Agl⁹]-dOTC	8.25	1.81		4.10
carba-1-OTC	0.89	1.67	0.89	0.95
carba-1,6-OTC	0.60	2.12		1.82
carba-1-dOTC	5.50		1.47	1.65
carba-6-dOTC	4.71		6.10	3.50
[Glu(Me)⁴]-OTC	0.99			0.83
[Glu(Me)⁴]-dOTC	2.20			1.05

[a] Abbreviations: OTC, oxytocin; dOTC, deaminooxytocin; Hpm, 2-hydroxy-3-mercaptopropionic acid; Agl, azaglycine. Carba-analogs cf.[5,8].

[b] Index of persistence for rat uterus[1]. Data from[5] used for computation.

[c] Rate constants from oil-immersion relaxation curves.

to the receptor compartment in the uterus.

Inactivation constants obtained from oil-immersion experiments reflect the actual rate of drug elimination from the receptor compartment. This process is relevant for the relaxation dynamics of the responding tissue. It is not astonishing that these constants correlate only very poorly with those of inactivation by minced tissue (Table I): the latter constants also comprise inactivation by enzymes from destroyed cells which, under normal circumstances, do not participate on target tissue inactivation. The inactivation rate in this instance is rather monotonic for all analogues, without major differences between them. On the other hand, the oil-immersion measurements indicate that analogue protected against N-terminal splitting (dOTC-analogues, [Hmp¹]-OTC[7]) are considerably more stable than the rest. Protection against S-S reduction (carba-analogues[8]) and against carboxamidopeptidase[9] ([Agl⁹]-OTC) has only a minor effect upon inactivation rate. Aminopeptidase splitting is apparently the rate determining inactivation mechanism in the rat uterus. Ratio of rate constants of corresponding OTC and dOTC analogues for target tissue inactivation is 2.2 to 15.3, whereas that for *in vivo* elimination from the receptor compartment, expressed as I_p (Table I), is around 1.6. Besides inactivation, several other processes are apparently participating in overall elimination from target tissue in *in vivo* conditions.

Supported by Swiss National Science Foundation, Grant 3.478-83.

References

1. Pliška, V. (1966) Arzneimittel-Forsch. 16, 886-893.
2. Kalsner, S., and Nickerson, M. (1968) Canad. J. Physiol. Pharmacol. 46, 719-730.
3. Pliška, V., and Rudinger, J. (1976) Clin. Endocrinol. 5, 73s-84s.
4. Pliška, V. (1969) Eur. J. Pharmacol. 5, 253-262.
5. Barth, T., Krejčí, I., Vaněčková, J., and Rychlík, I. (1974) Eur. J. Pharmacol. 25, 67-70.
6. Wanner, O., Crankshaw, D., and Pliška, V. (1977) Mol. Cel. Endocrinol. 6, 281-292.
7. Hope, D.B., and Wälti, M. (1971) Biochem. J. 125, 909-911; (1972) J. Chem. Soc. Perkin I, 1946-1950.
8. Pliška, V., Rudinger, J., Douša, T., and Cort, J.H. (1968) Am. J. Physiol. 215, 916-920.
9. Glass, J.D., Dubois, B.M., Schwartz, I.L., and Walter, R. (1970) Endocrinology 87, 730-737.

VASOPRESSIN ANTAGONIST ANALOGS CONTAINING α-METHYL AMINO ACIDS AT POSITION 4

Michael L. Moore, William F. Huffman, William M. Bryan, Joanne Silvestri and Hsin-Lan Chang

Department of Peptide Chemistry, Smith Kline & French Laboratories, Philadelphia, PA 19101.

Garland R. Marshall

Department of Pharmacology, Washington University School of Medicine, St.Louis, MO 63110.

F. Stassen, J. Stefankiewicz, L. Sulat, D. Schmidt, L. Kinter, J. McDonald and D. Ashton-Shue

Departments of Pharmacology and Molecular Pharmacology, Smith Kline & French Laboratories, Philadelphia, PA 19101.

Introduction

Alpha-methyl amino acids have been shown to impart well-defined conformational constraints to the peptide backbone.[1-3] It was of interest to us to examine the effects of an α-methyl amino acid on the cyclic hexapeptide moiety of the vasopressin antagonists shown in Figure 1.[4]

```
Pmp-X-Phe-Y-Asn-Cys-Pro-Arg-Z
 |               |
 S---------------S
```

Figure 1. Structure of vasopressin antagonists

Results and Discussion

Peptides were synthesized on BHA resin, cleaved from the resin with HF and then cyclized with dilute ferricyanide. Their biological activities were evaluated _in vitro_ for both receptor binding and inhibition of LVP-sensitive adenylate cyclase in pig renal medullary membranes.[5] The _in vivo_ activity was evaluated in the hydropenic rat.[6] The results are summarized in Table 1.

Regardless of the optical configuration at position 2 or the presence or absence of the carboxyl-terminal glycine, simple α-methyl amino acid substitutions at position 4 are compatible with biological activity, although there is a roughly five-fold decrease in potency. The cycloleucine substitution, however, behaves quite differently, giving analogs which are 100- to 1000-fold less potent. This is not an effect of deleting the ß-branch, since it has been previously shown[7] that a ß-branch is not essential at position 4 and the Aib analog does retain some activity.

Since cycloleucine resembles an α-methyl amino acid, it should produce a conformational effect on the peptide backbone comparable to an α-methyl amino acid. The biological data shown indicate that the constraint produced by α-methylvaline is compatible with the receptor-bound conformation, so the cycloleucine analogs would be expected to retain good biological activity on the basis of backbone conformational effects. The difference between cycloleucine and non-cyclic α-methyl amino acids is that in cycloleucine the side chain is held in a fixed relationship to the peptide backbone which puts steric bulk in a region which is not occupied in the non-cyclic α-methyl amino acids. This can be seen in figure 2 in which α-methylvaline and cycloleucine are shown in the same relative orientation.

This steric bulk must prevent the binding of an otherwise conformationally compatible molecule to the receptor, implying that the receptor makes a close contact to position 4 of the vasopressin antagonists. The presence of receptor bulk which would interfere with a cycloleucine side chain but not with an α-methylvaline side chain also suggests that the valine t rotamer (χ_1=180°) would similarly be sterically disallowed. Of the remaining gauche rotamers, the g$^-$ rotamer (χ_1= -60°) is the more favorable, suggesting that that orientation of the valine side chain is presented in the receptor bound conformation.

The steric volume of cycloleucine which is not in common with α-methylvaline can be calculated using the g$^-$ rotamer of the valine side chain. This volume must overlap with the steric bulk of the receptor, and thus a portion of the receptor topology can be mapped with respect to the peptide backbone at position 4. Determination of receptor topology will prove critical for the <u>de novo</u> design of non-peptide ligands which interact with peptide receptors, which is an important goal of the peptide drug design process.

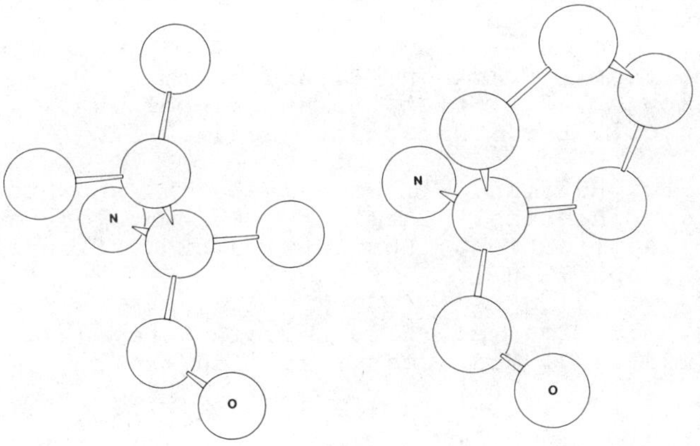

Figure 2. α-Methylvaline and cycloleucine

Table 1. Activity of Position 4 Analogs

X	Y	Z	K_b (nM)	K_i (nM)	ED_{300} (μg/kg)[8]
Tyr(Me)	Val	GlyNH2	88.	35.	98.
Tyr(Me)	α-MeVal	GlyNH2	490.	170.	61.
Tyr(Me)	Cle	GlyNH2	11,000.	n.d.	>300.
D-Tyr(Et)	Val	GlyNH2	12.	6.7	11.2
D-Tyr(Et)	α-MeVal	GlyNH2	311.	11.	18.
D-Tyr(Et)	Val	NH2	11.	4.5	9.2
D-Tyr(Et)	Aib	NH2	580.	140.	39.
D-Tyr(Et)	Cle	NH2	10,000.	n.d.	n.d.

References

1. Marshall, G. R. and Bosshard, H. E. (1972) Circ. Res., Suppl. II, 30-31, II-143-150.

2. Toniolo, C., Bonora, G. M., Bavoso, A., Benedetti, E., Di Blasio, B., Pavone, V. and Pedone, C. (1983) Biopolymers, 22, 205-215

3. Jung, G., Bosch, R., Katz, E., Schmitt, H., Voges, K. P. and Winter, W. (1983) Biopolymers, 22, 241-246.

4. Kruszynski, M., Lammek, B., Manning, M., Seto, J., Haldar, J. and Sawyer, W. H. (1980) J. Med. Chem., 23, 364-368.

5. Stassen, F. L., Erickson, R. W., Huffman, W. F., Stefankiewicz, J., Sulat, L. and Wiebelhaus, V. D. (1982) J. Pharmacol. Exp. Therapeut., 223, 50-54.

6. Kinter, L., Huffman, W., Wiebelhaus, V. and Stassen, F. (1984) in Diuretics: Chemistry, Pharmacology and Clinical Applications, Puschett, J. Ed., Elsevier, New York, NY, pp. 72-81.

7. Manning, M., Olma, A., Klis, W. A., Seto, J. and Sawyer, W. H. (1983) J. Med. Chem., 26, 1607-1613.

8. ED_{300} is the dose required to lower urine osmolality to 300 mOsm.

NOVEL VASOPRESSIN ANTAGONISTS LACKING THE PROLINE AND THE GLYCINE RESIDUES AND CONTAINING C-TERMINAL DIAMINOALKANES AND AMINO ALKYLGUANIDINES

James F. Callahan, Fadia E. Ali, William Bryan, Hsin-Lan Chang, Heidemarie Greene, William F. Huffman, Michael L. Moore, Joanne Silvestri and Nelson Yim

Department of Peptide Chemistry, Smith Kline and French Laboratories, Philadelphia, PA 19101

Grace Heckman, Lewis Kinter, Jeanne McDonald, Dulcie Schmidt, Darryl Shue, Frans Stassen and Lynn Sulat

Departments of Pharmacology and Molecular Pharmacology Smith Kline and French Laboratories, Philadelphia, PA 19101

Introduction

As part of our ongoing investigation to identify the minimum pharmacophore requirements for the vasopressin V_2-receptor antagonists, we decided to study the role of the tripeptide tail (Pro-Arg-Gly-NH_2) in antagonist activity. In the model proposed by Walter, et al.[1] for the biologically active conformation of vasopressin at the renal V_2-receptor, the entire tripeptide tail is necessary for full agonist activity and proline at position 7 plays a key role in the orientation of the tripeptide tail with respect to the cyclic hexapeptide ring as well as being an important binding element by virtue of its exposure as a corner residue in a ß-turn. We have, however, previously presented evidence that the pharmacophore requirement of the renal V_2-receptor differs for vasopressin agonists and antagonists[2]. Manning, et al.[3] as well as our group[4] have shown that potent vasopressin antagonists could be obtained by deletion of the carboxyl-terminal glycine residue leaving a terminal arginamide 1 and that antagonist analogs containing either D or L arginine at position 8 were essentially equipotent.[5] We have also found that proline at position 7 can be replaced by D-proline with good retention of biological activity.[6] These observations suggested that the terminal carboxamide group may not play a role in antagonist binding and also that the orientation of the tripeptide tail may not be as critical in antagonists as it is in agonists.

Results and Discussion

In order to further elucidate these points, we first synthesized compounds 3a,b and 4, where the terminal argininamide of 1 was replaced by an alkyl diamine or an aminoalkyl guanidine, respectively. Compounds 3a and 3b were synthesized by a combination of solid phase and solution synthesis. The proline acid 8 was synthesized on Merrifield resin, cleaved from the resin with anhydrous HF and cyclized under dilute conditions with aqueous potassium ferricyanide. After purification by flash chromatography (C-18 reversed phase), 8 was coupled in solution with the respective mono-protected diamine (DCC, HOBT), followed by deprotection (TFA) and purification by HPLC. The guanidino terminal peptide 4 was formed by treating 3a with excess aqueous O-methylisourea at pH 10 followed by purification by HPLC. Compounds 5 and 6[7] were prepared by solid phase synthesis on benzhydrylamine resin, cleaved with anhydrous HF, cyclized using dilute aqueous potassium ferricyanide, and purified using countercurrent distribution followed by gel filtration. Compound 7 was synthesized in an analogous way to 3 utilizing the corresponding cysteine acid 9.

These peptides were evaluated *in vitro* for binding (K_{bind}) and inhibition of LVP-sensitive adenylate cyclase (K_i) in porcine renal medullary preparation[5] and *in vivo* in a hydropenic rat model.[8] The results, summarized in

Table I. Biological Activities of Vasopressin Antagonists

```
Pmp-D-Tyr(Et)-Phe-Val-Asn-Cys-X
 |                           |
 S---------------------------S
```

		Pig		Rat
Cpd.	X	K_{bind} (nM)	K_i (nM)	ED_{300} (μg/kg)[9]
1	Pro-Arg-NH_2	12	4.5	9
2	Pro-Lys-NH_2	37	9.4	22
3a	Pro-$NH(CH_2)_4NH_2$	11	6.4	19
3b	Pro-$NH(CH_2)_5NH_2$	30	8.3	27
4	Pro-$NH(CH_2)_4NHC(=NH)NH_2$	14	6.0	22
5	Arg-NH_2	9	2.5	58
6	Lys-NH_2	26	5.8	59
7	$NH(CH_2)_5NH_2$	19	6.8	94
8	Pro-OH			
9	OH			

Table I, show that both the aminoalkyl analogs 3a,b and the guanidino analog 4 all retain very significant *in vivo* and *in vitro* activity when compared to 1 or 2. The des-proline analogs 5-7 retain very good *in vitro* activity compared to 1 and 2 but have somewhat reduced *in vivo* activity, which may reflect metabolic as well as posssible pharmacokinetic differences.

Taking into consideration the data presented above, we have shown that the terminal carboxamide is not necessary for antagonist activity and that proline at position 7 is not a critical element of the V_2-receptor antagonist pharmacophore. This data supports a minimum effective V_2-receptor antagonist pharmacophore which can be presented by a cyclic hexapeptide ring with a positively charged residue attached to the ring.

References

1. Walter, R., Smith, C., Mehta, P., Boonjarern, S., Arruda, J. and Kurtzman, N. (1977) Disturbances in Body Fluid Osmolality, Andreoli, T., Grantham, J., and Rector Jr., F., Eds., American Physiological Society, Bethesda, 1-36.

2. Moore, M., Huffman, W., Roberts, G., Rottschaefer, S., Sulat, L., Stefankiewicz, J. and Stassen, F. (1984) Biochem. Biophys. Res. Com. 121, 878-883.

3. Manning, M., Olma, A., Klis, W., Kolodziejcyk, A., Nawrocka, E., Misicka, A., Seto, J. and Sawyer, W. (1984) Nature 308, 652-653.

4. Huffman, W. and Moore, M. (1984) U.S. Patent No. 4,469,679.

5. Stassen, F., Erickson, R., Huffman, W., Stefankiewicz, J., Sulat, L. and Wiebelhaus, V. (1982) J. Pharmacol. Exp. Therapeutics 223, 50-54.

6. Stassen, F., Heckman, G., Schmidt, D., Stefankiewicz, J., Sulat, L., Huffman, W. and Kinter, L. (1984) Presented at the Conference on Vasopressin, Aspen, Colorado.

7. Ali, F. and Huffman, W. (1984) U.S. Patent Nos. 4,481,193 and 4,481,194.

8. Kinter, L., Huffman, W., Wiebelhaus, V. and Stassen, F. (1984) Diuretics: Chemistry, Pharmacology and Clinical Applications, Puschett, J. ed., Elsevier, New York, 72-81.

9. ED_{300} is the dose required to lower urine osmolality to 300 mOsm/kg in hydropenic rat model.

THE SYNTHESIS AND STRUCTURE-ACTIVITY STUDIES OF VASOPRESSIN ANTAGONISTS MODIFIED AT POSITIONS ONE AND TWO

Nelson C. F. Yim, Fadia E. Ali, William M. Bryan, William F. Huffman, Michael Moore, Joanne Silvestri, R. Erickson*, G. Heckman+, L. Kinter*, J. McDonald*, D. Schmidt+, D. Shue*, F. Stassen+, J. Stefankiewicz* and L. Sulat*

Depts. of Peptide Chemistry, Pharmacology* and Molecular Pharmacology +, Smith Kline and French Laboratories, Philadelphia, PA 19101.

Su-Sun Wang
Alpha Biochemicals, Inc., San Carlos, CA 94070

INTRODUCTION

In our attempt to determine the antagonist pharmacophore of potent peptide antagonists of vasopressin V_2-receptors, we have employed a broad spectrum of approaches. A portion of our effort was focused on the classical single amino acid substitution SAR approach. Some very interesting results were obtained in our investigation of positions one and two which are proposed to be key features of the peptide antagonists.[1] Manning and his co-workers reported[2] the first potent vasopressin antagonist by incorporating ß-mercapto-ß,ß-cyclopentamethylenepropionic acid (Pmp-OH) in position 1. From the literature the Pmp-OH in position 1 is apparently important for antagonists' activity.[1,3,4] However, the ring size requirement for the antagonism has never been reported.[5]

Using the methodology we developed for the synthesis of ß-(S-Benzylmercaptan)-ß,ß-cyclopentamethylenepropionic acid [Pmp(S-Bzl)-OH], we prepared analogs of various ring sizes from an appropriate alicyclic ketone as shown in Scheme 1.

These intermediates were incorporated in position 1 with general structure:

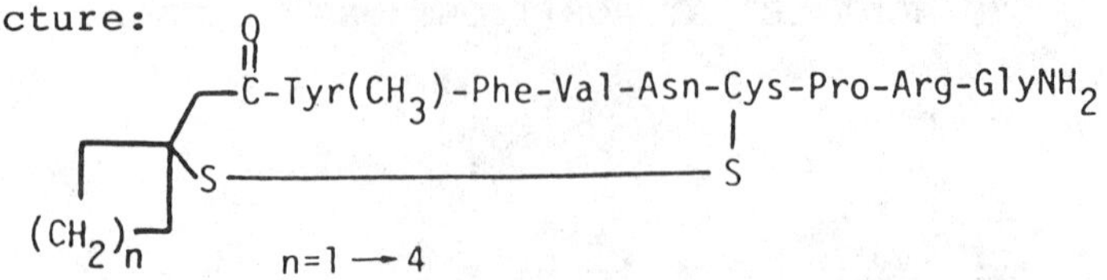

n=1 → 4

We have also independently prepared a number of analogs with D or L amino acids with aromatic, aliphatic, or heterocyclic side chains at position 2.

The biological activities of these peptides were examined.

Scheme 1

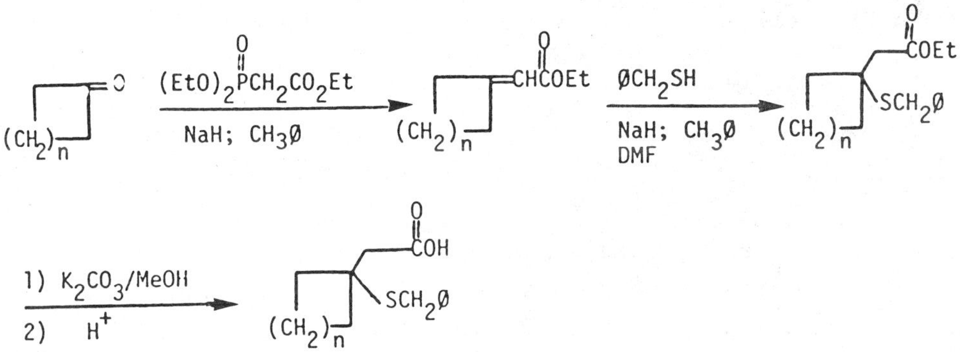

RESULTS AND CONCLUSIONS

All the peptides were prepared on chloromethylated polystyrene. The properly protected straight chain peptide was cleaved from the resin by ammonolysis, deprotected by treatment with sodium/liquid ammonia and subsequently cyclized with dilute ferricyanide. Their biological activities were evaluated in vitro for both receptor binding and inhibition of LVP-sensitive adenylate cyclase in the porcine medullary preparation[7] and in vivo in the hydropenic rat model.[8] The results are summarized in Table 1. We

Table 1. Biological Activities

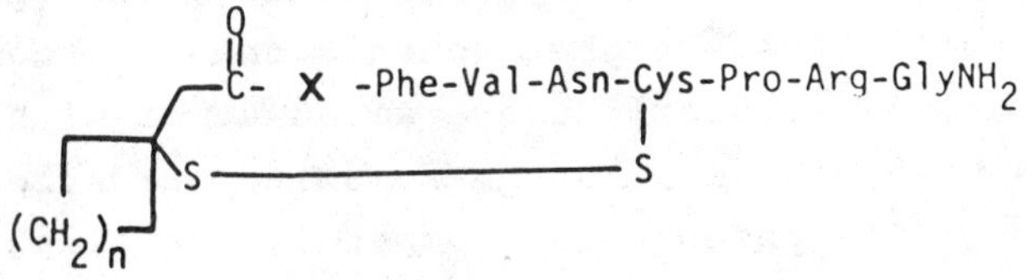

n	X	Pig K_b (nM)	Pig K_i (nM)	Other Species K_i	ED_{300}* µg/Kg
4	L-Tyr(Me)	290	100	-	w. ant.
3	L-Tyr(Me)	88	35	6.9^R; 29^H	97.5
2	L-Tyr(Me)	110	40	-	60
1	L-Tyr(Me)	660	220	-	6031
3	L-Tyr(Me)	88	35	6.9^R	97.5
3	D-Tyr(Me)	12	7.9	10^R	15.2±6.1
3	L-Phe(pEt)	14	54	-	-
3	L-Tyr(Et)	100	32	5.1^R; 25^H	25.8±8
3	D-Tyr(Et)	12	6.7	1.3^R; 4.6^H	11.2±5.3
3	L-Trp	2150	-	-	1800
3	D-Trp	130	63	-	13.8
3	L-Cha	1200	670	-	160
3	D-Cha	12000	520	180^H	11.5
3	L-Leu	5900	2300	84^R; 1800^H	109
3	D-Leu	11500	600	-	118
3	D-Phe	110	47	6^R	23.7±10
3	L-Phe	900	250	-	431
3	D-Phe	110	47	6^R	23.7
3	D-Phg	23000	1000	-	5000
3	D-Pba	170	110	-	19.3

Phg. = phenylglycine; Pba = Aminophenylbutyric Acid; Cha = Cyclohexylalanine; R = Rat; H = Human.
*ED_{300} is the dose required to lower urine osmolality to 300 mOsm.

observed that optimal antagonist potency was achieved with the cyclopentane or cyclohexane ring size in position 1. Either smaller or larger rings led to dramatically reduced activity. At position 2, we noticed several trends in relative biological activities. Generally, the substitution of D-amino acids gave a higher potency than the corresponding L-amino acids, particularly those amino acids with aromatic side chains. In addition, aromatic amino acids, whether D or L, were especially potent when compared with aliphatic or heterocyclic amino acids. Furthermore, the distance from the phenyl ring to the alpha carbon is extremely important as shown in compounds Phg^2, Pba^2 as compared with Phe^2. A species difference between rat and pig or human was observed. In particular, D- or L-alkyl amino acids gave antagonists which are quite potent in the rat, but markedly less so in the pig and human. The most potent substitution at position 2 in all species is the D-tyrosine alkyl ether or the isosteric D-(p-alkyl)-phenylalanine.

REFERENCES

1. Manning, M., Olma, A., Klis, W. Kolodziejczyk, A., Seta, J. and Sawyer, W. (1982) J. Med. Chem. 25, 45-50.
2. Lowbridge, J., Manning, M., Haldar, J. and Sawyer, W. (1978) J. Med. Chem. 21, 313-315.
3. Kruszynski, M., Lammek, G., Manning, M., Seto, J., Haldar, J. and Sawyer, W. (1980) J. Med. Chem. 23, 364-8
4. Manning, M., Nawrocka, E., Misicka, A., Olma, A., Klis, W., Seto, J. and Sawyer, W. (1984) J. Med. Chem. 27, 423-29.
5. A recent U.S. patent (1985, 4,491,577) claimed the cyclopentane as well as the cyclohexane ring at position 1.
6. Yim, N.C.F. and Huffman, W. (1983) Int. J. of Prot. and Pept. Res. 21, 568-70.
7. Stassen, F., Erickson, R., Huffman, W., Stefankiewcz, J., Sulat, L. and Wiebelhause, W. (1982) J. Pharmacol. Exp. Therapeut., 223, 50-54.
8. Kinter, L., Huffman, W., Wiebelhaus, W. and Stassen, F. (1984) in Diuretics: Chemistry, Pharmacology and Clinical Applications, Puschett, J. Ed., Elsevier, New York, N.Y. pp 72-81.

CONFORMATIONAL STUDY OF SOMATOSTATIN IN METHANOL SOLUTION.

Ellen M. van den Berg, Arnold W. Jans and Georges Van Binst.
ORGC, Vrije Universiteit Brussel, Pleinlaan 2, B1050 Brussel

Arison and Veber [1] studied Somatostatin in methanol solution and proposed Phe^6-Phe^{11} stacking and a proximity between Lys^9 and Trp^8. Our previous investigations in H_2O/D_2O to detect the proposed Phe^6-Phe^{11} interactions did not show the presence of such effects [2]. Starting from the known assignments in aqueous solution [3,4] we were able to assign the 500 MHz ^1H NMR spectrum in methanol-d4 by solvent titration. The assignments were confirmed by 2D correlation techniques. The important observations in these experiments are the following:

1°) the signals of α-Lys^4 and α-Lys^9 do not cross,

2°) the signals of γ-Lys^4 and γ-Lys^9 <u>do cross</u>,

3°) the signals of ε-Lys^4 and ε-Lys^9 <u>do cross</u>,

4°) the signals of α-Thr^{10} and α-Thr^{12} do not cross.

Furthermore:

one of the β-Cys^3 proton signals is shifted 0.67 ppm to low field in methanol with respect to water;

the α-Cys^3 proton signal is shifted 0.51 ppm to high field;

the β-proton signals of Asn^5 and Phe^6 remain the most upfield shifted β-protons;
the β-Ser^{13} and α-Gly^2 protons which are equivalent in water are now chemically unequivalent and appear respectively as twice a double doublet and as a double doublet.
In order to "freeze out" some conformations, we performed 2D NOE experiments at 263 K, which we found to be the limit below which line broadening occured, with t_m values = 340, 600, 800 and 1200 ms. The most important interresidue NOE's are the following:
$4Trp^8$ - αCys^{14}, $5Trp^8$ - βCys^{14}, $5Trp^8$ - $4Phe^7$, $2Phe^6$ - $2Phe^7$, βLys^4 -$7Trp^8$, γLys^4 - $7Trp^8$.

The main discrepancy with the previous publication of Arison and Veber [1] is the evidence that the γ-protons of Lys^4 show the most upfield signals instead of Lys^9, at 303 K as well as at 263 K. Furthermore, in methanol the signals of the α-protons are spread out over a range of 1.35 ppm (at 303 k) compared with 0.66 ppm for D_2O which is an indication for more conformational differenciation between them. The same occurs for the geminal β-protons which are more chemical unequivalent in methanol than in water. The observed upfield shift for the $2Phe^6$ and $3Phe^6$ protons is remarkably more pronounced in methanol. In a previous publication [4], we pointed out that only one set of minimum energy conformations was in agreement with all the parameters measured by NMR in water. We also already mentioned that a conformer of the proposed set B1 could give rise to shielding of Lys^4 by Trp^8.

The present measurements indicate a more pronounced proportion of this conformer in methanol solution as confirmed by the following interresidue NOE and chemical shift effects:

1°) Phe^6 and Phe^7 are in close proximity according to the important upfield shift of protons 2 and 3 of Phe^6 and NOE interactions between $2Phe^6$ and $2Phe^7$;

2°) proximity of the β, γ, δ and ε protons of Lys^4 to the indole ring of Trp^8 shown by the upfield shifts and the following NOE effects: $7Trp^8$ - βLys^4, $7Trp^8$ - γLys^4;

3°) particular orientation of the Lys^9 side chain: upfield shift of the α-proton, downfield shift of one of the β-protons which results in an important chemical unequivalence of these protons;

4°) at 303 K a very important chemical unequivalence of the β-protons of Cys^3 and Cys^{14} is observed. One signal of each β-proton pairs is shifted downfield. At 263 K, the effect is emphasized for Cys^{14} and disappears for Cys^3. These effects are due to the side chain orientation of an aromatic ring which has been identified as the Trp^8 by NOE interactions: $4Trp^8$ - αCys^{14} and $5Trp^8$ - βCys^{14};

5°) the αThr^{12} and βThr^{12} protons are deshielded by Phe^{11} or Phe^7; Thr^{10} undergoes the same influences as in water;

6°) from the temperature coefficients of the Thr^{10} NH

($\frac{\Delta\delta}{\Delta T} < 1.0\ 10^{-3}$) and from the J values ($^3J_{NH}$ - αLys^9 = 3.5 Hz), evidence is shown for an intramolecular H bond Thr^{10}-NH--O=C-Phe^7 stabilizing a β-turn at that level;
7°) the values of the Phe^{11} aromatic proton signals are normal ones and similar to their random coil values.
All these features confirm the presence of a conformer of set Bl as major conformation. No evidence has been shown for interactions between Phe^6 or Phe^7 and Phe^{11} although differentiation between Phe^6 and Phe^7 has not thoroughly been possible due to overlapping signals during the solvent titration.

References

1. Arison, B.H., Hirschmann, R., Palaveda, W.S., Brady, S.F., and Veber, D.F. (1981) Biochem. Biophys. Res. Comm. 100, 1148-1153.

2. Jans, A.W.H., Hallenga, K., Van Binst, G., Michel, A., Scarso, A., and Zanen, J. (1985) Biochim. Biophys. Acta 827, 447-452.

3. Hallenga, K., Van Binst, G., Scarso, A., Michel, A., Knappenberg, M., Dremier, C., Brison, J., and Dirkx, J. (1980) FEBS Letters 119, 47-52.

4. Knappenberg, M., Michel, A., Scarso, A., Brison, J., Zanan, J., Hallenga, K., Deschrijver, P., and Van Binst, G. (1982) Biochim. Biophys. Acta 700, 229-246.

CONFORMATION AND DESIGN OF CYCLIC SOMATOSTATIN ANALOGS WITH HIGH AFFINITY FOR MU-OPIATE RECEPTORS

John T. Pelton, Wayne L. Cody, and Victor J. Hruby
Department of Chemistry, University of Arizona
Tucson, Arizona 85721

Karoly Gulya, Lane D. Hirning, Henry I. Yamamura, and Thomas F. Burks
Department of Pharmacology, University of Arizona
Tucson, Arizona 85724

Introduction

Somatostatin, a cyclic tetradecapeptide, is a key regulatory hormone in several biological pathways. It is found in most, but not all organs and exhibits specific and selective functions, depending on its location. While in most systems the physiological significance of somatostatin is clear, in others its precise role is less well defined. The weak affinity of somatostatin for the opiate receptor is one such system. Recently, we have prepared a number of conformationally restricted analogs of somatostatin designed primarily to examine the features important for activity at the opiate receptor[1]. We wish to report here NMR spectral studies of our most potent, μ-opiate receptor selective analog, D-Phe-Cys-Tyr-D-Trp-Lys-Thr-Pen-Thr-NH_2.

Results and Discussion

The activities of somatostatin and several analogs at opiate receptors is summarized in Table I. The conformationally restricted analog, D-Phe-Cys-Tyr-D-Trp-Lys-Thr-Pen-Thr-NH_2 ([Cys^2, Tyr^3, Pen^7]NH_2), binds to mu-opiate receptors in the rat brain with high affinity (3.5 nM) and selectivity (IC_{50} δ/IC_{50} μ = 2030). In the guinea pig ileum assay, this analog acts as an opiate antagonist, shifting the dose-response curve of morphine

Table I. Biological Activity of Somatostatin and Analogs

Peptide	$[^3H]$Naloxone IC_{50} (nM)	$[^3H]$DPDPE[1] IC_{50} (nM)	Ke(GPI)[2] (nM)
Somatostatin	27,400 ± 4,200	16,400 ± 8,500	>1000
[Cys[2], Tyr[3], Pen[7]]NH$_2$	3.5 ± 0.2	7,100 ± 600	49
[Cys[2], Phe[3], Pen[7]]NH$_2$	9.9 ± 1.6	24,250 ± 3,100	41
SMS 201-995[3]	38 ± 60[4]		400

[1]DPDPE = [D-Pen[2], D-Pen[5]]Enkephalin; [2]Ke = Q/(DR-1), where Q = antagonist concentration and DR = dose ratio in antagonist-treated versus control preparations; [3]SMS 201-995 = D-Phe-Cys-Phe-D-Trp-Lys-Thr-Cys-Thr(ol); [4]value as reported in reference (2).

to the right (Ke = 49 nM). In the mouse vas deferens preparation, the adrenergic α_2 antagonist properties of somatostatin and our analogs give an inhibitory activity that is not reversed by opiate agonists or antagonists.

The flexibility of many 20-membered ring systems in aqueous solutions[3] led us to investigate the conformationally restricted, mu-opiate receptor selective analog D-Phe-Cys-Tyr-D-Trp-Lys-Thr-Pen-Thr-NH$_2$. In this peptide the geminal β,β-dimethyl group of penicillamine should restrict the dihedral angle of the disulfide bond and the 20-membered ring system, via transannular gem dimethyl effects. An examination of the NMR spectrum of this

Table II. ^1H NMR Parameters of the Amide and Cα Protons

Amino Acid	δ_{NH}	$^3J_{NH\text{-}C\alpha H}$	$-\Delta\delta/\Delta T$[1]	δ_α	$^3J_{\alpha\beta}$
D-Phe[1]				4.35	7.3|7.8
Cys[2]	----[2]			5.07	5.4|8.8
Tyr[3]	8.49	8.75	8.5	4.41	5.9|11.3
D-Trp[4]	8.56	7.25	9.8	4.68	6.4|7.8
Lys[5]	8.19	8.75	7.0	3.95	3.7|10.5
Thr[6]	8.15	10.50	4.8	4.42	4.4
Pen[7]	8.03	8.50	6.9	4.78	---
Thr[8]	8.62	5.50	8.9	4.38	6.4

[1]Amide proton temperature shifts in ppb/^{o}K; [2]Cys amide proton not found.

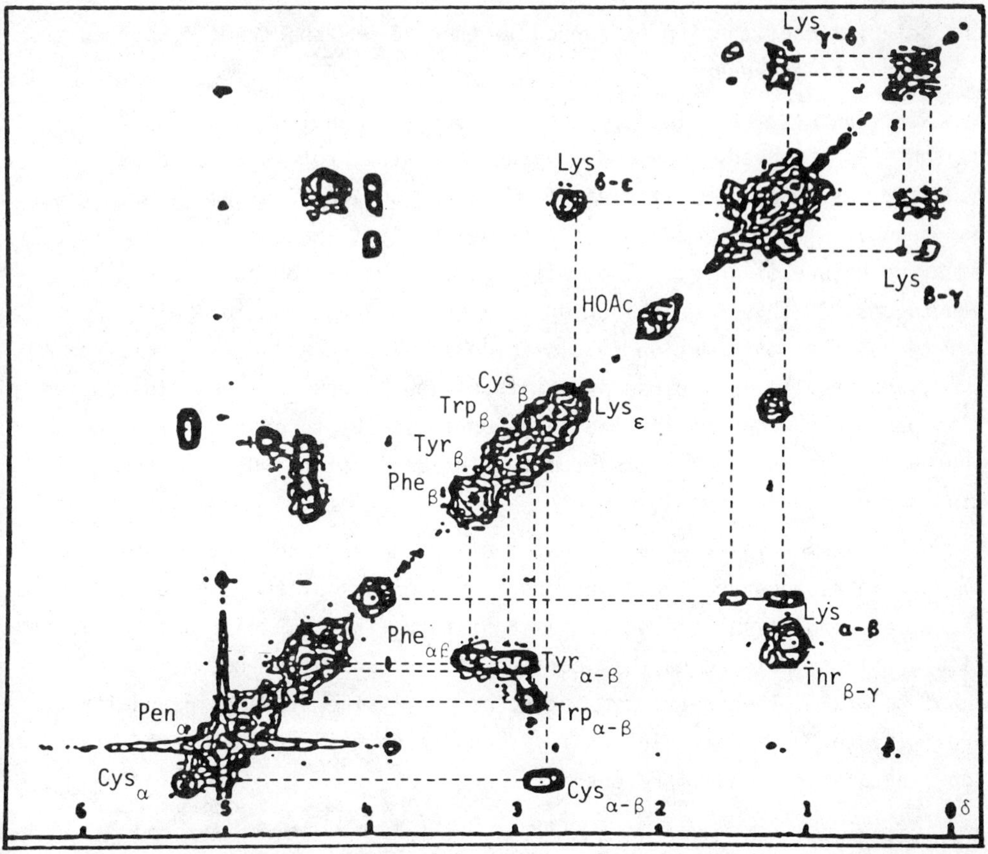

Fig. 1. The 0.0 to 6.5 ppm region of the ^1H COSY spectrum. The connectivities within each amino acid residue as well as the diastereotopic nature of the Lys β and γ methylene protons are readily apparent.

analog has confirmed these expectations.

Chemical shifts, temperature dependencies and coupling constants of the amide protons are summarized in Table II. Resonances were assigned with the aid of two-dimensional ^1H-^1H COSY experiments (Figure 1), and conventional one dimensional decoupling studies. Long-range correlations between the Cβ methylene protons and the C_2 protons of the aromatic residues were obtained from a COSY experiment incorporating a delay sequence[4]. Assignment of the phenylalanine resonances was subsequently confirmed through conventional pH titration experiments which, when employed with the C-terminal carboxylic acid analog, also led to an assignment of the individual threonine residues. Hetero-nuclear decoupling experiments, combined with the up-field position characteristic of the Trp Cβ carbon in the ^{13}C NMR spectrum, supported the assignment of these protons.

No evidence of any intra-molecular hydrogen bonding involving the amide protons was observed. The $\Delta\delta/\Delta T$ values of these protons were large and negative (as in the NMR studies of SMS 201-995[4], the Cys^2 amide proton was not found). The observed $^3J_{NH-C\alpha H}$ values reflect the conformationally constrained nature of the peptide, with four of the six observed coupling constants greater than 8 Hz. The $^3J_{NH-C\alpha H}$ values are compatible with the existence of a conformation in ϕ,ψ-space similar to a C_7 turn from Tyr^3 to Lys^5, and results in a close proximity of the D-Trp^4 and Lys^5 side chains. This proximity accounts for the large up-field shifts of the Lys β and γ proton resonances and results from the influence of the D-Trp^4 aromatic ring current.

Calculated rotamer populations about the Cα-Cβ bond of the individual amino acid residues reflect the constrained nature of the peptide with decreased populations of some rotamer states. The Lys^5 residue, in particular, is almost exclusively confined to the *gauche*(-) rotamer with χ_1 equal to 87%[5]. The restricted nature of the lysine side-chain and its proximity to the indole ring of tryptophan accounts for the marked diastereotopic nature of β and γ protons.

Acknowledgements

This work was supported, in part, by U.S. Public Health Service Grants AM 21085, HL 30956, MH 27257 and AM 06939

References

1. Pelton, John T., Gulya, Karoly, Hruby, Victor J., Duckles, Sue Piper, and Yamamura, Henry I. (1985) *Proc. Natl. Acad. Sci. USA*, *82*, 236-239

2. Maurer, R., Gaehwiler, B.H., Buescher, H.H., Hill, R.C., and Roemer, D. (1982) *Proc. Natl. Acad. Sci. USA*, *79*, 4815-4817

3. Hruby, Victor J., Mosberg, Henry I., Sawyer, Tomi K., Knittel, James J. Rockway, Todd W., Ormberg, James, and Darman, Paul (1983) *Biopolymers*, *22*, 517-530

4. Wynants, C., and Van Binst, G. (1984) *Biopolymers*, *23*, 1799-1804

5. DeLeeuw, F.A.A.M., and Altona, C. (1982) *Int. J. Pept. Protein Res.*, *20*, 120-125

SYNTHESIS AND EVALUATION OF ACTIVITIES OF OCTAPEPTIDE ANALOGS OF SOMATOSTATIN

Ren-Zhi Cai, Balazs Szoke, Dadin Fu, Tommie W. Redding, John Colaluca, I. Torres-Aleman and Andrew V. Schally
VA Medical Center and Tulane University Medical Center
New Orleans, Louisiana 70146

Introduction

Considerable progress[1] has been made in recent years in the synthesis of highly potent and selective analogs of somatostatin. One of the most important of these analogs is D-Phe-Cys-Phe-D-Trp-Lys-Thr-Cys-Thr-ol (SMS 201-995) synthesized by Bauer *et al.*[2]. This analog is 20-70 times more potent than somatostatin in inhibiting the secretion of GH *in vivo* and much more selective in suppressing the release of GH than that of insulin or glucagon. Meanwhile, a new cyclic hexapeptide, cyclo(N-Me-Ala-Tyr-D-Trp-Lys-Val-Phe), was reported by Veber *et al.*[3] in which the replacement of Phe-7 and Thr-10 by Tyr-7 and Val-10, respectively, resulted in a dramatic increase in potency for inhibition of insulin, glucagon and GH release. Recently nearly 200 octapeptide amide analogs of somatostatin related to compound SMS 201-995 have been synthesized by solid phase methods in our laboratory. This paper reports the synthesis and the evaluation of biological activities of some of these analogs after purification by HPLC. Two of these analogs, D-Phe-Cys-Tyr-D-Trp-Lys-Val-Cys-Thr-NH_2 and D-Phe-Cys-Tyr-D-Trp-Lys-Val-Cys-Trp-NH_2, have high potency and show prolonged action for inhibiting the secretion of GH.

Results and Discussion

Some of the analogs synthesized containing Phe-7 and Thr-10 or Tyr-7 and Val-10 and modifications in the N- and C-terminal residues are shown in Table 1. The biological activity was measured in vivo in Nembutal anesthetized rats and compared to that of somatostatin on the basis of 4-point assays.[4]

Table I Comparison of the Biological Activity on Inhibition of GH in vivo of the Series of Somatostatin Analogs Synthesized in our Laboratory with that of Somatostatin and the Analog of Bauer et al. (SMS 201-995).

Code Number	Structure	GH Inhibition
	Somatostatin-14	100
SMS 201-995	D-Phe-Cys-Phe-D-Trp-Lys-Thr-Cys-Thr-ol	7000*
	5 6 7 8 9 10 11 12	
RC-115-II-2H	D-Phe-Cys-Tyr-D-Trp-Lys-Val-Cys-Gly-NH_2	95
RC-138-2H	D-Phe-Cys-Tyr-D-Trp-Lys-Val-Cys-Ala-NH_2	1570
RC-114-2H	D-Phe-Cys-Tyr-D-Trp-Lys-Val-Cys-Seu-NH_2	4280
RC-150-2H	D-Phe-Cys-Tyr-D-Trp-Lys-Val-Cys-Abu-NH_2	1530
RC-121-2H	D-Phe-Cys-Tyr-D-Trp-Lys-Val-Cys-Thr-NH_2	11800
RC-113-2H	D-Phe-Cys-Tyr-D-Trp-Lys-Val-Cys-Tyr-NH_2	3020
RC-76-2H	D-Phe-Cys-Phe-D-Trp-Lys-Thr-Cys-Pro-NH_2	1070
RC-116-2H	D-Phe-Cys-Tyr-D-Trp-Lys-Val-Cys-Pro-NH_2	95
RC-102	D-Phe-Cys-Phe-D-Trp-Lys-Thr-Cys-Thr-NH_2	4000
RC-121	D-Phe-Cys-Tyr-D-Trp-Lys-Val-Cys-Thr-NH_2	19920
RC-159-II	D-Phe-Cys-Tyr-L-Trp-Lys-Val-Cys-Thr-NH_2	2380
RC-161	Ac-D-Phe-Cys-Tyr-D-Trp-Lys-Val-Cys-Thr-NH_2	5520
RC-102-II	Ac-Phe-Cys-Tyr-D-Trp-Lys-Val-Cys-Thr-NH_2	4600
RC-88	p-Cl-D-Phe-Cys-Phe-D-Trp-Lys-Thr-Cys-Thr-NH_2	2130
RC-88-II	p-Cl-D-Phe-Cys-Tyr-D-Trp-Lys-Val-Cys-Thr-NH_2	1740
RC-95-I	D-Phe-Cys-Phe-D-Trp-Lys-Thr-Cys-Trp-NH_2	5280
RC-160	D-Phe-Cys-Tyr-D-Trp-Lys-Val-Cys-Trp-NH_2	13460

* Data is taken from reference No. 2, Bauer et al.

Most of the analogs in Table I are much more potent than somatostatin. In octapeptides with N-terminal D-Phe and C-terminal $ThrNH_2$ or $TrpNH_2$ such as RC-102 and RC-95-I, it was possible to enhance the activity by incorporating Tyr-7 and Val-10 in place of Phe-7 and Thr-10, respectively, as in analags RC-121 and RC-160. However, the same substitutions in other analogs with different N- or C-terminal residues resulted in either no change (RC-88,RC-88-II) or a decrease (RC-76-2H, RC-116-2H) in the activity. These results indicate the importance of the C- and N-terminal residues in this series of somatostatin analogs. Analog RC-121 containing D-Trp-8 shows higher potency than its diastereomeric counterpart RC-159-II with L-Trp-8. This is in agreement with previous results[5] on somatostatin analogs containing 14 amino acids or a larger ring, but is at variance with findings on most of the cyclic hexapeptide analogs.[6] In general, the removal of the hydroxyl group from the C-terminal residue of the analogs decreases the activity (eg.RC-138-2H, RC-150-2H). This might indicate the presence of a hydrogen bond between this hydroxyl group of the receptor-bound hormone and the receptor. The incorporation of $ProNH_2$ at the C-terminus prevents the formation of a hydrogen bond not only intermolecularly, between the C-terminal residue and the receptor, but also intramolecularly, between the C- and N-terminal peptide bonds. The elimination of the conditions for the intermolecular hydrogen bond formation may account for the relatively low potency of analogs RC-76-2H and RC-116-2H, while the significant difference between the activities of these two compounds suggests that the intramolecular hydrogen bond between the terminal peptide bonds has a greater role in stabilizing the active conformation of the Tyr-7,Val-10 structure than in the case of the Phe-7,Thr-10 peptide.

Another interesting finding is that analogs RC-95-I and RC-160 containing $TrpNH_2$ show high potencies. The aromatic side chain of Trp-12 may be involved in building up a hydrophobic area through the stacking of aromatic rings, which may promote the formation of the active conformation. In this series of analogs, the most active analogs D-Phe-Cys-Tyr-D-Trp-Lys-Val-Cys-Thr-NH_2 (RC-121) and D-Phe-Cys-Tyr-D-Trp-Lys-Val-Cys-Trp-NH_2 (RC-160) are 200 times and 135 times more potent, respectively, than somatostatin-14 in tests for inhibition of growth hormone release. At a dose of 0.1 μg/100g administered subcutaneously, analog RC-121 significantly inhibited GH release for at least 3 hours, demonstrating a prolonged activity. In tests on the suppression of the release of insulin and glucagon in vivo, analog RC-121-2H was 3.9 and 7.4 times more potent, respectively, than somatostatin. These results are similar to those with SMS 201-995 which inhibits the secretion of GH much more selectively than insulin and glucagon release. In tests on inhibition of pentagastrin-induced gastric acid secretion in dogs, analogs RC-121 and RC-160 were found to be only 2-5 times more potent than SS-14. This again indicates a selectivity of action of this series of analogs. Some of these analogs possess antitumor activities. Treatment with these compounds caused a significant inhibition of MT/W9A mammary tumor growth in rats and increased the survival rate in mice bearing the Dunn osteosarcoma. Other endocrine, gastroenterological and oncological investigations on these analogs are in progress.

References

1. Veber,D.R., Freidinger,R.M., Schwenk-Perlow,D.S., Paleveda,W.J.,Jr., Holly,F.W., Strachan,R.G., Nutt,R.F., Arison,B.H., Homnick,C., Randall,W.C., Glitzer,M.S., Saperstein,R. and Hirschmann,R. (1981) Nature 292, 55-58.

2. Bauer,W., Briner,U., Doepfiner,W., Haller,R., Huguenin,R., Marbach,P., Petcher,T.J. and Pless,J. (1982) Life Sciences 31, 1133-1140.

3. Veber,D.F., Saperstein,R., Nutt,R.F., Freidinger,R.M., Brady,S.F., Curley,P., Perlow,D.S., Paleveda,W.J., Colton,C.D., Zacchei,A.G., Tocco,D.J., Hoff,D.R., Vandlen,R.L., Gerich,J.E., Hall,L., Mandarino,L., Cordes,E.H., Anderson,P.S. and Hirschmann,R. (1984) Life Sciences 34, 1371-1378.

4. Pugsley,L.I. (1946) Endocrinology 39,161-176.

5. a) Rivier,J., Brown,M. and Vale,W. (1975) Biochem.Biophys.Res.Commun. 65, 746-751;
   b) Meyers,C.A., Coy,D.H., Huang,W.Y., Schally,A.V. and Redding,T.W. (1978) Biochemistry 17, 2326-2331.

6. Nutt,R.F., Saperstein,R. and Veber,D.F. (1983) Peptides: Structure and Function, Proceedings of the Eighth American Peptide Symposium Hruby,V. and Rich,D. Eds., Pierce Chemical Company, Rockford,IL pp.345-348.

A NEW PROCEDURE FOR ^{3}H-LABELLING OF THE METHIONINE-CONTAINING PEPTIDES : APPLICATION TO THE LABELLING OF METHIONINE-AMIDE, SUBSTANCE P, NEUROKININ A AND NEUROKININ B

Gérard Chassaing, Solange Lavielle, Andrée Marquet

Laboratoire de Chimie Organique Biologique, CNRS UA 493 - 4, Place Jussieu 75230 Paris Cedex 05, France.

Roger Genet, Pierre Fromageot and Jean-Louis Morgat

Service de Biochimie, Département de Biologie, Centre d'Etudes Nucléaires de Saclay, 91191 Gif-sur-Yvette, France.

Introduction

The sulfide fonction of methionine is an attractive target for a specific tritium labelling of a peptide. The first approach used was the direct alkylation by tritiated methyl iodide with $AgClO_4$. It yielded a sulfonium compound which was demethylated by Na/NH_3 liq.[1,2]. This procedure required a reducing medium and led to the loss of half of the specific radioactivity. During our research on the pathways for the transformation of methionine, we have established that some synthetic intermediates represent excellent precursors of methionine[3,4]. In this paper we describe their use for tritium labelling.

Results and discussion

We have previously described the two-step conversion of methionine into (S-tert-butyl)-homocysteine which can be applied either to the free amino acid or to a methionine containing peptide[3]. A C-terminal methionine amide or carboxylate and a methionine included in a sequence can be transformed via the (S-tert-butyl)-homocysteine into one of the precursors (1, 2 or 3) shown on fig. 1. In this way, whatever the position of methionine is on a peptide, it can be easily tritiated.

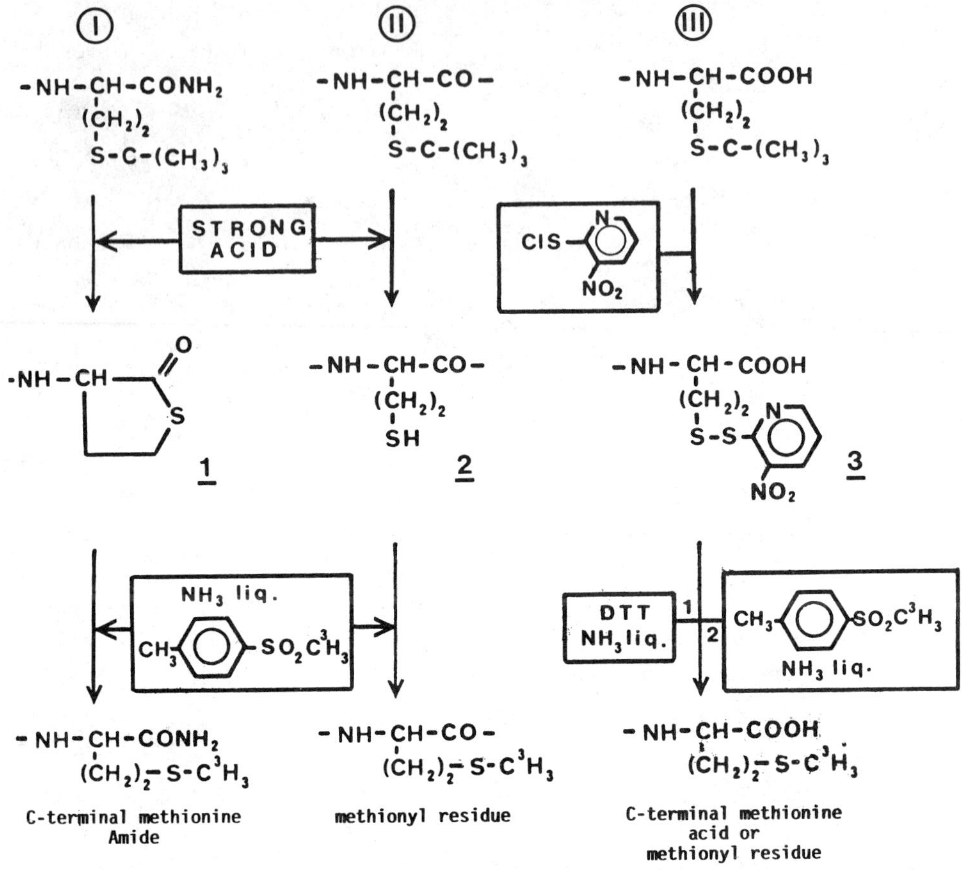

Pathway I was applied to the synthesis of the labelled methionine amide and C-terminal methionine amide peptides (SP, NKA, NKB). Pathway III was checked for the synthesis of tritiated Boc-Met. No example of pathway II is given here.

Methods

Tritium gas and tritiated methanol were obtained from le Commissariat à l'Energie Atomique (C.E.A., France). Purity control and reaction progress were analysed by HPLC (Waters, μ -Bondapak C-18, 10μ) or on TLC (Merck, Silicagel 60 F-254). Radioactivity on the thin layer chromatograms was detected by autoradiography on Kodirex films (Kodak X-O Mat,

X AR 5) or scanned with an automatic TLC linear analyzer (Berthold).

Synthesis of the precursors

The preparation and purification of the N-(tert-Butoxycarbonyl)-S-(3-nitro-2-pyridine sulfenyl)-homocysteine will be published somewhere else[5]. The three peptide precursors [Hcy^{11}]SP, [Hcy^{10}]NKA and [Hcy^{10}]NKB thiolactones have been synthesized by solid phase methodology, (the first residue being always N-α-Boc-S-(tert-butyl)-homocysteine, linked to a MBHA resin. All the other amino-acids, after suitable protection, were coupled by the dicyclohexylcarbodiimide-1-hydroxybenzotriazole method except for the Boc-Gln which was introduced as its p-nitrophenyl ester). After cleavage of the peptide from the MBHA-resin and concomitant removal of the protecting groups by HF, the resulting peptide was purified by partition chromatography on Sephadex G-25 with n-butanol-acetic acid-water (4:1:5), for [Hcy^{11}]SP , and for [Hcy^{10}]NKB thiolactones and n-butanol-2% trifluoroacetic acid (1:1) for [Hcy^{10}]NKA thiolactone.

Synthesis of ^{3}H-methyl-4-Tolysulfonate

To a mechanically stirred solution of 9 μmoles of ^{3}H-methanol (1 Ci, 110 Ci/mmole) in 50 μL of anhydrous THF, 18 μmoles of n-butyllithium in hexane was added. After removal of the solvents, 18 μmoles of tosyl chloride dissolved in 100 μL of dry THF were added and the mixture stirred for 1 hour. The crude alkylating reagent was purified by TLC, (solvent system : chloroform-benzene, (1:1), R_F:0.3). Autoradiochromatogram scanning performed on the crude material revealed a major peak eluted with 5 mL of chloroform and aliquoted (50-60 mCi in 500 μL of $CHCl_3$). This reagent can be stored in liquid nitrogen for at least six months. By radioimmunoassay performed on ^{3}H-substance P, the ^{3}H-methyl-4-tolylsulfonate showed a specific radioactivity close to 75 Cimmol.

^{3}H-labelling of C-terminal methionine amide residues (pathway I)

A mixture of 2 μmoles of homocysteine thiolactone (or [Hcy^{11}]SP, [Hcy^{10}]NKA, [Hcy^{10}]NKB thiolactones and 1 μmole of ^{3}H-methyl-4-tolylsulfonate was stirred at -60°C in 100 μL of dry liquid ammonia for 1 hour. After evaporation of liquid ammonia, the crude product was purified by HPLC, μ-Bondapak C-18 1.5 mL/min. percent of CH_3CN in 0.1% trifluoroacetic acid, ^{3}H-SP 23 min., iso. 21.5%; ^{3}H-NKA, 20.3 min., iso. 18% ;

^{3}H-NKB, 24.7 min., iso. 25%. After removal of CH_3CN and lyophilisation these tritiated peptides have been aliquoted, in water for ^{3}H-SP and ^{3}H-NKA and in 50% acetic acid for ^{3}H-NKB, in the presence of 0.2% mercaptoethanol and are kept in liquid nitrogen. The tritiated peptides (SP, NKA, NKB) have the same retention times that the unlabelled compounds.

^{3}H-labelling of C-terminal methionine (pathway III)

1 µmole of N-(tert-Butoxycarbonyl)-S-(3-nitro-2-pyridinesulfenyl)-L-homocysteine and 1 µmole of dithiothreitol was stirred at -60°C in 100 µL of dry liquid ammonia for 10 min. Then 0.5 µmole of ^{3}H-methyl-4-tolylsulfonate in 20 µL of $CHCl_3$ was added. The resulting solution was stirred for 1 hour. The radiopurity was checked on TLC and revealed a major peak comigrating with the cold reference (Silicagel, solvent system : $CHCl_3$-CH_3OH-CH_3COOH, 45:1.5:1, R_F : 0.59).

Guinea-pig ileum bioassay

The activity of the three labelled tachykinins (SP, NKA and NKB) was assayed on the isolated guinea-pig ileum[6]. Concentration-response curves were obtained using a cumulative dose-assay. The ED_{50} for the labelled peptides were SP: 0.75 10^{-9}M, NKB: 0.83 10^{-9}M and NKA 5.03 10^{-9}M, identical to those obtained for the unlabelled peptides.

References

1. Soucy, P. (1975) Rapport C.E.A.
2. Morgat, J.L. and Fromageot P. (1977) XIIe Rencontres Internationales de Chimie Thérapeutique, Nantes
3. Chassaing, G., Lavielle, S. and Marquet, A. (1983) J. Org. Chem., 48, 1757-1760
4. Chassaing, G., Lavielle, S., Julien, S. and Marquet, A. (1985) Tetrahedron Letters, 623-626
5. Ploux, O., Chassaing, G. and Marquet, A. (submitted)
6. Poujade, C., Lavielle, S., Torrens, Y. and Marquet, A. (1983) Int. J. Peptide Protein Res., 21, 254-257

DESIGN, SYNTHESIS AND BIOLOGICAL ACTIVITY OF TOPOLOGICALLY RELATED PARTIAL NON-PEPTIDIC PEPTIDOMIMETIC ANALOGS OF SUBSTANCE P.

Eli Rubini[1], Uri Wormser[2], Dina Levian-Teitelbaum[3], Ralph Laufer[2], Chaim Gilon[3], Zvi Selinger[2] and Michael Chorev[1].

Departments of Pharmaceutical[1], Biological[2] and Organic[3] Chemistry, The Hebrew University of Jerusalem, Jerusalem 91904, Israel.

The existence of receptors activated similarly by the opioid peptides and opiates, two types of natural products different in their chemical structure and biological sources, is a striking phenomenon.[1,2] The possible transformation of a biologically active peptidic structure into a non-peptidic peptidomimetically active compound is a tremendous challenge.[3] The potential associated with it may be the following: increased selectivity towards a specific receptor; narrowing the spectrum of biological activities; improved metabolic stability, improved transportation across membranes and bio-barriers such as the blood-brain barrier, more favourable pharmacokinetics and last but not least, tailor-made antagonism. As a result of all these, such a transformation may yield novel non-peptidic peptidomimetic drugs which are extensively searched for and can be of a great therapeutic potential.

The working hypothesis assumes that representative low energy conformations of a peptide form a typical surface which interacts with a specific and complementary surface of the receptor. The ultimate objective is to build a non-peptidic structure which will reproduce as close as possible this essential topology.

The key feature in this transformation is a stereochemically defined polyfunctional small ring which will serve as a matrix to carry the essential pharmacophores in a restricted spatial orientation. Replacements of peptidic bonds by different amide bond surrogates is used to acomplish the transformation of a peptidic structure to a non-peptidic one.[4]

The system selected by us to fulfill the above mentioned criteria is the 4-hydroxy-proline available as three isomers; trans-L, cis-L and cis-D. Substitution of these isomers by the essential substance P (SP) pharmacophores[5] i.e. pGlu-Phe on the imino function, Leu-Met-NH_2 on the carboxyl function and benzyl group on the hydroxy moiety substituting for the benzyl side chain of Phe^8 residue was examined by employing CPK models. Only the trans-L-Hyp system can reproduce closely the topology produced by a representative low energy conformation.[6] Glycine residue at position 9 was not needed to achieve this resemblance. Modelling was followed by synthesis employing classical solution techniques using Boc protection and mixed anhydride method for coupling. Table I summarizes the analogs prepared, some of their physical characterization and the potencies relative to the parent C-terminal SP related hexapeptide $[pGlu^6]SP_{6-11}$ as measured in an isolated guinea pig ileum (GPI) assay.

Table I. Characterization and Biological Activities of Hydroxyproline Analogs of $[pGlu^6]SP_{6-11}$.[a]

Analogs of SP_{6-11}	Relative Potency (%)	FAB-MS[b] $[m/z(M+H)^+]$	Mp. (^oC)	$[\alpha]^{25^oC}$ in MeOH (c)
1 $pGlu^6$	100	-	-	-
2 $pGlu^6$,$desGly^9$	1.2	667	230-234	-25.9(1.0)[c]
3 $pGlu^6$,trans-L-$Hyp(Bzl)^8$	2.0	780	70-75	-24.6(1.0)
4 $pGlu^6$,trans-L-$Hyp(Bzl)^8$,$desGly^9$	0.001	723	104-106	-71.0(0.5)
5 $pGlu^6$,cis-L-$Hyp(Bzl)^8$	0.00003	780	117-121	-28.3(1.0)
6 $pGlu^6$,cis-L-$Hyp(Bzl)^8$,$desGly^9$	0.00002	723	75-78	-27.8(1.0)
7 $pGlu^6$,cis-D-$Hyp(Bzl)^8$	0.00002	780	144-147	-19.7(1.0)
8 $pGlu^6$,cis-D-$Hyp(Bzl)^8$,$desGly^9$	0.000003	723	194-195	-36.9(1.0)

[a]All compounds gave the anticipated elemental and amino acid analysis and were pure on RP18-HPLC and TLC. [b]Beside the $(M+H)^+$ mass, pattern of fragmentation gave all the fragments corresponding to the sequence. [c]In DMF.

In all cases the hexapeptides 3, 5 and 7 were more active than the corresponding desGly^9 analogs namely 4, 6 and 8 which is in line with the decrease by two orders of magnitude in the potency of the normal desGly^9 analogs 2 compared to the parent hexapeptide [pGlu^6]SP$_{6-11}$ (1). The most active analogs in the Hyp series were those including the trans-L-Hyp(Bzl) residue, as was predicted from the modelling studies.

Although Gly^9 could be ommited to achieve topological resemblance, it was found to be important for biological activity (200 fold increase in potency of analog 3 and 4 in Table I). Following this observation and our previous finding that introduction of a Phe$^8\psi(CH_2O)$Gly^9 into the parent peptide 1 results in significant activity in the GPI assay (25% relative potency)[7], we have prepared the analog which combines both modifications. Scheme 1 summarizes the synthesis of [pGlu^6, trans-L-Hyp (Bzl)$^8\psi(CH_2O)$Gly^9]SP$_{6-11}$. The pseudopeptidic unit is prepared by the alkylation of the corresponding derivative of the 4-hydroxy-prolinol by ethyl bromoacetate in presence of sodium hydride and 18-crown-6. Incorporation of this non-peptidic pseudodipeptidic unit into the amino acid sequence is another step towards total transformation. As anticipated, this partial non-peptidic peptidomimetic analog was 5 fold more active than its corresponding peptidic analog [pGlu^6,trans-L-Hyp(Bzl)^8]SP$_{6-11}$(3) in the isolated GPI assay. In this analog we have maintained the regidity of the Hyp moiety but increased backbone flexibility in the Hyp(Bzl)^8-Gly^9 amide bond by replacing it with a CH_2-O function. In summary, replacement of two out of six amino acid residues, in a critical portion of the sequence, by a pseudodipeptidic unit following a novel conceptual working hypothesis yielded a potent peptidomimetic compound.

Biologically relevant topology may be roughly obtained through low energy conformations employing model building techniques such as CPK models. 4-Hydroxy-proline can serve as a matrix with the great advantage of being readily available. The static topological model may lack essential degrees of freedom important for an efficient receptor-agonist interaction, such flexibility may be attributed to the presence of Gly^9 in the C-terminal sequence of substance P. A non-rigid peptide bond surrogate, such as the methyleneoxy function, may yield the anticipated flexibility. We feel that our conceptual approach to non-peptidic peptidomimetic trans-

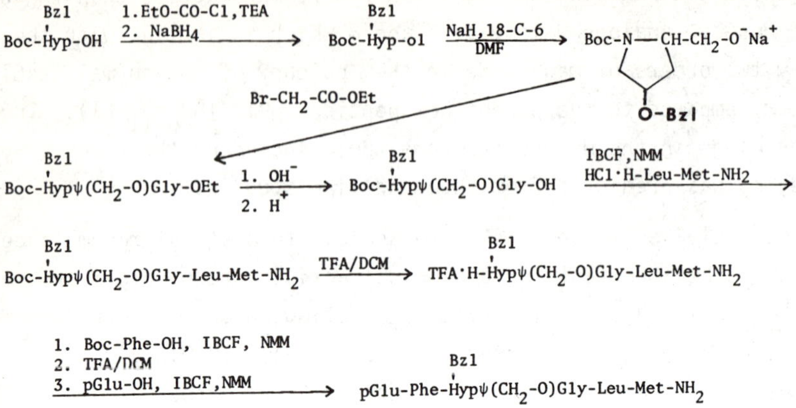

Scheme 1. Synthesis of [pGlu^6, trans-L-Hyp(Bzl)$^8\psi$(CH_2O)Gly^9]SP_{6-11}

formation is still in its infancy and will certainly enjoy farther elaboration and sophistication such as: computerized molecular graphics and molecular dynamics, tailor-made matrices and careful selection of amide bond surrogates to retain the essential hinges required for optimal agonist or antagonist-receptor interactions.

1. Schiller, P.W., Yam, C.F., and Lis, M. (1977) Biochem. 16, 1831-1838.
2. Hudson, D., Sharpe, R., and Szelke, M. (1980) Int. J. Pept. Prot. Res. 15, 122-129.
3. Farmer, P.S. (1979) in Drug Design Vol. X, Ariens, E.J. Ed., Academic Press, New York, pp. 121-143.
4. Spatola, A.F. (1983) in Chemistry and Biochemistry of Amino Acids, Peptide and Proteins, Vol. 7, Weinstein B., Ed., Marcel Dekker Inc., New York, pp. 267-357.
5. Sandberg, B.E.B., and Iversen, L.L. (1982) J. Med. Chem. 25, 1009-1015.
6. Manavalan, P. and Momany, F.A. (1982) Int. J. Pept. Prot. Res. 20, 351-365.
7. Rubini, E., Gilon, C., Levian-Teitelbaum, D., Selinger, Z., Weinstock-Rosin, M., and Chorev, M. (1984) in Peptides 1984, Ragnarsson, U., Ed. Almovist and Wiksell Int., Stockholm. pp. 337-340.

SYNTHESIS, CHARACTERIZATION AND BIOLOGICAL ACTIVITY OF KETO METHYLENE PSEUDOPEPTIDE ANALOGS RELATED TO THE C-TERMINAL HEXAPEPTIDE OF SUBSTANCE P.

A. Ewenson[1], R. Cohen[4], D. Levian-Teitelbaum[1], M. Chorev[2], R. Laufer[3], Z. Selinger[2] and C. Gilon[1].
Departments of Organic[1], Pharmaceutical[2] and Biological[3] Chemistry; Michlala Jerusalem[4] (College for Women, Bait Vagan, Jerusalem 91160), The Hebrew University of Jerusalem, Jerusalem 91904, ISRAEL.

Backbone modified peptide analogs are of great importance as tools to probe the contribution of a specific amide bond to the biological activity, conformation and metabolic stability of a certain compound. In the large array of amide bond surrogates, the keto-methylene modification presents the following salient features: a) it maintains the carbonyl function, thus providing hydrogen bond acceptor capacity and b) it increases conformational freedom[1]. As part of our ongoing research concerning backbone modified substance P (SP) analogs[2] we have been engaged in the synthesis of keto-methylene pseudopeptides related to SP. Only in few cases[3,4] has this modification been successfully applied to replace bonds other than AA'-Gly. The reason for this seems to be the intricate synthetic routes leading to such compounds. We have recently devised a novel and relatively simple method to obtain keto-methylene and α,β-dehydro keto-methylene isosteres of peptides having the general formulas H-AA'ψ($COCH_2$)AA"-OH and H-AA'ψ($COCH_2$)ΔAA'-OH.

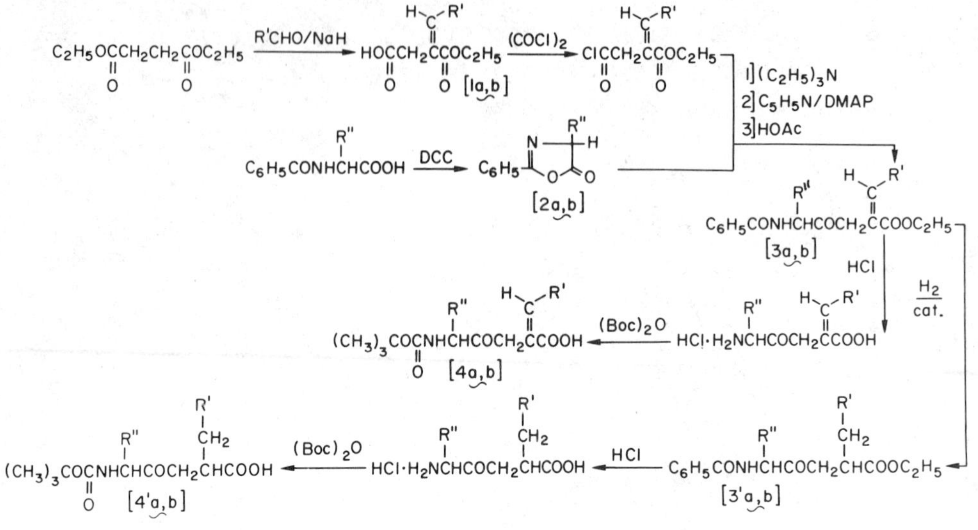

Scheme 1. Synthetic Scheme of $\psi(COCH_2)$ dipeptides.
a: R'=Bzl, R"-Ph; b: R'=i-Pr, R"=H.

Methods, Results and Discussion

The synthetic scheme (Scheme I) is based on a modified Dakin-West reaction, involving an N-Benzoyl amino acid-5-oxazolone (2a,b) with the appropriately α-substituted mono acyl chloride mono ethyl (or methyl) succinate obtained from a modified Stobbe condensation. The characterization of Intermediates and final products was performed by m.p., TLC, HPLC, IR, Elemental Analysis and FAB-MS (Table I). The coupling of such isosteres to give analogs of $[pGlu^6]SP_{6-11}$ is exemplified by the synthesis of the analog $[pGlu^6,(R,S)Phe^8\psi(COCH_2)Gly^9]$-$SP_{6-11}$, shown in Scheme 2. This compound was fully characterized (including amino acid analysis). It was a full agonist with a potency of 70% (relative to $[pGlu^6]SP_{6-11}$) in the guinea pig ileum assay. Furthermore, by a fast assay recently developed by us[5] for screening potential specific inhibitors of SP degradation, we found that this analog was a potent inhibitor for SP degradation with an IC_{50} of 18±2 μM.

Table I. Characterization of Typical Fully and Partially Protected Keto-Methylene and α,β-Dehydro Keto-Methylene Units.

Compound	R'	R"	mp(°C)	Formula		C	H	N
3a	Ph	Bzl	110-112	$C_{28}H_{27}NO_4$	(c)	76.17	6.16	3.17
					(f)	76.23	6.01	3.41
3'b	i-Pr	H	70-75	$C_{18}H_{25}NO_4$	(c)	67.71	7.83	4.38
					(f)	67.95	7.78	4.65
4a	Ph	Bzl	110-113	$C_{24}H_{27}NO_5$	(c)	70.40	6.65	3.42
					(f)	70.74	6.50	3.28
4'b	i-Pr	H	89-90	$C_{14}H_{25}NO_5$	(c)	58.52	8.77	4.87
					(f)	58.26	8.47	5.13

(c) Calc'd. (f) Found.

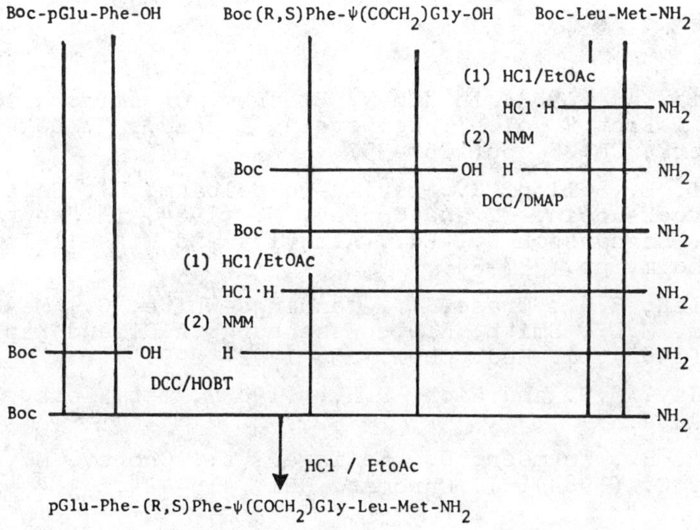

Scheme 2. Synthesis of $[pGlu^6,(R,S)Phe^8\psi(COCH_2)Gly^9]SP_{6-11}$

Conclusions

As shown by the biological results reported here and elsewhere[4], the keto-methylene modification represents a promising tool in the investigation of peptide actions at different levels. The synthetic scheme for the preparation of such analogs, outlined here, affords a relatively rapid and straightforward route to that end. Virtually any pseudodipeptidic unit (3) can be obtained in this fashion. One of the drawbacks in this scheme is that the product is a diastereomeric mixture. On the other hand, once separated the different stereoisomers could afford useful probes to assess spatial and conformational requirements of peptide-target interactions. We are currently engaged in the separation and stereochemical assignment of the isomers and in studies of their conformational implications.

References

1. Spatola, A. (1983) in "Chem. Biochem. of Amino Acids, Pept. & Prot." Vol. 7; Weinstein, B., Ed., M. Dekker, New York, Ch. 5, pp. 267-357.
2. Rubini, E., Gilon, C., Levian-Teitelbaum, D., Selinger, Z. Weinstock-Rosin, M. and Chorev, M. (1984) in "Peptides 1984", Ragnarsson U., Ed., Almqvist and Wiksell Intl. Stockholm, pp. 337-340.
3. Almqvist, R.G., Crase, J., Jennings-White, C., Meyer, R.F. Hoefle, M.L., Smith, R.D., Essenburg, A.D. and Kaplan, M.R. (1982), J. Med. Chem. 25, 1292-1299.
4. Holladay, M.W. and Rich, D.H., (1983), Tet. Lett. 24, 41 4401-4404.
5. Laufer, R., Wormser, U., Selinger, Z., Chorev, M., and Gilon, C. (1984), J. Chromat. 301, 415-424.

NEUROMEDINS: NOVEL NEUROPEPTIDES IDENTIFIED IN PORCINE SPINAL CORD

Naoto MINAMINO, Tetsuji SUDOH, Kenji KANGAWA & Hisayuki MATSUO
Department of Biochemistry, Miyazaki Medical College, Kiyotake, Miyazaki 889-16, Japan

The rapidly growing knowledge of neuropeptides has revealed the existence of a complex network of neural communication systems. In order to clarify the mechanism of peptidergic neurotransmission and neuromodulation, it is urgent that still unidentified neuropeptides be discovered. We have recently developed a systematic method for surveying unknown peptides by utilizing "non-specific" bioassay for the effects on the contractility of several smooth muscle preparations. By the use of this method coupled with high resolution HPLC, we have been performing the systematic search for unidentified neuropeptides in porcine spinal cord, and isolated a series of novel neuropeptides designated "NEUROMEDINS", including neuromedin K, L (kassinin-like)[1,2], neuromedin B, C (bombesin-like)[3,4] and neuromedin N (neurotensin-like)[5]. Amino acid sequences of these peptides were determined as shown below.

```
Gly-Asn-Leu-Trp-Ala-Thr-Gly-His-Phe-Met-NH2 : neuromedin B
Leu-Ser-Trp-Asp-Leu-Pro-Glu-Pro-Arg-Ser-
Arg-Ala-Gly-Lys-Ile-Arg-Val-His-Pro-Arg-
Gly-Asn-Leu-Trp-Ala-Thr-Gly-His-Phe-Met-NH2 : neuromedin B-30
                               Ala-Pro-
Leu-Ser-Trp-Asp-Leu-Pro-Glu-Pro-Arg-Ser-
Arg-Ala-Gly-Lys-Ile-Arg-Val-His-Pro-Arg-
Gly-Asn-Leu-Trp-Ala-Thr-Gly-His-Phe-Met-NH2 : neuromedin B-32

Gly-Asn-His-Trp-Ala-Val-Gly-His-Leu-Met-NH2 : neuromedin C

Asp-Met-His-Asp-Phe-Phe-Val-Gly-Leu-Met-NH2 : neuromedin K
His-Lys-Thr-Asp-Ser-Phe-Val-Gly-Leu-Met-NH2 : neuromedin L
```

Lys-Ile-Pro-Tyr-Ile-Leu : neuromedin N

Tyr-Phe-Leu-Phe-Arg-Pro-Arg-Asn-NH_2 : neuromedin U-8

Phe-Lys-Val-Asp-Glu-
Glu-Phe-Gln-Gly-Pro-Ile-Val-Ser-Gln-Asn-
Arg-Arg-Tyr-Phe-Leu-Phe-Arg-Pro-Arg-Asn-NH_2 : neuromedin U-25

In addition to the five neuromedins mentioned above, neuromedin B-30, B-32[6], U-8 and U-25[7] have recently been identified in porcine spinal cord and in porcine brain.

Methods

The outline of procedures for isolating neuromedins from porcine spinal cords is summarized in Fig. 1.

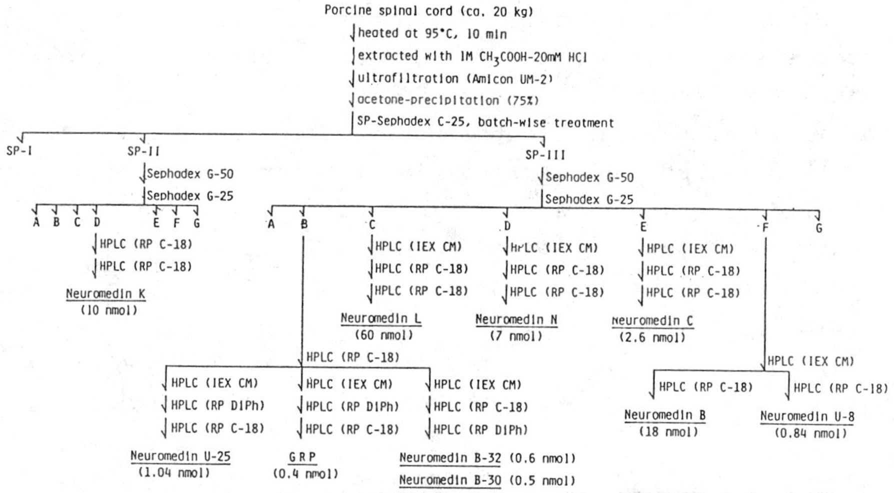

Fig. 1 Purification Procedures of Neuromedins

Results and Discussion

NEUROMEDIN B-30 AND B-32: By utilizing a specific radioimmunoassay for neuromedin B (NMB), we have isolated two novel "big" neuromedin B, designated neuromedin B-32 and B-30 (NMB-32 & -30), both of which were identified as N-terminally extended forms of NMB. The amino acid sequence of NMB-32 was determined to be as shown above, while NMB-30 was found to be an N-terminal two amino acids deleted form of NMB-32. Isolation of a family comprising NMB, B-30 and B-32 is indicative of their biosynthetic relationship, as observed in another mammalian family of bombesin-like peptides, comprising GRP and its C-terminal neuromedin C (NMC). As for conversion from "big" NMB to NMB, processing must take place after Pro-Arg, which is adjacent to the NMB unit in NMB-32 and NMB-30. Although the Pro-Arg structure is not a typical processing site, a similar relationship is found between GRP and NMC[4], suggesting that processing to NMB and NMC may proceed in a similar manner. Also in a series of human atrial natriuretic polypeptides (hANPs), α-hANP, the smallest hormone unit, is preceded by a Pro-Arg signal in the precursor molecule[8]. These facts indicate that the Pro-Arg structure is a specific signal for processing. The differences of distribution of NMB and NMC groups[9,10] suggest the possible existence of distinct precursors for two groups in mammalian bombesin-like peptide family.

NEUROMEDIN U-8 AND U-25: Two novel peptides eliciting a potent stimulant effect on the rat uterus smooth muscle have been purified also from porcine spinal cord. These peptides were designated as neuromedin U-8 (NMU-8: 8 residues long) and U-25 (NMU-25: 25 residues long) refering to their <u>uterus</u> stimulating activity. NMU-8 is a novel peptide with a C-terminal amide structure, while NMU-25 contains the NMU-8 sequence at its C-terminus, preceded by paired Arg residues. Although sequences highly related to NMU-8 or NMU-25 have never been found in the peptides so far identified, the C-terminal amide structure is a unique feature of peptides exhibiting hormonal or physiological activities, as observed in hypothalamic releasing factors and gastrointestinal hormones. A partial C-terminal sequence of Arg-Pro-Arg-X-$CONH_2$ occuring in NMU-8 and NMU-25 is also found in pancreatic polypeptide(PP). Thus, their potent uterus stimulating activity and hypertensive effect, as well as their unique C-terminal amide structure may indicate their specialized physiological function. The natural occurence of NMU-8 which is most likely processed from NMU-25 suggests that a similar processing may take place also in VIP/PHI family and PP at the site of paired basic residues to generate their respective truncated C-terminal peptide amides as endogenous entities, even though they have not so far been identified.

References

1) Kangawa, K., Minamino, N., Fukuda, A and Matsuo, H. (1983) Biochem. Biophys. Res. Commun. 114, 533-540

2) Minamino, N., Kangawa, K., Fukuda, A. and Matsuo, H. (1984) Neuropeptides 4, 157- 166

3) Minamino, N., Kangawa, K, and Matsuo, H. (1983) Biochem. Biophys. Res. Commun. 114, 541-548

4) Minamino, N., Kangawa, K., and Matsuo, H. (1984) Biochem. Biophys. Res. Commun. 119, 14-20

5) Minamino, N., Kangawa, K., and Matsuo, H. (1984) Biochem. Biophys. Res. Commun. 122, 542-549

6) Minamino, N., Sudoh, T., Kangawa, K., and Matsuo, H. (1985) Biochem. Biophys. Res. Commun., in press

7) Minamino, N., Kangawa, K., and Matsuo, H. (1985) Biochem. Biophys. Res. Commun., in press

8) Kangawa, K., Fukuda, A. and Matsuo, H. (1985) Nature 313, 393-400

9) Minamino, N., Masuda, H., Kangawa, K., and Matsuo, H. (1984) Biochem. Biophys. Res. Commun. 124, 731-738

10) Minamino, N., Kangawa, K., and Matsuo, H. (1984) Biochem. Biophys. Res. Commun. 124, 925-932

STRUCTURE-ACTIVITY STUDIES OF NEUROKININ A(α) AND B(β)*

Eisuke Munekata, Kazuki Kubo, Hisanori Tanaka and Fumio Osakada
Institute of Applied Biochemistry, University of Tsukuba, Ibaraki, Tsukuba, Japan

Neurokinin A (NKA) and B (NKB)[1)] are novel neuropeptides isolated from porcine spinal cord. The chemical features and pharmacological characteristics of both neurokinin peptides are closely related to the tachykinin families such as substance P (SP), physalaemin (PHY) and kassinin (KAS) as shown below:

NKA	His-Lys-Thr-Asp-Ser-Phe-Val-Gly-Leu-Met-NH_2
NKB	Asp-Met-His-Asp-Phe-Phe-Val-Gly-Leu-Met-NH_2
SP	Arg-Pro-Lys-Pro-Gln-Gln-Phe-Phe-Gly-Leu-Met-NH_2
PHY	pGlu-Ala-Asp-Pro-Asn-Lys-Phe-Tyr-Gly-Leu-Met-NH_2
KAS	Asp-Val-Pro-Lys-Ser-Asp-Gln-Phe-Val-Gly-Leu-Met-NH_2

SP has been regarded as the only member of tachykinin peptide since long time and well investigated from the viewpoints of peptidergic neurotransmitter in central and peripheral nerve systems. Studies on pharmacology and histochemistry of neurokinin peptides demonstrated that NKA and NKB are putative neurotransmitter like SP in mammals.[2-4)] NKA and NKB exhibit contracting activities on the smooth muscle preparations such as guinea pig ileum(GPI), rat vas deferens(RVD) and rat duodenum(RDN).

The receptor of tachykinin peptides are tentatively subclas-

*According to the recommendation of IUPHR-committee on August 1984 in London, we use A and B instead of α and β.

sfied into SP-P and SP-E type based on the susceptibility of the tissue to the peptide. The muscle of GPI is SP-P type and substance P is assigned to be endogenous agonist of it. On the other hand, the muscles of RVD and RDN are SP-E type but the endogenous agonist for it was unknown so far. Recently, several studies[5)] suggested that NKA and NKB might be endogenous agonist for SP-E in mammals. We have demonstrated also C-terminal heptapeptide derivatives of NKA and NKB retain nearly the same contracting activities on RVD and RDN with the native decapeptide although the activity on GPI decreases gradually as the shortening of peptide chain.

For the purpose of structure-activity studies, heptapeptide derivatives related to SP, NKA and B as shown below were prepared and the contractile activities were investigated comparatively.

Table 1. Synthetic analogs of NKA7,NKB7 and SP7.

Peptides		Amino acid sequences
DSFV :	NKA7 = NKA(4-10)	Asp-Ser-Phe-Val-Gly-Leu-Met-NH_2
DSFF :	Phe^7-NKA(4-10)	Asp-Ser-Phe-Phe-Gly-Leu-Met-NH_2
QSFV :	Gln^4-NKA(4-10)	Gln-Ser-Phe-Val-Gly-Leu-Met-NH_2
QSFF :	(Gln^4,Phe^7)-NKA(4-10)	Gln-Ser-Phe-Phe-Gly-Leu-Met-NH_2
DSAV :	Ala^5-NKA(4-10)	Asp-Ser-Ala-Val-Gly-Leu-Met-NH_2
DFFV :	NKB7 = NKB(4-10)	Asp-Phe-Phe-Val-Gly-Leu-Met-NH_2
DFFF :	Phe^7-NKB(4-10)	Asp-Phe-Phe-Phe-Gly-Leu-Met-NH_2
QFFV :	Gln^4-NKB(4-10)	Gln-Phe-Phe-Val-Gly-Leu-Met-NH_2
QFFF :	(Gln^4,Phe^7)-NKB(4-10)	Gln-Phe-Phe-Phe-Gly-Leu-Met-NH_2
QQFF :	SP7 = SP(5-11)	Gln-Gln-Phe-Phe-Gly-Leu-Met-NH_2
QQFV :	Val^8-SP(5-11)	Gln-Gln-Phe-Val-Gly-Leu-Met-NH_2
DQFV :	(Asp^5,Val^8)-SP(5-11)	Asp-Gln-Phe-Val-Gly-Leu-Met-NH_2
NKA-OH		His-Lys-Thr-Asp-Ser-Phe-Val-Gly-Leu-Met-OH

QSFF, QFFF and DQFV is also Ser^6-SP(5-11), Phe^6-SP(5-11) and kassinin(6-12),respectively.

Peptidederivatives listed in Table 1 were synthesized by solid phase method and purified by high performance liquid chromatography. The results of the pharmacological tests are shown in Table 2.

Table 2. Relative potencies and maximum effects of neurokinin and substance P heptapeptide derivatives on the guinea pig ileum and rat duodenum.

	G.P.I.		R.D.N.	
Peptides -GLM-NH_2	relative potency	maximum effect	relative potency	maximum effect
DSFV	11.4	90-100	100	100
DSFF	4.8	100	3.0	75-95
QSFV	4.5	100	2.5	85-100
QSFF	4.7	85-95	0.3	100
DSAV	< 0.01	n.d.	< 0.01	n.d.
DFFV	4.1	95-100	24.8	100
DFFF	6.6	90-100	0.7	60-75
QFFV	7.3	100	0.8	70-90
QFFF	4.3	85-100	0.8	60-75
QQFF	100	75-90	1.1	90-100
QQFV	6.5	90-100	2.0	80-100
DQFV	3.5	90-100	22.3	90-100
NKA-OH	0.3	80-90	0.2	60-80

EC_{50} of QQFF in the G.P.I. was 2.2 ± 0.3 nM.
EC_{50} of DSFV in the R.V.D. was 5.2 ± 0.6 nM.
G.P.I.; guinea pig ileum, R.D.N.; rat duodenum.
Each result was obtained from 4-7 determinations.

The pharmacological properties of the heptapeptide derivatives investigated in this study are summarized as follow:

1) tachykinin peptides are divided into two groups from the chemical structure, namely, in partial structure as below,

------ X - Y -Phe- Z -Gly-Leu-Met-NH_2

the peptides of group I have the sequence X=Gln or Asn and Z=aromatic amino acid, whereas the peptides group II have the sequence X=Asp and Z=aliphatic amino acid,

2) the peptides of group I (substance P, physalaemin) have high affinity to SP-P receptor and the peptides of group II (eledoisin, kassinin and neurokinins)possess specific affinity to SP-E receptor,

3) both Asp(or acidic amino acid in X) and Val(or aliphatic amino acid in Z) are at least essential to recognize the SP-E receptor.

Acknowledgements

This study was supported by Grant-in-aid for scientific research from the Ministerium of Education, Science and Culture, Japan, special research program on instinct and Ajinomoto Co., Ltd.

References

1) Kimura, S., Okada, M., Sugita, Y., Kanazawa, I., and Munekata E. (1983) Proc. Japan Acad. **59B**, 101-104.
2) Munekata, E., Okada, M., Kimura, S., Sugita, Y., Kanazawa, I., Matsudo, T., and Otsuka, M. (1984) Chem. Lett. 1013-1016.
3) Matsudo, T., Yanagisawa, M., Otsuka, M., Kanazawa, I., and Munekata, E. (1984) Neurosci. Res. **2**, 105-110.
4) Kanazawa, I., Ogawa, T., Kimura, S., and Munekata, E. (1984) Neurosci. Res. **2** 111-120.

TACHYKININ ANTAGONISTS AND MULTIPLE RECEPTORS

David E. Wright[*], Henry I. Jacoby[+], Roger W. Scott[+], and Jeffry L. Vaught[+]

Departments of Chemical[*] and Biological[+] Research, McNeil Pharmaceutical, Spring House, Pa 19477

Introduction

There is strong support for the existence of multiple receptor subtypes for neuropeptides as exemplified by the mammalian opioid peptides. An important aspect of the characterization of newly isolated neuropeptides is the delineation of receptor subtypes. Recently, two new tachykinin peptides, neuromedin K and substance K, were isolated from bovine spinal cords.[1-3] Both peptides have potencies similar to substance P on isolated guinea pig ileal strips.[4] However, in the mouse tachykinin-induced scratching test neuromedin K is about 30 times less potent than substance P.[5] Of interest to us is the neuropeptide neuromedin K which has a C-terminal region similar to substance P but whose N-terminal region is strikingly different. Neuromedin K has two aspartic acid residues while substance P has lysine and arginine residues in this region. This structural difference may suggest the potential existence of separate neuromedin K and substance P receptors. To examine the possibility of different tachykinin sub-receptors, a novel neuromedin K antagonist (D-Pro^2, D-Trp^{6,8}, Nle^{10})NK was synthesized. This new antagonist and the substance P antagonist (D-Pro^2, D-Trp^{7,9})SP were tested both centrally (mouse reciprocal hind-limb scratching) and peripherally (guinea pig ileum) for their ability to selectively antagonize the agonist actions of the mammalian tachykinins substance P, neuromedin K and substance K as well as the amphibian tachykinins eledoisin, kassinin, and physalaemin. The substance P antagonist (D-Arg^{1}, D-$Trp^{7,9}$, Leu^{11})SP also was tested against these tachykinin agonists in the guinea pig ileum.

Materials and Methods

The neuromedin K antagonist (D-Pro^{2}, D-$Trp^{6,8}$, Nle^{10})NK was synthesized by solid phase methodology on a 4-methylbenzhydrylamine resin. After the last amino acid was incorporated, the N-terminal Boc-group was removed, and the peptide was cleaved from the resin with HF. Purification was achieved by gel-filtration on a Bio-Gel P-2 column, and preparative HPLC used CN and C-8 columns. Analytical HPLC showed one peak and amino acid analysis gave the proper ratios. The substance P antagonists were purchased from Peninsula Laboratories.

The reciprocal hind-limb scratching responses were determined as described previously.[5] The pA_2 values were determined from multiple concentration-response curves using guinea pig ileal strips. Schild plots were constructed and pA_2 values were determined only for those concentration-response curves whose slopes were not significantly different from one.

Results and Discussion

The neuromedin K antagonist (D-Pro^{2}, D-$Trp^{6,8}$, Nle^{10})NK was a competitive antagonist in the guinea pig ileum of neuromedin K but not of the other two mammalian tachykinins substance P and substance K (Table I). Competitive antagonism was seen against all the amphibian tachykinins as well (not shown). The substance P antagonist (D-Pro^{2},

Table I. Mean pA_2 Values of Tachykinin Antagonists on the Guinea Pig Ileum

Antagonists[a]	NK	SK	SP
(D-Pro^2, D-Trp^{6,8}, Nle^{10})NK	5.55	SL 1[b]	SL 1
(D-Pro^2, D-Trp^{7,9})SP	6.70	SL 1	6.87
(D-Arg^1, D-Trp^{7,9}, Leu^{11})SP	SL 1	5.77	6.64

[a]Based on multiple dose response curves.
[b]SL 1 = Schild plot slope significantly less than 1.

D-Trp^{7,9})SP was a competitive antagonist of neuromedin K and substance P but not of substance K. The other substance P antagonist (D-Arg^1, D-Trp^{7,9}, Leu^{11})SP was a competitive antagonist of substance P and substance K but not of neuromedin K.

In the reciprocal hind-limb scratching test, (D-Pro^2, D-Trp^{6,8}, Nle^{10})NK was selective in antagonizing neuromedin K but was a non-selective and equipotent antagonist of the other mammalian (Table II) and amphibian tachykinins (not shown). This is in contrast to (D-Pro^2, D-Trp^{7,9})SP which was equipotent towards all the tachykinins. Also the neuromedin K antagonist did not antagonize bombesin or somatostatin-induced scratching thereby showing its selectivity towards tachykinins (not shown).

The occurance of different tachykinin peptides in mammalian tissues is suggestive for the potential existence of different types of tachykinin receptors. The new neuromedin K antagonist reported here shows selectivity towards neuromedin K both peripherally and centrally thereby suggesting a distinct neuromedin K receptor. Since centrally this receptor shows selectivity towards neuromedin K and not eledoisin or physalaemin, this receptor cannot be classified as an SP-P or SP-E type receptor. Also in the guinea pig

Table II. Effectiveness of Tachykinin Antagonists in Agonist-Induced Scratching

Agonist[b]	ID_{50} of Antagonists[a] (D-Pro^2, D-Trp^{6,8}, Nle^{10})NK	(D-Pro^2,D-Trp^{7,9})SP
NK (500 ng)	0.6 (0.3 - 1.2)	2.6 (0.8 - 4.8)
SK (100 ng)	5.9 (3.1 - 9.9)	8.2 (5.5 - 11.6)
SP (20 ng)	3.4 (2.3 - 5.2)	4.6 (2.9 - 6.9)

[a] ID_{50} = that dose of antagonist in ug which inhibits agonist-induced scratching by 50%. Values in parenthesis are 95% fiducial limits.

[b] A fixed dose of agonist (given in parenthesis) equaling an SD_{90} dose (that dose of agonist which causes 90% of the animals to scratch) was admininstered simultaneously with varying doses of antagonist.

ileum SP-P and SP-E receptor designations may not apply as evidenced by the tachykinin antagonist studies presented here. Thus there appears to be separate receptors for the three mammalian tachykinin peptides, and tachykinin receptor subtyping should be considered in this context.

References

1. Kimura, S., Okada, M., Sugita, Y., Kanazawa, I., and Munekata, E. (1983) Proc. Jap. Acad. Ser. B 59, 101-104.

2. Kangawa, K., Minamino, N., Fikuda, A., and Matsuo, H. (1983) Biochem. Biophys. Res. Comm. 114, 533-540.

3. Maggio, J.E., Snadberg, B.E.B., Bradley, C.V., Iverson, L.L., Santikarn, S., Williams, D.H., Hunter, J.C., and Hanley, M.R. (1983) Ir. J. Med. Sci. 152 (suppl. 1), 20-21.

4. Folkers, K., Lu, Y., and Rosell, S. (1984) Biochem. Biophys. Res. Comm. 118, 405-408.
5. Vaught, J. et al. (1984) Eur. J. Pharm. 103, 355-357.

DEVELOPMENT AND MODIFICATION OF COMPETITIVE ANTAGONISTS OF BRADYKININ

Raymond J. Vavrek and John M. Stewart

Department of Biochemistry,
University of Colorado Medical School
4200 East 9th Avenue, Denver, CO 80262

Until recently, no specific sequence-related antagonists of the classical _in vitro_ (isolated rat uterus (RUT) and guinea pig ileum (GPI)) or _in vivo_ (rat blood pressure (RBP)) effects of bradykinin (Arg-Pro-Pro-Gly-Phe-Ser-Pro-Phe-Arg; BK) had been described. We have now learned how to make BK competitive antagonists. The key modification was the replacement of the Pro residue at position 7 of BK with a D-Phe residue (see TABLE I)[1,2].

The weak BK-like agonist effect of [DPhe^7]-BK on the uterus (1%) is converted to antagonism of BK (pA_2 = 6.5) by the substitution of the Phe residues at positions 5 and 8 with the isosteric beta-2-thienyl-L-alanine (Thi). The antagonism of BK on the ileum is increased more than an order of magnitude (pA_2 = 6.3) by this substitution. Addition of basic residues at the N-terminal of [DPhe^7]-BK does not change its RUT or GPI effects, but addition of Lys-Lys or a DArg residue to the N-terminal of [Thi^{5,8}DPhe^7]-BK decreases somewhat both its RUT and GPI antagonist potency.

The substitution of a DPro residue for Pro at position 3 destroys agonist and antagonist activity of both [DPhe^7]-BK and [Thi^{5,8}DPhe^7]-BK. However, addition of a DArg residue to the N-terminal of both of these DPro^3-containing analogs produces moderate antagonism of BK action on the uterus (pA_2 = 4.6-5.2) without modifying their inactivity on ileum.

TABLE I. Activities of BK Analogs in Classical Smooth Muscle Assays

ANALOG	RUT	GPI
BK	100%	100%
[DPhe^7]-BK	1%	5.0*
[Thi^{5,8}]-BK	1000%	200%
[Thi^{5,8}DPhe^7]-BK	6.5*	6.3*
Lys-Lys-BK	35%	-
DArg-BK	143%	103%
Lys-Lys-[DPhe^7]-BK	0.3%	5.1*
DArg-[DPhe^7]-BK	0.1%	5.6*
DArg-[Thi^{5,8}DPhe^7]-BK	5.5*	6.1*
Lys-Lys-[Thi^{5,8}DPhe^7]-BK	6.0*	5.3*
[DPro^3]-BK	0.01%	0.02%
[DPro^{3}DPhe^7]-BK	0	0
[DPro^{3}Thi^{5,8}DPhe^7]-BK	0	0
DArg-[DPro^{3}DPhe^7]-BK	4.6*	0
DArg-[DPro^{3}Thi^{5,8}DPhe^7]-BK	5.2*	0

Assays performed according to Trautschold in Handbook of Experimental Pharmacology, Vol 25, EG Erdos, ed (Berlin-Heidelberg-NY) 1970. Agonist potency is relative to BK = 100. *Antagonist potency is listed as pA_2 values of Schild (Br. J. Pharmacol. 2:189, 1947)

Replacement of the Pro residue at positions 2 and/or 3 of [Thi^{5,8}DPhe^7]-BK with hydroxyproline (Hyp) generally decreases the antagonist potency in the GPI assay, with [Hyp^{2}Thi^{5,8}DPhe^7]-BK producing the best antagonism in this group of analogs (pA_2 = 6.10) (see TABLE II). Addition of Lys-Lys or DArg to this analog gives the most potent antagonists of BK on the ileum yet found (pA_2 = 6.57, 6.51).

TABLE II. Antagonism of BK in the GPI Assay by Hydroxyproline-containing Analogs

ANALOG	pA_2 ± SEM(N)
[$Thi^{5,8}DPhe^{7}$]-BK	6.26 ± 0.18(10)
[$Hyp^{2}Thi^{5,8}DPhe^{7}$]-BK	6.10 ± 0.32(5)
[$Hyp^{3}Thi^{5,8}DPhe^{7}$]-BK	5.67 ± 0.16(6)
[$Hyp^{2,3}Thi^{5,8}DPhe^{7}$]-BK	5.86 ± 0.21(5)
Lys-Lys-[$Hyp^{2}Thi^{5,8}DPhe^{7}$]-BK	6.57 ± 0.30(5)
Lys-Lys-[$Hyp^{3}Thi^{5,8}DPhe^{7}$]-BK	5.48 ± 0.24(5)
Lys-Lys-[$Hyp^{2,3}Thi^{5,8}DPhe^{7}$]-BK	5.56 ± 0.25(5)
DArg-[$Hyp^{2}Thi^{5,8}DPhe^{7}$]-BK	6.51 ± 0.06(6)
DArg-[$Hyp^{3}Thi^{5,8}DPhe^{7}$]-BK	6.28 ± 0.11(6)
DArg-[$Hyp^{2,3}Thi^{5,8}DPhe^{7}$]-BK	6.03 ± 0.14(6)

N = number of experiments

The specificity of these BK antagonists for the effects of BK-like kinins has been determined on the GPI for several antagonists. [$DPhe^{7}$]-BK, $DArg^{0}$-[$DPhe^{7}$]-BK, [$Thi^{5,8}DPhe^{7}$]-BK and $DArg^{0}$-[$Thi^{5,8}DPhe^{7}$]-BK antagonize the myotropic effect of BK, kallidin (Lys-BK) and Met-Lys-BK on the ileum, but do not affect the ileum response to angiotensin II or substance P. $DArg^{0}$-[$Thi^{5,8}DPhe^{7}$]-BK antagonizes the uterine effect of BK, kallidin and Met-Lys-BK, but not that of angiotensin II.

In the rat blood pressure assay infusions of solutions of [$DPhe^{7}$]-BK, [$Thi^{5,8}DPhe^{7}$]-BK or $DArg^{0}$-[$Thi^{5,8}DPhe^{7}$]-BK antagonize the depressor effect of BK in a reversible manner. Administration of a bolus mixture of BK plus antagonist by intraaortic or intravenous routes also produces inhibition of the BK depressor response.

Preliminary data indicate that the DPhe substitution at position 7 may not be unique in converting BK agonists to antagonists. Antagonism of the myotropic effect of BK has been seen with analogs having DThi or beta-2-pyridyl-D-Ala

(DPal) at position 7, with their $Thi^{5,8}$-substituted derivatives, and with derivatives having N-terminal extensions with basic amino acid residues.

Simultaneous substitution of both the Ser residue at position 6 and the Pro residue at position 7 with D-hydrophobic amino acid residues gives rise to another class of BK analogs with antagonist activity. These antagonists retain inhibitory activity when further modified with Thi residues at positions 5 and 8, with basic extensions at the N-terminal, and with Hyp substitutions for Pro in positions 2 and 3.

ACKNOWLEDGMENTS

This work has been supported by grant HL-26284 from the NHLBI-NIH. We thank Constance Young and Francis Shepperdson for performing the bioassays, and Virginia Sweeney for the amino acid analyses.

REFERENCES

1. Stewart, J.M. and Vavrek, R.J. In "Kinins 84", LM Greenbaum, Ed., Plenum Press, NY 1985 (in press).
2. Stewart, J.M. and Vavrek, R.J. Peptides 1985 (in press).

EFFECTS OF THE PEPTIDE BOND RETRO-INVERSION ON THE BIOLOGICAL ACTIVITY OF BRADYKININ POTENTIATING PEPTIDE 5a (BPP5a).

Antonio Groggia, Antonello Pessi, Alessandro Sisto,
Antonio S. Verdini, Antonino Virdia, Giuseppe C. Viscomi

ENIRICERCHE S.p.A., 00015 Monterotondo, Rome, Italy

Introduction

The retro-inversion of the Phe^3-Ala^4 peptide bond in Glp-Lys-Phe-Ala-Pro ($[Phe^3]$ BPP5a) to prevent cleavage by Angiotensin Converting Enzyme (ACE) produced two isomers equipotent to BPP5a as hypotensives in normotensive rats, while displaying rather weak ACE inhibition and no Bradykinin potentiation [1].

These unexpected results suggested that, due to the interchange of the carbonyl groups and the secondary nitrogens, a) the retro-inverso peptides did not retain all the crucial components for a strong interaction with the postulated active site of ACE and b) other modes of action than ACE and plasma bradykininases inhibition played a role in determining the biological activity.

To further investigate the effects of peptide bond retro-inversion on the biological activities of BPP5a analogues, we synthesized the retro-isomers of $[Phe^3]$ BPP5a listed in Table I.

Results and Discussion

The synthesis of the peptides was performed according to previously described procedures [1,2].

Peptides	I_{50} (μM)	Bradykinin Potentiation In Vitro*	Bradykinin Potentiation In Vivo**	Maximum BP decrease*** (mm Hg)
1 Glp-Lys-Trp-Ala-Pro-OH(Bpp5a)	0.1	+	+	15–15
2 Glp-Lys-Phe-Ala-Pro-OH	0.1	+	+	15–15
3 Glp-Lys-Phe-Ala-Pro(S-Ph)-OH	0.06	+ +	+ +	30–30
4 Glp-Lys-Phe-Ala-Pro(OBzl)-OH	1.1	+	+ –	25–20
5 Glp-Lys-gPhe-(S)mAla-Pro-OH	140	0	0	20–15
6 Glp-Lys-gPhe-(R)mAla-Pro-OH	290	0	0	10–10
7 Glp-Lys-gPhe-(S)mAla-Pro(S-Ph)-OH	1.1	0	0	20–15
8 Glp-Lys-gPhe-(R)mAla-Pro(S-Ph)-OH	70	0	0	5–5
9 Glp-gLys-(S)mPhe-Ala-Pro-OH	50	+ +	+	0
10 Glp-gLys-(R)mPhe-Ala-Pro-OH	130	+	+ –	0
11 gGlp-DLys-(S)mPhe-Ala-Pro-OH	85	0	0	0
12 gGlp-DLys-(R)mPhe-Ala-Pro-OH	280	0	0	0
13 gGlp-DLys-(S)mPhe-Ala-Pro(S-Ph)-OH	1.1	+ +	n. d.	0
14 gGlp-DLys-(R)mPhe-Ala-Pro(S-Ph) OH	9.0	+	n. d.	0
15 gGlp-DLys-(R,S)mPhe-Ala-Pro(OBzl)-OH	40	0	0	0
16 Glp-gLys-DPhe-(S)mAla-Pro-OH	300	0	0	15–15
17 Glp-gLys-DPhe-(R)mAla-Pro-OH	700	0	0	5–5
18 gGlp-DLys-DPhe-(S)mAla-Pro-OH	320	0	0	15–15
19 gGlp-DLys-DPhe-(R)mAla-Pro-OH	1000	0	0	5–0
20 gGlp-DLys-DPhe-(S)mAla-Pro(S-Ph)-OH	400	0	0	15–15
21 gGlp-DLys-DPhe-(R)mAla-Pro(S-Ph)-OH	1200	0	0	20–15
22 Bz-gPhe-(S)mAla-Pro-OH	400	0	0	10–10
23 Bz-gPhe-(R)mAla-Pro-OH	750	0	0	5–5
24 H-Ala-Pro(S-Ph)-OH	1.8	+ +	n.d.	20–20

* Isolated guinea pig ileum; peptide concentration = $5 \cdot 10^{-9}$ M

** Effect on the diastolic and systolic pressure following the hypotension induced by injection of 2.5 μg/Kg/i.v. of BradyKinin. Peptide dose: 0.180 mg/Kg/i.v.

*** Blood pressure decrease after injection of 0.180 mg/Kg/i.v. of peptide to anaesthetized normotensive rats.

Table I. Biological activity of BPP5a and BPP5a analogues.

As expected, the partially modified retro-inverso peptides 5-8 and 16-21 display complete resistance towards cleavage by ACE *in vitro* as compared with BPP5a (Figure 1). When tested against human plasma and rabbit lung, heart, spleen, brain and kidney homogenates only peptides 16-21 result stable to degradation (figure 2).

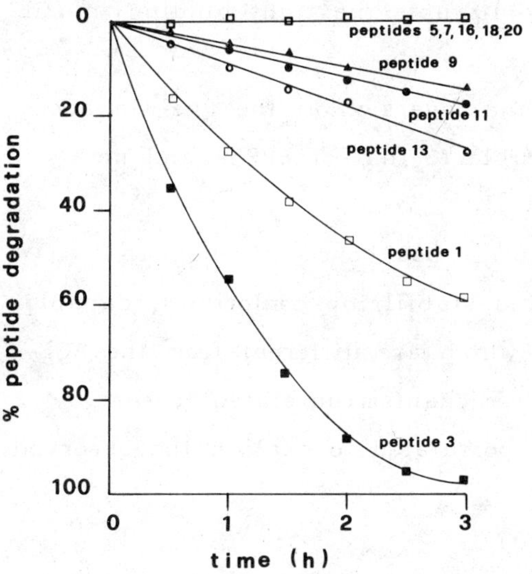

Fig. 1. Degradation of BPP5a and BPP5a analogues by ACE.

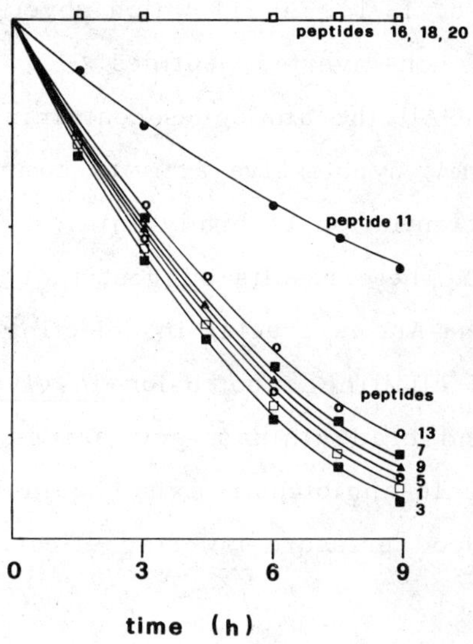

Fig. 2. Degradation of BPP5a and BPP5a analogues in human plasma.

ACE inhibitory activity shows a remarkable decrease not only when the scissile Phe-Ala bond is inverted (peptides 5-6, 16-19, 23-24) but also when the modification is adjacent to the Phe-Ala bond (peptides 9-12), confirming that BPP5a binding to ACE involves important interactions of the N-terminal residues with a region next to the obligatory binding site [3].

A thiophenyl group at the C^{γ} of the proline ring enhances the inhibitory activity (peptide 3), which is consistent with the hydrophobicity requirements of ACE for the C-terminal residues of substrates and inhibitors [4]; however, it is insufficient to compensate for retro-inversion at or adjacent to the Phe-Ala bond (peptides 7-8, 13-14, 20-21) which severely disrupts the tight binding to ACE of non-inverted peptides.

All the analogues containing the inversion at the Phe-Ala bond show hypotensive activity comparable to that of BPP5a and no potentiation of bradykinin.

These results suggest that peptide backbone inversion at Phe-Ala is crucial in selecting and stabilizing conformers, capable of eliciting hypotension in vivo, which are different from the ACE- and bradykininase-active ones. A mechanism unrelated to the renin-angiotensin axis should be postulated to explain the observed blood pressure lowering effect.

References

1. Verdini, A.S., and Viscomi, G.C. (1985) J. Chem. Soc. Perkin Trans. I, 697-701.

2. Pessi, A., Pinori, M., Verdini, A.S. and Viscomi, G.C. (1983) J. Chem. Soc. Chem. Commun., 195-197.

3. Ondetti, M.A., Rubin, R., and Cushman, D.W. (1977), Science, 196, 441-444.

4. Kim, D.H., Guinosso, C.J., Buzby, G.C., Herbst, D.R., Mc Caully, R.J., Wicks, T.L., and Wendt, R.L. (1983) J. Med. Chem., 26, 394-403.

DISCRIMINATION BETWEEN PARATHYROID HORMONE (PTH) AGONISTS AND ANTAGONISTS *IN VITRO*: MODIFICATION OF THE RENAL MEMBRANE ADENYLATE CYCLASE ASSAY TO REFLECT *IN VIVO* ACTIVITY OF PTH ANALOGS.

L. Caporale, P. DeHaven, J. Fisher, G. Tyler, and M. Rosenblatt
Merck Sharp & Dohme Research Laboratories, West Point, PA 19486

INTRODUCTION

PTH plays a critical role in calcium homeostasis.[1] Antagonists of PTH have potential utility in study of PTH mechanism of action and in treating disorders of calcium metabolism. The activity of a PTH analog, [Nle-8,Nle-18,Tyr-34]bPTH(3-34)amide, has been extensively evaluated. Although apparently a pure competitive antagonist in several *in vitro* assays, the 3-34 analogue has weak agonist activity *in vivo*. The availability of a reliable *in vitro* assay would greatly facilitate identification of more effective PTH antagonists and consume smaller amounts of peptide than *in vivo*. Therefore, we sought to develop an *in vitro* assay in which even the weak *in vivo* agonist properties of [Nle-8,Nle-18, Tyr-34]bPTH-(3-34)amide would be evident.

Here we report on the redesign of the *in vitro* cyclase assay for PTH to reveal agonism for this peptide. We (1) modified the method of cAMP detection to use HPLC with automatic injection (Fig. 1), then (2) altered incubation conditions in light of current models of hormone activation of of cyclase. These changes have not only made the assay more convenient, but, more importantly, have enabled us to demonstrate *in vitro* the agonism of 3-34.

RESULTS AND DISCUSSION

The cyclase assay conditions were altered with consideration for the important role of Mg^{++} in the activation of adenylate cyclase.[2] The enzyme

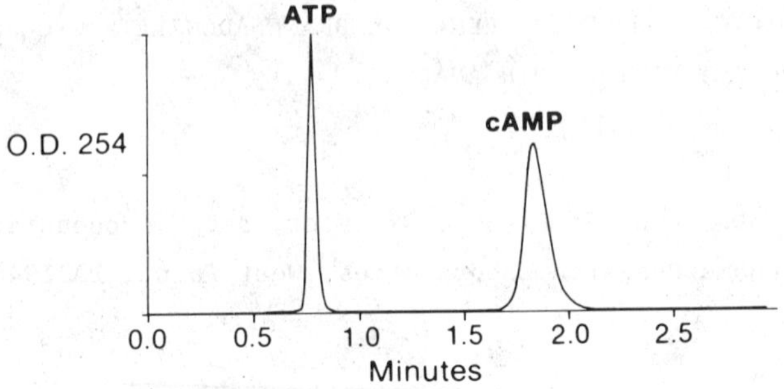

Fig. 1. HPLC separation of α-^{32}P-ATP from ^{32}P-cAMP. Samples are run isocratically in 20% methanol/80% 0.01 M $(NH_4)_3PO_4$, pH 6.0, on an 8 mm x 10 cm C_{18} μ Bondapak cartridge (Waters) with a flow rate of 6 ml/minute.

adenylate cyclase is now thought to be activated by binding to the α subunit of a "nucleotide regulatory" (N) protein bound to GTP. This free α subunit can be dissociated from the $N_{\alpha\beta\gamma}$ complex in the presence of Mg^{++} and GTP. This dissociation can occur in the absence of receptor at high Mg^{++} concentrations. Because high Mg^{++} levels increase dissociation of the N complex and activation of cyclase in the absence of hormone, cyclase stimulation is more hormone-dependent at low Mg^{++} concentrations.

Figure 2 illustrates the effects of the Mg^{++} concentration on fold-stimulation above basal for the full agonist [Nle-8,Nle-18,Tyr-34]bPTH-(1-34)amide. Although the actual quantity of cAMP produced is greater at high Mg^{++} concentrations (not shown), the fold-stimulation above basal is greatly enhanced at the lower Mg^{++} concentrations, as expected, because basal cyclase stimulation is more Mg^{++}-dependent than hormone-stimulated cyclase activity. Therefore, the use of low [Mg^{++}] should lower hormone-independent basal cyclase activity and thus permit detection of small amounts of cyclase stimulation by very weak agonists. In fact, as seen in Figure 3, the agonism of the 3-34 analogue is clearly revealed at low [Mg^{++}].

Thus, assays designed for increased hormone-dependency, rather than increased cAMP production, should provide a reliable in vitro measure of agonism and more accurately reflect relevant in vivo properties of PTH analogs, and could be applied, in principle, to other hormone systems.

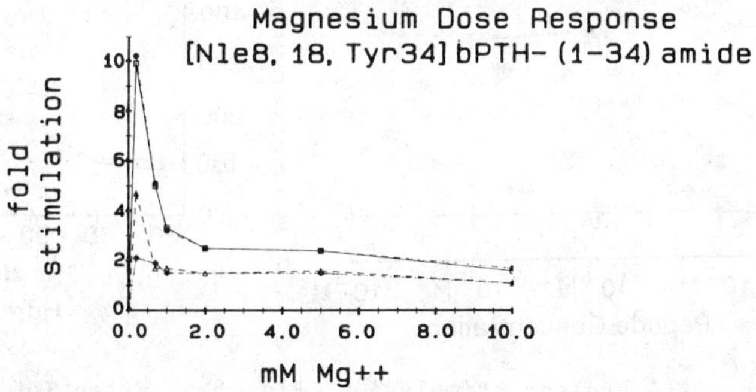

Fig. 2. Magnesium dose response. Fold stimulation above basal for [Nle-8,Nle-18,Tyr-34]bPTH-(1-34)amide at concentrations of —□— 10^{-5} M; --●-- 10^{-6} M; --△-- 10^{-7} M; -·-◆-·- 10^{-8} M.

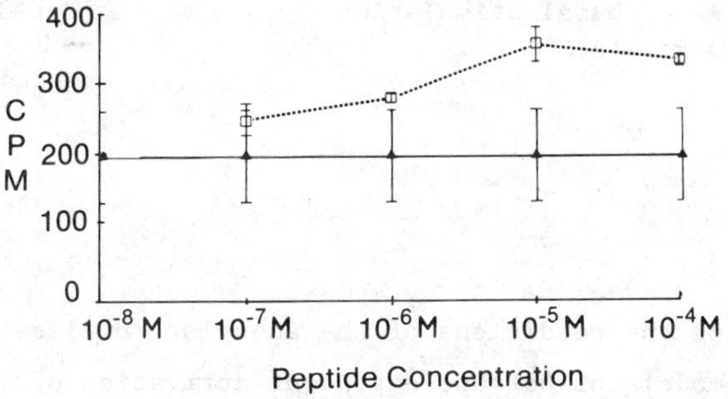

Fig. 3. Cyclase stimulation of bovine renal membranes by [Nle-8,Nle-18, Tyr-34]bPTH-(3-34)amide --□--; —▲— basal stimulation.

As a first test of the ability of the new in vitro cyclase assay to predict the in vivo behavior of the new PTH analogues, the activity of [Tyr-34]hPTH-(5-38) amide was investigated.

This compound is the first tested in a series investigating whether extending the peptide at the carboxyl terminus would stabilize a native conformation in this region, and thus increase receptor affinity. In addition because the 3-34 peptide is an agonist, while the 7-34 analogue has shown antagonism in vivo, it is also of interest whether truncation of the amino terminus to position 5 would result in agonism or antagonism. This compound is predicted to be an agonist by its ability to stimulate cAMP production by bovine renal membranes in vitro (Figure 4), and is indeed an agonist in vivo (Figure 5).

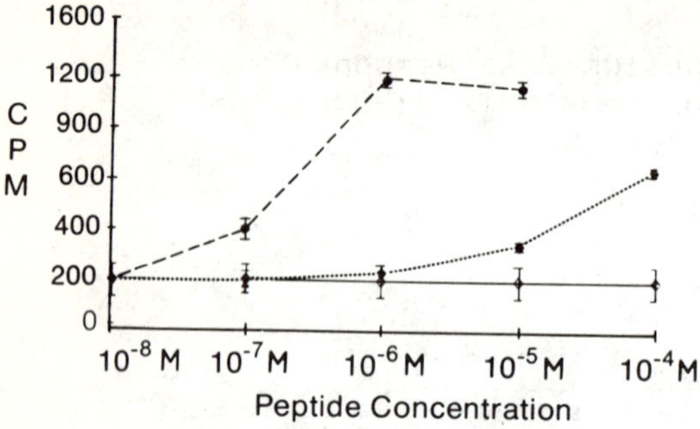

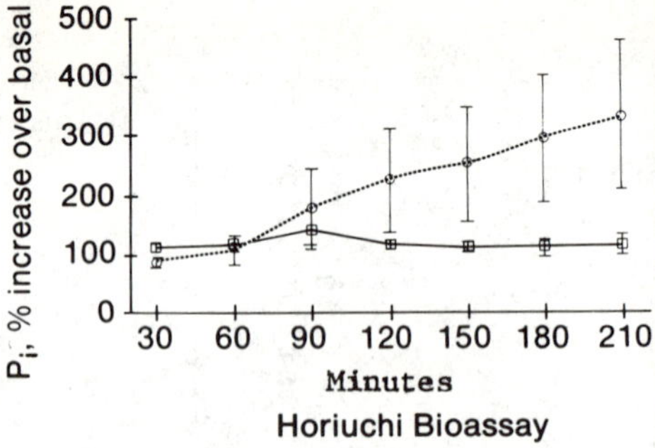

Fig. 4. *In vitro* cyclase stimulation of bovine renal membranes by [Nle-8,Nle-18,Tyr-34]bPTH-(1-34)amide -- ● --; [Tyr-34]hPTH-(5-38)amide ···◆···; basal stimulation — ◇ —.

Fig. 5. Effect of infusion of [Nle-8,Nle-18, Tyr-34]hPTH-(5-38)amide on urinary phosphate levels; ···o··· [Tyr-34]hPTH-(5-38)amide; — □ — perfusion vehicle only.

SUMMARY

By altering the conditions of the adenylate cyclase assay based upon current models of hormone-dependent activation of this enzyme, the assay now accurately reveals the *in vivo* agonism of this compound. The availability of a reliable *in vitro* method for predicting *in vivo* agonism of PTH analogues will greatly expedite the screening of compounds in a program to develop improved *in vivo* antagonists of PTH action. Furthermore, the principle of augmented hormone dependency applied to this assay may prove of general utility in designing bioassays for other cyclase-stimulating hormones.

REFERENCES

1. Habener, J., Rosenblatt, M., and Potts, J.T. Jr. (1984) Physiol. Rev. 64, 985-1053.

2. Iyengar, R. and Birnbaumer, L. (1982) Proc. Natl. Acad. Sci. U.S. 79, 5179-5183.

TRITIATION OF MELANIN CONCENTRATING HORMONE TO HIGH SPECIFIC ACTIVITY: SYNTHESIS OF ($^{3}H_{4}$)-NORVALINE^{3,6}-MCH

Alex N. Eberle, Klaus Neubert and Jürg Girard
Department of Research, University Hospital and University Children's Hospital, CH-4031 Basel, Switzerland

Albert Zeller
Radium-Chemie, CH-9053 Teufen, Switzerland

Introduction

Melanin concentrating hormone (MCH), a physiological antagonist of α-MSH in teleost fishes, induces pigment aggregation in their melanophores.[1] It has been isolated from salmon pituitaries and sequenced,[2] and its structure was confirmed by solid-phase[3,4] and classical synthesis.[5] A recent study[6] on MCH analogs revealed that oxidation or replacement of the two methionine residues hardly affects the potency of the hormone whereas iodination with chloramine T severely impairs its pigment-aggregating activity. Therefore radioiodinated MCH cannot be used for receptor studies. Consequently we prepared a MCH analog suitable for tritiation in which the two methionines were replaced by propargylglycine (Pra).

Peptide Synthesis

The synthesis of the linear, Cys^{5,14}(Acm)-protected heptadecapeptide (Figure 1) was performed on a composite polydimethylacrylamide-macroporous kieselguhr support, using Sheppard's[7]

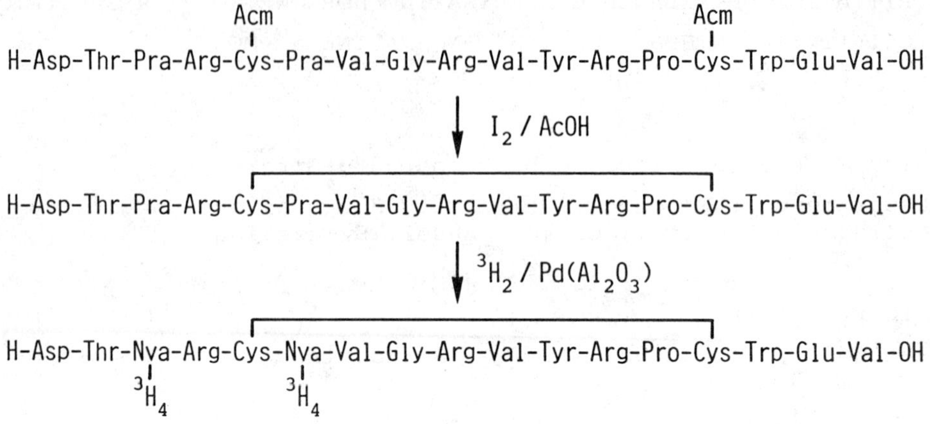

Fig. 1 Synthesis of tritiated [$Nva^{3,6}$]-MCH

continuous flow method. Fmoc-amino acid anhydrides were used throughout for acylation reactions; butyl esters and ethers and N_G-methoxytrimethylbenzenesulfonyl (Arg) were used for side-chain protection. The peptide was cleaved from the resin by a 6 hours treatment with trifluoroacetic acid/anisol/dithioethane 100:3.3:1 and then purified on CM-cellulose. Cyclization was carried out in glacial acetic acid by iodine oxidation,[8] followed by Sephadex LH-20 chromatography.

Tritiation

Since tritiations of peptides containing disulfide bridges are difficult to perform, we investigated the use of a new apparatus (Figure 2) with which tritium gas pressures of up to 1650 mbar can be reached. The catalyst was exposed to the tritium gas in dry form in the reaction vessel before it was pored into the peptide solution. Using an excess of catalyst, the tritiation was quantitative within a few minutes. The peptide was purified on Sephadex LH-20 and by reverse-phase HPLC.

Results and Discussion

Tritiation of the MCH derivative gave a compound with 220 Ci/mmol of tritium specifically incorporated. Non-specifically bound radioactivity could be removed quantitatively. The compound could be used as tracer in an MCH radioimmunoassay, resulting in a slight reduction in sensitivity as compared to iodinated tracer. In the bioassay using scale melanophores of

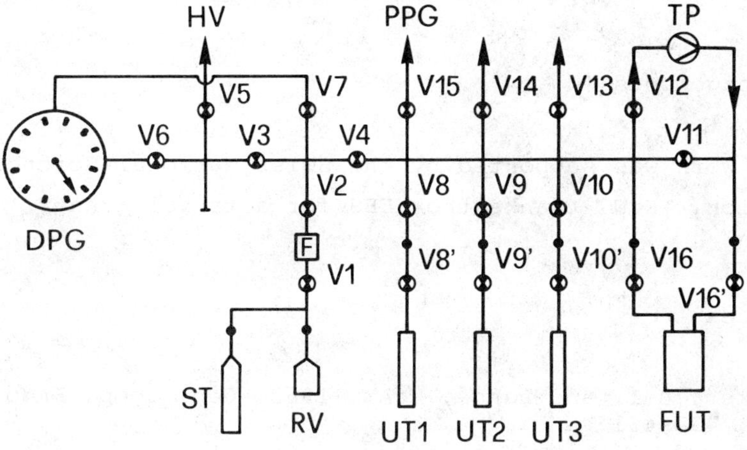

Fig. 2 Scheme of the tritiation apparatus. It consists of a fully metal-sealed manifold system made of welded stainless steel and is equipped with uranium traps for tritium storage. The reaction vessel (RV) is coupled to the unit by Swagelock PTFE ferrules. The solvent trap (ST) is mounted after the tritiation when the solvent from the reaction vessel is distilled off. A filter (F) connects the RV line with the rest of the unit. Various valves (V) separate/connect the uranium traps (UT), flow-through uranium trap (FUT), Toepler pump (TP), high-vacuum pump (HV), Pirani pressure gauge (PPG) and differential pressure gauge (DPG). Uranium is released by heating the UT to 400 - 500°C; the tritium gas pressure is adjusted to 1650 mbar (DPG). The reaction is terminated by freezing of the sample and adsorption of the gas onto a cold UT. FUT/Toepler pump remove traces of tritium gas left.

the Chinese grass carp, _Ctenopharyngodon idellus_,[6] the EC_{50} for half-maximal pigment aggregation was 200 pM; non-radioactive $[Nva^{3,6}]$-MCH and MCH had an EC_{50} of 170 pM and 63 pM, respectively. This means that the compound retained 32% of MCH-bioactivity and 85% of its non-radioactive analog.

In conclusion, tritiation of MCH at elevated gas pressure together with some specific manipulations yielded a radioactive MCH derivative which is suitable for receptor studies. This experiment could well serve as a model for other similar peptides.

Acknowledgements

This work was supported by the Swiss National Science Foundation. K.N. thanks the FEBS for a travel grant.

References

1. Baker, B.I. and Rance, T.A. (1983) Gen. Comp. Endocrinol. 50, 423-431.
2. Kawauchi, H., Kawazoe, I., Tsubokawa, M., Kishida, M., and Baker, B.I. (1983) Nature 305, 321-323.
3. Wilkes, B.C., Hruby, V.J., Sherbrooke, W.C., de L. Castrucci, A.M., and Hadley, M.E. (1984) Biochem. Biophys. Res. Commun. 122, 613-619.
4. Eberle, A.N., Atherton, E., Dryland, A., and Sheppard, R. C. (1985) J. Chem. Soc., in press.
5. Okamoto, K., Yasumura, K., Fujitani, K., Kiso, Y., Kawauchi, H., Kawazoe, I., and Yajima, H. (1984) Chem. Pharm. Bull. 32, 2963-2970.
6. Baker, B.I., Eberle, A.N., Baumann, J.B., Siegrist, W., and Girard, J. (1985) Peptides, in press.
7. Sheppard, R.C. (1983) Chemistry in Britain, pp. 402-413.
8. Kamber, B., Hartmann, A., Eisler, K., Riniker, B., Rink, H., Sieber, P., and Rittel, W. (1980) Helv. Chim. Acta 69, 899-915.

CONFORMATIONALLY RESTRICTED ANALOGUES OF PRO-LEU-GLY-NH_2

Rodney L. Johnson, Kuo Long Yu, and Irach Taraporewala
Department of Medicinal Chemistry, College of Pharmacy, University of Minnesota, Minneapolis, Minnesota 55455

Ram K. Mishra and G. Rajakumar
Department of Psychiatry and Neurosciences, McMaster University, Hamilton, Ontario L8N 3Z5, Canada

Introduction

The tripeptide L-prolyl-L-leucylglycinamide (PLG) has been shown to possess a number of neuropharmacological activities.[1] For example, PLG has been shown to potentiate the behavioral effects of L-DOPA and to selectively enhance the binding affinity of the dopamine agonist ^{3}H-apomorphine to dopamine receptors. PLG has also been shown to reverse haloperidol-induced supersensitivity of post-synaptic dopamine receptors. This pharmacological profile suggests that PLG may be exerting its neuropharmacological actions through the modulation of dopamine receptors within the central nervous system.

In order to gain a better understanding of PLG's biologically active conformation and its modulation of dopamine receptor sensitivity, we have undertaken the synthesis of a series of conformationally constrained analogues of this tripeptide.

Two approaches were employed to construct conformationally restricted analogues of PLG. In one approach, the γ- and δ-lactam residues developed by Freidinger et al.[2] were incorporated into the PLG molecule to give analogues 1-3. These analogues were synthesized in an attempt to mimic the

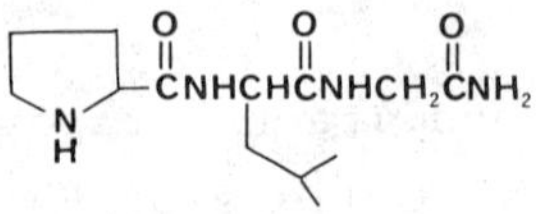

PLG

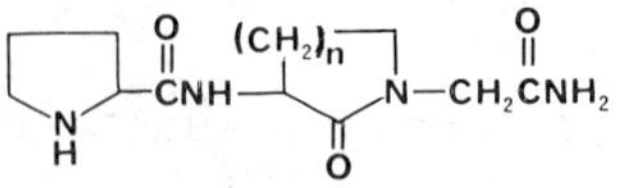

1, n=1
2, n=2

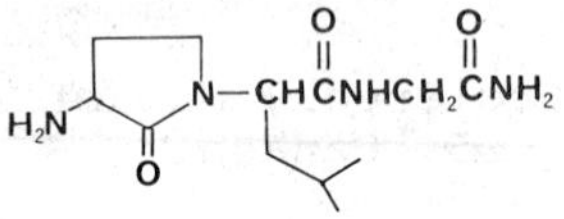

3

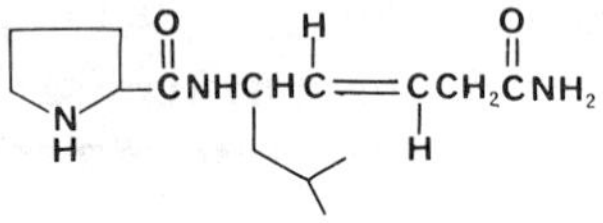

4

β- and γ turns that computational and spectroscopic studies have indicated are the preferred conformations of PLG.[3]

In the second approach, an olefinic analogue of PLG (4) was synthesized, wherein the Leu-Gly peptide bond was replaced with a trans-ethylenic bond. This analogue was made in an attempt to determine whether the Leu-Gly amide bond of PLG needs to be in a cis or trans configuration when the tripeptide interacts with its receptor. This question was of particular interest in view of the fact that a purported analogue of PLG, cyclo(Leu-Gly), has a cis Leu-Gly amide bond and possesses a pharmacological profile similar to that of PLG.

Results and Discussion

The synthesis of analogues 1 and 2 was carried out as outlined in Scheme 1. The γ- and δ-lactam residues that served as the respective starting materials were synthesized using the methods of Freidinger et al.[2] Analogue 3 was obtained using a modification of a literature procedure.[4]

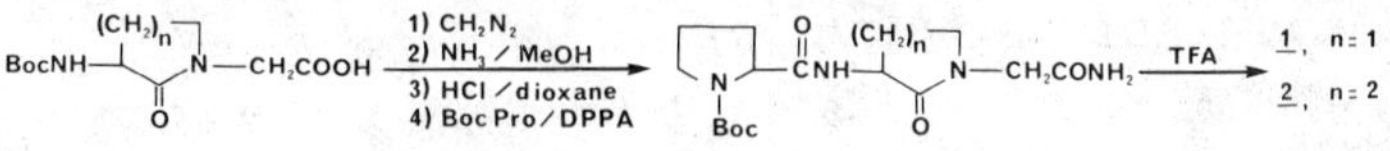

Scheme 1. Synthesis of the PLG lactam analogues 1 & 2.

The synthesis of 4 was carried out as pictured in Scheme 2. The olefinic dipeptide residue which served as the starting material was prepared as described previously.[5]

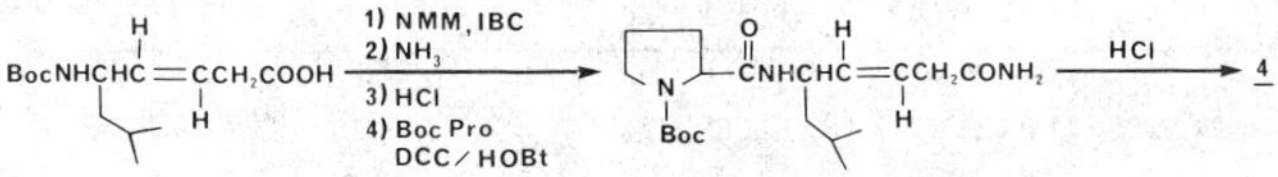

Scheme 2. Synthesis of ProLeuψ[E-CH=CH]Gly-NH_2

The PLG analogues 1-4 were tested for their effects on ^{3}H-PLG binding in a bovine striatal membrane preparation[6] and ^{3}H-ADTN (2-amino-6,7-dihydroxytetrahydronaphthalene) binding to rat striatum dopamine receptors[7] using previously described assay procedures. The results are depicted in Table I, and show that in their ability to displace ^{3}H-PLG from its binding sites, the olefinic analogue 4 was as effective as PLG. The γ-lactam analogue 1 was slightly less active, while lactam analogues 2 and 3 showed no activity at the dose tested. Analogues 1, 3, and 4 all showed a greater ability than PLG to enhance the binding of the dopamine agonist ^{3}H-ADTN to striatal dopamine receptors.

These results not only suggest that the Leu-Gly amide bond of PLG is in a trans configuration when PLG interacts with its receptor, but that the active conformation of PLG is one closely related to a β-turn in which the trans primary amide hydrogen atom of the glycinamide residue forms a hydrogen bond with the prolyl carbonyl group.

Table I. Effect of PLG and Its Analogues on ^{3}H-PLG and ^{3}H-ADTN Binding.

Compound (100 nM)	% Displacement of ^{3}H-PLG[a]	% Increase in ^{3}H-ADTN Binding[a]
1	30	35
2	0	9
3	0	29
4	41	36
PLG	40	22

[a]Average of four determinations.

Acknowledgement

This study was supported by NIH grant NS-20036.

References

1. Mishra, R.K., Chiu, S., Chiu, P., and Mishra, C.P. (1983) Meth. Find. Expt. Clin. Pharmacol. 5, 203-233.
2. Freidinger, R.M., Perlow, P.S., and Veber, D.F. (1982) J. Org. Chem. 47, 104-109.
3. Ralston, E., De Coen, J.-L., Walter, R. (1974) Proc. Nat. Acad. Sci. USA 71, 1142-1144.
4. Prochazka, Z., Lebl, M., Barth, T., Hlavacek, J., Trka, A., Budesinsky, M., and Jost, K. (1984) Coll. Czech. Chem. Commun. 49, 642-652.
5. Johnson, R. (1984) J. Med. Chem. 27, 1351-1354.
6. Chiu, S., Wong, Y., Ferris, J.A., Johnson, R.L., and Mishra, R.K. (1983) Pharmacol. Res. Commun. 15, 41-51.
7. Chiu, S., Paulose, C.J., and Mishra, R.K. (1981) Peptides 2, 105-111.

A NEUROPEPTIDE PROCTOLIN AND ITS ANALOGS: SYNTHESIS AND CARDIOEXCITATORY EFFECT ON INSECTS - PERIPLANETA AMERICANA L. AND TENEBRIO MOLITOR L.

Danuta Konopinska

Institute of Chemistry, Wroclaw University
Wroclaw, 50383 Poland

Wieslaw Sobotka

Institute of Organic Chemistry, Polish Academy of Sciences
Warszawa, 02691 Poland

Gregorz Rosinski, Andrzej Lesicki and Przemyslaw Sujak

Department of Animal Physiology, A. Mickiewicz University
Poznan, 60035 Poland

Introduction

The neuropeptide proctolin (Arg-Tyr-Leu-Pro-Thr) was isolated from the American cockroach (Periplaneta americana) by Brown and Starratt.[1,2] Proctolin has been shown to occur in species of six orders of insects and probably in some other arthropods.[4-7] Initially it was described as a neurotransmitter in the cockroach hindgut. It now appears that proctolin contracts the proctodeal and extensior tibia muscle and excites the heart of cockroach[4,5,8,9] and other invertebrates.[10,11]

The first proctolin synthesis was carried out by the classical method[11]. A solid phase procedure followed.[12] Among reported analogs[12,13] only [Phe-OMe]^2-proctolin (where the Tyr hydroxyl group was methylated) exhibited 278% proctolin activity[13] on cockroach proctodeal muscle. Proctolin analogs modified in position 2 of the peptide chain by other aromatic amino acid residues, D-Tyr and Phe,[13] or His and Trp,[12] preserved only 10-15% of proctolin activity. These results inspired us to undertake synthesis of the following proctolin analogs modified in position 2: Arg-Phe(p-NH_2)-Leu-Pro-Thr (I), Arg-Phe(p-NMe_2)-Leu-Pro-Thr (II), Arg-Phe(p-OMe)-Leu-Pro-Thr (III), Arg-Phe(p-NO_2)-Leu-Pro-Thr (IV) and Arg-Ala (β-4-methoxycyclohexyl)-Leu-Pro-Thr (V). Proctolin was prepared according to our method.[14] Synthesis of the analogs was carried out by conventional liquid-phase technique using N-hydroxysuccinimide esters or dicyclohexylcarbodiimide. α-Amino groups were protected by BOC. The N-terminal Arg was used as Z-Arg(NO_2)-OH or BOC-Arg(Tos)-OH derivatives. The C-terminal group of Thr was benzylated. Free peptides I-III and V were obtained by catalytic hydrogenation in presence of Pd/$BaSO_4$, whereas peptide IV and proctolin were deprotected with CF_3SO_3H in anisole. Peptides I-III and V were purified on Sephadex G-25 column and peptide IV and proctolin were purified using Dowex-IRC-50 and Sephadex G-25.

Results and Discussion

Biological effects of proctolin and its analogs I-V were investigated in a cardioexcitatory test according to Miller.[15] The test was performed on two insect species: cockroach (Periplaneta americana) and yellow mealworm (Tenebrio molitor). Proctolin and its analogs I-IV increased the frequency of mealworm and cockroach heartbeat (see Table) at the range of 10^{-9} to 10^{-8}M concentrations. Higher concentration of peptides (10^{-7}M) often provoked constant systole. Tenebrio molitor heart was stimulated by a 10-fold lower peptide concentration than Periplaneta americana heart.

The range of proctolin concentrations that stimulated the insect heart to more frequent beat was similar to that reported by the other authors.[4,15,16] According to the data

Table Cardioexcitatory Effect of Peptide I-V Relative to Proctolin (%) on Insects:

Compound	Periplaneta Americana at Conc. 2.1×10^{-8}M	Tenebrio Molitor at Conc. 2.6×10^{-9}M
Proctolin	100	100
I	380	183
II	100	320
III	390	155
IV	374	172
V	0	0

in the literature,[13] peptide III was more active than proctolin in a myotropic test on cockroach proctodeal muscle. We have found that peptide III, as well as the other analogs, I, II and IV, exerted higher activity on insect hearts than proctolin itself. Peptide V, in which the aromatic ring was replaced by 4-meth-oxycyclohexyl, was inactive in comparison with proctolin. Therefore, it may be inferred that replacement of the -OH group of Tyr in position 2 by $-NH_2$, $-NMe_2$, $-NO_2$ or -OMe increased stability of the analog in respect to degradation by specific proctolin enzymes.[17] Moreover, the high activity of peptide I-IV and lack of activity of the peptide V to some extent verified the Sullivan and Newcomb hypothesis[12] that biological function of proctolin depends on aromatic character of the amino acid residue in position 2 of the proctolin chain. Furthermore, our investigation seemed to indicate that the presence of polar group in the para position of the proctolin aromatic ring plays an important role in the binding process of the neuropeptide with its cellular receptor.

References

1. Brown, B.E., and Starratt, A.N. (1975) J. Insect Physiol. 21, 1879-1881.

2. Starratt, A.N., and Brown, B.E. (1977) Life Sci. 17, 1253-1256.

3. Brown, B.E. (1977) J. Insect Physiol. 23, 861-864.

4. Brown, B.E. (1977) Life Sci. 17, 1241-1252.

5. Adams, M.E., and O'Shea, M. (1983) Sciences 221, 286-289.

6. Bishop, C.A., Wina, J.J., and O'Shea, M. (1984) J. Neurosci. 4, 2001-2009.

7. Krawitz, E., Beltz, B., Glusman, S., Goy, M., Harris-Warrick, R., Johnson, M., Livingston, M. and Schwarz, Z. (1984) Pesticide Biochem. 22, 133-147.

8. Starratt, A.N., and Steele, R.W. (1984) Insect Biochem. 14, 97-102.

9. Miller, T.A. (1983) Endocrinol. Insects, R.A. Liss, Inc., New York, N.Y. pp. 101-107.

10. Watson III W.M., Augustin, G.J., Benson, J.A. and Sullivan, R.E. (1983) J. Exp. Biol. 103, 56-73.

11. Starratt, A.N., and Brown, B.E., (1977) Can. J. Chem. 55, 4238-4242.

12. Sullivan, R.E., and Newcomb, R.W. (1983) Peptides 3, 337-344.

13. Starratt, A.N. and Brown, B.E. (1979) Biochem. Biophys. Res. Comm. 90, 1125-1130.

14. Konopinska, D., Kazanowska, B., Sobotka, W. and Boguslawska-Jaworska, J. (1985) Experientia, in press.

15. Miller, T.A. (1979) Ann. Zool. 19, 77-89.

16. S-Rozsa, K., and Miller, T.A. (1981) Adv. Physiol. Sci. 22, 587-591.

17. Quistad, G.B., Adams, M.E., Scarborough, R.M., Carney, R.L., and Schooley, D.A. (1984) Life Sci. 34, 569-576.

IDENTIFICATION AND BIOLOGICAL ACTIVITY OF CHICKEN CALCITONIN I FROM ULTIMOBRANCHIAL GLANDS

Tamotsu Homma, Motoo Watanabe, Sachio Hirose and Akira Kanai
Mitsubishi Petrochemical Co., Ltd., 1315 Ami, Inashiki
Ibaraki 30003 JAPAN

Kenji Kangawa and Hisayuki Matsuo
Miyazaki Medical College, 5200 Kiyotake, Miyazaki 88916 JAPAN

Introduction

Calcitonin(CT) was first identified in rat thyroid gland as a hypocalcemic peptide hormone by Hirsch et al. in 1963.[1] In 1973, Nieto et al. reported the presence of the two forms of hypocalcemic activity in chicken ultimobranchial glands.[2] It is known that one of two chicken CTs elicits a hypocalcemic activity as potent as those of salmon I and eel CTs. The structure of chicken CT I and II has remained to be elucidated although only the amino acid composition of chicken CT I has been reported.

We report here the effective isolation of one of two-form chicken CTs by utilizing the radioimmunoassay(RIA) with the antiserum against eel CT and biological activities of chicken CT I determined by microsequencing and confirmed by synthesis.

Results and Discussion

The antiserum against eel CT was prepared in rabbits by immunization with antigenic conjugate of eel CT on thyroglobulin. The antiserum gave identical slopes of displacement of ^{125}I-labelled eel CT by chicken ultimobranchial extracts as by synthetic eel CT. This fact suggests that a closer structural similarity exists between chicken and eel CTs. It was found to be able to use the RIA for chicken CT assay in purification steps.

Frozen chicken ultimobranchial glands (ca. 179g) collected from 3384 chickens were boiled in the extraction solvent(1M acetic acid) for abolishing intrinsic proteolytic activity. Acid extracts (ca. 900ml) obtained after centrifugation were subjected to acetone precipitation at a concentration of 66%. After removal of the precipitates, the supernatant was evaporated to dryness. The dry materials were again dissolved in 1M acetic acid (330ml), and subjected on SP-Sephadex C-25 column. Successive elutions with 1M acetic acid, 2M pyridine and 2M pyridine-acetic acid (pH 5.0) afforded three respective fractions, SP-I, SP-II and SP-III. The immunoreactivity of chicken CT was observed in the fractions of SP-II and SP-III. The immunoreactivity in SP-III containing strongly basic peptides was found to be twice as much as that in SP-II. The higher immunoreactive SP-III was chosen as the material for the present purification.

Table I. Summary of the Purification of Chicken CT I

Steps	Weight (mg)	ir-CT[a] (ug)	Content of ir-CT (%)
Acid extract	13600	1157	0.0085
Acetone precipitation	-	1015	-
SP-Sephadex C-25(SP-III)	1350[b]	613	0.045
Sephadex G-50	20.8	485	-
HPLC(ion exchange)[d]	-	147	23.3
HPLC(reverse phase)[e]	0.137[c]	137	100

a) Immunoreactive CT

b) SP-II was not included.

c) Protein was determined by amino acid analysis.

d) TSK CM 2SW Column

e) Chemcosorb 5-ODS-H

The immunoreactive activity of each purification step is summarized in Table I. Finally 137 ug of immunoreactive CT was obtained from 3384 chickens.

The amino acid compositions of purified chicken CT I and fragment peptides produced by trypsine (T-1,2,3 and 4) and protease V-8 digestion (P-1 and 2) were elucidated. Chicken CT I contained one more Ala and one less Asx than eel CT. It was evident that half-cystine residues at the 1st and 7th positions formed a disulfide linkage in the chicken CT I, since the molecular weight of chicken CT I was ranging from 3,500 to 5,000 and one cystine could be found in the chicken CT I (taken value of amino acids: 32). Performic acid-oxidized chicken CT I and fragment peptides were separated by HPLC (reverse phase) and taken for sequence determination. The subsequence (1-14) of performic acid-oxidized chicken CT I and the sequences of T-1, T-2 and T-3 were determined by sequence analysis with dansyl-Edman method. T-4 was found to be the N-terminal peptide, since the amino acid composition of T-4 was identical with that of subsequence (1-11) of performic acid-oxidized chicken CT I. T-2 was found to be the second peptide from the N-terminus in chicken CT I, since the subsequence (12-14) of performic acid-oxidized chicken CT I was the same as the subsequence (1-3) of T-2. T-1 was identified to be the C-terminal peptide, since it contained a Pro-NH_2 and no basic residues (Arg and Lys) which should be formed to be the C-terminal residue of fragment peptides by tryptic digestion. Therefore, the order of the four tryptic peptides was established as T-4, T-2, T-3 and T-1 from the N-terminus of chicken CT I. Futhermore, ^{14}Gln, ^{15}Glu, ^{20}Gln and ^{26}Asp were determined from these results combined with the data of protease V-8 digestion, aminopeptidase M digestion and amino acid analysis. Finally, the complete structure of chicken CT I determined thus was shown as H-Cys-Ala-Ser-Leu-Ser-Thr-Cys-Val-Leu-Gly-Lys-Leu-Ser-Gln-Glu-Leu-His-Lys-Leu-Gln-Thr-Tyr-Pro-Arg-Thr-Asp-Val-Gly-Ala-Gly-Thr-ProNH$_2$.

Although a wide homology was found at the amino terminal region in the hormones except for chicken CT I, chicken CT I differed in two positions from eel CT, ^{2}Ala and ^{3}Ser in chicken CT I corresponded to ^{2}Ser and ^{3}Asn in eel CT. The 3rd amino acid from N-terminus of the CTs so far isolated was all Asn and Ser was only found in chicken CT I.

Hypocalcemic activity of chicken CT I thus purified was compared with those of synthetic chicken CT I and several synthetic CTs, bysubcutaneous injection into rats fasted, as summarized in Table II. Biological activity of hypocalcemic response was expressed as MRC unit according to porcine CT standard B. The sequence of chicken CT I was also confirmed by synthesis, since natural and synthetic chicken CT I were equipotent in hypocalcemic activity. Specific biological activity of chicken CT I (4500 MRCU/mg) was as potent as that of eel CT, which had been known to elicit the most potent activity, and was much more than that of salmon CT I. Thus, purified chicken CT I possessed one of the highest hypocalcemic response so far isolated.

Table II. Hypocalcemic Activities of Purified Chicken CT I and Various Synthetic CTs

Calcitonin		Hypocalcemic Activity (MRCU/mg)
Chicken CT I	natural	4500
Chicken CT I	synthetic	4500
Eel CT	synthetic	4500
Salmon CT I	synthetic	3500
Human CT	synthetic	270

References

1. Hirsch, P.F., Gauthier, G.F., and Munson, P.L. (1963) Endocrinology 73, 244- 252.

2. Nieto, A., Moya, F., and R-Candela, J.L. (1973) Biochem. Biophys. Acta, 332, 383-391.

STRUCTURE/FUNCTION STUDIES ON HUMAN RELAXIN

Paul D. Johnston, John Burnier, Sharon Chen, Dave Davis, Harvard Morehead, Melissa Remington, Martin Struble, Geoffrey Tregear* and Hugh Niall*

Genentech, Inc., 460 Pt. San Bruno Boulevard, South San Francisco, CA 94080

Howard Florey Institute, University of Melbourne, Parkville, Victoria, Australia, 3052*

Introduction

Relaxin is an ovarian peptide hormone of pregnancy, structurally related to insulin. Two human genes have been identified only one of which, designated gene-2 (h2)[1], is expressed in the ovary. The other form, gene-1 (h1), has been found only as a genomic clone. Figure 1 presents a comparison of relaxin sequences and human insulin. The exact processing of human relaxin is unknown; the estimate shown in Figure 1 is based on comparison to the known sequences of animal relaxins. The chain residue numbering is relative to the human gene-2 (h2) sequence; therefore it differs from that in reference.[2] The bioactive core (80% of full length) for pig relaxin has been shown to be A3-24,B5-24.[2] This core transposed to the human relaxins includes methionine 25. We describe in this communication the synthesis of several relaxin analogues which help define the active core of human relaxin and explore amino acid substitutions for the methionines in human B-chains. Several des-methionine B-chain analogues were found for both gene forms which lead to fully active chain recombined relaxins. These analogues are suitable for hybrid fusion expression in *E. coli* and chemical processing by CNBr.

Results and Discussion

```
                                 A-Chain
              1                 10                  20
h2 Relaxin    Q-L-Y-S-A-L-A-N-K-C-C-H-V-G-C-T-K-R-S-L-A-R-F-C
h1 Relaxin    R-P-Y-V-A-L-F-E-K-C-C-L-I-G-C-T-K-R-S-L-A-K-Y-C

p Relaxin         R-M-T-L-S-E-K-C-C-Q-V-G-C-I-R-K-D-I-A-R-L-C
r Relaxin     Q-S-G-A-L-L-S-E-Q-C-C-H-I-G-C-T-R-R-S-I-A-K-L-C
h Insulin             G-I-V-E-Q-C-C-T-S-I-C-S-L-Y-Q-L-E-N-Y-C-N

                                 B-Chain
                    1                 10                  20                  30
h2 Relaxin          D-S-W-M-E-E-V-I-K-L-C-G-R-E-L-V-R-A-Q-I-A-I-C-G-M-S-T-W-S-K-R-S-L
h1 Relaxin            K-W-K-D-D-V-I-K-L-C-G-R-E-L-V-R-A-Q-I-A-I-C-G-M-S-T-W-S-K-R-S-L

p Relaxin             Q-S-T-N-D-F-I-K-A-C-G-R-E-L-V-R-L-W-V-E-I-C-G-S-V-S-W-G-R-T-A-L
r Relaxin     R-V-S-E-E-W-M-D-Q-V-I-Q-V-C-G-R-G-Y-A-R-A-W-I-E-V-C-G-A- -S-V-G-R-L-A-L
h Insulin                   F-V-N-Q-H-L-C-G-S-H-L-V-E-A-L-Y-L-V-C-G-E-R-G-F-F-Y-T-P-K-T
```

Fig. 1. Sequence comparison of human relaxins with human insulin, porcine (p) and rat (r) relaxins. The numbering system is relative to h2 relaxin. The disulfides for the relaxins are analogous to insulin: A10-A15, A11-B11 and A24-B23. The A- and B-chain lengths are referred to outside parentheses, eg. A24B2-33 while amino acid substitutions are noted within the parentheses, eg. B2-33(K4,A25).

The individual chains of human relaxin and their analogues were synthesized by solid phase methods and purified by preparative reverse phase HPLC. Purification of the synthetic peptides was monitored by amino acid analysis, amino terminal sequencing and reverse phase HPLC. A modification of the *in vitro* chain recombination conditions for porcine relaxin[2] was used to produce the human relaxins.

Since gene-2 is the only relaxin expressed in the corpus luteum[1], we have primarily concentrated on analogues of h2 relaxin involving changes in the B-chain. The amino acid substitutions employed were Met4 to Lys4, the equivalent residue in h1 relaxin, and Met25 to Ala25, the equivalent residue in rat relaxin. Figure 2 shows a time course for the *in vitro* chain recombination of h2 A24 and B33(K4,A25). The reaction is stopped at optimal production of relaxin. All peaks have been collected and assayed by the *in vivo* murine pubic symphysis ligament (MPS) and the *in vitro* rat uterine contractility (RUC) bioassays. Two forms of active relaxin were isolated, elution times 22.5 and 23 minutes. The purified relaxins were characterized by amino acid analysis, amino terminal sequencing and reverse phase HPLC. A blocked amino-terminus of the later eluting relaxin's A-chain indicate that the difference probably involves the conversion of the amino terminal A-chain Gln (22.5 min) → pyro-Glu (23.0 min) in the later eluting

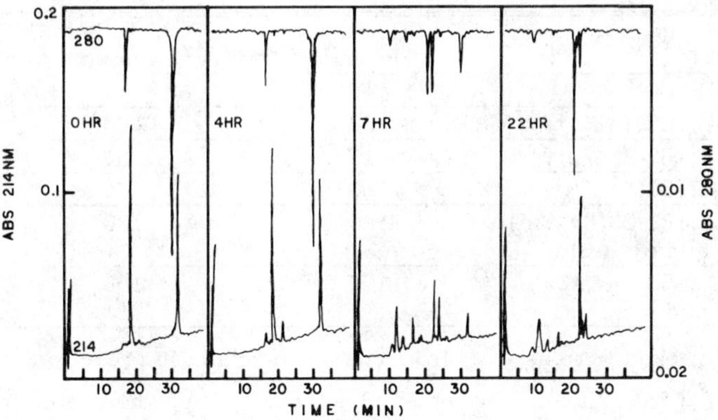

Fig. 2. HPLC time course profile for the *in vitro* chain recombination reaction of h2 A24 and B33(K4,A25). Chromatography was done using a Synchropak RP-C4 (4.6 x 250 mm; 300 A). A linear gradient of acetonitrile (15 → 60% in 50 minutes; flow 1 ml/min) in a 0.05% TFA, H2O buffer. Retention times: A-chain 18.6 min.; B-chain 31.9 min.; relaxin (Gln) 22.5 min.; relaxin (pyro-Glu) 23 min.

form. Conversion to pyro-Glu is also observed for natural animal relaxins. The other peaks observed throughout the course of the reaction have been identified as intramolecular oxidized forms of the A-chain (11.5 min to 18 min) and B-chain (24 min to 32 min) by amino acid analysis and low molecular weight polyacrylamide urea gels. The major species are monomeric or dimeric, and B-chain forms a larger amount of polymeric species than does A-chain. Comparison of the A214 nm/A280 nm ratios, Figure 2, as well as nonreducing urea gels indicate no formation of inactive soluble A-B hybrids. All forms of A- and B-chains including the mixed oligomers that precipitate out of the reaction can be reduced back to their original elution positions with 10 mM DTT.

Table I presents a comparison of the bioactivities of recombined relaxin analogues of human gene-1 and gene-2. Several major conclusions can be drawn from this data. The h1 B2-27 Met25 to Ala25 substitution does not affect the activity of the chain recombined relaxin, A24B2-27(A25). This is very interesting because des-methionine h1 A24B2-24 is inactive. The h2 B-chain des-methionine analogues using Met4 to Lys4 and Met25 to Ala25 lead to a fully active relaxin. There is no significant difference in biological activity between the Gln and pyro-Glu forms. The shortened B-

Table I. Comparative Bioactivities of Human Relaxin Analogues

Analogue	RUC Activity % Porcine	MPS Activity % Porcine
h1 A24B24*	0	0
h1 A24B27(A25)*	60	100
h2 A24B2-25*	73	30
h2 A24B33	100	100
h2 A24B2-33(K4,A25) Gln and pyro-Glu	100	100
h2 A24B33(K4,A25) Gln and pyro-Glu	100	100

* The species marked were assayed from chain recombination reactions without further purification of the relaxin. The reported activity has been corrected for recombination yield as assessed by analytical reverse phase HPLC of the final reaction mixture. The non-asterisked values were obtained from highly purified chain recombined relaxins and are indicative of absolute specific activity. Note that the Gln and pyro-Glu forms were assayed separately, even though they are reported together.

chain analogues did not substantially increase the chain recombination reaction yield; this result contrast that observed for porcine relaxin.[2] Unlike porcine relaxin[2], we observed by CD that no W28 dependent β-structure is found for either the shortened or full length h2 B-chains. In addition, shortening the B-chain amino-terminus by one residue left the biological activity unchanged.

References

1. Hudson, P., John, M., Crawford, R., Haralambidis, J., Scanlon, D., Gorman, J., Tregear, G.W., Shine, J., and Niall, H.D. (1984) EMBO Journal **3**, 2333-2339.

2. Tregear, G.W., Du, Y.C., Wang, K., Southwell, C., Jones, P., John, M., Gorman, J., Kemp, B., and Niall, H.D. (1983) in Biology of Relaxin and Its Role in the Human, Bigazzi, M., Greenwood, F.C., and Gasparri, F., Eds., Elsevier, N.Y., NY, pp. 42-55.

CIRCULAR DICHROISM, 2D-NMR AND BIOLOGICAL ACTIVITY OF THE DODECAPEPTIDE α-FACTOR FROM S. CEREVISIAE

Fred R. Naider and Ponniah Shenbagamurthi
College of Staten Island, C.U.N.Y. Staten Island, N.Y. 10301

Michelle S. Broido and Linda A. Hughes
Hunter College, C.U.N.Y. New York, N.Y. 10021

Susan Raths and Jeffrey M. Becker
University of Tennessee, Knoxville, Tenn. 37996

Sexual conjugation in Saccharomyces cerevisiae is controlled by diffusable peptides known as mating pheromones.[1] The synthesis, secondary structure and biological activity of amino acid analogs of the α-factor, Trp-His-Trp-Leu-Gln-Leu-Lys-Pro-Gly-Gln-Pro-Met-Tyr, is the subject of this communication.

Solid Phase Synthesis

Solid phase synthesis of the α-factor was carried out on a standard chloromethylated polystyrene resin and all residues were double coupled using either DCCI or DCCI-HOBt. Side chain protection included 2,6-Cl_2-Bzl for Tyr, 2-Cl-Z for Lys, Form for Trp, and Tos for His. Dimethylsulfide was used as a scavenger during all acidolysis steps. Final cleavage of the peptide was accomplished by both the low-high HF procedure[2] and the high HF procedure. Our experiments did not reveal any significant improvement with the low-high method. Although different impurities were observed on a gradient reversed phase HPLC (C_{18} support, $CH_3CN:H_2O:CF_3COOH$; 20-40% gradient), the impurities associated with the crude product from the low-high HF cleavage had K' values very near to those of the desired product. This increased the purification problem. Nevertheless the desired α-factor analog could be obtained using a one step purification on a Prep 500 reversed-phase column.[3] Final peptide was 96-98% pure as judged by analytical HPLC. Yields of product varied from 25-30%.

Biological Activity of desTrp^1,Cha^3,X^9-α-Factor Analogs

The presence of a Pro^8-Gly^9 sequence in the α-factor raises the possibility of a β-turn in this region of the peptide backbone. Analogs

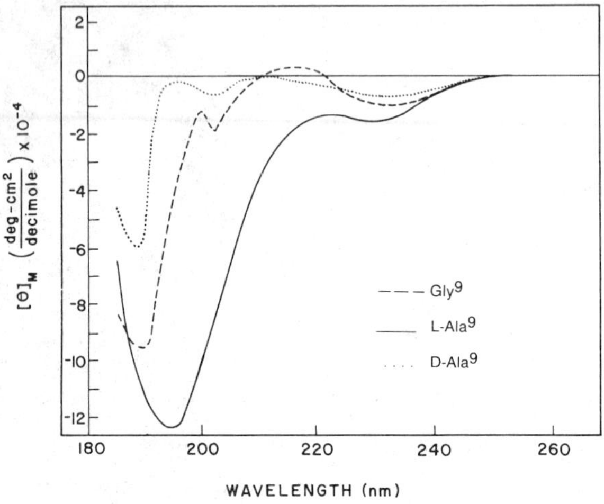

Fig. 1. CD of the desTrp^1, Cha^3,X^9 - α-Factor in Trifluoroethanol. Peptide concentration ~ 1 mg/ml.

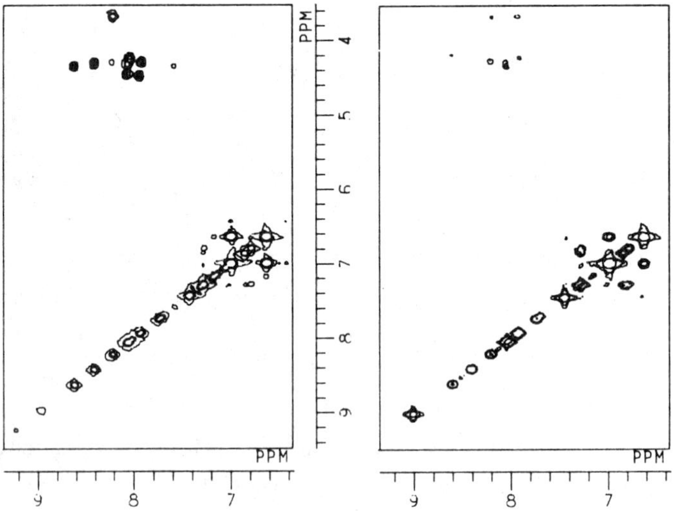

Fig. 2. COSY (left) and 2D NOE (right) spectra of the NH region of the desTrp^1,Cha^3,[α-^{2}H]Met^{12}-α-factor. Cross-peaks indicate J (COSY) or dipolar (NOE) connectivities.

of the desTrp^{1},Cha^{3}-α-factor in which Gly^{9} was replaced by L-Ala, D-Ala, L-Leu and D-Leu were prepared and purified using solution phase techniques.[3] The CD patterns of the alanine-containing analogs in trifluoroethanol fall into two different classes (Figure 1). Similar patterns were observed in desTrp^{1},Cha^{3},Gly^{9}- and desTrp^{1},Cha^{3},D-Ala^{9}-α-factors whereas the pattern for the desTrp^{1},Cha^{3}-L-Ala^{9}-pheromone was distinct. Although the patterns are not associable with ordered conformations they suggest one distribution of φ,ψ angles for the Gly^{9}- and D-Ala^{9} pheromones and a different distribution for the L-Ala^{9}-containing dodecapeptide. Similar results were found for the leucine-containing analogs. The biological activity of the position 9 analogs as judged by their ability to induce changes in the shape of S. cerevisiae MATa cells (shmoo assay) is summarized in Table 1. Both the D-Ala^{9} and Gly^{9} dodecapeptides induced aberrant morphologies at concentrations of 1.25 μg/ml. The D-Leu^{9} analog was also highly active. In contrast at concentrations of 200 μg/ml both the L-Ala^{9}- and L-Leu^{9}-analogs exhibited almost no activity.

Table 1. Biological Activity of desTrp^{1},Cha^{3},X^{9}-α-Factors[a]

X^{9}	Conc. (μg/ml)	% Aberrant	% Unbudded
Gly (natural residue)	1.25	89	82
L-Ala	200.00	23	27
D-Ala	1.25	80	83
L-Leu	200.00	27	32
D-Leu	5.00	70	63
None	-	0	30

[a] Assays were conducted using S. cerevisiae 2180-1A[3]

These results suggest that a Pro^{8}-Gly^{9} or a Pro^{8}-D-X^{9} sequence is tolerated whereas a Pro^{8}-L-X^{9} sequence greatly decreases the biological activity of α-factor. Together with the CD studies they support the hypothesis that the active pheromone assumes a Type II-β-turn in the Lys^{7}-Pro^{8}-Gly^{9}-Gln^{10} region of the pheromone.

2D-NMR Studies

Complete two-dimensional ^{1}H NMR analyses of α-factor and the desTrp[1], Cha[3]-dodecapeptide, dissolved in DMSO, have been performed. Selective deuteration of α-protons enabled assignment of NH resonances due to collapse of CHα-NH spin-spin (J) couplings;[4] correlated spectroscopy (COSY) which traces connectivities based on spin-spin couplings was used to assign the remainder of the protons in the molecule (Figure 2). The most efficacious means of obtaining structural information by NMR is through dipole-dipole connectivities, as manifested in nuclear Overhauser effects. 2D NOE (NOESY) experimental results gave no evidence for short internuclear separations other than those expected from primary structure. Greater than 90% of all proton resonances in the molecule are within 0.1 ppm of the resonance positions found for disordered tetrapeptides.[5] These data clearly indicate that alpha-factor peptides assume no structure in DMSO.

Acknowledgements

This work was supported by NIH Grants GM 22086 and GM 22087, NSF Grant PCM 8111745 and a grant from the Research Corporation.

References

1. Thorner, J. (1980) The Molecular Genetics of Development, Leighton, T. and Loomis, W.E., Eds., Academic Press, New York, pp. 119-178.

2. Tam, J.P., Heath, W.F., and Merrifield, R.B. (1983). J. Amer. Chem. Soc. 105, 6442-6455.

3. Shenbagamurthi, P., Baffi, R., Khan, S.A., Lipke, P., Pousman, C., Becker, J.M., and Naider, F. (1983) Biochemistry 22, 1298-1304.

4. Shenbagamurthi, P., Steinfeld, A.S., Khan, S.A., Becker, J.M., and Naider, F. (1983) Biopolymers 22, 815-820.

5. Wuthrich, K. (1976) NMR In Biological Research: Peptides and Proteins. Elsevier, New York, pp. 51-54.

CARBOXY-TERMINUS OF α-MELANOTROPIN: STRUCTURAL DETERMINANTS IMPORTANT FOR MELANOTROPIC ACTIVITY

Douglas J. Staples§, Tomi K. Sawyer§, Mac E. Hadley*, Michael H. Engel+, Anne E. deVaux§, Joseph A. Affholter§, Paul S. Darman**, Wayne L. Cody**, Brian C. Wilkes**, and Victor J. Hruby**

§Biopolymer Chemistry-Regulatory Peptide Research, The Upjohn Company, Kalamazoo, MI 49001; *Department of Anatomy and **Department of Chemistry, The University of Arizona, Tucson, AZ 85721; and +School of Geology and Geophysics, The University of Oklahoma, Norman, OK 73019.

INTRODUCTION

The regulatory role of α-melanotropin (Ac-Ser-Tyr-Ser-Met-Glu-His-Phe-Arg-Trp-Gly-Lys-Pro-Val-NH_2, α-MSH) on vertebrate pigmentation[1] was the first described and, perhaps, the most intensively studied physiological activity of this peptide hormone. Based on previous structure-activity studies, using the <u>in vitro</u> frog skin bioassay, several candidate "**active site**" sequences have been proposed[2,3] for α-MSH and include: His-Phe-Arg-Trp, Met-Glu-His-Phe-Arg-Trp-Gly, Phe-Arg, and Lys-Pro-Val-NH_2. Interestingly, the above putative "**active site**" sequences of α-MSH would, therefore, exemplify a peptide hormone having at least two distinct molecular regions (i.e., α-MSH_{4-10} and α-MSH_{11-13}) capable of independently eliciting intrinsic melanotropic activity. In this report, the physico-chemical and biological properties of several carboxy-terminal α-MSH fragment peptides (Figure 1) are described in order to ascertain whether, indeed, α-melanotropin is composed of one or more structurally-independent "**active-site**" sequences.

Figure 1. Primary sequences of α -MSH and several of its fragment peptides.

Ac-α-MSH_{11-13}-NH_2	Ac-Lys-Pro-Val-NH_2	**(I)**
Ac-α-MSH_{9-13}-NH_2	Ac-Trp-Gly-Lys-Pro-Val-NH_2	**(II)**
Ac-α-MSH_{7-13}-NH_2	Ac-Phe-Arg-Trp-Gly-Lys-Pro-Val-NH_2	**(III)**
Ac-α-MSH_{6-13}-NH_2	Ac-His-Phe-Arg-Trp-Gly-Lys-Pro-Val-NH_2	**(IV)**
Ac-α-MSH_{4-10}-NH_2	Ac-Met-Glu-His-Phe-Arg-Trp-Gly-NH_2	**(V)**
α-MSH	Ac-Ser-Tyr-Ser-Met-Glu-His-Phe-Arg-Trp-Gly-Lys-Pro-Val-NH_2	
	1 4 7 10 13	

RESULTS AND DISCUSSION

The α-MSH fragment peptides (compounds **I-IV**) were prepared by solid-phase peptide synthetic methods (Figure 2) similar to those previously reported[4]. The title compounds were purified by reverse-phase preparative HPLC (C_{18} Vydac column and a $CH_3CN/H_2O/0.2\%$TFA-based binary solvent system). The α-MSH and Ac-α-MSH_{4-10}-NH_2 were prepared as previously described[4]. The title compounds (I-IV) were characterized for their physico-chemical properties by amino acid analysis and fast atom bombardment-mass spectrometry (FAB-MS) as shown in Table I. In addition, compounds **I-III** were characterized by D/L amino acid analysis (Table II) to determine their stereochemical purity.

The compounds were tested for their relative potencies on the frog (Rana pipiens) skin bioassay as shown in Table III. In summary, the C-terminal 9-13 and 11-13 sequences of α-MSH did not possess melanotropic activity. Thus, the data do **not** support the hypothesis of a **two "active site"** model for α-MSH

Table I. Amino acid analysis and FAB-MS of compounds **I-IV**.

Compound	FAB-MS $(M\cdot+1)^+$ at m/z	Amino Acid Ratio							
		His	Phe	Arg	Trp	Gly	Lys	Pro	Val
I	384					0.99	1.00	1.00	1.00
II	627				0.85	1.01	1.00	0.99	1.01
III	930		1.05	0.96	1.01	1.01	1.00	0.99	1.01
IV	1067	0.96	0.99	0.94	1.05	1.03	0.99	1.01	1.01

Table II. D/L amino acid analysis[5,6] of compounds **I-III**.

Compound	D/L Amino Acid Value					
	Phe	Arg	Trp	Lys	Pro	Val
I				0.06	<0.01	<0.01
II			0.03	0.05	<0.01	0.01
III	0.02	<0.01	0.01	0.01	<0.01	0.01

bioactivity. However, the C-terminal heptapeptide (Ac-α-MSH_{7-13}-NH_2, **III**) exhibits full agonist activity, but low potency (5/100,000 relative to α-MSH). N-Terminal extension of the peptide by the incorporation of His to provide Ac-His-Phe-Arg-Trp-Gly-Lys-Pro-Val-NH_2 (**IV**) resulted in a 100-fold increase in potency. It is clear from our previous studies[1], however, that the C-terminal tripeptide sequence does contribute to the high melanotropic potency of α-MSH. For example, the central heptapeptide fragment, Ac-Met-Glu-His-Phe-Arg-Trp-Gly-NH_2, possesses only minimal activity (about 3/10,000) relative to α-MSH.

However, C-terminal extension of either Ac-α-MSH_{4-10}-NH_2 or Ac-[Cys[4], Cys[10]]-α-MSH_{4-10}-NH_2 to incorporate Lys-Pro-Val results in markedly increased (100- to 1000-fold) relative potencies for the resultant decapeptides.[1] The above data provides evidence to suggest a **one-"active site"** model for α-MSH and that Phe-Arg-Trp, the common tripeptide sequence of Ac-α-MSH_{4-10}-NH_2 and Ac-α-MSH_{7-13}-NH_2, may possess the minimal structural determinants important for melanotropic activity.

Table III. Relative melanotropic potencies in vitro of several α-MSH fragment peptides.

Compound	Relative Potency
Ac-α-MSH	1.0
Ac-α-MSH_{11-13}-NH_2	inactive[a]
Ac-α-MSH_{9-13}-NH_2	inactive[a]
Ac-α-MSH_{7-13}-NH_2	0.00005
Ac-α-MSH_{6-13}-NH_2	0.005
Ac-α-MSH_{4-10}-NH_2	0.0003

[a]No biological activity observed in the 10^{-5}-10^{-3} M [peptide] range.

REFERENCES AND NOTES

1. Hruby, V.J., Wilkes, B.L., Cody, W.C., Sawyer, T.K., and Hadley, M.E. (1984) in **Peptide and Protein Reviews**, Hearn, M.T.W., Ed., Marcel Dekker, Inc., New York, NY., pp. 1-69.

2. Otsuka, H. and Inouye, K. (1964) **Bull. Chem. Soc. Japan 37**, 1465-1471.

3. Eberle, A., and Schwyzer, R. (1976) in **Surface Membrane Receptors**, Bradshaw, R.A., Frazier, W.A., Merrell, R.C., Gottlieb, D.I., and Hogue-Angeletti, R.A., Eds., Plenum Publ. Corp., New York, NY, pp. 291-304.

4. Sawyer, T.K., Hruby, V.J., Draelos, M., Wilkes, B.C., and Hadley, M.E. (1982) **J. Med. Chem. 25**, 1022-1027.

5. Engel, M.H., and Hare, P.E. (1981) **Carnegie Inst. Wash. Year Book 81**, 394-397.

6. The D/L values obtained for compounds **I-III** generally reflect trace racemization which may occur during the peptide hydrolysis versus intrinsic stereochemical impurities within the intact title peptides.

INSULIN ANALOGUES BEARING UNNATURAL AROMATIC AMINO ACID SUBSTITUTIONS AT POSITION B25

Satoe H. Nakagawa, Matthew J. Sheetz and Howard S. Tager
Department of Biochemistry and Molecular Biology, The University of Chicago, Chicago, Illinois 60637

Introduction

Two out of three mutant human insulins isolated from unrelated diabetic patients were identified to have Leu for B25 Phe or Ser for B24 Phe substitutions[1-3]. Synthetic studies on [Leu^{B24}]-, [Leu^{B25}]-, [Ser^{B24}]-, and [Ser^{B25}]-insulin have shown that subsitution of B25 Phe with either residue reduces hormone potency markedly (to about 1% of that of the natural hormone), whereas substitution of B24 Phe reduces potency only moderately (to 10-20% of that of natural hormone)[4-6].

Crystallographic studies of 2Zn insulin or 4Zn insulin have presented accurate positions for the atoms of the corresponding insulin hexamers that are composed of three identical dimers, each of which contains two nonidentical monomers[7,8]. B25 Phe appears as part of a hydrophobic surface of the monomer, together with B12 Val, B16 Tyr, B24 Phe, and B26 Tyr. Moreover, these residues are nearly invariant and play important roles in monomer-monomer interactions. It is obvious from the comparison of the two monomers that the B25 Phe aromatic ring takes very different conformations in each case: in one molecule the ring packs into its own monomer; in the other molecule the ring protrudes from the surface.

Since the B25 Phe side chain has conformational flexibility in the Zn-insulin crystal and over-all structural changes are not seen by the substitution of B25 Phe with Leu or Ala, as studied by CD spectra[6,9], it is likely that the B25 Phe side chain plays an important role in interactions of the hormone with its receptor. By substituting B25 Phe with other aromatic amino acids (see Figure 1), we expect to obtain further information on the role of the B25 side chain in directing hormone interaction with its receptor.

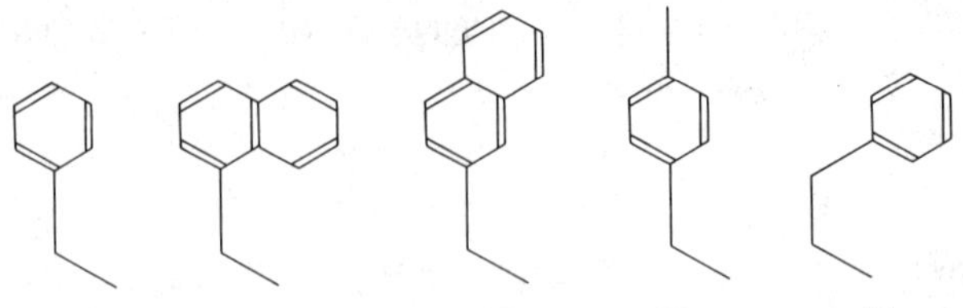

Fig. 1. Structures of B25 side chains of insulin and semi-synthetic insulin analogues.

Synthesis

The syntheses of four octapeptides (B23-30), Gly-Phe-X-Tyr-Thr-Pro-Lys(Boc)-Thr, where X is an unnatural amino acid to be substituted (I, 3-(1'-naphthyl)alanine; II, 3-(2'-naphthyl)alanine; III, p-methylphenylalanine; and IV, 2-amino-4-phenylbutyric acid) were carried out by Merrifield's solid-phase method to yield the corresponding Tfa-Gly-Phe-X-Tyr(Bzl)-Thr(Bzl)-Pro-Lys(ClZ)-Thr(Bzl)-resin esters. Cleavage of the peptides from the resin and removal of all protecting groups except the Tfa group were carried out according to the procedure reported by Anwer et al[10]. Introduction of the Boc group at the ε-amino group of Lys and removal of the Tfa group resulted in octapeptides used during enzyme-catalyzed semisynthesis[3].

The trypsin-catalyzed condensation between $(Boc)_2$-desoctapeptide (B23-30)-insulin prepared from porcine insulin and a synthetic octapeptide with an unnatural amino acid substitution at B25 were carried out by the procedure described by Inouye et al[6]. After removal of Boc groups, semi-synthetic insulin analogues were purified by repeated gel-permeation chromatography. The purity of each analogue was confirmed by reverse-phase HPLC and by amino acid analysis.

Results and Discussion

Binding affinities of the semi-synthetic insulin analogues were assessed by competition for ^{125}I-insulin binding to isolated canine hepatocytes (Figure 2). Biological activity was assessed by stimulation of glucose oxidation by isolated rat adipocytes[3]. The potencies of the analogues relative to porcine insulin (determined by the concentration of peptide required to cause half maximal response in binding inhibition or biological activity) are summerized in Table I. Naphthyl analogues, I and II, have identical mass, but differ in attachment of the bulky ring. Apparently, II is sterically more favored than I during the interaction of hormone with receptor. Analogue III, which has a methyl group at the p-position on the phenyl ring, showed a potency intermediate between the two naphthyl analogues. All three analogues bearing an aromatic ring at the β-carbon of the B25 side chain showed substantial binding and biological activities. However, analogue IV, which has an aromatic ring at the γ-carbon, showed markedly reduced binding and biological activity that is reminiscent of the low activities of [Leu^{B25}]- or [Ser^{B25}]insulin.

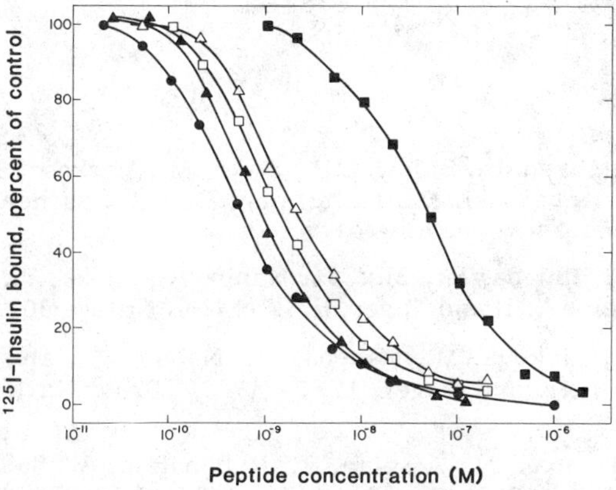

Fig. 2. Dose-response curves of insulin and semi-synthetic insulin analogues in inhibition of ^{125}I-insulin binding to canine hepatocytes: porcine insulin (●); analogue I (△); analogue II (▲); analogue III (□); analogue IV (■).

Table I. Binding and Biological Potencies of Semi-synthetic Insulin Analogues

Analogue	% of Insulin Potency	
	Binding	Glucose Oxidation
Porcine insulin	100	100
[Nal(1)^{B25}] human insulin (I)	24	17
[Nal(2)^{B25}] human insulin (II)	50	66
[pMePhe^{B25}] human insulin (III)	36	30
[Hph^{B25}] human insulin (IV)	1.1	1.1

Studies on the enhancement of the rate of dissociation of ^{125}I-insulin from canine hepatocytes showed that receptor occupancy by analogue II accelerates such dissociation, as does insulin, whereas analogue III does not (data not shown).

In conclusion, the B25 Phe aromatic ring in natural insulin apparently plays an important role in the binding of ligand to receptor. B25 Phe can be substituted with other amino acids bearing an aromatic ring at the β-carbon, although there are some restrictions in size or shape, without affecting over-all features of hormone interaction with its receptor. Nevertheless, an aromatic ring at the γ-carbon causes major perturbations in ligand-receptor interactions.

References

1. Tager, H.S., Given, B., Baldwin, P., Mako, M., Markese, J., Rubenstein, A.H., Olefsky, J., Kobayashi, M., Kolterman, O. and Poucher, R. (1979) Nature 281, 122-125.

2. Shoelson, S., Haneda, M., Blix, P., Nanjo, A., Sanke, T., Inouye, K., Steiner, D., Rubenstein, A.H. and Tager, H.S. (1983) Nature 302, 540-543.

3. Shoelson, S., Fickova, M., Haneda, M., Nahum. A., Musso. G., Kaiser, E.T., Rubenstein, A.H. and Tager, H.S. (1983) Proc. Natl. Acad. Sci. USA 80, 7390-7394.

4. Tager, H., Thomas, N., Assoian, R., Rubenstein, A., Saekow, M., Olefsky, J. and Kaiser, E.T. (1980) Proc. Natl. Acad. Sci. USA 77, 3181-3185.

5. Gattner, H.-G., Danho, W., Behn, C. and Zahn, H. (1980) Hoppe-Seyler's Z. Physiol. Chem. 361, 1135-1138.

6. Inouye, K., Watanabe, K., Tochino, Y., Kobayashi, M. and Shigeta, Y. (1981) Biopolymers 20, 1845-1858.

7. Blundell, J.T., Cutfield, J.F., Cutfield, S.M., Dodson, E.J., Dodson, G.G., Hodgkin, D.C., Mercola, D.A. and Vijayan, M. (1971) Nature 231, 506-511.

8. Bently, G., Dodson, E., Dodson, G., Hodgkin, D. and Mercola, D. (1976) Nature 261, 166-168.

9. Wollmer, A., Strassburger, W., Glatter, V., Dodson, G.G., McCall, M., Gattner, H.-G., Danho, W., Brandenburg, D. and Rittel, W. (1981) Hoppe-Seyler's Z. Physiol. Chem. 362, 581-591.

10. Anwer, M.K. and Spatola, A.F. (1983) J. Org. Chem. 48, 3503-3507.

SEMISYNTHETIC INSULIN ANALOGS: SUBSTITUTION FOR GLY-B23 SIGNIFICANTLY ALTERS THE ACTIVITY OF DESTETRAPEPTIDE-INSULIN

Mark W. Riemen*, John T. Hosoume, Nancy A. Hillyard, Michael P. Allen, and Frederick H. Carpenter (deceased)

Department of Biochemistry, University of California, Berkeley 94720.
*Current address: Dept. of Virus & Cell Biology Research, Merck Sharp & Dohme Research Laboratories, West Point, PA 19486

Introduction

The C-terminal region of the B-chain of insulin has been extensively studied. Most of this work has centered on the aromatic residues Phe^{B24}-Phe^{B25}-Tyr^{B26}. Modifications or deletions of these residues cause considerable alteration of biological activity. Inouye et al.[1,2] have semisynthetically prepared several analogs containing D-amino acids in this region. They reported the first semisynthetic insulin analog with supranormal activity, [DPhe^{B24}]-insulin. This derivative has 180% activity in receptor binding assays and 140% in hexose uptake assays. However, low coupling yields were observed during the preparation of these analogs. Inefficient coupling was caused by the proximity of D-amino acids to the site of bond formation[2]. Using semisynthetic methodology developed in our laboratory[3,4], we have prepared and biologically analyzed several derivatives of insulin in which Gly^{B23} and Phe^{B24} have been substituted with D-amino acids.

Results and Discussion

Synthetic tetrapeptides (for sequences see DOI-derivatives below) were prepared in solution using a stepwise (C to N) strategy. The di- and tripeptide intermediates and final tetrapeptide products were crystalized and characterized in the protected form. The protected peptides were homogeneous as indicated by melting point, optical rotation, elemental analysis and TLC (at least two systems). Two additional methods were used for characterization: a) modified[5] Manning and Moore[6] analysis for evaluation of optical character of amino acids; b) leucine amino peptidase digestion of the deprotected peptides. Both indicated that optical integrity was maintained throughout the synthetic procedure.

The following derivatives were synthesized from the semisynthetic intermediate, bis-Boc-desoctapeptide(B23-30)-insulin-phenylhydrazide (bis-Boc-DOI-NHNH-C_6H_5) and the appropriate synthetic peptide: DOI-(Ala-Phe-Phe-Tyr)$^{B23-26}$; DOI-(DAla-Phe-Phe-Tyr)$^{B23-26}$; DOI-(DAla-DPhe-Phe-Tyr)$^{B23-26}$; DOI-(Ala-DPhe-Phe-Tyr)$^{B23-26}$. Bis-Boc-DOI-NHNH-C_6H_5 was prepared, activated and coupled to the desired amino-deprotected peptide as described previously[4].

The efficiency of coupling of synthetic peptides to bis-Boc-DOI-NHNH-C_6H_5 was analyzed by polyacrylamide gel electrophoresis (PAGE)[4]. The analysis was done on the partially protected semisynthetic derivatives under alkaline conditions. Coupling yields of 85-90% were observed for DOI-(Ala-Phe-Phe-Tyr)$^{B23-26}$, DOI-(DAla-Phe-Phe-Tyr)$^{B23-26}$, and DOI-(DAla-DPhe-Phe-Tyr)$^{B23-26}$, while only 50% coupling efficiency was observed for DOI-(Ala-DPhe-Phe-Tyr)$^{B23-26}$. Initially, the partially protected semisynthetic insulin derviatives were purified using ion-exchange chromatography in 8M urea[4]. This led to low recovery (3.81mg, 8%) for DOI-(DAla-Phe-Phe-Tyr)$^{B23-26}$. Therefore, purification was attempted in 50% 2-PrOH[7]. For DOI-(DAla-DPhe-Phe-Tyr)$^{B23-26}$, this resulted in a superior yield (12.3mg, 18%). However, the yields of DOI-(Ala-Phe-Phe-Tyr)$^{B23-26}$ (3.9mg, 6%), and DOI-(Ala-DPhe-Phe-Tyr)$^{B23-26}$ (1.0mg, 2%) were still low.

PAGE was also used to evaluate the purity of the completely deprotected semisynthetic derivatives. Under both alkaline and acidic conditions, the four derivatives appeared homogeneous. Amino acid analysis of the analogs indicated that the final products contained the desired residues.

Table I. Biological Potency (% of Insulin Activity) of Insulin Analogs

Compound	Binding	Lipogenesis
Insulin	100	100
DOI-(Gly-Phe-Phe-Tyr)$^{B23-26}$	59	52
DOI-(DAla-Phe-Phe-Tyr)$^{B23-26}$	81	76
DOI-(DAla-DPhe-Phe-Tyr)$^{B23-26}$	19	16
DOI-(Ala-DPhe-Phe-Tyr)$^{B23-26}$	0.2	0.3
DOI-(Ala-Phe-Phe-Tyr)$^{B23-26}$	0.3	0.3
desoctapeptide(B23-30)-insulin(DOI)	0.2	0.1

The semisynthetic analogs were tested for biological activity in competitive binding and lipogenesis assays[4]. All derivatives produced typical dose response curves in both assays and stimulated full insulin-like response if administered in sufficient quantity. The biological potencies of these derivatives are included in table I. Also included in table I are activities of DOI-(Gly-Phe-Phe-Tyr)$^{B23-26}$ and DOI. DOI-(Gly-Phe-Phe-Tyr)$^{B23-26}$ (destetratpeptide-B27-30 insulin) was prepared and characterized previously[4]. It is the "parent" compound of the analogs described here. It was reassayed along with the D-amino acid containing derivatives for comparison. Biological potency (% of insulin activity) is determined as follows: $(C_i/C_d) \times 100$ = biological potency; where C_i = concentration of insulin required to stimulate half-maximal response and C_d = concentration of derivative required to stimulate half-maximal response.

Conclusions

Derivatives of insulin containing D-amino acid residues in positions B23-26 can be prepared in reasonable yields using bis-Boc-DOI-NHNH-C_6H_5 as a semisynthetic intermediate. This is probably the method of choice where the D-amino acid is to be added at postion B23.

The substitution of L-Ala for Gly at position B23 completely negates the increased biological activity observed upon readdition of the aromatic residues Phe-Phe-Tyr^{B24-26} to DOI. The derivative in which D-Ala is substituted for Gly at this position stimulates nearly full insulin-like activity in binding and lipogenesis assays. These data support the idea that there is a particular conformation at residue B23 which is required for full insulin activity. The inclusion of D-Phe at position B24 did not increase the activity of either the D-Ala of L-ala derivatives. The difference between the results reported here and those of Inouye et al[1,2] may be due to the fact that they constructed analogs containing the entire C-terminal B-chain octapeptide (DOI + octapeptide). The analogs discussed here are destetrapeptide insulin analogs (DOI + tetrapeptide).

References

1. Kobayashi, M., Ohgaku, S., Iwasaki, M., Maegawa, H., Shigeta, Y., & Inouye, K. (1982) Biochem. Biophys. Res. Comm., 107, 329-336.

2. Inouye, K., Watanabe, K., Tochino, Y., Kobayashi, M., & Shigeta, Y. (1983) In: Peptide Chemistry 1982, (S. Sakakibara, Ed.) pp. 277-282, Protein Research Foundation, Osaka.

3. Canova-Davis, E., & Carpenter, F.H., (1981) Biochemistry 20, 7053-7058.

4. Riemen, M.W., Pon, L.A., & Carpenter, F.H. (1983) Biochemistry 22, 1507-1515.

5. Mitchell, A.R., Kent, S.B.H., Chu, I.C., & Merrifield, R.B. (1978) Anal. Chem., 50, 637-640.

6. Manning, J.M. & Moore, S. (1968) J. Biol. Chem., 243, 5591-5597.

7. Jonczyk, A., Keffer, L.M., Naithani, V.K., Gattner, H.G., De Meyts, P., & Zahn, H. (1981) Hoppe-Seyler's Z. Physiol. Chem., 362, 557-561.

SYNTHESIS AND CARDIOVASCULAR ACTIVITY OF IMIDAZOLE-SUBSTITUTED ANALOGS OF TRH

Virender M. Labroo, Giora Feurerstein* and Louis A. Cohen
NIADDK, National Institutes of Health, Bethesda, MD 20205
* Uniformed Services University of Health Sciences, Bethesda, MD 20814

Introduction

TRH (PGlu-His-Pro-NH_2) is well known to have a wide variety of effects on CNS[1-2] and CVS[3-5] in addition to its well known endocrine effects of releasing thryotropin and prolactin[6-7]. We had previously found that 4-fluoro-Im-TRH (4-F-TRH) neither binds to rat pituitary cells *in vitro* nor releases prolactin from them[8]. When microinjected directly into rat hypothalmus, however, both TRH and 4-F-TRH, significantly increase blood pressure and heart rate of the animals[9]. In this report we describe the synthesis of 4-trifluoromethyl-Im-TRH(4-CF_3-TRH), 2-trifluoromethyl-Im-TRH (2-CF_3-TRH) and 4-nitro-Im-TRH (4-NO_2-TRH) and their effects on CVS in conscious rats after intravenous administration.

Synthesis

Boc-4-CF_3-His and Boc-2-CF_3-His were prepared from commercial Boc-His by direct photochemical trifluromethylation in MeOH in the presence of Et_3N. The mixture of position isomers was separated by flash chromatography over silica gel with AcOH : i-PrOH : $CHCl_3$, 0.5 : 4.5 : 45 as eluting solvent system (Scheme 1). Scheme 2 describes the synthesis of 4-CF_3-TRH and 2-CF_3-TRH from the corresponding Boc-amino acids. 4-NO_2-TRH was synthesized from 4-NO_2-His-OMe.HCl, as described in Scheme 3.

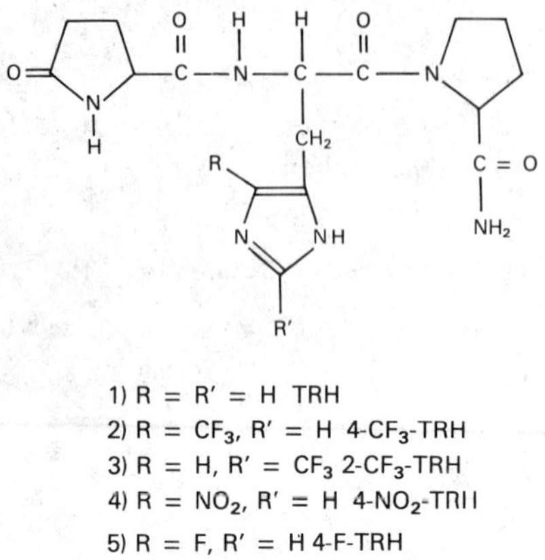

Fig. 1. Structures of TRH, 4-CF_3-TRH, 2-CF_3-TRH, 4-NO_2-TRH and 4-F-TRH.

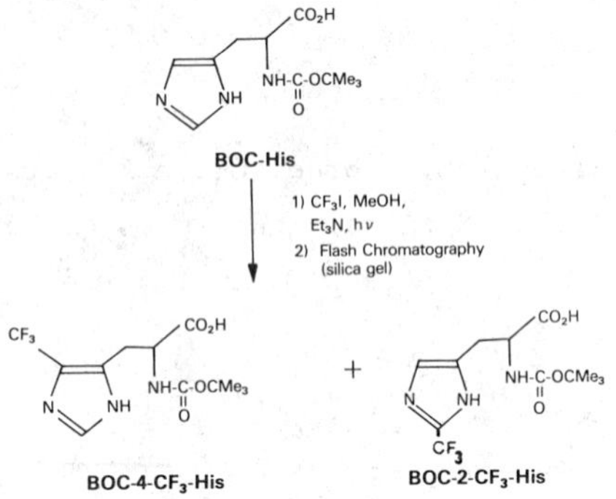

Scheme 1. Synthesis of Boc-4-CF_3-His and Boc-2-CF_3-HIS.

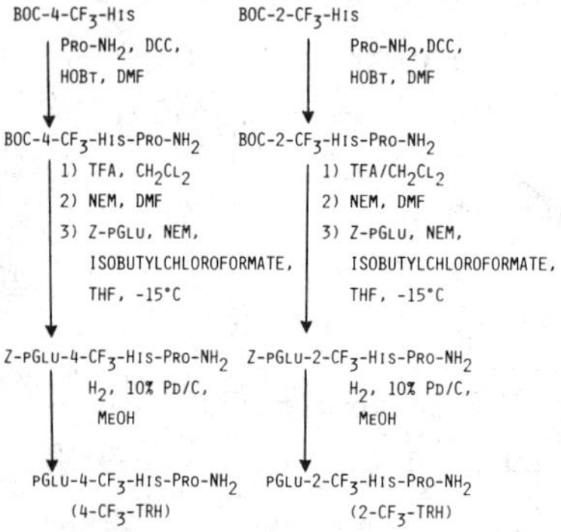

Scheme 2. Synthesis of 4-CF_3-TRH and 2-CF_3-TRH.

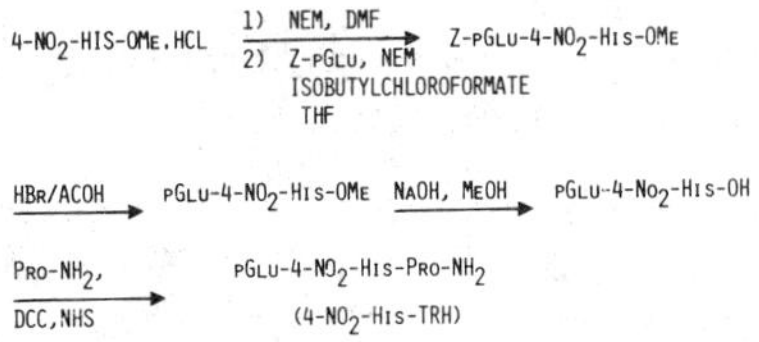

Scheme 3. Synthesis of 4-NO_2-TRH.

	n	ΔMAP	Δ PP	Δ H_r
TRH		mm Hg	mm Hg	Beats/Min
1mg/kg	6	23±5*	9±1*	91±19*
5mg/kg	6	20±2*	15±4*	73±9*
4-NO_2-TRH				
1mg/kg	8	17±2*	9±2*	85±11*
5mg/kg	8	20±3*	13±2*	105±17*
2-CF_3-TRH				
1mg/kg	9	19±4*	6±3*	85±15*
5mg/kg	9	27±3*	11±3*	50±15*
4-CF_3-TRH				
1mg/kg	11	23±2*	10±3*	69±11*
5mg/kg	10	16±3*	9±1*	78±13*

*P<0.001(Student t test), MAP = Mean Arterial Pressure. PP = Pulse Pressure, H_r = Heart Rate.

Table I. Effect of TRH and TRH-analogues on MAP and Heart Rate of conscious rats.

CARDIOVASCULAR ACTIVITY

Table I summarizes the effects of TRH, 4-NO_2-TRH, 2-CF_3-TRH and 4-CF_3-TRH on blood pressure and heart rate of conscious rats when the peptides were injected intravenously. The data shows that the three newly synthesized peptides significantly raise the blood pressure and heart rate in conscious rats and the increases are comparable to those produced by TRH under similar conditions. As previously described, 4-F-TRH also increases blood pressure and heart rate in conscious rats[9]. We had previously shown that 4-F-TRH and 2-CF_3-TRH do not bind to pituitary cells in vitro and do not release prolactin from them[9]. These results suggest that various chemical modifications of imidazole ring of TRH do not alter the cardiovascular effects of TRH, while they do inhibit binding to pituitary cells and the resultant endocrine effects. We are now studying the activity of these TRH analogs in releasing thyrotropin and prolactin when the peptides are administered intravenously, to establish if the ultimate clinical goal of a total separation of neuroendocrine and CVS activities can be achieved.

References

1. Bhargava, H.N. (1981) Life Science 28, 1261-1267.
2. Bhargava, H.N. (1981) Life Science 29, 1015-1020.
3. Koivusalo, F., Paakari, I., Leppaluoato, J. and 106, 83-86.
4. Tsay, B.L., and Lin, M.T. (1982) Neuroendocrinology 35, 173-177.
5. Farber, J.R., Connors, A.F., Gisolti, C.V., McCaffree, D.R., and Smith, R.M. (1981), Brain Research Bulletin 6, 13-17.
6. Meites, J. and Sonntang, W.E. (1981), Annual Reviews of Pharmacology and Toxicology 21, 295-322.
7. Tashjian, A.J., Barowsky, N.J. and Jensen, D.K. (1971), Biochem. Biophy. Res. Commun. 43, 516-523.
8. Labroo, V.M., Kirk, K.L., Cohen, L.A., Delbeke, D., and Dannies, P.S., (1983), Biochemical Biophysical Research Communications, 113, 581-585.
9. Feurenstein, G., Lazovsky, D., Cohen, L.A., Labroo, V.M., Kirk, K.L., Kopin, I.J. and Faden, A.I., (1984), Neuropeptides 4, 303-310.

Acknowledgement: This work was supported in part by the Uniformed Services University of Health Sciences, Protocol, #R09211.

SINGLE AMINO ACID DELETION ANALOGS OF HUMAN GROWTH HORMONE-RELEASING FACTOR

Toratane Munegumi, Andrew Baird, William Wehrenberg, Naoto Ueno and Nicholas Ling.
Laboratories for Neuroendocrinology, The Salk Institute, P.O. Box 85800, San Diego, CA. 92138.

Introduction

The regulation of pituitary growth hormone secretion is controlled primarily by two hypothalamo-hypophysiotropic factors, a growth hormone release-inhibiting factor characterized as somatostatin in 1973[1] and a growth hormone-releasing factor (GRF) recently characterized, first from two pancreas tumors obtained from acromegaly[2,3], and later from the hypothalamus[4]. Because growth hormone has been implicated in the development of microangiopathies in diabetic patients[5], our laboratories are interested in the preparation of antagonists to human growth hormone releasing factor (hGRF). The use of somatostatin to inhibit chronically the secretion of growth hormone in diabetic patients has been ruled out because of its multiple inhibitory functions in many endocrine cells. The alternative is to synthesize a competitive antagonist to GRF since the secretion of growth hormone is solely dependent on stimulation by this peptide. Furthermore, the action of GRF is very specific and its distribution in normal tissues is limited solely to the hypothalamus. Thus, if a potent antagonist to GRF can be found, the question of growth hormone-dependent complications in diabetic patients can be addressed.

For this purpose we first have to locate the minimal biologically active core fragment of hGRF so that fewer residues need to be modified in order to change it from an agonist to an antagonist. The location of the minimally active core region was accomplished by the synthesis of a series of amino-terminal deleted analogs of hGRF as well as another series of carboxy-terminal deleted analogs, which were then tested for their capacity to release growth hormone in monolayer cell culture prepared from adult male rat anterior pituitaries. Results of the bioassay showed that the minimal biologically active core of hGRF with full intrinsic activity comprised the fragment (3-21)[6]. Based on this finding, we have now prepared a series of single amino acid deletion analogs of hGRF(1-27)NH_2 to test which residue is responsible for imparting the molecule with the high intrinsic activity in the pituitary monolayer culture system. The (1-27) fragment was chosen because it has at least 10% of the activity of hGRF and brackets the active core region of the peptide.

Results and Discussion

hGRF(1-27)NH_2 and its single amino acid deleted fragments were synthesized by solid-phase methodology as previously described[6]. All the synthetic products yielded the correct amino acid composition after HCl hydrolysis and their purity ranged between 90% and 98% by reverse-phase high-performance liquid chromatography analysis (data not shown). Since hGRF contains a Leu^{22}-Leu^{23} sequence, only a single [des-Leu^{22}]hGRF(1-27)NH_2 fragment was prepared because this molecule is identical to [des-Leu^{23}]hGRF(1-27)NH_2. When these deletion analogs were tested in the male rat anterior pituitary culture system for their capacity to release growth hormone, most of them were found to have very low potency relative to hGRF(1-27)NH_2, as shown in Table 1. However, deletions after the 24th position generally do not decrease the activity as much as that resulting from deletions between the 1st and 23rd positions. This finding is concordant with the minimal biologically active core's being at the (3-21) region of the molecule. In addition, deletions at positions 8, 9, 12, 15 and 16 still yield fragments with substantial activity, whereas deletions at all the other positions up to residue 23 result in analogs with very low potency. This may reflect that residues other than 8, 9, 12, 15, 16, 24 and above are required for binding and/or activation, or that shortening of the peptide backbone by deletion at these positions is detrimental to the intrinsic activity of hGRF. In agreement with our finding, two other deletion analogs, hGRF(1-15)-(20-44) and hGRF(1-15)-

Table I. Relative Potencies of Single Amino Acid Deletion Analogs of hGRF(1-27)NH_2

	5 10 15 20 25	Relative Potencies
hGRF(1-27)NH_2	Y A D A I F T N S Y R K V L G Q L S A R K L L Q D I M	1
des	Y	0.00068
"	A	0.00018
"	D	0.00017
"	A	0.00006
"	I	0.00009
"	F	0.00011
"	T	0.00064
"	N	0.00511
"	S	0.00349
"	Y	0.00025
"	R	0.00012
"	K	0.00113
"	V	0.00004
"	L	0.00005
"	G	0.00639
"	Q	0.00770
"	L	0.00015
"	S	0.00005
"	A	0.00028
"	R	0.00001
"	K	0.00003
"	L	0.00009
"	Q	0.00336
"	D	0.0756
"	I	0.0335

(18-44), were also reported to be inactive[7]. Furthermore, no antagonistic activity was found in any of the less potent single amino acid deleted hGRF(1-27)NH_2 analogs because none of these can block the activity of hGRF, even at concentrations of up to 10^{-5}M (data not shown). These results suggest that, in order to convert GRF to its antagonistic form, it may be necessary to modify more than one residue.

Acknowledgements

We wish to express our appreciation to D. Angeles, T.-C. Chiang, T. Durkin, R. Klepper, B. Rathbun, M. Regno and R. Schroeder for their excellent technical assistance. The GRF research project was supported by program grants from NIH (HD-09690 and AM-18811) and the Robert J. and Helen C. Kleberg Foundation.

References

1. Brazeau, P., Vale, W., Burgus, R., Ling, N., Butcher, M., Rivier, J. and Guillemin, R. (1973) Science 179, 77-79.

2. Guillemin, R., Brazeau, P., Böhlen, P., Esch, F., Ling, N. and Wehrenberg, W. (1982) Science 218, 585-587.

3. Rivier, J., Spiess, J., Thorner, M. and Vale, W. (1982) Nature 300, 276-278.

4. Ling, N., Esch, F., Böhlen, P., Brazeau, P., Wehrenberg, W. and Guillemin, R. (1984) Proc. Natl. Acad. Sci. USA 81, 4302-4306.

5. Gerich, J. (1984) New England J. Med. 310, 848-850.

6. Ling, N., Baird, A., Wehrenberg, W., Ueno, N., Munegumi, T. and Brazeau, P. (1984) Biochem. Biophys. Res. Commun. 123, 854-861.

7. Blake, J., Westphal, M., and Li, C.H. (1984) Int. J. Peptide Protein Res. 24, 498-504.

EFFECT OF NATURAL AND SYNTHETIC PEPTIDES ON APLYSIA NEURONS

M. K. Rock, S. B. Shope, J. E. Blankenship

Marine Biomedical Institute
University of Texas Medical Branch
Galveston, Texas 77550

D. H. Schlesinger
Department of Medicine and Cell Biology
New York University School of Medicine
New York, New York 10016

Introduction

The neuroendocrine bag cells of the abdominal ganglion and the atrial gland of the reproductive tract of the marine mollusc *Aplysia* contain a family of peptides that are structurally and functionally related and appear to be involved in the regulation of egg laying behavior. Alpha bag cell peptide (α-BCP) is cleaved from a large precursor molecule as a 9-amino acid residue of relatively low biological activity; it is further cleaved at the C-terminus to an 8- and then a 7-amino acid peptide. The 8- and 7-residue peptides have two effects: they cause an inhibition of the left upper quadrant (LUQ) neurons (L2-L6) of the abdominal ganglion and also can depolarize or activate the bag cell themselves, inducing and/or enhancing and afterdischarge[2].

Three atrial gland peptides have been isolated and sequenced: A, B peptides and egg releasing hormone (ERH)[3-6]. All three peptides as well as α-BCP_{1-8} activate the bag cells, thereby indirectly causing egg laying. However, only ERH acts directly on the ovotestis to cause egg release.

α-BCP shares a great deal of sequence homology with peptides A. B and ERH: six of the eight residues of α-BCP_{1-8} are identical with residues 27-32 of A/B/ERH (Fig. 1). We undertook the present study to determine a) whether the atrial gland peptides also shared α-BCP's ability of inhibiting the LUQ cells; b) whether BCP and the atrial gland peptides are able to mimic the effect of bag cell after discharge on neuron R2 inhibition[7] and c) the structural requirements of BCP and atrial gland peptides for biological activity.

Methods

Peptide synthesis - Syntheses were performed according to the solid phase methodology of Merrifield[8] on a Vega model 250C automated peptide synthesizer. The peptides were then cleaved from the support and de-protected with HF in the presence of anisole (0°C, 1 hr) and were desalted and purified on Sephadex G-25 column (100 x 0.9 cm) equilibrated with 1% acetic acid. They were then characterized by amino acid analysis following acid hydrolysis (5.7N HCl, 24 hours, 110°C) and were evaluated for homogeneity by thin-layer chromatography (butanol: acetic acid: pyridine: water; 15: 3:10:12) and automated N-terminal peptide sequence analysis[9].

Experiments were performed on mature (200-600) g) *A. californica* as described (in preparation Rock et al., J. Neurobiology). Four methods of application of peptides (crude atrial gland extract from *A. californica*[6] and synthetic α BCP^{2}_{1-8} and $A/B/ERH^{3-6}_{26-34}$ were used: superfusion into the bathing solution, arterial perfusion, pressure ejection through micropipettes, and stimulation of the bag cells to afterdischarge. Peptides were dissolved in a peptidase inhibitor/ASW solution. In superfusion and perfusion experiments, the inhibitor solution preceded the peptide solution.

Results and Discussion

Homogeneity of synthetic peptides - All peptides were in excess of 97% homogeneous by N-terminal sequencing and only one spot was observed on thin-layer chromatography.

Effect of peptides on left upper quadrant cells - When the bag cells were activated during intracellular recording of the LUQ cells, inhibition of the LUQ cells ensued within 1 min after the onset of the afterdischarge (Fig. 2A). When synthetic α-BCP_{1-8} (10^{-4}M) was superfused, arterially perfused or pressure ejected through micropipettes onto somata (Fig. 2C), it always caused an inhibition of the LUQ cells.

We have examined the influence of crude atrial gland extract and B_{26-34} on the LUQ cells. Crude extract had a slight excitatory effect on the LUQ cells (Fig. 2E), whereas B_{26-34} had either no noticeable effect on the LUQ cells when applied at concentrations that were quite effective for the similar α-BCP_{1-8}, or B_{26-34} had an inhibitory effect on the LUQ cells when applied by pressure ejection at higher concentrations (Fig. 2F).

Effects of peptides on R2 - Bag cell afterdischarge leads to an inhibition of the giant cell R2[7] a result we consistently observed (Fig. 2B). Since α-BCP_{1-8} is released by a bag cell afterdischarge, it might account for the inhibition of R2 following an afterdischarge. Synthetic α-BCP_{1-8} at 10^{-4} at 10^{-4}M produced inhibition of R2 when superfused, arterially perfused or when pressure ejected (Fig. 2D).

On the other hand, superfusion of crude atrial gland extract had a long and powerful excitatory effect on cell R2 (Figure 2E). It is not clear which of the atrial gland peptides is responsible for this excitatory effect. However, we assume it is peptide, but not B_{26-34} because exposure of the extract to a non-specific protease (Sigma Type XIV) abolishes the excitatory effect effect and B_{26-34} inhibited (Fig. 2G) rather than excited R2. These results indicate that at similar concentrations B_{26-34} mimicks the effect that α-BCP_{1-8} has on R2. The comparable effects of α-BCP_{1-8} and B_{26-34} on R2 are unlike the effects these peptides have on the LUQ cells at which BCP_{1-8} exhibits a more powerful inhibitory influence.

The results indicate that peptides associated with the atrial gland often have different effects upon identified abdominal ganglion neurons than do the peptides of the bag cells of A. californica.

This research was supported by NIH: NS-11255, NS-07185(SBS),

NS-18109(DHS); NSF: PCM 82-15185(JEB).

References

1. Rothman, B. S., Mayeri, E., Brown, R. O., Yuan P.M., ad Shively, J. E. (1983) Proc. Natl. Acad. Sci. USA. 80: 5733-5757.
2. Strumwasser, F., Kaczmarek, L. K., Jennings, K. R., and Chiu, A. Y. (1981) in Neurosecretion: Molecules, Cells, Systems. Farner, D. S. and Lederis, K., Eds., Plenum Press, New York, pp. 249-268.
3. Heller, E., Kaczmarek, L. K., Hunkapillar, M. W., Hood, L. E., and Strumwasser, F. (1980) Proc. Natl. Acad. Sci. USA. 77: 2328-2332.
4. Chiu, A. Y.,Hunkapiller, M. W., Heller, E., Stuart, D. K., Hood, L. E. and Strumwasser, F. (1979) Proc. Natl. Acad. Sci. USA. 76: 6656-6660.
5. Blankenship, J. E., Rock, M. K., Fuller, J. M., and Schlesinger, D. H. (1984) Fed. Proc. 43: 927.
6. Schlesinger, D. H., Babirak, S. B., and Blankenship, J. E. (1981) in Symposium on Neurohypophysial Peptide Hormones and Other Biologically Active Peptides. Schlesinger, D. H., Ed., Elsevier/North-Holland Biomedical Press, Amsterdam, pp. 137-150.
7. Mayeri, E., Brownell, P., and Branton, W. D. (1979) J. Neurophysiol. 42: 1185-1197.
8. Merrifield, R. B. (1963) J. Am. Chem. Soc., 85: 2149-2154.
9. Schlesinger, D. H. (1983) Methods in Enzymology 91: 494-502.

	1	2	3	4	5	6	7	8	9	10	11	12	13	14	15	16	17
α-BCP																	
B/A Peptides	ala	val	lys	ser/leu	ser	ser	tyr/asp	glu/gly	lys/asn	tyr	pro	phe	asp	leu	ser	lys	glu
ERH	ile	ser/val	ile	val	ser	leu	phe	lys	ala	ile	thr	asp	met	leu	leu	thr	glu

	18	19	20	21	22	23	24	25	26 (1)	27 (2)	28 (3)	29 (4)	30 (5)	31 (6)	32 (7)	33 (8)	34 (9)
α-BCP									ala	pro	arg	leu	arg	phe	tyr	ser	leu
B/A Peptides	asp	gly	ala	gln	pro	tyr	phe	met	thr	pro	arg	leu	arg	phe	tyr	pro	ile-NH_2
ERH	gln	ile	tyr	ala	asn	tyr	phe	ser	thr	pro	arg	leu	arg	phe	tyr	pro	ile

Fig. 1: The primary structure of reproductively related peptides in *Aplysia californica.*

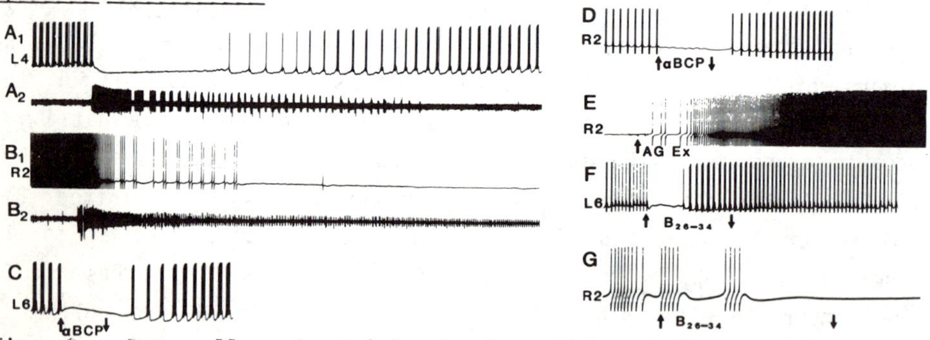

Fig. 2: Bag cell and atrial gland peptides influence LUQ and R2 neuronal activity. A bag cell afterdischarge recorded extracellularly (A_2, B_2) inhibited L4 (A_1) and R2 (B_1). Pressure ejection of α-BCP_{1-8} (10^{-4}M) directly onto soma inhibited L6 (C) and R2 (D); pressure pules, 5 sec on/5 sec off, begin at up arrow, end at down arrow, throughout. (E) Superfusion of atrial gland extract (AG Ex) at arrow strongly excites R2. Pressure ejection of B_{26-34} (5 x 10^{-4} M) directly onto soma inhibited L6 (F) and R2 (G); pulses as above. Intracellular action potential amplitudes ranged from 60 to 80 mV.

METABOLIC PROFILING OF PITUITARY PEPTIDES BY A COMBINATION OF RP-HPLC, RADIORECEPTOR ASSAY, RADIOIMMUNOASSAY, AND MASS SPECTROMETRY

Dominic M. Desiderio[1,2], Hisayoshi Takeshita[2], Hiroshi Onishi[2], Genevieve H. Fridland[2], Chhabil Dass[2], Peter Tinsley[2] and John Killmar[2]

Dept. of Neurology[1] and the Charles B. Stout Neuroscience Mass Spectrometry Laboratory[2], University of Tennessee Center for the Health Sciences, 800 Madison Avenue, Memphis, TN 38163

Introduction

This manuscript describes the development and use of a comprehensive analytical scheme to analyze endogenous peptides found in HPLC-purified pituitary extracts[1]. Various analytical methods have been utilized to achieve this goal, but the molecular specificity of some of those methods is relatively limited. While RP-HPLC, radioimmunoassay (RIA), radioreceptor assay (RRA), and bioassay (BA) all suffer different limitations to achieving that goal, mass spectrometry (MS) has the unique potential to provide the structure of the peptide being measured. On one hand, the protonated molecular ion, $(M+H)^+$, of a peptide can be formed; but it must be remembered that this ion does not readily reveal the amino acid sequence information. On the other hand, linked-field scanning MS methods can be utilized to provide amino acid sequence-determining information.

This program aims towards monitoring in normal and stressed biological systems several peptidergic pathways and individual peptides, and the metabolic relationships amongst pathways and/or individual peptides. The variety of human tissues studied includes tooth pulp, pituitary, various brain regions, ectopic and pituitary tumors, and cerebrospinal fluid. The objective is to elucidate the molecular mechanisms involved in pain, stress, and drug addiction.

This report combines the salient features of our work in the use of HPLC, RIA, RRA, and MS to measure endogenous peptides.

Materials and Methods

Tissue is homogenized and proteins are precipitated; gradient RP-HPLC is performed and 90 fractions collected; a canine limbic system synaptosomal or P2 preparation is utilized for RRA utilizing the tritiated ligands etorphin, D-^{2}ala, D-^{5}leu-leucine enkephalin (DADL), dihydromorphine, or other appropriate radioligand; commercial RIA kits are purchased (Immunonuclear, Stillwater, MN); and fast atom bombardment (FAB-MS) utilizes either a Finnigan MAT 731 or VG 7070E-HF/11-250 mass spectrometer. For the MS data, either the $(M+H)^+$ ion can be utilized to corroborate the potential presence of a peptide, or the $(M+H)^+$ is subjected to collision activated dissociation (CAD) processes and/or linked-field scanning in the B/E mode to provide a unique amino acid sequence-determining fragment ion to increase significantly the molecular specificity of the measurement process[1]. A B/E scan collects all of the fragment ions produced from a selected precursor ion. Stable isotope (^{18}O)-labeled peptide internal standards are utilized.

Results and Discussion

The gradient RP-HPLC chromatogram of the peptide-rich fraction derived from a canine pituitary homogenate is shown in Figure 1, where it can be seen that synthetic peptides elute at the positions denoted by the arrows along the top of the chromatogram, the peptide bond UV absorption at 200 nm is indicated by the curve, the gradient (0-30%) of the organic modifier acetonitrile is indicated, triethylamine-formic acid is the volatile buffer, 90-fractions are collected by a fraction collector and are represented by the numbers along the bottom-axis, and the ^{3}H-etorphin radioligand is used in conjunction with the receptor preparation noted above, where the interaction of the endogenous peptide ligand with that receptor preparation is indicated by the height of the hatched box. Because methionine enkephalin is used for the calibration curve, the data listed along the left-hand axis are represented as picomoles of receptoractive (ra-) methionine enkephalin equivalents mg^{-1} protein. The term "receptoractive" is used until structural elucidation of individual peptides is performed.

A FAB mass spectrum was taken of each and every one of the 90-fractions indicated in Figure 1. Table I collects the analytical data obtained from the variety of analytical methods. Rather than give complete descriptive analytical data for each system, the purpose of Table I is to indicate in one glance the comprehensive nature of this type of analytical technique. For example, the three target peptides for our current studies include methionine enkephalin, leucine enkephalin, and substance P. These peptides have been studied with RIA and RRA, with development of a substance P RRA now in progress. Furthermore, each of the synthetic peptide does indeed co-elute at the indicated fraction number and an appropriate $(M+H)^+$ ion is found for each peptide. These analytical data are significantly corroborated by the presence of the molecular ion; and in the case of methionine enkephalin and leucine enkephalin, measurement utilizing a unique amino acid sequence-determining fragment ion (-Gly-Phe-Met and -Gly-Phe-Leu), respectively[1].

Conclusions

This comprehensive one-laboratory analytical scheme has been applied to analysis of the peptide-rich fraction from the canine pituitary tissue homogenate. While RIA and RRA are sufficiently sensitive to quantify endogenous amounts of peptides, peptide structures cannot be assigned to those measurements. On the other hand, MS, while having more limited sensitivity relative to RIA and RRA, does provide either corroborating molecular ion or unambiguous amino acid sequence-determining fragment ion information. It must be remembered that molecular ions do not unambiguously establish the amino acid sequence of a peptide (N amino acids yield N! different peptides). In some favorable cases, it is possible to fragment the molecular ion in transit in the mass spectrometer and provide a unique amino acid sequence-determining fragment ion that correlates only to the peptide of interest, and furthermore to use an ^{18}O-labeled internal standard for quantification purposes[1].

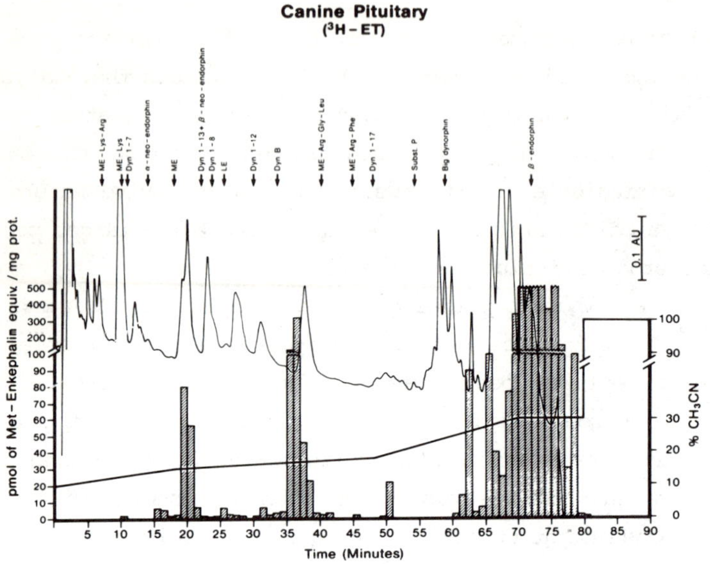

Fig. 1. RP-HPLC chromatogram of canine pituitary.

Table I. Analytical Data for Several Selected RP-HPLC Fractions

HPLC Fraction No.	FAB-MS			RIA	RRA	Coeluting Peptide
	$(M+H)^+$	B/E	HRMS			
3	317	-	-	-	-	-
8	$[400_n+H]$	yes	yes	-	-	LC "bleed"
10	702	-	-	-	-	YGGFMK
19	574	yes (SIM)	-	yes	yes	YGGFM
23	556	yes (SIM)	-	yes	yes	YGGFL
26	1,007	-	-	-	-	-
54	1,348	-	-	yes	"yes"	Substance P

References

1. Desiderio, D.M. (1984) Analysis of Neuropeptides by Liquid Chromatography and Mass Spectrometry, Elsevier, Amsterdam, 235 pp.

USE OF SYNTHETIC FRAGMENTS TO STUDY ENZYMATIC PROCESSING AND MOLECULAR ORGANIZATION OF NEUROPEPTIDE BIOSYNTHETIC PRECURSORS

Rao S. Rapaka, Shoji Ando, Tatsuhiko Kanmera, Ted H. Miller and Irwin M. Chaiken

Molecular, Cellular, and Nutritional Endocrinology Branch, NIADDK
National Institutes of Health, Bethesda, MD 20205 USA

Introduction

Active neuropeptides are derived from biosynthetic precursors by enzymatic processing involving endo- and exoproteolytic steps, and often C-terminal amidation[1-5]. Although current understanding of precursor structures is limited largely to the linear constructs deduced from DNA sequencing (Figure 1), it is likely that these molecules achieve at least locally ordered conformations and that enzymatic processing is controlled not only by local sequences in the precursors but also by the conformational order of these molecules.

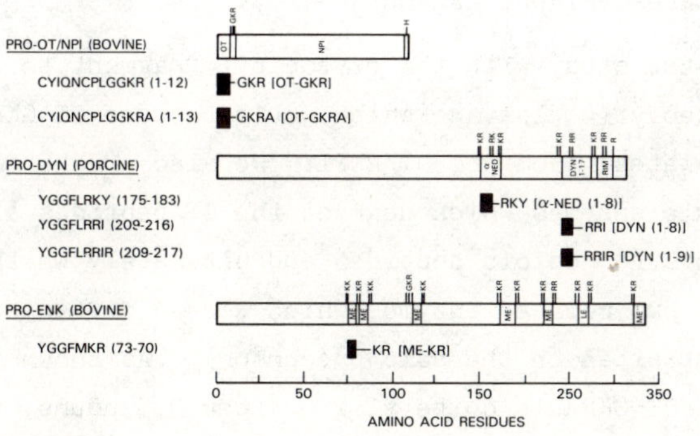

Fig. 1. Primary structures of pro-OT/NPI (prooxyphysin), pro-DYN (prodynorphin), pro-ENK (proenkephalin) and synthetic peptides used as substrates for enzymatic processing. Abbreviations: OT, oxytocin; NPI, neurophysin I; α-NED, α-neoendorphin; DYN, dynorphin, RIM, rimorphin; ME, [Met[5]]enkephalin; LE, [Leu[5]]enkephalin.

To evaluate the contribution of sequence-dependent and conformational elements to processing mechanisms, we are examining enzymatic conversion of synthetic peptide substrates and of larger semisynthetic precursor constructs derived from these peptides. Several synthetic peptides were obtained (Figure 1) which contain sequences of neurohypophysial and opioid hormones combined with putative sites for dibasic endoproteases, carboxypeptidase B and amidating enzyme. The OT-peptides were prepared by solid phase peptide synthesis as before[6]. The opioid peptides were from Peninsula Laboratories (Belmont, CA). In addition, a semisynthetic pro-OT/BNPI analogue was produced by coupling OT-GKR and bovine NPI[7].

Results and Discussion

Peptides as Substrates to Study Processing Reactions - A major purpose of using synthetic precursor fragments as processing substrates was to detect and characterize the set of enzymes which produce the neuroendocrine peptides. OT-GKR is degraded sequentially by CPase B from specific posterior pituitary neurosecretory granule subfractions[6]. Both CPase B steps, Arg cleavage followed by Lys cleavage, have pH optima close to 5.5-6.0 and are stimulated by Co^{++}. Subsequent amidation of OT-G to yield OT (Gly 9 amidated) is slow but can be detected with fresh granule lysates (Figure 2A and Ref. 6).

We extended the study with the oxytocinyl fragment to detect the putative endoproteolytic cleavage which would produce OT-GKR (or other possible intermediates) from pro-OT/BNPI. We also sought to determine whether the granule enzymes which degrade the OT peptides also degrade precursor fragments for opioid peptides and ultimately whether it is the same or different-but-related enzymes which cleave different precursors (or even different sites on the same precursor). As shown in Figure 3C, the conversion of OT-GKRA is quite slow. After 3.5 hours, small amounts of OT-GKR, OT-GK, OT-G and OT all are detected. The data suggest a model in which initial cleavage of OT-GKRA to produce OT-GKR (by a dibasic protease) is slow, that the rates of CPase B are faster, and thus that the initial product OT-GKR does not accumulate. Nonetheless, the results provide preliminary evidence that all of the enzymatic activities - dibasic endoprotease (here acting as an exoprotease), CPase B, and amidating enzyme - are contained in the neurosecretory granules and that these enzymes act in sequence to produce OT. Sequential CPase conversion also was observed for ME-KR (Figure 1) to ME.

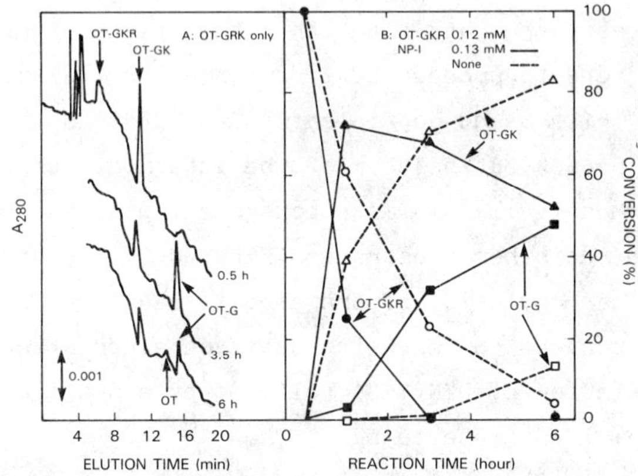

Fig. 2. (A) Enzymatic processing of OT-GKR. OT-GKR (12-70 nmol) incubated at 37°C with 20 µl granule lysate in 80 µl of 0.2 M sodium acetate buffer, pH 5.5, containing 1 mM $CoCl_2$. Aliquots fractionated by CN-propylsilyl RP-HPLC (0.46 x 25 cm) using a linear gradient, from 90% TEAP/10% CH_3CN at 0 time to 75% TEAP/25% CH_3CN at 20 min to 30% TEAP/70% CH_3CN at 30 min. Flow rate: 0.8 ml/min. TEAP: 67 mM triethylammonium phosphate, pH 3. (B) Effect of NP on OT-GKR processing. Processing reaction and HPLC separation as in A.

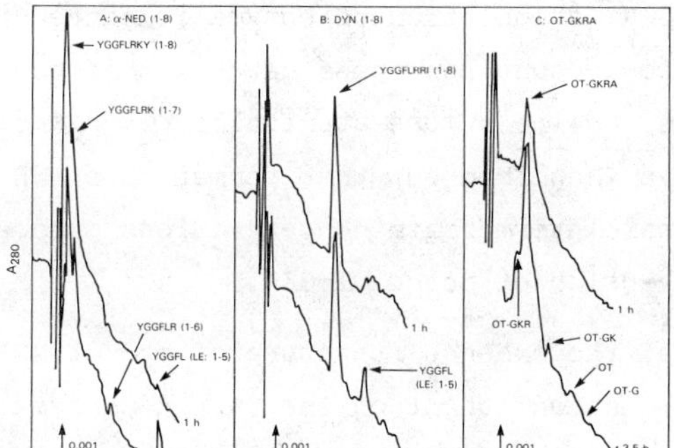

Fig. 3. Enzymatic processing of (A) α-NED(1-8), 30 nmoles (B) DYN(1-8), 20 nmoles (C) OT-GKRA, 42 nmoles. Other reaction conditions as for Figure 2A.

α-NED(1-8) and DYN(1-8) also are converted to mature neuropeptides by granule enzymes. For α-NED(1-8), Figure 3A, cleavage C-terminal to the dibasic sequence is relatively rapid, and intermediates 1-7 and 1-6 are observed in addition to final α-NED(1-5), or [Leu^5]enkephalin (LE). For DYN(1-8), Figure 3B, LE is formed but little of 1-7 and 1-6 are seen. In this case, the data fit with either (i) a single cleavage at the Leu 5-Arg 6 bond to produce LE in one step or (ii) sufficiently slow cleavage at Arg 7-Ile 8 and sufficiently fast CPase B cleavages that only the final product but neither of the intermediates is accumulated significantly. Model (ii) is more consistent with the conversion patterns found with the other substrates (Figures 2, 3), namely initial cleavage on the carboxyl side of the dibasic sequence followed by further sequential conversion, and with data, obtained with DYN(1-9), showing rapid accumulation of DYN(1-8) followed by a small but observable accumulation of 1-7 en route to LE.

The Impact of Ordered Conformation on Enzymatic Processing - An observation supporting the likely contribution of molecular order in processing has been made by the effect of neurophysin on the enzymatic conversion of OT-GKR. We previously found that addition of neurophysin to OT-GKR at high concentrations (0.7 mM) can cause partial suppression of CPase B conversion[6]. In contrast, Figure 2B, with lower concentrations of NP and OT-GKR, stimulation of enzymatic conversion occurs. A working hypothesis to account for these data is that the OT-GKR in 1:1 peptide-NP complexes attains a more sterically favored state for enzymatic conversion than when free but that formation of NP dimers at higher NP and hormonal intermediate concentrations suppresses this more-favored rate of conversion of bound OT-GKR.

Given the above, the ordered structure of pro-OT/NPI and processing intermediates likely has an effect on the rate of enzymatic conversion of precursors to mature OT and NP. This possibility makes it important ultimately to evaluate processing reactions with intact or close-to-intact precursors. In our own studies, preliminary data suggest that semisynthetic pro-OT/BNPI, which is folded into a conformation in which the OT domain binds intramolecularly with the NPI domain[7], is processed by neurosecretory granule lysate to form hormone-containing peptides. We presently are evaluating the nature of this conversion.

Conclusion - Synthetic peptides containing processing regions have proven helpful to detect and characterize enzymatic reactions which produce biologically active neuroendocrine hormones from biosynthetic precursors. While such studies show the impact of amino acid sequence on enzymatic rates, these peptides also help to visualize the contribution of ordered conformation to regulate processing. The availability of close-to-intact semisynthetic precursors promises to be useful to better define the interplay of conformation and local sequence in determining production of neuroendocrine peptides.

References

1. Mains, R.E., Eipper, B.A., Glembotski, C.C., and Dores, R.M. (1983) Trends in Neurosciences 6, 229-235.
2. Douglass, J., Civelli, O., and Herbert, E. (1984) Annu. Rev. Biochem. 53, 665-715.
3. Loh, Y.P., Brownstein, M.S., and Gainer, H. (1984) Annu. Rev. Neurosci. 7, 189-222.
4. Marks, N., Benuck, M., and Berg, M.S., in: Medicinal Chemistry and Molecular Pharmacology of Opioid Peptides, R.S. Rapaka and R.L. Hawks, Eds., NIDA Press, in press.
5. Chaiken, I.M., Kanmera, T., and Sequeira, R.P., ibid., in press.
6. Kanmera, T., and Chaiken, I.M. (1985a) J. Biol. Chem., in press.
7. Kanmera, T., and Chaiken, I.M. (1985b) J. Biol. Chem., in press.

MOLECULAR CORRELATIONS OF TASTE INCLUDING A NEW CLASS OF AMINO ACID BASED SWEETENERS

M. Goodman, J. Bland, J. Tsang and J. Coddington

Department of Chemistry, University of California, San Diego
La Jolla, California 92093

P.A. Temussi, T. Tancredi and F. Lelj

Istituto Chimico, Universita di Napoli, Via Mezzocannone 4
Naples, Italy

W.D. Fuller and M.S. Verlander

BioResearch, Chemical Technology Division
San Diego, California 92121

The discovery of sweet and bitter peptides has led to studies of their structure-taste relationships. The sweet and bitter tastes are elicited by molecules of widely different classes of chemical structures. Over the past decade mapping of the topological features of the taste receptors has been carried out.[1,2] The sweet and bitter tastes are closely related as can be seen by the following:

(1) Many peptide derivatives change their taste from sweet to bitter when the configuration of one chiral center is altered.[3]

(2) For the series of trifluoroacetyl-L-aspartyl anilides the taste changes from sweet to bitter when the position of electron-withdrawing substituents such as halogen, cyano or nitro is altered from the para to the meta.[4]

(3) The taste of L-aspartyl α-aminocycloalkane carboxylic acid methyl ester changes from sweet to bitter as the ring size increases from 5 to 6.[5]

These observations can be explained on the basis of two models: (1) two similar receptors having the same flat, elongated cavities or (2) one receptor with two modes of interaction. In both models one end of the cavity has a hydrogen bonding donor and acceptor (i.e. the AH-B structure of Schallenberger[6]) while the other end contains hydrophobic interacting groups. The validity of the models is demonstrated by the fact that 4-bromosaccharin is both sweet and bitter since it fits into the dimensions of aspartame used to define the outline of the sweet and bitter receptors. In addition, the sulfononaphthimidazoles fit only the sweet receptor.[7]

The sweet-tasting N-aryl ureido derivatives of aspartame and related dipeptides provide further insight into the topology of the taste receptors.[8] The para-substituents on the aryl ring of the ureido group demonstrate that strongly electron-withdrawing groups are required for the sweet taste.[8] Recently we synthesized the p-, m-, and o-methyl ureidyl aspartame and found that all isomers are tasteless. We also modified the aryl ring of the ureido function by substituting nitro and cyano groups at the meta and ortho positions and found the meta derivatives bitter while ortho are tasteless. This is in agreement with our previously reported observations on meta- and ortho-nitro substituted trifluoroacetyl-L-aspartyl anilides.[4]

Since the trifluoro-L-aspartyl anilides are much sweeter than the corresponding zwitterions,[9] we chose to combine the ureido and anilide functions into a single molecule. The para-, meta-, and ortho-substituted ureidyl-L-aspartyl para'-, meta'-, and ortho'-substituted anilides (**1**) were prepared. The results show that only the para-ureido, para'-anilide is sweet. When either group is meta-substituted (para, meta' or meta, para') the compound is very bitter. When the substitutions are para, ortho', the tastant is also bitter while ortho, para' is tasteless.

The taste relationships for these ureidyl L-aspartyl anilides fit the proposed models of the receptor. The aryl

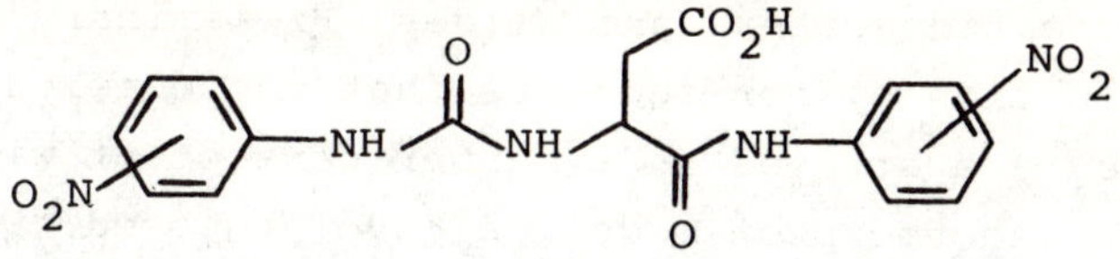

1*

*Ortho, meta, para refer to substitution on the ureido function, while ortho', meta' and para' refer to substitution on the anilide group.

ring of the ureido group assumes a perpendicular array to the plane of the flat L-aspartyl anilide structure. Substitutions at either meta or meta' position lead to projections into the bitter zone of the receptor while the ortho-substitution on the ureido-aryl grouping interferes with the AH-B interactions.

Lastly we wish to report a new class of sweeteners based on the partial retro-inverso modification. Some years ago we reported the results of the reversal of the peptide bond in DL-aminomalonyl L-phenylalanine methyl ester.[10] The product (RS)-α-aminoglycyl-(RS)-2-benzylmalonyl methyl ester was unstable and tasteless.[10] We have now extended our studies to acylated L-aspartyl-1,1-diaminoalkanes.[11] These derivatives 3 were prepared by reversing the amide bond in the dipeptide amides 2.[12] The syntheses will be reported[12] elsewhere.

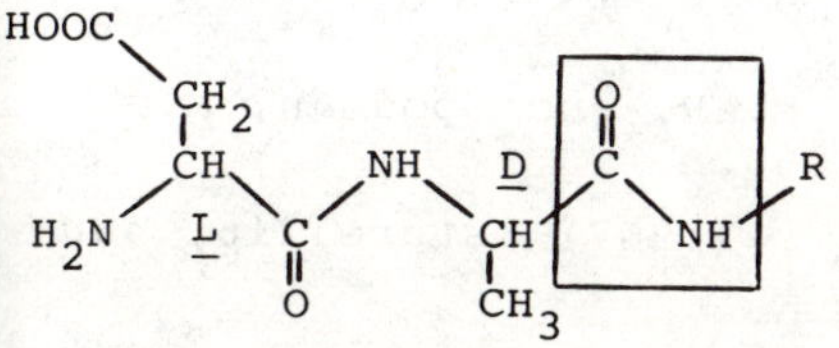

2

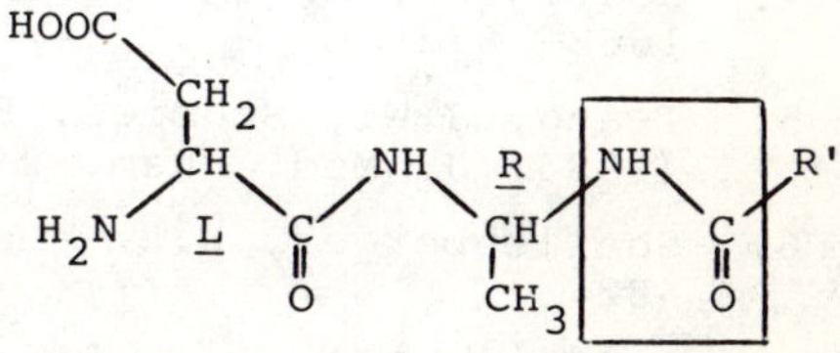

3

The taste characteristics of these compounds are strikingly similar in quality to sucrose and their intensity of sweetness depends on the nature of the group R' of the carboxylic acid used to acylate the 1,1-diaminoalkane. Taste intensities varied from 75-100 times sucrose for R' = t-butyl or cyclopentyl to 800-1000 times sucrose for R' = 2,2,5,5-tetramethylcyclopentyl. Surprisingly, sweetness was not dependent on the chirality of the 1,1-diaminoalkane residue, in contrast with the analogous dipeptide esters and amides. Preliminary conformational evaluation by nmr suggests that these molecules are quite rigid. They are arrayed in a manner consistent with the requirements for the receptor model we have proposed. The compounds are extremely stable towards hydrolysis and cannot form diketopiperazines. Thus, through use of a fundamental structural modification, we have designed a new and useful class of sweeteners.

The taste of the different classes of compounds discussed in this paper are consistent with receptor models we have proposed. Further studies are currently in progress.

References

1. Temussi, P.A., Lelj, F. and Tancredi, T. (1978) J. Med. Chem. 21, 1154-1158.
2. Temussi, P.A., Lelj, F., Tancredi, T., Castiglione-Morelli, M.A., and Pastore, A. (1984) Int. J. Quantum Chem. XXVI, 889-906.
3. Mazur, R.H., Schlatter, J.M. and Goldkamp, A.H. (1969) J. Amer. Chem. Soc. 91, 2684-2691.
4. Rodriguez, M. and Goodman, M. (1984) J. Med. Chem. 27, 1668-1672.
5. Tsang, J.W., Schmied, B., Nyfeler, R. and Goodman, M. (1984) J. Med. Chem. 27, 1663-1668.
6. Shallenberger, R.S. and Acree, T. (1967) Nature 216, 480-482.
7. Temussi, P.A., unpublished results.
8. Norte, C. and Tinti, J.-M., Eur. Pat. Appl. EP 0107,597 (Cl. C07103/52), 15 Sept. 1983.
9. Lapidus, M. and Sweeney, M. (1973) J. Med. Chem. 16, 163-166.
10. MacDonald, S.A., Willson, C.G., Chorev, M., Vernacchia, F.S. and Goodman, M. (1980) J. Med. Chem. 23, 413-420.
11. Verlander, M.S., Fuller, W.D. and Goodman, M. Eur. Pat. Appl. EP 128,654 (Cl. C07C103/737), 13 June 1983.
12. Brennan, T.M. and Hendrick, M.E. Eur. Pat. Appl. EP 34,876 (Cl. C07C103/52) 2 Sept. 1981.

"TRANSITION STATE" SUBSTITUTED RENIN INHIBITORY PEPTIDES: STRUCTURE-CONFORMATION-ACTIVITY STUDIES ON N^{IN}-FORMYL-TRP AND TRP MODIFIED CONGENERS

Tomi K. Sawyer[δ], Donald T. Pals*, Clark W. Smith[δ], Hossain H. Saneii*, Dennis E. Epps[‡], David J. Duchamp[‡], Jackson B. Hester*, Ruth E. TenBrink*, Douglas J. Staples[δ], Anne E. deVaux[δ], Joseph A. Affholter[δ], Gerald F. Skala[δ], Warren M. Kati*, Judy A. Lawson*, Margaret R. Schuette*, Bharat V. Kamdar* and D. Edward Emmert*

[δ]Biopolymer Chemistry-Regulatory Peptide Research, *Cardiovascular Diseases Research, and [‡]Physical and Analytical Chemistry, The Upjohn Company, Kalamazoo, Michigan 49001

William D. Carlson and Mark Handsumacher
Cardiac Unit, Massachusetts General Hospital, Boston, Massachusetts 02114

Introduction

Human renin is the rate-limiting enzyme of the well known renin-angiotensin converting enzyme (ACE) cascade as illustrated in Figure 1. Angiotensin II is the biologically-active product of the renin-ACE cascade and its effects at various target tissues include: (1) vasoconstriction at vascular smooth muscle: (2) aldosterone secretion at the adrenal cortex; (3) catecholamine secretion at the adrenal medulla; (4) NaCl retention at the kidney; (5) H_2O absorption in the gut; and (6) stimulation of thirst behavior, NaCl appetite and vasopressin secretion in the brain. Pharmacological intervention of the biosynthesis of angiotensin II has been a challenging area of research as directed towards the development of potential antihypertensive drugs[1,2].

The inhibition of renin by substrate-based competitive antagonists specific against this aspartic acid protease has been previously hallmarked by several structure-activity investigations[3,11]. In retrospect, the systematic structure-activity studies of Burton and co-workers[5-7] first led towards the development of moderately potent (i.e., having IC_{50} values in the 10^{-5}M to 10^{-6}M range)

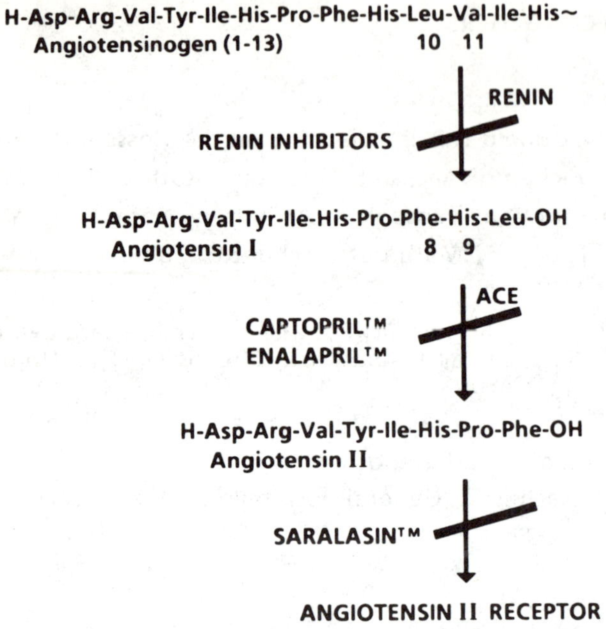

Fig. 1. Selected inhibitors of the renin-ACE cascade

compounds of which the decapeptide, known as RIP (Figure 2), became the first substrate-based renin inhibitor active *in vivo* (in sodium-deplete monkeys) in which it effected blood pressure lowering. Subsequently, major advances in the development of more potent renin inhibitors have been achieved by Szelke and co-workers[8,9] as well as Boger and co-workers[10,11] via the incorporation of "transition-state" dipeptide isosteres[12] (e.g., Pheψ(CH$_2$NH)Phe, Leuψ(CH$_2$NH)-Val, Leuψ(CH[OH]CH$_2$)Val and Statine) as replacements for Leu-Val at the cleavage site (P_1-P_1') of human substrate (Figure 2). These results have provided an excellent database for further comparative structure-conformation-activity studies of renin-inhibitor interactions. In this report, we describe our initial studies on renin inhibitors based on a multidisciplinary strategy which included the design and synthesis of analogues of RIP, structure-activity *in vitro* and enzyme kinetic studies, molecular modeling and fluorescence spectroscopy studies on selected RIP congeners.

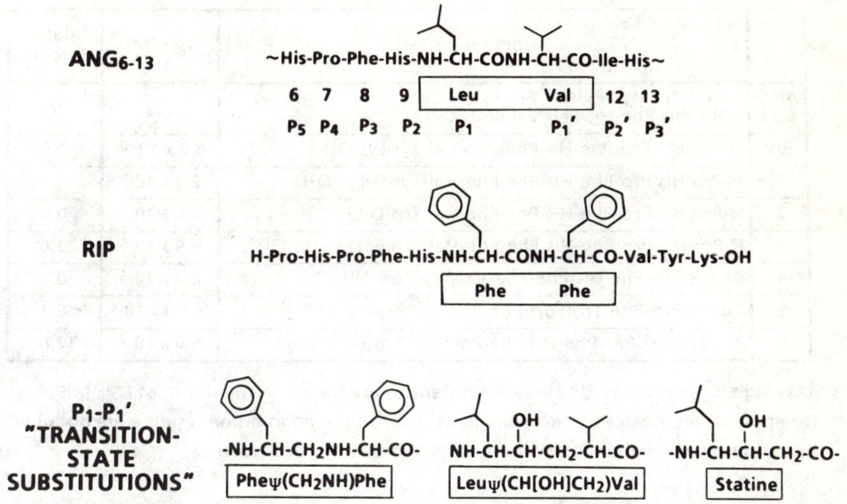

Fig. 2. Chemical structures of human renin substrate(6-13), Leu^{10}-Val^{11} modifications and Burton's "renin inhibitory peptide (RIP)".

Results and Discussion

The design and synthesis of human renin inhibitors was initially based on the decapeptide, RIP, and all analogues were synthesized by solid-phase chemistry methods, purified to homogeneity by preparative reverse-phase C-18 HPLC, and characterized by amino acid analysis, analytical HPLC and FAB-mass spectrometry (experimental details will be reported elsewhere). We first examined the P_3' Tyr residue by substitution with Trp and N^{in}-formyl-Trp (Trp(For)) and found that the Trp(For)-modified RIP analogue (compound **3**, Table I) was significantly more potent (>5-fold) than its Trp-modified correlate (compound **2**). In human angiotensinogen the P_3' residue is His, therefore, we decided to examine the effects of Trp(For)-substitutions at the P_2 His and P_5 His residues. Noteworthy was the bis-Trp(For)-modified octapeptide, compound **6**, which was significantly more potent (~12-fold) than RIP. In fact, the Ac-Trp(For)-Pro-Phe-His-Phe-Phe-Val-Trp(For)-NH_2 may be the most potent RIP analogue known having Phe-Phe at the P_1-P_1' cleavage site. Compound **6** was

Table I. Human renin inhibitory peptide analogues having Phe-Phe at the P_1-P_1' cleavage site.

	Compound	IC_{50} (M)	Relative Potency
ANG 6-13	P_5 P_4 P_3 P_2 P_1 P_1' P_2' P_3' ~His-Pro-Phe-His-Leu-Val-Ile-His~	***	***
RIP	H-Pro-His-Pro-Phe-His-Phe-Phe-Val-Tyr-Lys-OH	$8.0 \times 10^{-6\delta}$	1.0^{δ}
1	H-Pro-His-Pro-Phe-His-Phe-Phe-Val-Tyr-D-Lys-OH	2.5×10^{-5}	1.1
2	H-Pro-His-Pro-Phe-His-Phe-Phe-Val-Trp-D-Lys-OH	4.4×10^{-5}	0.4
3	H-Pro-His-Pro-Phe-His-Phe-Phe-Val-Trp(For)-D-Lys-OH	6.9×10^{-6}	2.6
4	Ac-His-Pro-Phe-His-Phe-Phe-Val-Trp(For)-NH_2	8.3×10^{-6}	0.7
5	Ac-His-Pro-Phe-Trp(For)-Phe-Phe-Val-Trp(For)-NH_2	$>1.0 \times 10^{-4}$	<0.1
6	Ac-Trp(For)-Pro-Phe-His-Phe-Phe-Val-Trp(For)-NH_2	5.0×10^{-7}	12.0

δAverage IC_{50} value for RIP. Relative potencies are based on the ration of IC_{50} (RIP)/IC_{50} (compound) on the day of the experiment. Human renin inhibition results were obtained by standard radioimmunoassay techniques as previously described.[13]

also shown (Figure 3) to be a competitive inhibitor (Lineweaver-Burke analysis) and it effected a Ki of 2.7×10^{-8}M.

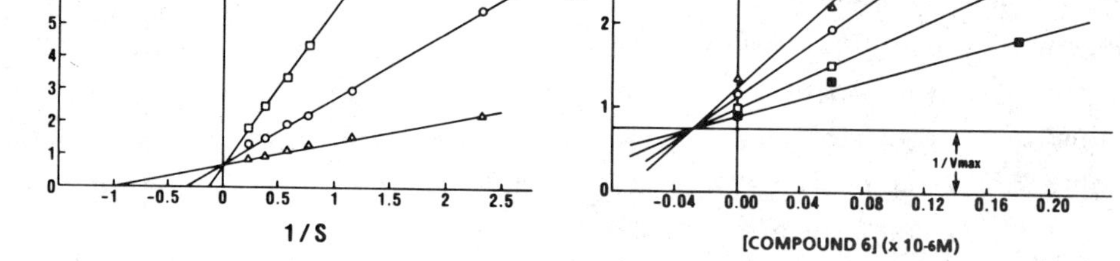

Fig. 3. Competitive inhibition of human renin-ANG reaction by compound **6** (left panel) and Ki determination for compound **6** (right panel).

Molecular modeling of Ac-Trp(For)-Pro-Phe-His-Phe-Phe-Val-Trp(For)-NH_2 (compound **6**) was then performed using the computer-generated hypothetical 3-D structure of human renin as previously described by Carlson *et al.*[14]. As shown in Figure 5, compound **6** was docked into the "active site" of the human renin model[15] in a rational manner to achieve energetically favorable intermolecular interactions within the ligand-enzyme complex. The peptide backbone of compound **6** exists in a twist-like secondary structure and the amino acid side-chain moieties at the P_5, P_3 and P_1 sites (i.e., Trp(For), Phe and Phe) were observed to form a hydrophobic cluster as based on its van der waals surface display representation. Accordingly, the complementary surface (i.e. S_5, S_3 and S_1) of the Carlson-Karplus human renin model to these substituents of compound **6** is primarily composed of aromatic residues within the C-terminal lobe of the enzyme that lie in the "active site" groove.

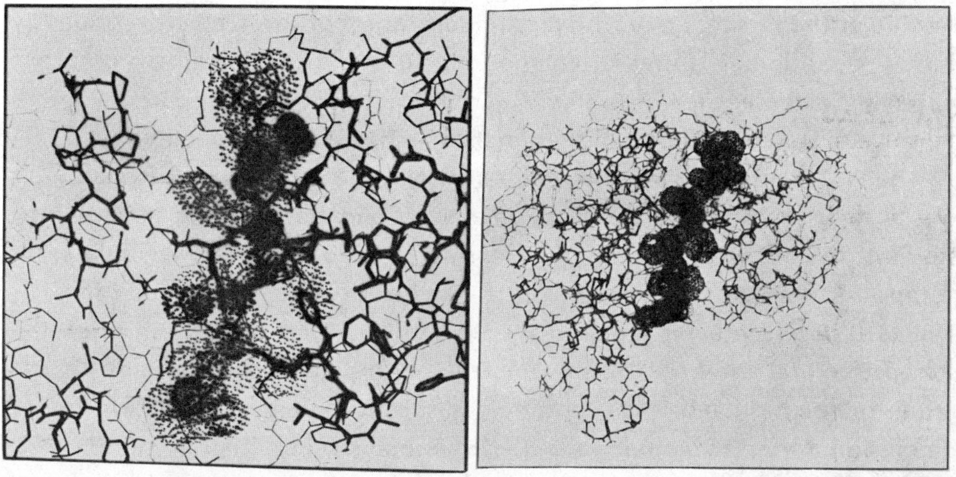

Fig. 5. (a) A computer graphics representation of compound **6** (van der waals surface display) docked into the "active site" of the Carlson-Karplus human renin 3-D structural model (ball-and-stick display).
(b) Computer graphics representation of the compound **6**-human renin complex (similar display as before).

Based on the octapeptide template of compound **6**, we next examined P_1-P_1' Pheψ(CH_2NH)Phe-substituted "transition state" RIP analogues (Table II). The bis-Trp(For))-modified congener, compound **9**, was highly potent (>1000-fold) versus RIP and significantly more potent (>15-fold) than its

Table II. Human renin inhibitory peptide analogues having Pheψ(CH_2NH)Phe at the P_1-P_1' cleavage site.

	Compound	IC_{50} (M)	Relative Potency
ANG 6-13	P_5 P_4 P_3 P_2 P_1 P_1' P_2' P_3' ~His-Pro-Phe-His-Leu-Val-Ile-His~	***	***
RIP	H-Pro-His-Pro-Phe-His-Phe-Phe-Val-Tyr-Lys-OH	$8.0 \times 10^{-6\delta}$	1.0^{δ}
7	H-Pro-His-Pro-Phe-His-Pheψ(CH_2NH)Phe-Val-Tyr-D-Lys-OH	1.7×10^{-7}	80.0
8	Ac-His-Pro-Phe-His-Pheψ(CH_2NH)Phe-Val-Trp(For)-NH_2	1.0×10^{-7}	60.0
9	Ac-Trp(For)-Pro-Phe-His-Pheψ(CH_2NH)Phe-Val-Trp(For)-NH_2	6.0×10^{-9}	1050.0
10	Ac Trp(For)-Pro-Phe-His-Pheψ(CH_2NH)Phe-Val-D-Trp(For)-NH_2	4.0×10^{-8}	200.0
11	Ac-D-Trp(For)-Pro-Phe-His-Pheψ(CH_2NH)Phe-Val-D-Trp(For)-NH_2	6.3×10^{-6}	2.0

δRefer to Table I for further details.

Pheψ(CH_2NH)Phe-substituted correlate, compound **8**, which incorporated a His residue at the P_5 site. Stereochemical modification of the Trp(For) residues by their D-enantiomers (refer to compounds **10** and **11**) effected a significant potency decrease of which the P_5 Trp(For) moiety was apparently more sensitive to stereotransformation than the P_3' Trp(For) functionality.

The P_5 Trp(For) residue of the Pheψ(CH_2NH)Phe-substituted RIP analogues was further investigated by structure-activity and fluorescence spectroscopy studies. This was accomplished using the octapeptide template of compound **12** (Ac-Trp(For)-Pro-Phe-His-Pheψ(CH_2NH)Phe-Val-Tyr-NH_2, Table III) in which a Tyr replaced the P_3' Trp(For) functionality. In brief, the stereostructural integrity of the P_5 Trp(For) moiety in compound **12** was requisite for high biological potency as substitution by D-Trp(For), Trp or ommission (refer to compounds **13-16**, Table III) resulted in significantly reduced potencies. Based on the P_4-P_3' heptapeptide template of compound **16**, the N-terminal P_4 Pro residue was shown to be effectively substituted by a t-butylacetyl (Tba) moiety. Subsequent modification of the P_3 Phe residue of compound **17** by Trp(For) or D-Trp(For) resulted in significant decreases (40- and 100-fold for compounds **18** and **19**, respectively) in biological potency. A similar effect for P_3 Phe substitution by Trp(For) was observed for the octapeptide, compound **20**, as compared to compound **12**. Overall, these data illustrate the regiospecificity of Trp(For)-modification of RIP analogues for effecting enhanced renin inhibitory potency to be at the P_5 site.

Table III. Human renin inhibitory peptide analogues having Pheψ(CH_2NH)Phe at the P_1-P_1' cleavage site.

	Compound	IC_{50} (M)	Relative Potency
ANG 6-13	P_5 P_4 P_3 P_2 P_1 P_1' P_2' P_3' ~His-Pro-Phe-His-Leu-Val-Ile-His~	***	***
RIP	H-Pro-His-Pro-Phe-His-Phe-Phe-Val-Tyr-Lys-OH	8.0×10^{-6} [δ]	1.0[δ]
12	Ac-Trp(For)-Pro-Phe-His-Pheψ(CH_2NH)Phe-Val-Tyr-NH_2	3.0×10^{-9}	2700.0
13	Ac-D-Trp(For)-Pro-Phe-His-Pheψ(CH_2NH)Phe-Val-Tyr-NH_2	1.9×10^{-7}	40.0
14	H-D-Trp(For)-Pro-Phe-His-Pheψ(CH_2NH)Phe-Val-Tyr-NH_2	2.0×10^{-7}	40.0
15	Ac-Trp-Pro-Phe-His-Pheψ(CH_2NH)Phe-Val-Tyr-NH_2	1.5×10^{-8}	530.0
16	Ac-Pro-Phe-His-Pheψ(CH_2NH)Phe-Val-Tyr-NH_2	7.9×10^{-8}	110.0
17	Tba-Phe-His-Pheψ(CH_2NH)Phe-Val-Tyr-NH_2	1.9×10^{-7}	40.0
18	Tba-Trp(For)-His-Pheψ(CH_2NH)Phe-Val-Tyr-NH_2	6.1×10^{-6}	1.0
19	Tba-D-Trp(For)-His-Pheψ(CH_2NH)Phe-Val-Tyr-NH_2	$\sim 1.0 \times 10^{-5}$	~0.3
20	Ac-Trp(For)-Pro-Trp(For)-His-Pheψ(CH_2NH)Phe-Val-Tyr-NH_2	4.6×10^{-8}	240.0

δRefer to Table I for further details.

Fluorescence spectroscopic analysis of (Table IV) two P_1-P_1' Pheψ(CH_2NH)Phe-substituted RIP analogues, compounds **12** and **15**, was then performed by methods similar to those previously described[16] for solution-phase conformational studies on peptide analogues of angiotensin II, cholecystokinin, enkephalin, and somatostatin. Determination of the average intramolecular distance between the P_3' phenol side-chain (donor) and the P_5 indole side-chain (acceptor) for each compound (concentrations at 20-30 x 10^{-6}M in H_2O) was made by evaluation of singlet-singlet energy transfer between donor and acceptor on basis of the Förster equation. In brief, the average

Table IV. Fluorescence spectroscopy of P_1-P_1' Pheψ(CH_2NH)Phe-substituted RIP analogues having P_5 Trp(For)/Trp and P_3' Tyr modifications.

Compound	$\phi_{Trp(For)}$	ϕ_{Trp}	ϕ_{Tyr}	E	Ro (Å)	r (Å)
12[δ]	0.010 ± 0.002	***	0.024 ± 0.003	0.611 ± 0.042	16.6	14.2 ± 0.3
15[δ]	***	0.053 ± 0.014	0.032 ± 0.002	0.470 ± 0.031	12.4	11.2 ± 0.1

δFor structures of compounds **12** and **15**, refer to Table III above. Förster critical distance; and r, intramolecular P_5 to P_3' amino acid side-chain distances. The experimental details of this study will be reported elsewhere.

intramolecular distance between the P_5 Trp(For) and P_3' Tyr side-chain functionalities of compound **12** was 14.2Å which suggested the existence of a reverse-turn type secondary structure within the P_4-P_2' hexapeptide sequence. For compound **15**, the average intramolecular distance between the P_5 Trp and P_3' Tyr side-chain functionalities was 11.2Å which again implicated potential conformational preferences of the peptide backbone to effect spatial proximity of these two primary sequence-distal amino acid residues.

Both Statine and Leuψ(CH[OH]CH_2)Val-modified "transition-state"-substituted analogues were evaluated for their comparative renin inhibitory potencies (Table V). In both of these dipeptide isosteres, the chirality of the hydroxymethylene (CH(OH)) carbon atom was of the S-configuration as previously reported[11] to be requisite for providing high potency. In brief, the Trp(For)-substituted octapeptide, compound **6**, provided a superior template for the incorporation of the P_1-P_1' Statine or Leuψ(CH[OH]CH_2)Val dipeptide isosteres. The resultant analogues, compounds **24** and **25** were superpotent ($IC_{50} \approx 5 \times 10^{-10}$M) renin inhibitors. Similar to previous studies by Boger[11], we found that the t-butyloxycarbonyl (Boc) moiety was an effective replacement for the N-terminal Pro-His-Pro tripeptidyl moiety (refer to compounds **22** and **23**).

Table V. Human renin inhibitory peptide analogues having Leuψ(CH[OH]CH_2)Val at the P_1-P_1' cleavage site.

	Compound	IC_{50} (M)	Relative Potency
ANG 6-13	P_5 P_4 P_3 P_2 P_1 P_1' P_2' P_3' ~His-Pro-Phe-His-Leu-Val-Ile-His~	***	***
RIP	H-Pro-His-Pro-Phe-His-Phe-Phe-Val-Tyr-Lys-OH	8.0×10^{-6} δ	1.0δ
21	H-Pro-His-Pro-Phe-His-Phe-Phe-Ile-His-D-Lys-OH	1.7×10^{-4}	0.1
22	H-Pro-His-Pro-Phe-His-Statine-Ile-His-D-Lys-OH	1.0×10^{-8}	180.0
23	Boc-Phe-His-Statine-Ile-Phe-NH_2	2.6×10^{-8}	780.0
24	Ac-Trp(For)-Pro-Phe-His-Statine-Val-Trp(For)-NH_2	3.8×10^{-10}	20,000.0
25	Ac-Trp(For)-Pro-Phe-His-Leuψ(CH[OH]CH_2)Val-Val-Trp(For)-NH_2	6.1×10^{-10}	10,300.0

δRefer to Table I for further details.

Finally, we prepared several [Cys, Cys]-substituted cyclic derivatives to test for their comparative bioactivity relationships (Table VI) and examine by molecular modeling. The P_5, P_1-P_1' and P_3' residues were conserved, and

Cys, Cys-substitutions were systematically made such that the P_1-P_1' cleavage site modifications were included within the resultant disulfide-bridged ring structures. The most potent cyclic P_1-P_1' Phe-Phe-substituted RIP analogue was Ac-Trp(For)-Cys-Phe-His-Phe-Phe-Cys-Trp(For)-NH_2 (compound **28**) which was more than 40-fold more potent than RIP. Compound **28** contains a 20-membered ring structure flanked N- and C-terminally by exocyclic Trp(For) residues, and molecular modeling (data not shown) indicated a reverse-turn conformation existing within the P_3-P_1' tetrapeptide sequence. Substitution of the P_1-P_1' dipeptide isosteres Pheψ(CH_2NH)Phe and Statine for Phe-Phe in compound **28** did not effect increased renin inhibitory potency.

Table VI. Cyclic, conformationally-restricted human renin inhibitory peptide analogues having Phe-Phe, Pheψ(CH_2NH)Phe or Statine at the P_1-P_1' cleavage site.

	Compound	IC_{50} (M)	Relative Potency
ANG 6-13	P_5 P_4 P_3 P_2 P_1 P_1' P_2' P_3' ~His-Pro-Phe-His-Leu-Val-Ile-His~	***	***
RIP	H-Pro-His-Pro-Phe-His-Phe-Phe-Val-Tyr-Lys-OH	$8.0 \times 10^{-6\delta}$	1.0^{δ}
26	Ac-His-Pro-Phe-Cys-Phe-Phe-Cys-Trp(For)-NH_2	2.9×10^{-6}	3.0
27	Ac-Trp(For)-Pro-Cys-His-Phe-Phe-Cys-Trp(For)-NH_2	$>1.0 \times 10^{-4}$	<0.1
28	Ac-Trp(For)-Cys-Phe-His-Phe-Phe-Cys-Trp(For)-NH_2	1.9×10^{-7}	44.0
29	Ac-Trp(For)-Pro-Phe-Cys-Pheψ(CH_2NH)Phe-Cys-Trp(For)-NH_2	$>1.0 \times 10^{-5}$	~0.1
30	Ac-Trp(For)-Pro-Phe-Cys-Statine-Cys-Trp(For)-NH_2	1.5×10^{-6}	5.0
31	Ac-Trp(For)-Cys-Phe-His-Pheψ(CH_2NH)Phe-Cys-Trp(For)-NH_2	7.6×10^{-7}	11.0
32	Ac-Trp(For)-Cys-Phe-His-Statine-Cys-Trp(For)-NH_2	4.3×10^{-6}	2.0

δRefer to Table I for further details.

In summary, we have descibed the comparative structure-activity relationships of several different P_1-P_1'-modified RIP analogues. The Trp(For) residue was shown to effect increased renin inhibitory potency in vitro, and this structural modification is stereospecific (L isomer preference) and regiospecific (P_5 site preference). In fact, the P_5 Trp(For)-substituted compounds noted above may provide the first examples of Trp(For)-

modification into a peptide in which significant biological potency enhancement may be unambiguously attributed to this heteroaromatic and highly lipophilic aminoacyl derivative. Results obtained from fluorescence spectroscopy, molecular modeling, and the bioactivity relationships of [Cys, Cys]-substituted RIP analogues provide evidence to suggest reverse-turn type conformational features ocurring within the P_3-P_1' tetrapeptide sequence.

References and Notes

1. Geiger, R. (1984) **Arzneimittel-Forschung/Drug Res. 35**, 1386-1391.
2. Gantem, D., Unger, Th., and Lang, R.E. (1984) **Arzneimittel-Forschung/Drug Res. 35**, 1391-1398.
3. Kokubu, T., Hiwada, K. Ito, T., Ueda, E., Yamamura, T., Mizoguchi, T., Shigezane, K. (1973) **Biochem. Pharmacol. 22**, 3217-.
4. Saketa, Y., and Inagami, T. (1975) **Biochemistry 14**, 3188-.
5. Poulson, K., Haber, E., and Burton, J. (1973) **Biochemistry 12**, 3877-3882.
6. Burton, J., Quinn, T., and Poulesen, K. (1979) in **Peptides: Synthesis-Structure-Function**, Rich, D.R. and Gross, E., Eds., Pierce Chemical Co., Rockford Il., pp. 447-449.
7. Burton, J., Cody, R.J., Herd, J.A., and Haber, E. (1980) **Proc. Natl. Acad. Sci. USA 77**, 5476-5479.
8. Szelke, M., Leckie, B., Hallett, A., Jones, D.M., Sueiras, J., Atrash, B., and Lever, A.F. (1982) **Nature 299**, 555-557.
9. Szelke, M., Jones, D.M., Atrash, B., Hallett, A., and Leckie, B.J. (1983) in **Peptides: Structure and Function**, Rich, D.H. and Hruby, V.J., Eds., Pierce Chemical Co., Rockford, Il., pp. 579-582.
10. Boger, J., Lohr, N.S., Ulm, E.H., Poe, M., Blaine, E.H., Fanelli, G.M., Lin, T.-Y.,Payne, L.S., Schorn, T.W., LaMont, B.I., Vassil, T.C., Stabilito, I.I., and Veber, D.F. (1983) **Nature 303**, 81-84.
11. Boger, J. (1983) in **Peptides: Structure and Function**, loc. cit., pp. 569-578.
12. The term "transition-state" has been used to implicate the P_1-P_1' sisscile amide bond isosteres of CH_2NH and $CH[OH]CH_2$ as possible tetrahedral mimics of the substrate P_1-P_1' CONH transition-state produced during the renin-ANG cleavage reaction. The concept of synthetic transition-state mimics has been recently reviewed (Rich, D.H. [1985] **J. Med. Chem. 28**, 263-273), and it was suggested that in the case of the P_1-P_1' statine-substituted renin inhibitors the term "collected substrate" may be a more accurate molecular descriptor.
13. Pals, D.T., deGraaf, G.L., Kati, W.M., Lawson, J.A., Smith, C.W., and Skala, G.F. (1985) **Clin. Exp. Hypertension A7**, 105-121.
14. Carlson, W., Karplus, M., and Haber, E. (1985) **Hypertension, 7**, 13-26.
15. Another computer graphics model of human renin has also been reported (Sibanda, B.L., Blundell, T., Hobart, P.M., Fogliano, M., Bindra, J.S., Dominy, B.W., and Chirgwin, J.M. [1984] **Fed. Eur. Biochem. Soc.** 174, 102-111).
16. Schiller, P.W., Abraham, N., Bellini, F., and Immer, H. (1981) in **Peptides: Synthesis-Structure-Function**, loc. cit., pp. 367-370 (and references therein)..

POTENT INHIBITORS OF HOG AND HUMAN RENIN CONTAINING AN AMINO ALCOHOL DIPEPTIDE SURROGATE

Denis E. Ryono, Charles A. Free, Richard Neubeck, Sylvia G. Samaniego, Jollie D. Godfrey and Edward W. Petrillo, Jr.

The Squibb Institute for Medical Research, P.O. Box 4000, Princeton, New Jersey, 08540

Renin research continues to attract efforts directed toward the preparation of novel inhibitors. We chose as a point of departure the general aspartic protease inhibitor, pepstatin A[1], and sought to extend this well-known example of a transition state analog inhibitor[2] by incorporating another of the heteroatom elements, a secondary amine function, which comprises the putative tetrahedral intermediate of amide bond hydrolysis (Figure 1). We placed the resulting amino alcohol dipeptide surrogate into various renin substrate analogs in order to: (1) assess the affinity of the amino alcohol moiety for the enzyme active site; and (2) identify the most effective regions of binding with regard to the various subsites of substrate/inhibitor interactions.

Fig. 1 The amino alcohol dipeptide surrogate.

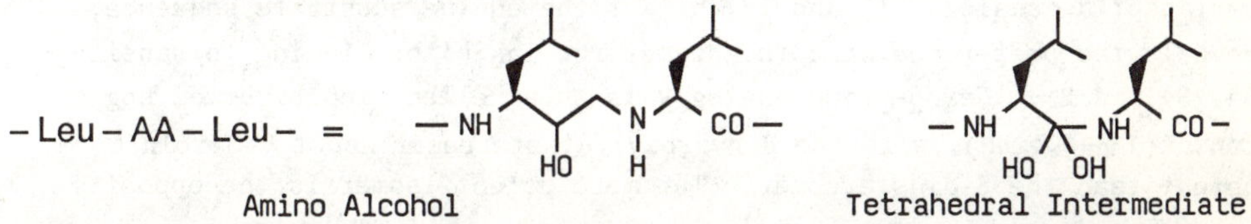

For the preparation of equine sequence (Leu,10-Leu,11) amino alcohol analogs we employed the key synthetic intermediate <u>1</u>[3], shown in Scheme 1 in the synthesis of a pentapeptide amino alcohol analog. Using the differential protection afforded by the Boc, Z and TMSE ester protecting groups, a sequence of C-terminal followed by N-terminal elaboration was used to prepare amino alcohols of any desired composition. The temporary hydroxyl protection by the oxazolidine ring present in <u>1</u> allowed carboxyl activation to proceed without lactonization. With benzyloxymethyl (BOM) protection of the histidine imidazole ring[4], final deprotection involved BOM and Z group removal by hydrogenolysis using $Pd(OH)_2$-C catalyst with added HCl to suppress reductive methylation of the amino alcohol secondary amine.

Scheme 1. Synthesis of a pentapeptide amino alcohol analog, Boc-His-Leu-AA-Leu-Val-Phe-OMe.

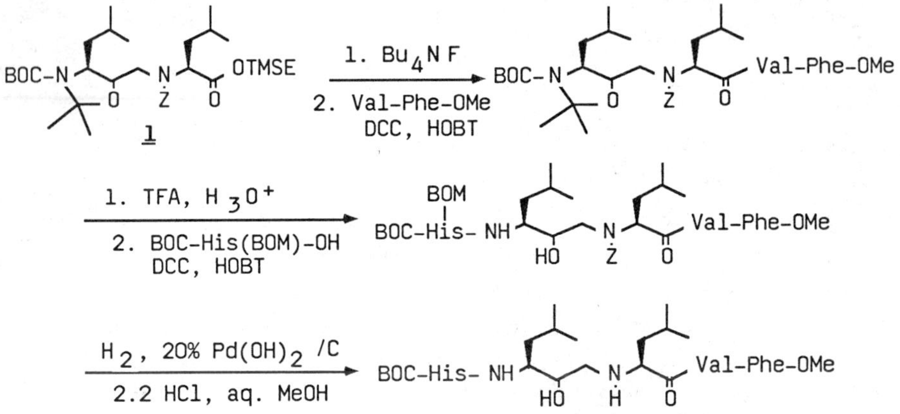

Table I summarizes the inhibitory potencies of various amino alcohols against renins from hog and human kidney, measured by RIA of angiotensin I generated at pH 7.0 from the respective substrates, rat and human angiotensinogen. A comparison of tetrapeptide analogs I and II and pentapeptide analogs III and IV, all of the equine substrate sequence, reveals the preference of both enzymes for inhibitor binding to subsites S_3, S_2 and S_1. Hexapeptide analog V is an excellent inhibitor of hog renin (I_{50}=4.4 nM), with the R hydroxyl diastereomer about 2-3 fold more potent than the S diastereomer. The more potent isomer is the opposite of that found in the related statine[5] and "hydroxy isostere"[6] classes of renin inhibitors. This relationship holds for many of the amino alcohol analogs that we have prepared, thus distinguishing them from the previous two groups of hydroxyl-based renin inhibitors. In one or both of the amino alcohol isomers it is conceivable that the secondary amine is involved in specific binding interactions with the enzyme active site. Removal of the Boc group (VI) and deletion of the hydroxyl (VII) both result in substantial loss of potency against the two enzymes.

Table I. Hog and Human Renin Inhibition of Amino Alcohol Analogs

			I_{50}(μM): Hog Renin	Human Renin
I	R	Boc-Leu-AA-Leu-Val-Phe-OMe	280	>1000
II	R	Boc-Phe-His-Leu-AA-Leu-NH_2	13	13
III	R	Boc-His-Leu-AA-Leu-Val-Phe-OMe	12	>1000
IV	R	Boc-Phe-His-Leu-AA-Leu-Val-OMe	0.40	12
V	R	Boc-Phe-His-Leu-AA-Leu-Val-Phe-OMe	0.0044	0.55
V	S	Boc-Phe-His-Leu-AA-Leu-Val-Phe-OMe	0.013	1.9
VI	R	Phe-His-Leu-AA-Leu-Val-Phe-OMe	0.089	180
VI	S	Phe-His-Leu-AA-Leu-Val-Phe-OMe	0.21	280
VII		Boc-Phe-His-Leu-AA*-Leu-Val-Phe-OMe (*des-OH)	0.067	67
VIII	RS	Boc-Phe-His-Leu-AA-Val-Ile-His-OMe	0.14	0.15
IX	RS	Boc-Phe-His-Leu-AA-Gly-Val-Phe-OMe	1.0	0.12
X	RS	Boc-Phe-His-Leu-AA-Gly-Leu-Phe-OMe	2.2	0.30
XI	RS	Boc-Phe-His-Leu-AA-Gly-Ile-His-OMe	13	0.031

Although excellent hog renin inhibitors, analogs V were less effective inhibitors of human renin. The human sequence analog VIII gives improved potency, although less than that expected based on results from related inhibitors.[5,6] Compounds IX, X and XI explore the effect of deleting the alkyl sidechains of the P_1' groups of the Leu-AA-Leu and Leu-AA-Val analogs. Analog IX, a Leu-AA-Gly modification of the hexapeptide analogs V, is a poorer hog renin inhibitor by about 200-fold. However, IX does show definite improvement against human renin versus the two isomers of V. In compound X, the glycine modification with a terminal Leu-Phe-OMe dipeptide, human renin potency falls between analogs V and IX. Finally, amino alcohol XI, combining glycine modification with the human sequence terminal dipeptide, Ile-His-OMe, is a very good inhibitor of human renin (I_{50}=31 nM). In XI the hog renin inhibition falls off (I_{50}=13000 nM), indicating high specificity for the human enzyme. Relief of deleterious steric interactions is probably responsible for the better human renin inhibition displayed by the Leu-AA-Gly compounds. Compounds V and XI are specific inhibitors of renin. Against pepsin (pH 1.6) and cathepsin D (pH 3.0), the R isomer of V gives I_{50} values of 450 and 7.0 μM, respectively, while the corresponding values for the S isomer are >500 and 100 μM. Amino alcohol XI is essentially inactive against both enzymes (I_{50} >500 μM).

SUMMARY

Analogs incorporating an amino alcohol dipeptide surrogate in place of the scissile dipeptide are potent inhibitors of renin. The most effective inhibitors encompassed interactions with S_3, S_2, and S_1 enzyme subsites, and are of the opposite hydroxyl configuration as that found for the related statine and "hydroxy isostere" renin inhibitors. Amino alcohol analogs designed for potent inhibition of hog and human renin are much less inhibitory towards the related aspartic proteases, pepsin and cathepsin D. The successful application of the amino alcohol dipeptide surrogate to renin inhibition paralleled its utilization at Squibb in the development of novel inhibitors of angiotensin converting enzyme,[7] and the potential remains for extension to other therapeutically relevant peptidases.

REFERENCES

1. Umezawa, H., Aoyagi, T., Morishima, H., Hamada, H., Takeuchi, T. (1970) J. Antibiot., 23, 259-262.

2. Lienhard, G. E. (1973) Science, 180, 149.

3. The synthesis and stereochemical assignments of amino alcohol intermediate 1 will be reported elsewhere.

4. Brown, T., Jones, J. H., Richards, J. D. (1982) J. Chem. Soc. Perkin Trans. I, 1553-1561.

5. Boger, J., Lohr, N. S., Ulm, E. H., Poe, M., Blaine, E. H., Fanelli, G. M., Lin, T-Y., Payne, L. S., Schorn, T. W., LaMont, B. I., Vassil, T. C., Sabilito, I. I., Veber, D. F., Rich, D. H., Boparai, A. S. (1983) Nature, 303, 81-84.

6. Szelke, M., Jones, D. M., Atrash, B., Hallett, A., Leckie, B. (1983) "Peptides: Structure and Function. Proceedings of the Eighth American Peptide Symposium", Hruby, V. J., Rich, D. H. Eds., Pierce Chemical Co., Rockford, Illinois, pp 579-582.

7. Gordon, E. M., Godfrey, J. D., Pluscec, J., Von Langen, D. and Natarajan, S. (1985), Biochem. Biophys. Res. Commun., 126, 419-426.

A STEREOCONTROLLED SYNTHESIS OF HYDROXYETHYLENE DIPEPTIDE ISOSTERES USING NOVEL, CHIRAL AMINOALKYL EPOXIDES; NEW RENIN INHIBITOR ANALOGS.

Ben E. Evans, Kenneth E. Rittle, Edgar H. Ulm and Daniel F. Veber

Merck Sharp & Dohme Research Laboratories, West Point, PA 19486

James P. Springer and Martin Poe

Merck Sharp & Dohme Research Laboratories, Rahway, NJ 07065

In connection with an investigation of renin inhibitors containing non-hydrolyzable isosteres in place of specific peptide bonds, we required a synthesis of the hydroxyethylene dipeptide isostere unit **1** which could provide the individual

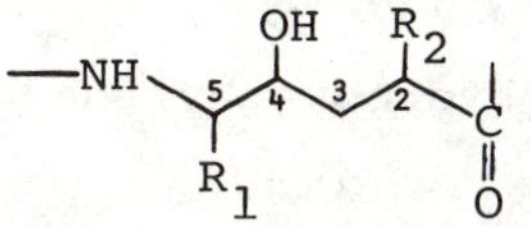

1

diastereomers of this unit with control of stereochemistry at all three chiral centers and with ample latitude for independent variation of substituents R_1 and R_2.

The previously unreported epoxide **3** (Scheme 1) was prepared not only in good yield, but with virtually complete retention of chirality by reaction of Boc-L-phenylalanine aldehyde **2**[1] with dimethylsulfonium methylide. The epoxide **3** was produced as a chromatographically separable mixture of two isomers, the 2R,3S("threo") and 2S,3S("erythro"). The two were distinguishable by NMR (C_2 proton: 4.1δ-isomer A, 3.7δ-isomer B), and their absolute stereochemistries were established by X-ray crystallographic analysis of one isomer (4.1δ=threo=2R,3S). Chiral purity was verified (>95%) by

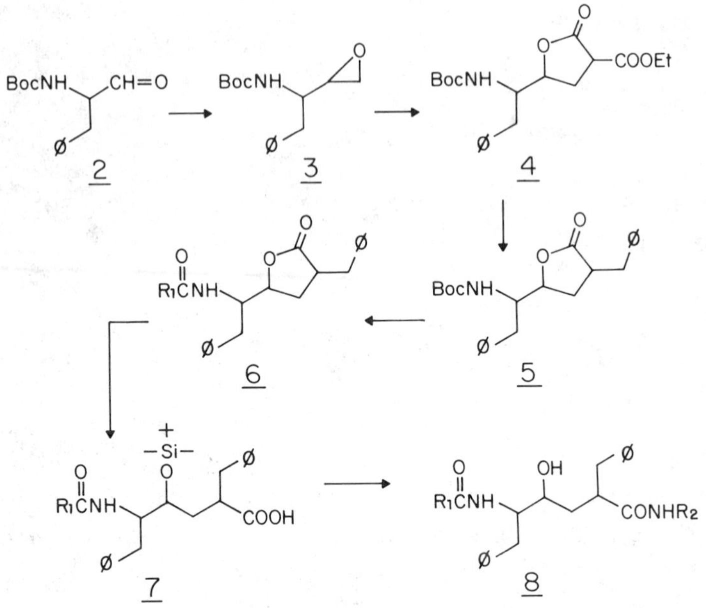

Scheme 1. Synthesis of hydroxyethylene dipeptide isosteres.

identification (360 MHz NMR) of a single carbinolamine upon treatment of the chiral erythro (2S,3S) epoxide with d-(+)-α-methylbenzylamine. Racemic erythro (2S,3S + 2R,3R) epoxide from racemic Boc-phenylalanine aldehyde was used as a control, yielding a pair of carbinolamine diastereomers. Each of the four isomers of epoxide **3** could thus be obtained with known absolute stereochemistry.

With diethylmalonate and sodium ethoxide, the individual epoxides could each be converted to a pair of diastereomers (C_2) of the lactone **4**. Alkylation followed by hydrolysis and decarboxylation gave the lactone **5**, an amine-, hydroxy-, and acid-protected form of the hydroxyethylene dipeptide isostere **1**.

Conventional deblocking and coupling techniques allowed elaboration of the N-terminus of lactone **5** to the desired configuration (**6**). For elaboration of the C-terminus, the lactone ring was first opened with base, and the resulting salt disilated then selectively monodesilated according to the procedure of Corey and Venkateswarlu[2] to give the hydroxy-protected acid **7**. Coupling by conventional tech-

Table I. Renin Inhibitory Peptides Containing Hydroxyethylene Dipeptide Isosteres

	IC_{50} (mM) Hog Renin	K_i (mM) Purified Human Kidney Renin[4]
Boc-Phe-Phe-Sta-Leu-Phe-NH_2[3,a]	0.091	0.0048
Boc-Phe-Phe-Phe$\overline{\text{OH}}$Phe-Leu-Phe-NH_2[b]		
2R(S), 4R,5S	>300	>0.3
2S(R), 4R,5S	>300	>0.3
2R(S), 4S,5S	>300	>0.3
2S(R), 4S,5S	0.082	0.068
Boc-Phe-Phe-Sta-Leu-$NHCH_2$Ø	0.017	0.043
Boc-Phe$\overline{\text{OH}}$Phe-Sta-Leu-$NHCH_2$Ø		
2R(S), 4S,5S	0.63	1.4
2S(R), 4S,5S	0.12	0.16
2R(S), 4R,5S	0.27	0.24
65/35 2R + 2S, 4R,5S	0.35	0.012
Boc-Phe$\overline{\text{OH}}$Phe-ACHPA-Leu-$NHCH_2$(m-H_2NCH_2Ø)[c]		
2R(S), 4S,5S	2.4	2.2
2S(R), 4S,5S	0.027	0.048
2R(S), 4R,5S	1	1.0
2S(R), 4R,5S	0.045	0.038

[a]Sta=statine, 3S,4S-4-amino-3-hydroxy-6-methylheptanoic acid. [b]Phe$\overline{\text{OH}}$Phe=hydroxyethylene analog of Phe-Phe. [c]ACHPA=3S,4S-4-amino-5-cyclohexyl-3-hydroxypentanoic acid.

niques served to elaborate the C-terminus, and final OH deprotection provided the completed peptides **8**, each as a pair of diastereomers (C_2). These were separated by column chromatography or by HPLC. Stereochemistry at C_2 has not been established unambiguously.

Renin inhibitory peptides prepared by this route are listed in Table I. Several of these compounds serve as effective renin inhibitors when compared with the parent peptide.

The methods presented in this paper provide access to the hydroxyethylene dipeptide isostere unit **1** in each of its stereochemical modifications. Chirality at C_5 and the substituent R_1 are set by the choice of starting amino acid. Chiralities at C_2 and C_4 are selected by separation of diastereomers, and the substituent R_2 is subject only to the limitations of a malonate alkylation and the availability of a suitable alkylating agent. Peptides incorporating the unit **1** at various positions are effective inhibitors of both purified human kidney renin[4] and crude hog renin.

This work also introduces the chiral epoxide **3**. All four individual diastereomers of the epoxide are accessible, and the absolute stereochemistry at both chiral centers may be ascertained. The new γ-substituted-γ-lactones **4**, **5**, and **6** are also described.

References

1. Rittle, K.E., Homnick, C.F., Ponticello, G.S., and Evans, B.E. (1982) J. Org. Chem. 47, 3016-3018.
2. Corey, E. J. and Venkateswarlu, A. (1972) J. Am. Chem. Soc. 94, 6190-6191.
3. Bock, M.G., DiPardo, R.M., Evans, B.E., Rittle, K.E., Boger, J.S., Freidinger, R.M., and Veber, D.F. (1985) Chem. Commun. 109-110.
4. Slater, E.E. and Strout, H.V. (1981) J. Biol. Chem. 256, 8164-8171.

RENIN INHIBITORS. TRIPEPTIDE AND TETRAPEPTIDE COMPETITIVE INHIBITORS CONTAINING A NOVEL ANALOG OF STATINE

Joshua Boger, Linda S. Payne, Debra S. Perlow, Martin Poe,[+] Edward H. Blaine, Edgar H. Ulm, Terry W. Schorn, Bruce I. LaMont, Mark G. Bock, Roger M. Freidinger, Ben E. Evans, and Daniel F. Veber

Merck Sharp & Dohme Research Laboratories, West Point, Pennsylvania 19486 and [+]Rahway, New Jersey 07065

Inhibitors of renin based upon the substrate sequence have been reported from several laboratories and may offer an attractive alternative to inhibition of angiotensin converting enzyme for the treatment of hypertension.[1] We have described inhibitors of renin, with potencies around 10 nM ($K_i = 10^{-8}$M), which are analogs of the minimum substrate octapeptide (His-Pro-Phe-His-Leu-Leu-Val-Tyr, angiotensinogen sequence 6-13) and which contain the transition-state (or intermediate) analog statine, (3S,4S)-4-amino-3-hydroxy-6-methyl-heptanoic acid (=Sta), which we hypothesize to be an analog of the scissile dipeptide.[2]

Molecular modelling studies on a renin-related enzyme suggested that additional binding interactions might be realized by extension of the Sta side chain further into its hypothesized binding pocket in renin.[3] Replacement of the isobutyl side chain with cyclohexylmethyl was predicted to be ideal, while replacement with benzyl might sacrifice specific interactions which renin might require from the Leu-type side chain methyl groups. These predictions were confirmed (Table I) in the heptapeptide series **1-3**, with the cyclohexylmethyl side chain analog of Sta, ACHPA, leading to an increase in human renin inhibition, versus Sta, of 50-75 fold.

It was found that the heptapeptide sequence could be shortened at the N-terminus to pentapeptide renin inhibitors, such as **4-6**, with only a 3- or 4-fold loss in human renin inhibitory potency. Attempts to shorten the inhibitors further at the N-terminus generally gave poor inhibitors of plasma renin, such as **7**, although the low K_i for purified human kidney renin was encouraging. This phenomenon, in which small renin inhibitors give disparate inhibition values using purified renin enzyme versus renin in plasma, has led to the hypothesis of a non-renin plasma component capable of binding certain renin inhibitors with high efficiency (B. E. Evans, K. E. Rittle, M. G. Bock, C. D. Bennett, R. M. DiPardo, J. Boger, M. Poe, E. H. Ulm, B. I. LaMont, G. M. Fanelli, and D. F. Veber, unpublished results). This effect seems to be ameliorated by addition of charged groups to these small, generally hydrophobic, inhibitors. Renin substrate analogs have been reported from these laboratories in which the C-terminal Phe residue in Boc-Phe-His-Sta/-ACHPA-Leu-Phe-NH_2 has been replaced with various solubilizing substituents, including m-aminomethylbenzylamide (m-AMBA, as in **8**), m-guanidinylmethylbenzylamide (m-GMBA), and 2- and 4-pyridylmethylamides.[4]

As an extension of that work, we have prepared the tetra-and tripeptide renin inhibitors **8-18** shown in Table I. Tetrapeptides **8-10**, containing M-AMBA and blocked at the N-terminus with various alkyl urethanes, show that the larger Boc-group is preferred in this series over smaller urethanes, based on the human plasma renin inhibition (IC_{50}) values, in marked contrast to the results seen with the pentapeptide inhibitors **4-6**, where the smaller ethyloxycarbonyl (Etoc) group is preferred. Extremely potent analogs, such as **11**, with an acid stable N-terminal Etoc, can be obtained with C-terminal groups other than m-AMBA, such as the free carboxylate-containing AHPPA, the phenylalanine-derived Sta analog. The subtle relationships between the optimal N- and C-terminae in these tetra- and penta-peptide inhibitors merit additional investigation.

Table I. Renin Inhibition by Substrate Analogs Containing Statine, AHPPA, and ACHPA[a]

Number	Compound	Renin Inhibition, 10^{-9}M Human Kidney, K_i	Human Plasma, IC_{50}	Dog Plasma, IC_{50}
	6,7 8 9 10,11 12 13			
1	IHP-Phe-His -Sta- Leu-Phe-NH_2	8.8[b]	13[b]	32
2	IHP-Phe-His-AHPPA-Leu-Phe-NH_2	18	2.2	100
3	IHP-Phe-His-ACHPA-Leu-Phe-NH_2	0.16[b]	0.17[b]	1.9[b]
4	Boc-Phe-His-ACHPA-Leu-Phe-NH_2	0.88	2.2	13
5	Poc-Phe-His-ACHPA-Leu-Phe-NH_2	1.1	5.1	20
6	Etoc-Phe-His-ACHPA-Leu-Phe-NH_2	0.52	0.76	9.5
7	POA-His-ACHPA-Leu-Phe-NH_2	14	1000	>1000[c]
8	Boc-Phe-His-ACHPA-Leu-m-AMBA	0.48	0.26[b]	3.3
9	Poc-Phe-His-ACHPA-Leu-m-AMBA	0.21	0.77	8.0
10	Etoc-Phe-His-ACHPA-Leu-m-AMBA	1.0	4.8	5.6
11	Etoc-Phe-His-ACHPA-Leu-AHPPA	0.057	2.2	1.1
12	Boc-Phe-His-ACHPA-Ile-$NHCH_2\phi$	0.080	2.6[b]	78
13	Boc-Phe-His-ACHPA-Ile-$NHCH_2$-2-Pyr	0.026	0.83[b]	8.9[b]
14	POA-His-ACHPA-Ile-$NHCH_2\phi$	14	>1000[c]	>1000[c]
15	POA-His-ACHPA-Ile-$NHCH_2$-2-Pyr	21	120	>1000[c]
16	POA-His-ACHPA-Leu-$NHCH_2$-4-Pyr	4.1	130	94
17	POA-His-ACHPA-Leu-m-AMBA	6.1	33	850
18	POA-His-ACHPA-Leu-m-GMBA	3.1	48[b]	340

[a]**IHP**, isovaleryl-His-Pro-; **Sta**, Statine; **AHPPA**, (3S,4S)-4-amino-3-hydroxy-5-phenylpentanoic acid; **ACHPA**, (3S,4S)-4-amino-3-hydroxy-5-cyclohexylpentanoic acid; **Poc**, i-propyloxycarbonyl; **Etoc**, ethyloxycarbonyl; **POA**, phenoxyacetyl; **Pyr**, pyridyl; **m-AMBA**, m-aminomethylbenzylamide; **m-GMBA**, m-guanidinylmethylbenzylamide. Compounds characterized by TLC, HPLC, NMR, FAB-MS, and amino acid analysis. Human kidney renin, pH 7.2, 37°C. Plasma renins, 7.4, 37°C. [b]Average of multiple determinations. [c]Inhibition <40% at 1000 nM.

The extremely potent tetrapeptide renin inhibitors **12** and **13**, both containing the human-renin specific Ile-12 substitution, show a large disparity between purified human kidney renin (K_i) and human plasma renin (I_{50}) determinations, with **13** being one of the most potent human kidney renin competitive inhibitors yet described (K_i=2.6 x 10^{-11}M) yet of only nanomolar potency when measured in plasma, despite the solubilizing 2-pyridyl group. In tripeptide inhibitors using phenoxyacetyl as a Phe-8 replacement, replacement of C-terminal benzylamide, as in **14**, with 2- or 4-pyridylmethylamide, as in **15** and **16**, gives renin inhibitors of modest potency in both plasma and pure enzyme assays. Replacement of the pyridyl C-terminae with the more solubilizing m-aminomethyl or m-guanidinylmethyl groups, as in **17** and **18**, narrows the disparity somewhat, giving renin inhibitors of nanomolar potency versus human kidney renin which retain much of that potency in human plasma assays. The in vivo evaluation of these small renin inhibitors is a subject for further investigation.

References

1. Boger, J. (1985) Annu. Rep. Med. Chem., in press.
2. Boger, J., Lohr, N.S., Ulm, E.H., Poe, M., Blaine, E.H., Fanelli, G.M., Lin, T.-Y., Payne, L.S., Schorn, T.W., LaMont, B.I., Vassil, T.C., Stabilito, I.I., Veber, D.F., Rich, D.H., and Boparai, A.S. (1983) Nature **303**, 81-84.
3. Boger, J., Payne, L.S., Perlow, D.S., Lohr, N.S., Poe, M., Blaine, E.H., Ulm, E.H., Schorn, T.W., LaMont, B.I., Lin, T.-Y., Kawai, M., Rich, D.H., and Veber, D.F. (1985) J. Med. Chem., in press.
4. Bock, M.G., DiPardo, R.M., Evans, B.E., Freidinger, R.M., Whitter, W.W., Payne, L.S., Boger, J., Ulm, E.H., Blaine, E.H., Veber, D.F., and Poe, M., this volume.

RENIN INHIBITORS. SYNTHESIS AND BIOLOGICAL ACTIVITY OF STATINE- AND ACHPA-CONTAINING PEPTIDES HAVING POLAR END GROUPS

Mark G. Bock, Robert M. DiPardo, Ben E. Evans, Roger M. Freidinger, Willie L. Whitter, Linda S. Payne, Joshua Boger, Edgar H. Ulm, Edward H. Blaine, and Daniel F. Veber

Merck Sharp & Dohme Research Laboratories, West Point, PA 19486

Martin Poe

Merck Sharp & Dohme Research Laboratories, Rahway, NJ 07065

The design and synthesis of renin substrate analogs has led to peptidal renin inhibitors achieving potency levels in the nanomolar range.[1] Yet, despite these successes, important questions remain to be answered before medicinally useful agents result.

The high renin inhibitor potency of Boc-Phe-His-Sta-Leu-4-amido-1-benzylpiperidine, **2** (K_I (human kidney renin) 6.4 x 10^{-11}) synthesized previously in this laboratory, demonstrated the practicability of incorporating non-peptidal segments at the carboxy terminus of the pentapeptide renin inhibitor Boc-Phe-His-Sta-Leu-Phe-NH_2, **1**. Results from this study suggested that further modifications along these lines would provide an approach with which unresolved issues concerning the efficacy of peptide renin inhibitors, such as their solubility in physiological media and their duration of action, could be addressed.

Accordingly, using standard methodology, we prepared a series of tetrapeptides systematically substituted at the C-terminus with readily available polar end groups and containing the amino acids 4S-amino-3S-hydroxy-6-methylheptanoic acid (Sta) or 4S-amino-3S-hydroxy-5-cyclohexylpentanoic acid (ACHPA).[2] These are tabulated along with aqueous solubility measurements in Table I. Relative to our standards, **1** and **2**, we were successful in increasing the solubility of these potent renin inhibitors over a range of one hundred to more than one thousand fold. Importantly, this was accomplished without significant increase in molecular weight and/or peptide chain length.

Compounds **1-14** were tested *in vitro* and *in vivo* for their ability to inhibit porcine, canine, and human (plasma and purified kidney) renins (Table II). A considerable range of activities was observed among the compounds tested in the four assays. However, the renin inhibitors **3-14** containing polar end groups were uniformly more effective in inhibiting human renins than porcine or canine renins. Most importantly, the addition of polar end groups appears to be one solution to a problem encountered with small (<6 amino acids) peptidal renin inhibitors, e.g. **2**, containing hydrophobic C-termini.[3] The enhanced polarity at the C terminus of these compounds overcomes the loss of potency which is observed for the compounds lacking polar end groups[3] in the presence of plasma: e.g. compare **2** with **5**, **9**, **10**, **11**, and **12**. In all cases studied, substitution of the more lipophilic ACHPA for Sta led to enhanced potency. This trend prevailed even vs human plasma renin, despite the relatively lower solubility of the ACHPA-containing compounds. We conclude, therefore, that a direct relationship between *in vivo* potency and aqueous solubility does not exist with these compounds.

A further advance was realized with the synthesis of **8** and **9**. Both compounds displayed activity *in vivo* when administered i.v. to sodium deficient dogs. At a dose of 2.1 mg/kg for **8** and 2.2 mg/kg for **9**, plasma renin activity was suppressed for 6.9 hr and 5.7 hr, respectively.

Table I. Physical Properties of Statine- and ACHPA-Containing Tetrapeptide Substrate Analogs

Boc-L-Phe-L-His-X-L-Leu-R

Compd[a]	X	R	Solubility (μM)[b] pH 7.4 Buffer	H_2O
1	Sta	Phe-NH_2	<20	<20
2	Sta	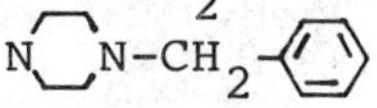	<10	<10
3	Sta	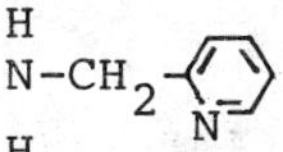	180	380
4	Sta	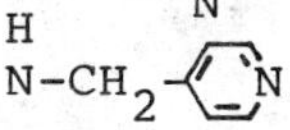	340	440
5	ACHPA	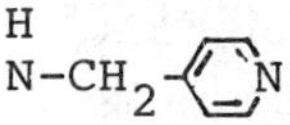	<10	—
6	Sta	N(H)-CH_2CH_2-NH_2	4,900	2,600
7	Sta	N(H)-CH_2-CH_2-N=C(NH_2)(NH_2) •HNO_3	11,000	24,000
8	Sta	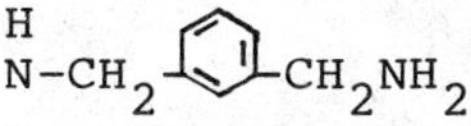	1,400	880
9	ACHPA	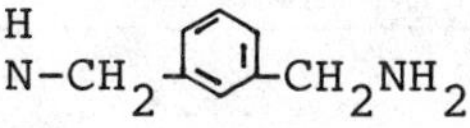	20	20
10	Sta	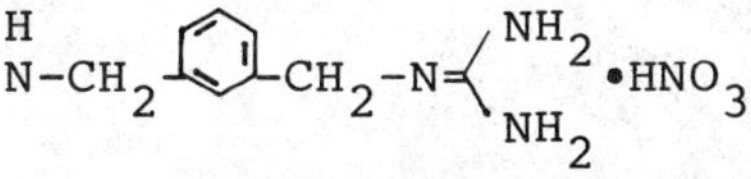	930	3,000
11	ACHPA	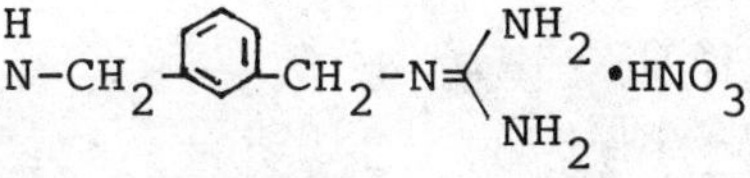	40	>250
12	Sta	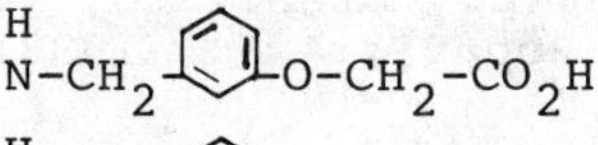	1,800	270
13	ACHPA	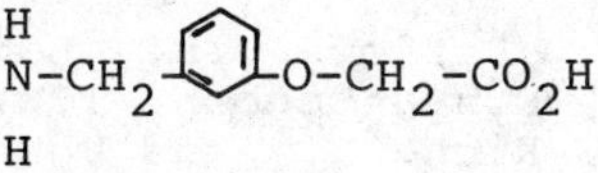	210	30
14	Sta	N(H)-CH_2CH_2-SO_3H	>11,000	>11,000

[a]Amino acid analyses were within 5% of the theoretical value.
[b]Single determination.

Table II. Inhibition of Renins by Statine- and ACHPA-Containing Tetrapeptide Substrate Analogs

Compd	Human Kidney K_I (M)	Human Plasma IC_{50} (M)	Dog Plasma IC_{50} (M)	Porcine Kidney IC_{50} (M)
1	$1.9x10^{-7}$	$1.4x10^{-7}$	$2.9x10^{-7}$	$6.3x10^{-8}$
2	$6.4x10^{-11}$	$2.5x10^{-8}$	$3.1x10^{-7}$	$2.8x10^{-7}$
3	$2.5x10^{-9}$	$5.0x10^{-8}$	$9.2x10^{-8}$	$2.4x10^{-7}$
4	$9.0x10^{-9}$	—	$2.0x10^{-8}$	$2.1x10^{-7}$
5	$9.6x10^{-10}$	$4.7x10^{-11}$	$9.9x10^{-10}$	$6.7x10^{-9}$
6	$1.5x10^{-8}$	$1.4x10^{-7}$	32% (10^{-6})	31% (10^{-6})
7	$4.7x10^{-9}$	$6.4x10^{-8}$	32% (10^{-6})	$7.1x10^{-7}$
8	$5.3x10^{-10}$	$9.0x10^{-9}$	$1.1x10^{-7}$	$5.4x10^{-8}$
9	$4.8x10^{-10}$	$7.6x10^{-11}$	$3.3x10^{-9}$	$8.3x10^{-9}$
10	$1.3x10^{-9}$	$9.0x10^{-9}$	$4.4x10^{-8}$	$\sim 2x10^{-8}$
11	$4.6x10^{-10}$	$1.0x10^{-10}$	$5.9x10^{-10}$	$9.0x10^{-9}$
12	$4.8x10^{-8}$	$6.8x10^{-9}$	$1.4x10^{-8}$	$4.3x10^{-8}$
13	$3.5x10^{-10}$	$3.8x10^{-9}$	$1.3x10^{-9}$	—
14	$7.6x10^{-10}$	$9.2x10^{-8}$	$3.5x10^{-7}$	0 (10^{-7})

References

1. Boger, J., Lohr, N.S., Ulm, E.H., Poe, M., Blaine, E.H., Fanelli, G.M., Lin, T.-Y., Payne, L.S., Schorn, T.W., LaMont, B.I., Vassil, T.C., Stabilito, I.I., Veber, D.F., Rich, D.H., and Bopari, A.S. (1983) Nature **303**, 81-84.

2. Boger, J., Payne, L.S., Perlow, D.S., Lohr, N.S., Poe, M., Blaine, E.H., Ulm, E.H., Schorn, T.W., LaMont, B.I., Lin, T.-Y., Kawai, M., Rich, D., and Veber, D.F. J. Med. Chem., in press.

3. Evans, B.E., Rittle, K.E., Bock, M.G., Bennett, C.D., DiPardo, R.M., Boger, J., Poe, M., Ulm, E.H., LaMont, B.I., Fanelli, G.M., and Veber, D.F., submitted.

RENIN INHIBITION BY LINEAR AND CONFORMATIONALLY RESTRICTED ANALOGS OF RENIN SUBSTRATE

Clovis R. Nakaie, Jorge L. Pesquero, Maria C. F. Oliveira, Luiz Juliano and Antonio C. M. Paiva

Department of Biophysics, Escola Paulista de Medicina, Caixa Postal 20.388, 04034 São Paulo, SP, Brazil

Introduction

Renin is a special aspartyl protease with neutral pH optimum and very restricted specificity for the Leu-Leu or Leu-Val bond at the 10-11 position of the substrate sequence. Studies of the secondary specificity, done with small peptide inhibitors, have demonstrated that an aromatic residue at subsite S_3 is essential for good recognition by renin, and that histidine at S_2 is not critical.[1,2]

We have previously hypothesized that a beta-turn involving residues 6-9 (His-Pro-Phe-His) of renin synthetic substrate[3] and of renin inhibitors containing this sequence[4,5] might be important for their binding to the enzyme. To further characterize the possible role of peptide chain conformation as a secondary requirement for binding of inhibitors of this class to renin, we have synthesized a series of linear and cyclic analogs of the angiotensinogen-(6-11) sequence and assayed them as potential renin inhibitors. The pK values of titratable groups of these peptides were also determined in order to obtain some information about the conformational state of these molecules. The peptide syntheses, renin assays and titrations were done as previously described.[3-5]

Results and Discussion

Compound 1 (Table I), containing the natural sequence of equine angiotensinogen-(6-11), had no renin inhibitory activity, but compound 2, in which the C-terminal Leu-Leu segment was replaced by Pro-Phe, was a competitive inhibitor of the enzyme. Compounds 3 and 4 are lower homologs of compound 2, in which the N-terminal His residue and His-Pro sequence were respectively absent. The K_i value for compound 3 is one order of magnitude higher than that for compound 2 and compound 4 was not an inhibitor. Since compound 5, in which

Table I. Inhibitory Constants and pK Values

Compounds		K_i (M)	pK values			
No.	Sequence	$x10^{-5}$	COOH	Imidazoles		Amino
	6 7 8 9 10 11					
1	His-Pro-Phe-His-Leu-Leu	ni	3.70	5.30	6.55	7.55
2*	His-Pro-Phe-His-Pro-Phe	4.9	3.50	5.60	6.95	8.05
3	Pro-Phe-His-Pro-Phe	14.8	3.62	-	6.59	8.70
4	Phe-His-Pro-Phe	ni	3.35	-	6.35	7.35
5	His-Pro-Phe-His-Pro	ni	3.45	5.30	6.98	8.18
6*	His-Pro-Pro-His-Pro-Phe	ni	3.18	5.38	6.79	8.30
7*	His-Pro-Phe-His-Pro-Phe-Hz	9.1	-	5.20	6.43	7.69
8	His-Pro-Phe-His-Pro-Phe	5.7	-	5.90	7.10	-
9	Acp-His-Pro-Phe-His-Pro-Phe	7.1	-	5.60	6.15	-
10	Cys-His-Pro-Phe-His-Cys-NH_2	0.7	-	5.50	6.25	7.55
11	Cys-His-Pro-Phe-His-Cys	ni	3.49	5.50	6.69	9.48

Acp, epsilon-aminocaproyl; Hz, hydrazide; ni, no inhibition
*Reference 5

the C-terminal Phe residue was absent, was also not an inhibitor, we conclude that Pro-Phe-His-Pro-Phe is the minimum sequence able to bind significantly to renin.

The pK values obtained for titratable groups in the linear peptides (Table I) indicate that in compound 2 the C-terminal carboxylate group is in closer proximity to the alpha-ammonium group and to the imidazolium group of His^6 than would be expected in a random conformation, since the pK values of the latter groups are ca. 0.3 to 0.4 units higher in compound 2 than in compound 7, in which the carboxyl group is blocked. A similar electrostatic effect, but only with the alpha-ammonium group, is observed with compounds 5 and 6, whereas no interaction between the carboxylate and either the ammonium or the imidazolium groups is seen in compound 1. These data suggest that compound 2 presents a bend that is different from those favoured in compounds 5 and 6. It is interesting to note that no inhibitory activity was found in compounds 5 and 6, as well as in compound 1, in which the interaction between carboxylate and ammonium or imidazolium groups was not detected by the titration results.

Our results indicate that the inhibitory peptides containing the His-Pro-Phe-His sequence bind to subsites S_2 - S_5 of renin´s active center as long as the residues in positions 10 and 11 are present to favour binding of inhibitor to enzyme. The finding that compound 2, but not compound 1, is an inhibitor, might be due to the Pro-Phe segment of the former introducing a bend in the molecule that would avoid unfavourable interactions with the enzyme. This hypothesis is supported by the observation that cyclic compound 8, in which two beta-turns are present,[6] and compound 9, were as efective inhibitors as the corresponding linear compound 2, indicating that the conformational constraint imposed by cyclization does not affect the binding ability of the molecule.

The stabilization of the beta-turn conformation of the

His-Pro-Phe-His segment by a disulfide bridge (compound 10) reduced the K_i value by one order of magnitude in relation to compound 2, compound 10 being the best inhibitor in the present series. The analog of this compound in which the C-terminal carboxyl group was free (compound 11) was not an inhibitor. We have also synthesized and assayed (Cys^5,Cys^{10})-angiotensin I, wich also had no inhibitory activity. This is probably related to an unfavourable interaction between the free carboxylate groups in the peptides and in the enzyme´s active center. The present results give further support to the idea that a beta-turn-like structure involving the His-Pro-Phe-His region of angiotensinogen, and of the competitive inhibitors containing this sequence, may be regarded as a possible "binding conformation".

This work was supported by grants from Fundação de Amparo à Pesquisa do Estado de São Paulo, Financiadora de Estudos e Projetos and Conselho Nacional de Pesquisas, Brazil.

References

1. Fehrentz, J.-A., Heitz, A., Castro, B., Cazanbon, C., and Nisato, D. (1984) FEBS-Letters 167, 273-276.

2. Kokubo, T., Hiwada, K., Murakami, E., Imamura, Y., Matsueda, R., Yabe, Y., Koike, H., and Iijima, Y. (1985) Hypertension, Suppl I 7, 8-11.

3. Oliveira, M. C. F., Juliano, L., and Paiva, A. C. M. (1977) Biochemistry 16, 2606-2611.

4. Nakaie, C. R., Oliveira, M. C. F. , Juliano, L., and Paiva, A. C. M. (1982) Biochem. J. 205, 43-47.

5. Nakaie, C. R., Oliveira, M. C. F., Juliano, L., Pesquero, J. L., and Paiva, A. C. M. (1983) Peptides. Structure and Function, Proc. 8th. Amer. Pep. Symp., Hruby, V., and Rich, D. H., eds. Pierce Chem. Co., Ill, pp. 595-598.

6. Liepina, I., Nikiforovich, G. V., and Paiva, A. C. M. (1984) Biochem. Biophys. Res. Comm. 122, 700-705.

RENIN INHIBITORS CONTAINING THE NOVEL AMINO-ACID 3-AMINO-DEOXYSTATINE

Michael Jones, Javier Sueiras-Diaz and Michael Szelke

Department of Chemical Pathology, Royal Postgraduate Medical School, London, W12 OHS, England

Brenda Leckie and Sheila Beattie

MRC Blood Pressure Unit, Western Infirmary, Glasgow, G11 6NT, Scotland

The pepstatins 1, isolated from actinomyces, are general inhibitors of aspartic proteinases. Their inhibitory activity is due to the unusual amino acid statine (Sta, 2a) which mimics the tetrahedral transition state 3 formed during amide hydrolysis[1]. Transition state theory predicts that the enzyme will have a greater binding affinity for the transition state 3 (or for stereochemically similar structures) than for the ground state of the substrate[2].

Potent inhibitors of renin have been synthesised in several laboratories[3-5] by incorporating Sta as a transition state analogue in place of the scissile dipeptide in the N-terminal sequence of angiotensinogen. We have shown previously that the hydroxy-ethylene moiety 4 which, unlike Sta, is isosteric with 3, provides even more potent renin inhibitors[6] and so far represents the closest approximation to the transition state. A pre-requisite of tight binding by Sta-containing inhibitors to pepsin is the S-configuration at the carbon bearing the hydroxyl[7]. Crystallographic studies with other aspartic proteinases indicate that the hydroxyl group of Sta displaces a water molecule present at the active site and then hydrogen bonds to the aspartic carboxyls[8].

R-Val-Val-Sta-Ala-Sta-OH 1

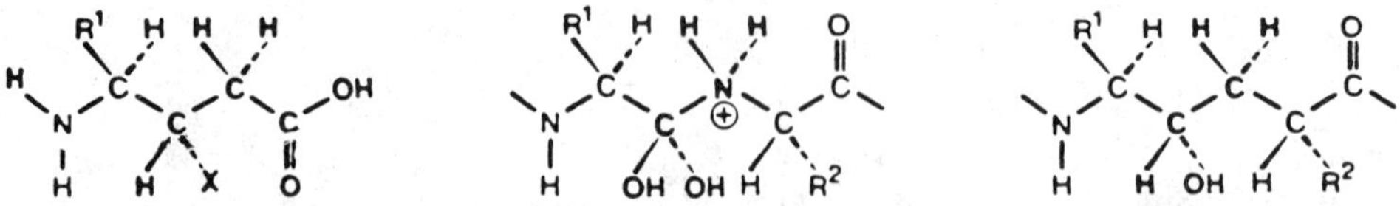

2a X = OH
2b X = NH_2

3

4

Replacement of the hydroxyl in Sta with a primary amino group yields the novel amino acid 3,4-diamino-6-methylheptanoic acid (3S-amino-deoxystatine Ads, 2b). Electrostatic interaction between the basic group of 2b and the aspartic carboxyls of the enzyme is expected to result in tighter binding. We present here the synthesis from L-leucine of the diastereomeric 3R,4S and 3S,4S-amino-deoxystatines (Scheme 1) and their comparison with Sta in the known renin inhibitory N-terminal sequence of human angiotensinogen (Table I). Configuration at C3 in 5 was determined by X-ray crystallography of the amide obtained by acylating R(+)-1-methyl-benzylamine with the faster moving (tlc, silica) diastereomer of BocAds(Z)OH. The synthesis (Scheme 1) was repeated with Boc-D-LeuOH to give 3R,4R- and 3S,4R-BocAds(Z)OH 5. After removal of the Boc-protecting group, the 3S,4S and 3R,4R isomers of 5 were derivatised

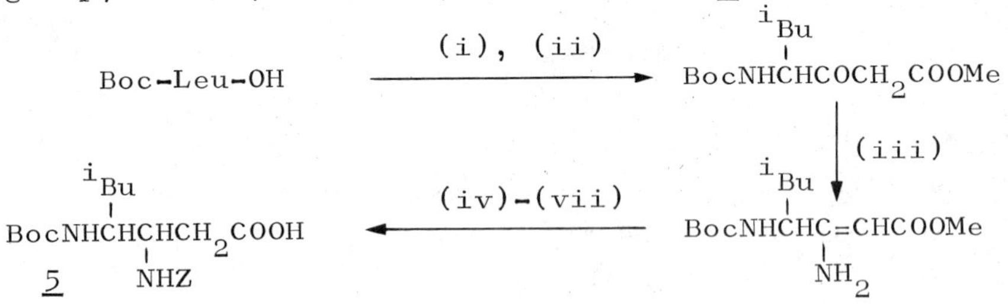

(i) $COIm_2$ (ii) $(MeOOCCH_2COO^-)_2Mg^{++}$ (iii) NH_4OAc-MeOH
(iv) $NaCNBH_3$ (v) ZONSu (vi) separation of diastereomers
(vii) NaOH-MeOH

Scheme 1. Synthesis of BocAds(Z)OH 5.

Table I. Inhibition of Plasma Renin[9] by Compounds 6 - 23

No.	Structure P_5 P_4 P_3 P_2 P_1 P'_1 P'_2 P'_3	IC_{50} (uM) human	IC_{50} (uM) rat
6	Boc-His-Pro-Phe-His--Sta-----Ile-His-NH_2	0.026	7.3
7	*RAds	0.028	0.26
8	*SAds	0.09	1.3
9	Sta-Val	0.007	0.5
10	RAds-Val	0.027	0.012
11	SAds-Val	0.22	0.42
12	Boc-Pro-Phe-His--Sta-----Ile-His-NH_2	1.6	>13.0
13	RAds	0.31	3.0
14	SAds	1.4	4.4
15	Sta-Val	0.61	7.2
16	RAds-Val	0.56	0.16
17	SAds-Val	3.5	3.7
18	Boc-Phe-His--Sta-----Ile-His-NH_2	0.32	110.0
19	RAds	0.1	1.0
20	SAds	0.44	4.9
21	Sta-Val	0.055	3.2
22	RAds-Val	0.11	0.11
23	SAds-Val	1.9	2.6

*R designates 3R,4S-Ads, S the 3S,4S-diastereomer

with 2,3,4,6-tetra-O-acetyl-β-D-glucopyranosyl-isothiocyanate. The resulting thioureas were readily separated by hplc and thus enabled us to determine that racemisation at C4 during the synthesis of 5 was less than 5%. Both the 3R, 4S and 3S,4S-isomers of 5 were incorporated into N-terminal fragments of human angiotensinogen by solid-phase synthesis.

Overall, Ads is a comparable replacement for 3S-Sta where inhibition of human renin is concerned, but is consistently superior to 3S-Sta against rat renin. Compound 10 is the most potent inhibitor of rat renin known at present. While inhibitors of pepsin containing 3R-Sta are several orders of magnitude less potent that the corresponding

3S-analogues, 3R,4S-Ads yields more potent inhibitors of both human and rat renin than the 3S,4S-epimer. In order to explore further the differences in binding of Sta and Ads, we are also investigating inhibitors with a 3R-Sta substitution. Although Sta has come to be accepted as isosteric replacement for a dipeptide unit rather than for a single amino acid residue[8], in our series it has proved to be more effective in the latter role: compounds 9, 15 and 21 are all significantly more potent than their counterparts 6, 12 and 18 lacking the P_1' residue.

We wish to thank Ferring Pharmaceuticals Ltd., England and Ferring AB, Sweden for financial support, Mrs. P. A. Windebank for amino acid analyses and Dr. S. Neidle for the X-ray crystallographic determinations.

References

1. Marciniszyn, J.,Jr., Hartsuck, J.A. and Tang, J. (1976) J. Biol. Chem. 251, 7088-7094.

2. Lienhard, G.E. (1973) Science 180, 149-154.

3. Boger, J. (1983) in Peptides: Structure and Function, Hruby, V.J. and Rich, D.H., Eds., Pierce Chemical Co., Rockford, IL. pp.569-578.

4. Evin, G., Castro, B., Devin, J., Menard, J., Corvol, P. Guegan, R., Diaz, J., Damarne, H., Cazaubon, C., and Gagnol, J.-P. (1983) in Peptides: Structure and Function, Hruby, V.J. and Rich, D.H., Eds., Pierce Chemical Co., Rockford, IL. pp.583-586.

5. Hofbauer, K.G. and Wood, J.M. (1985) Trends in Pharmacol. Sci., 6, 173-177.

6. Szelke, M., Jones, D.M., Atrash, B., Hallett, A. and Leckie, B.J. (1983) in Peptides: Structure and Function Hruby, V.J. and Rich, D.H., Eds., Pierce Chemical Co., Rockford, IL. pp.579-582.

7. Rich, D.H., Sun, E.T.O. and Ulm, E. (1980) J. Med. Chem. 23, 27-33.

8. Rich, D.H. (1985) J. Med. Chem. 28, 263-273.

9. Millar, J.A., Leckie, B.J., Morton, J.J., Jordan, J. and Tree, M. (1980) Clin. Chim. Acta 101, 5-15.

CRYSTALLOGRAPHIC ANALYSIS OF A PEPSTATIN ANALOGUE : BOC"LEU=LEU"-STA-ALA-STA-OMe

Gilles Précigoux, Elisabeth Ouvrard and Serge Geoffre

Laboratoire de Cristallographie, associé au CNRS, Université de Bordeaux I,
351, Cours de la Libération, 33405 Talence Cedex (France)

Introduction

X-ray diffraction investigations of renin substrates and renin inhibitors can provide accurate informations about molecular conformations and molecular interactions. However they are static informations observed in a specific surroundings and in the solid state. Due to the high flexibility of linear oligopeptides in solution and to the lack of conformation-affinity relationships studies, in the peptidic field, the greatest care must be taken in using these results. In the present study, the validity of the observed conformations in the crystals, as three-dimensional basis for substrate binding, can be partly tested by comparison with results from crystallographic analysis of pepstatin (or analogue) binding to aspartyl proteinases[1,2] and from computer graphics modelling of human renin[3].

The paper reports the X-ray structure of a pepstatin analogue Boc - "Leu-Leu"-Sta-Ala-Sta-OMe, a pentapeptide with a modified Leu-Leu peptide bond (Figure 1).

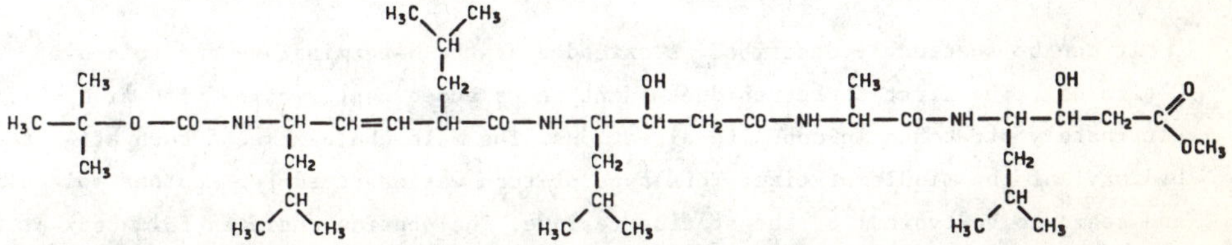

Fig.1 : Structure of the title compound.

The conformation is discussed in terms of other conformations observed for substrate analogue (Leu-Leu-Val-Tyr-OMe[4]) and for pepstatin analogue binding to penicillopepsin.

Results and Discussion

The molecule crystallizes as monohydrate in the orthorhombic space group $P2_12_12_1$ with a = 11.399(4), b = 15.073(3), c = 26.893(5) Å and Z = 4. The structure was solved by means of the direct methods program MULTAN[5] and refined to a residual of 0.06. Because of the limited number of observed data (1674) it was not possible to refined hydrogen atoms parameters. The conformation of the peptide molecule is shown in Figure 2.

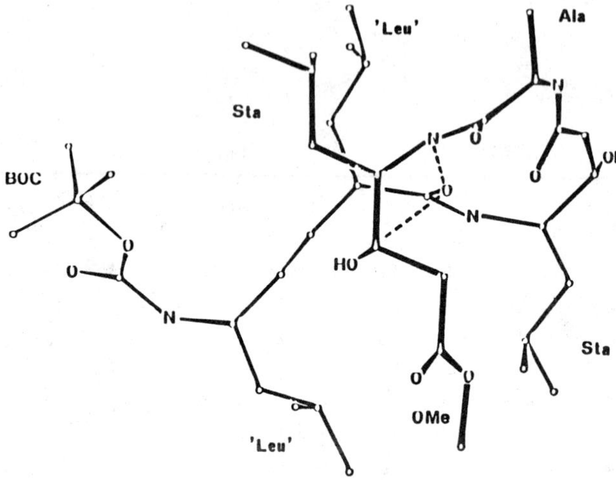

Fig.2 : Observed X-ray conformation for Boc-"Leu=Leu"-Sta-Ala-Sta-OMe

It can be succinctly described as extended at the N-terminal end and folded at the C-terminal. The first three residues adopt an extended conformation with side chains alternately stretched in opposite directions. The main chain makes a turn after the hydroxyl of the middle statine. This type of turn was observed for another molecule[6] and seems to be typical of the statine residue. The peptide chain is fold back at the Sta and Ala residues to form a twelve membered ring with an intramolecular hydrogen bond between the carbonyl of "Leu" and the nitrogen atom of the last statine residue (2.98 Å).For the present crystal structure, a second hydrogen bond is observed between the carbonyl of "Leu" and the last hydroxyl oxygen (2.77 Å). By analogy with the classical O_i-N_{i+3} turn, such a Sta-turn is of type I.

In figure 3, the molecule is compared with a pepstatin analogue, Iva-Val-Val-Sta-OEt in the conformation binding to the penicillopepsin and with a substrate analogue, Leu-Leu-Val-Tyr-OMe.

The comparison between the pepstatin analogues shows very similar conformations. The first four residues (residues-like) fill the same regions and the first hydroxyl

Boc-"Leu=Leu"-Sta-Ala-Sta-OMe

Leu-Leu-Val-Tyr-OMe[4]

Iva-Val-Val-Sta-OEt[1]

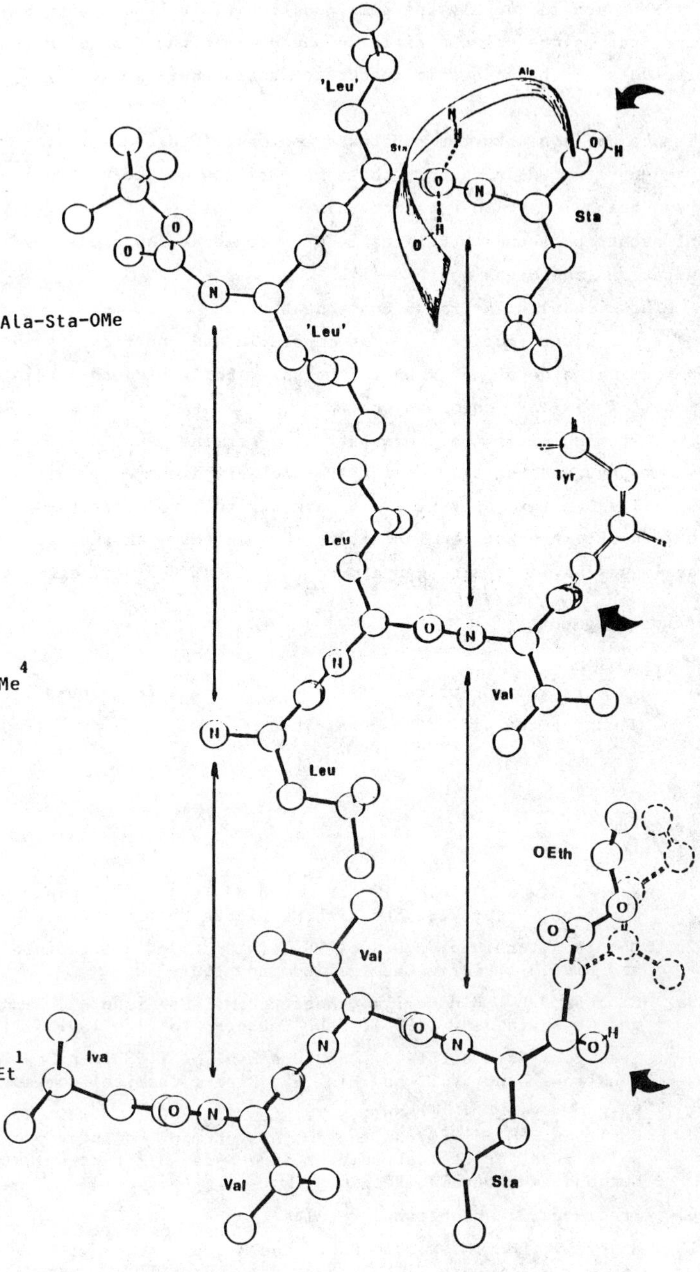

Fig.3 : Conformations of pepstatin analogues and substrate analogue viewed perpendicular to the mean plane of the first three residues.

oxygens are in roughly the same position. This result suggests that the "Leu","Leu" and Sta residues of the title compound mimic the binding of the Phe, His and Leu residues of the endogenic substrate (angiotensinogen 8-10).

The comparison with the substrate anologue (Leu-Leu-Val-Tyr-OMe) shows the hydroxyl oxygen situated in roughly the same position as the carbonyl oxygen of a peptide bond belonging to an extended main chain.This point leads to the hypothetical model of substrate bound to the enzyme with at least four residues in an extended configuration (angiotensinogen 8-11).

These results incite to conclude that :

- the first four residues of pepstatin and the first residues situated on the N-terminal side of the substrate sissile bond, bind the renin with minimum adjustements of similar conformations.
- The inhibitor and the substrate C-terminal ends do not bind the renin in a similar way, and the side chains are probably directed towards different pockets.
- The Sta-turn of type I may have the role to correctly position the hydroxyl oxygen of the pepstatin inhibitor. This agrees with the fact that the inhibitor analogues having a nitrogen atom N_{i+3} present the best affinity to the renin[7,8].

Acknowledgments

This work was supported in part by a grant from Roussel-Uclaf. We are grateful to Dr Gasc for the gift of the samples.

References

1. James, M.N.G., Sielecki, A.R., and Moult,J. (1983) in Peptides : Structure and function, Hruby,V.J. and Rich,H. Pierce Chemical Company, 521-530.

2. Bott,R.R. and Davies,D.R. (1983) in Peptides : Structure and function, Hruby,V.J. and Rich,H., Pierce Chemical Company, 531-540.

3. Sibanda,B.L., Blundell,T., Hobart,P.M., Fogliano,M., Bindra,J.S., Dominy,B.W. and Chirgwin,J.M. (1984). Febs Letters, Vol.174.1, 103-111.

4. Precigoux,G., Geoffre,S., and Hospital,M. (1983) in Peptides. Structure and function, Hruby,V.T. and Rich,H., Pierce Chemical Company, 599-602.

5. Main,P., Fiske,S.J., Hull,S.E., Lessinger,L., Germain,G., Declercq,J.P. and Woolfson,M.M. (1980) Multan 80. A system of computer programs for the automatic solution of crystal structures from X-ray Diffraction Data.Univs. of York, England and Louvain, Belgique.

6. Precigoux,G. Unpublished results.

7. Boger,J., Lohr,N.S., Ulm,E.H., Poe,M., Blaine,E.H., Fanelli,G.M., Lin,T-Y., Payne,L.S., Schorn,T.W., Lamont,B.I., Vassil,T.C., Stabilito,I.I., Veber,D.F. Rich,D.H., and Bopari,A.S. (1983), Nature, Vol.303,81-84.

8. Rich,D.H. and Sun,E.T.O. (1980) J.Med.Chem. 23,27-33.

TRANSPORT OF AN INTACT NONAPEPTIDE RENIN INHIBITOR ACROSS ADULT JEJUNUM

Kazuo Takaori

Laboratory of Cellular & Molecular Research, Massachusetts General Hospital, Boston, Massachusetts 02114

Mark Donowitz

Department of Medicine, New England Medical Center, Boston, Massachusetts 02111

James Burton

Department of Medicine, University Hospital, Boston, Massachusetts 02118

Introduction

The absorption of intact di- and tripeptides from adult intestine has been repeatedly demonstrated; little however, is known about the uptake of larger peptides. Studies in which peptides administered orally produce the expected physiologic response imply absorption of intact material[1]. The lack of any quantitative measure of this process or correlation with the physical properties of the peptide prevents the rational design of orally active compounds. In the present study a lipophilic nonapeptide renin inhibitor was shown to cross adult rabbit jejunum in intact form at an appreciable rate.

Materials and Methods

Labeled RI-61, Pro-His-Pro-Phe-His-Leu-Phe-Val-[^{3}H]Phe, was prepared using reactions outlined in Figure 1.

$[^3H]$Phe (10 mCi) —$(Boc)_2O$→ Boc-$[^3H]$Phe —$CsHCO_3$→

Boc-$[^3H]$PheO$^-$Cs$^+$ —$ClCH_2Pl$→ Boc-$[^3H]$Phe-OPl (250 mg)

—2. HF, 3. HPLC→ Pro-His-Pro-Phe-His-Leu-Phe-Val-$[^3H]$Phe (1.7 mCi, 17%)

Figure 1. Synthesis of labeled RI-61 (100 Ci Mol^{-1})

All potetial labeled proteolytes were synthesized in 50 μM quantities by solid phase peptide synthesis, purified to homogeneity by HPLC and fully characterized.

An HPLC system for separation of intact nonapeptide from potential cleavage products was developed (Figure 2) using previously described apparatus[2].

For transport studies, the serosa and muscularis propria were removed from adult rabbit jejunum which was then mounted in an Ussing chamber in a non-short-circuited state, maintained at 37° and gassed with 95% O_2-5% CO_2. Labeled RI-61 (50 M, 100 Ci/Mol) was added to the mucosal chamber and 1.0 mL aliquots of withdrawn from the mucosal or serosal chamber at various times. 50 μL aliquots were counted and the remainder of the sample lyophilized. A mixture of unlabeled proteolytes in water-0.2% CF_3COOH was added to the residue and this solution fractionated by HPLC.

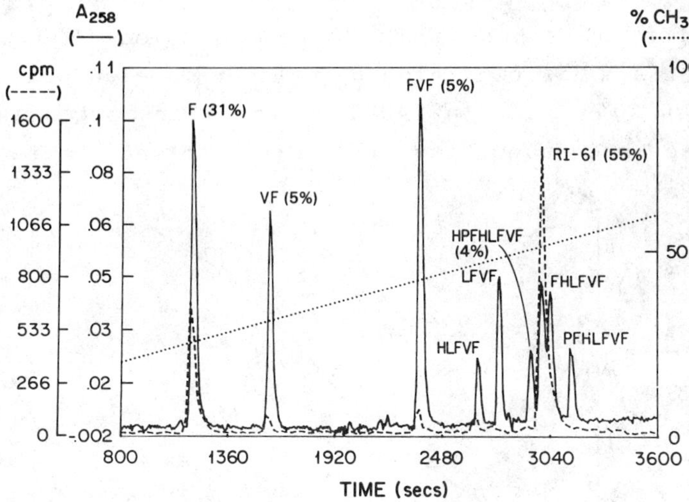

Figure 2. HPLC separation of RI-61 proteolytes after exposure of labeled peptide to the mucosal surface of rabbit jejunum for 5 minutes in the absence of inhibitors.

Results and Discussion

[^{3}H]RI-61 (50 μM) was added to the mucosal bath and samples taken periodically for analysis by HPLC. Degradation of RI-61 occurred rapidly; 45% of the RI-61 was intact at 5 minutes while none remained after 30 minutes. Phenylalanine was the major catabolite indicating that most degradation occurred at the C-terminus. Addition of phosphoramidon (30 μM) and pepstatylglutamic acid (10 μM) slowed breakdown of RI-61; 15% of the starting material remaining in the mucosal bath after 30 minutes. As shown in Figure 3a label appeared in the serosal bath at a constant rate after 15 minutes. More than 90% of the transported material was intact nonapeptide (Fig. 3b). The rate of transport of RI-61 (215 $\pm$ 76 pMol cm^{-2} hr^{-1}) was 4.1 % that observed for the rate of transport of 3-O-methyl glucose (5.2 $\pm$ 0.6 μMol cm^{-2} hr^{-1}). In the absence of sodium passage of the peptide from the mucosal to serosal baths remained constant (246 pMol cm^{-2} hr^{-1}) and is 31% of that observed for transport of the glucose analog (800 pMol cm^{-2} hr^{-1}) under similar conditions. Neither labeled polyethyleneglycol (900 D) or inulin (4000 D) are transported indicating that passage of the peptide (1146 D) is not an artifact.

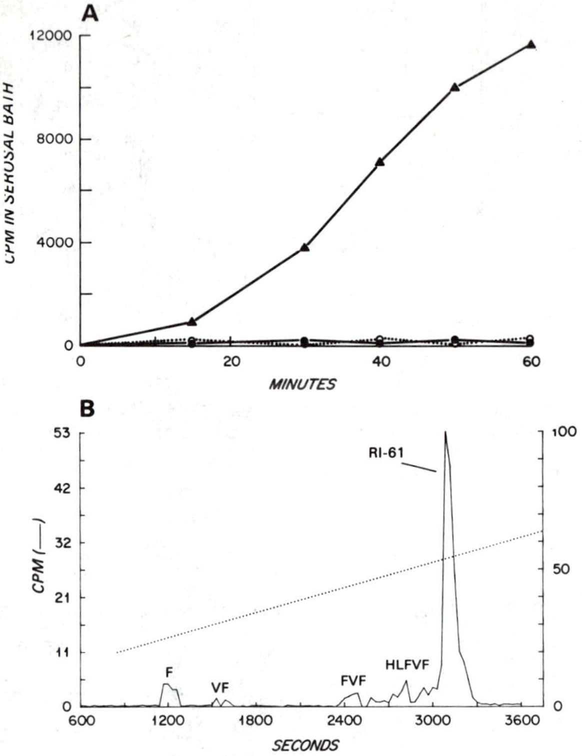

Figure 3. (a) Transport of RI-61 (▲——▲), polyethyleneglycol (●——●) and inulin (○——○); (b) Identification of labeled material in the mucosal bath as intact RI-61.

References

1. Gardner, M.L.G. (1983) Biochem. Soc. Trans. 11, 810-813.

2. Soparkar, C.N.S., Malison, R.T., Burton, J (1985) J. Chrom. 337, 110-114.

STATINE-CONTAINING RENIN INHIBITORY PEPTIDES: HEMODYNAMIC EFFECTS IN THE PRIMATE

Kwan Y. Hui, Michael S. Bernatowicz, Randall M. Zusman, William Carlson and Edgar Haber
Massachusetts General Hospital, Boston, MA 02114

L. Howard Hartley
New England Regional Primate Research Center, Southborough, MA 01772

Compounds that could selectively inhibit renin, and be used in clinical investigation, would be very valuable in defining the role of the renin-angiotensin system in the genesis of essential hypertension. Utilizing both structure-activity relationships of published compounds and molecular modelling to direct our work, we have designed and synthesized a number of statine-containing substrate analogs with binding constants in the same range as those reported by Boger et al.[1] and Szelke et al.[2]

Table I. Statine-containing Renin Inhibitory Peptides

Peptide	Sequence	IC50 (10^{-9}M)
R-Pep-19	His-Pro-Phe-His-Statine-Val-Tyr-Lys	400
R-Pep-20	Pro-His-Pro-Phe-His-Statine-Val-Tyr-Lys	10
R-Pep-21	Pro-His-Pro-Phe-His-Statine-Ile-His-Lys	8
R-Pep-24	Ac-Pro-His-Pro-Phe-His-Statine-Ile-His-Lys	20
R-Pep-25	Pro-Phe-His-Statine-Ile-His-Lys	900
R-Pep-26	Ac-Pro-Phe-His-Statine-Ile-His-Lys	40
R-Pep-27	Pro-His-Pro-Phe-His-Statine-Ile-Phe-amide	2
R-Pep-28	Pro-Phe-His-Statine-Ile-Phe-amide	220
R-Pep-29	Pro-His-Pro-Phe-Phe-Statine-Ile-Phe-amide	30
R-Pep-30	Pro-Phe-Phe-Statine-Ile-Phe-amide	350

The statine-containing peptides were synthesized by the Merrifield solid phase method. Their sequences and in vitro IC50 values are listed in Table I. IC50 values were determined by incubating high-renin human plasma in the presence of a series of concentrations of peptide at pH 7.4, and then by measuring the rate of AI generation by radioimmunoassay.

Dog, rabbit and rat were also examined with the three most potent analogs, R-Pep-20, -21 and -27, to define species specificity. Results are presented in Table II.

Table II. Species Specificity of Renin Inhibitory Peptides

	IC50 (M)			
Peptide	Human	Dog	Rabbit	Rat
R-Pep-20	1×10^{-8}	4×10^{-8}	1×10^{-7}	1×10^{-5}
R-Pep-21	8×10^{-9}	5×10^{-7}	2×10^{-6}	$> 10^{-4}$
R-Pep-27	2×10^{-9}	5×10^{-8}	2×10^{-7}	$> 10^{-5}$

IC50 values were determined as above.

R-Pep-21 is highly species specific, a similarity that it shares with the peptide H.142 described by Tree et al.[3] The close homology that these peptides bear to human substrate probably accounts for this result.

R-Pep-27 inhibits human renin most potently. This peptide was studied in conscious adult primates (Macaca fascicularis) chronically instrumented with arterial and venous catheters. The animals were either sodium-replete or sodium-depleted with a low salt diet followed by an intramuscular injection of furosemide. The peptide was infused in dosages of 1 to 16 ug/kg/min over 10 min periods. Electrically integrated mean arterial pressure was continuously recorded. Venous blood was withdrawn occasionally to assay plasma renin activity. Following the R-Pep-27 studies, Captopril (400 ug/kg) was administered IV as a bolus injection. Table III summarizes the effects of these agents on heart rate, mean arterial pressure (MAP) and plasma renin activity (PRA).

Table III. Effects of R-Pep-27 and Captopril on Hemodynamics and Plasma Renin Activity in the Primate

Na Balance	Drug	Dose ug/kg/ min	Change in Heart Rate	Change in MAP (mm Hg)	Change in PRA (ng/ml/h)
Replete	R-Pep-27	16		$+1.8 \pm 6.9$	-1.9 ± 1.8
Depleted	R-Pep-27	1	$+2 \pm 17$	-5 ± 4.6	
	"	4	-4 ± 20	-14.5 ± 7.0	
	"	16	-4 ± 17[a]	-14.2 ± 5.1[b]	-22.8 ± 15.4[c]
	Captopril	400*	-10 ± 25	-16.8 ± 10[d]	
	*(ug/kg IV bolus)				

Normal primate prior to study: Heart rate = 185; MAP = 89.7 ± 17.6; PRA = 2.26 ± 1.8

Na depleted primate prior to study: Heart rate = 192 ± 18; MAP = 98.2 ± 16.1; PRA = 33.1 ± 23.5[e]

[a] ns in relation to controls prior to drug administration at the same Na state

[b] $p < 0.003$ in relation to Na replete controls after drug administration

[c] correlation of change in MAP with change in PRA, $r = 0.63$; $n = 8$; $p < 0.05$

[d] ns in relation to administration of R-Pep-27 at 16 ug/kg/min

[e] $p < 0.04$ in relation to Na replete controls prior to drug administration

Thus R-Pep-27 is a potent hypotensive agent in the Na-depleted, but not Na-replete, primate; the effective dose being as low as 1 ug/kg/min. At a dose of 4 ug/kg/min, the hypotensive effect was comparable to that of Captopril given at 400 ug/kg as a bolus injection. No further hypotension occurs at doses up to 100 times the minimal hypotensive dose (data not shown). There was no significant effect on the heart rate at any of the dosages studied. As anticipated, plasma renin activity decreased in parallel with blood pressure. In the same species, the effective hypotensive dose of R-Pep-27 is approximately 0.001 times that of RIP,[4] though the marked hypotension seen at higher doses of RIP was not observed. While not compared in the same species, R-Pep-27 appears to be more potent than H.142, H.189[3] and SCRIP.[5] It is hoped that further study of R-Pep-27 will show it to be a selective agent for the study of the renin-angiotensin system.

We gratefully acknowledge financial support from The Upjohn Company and Reynolds Industries, Inc.

References

1. Boger, J., Lohr, N., Ulm, E.H., Poe, M., Blaine, E.H., Fanelli, G.M., Lim, T.Y., Payne, L.S., Schom, T.W., LaMont, B.I., Vassil, T.C., Stabilto, I.I., Veber, D.F., Rich, D.H. and Bopari, A.S. (1983) Nature 303, 81-84.

2. Szelke, M., Leckie, B., Hallett, A., Jones, D.M., Sueiras, J., Atrash, B. and Lever, A.F. (1982) Nature 299, 555-557.

3. Tree, M., Atrash, B., Donovan, B., Gamble, J., Hallett, A., Hughes, M., Jones, D.M., Leckie, B., Lever, A.F., Morton, J.J. and Szelke, M. (1983) J. Hypertens. 1, 399-403.

4. Burton, J., Cody, R.J., Herd, J.A. and Haber, E. (1980) Proc. Natl. Acad. Sci. USA 77, 5476-5479.

5. Blaine, E.H., Schorn, T.W. and Boger, J. (1984) Hypertension (Suppl I) 6, I111- I118.

MAPPING OF THE HUMAN RENIN "PRO" SEGMENT. COMPARISON OF IMMUNOGENIC AND INHIBITORY PROPERTIES OF PEPTIDES FRAGMENTS

Jacob Bouhnik, Frédéric Cumin, François-Xavier Galen, Geneviève Evin, Joel Ménard and Pierre Corvol

INSERM U36, 17 rue du Fer-à-Moulin, 75005 Paris, France

Isabelle Gaillard, Dominique Simon, Bernard Pau, Dino Nisato and Pierre Gagnol

Centre de Recherches Clin-Midy-SANOFI, rue du Professeur Blayac, 34082 Montpellier, France

Jean-Alain Fehrentz, René Seyer and Bertrand Castro

Centre CNRS-INSERM de Pharmacologie-Endocrinologie, rue de la Cardonille, BP 5055, 34033 Montpellier, France

Recently, human inactive renin[1] has been proved to be identical to the active renin biochemical precursor (prorenin[2]) ; these studies were based on the use of antisera directed against prosegments fragments[3 4]. Here we present some additional direct evidences using monoclonal antibodies and an attempt of rationalization of the roles played by the different regions of the prosegment.

The primary structure of the prosegment[5] is given in Fig. 1. Five peptides corresponding to different regions of the prorenin prosegment have been synthetized with the exception of the peptides located either at the N-terminal [1-8] or the C-terminal [41-46] ends : Y[9-15], Y[9-20], [16-20], Y[21-30] and Y[28-40]. The syntheses were performed by solution techniques and will be detailed elsewhere.

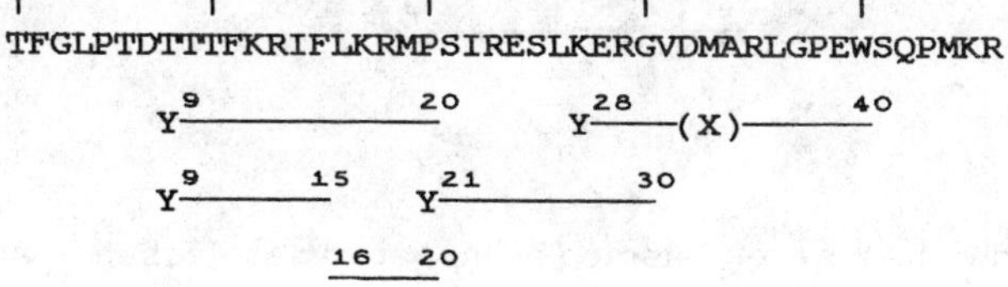

Figure 1

These synthetic peptides made it possible for us to study and to compare the relative inhibitory potency and immunological properties of the different portions of the prosegment. Some of the synthetized peptides bear an extra tyrosine at the N-terminus for radiochemical detection convenience ; Met_{33} was replaced with a more handy norleucyl residue.

The five peptides were tested for human plasma renin inhibition ; two of them, Y[21-30] and Y[28-40], were coupled to proteic carriers in order to get antigens for immunisations ; injections to rabbits were used to produce polyclonal antisera ; mice were used to rise monoclonal antibodies.

Experimental.

H. R. Biozzi mice were immunized with both BSA-supported Y[21-30] and Y[28-40] ; antibodies appearance were monitored with radiolabeled haptens. After three booster injections, spleen cells from a selected mouse were fused with P3-X63-Ag8-653 myeloma cells as described by Di Pauli *et al.*[6]. The production of antibodies in the supernatant of hybridoma cultures was detected by their ability to bind radiolabeled haptens. 12 hybridoma against Y[21-30] and one against Y[28-40] were obtained.

These monoclonal antibodies were assayed against inactive renin (from a chorion culture) and trypsin activated inactive renin using a sandwich technique (Fig. 2). $F(ab')_2$ fragment from F55 3E8 anti-renin antibody[7], binding inactive renin as well as active renin was adsorbed on assay wells, incubated with inactive or active renin and then with the tested monoclonal antibody.

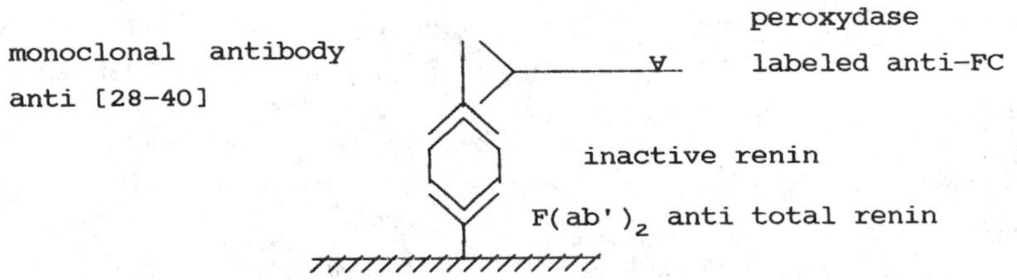

Figure 2

Successful sandwiches were detected with peroxydase labeled anti-mouse IgG-FC antibodies. The only monoclonal antibody against Y[28-40] recognized inactive renin, thus none out of six monoclonal antibodies against Y[21-30] recognized it. None of them recognized trypsin activated renin.

Human high renin plasma from selected hypertensive patients was incubated with or without peptides at various concentrations (10^{-6} to 10^{-3}) at pH 5.7 ; phenylmethylsulfonyl fluoride was used in order to prevent angiotensin I to angiotensin II conversion. The reaction was stopped by chilling and angiotensin I was measured by radioimmunoassay as previously described[8]. In order to provide a sound comparison between the peptides and owing to the weak inhibitory potency of the synthetic peptides, the results were obtained as inhibition ratios for the same concentration of inhibitors (350 μM). The results are summarized in the Table.

Peptides	% Inhibition of PRA at 350 μM	Monoclonal antibodies recognition of	
		Renin	Prorenin
Y[9-15]	11	ND	ND
Y[9-20]	57	ND	ND
[16-20]	65	ND	ND
Y[21-30]	13	0	0
Y[28-40]	27	0	+

Discussion.

The results obtained with the monoclonal antibody against Y[28-40] provide a new, direct, strong evidence for the exact likeness between inactive renin and at least a part of the biosynthetic precursor prorenin. On the other hand, the inhibition data give an insight on the possible affinity of the different peptides for a region close to the active site, the cleft or the flap of the enzyme. (It is to be noticed that human active renin inhibition by the [16-20] peptide is about 1/10 th of that obtained on mouse submaxillary renin with the homologous fragment[9]; this difference was reduced to 1/6 in an inhibition assay using pure renin and pure angiotensinogen for both species instead of plasmas[10]).

These results support a tentative modelisation of the tertiary structure of the prosegment. It can be suggested that the [28-40] C-terminal part of the prosegment lies at a well exposed position at the surface of the protein. The following [21-30] intermediate section of the prosegment is "neutral" and plays neither inhibiting nor immunoreactive role. It could be seen as a connecting section, slightly buried either under the protein surface or in a furrow. This part would not interact with either the cleft or the active site. Finally the [9-20] and especially the [16–20] section corresponds to the more affine peptide and probably deeply dives into the catalytic cleft insuring the inactivation of the proenzyme at neutral pH. We suggest that reversible activation of inactive renin[9] at low pH could be explained by an unfolding of the prosegment due to the breaking of the elctrostatic interaction between the positively charged protonated Lys_{17} and Arg_{18} residues and some carboxylate residues, loosing their negative charge by protonation into the active site.

References

1. Acker, G.M., Galen, F.-X., Devaux, C., Foote, S., Papernik, E., Pesty, A., Ménard, J. and Corvol, P. (1982) J. Clin. Endocrinol. Metab., 55, 902.

2. Sealey, J.E., Atlas, S.A. and Laragh, J.H. (1983) Federation Proceedings, 42, 2681.

3. Bouhnik, J., Fehrentz, J.A., Galen, F.-X., Seyer, R., Evin, G., Castro, B., Ménard, J., and Corvol, P. (1985) J. Clin. Endocrinol. Metab., 60, 399.

4. Kim, S.-J., Hirose, S., Miyazaki, H., Ueno, N., Higashimori, K., Morigana, S., Kimura, T., Sakakibara, S., and Murakami, K. (1985) Biochem. Biophys. Res. Comm. 126, 641.

5. Imai, T., Miyazaki, H., Hirose, S., Hori, H., Hayashi, T., Kageyama, R., Ohkubo, M, Nakanishi, S., and Murakami, K. (1983) Proc. Natl. Acad. Sci. USA 80, 7405.

6. Di Pauli, R., and Raschke, W.C. (1978) in Current Topics in Microbiology and Immunology, Melchers, F., Potte, M., and Warner, N.L., editors, Springer-Verlag, Berlin, Vol. 81 p 37-39.

7. Galen, F.-X., Devaux, C., Atlas, S., Guyenne, T., Ménard, J., Corvol, P., Simon, D., Cazaubon, C., Richer, P., Badouaille, G., Richaud, J.P., Gros, P., and Pau, B. (1984) J. Clin. Invest. 74, 723.

8. Ménard, J., and Catt, K.J. (1972) Endocrinology 90, 389.

9. Evin, G., Devin, J., Castro, B., Ménard, J., and Corvol, P. (1984) Proc. Natl. Acad. Sci. USA, 81, 48.

10. Cumin, F., Evin, G., Fehrentz, J.-A., Seyer, R., Castro, B., Ménard, J., and Corvol, P. (1985) under press.

SUBSTRATES FOR THE ASPARTYL PROTEINASES

Ben M. Dunn, Melba Jimenez, Benne Parten, and Jeffrey R. Weidner

Department of Biochemistry and Molecular Biology,
University of Florida College of Medicine,
Box J-245, Gainesville, Florida 32610

John Kay, Martin Valler, and Carole Rolph,

Department of Biochemistry, University College,
Cardiff CF1 1XL, Wales, U.K.

Introduction

Recent studies of the biochemistry, structure, and function of the Aspartyl Proteinase family of enzymes have emphasized the similarities in this group of proteins[1,2]. However, the various members of this class of proteinase are designed to operate at slightly different pH optima and cover a range of specificities from very broad for the gastric digestive enzymes such as the pepsins to the exquisitely specific in the case of renin. We have designed a series of new peptide substrates to probe the nature of binding interactions at the active sites of members of this family[3,4]. These peptides are large enough to fill up the active site cleft and, therefore, bind in a unique productive orientation. The peptide bond of the substrates cleaved by these enzymes is in all cases a -Phe-(NO_2)Phe- bond, where (NO_2)Phe represents p-nitro-L-phenylalanine. We have prepared derivatives of an octapeptide sequence containing the sensitive bond plus variation in the residues that flank that locus[2,4]. This report presents new data on hydrolysis of these substrates by a series of Aspartyl Proteinases.

Experimental

Peptides were synthesized by solid phase methods using the DCC-HOBT method[5]. Cleavage was done with anhydrous HF and the peptides were extracted with 50% acetic acid/water. Purification included an ion exchange step on CM-Sepharose and HPLC on microbondapak C_{18} columns. In some cases, HPLC of peaks from ion exchange demonstrated homogeneity but normally, further purification by HPLC was required. In all cases, the final product gave an amino acid composition as expected.

Enzyme kinetics were performed as previously described[3] with the decrease in absorbance at 300 nm followed to obtain initial rates of cleavage. In all cases, a sample of substrate was hydrolyzed to completion and the peptide products isolated by HPLC, hydrolyzed with 6 M HCl and analyzed. These experiments demonstrated that cleavage was between the Phe and (NO_2)Phe bond.

Results and Discussion

We previously reported[4] results obtained for the pig pepsin catalyzed hydrolysis of six peptide substrates in the series represented by I, where P3 represents a variable amino acid occupy-

I Lys-Pro-P3-Glu-Phe-(NO_2)Phe-Arg-Leu

ing that position. In Table I we present results for six new peptide substrates with two further variations in the P3 position and with five variations in the P2 position originally occupied by Glu in structure I.

In this series, the highest kcat is 122 sec^{-1}. The Km value observed for P3=Val and P2=Asn of 0.014 μM is among the lowest observed for pig pepsin. This suggests that we may have optimized binding interactions at those sites. This may, however, contribute to the lower value of kcat observed for that substrate, since

Table I. Kinetic Parameters for Pig Pepsin Catalyzed Hydrolysis of Substrates, pH 3.5, 37°, 0.1 M sodium formate

P3 residue	P2 residue	kcat, sec^{-1}	Km, mM
Gly	Glu	25	0.063
Asp	Glu	34	0.078
Thr	Val	76	0.151
Val	Val	12	0.049
Val	Asp	71	0.058
Val	Asn	25	0.014
Val	Ser	55	0.025
Val	Glu	122	0.079

optimal catalysis may require either strain at some point in the reaction or efficient transfer of binding energy to the catalytic steps or both.

To illustrate the versatility of these new substrates data has been collected for hydrolysis of many of these by a variety of members of the Aspartic Proteinase family. Table II presents results for the substrate Lys-Pro-Pro-Glu-Phe-(NO_2)Phe-Arg-Leu.

Table II. Kinetic data for hydrolysis of Lys-Pro-Pro-Glu-Phe-(NO_2)Phe-Arg-Leu by representative Aspartic Proteinases

enyme	kcat, sec^{-1}	Km, mM
pig pepsin[ref 4]	218	0.19
Human Cathepsin D	25	1.08
Yeast Proteinase A	14	0.16
Mucor Pusillus A.P.	2	0.07
Rhizopus chinensis A.P.	2	0.03
Endothia Parasitica A.P.	7	0.004

A large variation in the observed catalytic parameters is seen in Table II. In general, the mammalian enzymes show higher kcat values but poorer Km values than the microbial enzymes. The point of cleavage by all these enzymes is the same as shown by product analysis. Thus, it should be emphasized that, although this family of enzymes displays homologous structures and generally similar specificity, there is considerable variation in the catalytic efficiency. This must arise from subtle differences in binding in the active site crevice. These details may best be studied by a combination of kinetic and structural probes[6].

Acknowledgement: This work was supported by a grant from the NIH (AM18865) and grants from the SERC of Great Britian (GR/C42521 and 42538).

References

1. Kay, J. (1985) in Aspartic Proteinases and Their Inhibitors, Kostka, V., Ed., Walter de Gruyter, Berlin, in press.

2. Dunn, B.M., Parten, B., Jimenez, M., Rolph, C.E., Valler, M., and Kay, J. (1985) in Aspartic Proteinases and Their Inhibitors, Kostka, V., Ed., Walter de Gruyter, Berlin, in press.

3. Dunn, B.M., Kammerman, B., and McCurry, K.R. (1984) Anal. Biochem., 138, 68-73.

4. Dunn, B.M., deLucy, P., Magazine, H., Parten, B. and Jimenez, M. (1983) in Peptides: Structure and Function, Hruby, V. and Rich, D.H., Eds., Pierce Chemical Co., Rockford, Ill. 603-606.

5. Konig, W., and Geiger, R. (1970) Chem. Ber. 103, 788-798, 2024-2033, and 2034-2040.

6. Hallett, A., Jones, D.M., Atrash, B., Szelke, M., Leckie, B.J., Beattie, S., Dunn, B.M., Valler, M.J., Rolph, C.E., Kay, J., Foundling, S.I., Wood, S.P., Pearl, L.H., Watson, F.E., and Blundell, T.L. (1985) in Aspartic Proteinases and Their Inhibitors, Kostka, V., Ed., Walter de Gruyter, Berlin, in press.

COMPARISON OF THE STRUCTURES OF INHIBITORS AND A SUBSTRATE ANALOG OF ANGIOTENSIN CONVERTING ENZYME

Alice Vrielink and Penelope W. Codding

Departments of Chemistry and Pharmacology and Therapeutics, University of Calgary, Calgary, Alberta, Canada, T2N 1N4

Introduction. Effective inhibitors of angiotensin converting enzyme (ACE) are developed to have several binding sites to the enzyme. One set of molecules with a hydrophobic moiety are the mercaptopropanoylindoline-2-carboxylic acids developed by Wyeth Laboratories[1]. The structures of two compounds: WY44221 and its benzoylthio derivative, WY44088, were determined by single crystal X-ray diffraction. Also, the structure of a substrate analog, hip-L-his-L-leu, was determined for comparisons with the inhibitors.

Experimental. All three compounds crystallized in the orthorhombic space group, $P2_12_12_1$, with 4 molecules in the unit cell. The substrate analogue crystals contained five water molecules per substrate. All crystal data were collected at room temperature. Lorentz and polarization corrections were applied to each data set and all three structures were solved by direct methods. Crystal data and refinement descriptors are in Table I.

Table I. The Crystal Data for Three ACE Inhibitors

	WY44221	WY44088	Hip-His-Leu
a(Å)	9.981(3)	7.4867(4)	10.055(2)
b(Å)	10.343(4)	11.5461(6)	14.975(5)
c(Å)	12.801(6)	22.006(2)	17.825(5)
ρgcm^{-3}	1.339	1.291	1.069
$\Theta max(°)$	22.5	66	22.5
R	0.070	0.093	0.054
R_w	0.074	0.073	0.059

Results and Discussion. The two mercapto-indoline compounds have similar conformations in the crystalline state. The peptide bond next to the proline residue is *cis* in each compound as can be seen in Figure 1 (a and b). The torsion angles are -14° for WY44221 and -1° for the phenyl derivative. In each structure, the *cis* conformation is stabilized by an intermolecular hydrogen bond from the terminal carboxyl group to the carbonyl oxygen atom of the peptide bond in another molecule. In the sulfhydryl compound, the mercaptopropanoyl chain is in an extended conformation with a torsion angle from the carbon atom of the peptide bond to the S atom of -175°; in the phenyl derivative, the sulfur atom is gauche to the carbon atom of the peptide bond.

The structure of the substrate analogue is extended with all peptide bonds *trans*. This conformation facilitates the hydrogen bonding pattern shown in Figure 1c. The five water molecules form a hydration coat about the peptide. All of the potential hydrogen bonding groups of the peptide are used in this packing arrangement. These interactions model the multiple interactions with the ACE active site necessary for a high affinity ligand.

The striking difference between the structures of the indoline inhibitors and that of captopril[2] is the torsion angle of the peptide bond. In captopril and the substrate analogue, the

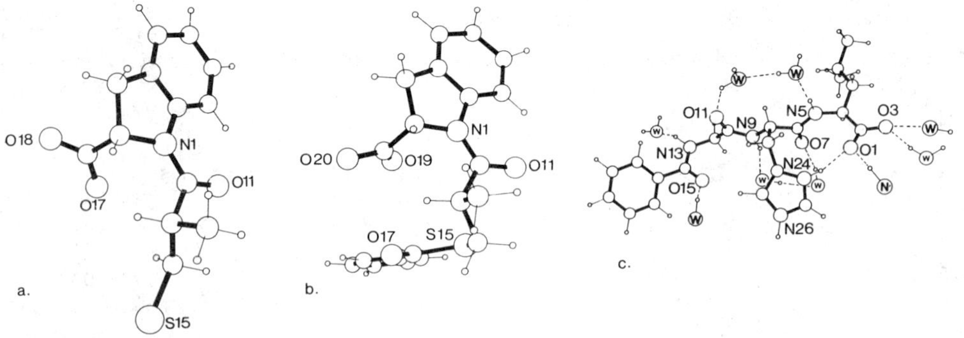

Fig. 1 a. Structure of WY44221, b. structure of WY44088, and c. structure of hip-L-his-L-leu showing the hydrogen bonding network to neighboring water molecules.

peptide bond is *trans* while in the Wyeth compounds this bond is *cis*. Recent molecular mechanics calculations on the low energy conformations of ACE inhibitors[3] indicate that the difference between the *cis* and *trans* conformers of WY44221 is small; the authors conclude that the *trans* form of ACE inhibitors is the active conformation.

Since the structure of ACE is unavailable, other zinc containing enzymes must be used as models for the ACE active site. One such enzyme is thermolysin; the crystal structure of thermolysin with a mercaptan inhibitor has been reported[4]. The crystallographically determined structure of WY44221 was superimposed on the positions of the thermolysin inhibitor, BAG. Figure 2 shows the ACE inhibitor in the thermolysin active site. It is evident that a better fit would be obtained with the *trans* conformer. In contrast, a superposition of the crystal structure of captopril on the BAG coordinates produces a reasonable fit. The position of captopril in the active site can be improved with a empirical energy calculation using the program YETI[5] to optimize the inhibitor-enzyme interactions. Figure 3 shows the resulting fit of captopril in the active site of thermolysin. The energy calculation places the carbonyl oxygen atom of the peptide bond in captopril near Arg 203 and forms a hydrogen bond between the proline N atom of captopril and the side chain of Asp 112.

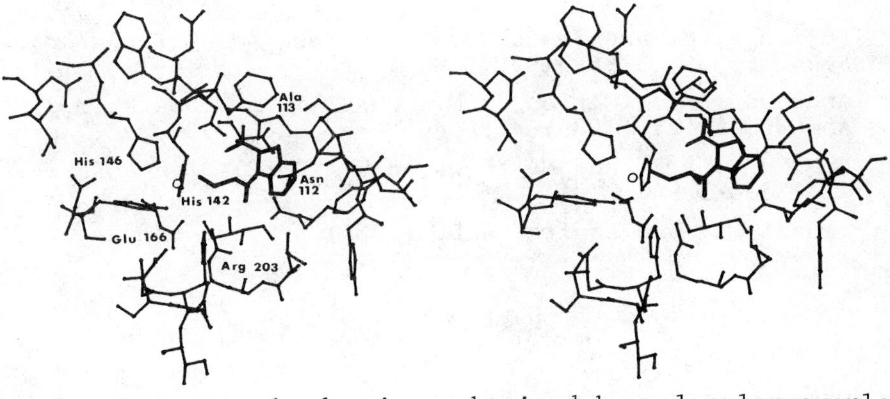

Fig. 2 Stereoscopic drawing, obtained by molecular overlap, of the structure of WY44221 in the active site of thermolysin.

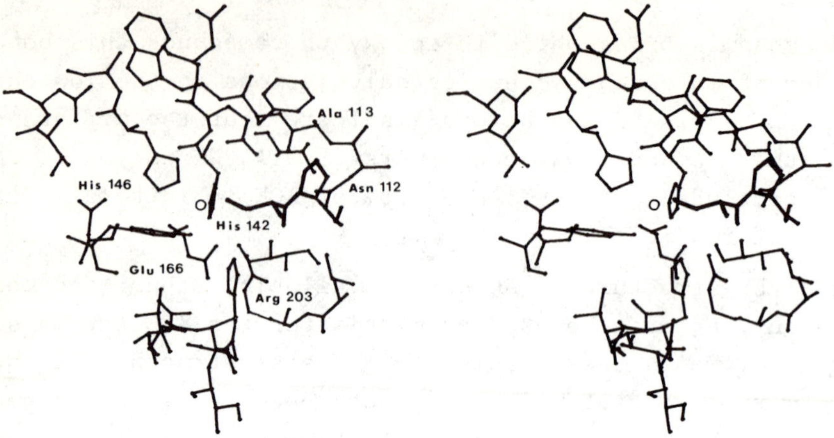

Fig. 3 Stereoscopic drawing of the structure of captopril, positioned by energy minimization, in the active site of thermolysin.

This combination of crystal structure analysis and molecular mechanics exploration of alternate conformations appears to be a fruitful way to examine the binding features necessary for ACE inhibition.

References

1. Kim, D.H., Guinosso, C.J., Buzby, G.C., Herbst, D.R., McCaully, R.J., Wicks, T.C. and Wendt, R.L. (1983) *J. Med. Chem.*, 26, 394-403.
2. Fujinaga, M. and James, M.N.G. (1980) *Acta Crystallogr. Sect. B*, 36, 3196-3199.
3. Andrews, P.R., Carson, J.M., Caselli, A., Spark, M.J. and Woods, R. (1985) *J. Med. Chem.*, 28, 393-399.
4. Monzinga, A.F. and Matthews, B.W. (1982) *Biochemistry*, 21, 3390-3394.
5. Vedani, A. and Dunitz, J.D. (1985) private communication.

LARGE SCALE N-CARBOXYANHYDRIDE PREPARATION OF ALA-PRO AND N_{ε}-(TFA)-LYS-PRO: SYNTHESIS OF ACE INHIBITORS

Thomas J. Blacklock*, John W. Butcher, and Richard F. Shuman,
Merck Sharp & Dohme Research Laboratories, Rahway, NJ 07065

The ACE inhibitors enalapril (2a, MK-421)[1] and lisinopril (2b, MK-521)[1] are needed in large scale for clinical trials. We have developed a general synthesis applicable to both compounds. The standard methods for peptide bond formation involving conventional protecting groups can become lengthy and uneconomical. Towards this end we developed a reductive amination procedure whereby Schiff base formation, when succeeded by catalytic hydrogenation in ethanol over Raney nickel, affords 2a and 2b in 80-90% with high diastereoselectivity (2a, SSS:RSS 87:13), (2b, SSS:RSS 95:5). See Wyvratt, et.al.[2] for a discussion of other methods.

The requisite dipeptides ala-pro (1a) and N_{ε}-(Tfa)-lys-pro (1b), were prepared in kilogram quantities via N-carboxyanhydride (NCA) chemistry. The economic advantage to such an approach was readily apparent as both NCA formation and subsequent condensation proved amenable to scaleup.

Alanine NCA (6a) and N_{ε}-(Tfa)-lysine NCA (6b) were prepared in 95% and 97% yields, respectively, via an optimized Fuchs-

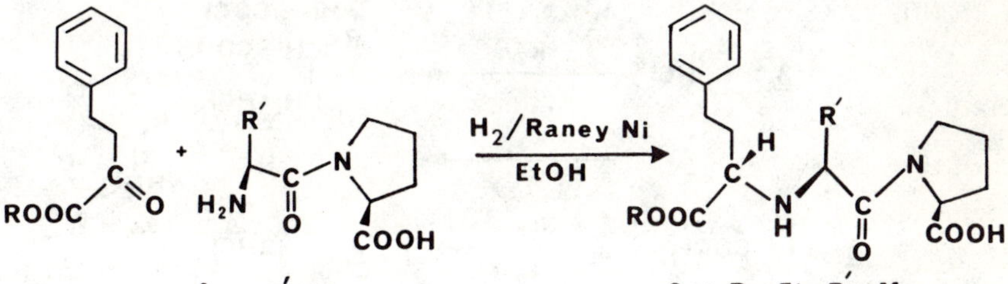

1a) R′=Me
1b) R′=$CF_3CONH(CH_2)_4$

2a) R=Et, R′=Me
2b) R=H, R′=$H_2N(CH_2)_4$

Farthing procedure (Scheme 1).[3] [The Tfa-lysine precursor to 6b was prepared in 85% yield from ethyl trifluoroacetate and lysine in aqueous sodium hydroxide. We found ethyl trifluoroacetate useful as an alternative to trifluoroacetic anhydride or to S-ethyl trifluorothioacetate for the derivatization.] In a manner similar to that recommended by Goodman[4] the appropriate amino acid was added to a preformed solution of phosgene (4M) in THF. This removed phosgene as a rate limiting reagent and any thermal contribution from the heat of solvation of phosgene with THF was eliminated. The reactions were complete within 4 hours with minimal HCl-induced THF cleavage. Upon addition of an amino acid to the phosgene/THF solution at 15-20°C, a rapid, self-limiting exotherm occurred due to initial formation of 4 (k_1) followed immediately by formation of 5 (k_2). Thus, almost instantaneously, between 33% and 50% of the amino acid was converted to 4, the amount depending on k_3. The reaction was then allowed to continue until complete dissolution (reaction) of the remaining amino acid hydrochloride had occurred (k_4). Traditionally this observation has been used to signify the end of reaction. Instead, we found that ring closure of 4 to the NCA (k_3) at 30°C or below was indeed much slower than expected and, in some cases, k_3 was observed to be proportional to concentration (this phenomenon is currently under further investigation). Accordingly, once amino acid

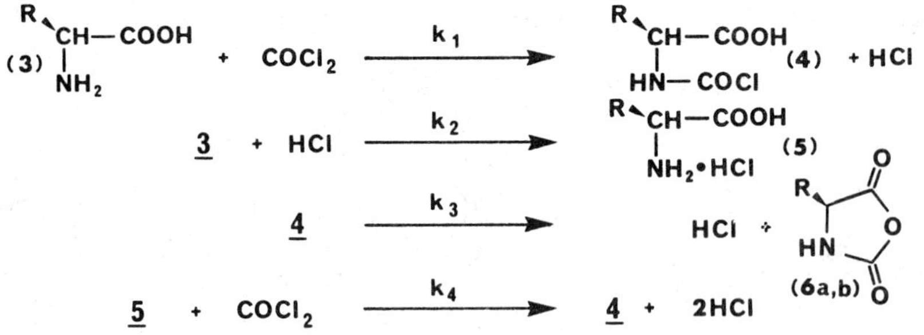

Scheme 1. Simplified Mechanistic Pathway to NCA Formation

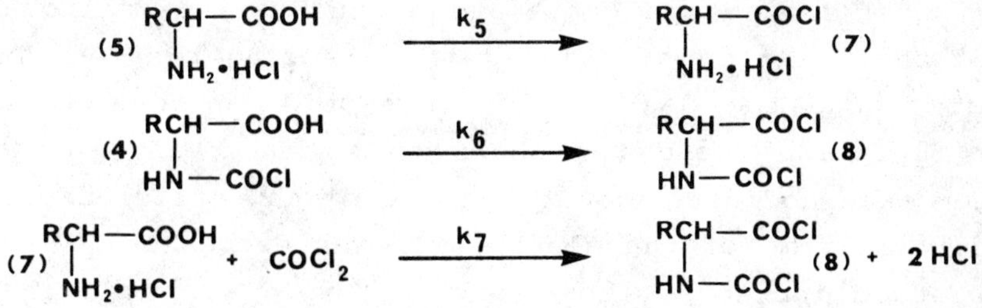

Scheme 2. DMF/Phosgene - Catalyzed Pathway to an N-Chlorocarbonyl Acid Chloride during NCA Formation

hydrochloride was no longer evident, the reaction was concentrated to remove excess phosgene and HCl, and aged to allow for completion. Subsequent dilution with THF followed by concentration removed most of the remaining HCl to afford an NCA solution sufficiently pure for direct use.

Solvent purity for the NCA reaction was found to be critical. Dimethylformamide (DMF), a common low level contaminant in many bulk solvents, was found to catalyze a significant reaction loss to the corresponding N-chlorocarbonyl amino acid chloride 8 (Scheme 2). This product (8) was identified by ^{13}C nmr and presumably arises from a Vilsmeier-Haack - type reaction. At levels of DMF as low as 1 ppt (parts per thousand) a nearly quantitative yield of 8 was obtained. This side reaction seriously affects the yield of an NCA if k_5 and/or k_6 effectively compete with k_3. Once any acid chloride has formed subsequent NCA formation from 7 or 8 is precluded. Whether k_5 or k_7 is rate determining depends primarily on the relationship of k_1 to k_3. If $k_1 \gg k_3$, then k_6 is the primary determinant of 8. Conversely, if $k_3 \gg k_1$ then k_5 must account for the formation of 8.

The general method for peptide bond formation reported by Iwakura et.al.,[5,6] was adopted for the preparation of 1a and 1b. However, we discovered that when alanine NCA was coupled with sodium prolinate in a sodium carbonate-buffered water/acetonitrile mixture, the product mixtures either froze and/or crystallized and contained as much as 40% oligomer(s). Warming the reaction alleviated the freezing problem, but it did not reduce the degree of overreaction. At a 0.1 molar scale, dipeptide yield was shown to be inversely related to addition time and directly related to mixing efficiency. When the addition time of the alanine NCA was decreased, the yield of dipeptide increased from about 65% to 90%. This increase however was greatly attenuated when the reaction was carried out at a concentration approaching 1 molar. A dramatic effect came from the discovery that when both the buffer and counter cation was changed from sodium to potassium the reaction could be run at higher concentrations and at 0°C without solubility problems. The yield of ala-pro exceeded 90%. Subsequently, a distinct correlation among the group I metals vs. yield of dipeptide was noted. Under identical conditions, including concentration, yields progressively increased as the Group I counterion increased in atomic number, i.e., the highest yield of 93% was obtained when the cesium cation was employed.[7]

Tfa-lys-pro was obtained in similar fashion (94% yield).[8]

References

1. Harris, E.E., Patchett, A.A., Tristram, E.W., and Wyvratt, M.J. (1982). US Pat. #4,374,829.

2. Wyvratt, M.J., Tristram, E.W., Ikeler, T.J., Lohr, N., Joshua, H., Springer, J.P., Arison, B., and Patchett, A.A. (1984). J. Org. Chem. **49**, 2816-2819.

3. Farthing, A. (1950). J. Chem. Soc. 3222-3229.

4. Fuller, W.D., Verlander, M.S., and Goodman, M. (1976). Biopolymers **15**, 1869-1871.

5. Iwakura, I., Uno, K., Oya, M., and Katakai, R. (1970). Biopolymers **9**, 1419-1427.

6. Katakai, R., Oya, M., Uno, K., and Iwakura, Y. (1971). Biopolymers **10**, 2199-2208.

7. Blacklock, T.J. and Shuman, R.F. US Pat. #4,510,083 Pub. 4/9/85. Eur. Pat. #127108, Pub. 12/5/84.

8. Blacklock, T.J. and Shuman, R.F.; U.S. Pat. Pending

KETOMETHYLDIPEPTIDES - A NEW CLASS OF ANGIOTENSIN CONVERTING ENZYME INHIBITORS

S. Natarajan, E. M. Gordon, E. F. Sabo, J. D. Godfrey, H. N. Weller, Jelka Pluscec, M. B. Rom, D. W. Cushman, J. M. DeForrest and J. R. Powell
The Squibb Institute for Medical Research, Princeton, NJ 08540.

INTRODUCTION:

Ketomethylene tripeptide analogs such as 1 and 2 constitute a novel class of angiotensin converting enzyme (ACE) Inhibitors.[1,2] Although compound 1 is a potent ACE inhibitor *in vitro*, its potency is not expressed *in vivo*.[2] On the other hand, an analog of compound 1 which incorporates a P_1' methyl residue (2) is not only a potent ACE inhibitor *in vitro* but also expresses its intrinsic potency *in vivo*.[3] We recently described the design and *in vitro* structure-activity relationships of a new class of ACE inhibitors which we termed "ketomethyldipeptides."[4,5] These compounds which are modelled after the tripeptide substrates (i.e., N-benzoyl-Phe-Ala-Pro), incorporate a ketone carbonyl and a secondary amine linked by a methylene unit ($-CO-CH_2-NH-$) as a surrogate for the scissile amide grouping. In the present communication, we report on the *in vivo* activity of members of this new series in which P_1' and P_1 residues have been varied.

RESULTS:

Ketomethyl dipeptide 3 was found to be a potent inhibitor of ACE *in vitro* (I_{50}=6 nM).[6,7] When this substance was administered intravenously[8] to normotensive rats, it inhibited the pressor response induced by exogenously administered angiotensin I. Encouraged by this result, we next examined the consequence of varying the P_1' residue (-Ala-) of compound 3 while maintaining the balance of the inhibitor molecule constant.

Table I: P_1'-Residue Modification

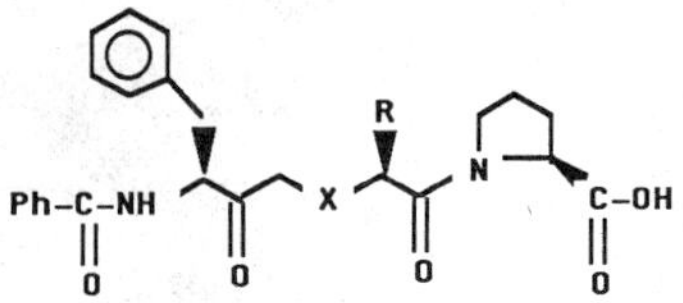

	X	R	I_{50}(nM)
1	--	H	12
2	--	CH_3	3
3	NH	CH_3	6
4	NH	H	4
5	NH	$(CH_2)_3-NH_2$	5
6	NH	$(CH_2)_4-NH_2$	9
7	NH	CH_2-Ph	10

Table II: P_1-Residue Modification

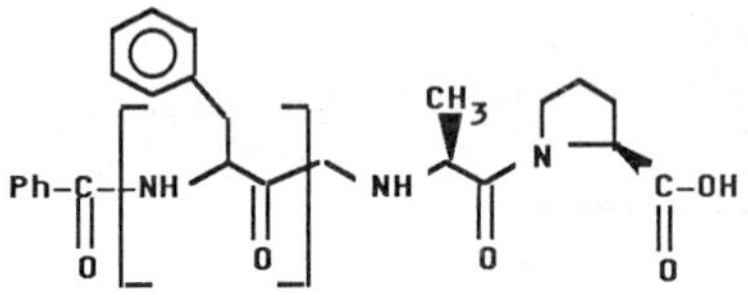

	[RESIDUE]*	I_{50}(nM)
3a	Phe	12
8	3-Pyridylala	7
9	Trp	8
10	Tyr	10
11	Nle	27
12	Lys	92

* 1:1 mixture of diastereoisomers at bracketed position

Inspection of the *in vitro* I_{50} values reveals that a good deal of tolerance exists at the enzyme S_1' subsite for enzyme-inhibitor interaction in this class of inhibitors. Analogs containing Gly(4), Orn (5), Lys(6) or Phe(7) at this position exhibited potent inhibition. Compound 4 is the most active ketomethyldipeptide prepared in this series, a somewhat surprising finding, since other major ACE inhibitor classes suffer substantial loss of inhibitory potency when the methyl group corresponding to the penultimate alanine side chain is removed.[9]

Although these analogs exhibited potent ACE inhibition *in vitro*, their inhibition of the pressor response to angiotensin I after I.V. administration was quite varied (Figure 1). Compounds 4, 5 and 7 were less active than 3 by varying levels while compound 6 exhibited maximal AI inhibition comparable to 3.

% MAXIMAL INHIBITION OF AI PRESSOR RESPONSE IN NORMOTENSIVE CONSCIOUS RATS AFTER I.V. ADMINISTRATION

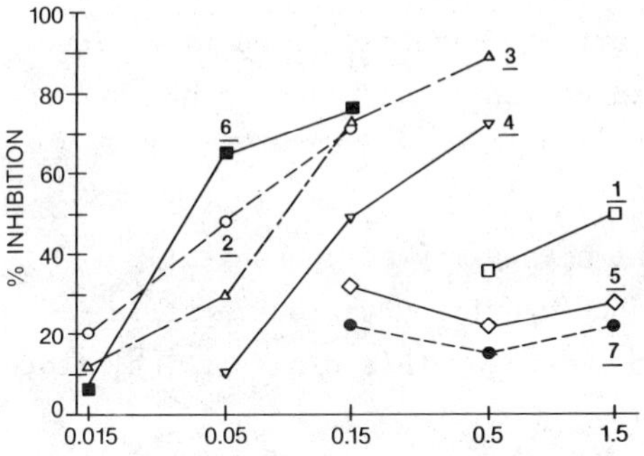

FIG. 1. EFFECT OF P_1' RESIDUE MODIFICATION

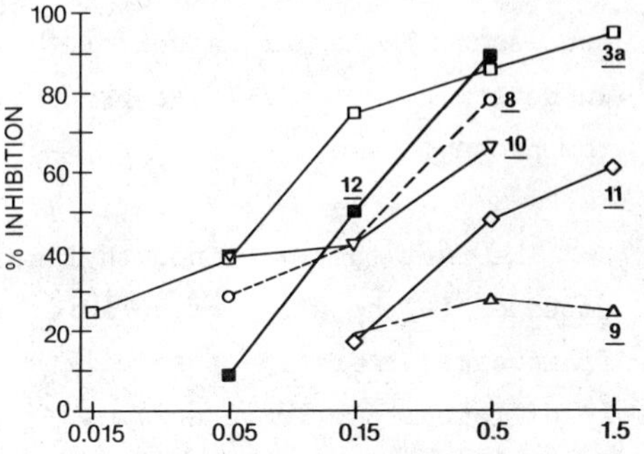

FIG. 2. EFFECT OF P_1 RESIDUE MODIFICATION

We next synthesized and examined analogs (8-12) to determine the effect of varying P_1 (-Phe-) residue of compound 3a while keeping the rest of the molecule same. This study resulted in a series of inhibitors with differing levels of intrinsic potency. Although compounds 8, 9 and 10 were almost equipotent with compound 3a in the *in vitro* assay, they were less active when tested *in vivo* by the i.v. route (Figure 2). Compound 11 was less active than 3a both *in vitro* and *in vivo*. On the other hand, substance 12 which was eight-fold less potent than 3a in the *in vitro* assay, was only slightly less active *in vivo*. When compounds 3 and 6 were administered to normotensive rats by the oral route, they proved to be poor inhibitors of AI pressor response.

CONCLUSIONS:

In conclusion, our findings show that ketomethyldipeptides are potent inhibitors of ACE. Structure-activity relationships in this series indicate that, while a good deal of structural latitude is permitted for P_1' and P_1 residues of these molecules for retention of potent *in vitro* activity, these residues play a key role in determining their *in vivo* activity. Though it is possible that factors such as transport and metabolism could play an important role in these processes, the exact nature of the participation of P_1' and P_1 residues in determining the *in vivo* properties of these molecules is not known at the present time.

The fact that α-aminomethylketone moiety may be conveniently embedded in the place of an amide grouping within various peptide frameworks offers the possibility of extending this class of inhibitors to different peptidase enzymes.

REFERENCES:

1. Almquist, R. G., Chao, W., Ellis, M. E. and Johnson, H. L. (1980) J. Med. Chem., 23, 1392-1398.
2. Meyer, R. F., Nicolaides, E. D., Tinney, F. J., Lunney, E. A., Holmes, A., Hoefle, M. L., Smith, R. D., Essenburg, A. D., Kaplan, H. R., and Almquist, R. G. (1981) J. Med. Chem., 24, 964-969.
3. Almquist, R. G., Crase, J., Jennings-White, D., Meyer, R. F., Hoefle, M. L., Smith, R. D., Essenburg, A. D. and Kaplan, H. R., (1982) J. Med. Chem., 25, 1292-1299.
4. Natarajan, S., Gordon, E. M., Sabo, E. F., Godfrey, J. D., Weller, H. N., Pluscec, Jelka, Rom, M. B. and Cushman, D. W. (1984) Biochem. Biophys. Res. Commun., 124, 141-147.
5. Gordon, E. M., Natarajan, S., Pluscec, Jelka, Weller, H. N., Godfrey, J. D., Rom, M. B., Sabo, E. F., Engebrecht, J. and Cushman, D. W., (1984) Biochem. Biophys. Res. Commun., 124, 148-155.
6. I_{50} values were determined by the method of Cushman, D. W. and Cheung, H. S. (1971) Biochem. Pharmacol., 20, 1637-1648.
7. We have briefly outlined the synthesis of this class of compounds earlier (Ref. 4). Details of their synthesis will be reported elsewhere.
8. Bianchi, A., Evans, D. B., Cobb, M., Peschka, M. T., Schaeffer, T. R. and Laffan, R. J. (1973) Eur. J. Pharmacol., 23, 90-96.
9. Petrillo, Jr., E. W. and Ondetti, M. A. (1982) Medicinal Research Reviews, 2, 1-41.

KETOMETHYLENE PENTAPEPTIDE ANALOGS AS ORALLY ABSORBED ANGIOTENSIN CONVERTING ENZYME INHIBITORS

Ronald G. Almquist, Clive Jennings-White, Wan-ru Chao, Tom Steeger, Kevin P. Wheeler, and Chozo Mitoma

SRI International, 333 Ravenswood Avenue, Menlo Park, CA 94025

Introduction

The ketomethylene tripeptide analog 5(S)-benzamido-4-oxo-6-phenylhexanoyl-L-proline (ketoACE) is a potent <u>in vitro</u> angiotensin converting enzyme (ACE) inhibitor, I_{50} = 70 nM.[1] Unfortunately this compound has poor activity as a blood pressure lowering agent when given either orally or intravenously to renal hypertensive rats.[2]

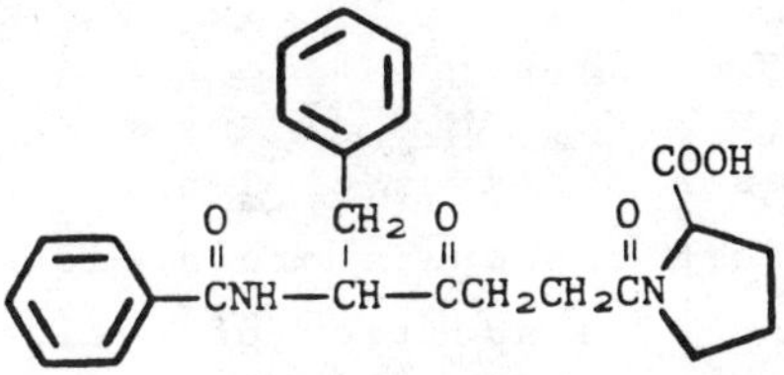

KetoACE

One possible method to improve the antihypertensive activity of ketoACE in the rat would be to develop analogs of it that have greatly increased ACE inhibition activity. An examination of the ACE inhibition data on various peptide analogs of the original snake venom peptide ACE inhibitors, indicates that the pentapeptide pGlu-Lys-Phe-Ala-Pro is almost 30 times more potent than the tripeptide Phe-Ala-Pro.[3] By analogy, extending ketoACE to a pentapeptide might greatly increase its ACE inhibition activity. It is also known that replacing pyroglutamic acid in pGlu-Lys-Phe-Ala-Pro with cyclobutane carboxylic acid (Cbc) yields a pentapeptide analog with slightly better ACE inhibition activity.[4] Since Cbc would also be expected to be more stable to peptidase degradation than pGlu, it was used in the pentapeptide analogs that we have prepared. The two analogs chosen for synthesis, 1 and 2 are analogs of Cbc-Lys-Phe-Gly-Pro and Cbc-Lys-Phe-Ala-Pro respectively in which the amide linkages connecting Phe-Gly and Phe-Ala have been replaced with ketomethylene linkages.

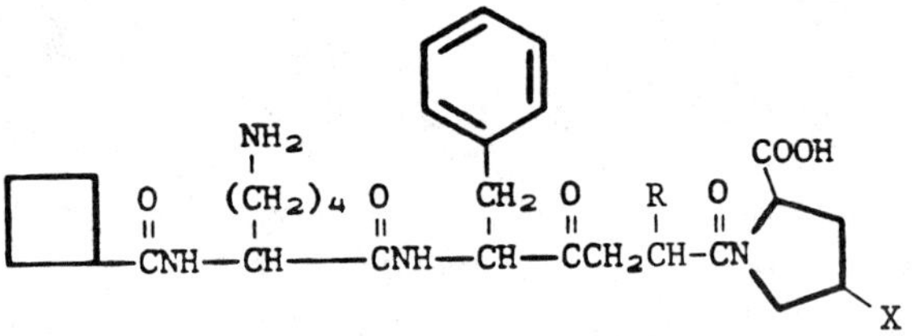

1; R = X = H
2; R = CH_3, X = H
[^{3}H]-1; R = H, X = ^{3}H

Biological Activity

The porcine plasma ACE I_{50}s of 1 and 2 are 7.0 nM and 3.0 nM respectively. Both compounds are at least 10 times more potent than ketoACE. The addition of cyclobutane carbonyl-L-lysine to the amino terminus of ketoACE must provide additional binding points for 1 and 2 to the angiotensin converting enzyme.

Both 1 and 2 were tested for their ability to lower blood pressure following oral administration in renal hypertensive rats of 3.0 and 6.4 mg/kg, respectively. Neither compound lowered blood pressure to a statistically significant extent at the doses tested.

Metabolism Studies

In order to gain a better understanding of the reasons for the low antihypertensive activity of 1 and 2, a radiolabeled derivative of 1 was synthesized in which the 4-position in the proline ring was tritiated. Studies of $[^3H]$-1 in rats yielded the following results. $[^3H]$-1 is well absorbed after oral administration in rats in which approximately 50% and 20% of an oral dose appears in the bile and urine, respectively, over a 24 h period. Forty percent of this oral dose appears in the bile within the first 8 h following dosing. It was also found that the blood half life of $[^3H]$-1 is only 24 min following intravenous injection.

Another interesting finding was that the pentapeptide analog $[^3H]$-1 was metabolically stable. TLC autoradiography of the 1, 2, 3 and 4 h bile aliquots and the 24 h urine revealed only a single spot corresponding to $[^3H]$-1 for each sample. In addition, treatment of 1 with trypsin caused only 20-30% decomposition of 1 as seen by TLC after 5 h. This data indicates that the lysyl-phenylalanine amide bond in $[^3H]$-1 is not significantly cleaved after oral administration.

Conclusions

The stability of $[^3H]$-1 and 1 to oral administration and trypsin cleavage indicates that insertion of a ketomethylene linkage in place of the normal amide linkage connecting Phe^3-Gly^4 not only prevents the cleavage of Phe from Gly but also greatly increases the stability of the adjacent Lys^2-Phe^3 peptide bond to peptidase cleavage. Peptidases capable of cleaving such a Lys-Phe bond may have a strong requirement for hydrogen bonding with the NH group of the next amide bond on the C-terminal side, an NH group not present in 1 or $[^3H]$-1.

At the present time the best explanation for the lack of antihypertensive activity after oral administration of 1 is that a blood level of 1 necessary to cause significant blood pressure lowering cannot be acheived due to rapid excretion of 1 into the bile and urine. The absorption of [^{3}H]-1 after oral dosing when based on biliary excretion values over time occurs at a fairly constant rate over the first 5 hours and continues at a slower rate over the next 19 hours. Although the blood half life of [^{3}H]-1 after intravenous administration is 24 minutes, the blood levels acheived after oral administration are much lower. Unlike the intravenous case in which a large bolus of [^{3}H]-1 is injected directly into the blood, a much smaller amount of orally administered [^{3}H]-1 is absorbed from the GI tract into the blood over a given time period and this small amount can rapidly be excreted by the liver and kidneys.

References

1. Almquist, R.G., Chao, W.-R., Ellis, M.W., and Johnson, H.L. (1980) *J. Med. Chem.* **23**, 1392-1398.
2. Meyer, R.F., Nicolaides, E.D., Tinney, F.J., Lunney, E.A., Holmes, A., Hoefle, M.L., Smith, R.D., Essenburg, A.D., Kaplan, H.R., and Almquist, R.G. (1981) *J. Med. Chem.* **24**, 964-969.
3. Cushman, D.W., Pluscec, J., Williams, N.J., Weaver, E.R., Sabo, E.F., Kocy, O., Cheung, H.S., and Ondetti, M.A. (1973) *Experientia* **29**, 1032-1035.
4. Ondetti, M.A., Pluscec, J., Weaver, E.R., Williams, N., Sabo, E.F., Kocy, O. (1972) Chemistry and Biology of Peptides, Meienhofer, J. Ed., Ann Arbor Science: Ann Arbor, pp. 525-531.

A NEW APPROACH TO THE SYNTHESIS OF PEPTIDE ALDEHYDE INHIBITORS.

Ian J. Galpin, Robert J. Beynon*, Janet Chetland, Mark T. Mulligan* and Graham A. Place*.

The Robert Robinson Laboratories and Department of Biochemistry*, University of Liverpool, P.O.Box 147, Liverpool, L69 3BX, England.

The peptide aldehydes initially discovered by Umezawa[1] are potent inhibitors of serine and cysteine proteinases. Of interest to us is chymostatin, which has the ability to inhibit chymotrypsin and related enzymes and to supress protein breakdown in muscle. A number of inhibitors of the general type (1) have been synthesised[2] where X = Leu, Ile or Val,

Z.Arg.X.Phe.H (1)

and the biological activity of these compounds have been evaluated.[2-4]

The general method of synthesis involves protection of the phenylalanine aldehydes as its semicarbazone, allowing chain extension by DCCI/HOBt coupling to give the tripeptide aldehyde semicarbazones. These were then deprotected by treatment with formaldehyde in MEOH.

In order to improve and streamline the synthesis we have investigated the use of a solid phase approach to the synthesis of these compounds. In this work we have examined the use of a polymeric semicarbazide hydrochloride derived from poly[methylene(polyphenyl isocyanate)]. The polymeric semicarbazide hydrochloride was prepared by reaction of the polymeric isocyanate with ^{t}Butyl carbazate. The semicarbazide hydrochloride was obtained after BOC removal with HCl in ethyl acetate, the direct use of hydrazine hydrate led to the formation of a highly cross-linked insoluble polymer. Benzyloxycarbonyl phenylalanine aldehyde was then prepared by the usual method involving reduction of the corresponding methyl ester with di-isobutyl lithium aluminium hydride at low temperature.[2] The resulting aldehyde was reacted with the polymeric semicarbazide hydrochloride in the presence of sodium acetate in a manner similar to that used in solution synthesis. It is advantageous that the polymer bound semicarbazone of the protected phenylalanine is soluble in DMF and other dipolar aprotic solvents. Thus, it may readily be precipitated by the addition of water or organic solvents such as ether, allowing purification by reprecipitation and washing.

The polymeric semicarbazone which was obtained in high yield was then hydrogenolysed in the presence of p-toluene sulphonic acid using DMF as the solvent. The hydrogenolysis with 10% palladium on charcoal, was rather slow and may be attributed to the polymeric nature of the substrate. The chain was then extended by coupling with the appropriate benzyloxycarbonyl (Z) or fluorenylmethyloxycarbonyl (Fmoc) amino acid. Activation in all cases was provided by the formation of a diphenyl phosphinic mixed anhydride[5] as it has been demonstrated that these intermediates have considerable utility in solid phase synthesis[6] due to their high stability to disproportionation and regioselective opening by the attacking amino component. After formation of the mixed anhydride at 0° complete acylation was achieved using a six-fold excess of the mixed anhydride. The protected dipeptide semicarbazone was then isolated by a washing procedure allowing excess reagents and impurities to be removed from the resin bound protected dipeptide semicarbazone. Some characterisation of these intermediates was carried out by proton N.M.R. using DMSO as the solvent, although good resolution was difficult to obtain due to the polymeric nature of these compounds.

Chain extension was then carried out using two approaches depending on the protecting group which had been chosen at the dipeptide stage. In the case of the benzyloxycarbonyl group, hydrogenolysis in the presence of p-toluene sulphonic acid could be used, but cleavage was again very slow. The use of the Fmoc protecting group, however, allowed very rapid cleavage of the amino protecting group by treatment with piperidine in DMF, the reaction being complete in a few minutes. Coupling with an amino-protected third residue was then carried out in a similar manner, again using a diphenyl-phosphinic mixed anhydride for activation. Following cleavage from the resin with 0.5M HCl/formaldehyde/DMF the final purification prior to biological testing was carried out by gel filtration in DMF using Sephadex LH20. The range of compounds prepared is indicated in Table I and new compounds were characterised by fast atom bombardment spectrometry[7] and by proton N.M.R.

TABLE I The effect of chymostatin and analogues on the hydrolysis of N-glutaryl Phe-4-methylcoumarylamide by chymotrypsin and a mouse skeletal muscle preparation[4] IC_{50} (μM).

INHIBITOR		CHYMOTRYPSIN	DIAPHRAGM
(1)	Z.Arg.Leu.Phe.H	0.23	33
(2)	Z.Arg.Ile.Phe.H	0.72	42
(3)	Z.Arg.Val.Phe.H	0.32	32
(4)	Z.Nle.Leu.Phe.H	0.35	50
(5)	Z.Nle.Ile.Phe.H	25	74
(6)	Z.Nle.Val.Phe.H	27	>100
	Chymostatin	0.14	25

The arginyl and norleucyl series of tripeptides were examined for inhibition of the hydrolysis of N-glutaryl Phe 4-methylcoumarylamide by chymotrypsin or a mouse skeletal muscle preparation.[4]

In general, substitution of the basic residue by a neutral residue at position one diminishes the inhibitory potency of the compounds towards chymotrypsin. The exception of the leucine-containing pair (1) and (4) suggest that leucine in this position leads to a particularly favourable interaction that can overcome a generally unfavourable substitution of arginine by norleucine. The differences are much less pronounced although qualitatively similar with the muscle preparation. At present, it is not clear that optimisation of the inhibitor structure towards chymotrypsin will lead to improved inhibitors of muscle proteinases, however a solid phase based synthesis should allow a more rapid investigation of these compounds.

Finally, we would like to thank the M.R.C., the S.E.R.C., and the Muscular Dystrophy Group of Great Britain and Northern Ireland for support.

REFERENCES

1. Umezawa, H., (1982). *Ann.Rev.Microbiol.*, 36, 75 - 99.

2. Galpin, I.J., Wilby, A.H., Place, G.A., and Beynon, R.J., (1984). *Int.J.Peptide Protein Res.*, 23, 477 - 486.

3. Grinde, B., Galpin, I.J., Wilby, A.H., and Beynon, R.J., (1983). *J.Biol.Chem.*, 258, 10821 - 10823.

4. Mulligan, M.T., Galpin, I.J., Wilby, A.H., and Beynon, R.J., (1985). *Biochem.J.*, 230, in press.

5. Jackson, A.G., Kenner, G.W., Moore, G.A., Ramage, R., and Thorpe, W.D., (1976). *Tetrahedron Letts.*, 3627 - 3630.

6. Galpin, I.J., and Robinson, A.E., (1984). *Tetrahedron*, 40, 627 - 634.

7. Rose, M.E., Prescott, M.C., Wilby, A.H., and Galpin, I.J., (1984). *Biomedical Mass Spectrometry*, 11, 10 - 23.

FLUOROPHOSPHORAMIDE-CONTAINING PEPTIDE ANALOGS AS IRREVERSIBLE PEPTIDASE INHIBITORS

Michael Calandra
Department of Chemistry, Rutgers University
New Brunswick, New Jersey 08903

David C. Roberts
MITRE Corporation, McLean, Virginia 22102-3481

Introduction

A recently developed and highly successful approach to the design of drugs focuses upon substances which can serve as inhibitors of those enzymes which mediate the formation and breakdown of peptide hormones. Irreversible inhibitors are of particular interest since they would be expected to exhibit highly potent and prolonged activity *in vivo*.

The first peptide-derived fluorophosphonates were reported by Lamden and Bartlett[1]. These were constructed by incorporating α-aminophosphonic acid residues into short peptide-like chains, and were efficient inactivators of chymotrypsin and elastase. Rapid hydrolysis prevented detailed kinetic analysis, and would probably limit *in vivo* applicability of these inhibitors.

Our approach has been to avoid the use of exotic starting materials by replacing the α-carbon of the modified amino acid residue with nitrogen, increasing ease of synthesis and stability to aqueous hydrolysis. A series of simple substrate analogs, and also a series of inhibitors of more extended structure, have been prepared and evaluated. The latter were designed so as to incorporate features identified as conferring high specificity to substrates of chymotrypsin (-Leu-Phe-/-Gly- and -Val-Phe-/-Gly sequences[2,3]) and elastase (-Ala-Pro-Nva-/-X sequences[4]). No attempts were made to deal with epimerism at phosphorus; the compounds reported here all behaved kinetically as single entities.

Methodology

(1) Synthesis of inhibitors.

(a) Substitution chemistry of phosphoryl halides[5].

$$POCl_2F \xrightarrow[-78^o,\ <1\ hr]{R_2NH} R_2N\text{-}POClF \xrightarrow[0\text{-}25^o,\ \frac{1}{2}\text{-}24\ hr]{R'NH_2 \quad Et_3N} \text{1b-e, 4,5,6}$$

Compound 2b was made similarly from $PSCl_2F$. Compounds 2a and 2c were obtained using similar chemistry via EtOPSClF, and 1a was obtained from EtOPOClF. Compounds of the following types proved too unstable to isolate or use as intermediates:

$$RNH\text{-}POClF \qquad PhCH_2OPOClF \qquad RCONHNH\underset{F}{P}(O)\text{-}OR\ (-NR_2)$$

(b) Alkyl fluorophosphonic acid derivatives. These were investigated for comparative purposes, and were prepared by a sequence involving Arbutsov reaction of the appropriate alkyl halide[6] followed by replacement of OEt with Cl by PCl_5[7] and displacement of Cl using KF in CH_3CN[8].

(2) Kinetics.

(a) Enzyme inactivation. Inhibitor incubation steps and chymotrypsin assays[9] were performed in a pH 7.8 TRIS buffer. Data were treated according to Kitz and Wilson[10].

(b) Hydrolysis kinetics. Studies were carried out using $CO_3^{=}$-HCO_3^- buffers, following release of F^- with a fluoride-sensitive electrode previously calibrated with known fluoride solutions of various concentrations in the same buffer. Pseudo first order kinetics were observed in all cases at initial stages of hydrolysis.

Table I. Structures and Kinetics Results for Inhibitors

Compd.	Structure	$k^{\dagger}_{inact}$, $M^{-1}min^{-1}$ x10^3	Hydrolysis $t_{1/2}$, min. (pH)
1a	$PhCH_2N(CH_3)P(O)(F)OEt$	17.5, 13.0 (3.3E-4; 4.34)	0.04 (10) 0.07 (11)
1b	$PhCH_2N(CH_3)P(O)(F)NHEt$	0.55, 1.43 (2.5E-4; 0.036)	3.5 (13)
1c	$PhCH_2N(CH_3)P(O)(F)GlyOEt$	4.0	0.09 (9)
1d	$PhCH_2N(CH_3)P(O)(F)GlyNHBu$	0.38, 0.31 (1.9E-2; 7.1)	0.40 (9)
1e	$PhCH_2N(CH_3)P(O)(F)GlyGlyOEt$	-	-
2a	$PhCH_2N(CH_3)P(S)(F)OEt$	-	-
2b	$PhCH_2N(CH_3)P(S)(F)NHEt$	-	-
2c	$PhCH_2N(NHAc)P(S)(F)OEt$	-	-
3a	$PhCH_2CH_2P(O)(F)OEt$	(large)	0.18 (9)
3b	$PhCH_2N(CH_3)P(O)(F)CH_2CH_2COOEt$	14.0	-
4a	$nPrNHP(O)(F)N(Me)CH_2COOEt$	-	-
4b	$iBuNHP(O)(F)N(Me)CH_2COOEt$	0.31, 0.37 (8E-5; 0.03)	0.18 (9)
5a	iBuN(Me)P(O)(F)GlyOEt	-	0.35 (10)
5b	$\overline{(CH_2)_4N}$-P(O)(F)GlyOEt	-	0.11 (9) 0.20 (10)
6a	$iPrNHCOCH_2N(CH_2Ph)P(O)(F)N(Me)CH_2COOEt$	0.038	-
6b	$iPrNHCOCH_2N(iBu)P(O)(F)N(Me)CH_2COOEt$	0.096	-
6c	$\overline{(CH_2)_4N}COCH_2N(nPr)P(O)(F)N(Me)CH_2COOEt$	0.063	-
6d	$\overline{(CH_2)_4N}CO\overline{CH(CH_2)_3N}P(O)(F)N(Me)CH_2COOEt$	-	-

†For chymotrypsin. Values in parentheses are, respectively, K_I (M) and k_2 (min^{-1}).

Results and Discussion (refer to Table I)

We have successfully demonstrated that (1) the phosphoryl fluoride function is compatible with large peptide-like structures, and lends itself to wide structural variation; (2) such compounds are stable enough to hydrolysis as to allow full kinetic evaluation (P=S compounds appear from preliminary observations to be 10-100 times yet more stable); and (3) compounds of this type can serve as moderately effective inhibitors of chymotrypsin. The effects of N- versus C- substitution on phosphorus, and of adjacent N-methyl groups, in slowing the inactivation step, may be deduced from the results.

References

1. Lamden, L. and Bartlett, P. (1983). Biochem. Biophys. Res. Comm. 112, 1085-1090.
2. Kurachi, K., Powers, J., and Wilcox, P. (1973). Biochemistry 12, 771-777.
3. Segal, D. (1972). Biochemistry 11, 349-356.
4. Powers, J. (1984). Biochemistry 23, 2995-3002.
5. Sasse, K. (1964). In Houben-Weyl Methoden der Organischen Chemie, Vol. XII/2, pp. 377-475. Thieme, Stuttgart.
6. Bhattacharya, A. and Thyagarajan, G. (1981). Chem. Rev. 81, 415-430.
7. Razumov, A., Mukhacheva, O., and Markovich, E. (1957). Zh. Obshch. Khim. 27, 2389.
8. Saunders, B. (1949). J. Chem. Soc., 2921.
9. Hummel, B. C. (1959). Canad. J. Biochem. Physiol. 37, 1393-1399.
10. Kitz, R. and Wilson, I. B. (1962). J. Biol. Chem. 237, 3245-3249.

SNTHESIS OF KELATORPHAN, A FULL ENKEPHALIN DEGRADING ENZYME INHIBITOR AND OF A TRIATIATED PROBE FOR STUDIES ON ENKEPHALINASE.

Marie-Claude Fournié-Zaluski, Romaine Bouboutou, Gilles Waksman and Bernard. P. Roques

Département de chimie organique (UA498 CNRS, U266INSERM) UER des sciences Pharmaceutiques et Biologiques, 4 avenue de l'Observatoire 75006 Paris France.

Richard Bessèlièvre

CEN Saclay 91190 Gif/Yvette France.

Introduction

It has been shown that the selectivity of substrates or inhibitors for metallopeptidases is related to specific interactions between the side chains of these molecules and enzyme subsites. However the increased affinity of inhibitors, related to their strong metal chelating ability, is generally associated with a decreased selectivity. For instance thiorphan (1) designed for selective inhibition of enkephalinase is also a relatively good ACE inhibitor. Taking this feature into account, we have attempted to synthesize a single compound able to inhibit the three enzymatic activities which degradate enkephalins (2) : aminopeptidase, dipeptidylaminopeptidase and enkephalinase. For this purpose, an hydroxamate group, able to behave as a bidentate ligand for the metal atom was introduced on a peptide backbone bearing the subsite specificity for enkephalinase.

Results and discussion

Compounds of the general formula HN(OH)-$COCH_2CH(CH_2Ø)$-CONH-CH(R)COOH were synthesized. A first synthetic way, starting from β-methylbenzylsuccinate <u>1</u> was used. This compound was obtained from diethylsuccinate by a method derived from that of Cohen and Milovanovié (3). However this way was not unambigous, since the alkaline hydrolysis of compound 2 induced the intermediate formation of the N- substituted benzylsuccinimide 3 which led to a mixture of isomers 4 and 5 by alkaline cleavage of the cyclic succinimide (scheme 1).

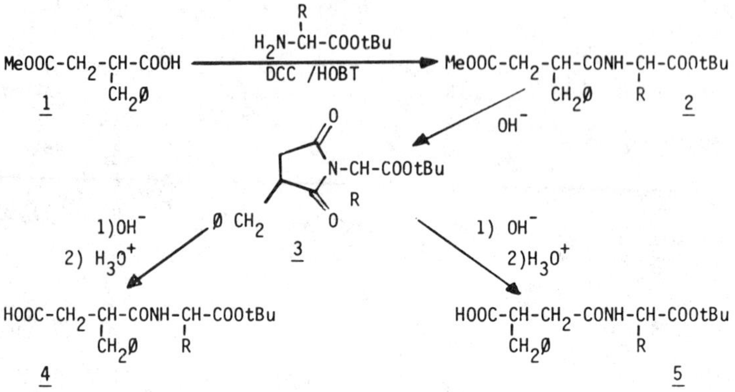

Scheme 1

This type of rearrangement was already described for methylsuccinic derivatives (4). To overcome this difficulty, a benzylidene moiety was maintened during all steps of the synthesis (scheme 2).

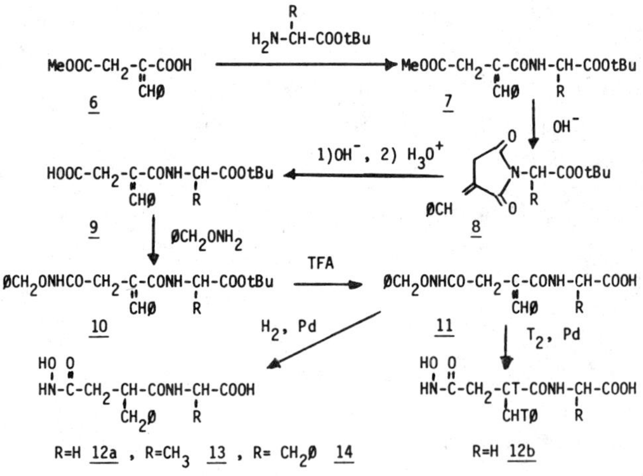

Scheme 2

The alkaline hydrolysis of 7 gave only the expected isomer 9 since the cyclic benzylidene succinimide 8 which can be intermediatly obtained, is specifically cleaved at the level of the less hampered amide bond. After deprotection of the t.butylester the final products were obtained by simultaneous deprotection of the hydroxamate group and saturation of the benzylidene moiety, by catalytic hydrogenation. The double bond reduction in compounds 11 gave compounds 13 and14 as diastereoisomeric mixture of RS and SS isomers, which were separated by chromatography. The inhibitory potency of these hydroxamates was studied on the three enkephalin degrading enzymes, purified from rat brain as described (5). As expected, hydroxamates 12, 13 and 14 are good inhibitors of the three enzymatic activities (table 1).On enkephalinase, all these compounds exhibit IC_{50} values in the nanomolar range, and the inhibitory potency increases with the hydrophobicity of the C-terminal aminoacid. As already observed in the case of thiorphan the absolute configuration of the succinyl asymmetric carbon is ineffective for enkephalinase recognition, while it is a discriminative factor for the two other enzymes.

			IC_{50} (M)		
			ENKEPHALINASE	AMINOPEPTIDASE M	DIPEPTIDYLAMINO PEPTIDASE
12	HO O HN-C-CH_2-CH-CONH-CH_2-COOH CH_2Ø	(R + S)	$5.6 \pm 0.6 \times 10^{-9}$	$2.2 \pm 0.2 \times 10^{-6}$	$2.1 \pm 0.5 \times 10^{-8}$
13	HO O HN-C-CH_2-CH-CONH- CH- COOH CH_2Ø CH_3	(R,S)	$1.7 \pm 0.6 \times 10^{-9}$	$3.8 \pm 0.5 \times 10^{-7}$	$0.9 \pm 0.1 \times 10^{-9}$
		(S,S)	$1.8 \pm 0.5 \times 10^{-9}$	$2.9 \pm 0.5 \times 10^{-5}$	$1.0 \pm 0.5 \times 10^{-7}$
14	HO O HN-C-CH_2-CH-CONH- CH- COOH CH_2Ø CH_2Ø	(R,S)	$3.4 \pm 0.7 \times 10^{-10}$	$7.2 \pm 0.3 \times 10^{-7}$	$2.5 \pm 0.3 \times 10^{-9}$
		(S,S)	$5.6 \pm 0.5 \times 10^{-10}$	$7.4 \pm 0.5 \times 10^{-5}$	$3.5 \pm 0.7 \times 10^{-7}$

Table 1. Inhibitory potency of hydroxamates on the three enkephalin degrading enzymes.

The RS isomer of compound 13 designated kelatorphan is, in this series,the best inhibitor of the three enkephalin degrading enzymes. The ability of kelatorphan to protect in vivo endogenous enkephalins was shown by its analgesic properties (fig. 1) . Kelalorphan is, at least, as efficient as the association of bestatin, (unspecific aminopeptidase inhibitor) and thiorphan(enkephalinase inhibitor). Interestingly, further addition of bestatin to kelatorphan does not improve the antinociceptive effect. These results evidence for the first time the selective involvement of a "kelatorphan-sensitive aminopeptidase" (likely aminopeptidase M) in the metabolism of enkephalins.

Furthermore compound 12 which is the most selective enkephalinase inhibitor in this series was tritiated as shown on scheme 2, and the tritiated analogue 12b allowed the visualization of enkephalinase in rat brain by autoradiographic measurements (fig. 2). The distribution of enkephalinase resembles that of opioïd receptors and firmly establishes the physiological implication of enkephalinase in enkephalin metabolism (6).

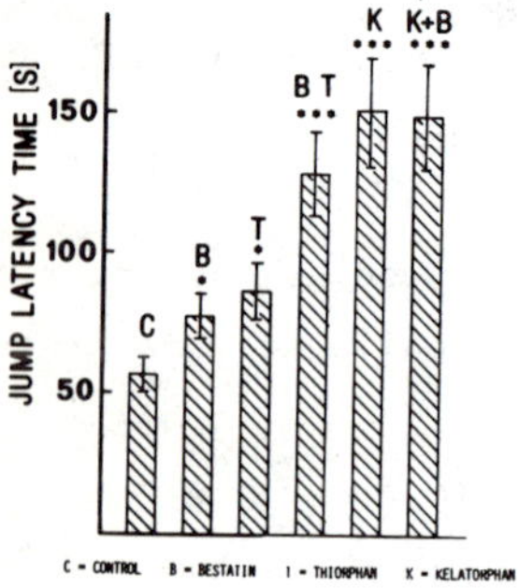

Figure 1. Analgesic effect of kelatorphan (hot plate test).

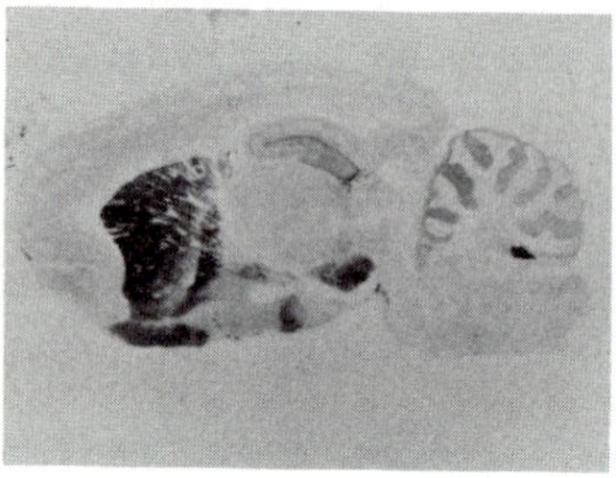

Figure 2. Visualization of enkephalinase in rat brain.

References

1. Roques B.P., Fournié-Zaluski M.C., Soroca E., Lecomte J.M., Malfroy B., Llorens C., and Schwartz J.C. (1980) Nature 288 286-288.
2. Roques B.P., and Fournié-Zaluski M.C., in "The medicinal chemistry and molecular pharmacology of opioïd peptides and the opiates". NIDA research monograph, R.S. Rapaka Ed. Washington (in press).
3. Cohen S.G. and Milovanović A. (1968) J. Amer. Chem. Soc. 90 3495-3502.
4. Hancock J.E.H and Linstead R.P. (1953) J. Org. Chem. 3490-3496.
5. Fournié-Zaluski M.C., Couland A., Bouboutou R., Chaillet P., Devin J., Waksman G., Costentin J and Roques B.P. (1985) J. Med. Chem. (in press)
6. Waksman G., Bouboutou R., Devin J., Bourgoin S., Cesselin F., Fournié-Zaluski M.C. and Roques B.P. (1985) Eur.J. Pharm. (in press).

THE SOLID PHASE SYNTHESIS OF TRYPSIN INHIBITORS CMTI I AND CMTI III FROM SQUASH SEEDS (*Cucurbita maxima*)

Gotfryd Kupryszewski and Krzysztof Rolka
Institute of Chemistry, University of Gdansk
80952 Gdansk, Poland

Ulf Ragnarsson
Institute of Biochemistry, Biomedical Center, University of Uppsala, 75123 Uppsala, Sweden

Tadeusz Wilusz and Antoni Polanowski
Institute of Biochemistry, University of Wroclaw
50137 Wroclaw, Poland

Introduction

Two polypeptide inhibitors of trypsin or of Hageman factor were independently isolated from squash seeds at the University of Wroclaw, Poland [1] and from pumpkin seeds by a group at the U.S. National Institutes of Health [2]. Both inhibitors CMTI I and CMTI III were sequenced at the Purdue University, West Lafayette [3].

The molecule of CMTI I, as well as that of CMTI III consists of 29 amino acid residues cross-linked by three disulphide bridges. The sequences of both inhibitors differ in only one position, Glu^9 in CMTI I being replaced by Lys^9 in CMTI III [3].

CMTI I: Arg-Val-Cys-Pro-Arg-Ile-Leu-Met-Glu-Cys-Lys-Lys-Asp-Ser-Asp-Cys-Leu-Ala-Glu-Cys-Val-Cys-Leu-Glu-His-Gly-Tyr-Cys-Gly.

CMTI III: Arg-Val-Cys-Pro-Arg-Ile-Leu-Met-Lys-Cys-Lys-Lys-Asp-Ser-Asp-Cys-Leu-Ala-Glu-Cys-Val-Cys-Leu-Glu-His-Gly-Tyr-Cys-Gly.

It was established that Arg^5-Ile^6 is the reactive side peptide bond in both inhibitors [3,4]. The interaction of CMTI I and CMTI III with trypsin is accompanied with cleavage of this peptide bond, resulting in the modified inhibitors CMTI I* and CMTI III* [1]. The molecules of modified inhibitors consist of two peptide chains held together by one disulphide bridge[4].

Results and Discussion

The inhibitors CMTI I and CMTI III were synthesized by the solid phase procedure. The chloromethyl resin with the capacity 0.22 meq/g was prepared according to Gisin [5]. All deblocking, rinsing and coupling steps were carried out using an automatic peptide synthesizer (Beckman, Model 990). N-Boc-protection was used through the synthesis with the following side-chain protecting groups: Arg(NO_2), Cys(Acm), Glu(OBzl), His(Tos), Lys(2-Cl-Z), Ser(Bzl) and Tyr(2,6-Cl-Bzl). Double couplings in CH_2Cl_2 were performed using DCC. Deblockings were achieved using 33% TFA/CH_2Cl_2 with 2% dimethylsulphide. 10% TEA/CH_2Cl_2 was used for neutralization. After completing the synthesis, the Acm-peptides were cleaved from the resin with liquid HF at $0^{o}C$ in the presence of anisole. The Acm-peptides were purified on Sephadex G-25F column. Both peptides gave the expected composition on amino acid analysis. Then the Acm-groups were removed with mercuric acetate. The reduced peptides were desalted on a Sephadex G-10 column and reacted with a mixture of oxidized and reduced gluthathione in Tris-HCl buffer (pH 8.5). The oxidized peptides were purified with 10% yield on a column of immobilized anhydrotrypsin.

In the assay according to Erlanger et al [6], the synthetic oxidized peptides had the same antitrypsin activity as native CMTI I and CMTI III respectively. Both synthetic peptides were modified with catalytic amounts of trypsin to the CMTI I* and CMTI III* respectively.

The synthetic oxidized peptides gave the identical electrophoretic and immunological [7] patterns as native inhibitors CMTI I and CMTI III respectively, as well as similar UV and CD spectra. Amino acid analyses of the isolated synthetic material were in agreement with theory.

Acknowledgements

This work was supported by the Swedish Natural Science Research Council (K-GF 3020-107), the Polish Academy of Sciences (MR I 12.1.6.1) and the Polish Ministry of Science and Higher Education (R 1.9).

References

1. Polanowski, A., Wilusz, T., Nienartowicz B., Cieslar, E., Slominska, A., and Nowak, K. (1980) Acta Biochim. Polon. 27, 371-381.

2. Hojima, Y., Pierce, J.V., and Pisano, J.J. (1982) Biochemistry 21, 3741-3746.

3. Wilusz, T., Wieczorek, M., Polanowski, A., Denton, A., Cook, J., and Laskowski, M. Jr. (1983) Hoppe-Seyler's Z. Physiol. Chem. 364, 93-95.

4. Nowak, K., Slominska, A., Polanowski, A., Wieczorek, M., and Wilusz, T. (1981) Hoppe-Seyler's Z. Physiol. Chem. 362, 1017-1019.

5. Gisin, B.F. (1975) Helv. Chim. Acta 56, 1476-1482.

6. Erlanger, B.F., Kokowsky, N., and Cohen, W., (1961) Arch. Biochem. Biophys. 95, 271-278.

7. Szewczuk, A., unpublished data.

INCORPORATION OF COORDINATING AMIDE BOND SURROGATES WITHIN MAMMALIAN COLLAGENASE INHIBITORS

Robert B. Miller, Arno F. Spatola, and Robert D. Gray

Departments of Chemistry and Biochemistry, University of Louisville, Louisville, KY 40292

Introduction

The mammalian collagenases are a group of structurally distinct but chemically similar zinc proteases which function to initiate the degradation of collagen.[1] Collagenases have been implicated in a number of tissue-destructive diseases, including rheumatoid arthritis, corneal ulceration, and in periodontal disease.[2] Recently, the role of collagenases in the metastases of invasive tumors has come to light.[3] A collagenase specific for basement membrane Type IV collagen has been shown to be produced by metastatic tumor cells, and to play an active role in the invasiveness of these cells.[4] We have thus begun a research program to design and prepare peptide-based collagenase inhibitors with the eventual goal of the inhibition of tumor metastases. The eventual clinical implication becomes quite apparent when statistics show that 50% of the patients that die from malignant cancers do so from the secondary tumor lesions rather than the primary tumor.[5]

Results and Discussion

We sought to design an inhibitor that would incorporate a coordinating amide bond surrogate into a molecule that utilizes binding specificities on both sides of the cleaved peptide bond (Figure 1). The collagen α(I) substrate, the structure of which has been fully determined, was chosen as the model on which to base the structures of the inhibitors (Figure 2). The Glyψ[CH(SH)CH_2]Gly pseudodipeptide ($NHCH_2CH(SH)CH_2CH_2COOH$)

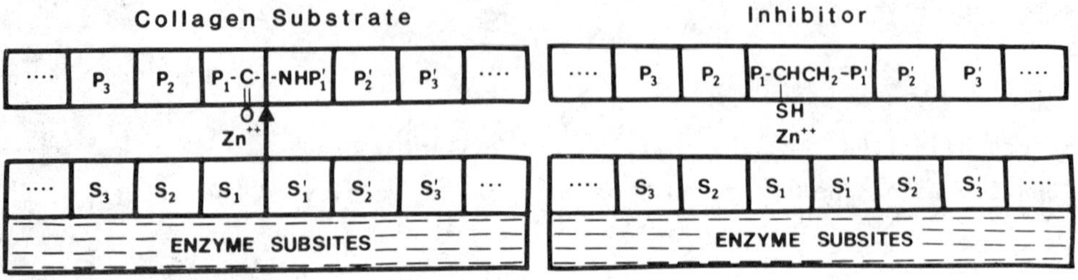

Fig. 1. Proposed interaction of collagen substrate and inhibitor with collagenase.

771 772 773 774 775 776 777 778 779 780 781

-Pro-Gly-Pro-Gln-Gly-Ile(Leu)-Ala-Gly-Gln-Arg-Gly-

↑

Cleaved Bond

Fig. 2. Structure of collagen α(I) immediately surrounding the cleavage site.

was the first synthetic target attempted in order to avoid the complication of 2 asymmetric centers. The synthesis of the unit that is actually incorporated into the inhibitor is shown in Scheme 1.

The synthesized unit is very versatile in the synthesis of the collagenase inhibitors. Orthogonal protecting groups were chosen; the Boc-protecting group is cleaved with mild acid, the methyl ester with mild basic hydrolysis, and the S-benzyl function cleaved with sodium/ammonia. Likewise, synthesis of pseudopeptide units containing funtionalities that interact with S_1' subsites are readily prepared from easily obtained precursors.[6-8] We thus have a synthetic entry to the preparation of pseudopeptides that contains a zinc coordinating moiety and utilizes both S and S' binding subsites.

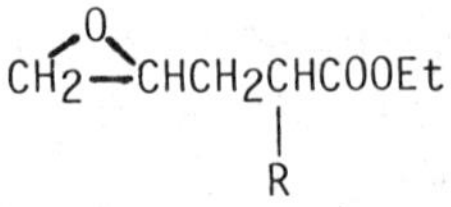

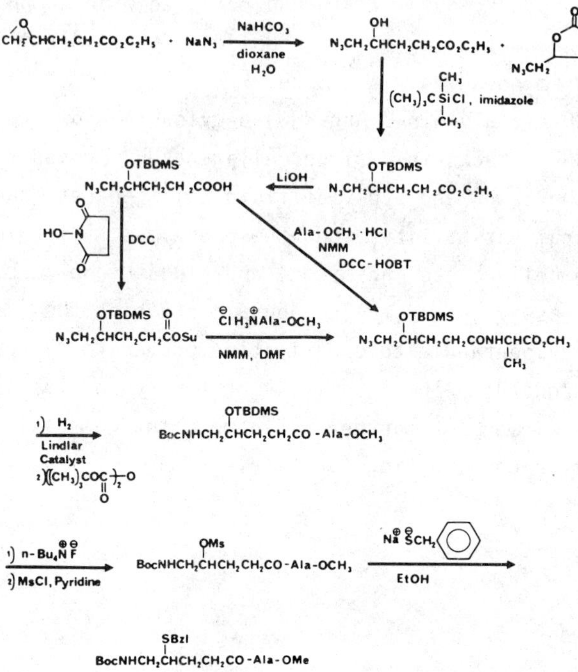

Scheme 1. Synthesis of the pseudopeptide fragment that is incorporated into the inhibitor peptide chain.

Table I. Inhibitory Activity of Collagenase Analogs Containing the Surrogate Amide Bond Replacement

Compound	IC$_{50}$(mM)[a]
Boc-Gln-CO-NHCH$_2$CH$_2$SH	1.5
HSCH$_2$CH(CH$_2$CH(CH$_3$)$_2$)CO-Ala-OEt	0.040[b]
BocNHCH$_2$CH(OH)CH$_2$CH$_2$CO-AlaOMe	> 8.0
Cysteine	2.5
Glutathione	~ 5.0

[a]Numbers represent inhibition of rabbit tumor V2 enzyme toward rat tail collagenase; similar results were obtained with pig synovial fluid.
[b]Mixture of diastereomers.

The inhibitor molecules were evaluated with collagenases derived from rabbit V2 tumor and pig synovial fluid and compared with compounds having only a sulfur containing moiety coupled to the N-terminal and C-terminal subsite amino acids. A compound ($HSCH(CH_2CH(CH_3)_2CO$-Ala-Gly-Gln-D-Arg-NH_2) previously shown to be the best inhibitor of a tadpole backskin-derived collagenase,[9] proved to be less effective toward the tumor and synovial fluid collagenases, thereby suggesting structural variability among these enzymes. Cysteine (formerly the best mammalian collagenase inhibitor) and glutathione were tested as standards. In our assays, only the compound containing the 2-methylthio-4-methylpentanoic acid derivative proved to be active compared to the standards. The inactivity of the hydroxy pseudopeptide analog points to the need for further refinement in the subsite binding sites and a good zinc-coordinating function.

References

1. Harper, E. (1980) Ann. Rev. Biochem. 49, 1063-1078.
2. Collagenase in Normal and Pathological and Connective Tissues, Woolley, D. E., and Evanson, J. M., Eds. (1980), Wiley-Interscience.
3. Mullins, D. E., and Rohrlich, S. T. (1983), Biochimica et Biophysica Acta 695, 177-214.
4. Liotta, L. A., Thorgeirsson, U. P., and Garbisa, S. (1982), Cancer Metastases Reviews 1, 277-288.
5. Rawls, R. L. (1985), Chemical and Engineering News, Feb. 25, 10-17.
6. Cregge, R. J., Herrmann, J. L., Lee, C. S., Richman, J. E., and Schlessinger, R. H. (1973), Tetrahedron Letters, 2425-2428.
7. Riley, R. G. and Silverstein, R. M. (1974), Tetrahedron 30, 1171-1174.
8. Kurth, M. J., Decker, O. H. W., Hope, H., and Yanuck, M. D. (1985), J. Am. Chem. Soc. 107, 443-448.
9. Gray, R. D., Saneii, H. H., and Spatola, A. F. (1981), Biochem. Biophys. Res. Commun. 101, 1251-1258.

INHIBITION OF HUMAN LEUKOCYTE ELASTASE, PORCINE PANCREATIC ELASTASE AND CATHEPSIN G BY PEPTIDE KETONES

Hitoshi Hori, Akira Yasutake, Yoshihiro Minematsu, and James C. Powers

School of Chemistry, Georgia Institute of Technology, Atlanta, GA 30332

Introduction

Serine proteases have previously been shown to be inhibited by peptide aldehydes. Elastase is inhibited by synthetic peptide aldehydes such as Ac-Pro-Ala-Pro-Ala-H, trypsin-like enzymes are inhibited by the fermentation products leupeptin and antipain, and chymotrypsin is inhibited by chymostatin. Peptide methyl ketones are generally much poorer inhibitors than the corresponding aldehydes. One ketone which is a very effective serine protease inhibitor is 4-amidinophenylpyruvate (APPA).[1] This inhibits trypsin and thrombin with K_I values of 0.004 mM and 0.008 mM respectively. Since it appeared that the increased affinity of APPA for trypsin was due both to increased reactivity of the ketone carbonyl group to nucleophilic addition because of the adjacent electronegative carboxyl group and to additional interactions between the enzyme and the carboxyl group of the inhibitor, we decided to investigate extended ketone structures as inhibitors of elastase. In this study, we describe the synthesis of a new series of peptide ketones and peptide α-ketoesters (Figure 1) and report the development of several potent reversible inhibitors for elastase and other serine proteases.

$R'CO-NH-CH(R)-CO-CH_2-CH_2CO-X$ $R'CO-NH-CH(R)-CO-CO-OR''$

Figure 1. Structures of peptide ketones (left) and peptide α-ketoesters (right) synthesized. R = alkyl or benzyl; R' = phenyl, aminoacyl, or peptidyl; R'' = alkyl or benzyl; X = alkoxy or amino acid residues.

Results and Discussion

The peptide ketones were synthesized by reaction of acyl amino acids with 3-carbomethoxypropionyl chloride in a modified Dakin-West reaction to give 5-acylamino-4-ketocarboxylic acid esters in 50-70% yield.[2] Hydrolysis of the ester and coupling with amino acid derivatives gave extended peptide structures which can interact with subsites on both sides of the catalytic residues. The peptide α-ketoesters were synthesized from acyl amino acids and peptides in 10-45% yields by a Dakin-West reaction using alkyl or benzyl oxalyl chloride.[3] The products obtained from the Dakin-West reactions are mixtures of diastereomers in which the amino acid residue which forms the ketone or ketoester residue is racemic.

The inhibition of HL elastase, PP elastase, and cathepsin G was studied at pH 7.5 and 25° C and the results are shown in Table I. The two elastases were inhibited competitively by both types of compounds with the α-ketoesters being much more potent. The best inhibitors were Z-Ala-Ala-DL-Abu-COOEt (Abu = 2-aminobutanoic acid), the corresponding benzyl ester, and Z-Ala-Ala-Ala-DL-Ala-COOEt. In cases where comparisions were made, the corresponding α-ketoacids were poorer inhibitors than the esters. Several derivatives containing Phe residues were good inhibitors of cathepsin G and chymotrypsin. These include Bz-DL-Phe-COOEt (K_I = 0.058 and 0.00028 mM with Cat G and chymotrypsin respectively) and MeO-Suc-Val-Pro-DL-Phe-COOMe (0.0011 and 0.00026 mM).

Table I. Inhibition of Human Leukocyte Elastase (HLE), Porcine Pancreatic Elastase (PPE), and Cathepsin G (Cat G) by Peptide Ketoesters and Ketoacids.

	K_I (mM) HLE	PPE	Cat G
Bz-DL-Ala-CH_2CH_2COOMe	3.4	4.2	2.2
Bz-DL-Abu-CH_2CH_2COOMe	3.5	5.7	2.1
Bz-DL-Ala-CH_2CH_2COOH	NI	NI	3.6
Bz-DL-Abu-CH_2CH_2COOH	14	43	4.7
Bz-DL-Nle-CH_2CH_2COOMe	3.7	2.4	1.9
Bz-DL-Nle-CH_2CH_2CO-Val-OMe	0.079	0.77	1.8
Bz-DL-Ala-COOEt	0.64	0.59	
B -DL-Ala-COOH	3.1	3.2	
Z-Ala-Ala-DL-Abu-COOEt	0.00012	0.00015	
Z-Ala-Ala-DL-Abu-COOBzl	0.00009	0.00008	
Z-Ala-Ala-Ala-DL-Ala-COOEt	0.0003	0.00014	

0.1 M Hepes, pH 7.5, 0.5 M NaCl, 9.0-9.8% DMSO at 25° C. NI = no inhibition

The mechanism of inhibition most likely involves interaction of the active site serine residue of the serine protease with the ketone carbonyl group of the inhibitor to form a hemiketal structure which resembles the tetrahedral intermediate involved in peptide bond hydrolysis (Figure 2). This model is based on the refined x-ray crystal structure of the complex formed by bovine trypsin and APPA.[4] The amidinophenyl group is located in the primary specificity pocket of trypsin and the active site serine has added to the 2-carbonyl group in APPA to give a tetrahedral structure. The oxyanion is stabilized by hydrogen bonding with groups in the oxyanion hole. A unique feature of this structure is the hydrogen bonding observed between the carboxylate oxygen and the serine oxygen and the NH of histidine-57. We expected that the negative charge on the carboxylate would significantly contribute to the binding energy. Our observation that α-ketoesters were much better inhibitors than the corresponding acids indicates that additional interactions with the S_1' subsite and hydrogen bonding of the ester oxygen with the histidine N-H can result in signicant binding energy in the case of elastase.

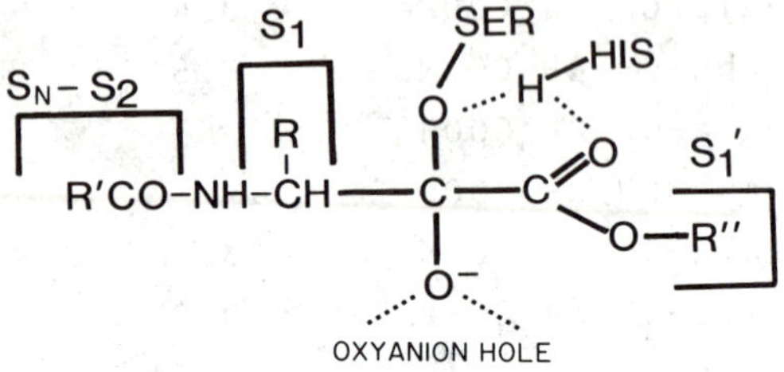

Figure 2. Proposed Model for the Interaction of a Peptide α-Ketoester with the Active Site of Elastase.

HL elastase is thought to be responsible for the destruction of lung connective tissue elastin which is observed in pulmonary emphysema. The inhibitors reported here may be useful in therapy and are currently being tested in animals. This research was supported by a grant (HL 29307) from the National Institutes of Health. We are grateful to Dr. James Travis and his research group at the U. of GA for supplying the HL elastase and cathepsin G used in this study.

References

1. Walsmann, P., Markwardt, F., Richter, P., Sturzebecher, J., Wagner, G., and Landmann, H. (1974) Pharmazie 29, 333-336.
2. Meyer, R.F., Nicolaides, E.D., Tinney, F.J., Luney, E.A., Holmes, A., Hoefle, M.L., Smith, R.D., Essenburg, A.D., Kaplan, H.R., and Almquist, R.G. (1981) J. Med. Chem. 24, 964-969.
3. Charles, I., Latham, D.W.S., Hartley, D., Oxford, A.W., and Scopes, D.I.C. (1980) J. Chem. Soc. Perkin I, 1139-1146.
4. Walter, J., and Bode, W. (1983) Hoppe-Seyler's Z. Physiol. Chem. 364, 949-959

STRUCTURE-ACTIVITY RELATIONSHIPS IN α-CONOTOXINS: A MODEL

William R. Gray, David M. Middlemas, Regina Zeikus, Baldomero M. Olivera and Lourdes J. Cruz

Department of Biology, University of Utah
Salt Lake City, Utah 84112

Introduction

The genus Conus comprises approximately 300 species of predatory snails. Among them are many that paralyze fish, and several human deaths have resulted from stings by the large *C. geographus*. We have purified a remarkable variety of peptides from the venom, properties of some of which are given in Table I. There are at least three classes of paralytic toxins which attack successive stages in muscle excitation: ω block release of acetylcholine (ACh) from nerve terminals[1], α inhibit activation of the receptor by ACh[2 3], and μ block transmission of the signal along muscle membrane[4]. We have argued[1] that a set of this type may be more potent than any one toxin. Other peptides are active intracerebrally (i.c.), but their physiological functions are unknown.

One notable feature is the presence of modified amino acids, including Hypro and γ-carboxyglutamate[5], both of which are linked to Ca metabolism in vertebrates. Most of the peptides are amidated at the C-terminus, and many have disulfides.

This paper is concerned mainly with the α-conotoxins GI and MI from *C. geographus* and *C. magus* respectively.

Table I. Some Major Peptides from Conus geographus Venom

Class(exx)	Target	Effect	Refs
α(GI, MI)	Acetylcholine receptor	Paralysis (i.p.)	2,3
E C C N R A C G R H Y S C*			
G R C C H P A C G K N Y S C*			
μ (GIIIA)	Muscle Na Channel	Paralysis (i.p.)	4,10
R D C C T Hy Hy K K C K D R Q C K Hy Q R C C A*			
ω_1(GVIA)	Presynaptic Ca Channel	Paralysis (i.p.,fish) Shaking (i.c., mice)	1
C K S Hy G S S C S Hy T S Y N C C R S C N Hy Y T K R C Y*			
ω_2(GVIIA)	Presynaptic Ca Channel	Shaking (i.c., mice)	1
C K S Hy G T Hy C S R G M R D C C T S C L L Y S N K C R R Y˙			
GV	Unknown	"Sleep" (i.c., mice)	5
G E γ γ L Q γ N Q γ L I R γ K S N*			

* indicates amidated C-terminus; ˙ status unknown

Structure-Function Relationships in α-Conotoxins

Once it was shown that the α-conotoxins block the nicotinic AChR, and that they compete directly with α-bungarotoxin, we explored their possible relationships to other AChR antagonists. These include alkaloids such as d-tubocurarine, as well as the snake α-neurotoxins which are much larger than α-conotoxins.

Amino acid sequence comparison shows no direct homology between the snail and snake toxins. Since Hider and Dufton[6] proposed a functional resemblance between the alkaloids and the "active tip" of short neurotoxins from snakes, we investigated whether GI could assume a conformation similar to that of the active tip of erabutoxin b, whose crystal structure was known. First we built a CPK model of the snake toxin, based on the coordinates from Kimball _et al_[7], and then tested GI for models that might resemble it.

For the α-conotoxins the disulphide bridges impose major constrains on structure, as there are two in only 13 amino acid residues. Our direct analysis[2], which was confirmed by synthesis, showed that bridges link Cys2-Cys7 and Cys3-Cys13 in GI. Independent proof came from Nishiuchi and Sakakibara's elegant synthesis of all three disulphide isomers.[8]

Calculation of β-turn potentials by Chou and Fasman's method[9] gave the results included in Table II for GI and MI. There is a high probability of turns initiating at Asn/His4, Gly8, and His/Asn10. We concentrated on a model for GI having turns starting at Asn4 and Gly8, because we were readily able to dispose the side-chains so that a close resemblance to erabutoxin could be obtained.

Table II. Sequences and Turn Potentials* for α-Conotoxins

	0	1	2	3	4	5	6	7	8	9	10	11	12	13	
GI		E	C	C	N	P	A	C	G	R	H	Y	S	C	- NH2
β		3	10	2	22	14	5	7	13	4	15	.	.	.	
MI	G	R	C	C	H	P	A	C	G	K	N	Y	S	C	- NH2
β	16	2	5	1	19	14	6	8	28	6	17	.	.	.	

* Values are expressed as Pt values x 10^5 (Chou and Fasman)[9]

Description of GI Model: Relation to Erabutoxin

The chain forms a tightly-bridged double loop. One reverse turn, made from -Asn.Pro.Ala.Cys-, is folded back to form a snug pocket for the Tyr side-chain. The second turn, -Gly.Arg.His. Tyr-, is brought back next to the amino terminus. All charged residues are thus clustered and there is a relatively hydrophobic rear surface. An H-bonded salt bridge was made between Glu1 and Arg9, to mimic more closely the crystal structure of erabutoxin[7] - but see below. A highly schematic drawing of the chain fold is shown in Fig. 1.

An end-on view of the conotoxin model is shown in Fig. 2, with a similar view of the "active tip" of the erabutoxin

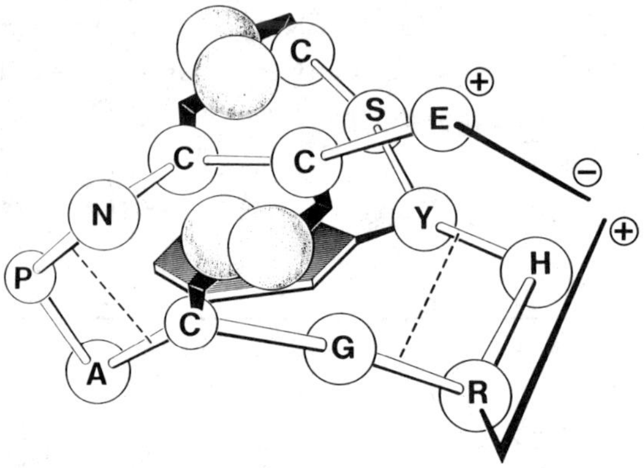

Fig. 1. Schematic chain-fold for conotoxin GI. Approx. main chain torsion angles (ϕ, ψ) are as follows: Glu1 (nd, -40); Cys2 (-150,130); Cys3 (-120,-60); Asn4 (-180,-70); Pro5 (-60, 110); Ala6 (60, 40); Cys7 (-60,160); Gly8 (150,-150); Arg9 (-60,130); His10 (60,30); Tyr11 (-60,150); Ser12 (-120, 50); Cys13 (70,150). Sulphur atoms of Cys are stippled; side chains of Tyr,Glu, and Arg are indicated schematically; possible H-bonds across the reverse turns are shown by dotted lines.

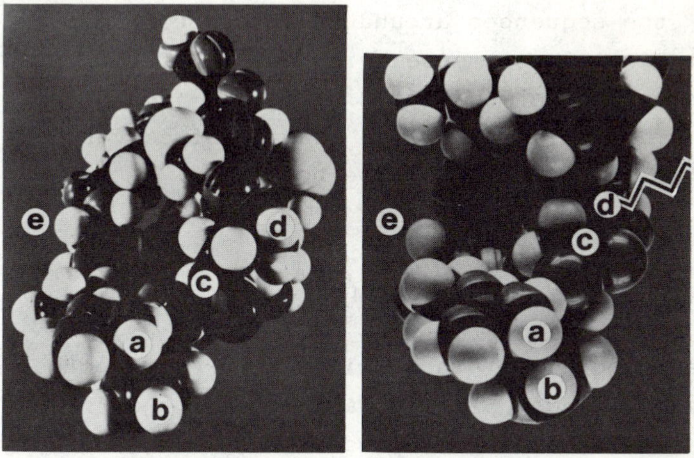

GI **Erabutoxin**

Fig. 2. CPK models of conotoxin GI (A) and active tip of erabutoxin (B). Letters indicate the following features: a, His10 (A), His32 (B); b, Arg9 (A), Arg33 (B); c, Glu1 (A), Asp31 (B); d, α-amino (A), Lys47 (B); e, Gly8 (A), Gly34 (B). Lys47 is not physically present in the model.

model. There is immediately apparent a strong resemblance between the two, which is emphasized by our having replaced Phe32 by His in erabutoxin: this is by far the most common amino acid in this position in the short α-neurotoxins from snakes.[11]

The general disposition of amino acid side-chains is similar in the two models, with charged residues on one side and the other side more hydrophobic. Two of the three "functionally invariant" residues of erabutoxin (Arg33 and Gly34) have direct counterparts in Arg9 and Gly8. All the natural α-conotoxins can take up this same conformation without distortion of the basic chain fold.

A striking feature of the models of the two toxins is that very similar structures can be obtained from non-homologous sequences. Behind this seeming contradiction is a curious pattern: the sequences around the proposed binding loop are related in an inverse sense, with -Gly.Arg.His.Tyr.Ser- in the conotoxin replacing -Ser.Asp.His.Arg.Gly- in the snake toxins. Presumably this is an example of convergent evolution: we call it "reverse homology".

Structure-Activity Relationships

In relating GI to erabutoxin we link it to other groups of cholinergic ligands. Pauling and Petcher[12] carried out detailed model-building studies of low-molecular weight compounds of this type, based on x-ray structures. They proposed that an "ideal" antagonist should be rigid and contain two cationic groups separated by about 11A, the intervening region being lipophilic. Very flexible molecules which have these features, such as decamethonium, may bind successfully but still allow conformational change of the receptor: they may thus act as agonists rather than antagonists.

Hider and Dufton[6] made the interesting proposal that the active tip of erabutoxin brings together a number of chemical features similar to those present in nicotinic antagonists such as curare and the calabash alkaloids. In erabutoxin the cationic roles are played by the side chains of Arg33 and Lys47 of the adjacent loop. The lipophilic region is provided by Trp29 and possibly Phe(His)32.

The α-conotoxins provide a novel set of molecules with which to test these ideas. Table III shows the proposed functional significance of the various parts of GI. In the following discussion of variants we assign the relative potency (P) of GI as 1.00, and compare other peptides to it. Since Nishiuchi and Sakakibara[13] obtained different relative activities than we did for GI and MI (1.3 rather than 2.4), comparisons of MI variants are given relative to MI itself. The variants include: natural GIA, GII, and MI[2,3], chemically modified GI[2]; and synthetic analogues of GI and MI.[13]

Table III. Functional Assignments for Conotoxin GI

Disulphides		RESIDUE	PROPOSED FUNCTION	
		+ NH_3	Cationic group for binding to AChR	
	1	Glu		
	2	Cys		
	3	Cys		
	4	Asn	Strong beta turn potential "Non-binding loop"	
	5	Pro		
	6	Ala		
	7	Cys		
	8	Gly	Sterically free	Turn: "Binding loop"
	9	+ Arg	Cationic site	
	10	·His	Indifferent	
	11	Tyr	Bulk to fill "back pocket"	
	12	Ser		
	13	Cys		
		$CO.NH_2$	Relatively indifferent	

Cationic groups. In GI we propose that the α-amino and the side chain of Arg9 play the roles of the paired cationic groups. As depicted in Fig. 2 the groups are separated by about 11A. Removal of Glu1 by Edman degradation leaves a positive charge on the newly exposed amino of Cys2, which is still at an acceptable distance. Such a derivative was found to be fully active (P = 1.0)[2], showing clearly that the putative salt bridge involving the Glu side chain is completely unnecessary.[2] Additional positive charges at the amino end result in increased potency relative to GI: GI

(Glu(+-), P=1.0); MI (Gly.Arg(++), P=2.4)[3]; desGly-MI (Arg(++), P > 1.7)[13]; Arg1-MI (Arg.Arg(+++) , P > 1.7).[13] Removal of the charged group results in substantial of activity: α-TNP-GI (TNP-Glu(-), P = 0)[2]; replacement of Gly.Arg.Cys- in MI by mercaptopropionyl decreases activity ten-fold, despite retaining correct disulphide arrangement.[13]

The other cationic group (Arg9) can be replaced by Lys in GII and MI, and the Lys of the latter can be replaced by Orn with little change of potency.[13] Nle9 conotoxin MI has only 30% of the activity of the native toxin[13]; however, there are two charges on the Gly.Arg- of the amino terminus, and these can be extended to a separation of about 10A.

Other modifications which decrease overall positive charge also decrease potency, as shown by two synthetic analogues that lack the carboxy-terminal amide; desamido-MI (30% of native activity), and des-(Gly.Arg) desamido-MI (3% of native activity).

Conformational Determinants

The two disulphide bridges are of critical importance, as shown by the effects of reduction, of wrong pairing, and of derivatization: all these result in loss of activity.[2 8 13]

Exchanging Phe for His4, or Gly for Pro5, reduces the β-turn potential about three-fold for the first loop. Activity was substantially reduced when these MI analogues were tested.[13]

In our model only Gly or a D-amino acid are acceptable at position 8; substitution by L-Ala drastically inactivates MI.[13] All natural α-conotoxins have Gly at this position.[2 3]

<u>Reverse side of binding loop</u>. We do not consider His10 and Tyr11 to be directly involved in binding, as they are on the opposite face of the molecule to the cationic groups. They can tolerate major changes. His10 has been replaced by Asn[3],

Ala[13], Gln[13], monoiodo- and diiodo-His[2], with relatively minor effects. Likewise, Tyr11 can be substituted by Phe, monoiodo-, and diiodo-Tyr. The toxin with both His and Tyr fully iodinated (tetraiodo-GI) still retains about 50% of its potency, despite being > 25% by weight of iodine[2]. However, exchanging Tyr11 by D-Tyr abolishes activity.[13] This could be due to major conformational change, or to direct interference with binding, as the D-Tyr side chain is brought onto the proposed binding face of the molecule.

Conclusion. While recognizing the limitations of "free" model-building, we feel that the backbone of the α-conotoxins is under rather tight constraints due to the disulphides. This is not necessarily true of some of the side chains, especially the hydrophilic ones. The toxins provide an unusually good system with which to explore structure activity relationships.

Acknowledgements

It is a pleasure to acknowledge the collaboration with Drs. J. Rivier, R. C. Sheppard, and E. Atherton. The work has been supported by USPHS grants GM 22737 and GM 34913.

References

1. Olivera, B.M., McIntosh, J.M., Cruz, L.J., Luque, F.A. and Gray, W.R. (1984). Biochemistry 23:5087; Olivera, B.M., Gray, W.R., Zeikus, R., McIntosh, J.M., Varga, J., Rivier, J., de Santos, V. and Cruz, L.J. (1985). Science, submitted.

2. Gray, W.R., Luque, F.A., Galyean, R., Atherton, E., Sheppard, R.C., Stone, B.L., Reyes, A., Alford, J., McIntosh, M., Olivera, B.M., Cruz, L.J. and Rivier, J. (1984) Biochemistry 23:2796.

3. McIntosh, M., Cruz, L.J., Hunkapiller, M.W., Gray, W.R. and Olivera, B.M. (1982) Arch. Biochem. Biophys. 218:329-334.

4. Cruz, L.J., Gray, W.R., Olivera, B.M., Zeikus, R.D., Kerr, L., Yoshikami, D. and Moczydlowski, E. (1985) J. Biol. Chem., in the press.

5. McIntosh, J.M., Olivera, B.M., Cruz, L.J. and Gray, W.R. (1984) J. Biol. Chem. 259:23.

6. Hider, R.C. and Dufton, M.J. (1979) in "Natural Toxins", eds. Eaker, D. and Wadstrom, T. Pergamon, pp. 515-522.

7. Kimball, M.R., Sato, A., Richardson, J.S., Rosen, L.S. and Low, B.W. (1979) Bioch. Bioph. Res. Comm. 88:950-959.

8. Nishiuchi, Y. and Sakakibara, S. (1982) FEBS Lett. 148:260-262.

9. Chou, P.Y. and Fasman, G.D. (1979) Biophys J. 26:367.

10. Sato, S., Nakamura, H., Ohizumi, Y., Kobayashi, J. and Hirata, Y. (1983) FEBS Lett. 155:2.

11. Karlsson, E. (1979) Handbook of Experimental Pharmacology 52:159-204.

12. Pauling, P. and Petcher, T.J. (1973) Chem-Biol. Interactions 6:351.

13. Nishiuchi, Y. and Sakakibara, S. (1985) in "Peptides 1984", in the press.

SNAKE VENOM NEUROTOXINS: CONFORMATION AND TOPOGRAPHY OF THEIR BINDING TO NICOTINIC ACETYLCHOLINE RECEPTOR

Victor I. Tsetlin
Shemyakin Institute of Bioorganic Chemistry, USSR Academy of Sciences, ul. Miklukho-Maklaya 16/10, Moscow

Introduction

At present, spatial structures in the crystal state and in solution are known for several representatives of the so-called short-chain (60-62 residues, 4 disulfides) and long-chain (66-74 residues, 5 disulfides) neurotoxins of postsynaptic action (see reviews[1-3]). The target of these neurotoxins - nicotinic acetylcholine receptor - is also well characterized[4]. The purpose of the present report is to summarize briefly the studies on the solution conformation of neurotoxins and their binding to the acetylcholine receptor carried out at our Institute (main references are given in the review[3]). These two problems are closely related and to solve them we relied upon a fine ajustment of spectroscopic and chemical approaches. Namely, the derivatives containing spin or fluorescence labels in the specified positions were prepared to shed light on the neurotoxin conformation. It was verified that incorporation of labels does not considerably perturb the spatial structure of neurotoxins.

The availability of spin and fluorescence labeled analogues made possible experimental studies of the neurotoxin-receptor topography using EPR and fluorescence spectroscopy. The knowledge of conformational properties of these analogues and, in particular, of the reporter group microenvironment was essential for meaningful interpretation of the binding studies.

Another means of elucidating the spatial structure of neurotoxin-receptor complexes relied upon photoinduced cross-linking between the acetylcholine receptor and a series of neurotoxins bearing photoactivable labels. Collectively, EPR, fluorescence and photolabeling studies allowed the delineation of the "binding area" in neurotoxins and shed light on the disposition of neurotoxin molecules in the binding site of the receptor.

Conformational Studies

Chemical modification. Analysis of spatial structure - biological activity relationships required an appropriate series of naturally occurring homologous neurotoxins and their chemically modified derivatives[5]. A large array of neurotoxin II *N.naja oxiana* derivatives have been prepared via amino group modification (Table I). In order to obtain a series of singly labeled compounds, mild acylating reagents were used at low reagent-protein ratio, and the reaction mixture was subsequently resolved by ion exchange chromatography. Thus, 5 acetylated derivatives were prepared, and their trifluoroacetylation was a key step in the ^{19}F NMR signal assignments for totally trifluoroacetylated neurotoxin II[6].

Another group contains either extrinsic fluorescence labels (Dns) or the groupings (TNP, SL) quenching fluorescence of intrinsic fluorophores. Singly modified neurotoxin II derivatives with grafted spin labels were obtained by reaction with N-hydroxysuccinimidyl 2,2,6,6-tetramethyl-4-carboxymethylpiperidine-1-oxyl in 6M Gu·HCl[7,8]. At optimal reagent-protein ratio it was possible to achieve the predominant formation of di-spin-labeled neurotoxins[9]. Neurotoxin II reactions with 2,2,6,6-tetramethyl-4-(2-iodacetamido)-piperidyl-1-oxyl or with 4-amino-2,2,6,6-tetramethylpiperidyl-1-oxyl in the presence of water-soluble carbodimide gave rise to complex mixtures from which we could isolate the derivatives bearing the spin labels at His 32 and Glu 2, respectively.

A number of spin and fluorescence labeled long-chain neurotoxins were also prepared[8,10]. Noteworthy, considerable differences were observed in the reactivity of residues occupying identical positions in various neurotoxins that may be relevant to different efficiency of the receptor binding reported in literature for the short-chain and long-chain neurotoxins.

Table I. Neurotoxin II Chemically Modified Derivatives*

Leu 1	Lys 15	Lys 26	Lys 27	Lys 45	Lys 47	Glu 2	His 31
	Ac	Ac	Ac	Ac	Ac		
(CF_3CO,	Ac,	CF_3CO,	CF_3CO,	CF_3CO,	CF_3CO)		
(CF_3CO,	CF_3CO,	CF_3CO,	Ac,	CF_3CO,	CF_3CO)		
(CF_3CO,	CF_3CO,	CF_3CO,	CF_3CO,	CF_3CO,	CF_3CO)		
(CF_3CO,	CF_3CO,	CF_3CO,	SL**	CF_3CO,	CF_3CO)		
Dns	Dns	Dns	Dns	Dns	Dns		
			TNP				
SL	SL	SL	SL	SL	SL	SL''	SL'''
	(SL',	SL')	(SL',	SL')			
		(SL',	SL')	(SL',	SL')		
N_3Bz	N_3Bz	N_3Bz	N_3Bz	N_3Bz	N_3Bz		

*Label position in singly modified derivatives is shown. The parentheses indicate simultaneous labeling of several residues numbered according to homology series[1].

**SL = O-N 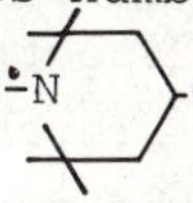 CH_2CO; SL' = O-N 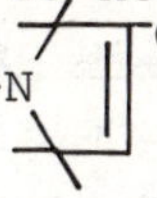 CO; SL'' = O-N 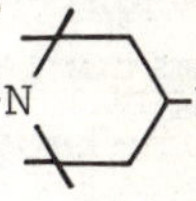 NH; SL''' = O-N 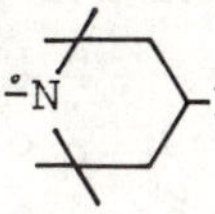 $NHCOCH_2$; N_3Bz = N_3  CO; TNP - trinitrophenyl.

The neurotoxin II reaction with N-hydroxysuccinimidyl *p*-azidobenzoate afforded 6 singly modified photoactivable derivatives. A crucial step in localizing the labels was mild reduction of the azido group and disulfide bridges[11].

NMR investigation of neurotoxin II *N.naja oxiana*. Earlier the resonances in the aromatic region of the ^{1}H NMR spectra were assigned and the microenvironment of several functionally important residues was elucidated[5]. A considerable number of slowly exchanging amide hydrogens were detected. Along with differential scanning calorimetry data[12] and CD curves recorded over a wide range of temperature, pH, and solvent polarity[2,6] it evidences in favor of a rigid spatial structure.

Later all the CH_3-signals were resolved, assigned to amino acid types or to a particular position in the polypeptide sequence. A network of intramolecular interactions between the specified residues was traced via Nuclear Overhauser effects or titratable group influence on the chemical shifts[13].

^{1}H NMR studies on the neurotoxin II derivatives with spin labels at Lys 26, Lys 27, Lys 45 or Lys 47 provided a set of distances from the spin label unpaired electron to spatially close protons. Thus, nearness of side chains belonging to residues remote in the amino acid sequence was disclosed.

All resonances in the ^{19}F NMR spectrum of hexa(trifluoroacetyl)neurotoxin II were assigned and the proximity of the Lys 27 and Lys 47 side chains was surmized from the mutual influence of fluorine atoms[6]. A possibility of determining intramolecular distances basing on ^{19}F signal broadening by covalently attached spin labels was demonstrated using a neurotoxin II derivative with the spin label at Lys 27 and trifluoroacetyl groups at the remaining amino groups[14].

EPR studies on spin labeled neurotoxins. EPR spectra of mono-spin-labeled derivatives displayed marked differences in mobility of labels located in distinct sites of the neurotoxin molecule[7]. By varying temperature and viscosity of the medium[15] the protein environment was shown to impose the most pronounced restrictions on the motional freedom of the spin labels at positions Lys 26 and Lys 27. Spin label exposure on the neurotoxin surface was assessed by a paramagnetic probe technique[7,8]. EPR studies of di-spin-labeled neurotoxin II derivatives at 77K provided a set of inter-spin distances and thus shed light on the spatial arrangement of lysine residues[9].

Fluorescence studies. These pursued a dual aim: to verify the conformational "nativity" of modified derivatives, and to characterize the microenvironment of tryptophan residues[10]. Availability of specifically labeled neurotoxins allowed us to exploit more fully the possibilities of fluorescence spectroscopy. For example, analysis of pH-dependences and fluorescence-quenching effects of covalently bound spin labels provided the distances from the spin labels to Trp 28 and Trp 29 side chains.

Comparison of short-chain and long-chain neurotoxins. Intramolecular distances measured by NMR, EPR and fluorescence spectroscopy underlay reconstruction of the polypeptide backbone folding in neurotoxin II. The solution conformation is similar to those of other homologous short-chain neurotoxins and to the X-ray structure of erabutoxin *b* (cf. review[2]). Although not accurate as X-ray analysis, solution studies allowed us to elucidate in detail the topology of certain regions of the neurotoxin surface, reactivity of functional groups and dynamic features.

Specific labels facilitated signal assignments in the NMR spectra of the long-chain neurotoxins. Analysis of the CD, fluorescence and ^{1}H NMR pH-dependences demonstrated that protonation of the His 22 imidazole ring entails a change in its spatial disposition, thereby affecting the microenvironment of some residues[8,16]. ^{1}H NMR analysis of toxin 3 *N.naja siamensis* and its monoacylated analogues provided the pK values for lysine residues. Several intramolecular interactions were disclosed by ^{1}H NMR studies of toxin 3 and neurotoxin I spin labeled at Lys 27 and His 71, respectively.

The long-chain neurotoxins manifested considerable flexibility of certain segments which might hamper reconstruction of their conformation basing on the identified intramolecular "contacts". However, application of two-dimensional NMR spectroscopy to toxin 3 made possible quite a detailed description of the solution conformation[16]. The short- and long-chain neurotoxins appear to have a similar overall conformation[2,3].

For better understanding of neurotoxin - receptor interactions it is essential not only to define common features of various neurotoxins, but also to specify even minor differences between the solution and crystal structures, or between distinct neurotoxins in solution. Another challenging task is to identify those interactions which, along with the disulfide bridges, ensure high stability of the neurotoxin spatial structure. For instance, ^{1}H NMR and CD studies on native and modified neurotoxin II in aqueous and organic media demonstrated that ionic interactions between amino group(s) and a carboxylic group with $pK<2$ (apparently, C-terminal Asn 62) stabilized the neurotoxin II conformation[6,13].

Neurotoxin - Acetylcholine Receptor Interaction

Binding studies were performed using *Torpedo marmorata* acetylcholine receptor purified by affinity chromatography on neurotoxin-Sepharose 4B in Triton X-100. When neurotoxin II with dansyl group at Lys 47 interacts with the receptor, a marked enhancement of the dansyl fluorescence and a blue shift of the emission maximum are observed[7]. These changes indicate that the label senses more hydrophobic environment in the receptor complex. The energy transfer from the tryptophan residue(s) of the receptor onto a dansyl group was also detected[8]. Similar spectral changes were observed with neurotoxin II derivatives dansylated at Lys 26 or Lys 27. If dansyl groups are attached to Leu 1 or Lys 15, the receptor does not affect their microenvironment.

Stoichiometry of the toxin-receptor complexes, 2:1, was determined at the titration of the acetylcholine receptor with dansylated derivatives of neurotoxin II. Hexa(trifluoroacetyl)-

neurotoxin II was discovered to compete for only one of the two neurotoxin binding sites in the acetylcholine receptor[7].

Acetylcholine receptor interaction with mono-spin-labeled derivatives of neurotoxin II was monitored by EPR spectroscopy[7]. Mobilities of the labels in certain positions (e.g. Lys 45) were not restricted in the toxin-receptor complexes, whereas limitations were clearly seen for other groups (e.g. Glu 2, Lys 27, His 32). Application of paramagnetic probes (ferricyanide and nickel ions) allowed us to distinguish whether a spin label was buried or exposed on the receptor surface. These experiments also suggested the presence of a positive charge in the receptor nearby the Leu 1 spin label, and negative charges close to bound Glu 2 and His 32 spin labels[8].

Fluorescence and EPR studies provided experimental evidence for multi-point binding of neurotoxin II to the acetylcholine receptor. In case of long-chain neurotoxins we did not have a series of derivatives with different labels at the same positions of the polypeptide chain and our knowledge of their binding topography is more fragmentary. For example, spin label at His 71 and dansyl groups at Lys 39 and Lys 60 of neurotoxin I took no part in binding, while the spin label at Lys 27 in toxin 3 was screened from the outer medium by the receptor groups[8,10]. Binding of long and short neurotoxins seems to be of a similar type[16], and available data are insufficient to relate conformational differences between them with the different kinetics of the receptor binding.

Six *p*-azidobenzoyl derivatives of neurotoxin II (Table I) bind efficiently to the acetylcholine receptor. Irradiation of the toxin labeled at Lys 45 failed to produce any toxin-receptor cross-links, whereas the toxins modified in positions Leu 1, Lys 15, Lys 26, Lys 27 and Lys 47 formed the covalent complexes with the receptor[8,11]. The neurotoxin labeled at Lys 26 manifested unusual properties: formation of covalent bonds induced in the acetylcholine receptor a conformational rearrangement and it acquired the capacity to bind up to three neurotoxin molecules, in addition to normal 2:1 stoichiometry. Noteworthy, only the short-chain neurotoxins but not the long ones can bind to these "secondary" sites[17].

The complementary photolabeling and spectroscopic approaches showed binding to involve a large area of the neurotoxin molecule. It gave us grounds to suggest that two neurotoxin molecules cannot be totally accommodated within the boundaries of the two α-subunits in a pentameric ($\alpha_2\beta\gamma\delta$) receptor complex, but should also contact the other subunits[8].

Indeed, SDS polyacrylamide gel electrophoresis demonstrated labeling of α,β,γ or δ-subunit, depending on the position of [^{14}C]-*p*-azidobenzoyl group in the neurotoxin II molecule[18]. For instance, with the label at Lys 27 the radioactivity was predominantly incorporated into the δ-subunit.

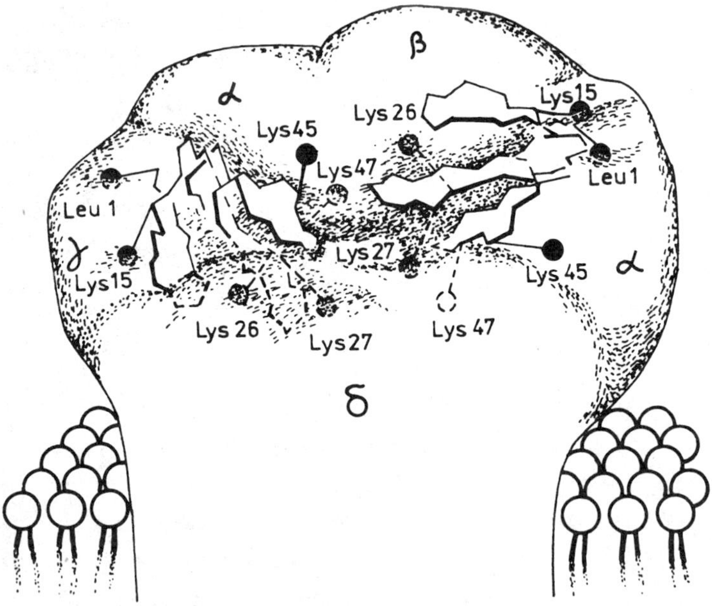

Fig. 1. Model for neurotoxin II disposition in the binding site of acetylcholine receptor.

Photolabeling under the conditions when one binding site was protected by hexa(trifluoroacetyl)neurotoxin II ensured the distinction of the two binding sites in terms of their contacts with the specified regions of the neurotoxin molecule. The obtained results allowed us to propose a model for the arrangement of neurotoxins in the binding site of the acetylcholine receptor (Figure 1).

Recent literature points to a dominant contribution of α-subunits into the neurotoxin binding. Therefore, some of the contacts between the receptor and labeled neurotoxins might not play a significant functional role. However, reconstruction of the whole network of contacts - at first at the level of the distinct subunits and then in terms of their specified amino acid residues - should provide valuable information on the acetylcholine receptor exposed domain, including the neurotoxin binding site and its nearest surroundings.

Acknowledgments

The participation of Drs. A. Arseniev, Yu. Utkin, E. Karlsson (Uppsala) A. Surin, K. Pluzhnikov, V. Pashkov, V. Kondakov and A. Karelin in the project is acknowledged. The author is grateful to Professors Yu. Ovchinnikov, V. Ivanov and V. Bystrov for many fruitful discussions.

References

1. Karlsson, E. (1979) in Handbook of Experimental Pharmacology, vol. 52, Lee, C.Y., Ed., Springer-Verlag, Berlin-Heidelberg, pp. 159-212.
2. Dufton, M.J., and Hider, R.C. (1983) CRC Crit. Rev. Biochem. 14, 113-171.
3. Ovchinnikov, Yu.A. (1984) Pure & Appl. Chem. 56, 1049-1068.
4. Popot, J.L., and Changeux, J.-P. (1984) Physiol. Rev. 64, 1162-1239.

5. Arseniev, A.S., Balashova, T.A., Utkin, Yu.N., Tsetlin,V.I., Bystrov, V.F., Ivanov, V.T., and Ovchinnikov, Yu.A. (1976) Eur. J. Biochem. 71, 595-606.

6. Tsetlin, V.I., Arseniev, A.S., Utkin, Yu.N., Gurevich, A.Z., Senyavina L.B., Bystrov, V.F., Ivanov, V.T., and Ovchinnikov, Yu.A. (1979) Eur. J. Biochem. 94, 337-346.

7. Tsetlin, V.I., Karlsson, E., Arseniev, A.S., Utkin, Yu.N., Surin, A.M., Pashkov, V.S., Pluzhnikov, K.A., Ivanov, V.T., Bystrov, V.F., and Ovchinnikov, Yu.A. (1979) FEBS Letters 106, 47-52.

8. Tsetlin, V.I., Karlsson, E., Utkin, Yu.N., Pluzhnikov,K.A., Arseniev, A.S., Surin, A.M., Kondakov, V.I., Bystrov, V.F., Ivanov, V.T., and Ovchinnikov, Yu.A. (1982) Toxicon 20, 83-93.

9. Utkin, Yu.N., Pashkov, V.S., Pluzhnikov, K.A., Kuryatov,A.B., Arseniev, A.S., Tsetlin, V.I., Bystrov, V.F., and Ivanov,V.T. (1983) Bioorg. Khim. (USSR) 9, 437-449.

10. Surin, A., Utkin, Yu., Pluzhnikov, K., and Tsetlin, V. (1983) in Toxins as Tools in Neurochemistry, Hucho, F., and Ovchinnikov Yu.A., Eds., Walter de Gruyter, Berlin-N.Y., pp. 173-191.

11. Pluzhnikov, K.A., Karelin, A.A., Utkin, Yu.N., Tsetlin,V.I., and Ivanov, V.T. (1982) Bioorg. Khim, (USSR) 8, 905-913.

12. Khechinashvili, N.N., and Tsetlin, V.I. (1984) Mol. Biol. (USSR) 18, 786-791.

13. Pashkov, V.S., Pluzhnikov, K.A., Utkin, Yu.N., Hintsche, R., Arseniev, A.S., Tsetlin, V.I., Ivanov, V.T., and Bystrov, V.F. (1983) Bioorg. Khim. (USSR) 9, 1181-1219.

14. Arseniev, A.S., Utkin, Yu.N., Pashkov, V.S., Tsetlin, V.I., Ivanov, V.T., Bystrov, V.F., and Ovchinnikov, Yu.A. (1981) FEBS Letters 136, 269-274.

15. Timofeev, V.P., and Tsetlin, V.I. (1983) Biophys. Struct. Mech. 10, 93-108.

16. Bystrov, V.F., Tsetlin, V.I., Karlsson, E., Pashkov, V.S., Utkin, Yu.N., Kondakov, V.I., Pluzhnikov, K.A., Arseniev, A.S., Ivanov, V.T., and Ovchinnikov, Yu.A. (1983) in Toxins as Tools in Neurochemistry, Hucho, F. and Ovchinnikov, Yu.A., Eds., Walter de Gruyter, Berlin-N.Y., pp. 193-233.

17. Tsetlin, V.I., Pluzhnikov, K.A., Karelin, A.A. (1984) 16th FEBS Meeting, Abstracts, p. 326.

18. Tsetlin, V., Pluzhnikov, K., Karelin, A., and Ivanov V. (1983) in Toxins as Tools in Neurochemistry, Hucho, F., and Ovchinnikov, Yu.A., Eds., Walter de Gruyter, Berlin-N.Y., pp. 159-169.

A PRELIMINARY STUDY OF THE RING-CURRENT INDUCED SHIFTS IN THE NMR SPECTRUM OF THE VARIANT-3 NEUROTOXIN FROM CENTRUROIDES SCULPTURATUS EWING

David G. Nettesheim, Charles E. Bugg and N. Rama Krishna

Comprehensive Cancer Center and the Department of Biochemistry
University of Alabama at Birmingham, University Station,
Birmingham, AL 35294

Dean D. Watt

Department of Biochemistry, Creighton University, Omaha, NB
68178

Introduction

The venom from the scorpion Centruroides sculpturatus Ewing contains a large number of small basic proteins which are responsible for the toxic effects of the venom. These proteins bind to the sodium channels of excitable membranes to enhance the sodium inward current and to delay sodium current inactivation.[1] They also exhibit target specificity. As part of a long range program on the characterization of the conformation activity relationships among these neurotoxins, we have initiated a series of experiments aimed at a detailed solution phase structural characterization of the variant-3 (CsE-v3) neurotoxin (M.W. $\simeq$ 7,000) from the Centruroides venom. In this paper, we present a preliminary comparison of the experimental NMR spectrum with the theoretical spectrum based on ring-current calculations[2] using the x-ray crystallographic structure for this molecule.[3]

Results and Discussion

The amino acid sequence for this toxin, shown in Figure 1, is characterized by six tyrosines, one phenylalanine and one

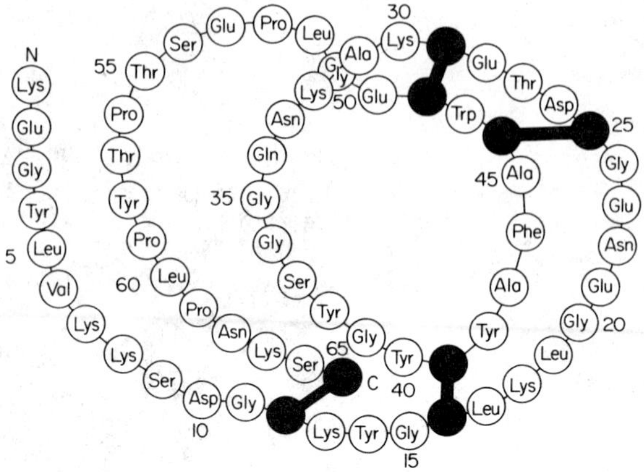

Figure 1. The amino acid sequence for the CsE-v3 toxin

tryptophan. A preliminary 400 MHz ^{1}H NMR study of this toxin has been described earlier.[4] The ^{1}H NMR spectrum has been found to show substantial deviations from the random coil spectrum due to the presence of ring-current interactions. The ring-current induced shift δ_r (in ppm) due to a single aromatic ring can be expressed as $\delta_r x 10^{-6} = i\,BG\,(\rho, z, \phi)$ where i is the ring-current intensity factor, G is the spatial term relating the cylindrical coordinates ρ, z and ϕ of a proton relative to the ring plane and ring centers, and B is a constant of proportionality. A detailed discussion of the three main approaches in ring-current calculations viz, the dipolar, Johnson-Bovey, and the Haigh-Maillion equations, and a detailed discussion of the methodology together with specific examples are contained in the authoritative review by Perkins.[2]

For the ring-current calculations, the coordinates from a 1.5 A^o data set obtained from the orthorhombic CsE-v3 toxin were used after refinement by the Hendrickson-Konnert method. The theoretical NMR spectrum based on the Johnson-Bovey equation[2] was calculated for the static structure using the program RCCAL written by Perkins. Figure 2 shows a comparison of the experimental spectrum with the spectrum predicted for the

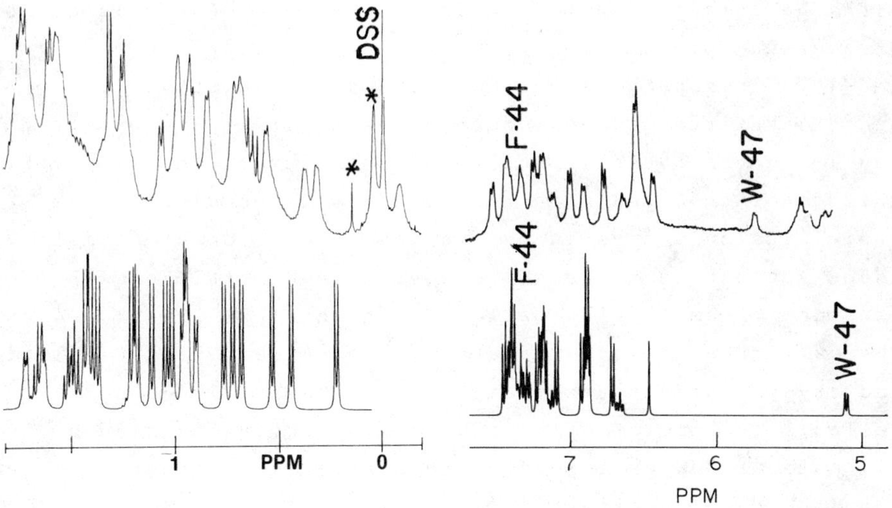

Figures 2 (left) and 3 (right). Comparison of the experimental (top, at pH = 6.8) and calculated (bottom) NMR spectra for the $C\underline{H}_3$ and the aromatic $C\underline{H}$ resonances.

methyl group containing residues (Leu, Val, Thr, and Ala) in the CsE-v3. Figure 3 shows a comparison of the aromatic CH resonance spectra. We have previously reported[4] that the Trp-47 in CsE-v3 shows an unusual spectrum with large upfield shifts (e.g., the $C_4\underline{H}$ resonance at 5.8 ppm in Figure 3 is upfield shifted by as much as 1.7 ppm from its random coil value), and based on NOE experiments we suggested that ring current interactions with a proximal tyrosine might be responsible for these shifts. This suggestion is confirmed by the present calculations which also predict the proximal tyrosine to be Tyr-4. There is a qualitative agreement between experiment and theory for Phe-44 and Trp-47. The assignments for the tyrosines are in progress and hence we could not yet judge the agreement for each of these at this point. Comparison of the spectra in Figure 2 shows a qualitative agreement for many of the CH_3 resonance positions. For example, two resonances tentatively assigned to A-45 and A-43 by 2D-NMR methods appear at 1.3 and 1.06 ppm respectively, in qualitative agreement with the calculations. Similarly, ring-current calculations predict the upfield CH_3 resonance at -0.08 ppm to be from Leu-19. This peak also shows 2D-NOESY connectivities to A-43 and F-44 NH resonances. Such connectivities are expected for Leu-19 on the basis of the x-ray structure. The ring current shifts for the remaining resonances will be discussed elsewhere. Calculations incorporating the motion of aromatic residues suggest that the NMR spectrum is compatible with the presence of small fluctuational motions ($\pm 20^{o}$) of Tyr-4 and Phe-44 in the CsE-v3. A detailed comparison of the solution and crystal structures must, however, await independent and reliable assignment of resonances for several residues. 2-Dimensional NMR experiments are currently in progress to accomplish this objective.

Acknowledgments

Support of this work by NIH grants NS-19183, CA-13148 and ES-02467 is gratefully acknowledged. N.R. Krishna is a Leukemia Society of America Scholar. The authors are indebted to Dr. Stephen J. Perkins for generously supplying a copy of his program RCCAL and for his advice on its use.

References

1. Watt, D.D., and Simard, J.M., (1984) J. Toxicol. Toxin Reviews, 3 181-221.
2. Perkins, S.J., (1982) in "Biological Magnetic Resonance" Vol. 4., Eds., Berliner, L.J., and Reuben, J., Plenum Press, New York, pp. 193-336.
3. Fontecilla-Camps, J.C., Almassy, R.J., Suddath, F.L., Watt, D.D., and Bugg, C.E., (1980) Proc. Natl. Acad. Sci. USA, 77 6496-6500.
4. Krishna, N.R., Bugg, C.E., Stephens, R.L., and Watt, D.D., (1983) J. Biomol. Struct. Dyna. 1 829-842.

CHEMICAL SYNTHESIS OF HEAT-STABLE ENTEROTOXINS PRODUCED BY ENTERIC BACTERIA —STRUCTURE AND BIOLOGICAL ACTIVITY—

Saburo Aimoto, Shoko Yoshimura, Yuji Hidaka, Haruo Ikemura and Yasutsugu Shimonishi

Institute for Protein Research, Osaka University, Yamada-oka, Suita, Osaka 565 Japan

Tae Takeda, Michiko Arita and Toshio Miwatani

Research Institute for Microbial Diseases, Osaka University, Yamada-oka, Suita, Osaka 565, Japan

Yoshifumi Takeda

The Institute of Medical Science, The University of Tokyo, Shirokanedai, Minato-ku, Tokyo 108 Japan

Introduction

Three different heat-stable enterotoxins (STs) were isolated from enterotoxigenic Escherichia coli strain SK-1, 18D and Yersinia enterocolitica, and their primary structures were determined as shown in Figure 1.[1 2 3] They have a similar region in their primary structures each other and the region has high contents of cystine residues.

E. coli STh [1]	NSSNYCCELCCNPACTGCY
E. coli STp [2]	NTFYCCELCCNPACAGCY
Yersinia ST [3]	QACDPPSPPAEVSSDWDCCDVCCNPACAGC

Fig. 1 Primary structures of E. coli STh, STp and Yersinia ST

In our previous papers [4 5 6] we reported the chemical synthesis of E. coli STs and the determination of the sequences essential for the expression of the toxic activity. In the present paper the chemical synthesis and the toxic activity of the shorter analogues of Yersinia enterocolitia ST (Yersinia ST) will be reported and the relation between structure and biological properties among E. coli STs and Yersinia ST will be discussed.

Result and Discussion

All the protected peptides were synthesized by a solution method. Coupling of peptide segments was carried out by an azide method. 4-Methylbenzyl group was used for the protection of SH-groups of Cys residues. The protected peptide obtained was converted to the native form after HF-treatment followed by air-oxidation. The air-oxidized peptide mixture was separated on reverse phase HPLC.

Each fraction was identified by amino acid analysis and fast atom bombardment mass spectometry. Further identification was carried out by comparing the synthetic ST with native ST on HPLC. In the case of the shorter analogue of Yersinia ST it was compared with the Edman degradation product of native Yersinia ST[13-30] on HPLC.

Each of the synthetic ST thus identified was also examined the cross-reactivities with a monoclonal or a polyclonal anti-STh antibody. These antibodies neutralized the toxicity of the synthetic ST which was expected to be neutralized. This fact suggests the structural homology of the synthetic ST to that of the native one. In most cases the main product in the air-oxidized peptide mixture was a right product.

The toxic activities and the immunological properties of synthetic STs are summarized in. Table I. Minimum

Table I Biological and Immunological Properties of Synthetic STs

No	Name	Amino Acid Sequence	MED(pmol)	PCA	MCA
		E. coli STh and Its Analogues			
		1 10 19			
1	*E. coli* STh[1-19]	Asn-Ser-Ser-Asn-Tyr-Cys-Cys-Glu-Leu-Cys-Cys-Asn-Pro-Ala-Cys-Thr-Gly-Cys-Tyr	0.4	+	+
2	*E. coli* STh[6-19]	Cys-Cys-Glu-Leu-Cys-Cys-Asn-Pro-Ala-Cys-Thr-Gly-Cys-Tyr	0.4	+	+
3	*E. coli* STh[6-18]	Cys-Cys-Glu-Leu-Cys-Cys-Asn-Pro-Ala-Cys-Thr-Gly-Cys	0.4	+	-
		E. coli STp and Its Analogues			
		1 10 18			
4	*E. coli* STp[1-18]	Asn-Thr-Phe-Tyr-Cys-Cys-Glu-Leu-Cys-Cys-Asn-Pro-Ala-Cys-Ala-Gly-Cys-Tyr	0.5	+	+
5	*E. coli* STp[5-18]	Cys-Cys-Glu-Leu-Cys-Cys-Asn-Pro-Ala-Cys-Ala-Gly-Cys-Tyr	0.6-0.7	+	+
6	*E. coli* STp[5-17]	Cys-Cys-Glu-Leu-Cys-Cys-Asn-Pro-Ala-Cys-Ala-Gly-Cys	0.5	+	+
		Yersinia ST Analogues			
		13 20 30			
7	*Yersinia* ST[13-30]	Ser-Ser-Asp-Trp-Asp-Cys-Cys-Asp-Val-Cys-Cys-Asn-Pro-Ala-Cys-Ala-Gly-Cys	14	w	-
8	*Yersinia* ST[14-30]	Ser-Asp-Trp-Asp-Cys-Cys-Asp-Val-Cys-Cys-Asn-Pro-Ala-Cys-Ala-Gly-Cys	14	w	-
9	*Yersinia* ST[15-30]	Asp-Trp-Asp-Cys-Cys-Asp-Val-Cys-Cys-Asn-Pro-Ala-Cys-Ala-Gly-Cys	15	w	-
10	*Yersinia* ST[16-30]	Trp-Asp-Cys-Cys-Asp-Val-Cys-Cys-Asn-Pro-Ala-Cys-Ala-Gly-Cys	16	w	-
11	*Yersinia* ST[17-30]	Asp-Cys-Cys-Asp-Val-Cys-Cys-Asn-Pro-Ala-Cys-Ala-Gly-Cys	9.1	w	-
12	*Yersinia* ST[18-30]	Cys-Cys-Asp-Val-Cys-Cys-Asn-Pro-Ala-Cys-Ala-Gly-Cys	10	w	-
13	*Yersinia* ST[18-30]-Tyr	Cys-Cys-Asp-Val-Cys-Cys-Asn-Pro-Ala-Cys-Ala-Gly-Cys-Tyr	4.4	w	w
		Chimeric STs			
14		Cys-Cys-Asp-Val-Cys-Cys-Asn-Pro-Ala-Cys-Thr-Gly-Cys	9.7	w	-
15		Cys-Cys-Asp-Val-Cys-Cys-Asn-Pro-Ala-Cys-Thr-Gly-Cys-Tyr	2.2	w	w
16		Cys-Cys-Asp-Leu-Cys-Cys-Asn-Pro-Ala-Cys-Ala-Gly-Cys-Tyr	1.4		+
17		Cys-Cys-Gln-Leu-Cys-Cys-Asn-Pro-Ala-Cys-Thr-Gly-Cys-Tyr	2.8		+

MED: minimum effective dose, PCA: anti-STh polyclonal antibody, MCA: anti-STh monoclonal antibody, +: neutralized, -: not neutralized, w: weakly neutralized.

effective dose (MED) was determined by a suckling mouse test according to the method of Takeda _et al_.[7]

The essential part for the expression of the toxic activity in _Yersinia_ ST was determined to be in the region with high similarity to that of _E. coli_ STs. That is corresponding to _Yersinia_ ST[17-30]. The toxic activity of _Yersinia_ ST was lower than those of _E. coli_ STs. The replacement of -Glu-Leu- residue to -Asp-Val- residue seems to be responsible for the low toxic activity. The C-terminal Tyr residue in _E. coli_ STs did not affect its toxic activity, but the Tyr residue coupled to the C-terminal of _Yersinia_ ST[18-30], that is ST No 13, increased the toxic activity. The shorter analogues of _Yersinia_ ST showed higher heat-stability than that of _E. coli_ STh[1-19] or STp[1-18].

References

1. Aimoto, S., Takao, T., Shimonishi, Y., Hara, S., Takeda, T., Takeda, Y., and Miwatani, T. (1982) Eur. J. Biochem., 129, 257-263 .
2. Takao, T., Hitouji, T., Aimoto, S., Shimonishi, Y., Hara, S., Takeda, T., Takeda, Y. and Miwatani, T. (1983) FEBS Lett., 152, 1-5.
3. Takao, T., Tominaga, N., Shimonishi, Y., Hara, S., Inoue, T. and Miyama, A. (1984) Biochem. Biophys. Res. Commun., 125, 845-851.
4. Ikemura, H., Yoshimura, S., Aimoto, S., Shimonishi, Y., Takeda, T., Takeda, Y. and Miwatani, T. (1983) Chem. Lett., 101-104.
5. Yoshimura, S., Miki, M., Ikemura, H., Aimoto, S., Shimonishi, Y., Takeda, T., Takeda, Y. and Miwatani, T. (1984) Bull. Chem. Soc. Jpn.,57, 125-133.
6. Yoshimura, S., Miki, M., Ikemura, H., Watanabe, H., Aimoto, S., Shimonishi, Y., Hara, S., Takeda, T., Miwatani, T., and Takeda, Y., (1985) FEBS Lett,. 181, 138-142.
7. Takeda, Y., Takeda, T., Yano, T., Yamamoto, K. and Miwatani, T. (1979) Infect. Immun. 25, 978-985.

PHOTOACTIVABLE APAMIN DERIVATIVES AS K^+ CHANNEL PROBES.

Labbé-Jullié, C., Seagar, M., Granier, C., Couraud, F. and Van Rietschoten, J.

INSERM U 172, CNRS UA 553, Laboratoire de Biochimie, Faculté de Médecine Secteur Nord, 13326 Marseille Cedex 15, France.

Apamin is an 18-peptide isolated from bee-venom which shows central neurotoxicity. Interactions of apamin with the slow Ca^{++}-dependant K^+ channel have been characterized by electrophysiology (1, 2, 3). High affinity binding was demonstrated of mono ^{125}I iodo-apamin (^{125}I-apamin) to intact primary cultured rat neurons, which could be correlated to an inhibition of $^{86}Rb^+$ efflux (4). Recently, photoaffinity labeling of neuronal membranes with arylazide-^{125}I-apamin was performed (5). For these experiments, the derivative was prepared in the following way : ^{125}I-apamin (4) was reacted with 5 to 10 excess of 4-azido-2-nitro-phenylaminoacetylsuccinimidyl ester (ANPAA-OSu was a gift of Dr. K. Angelides, University of Florida, Gainesville, and has been usefull in the derivatization of scorpion toxins (6)) for three hours at pH 8.5 in presence of BSA (1 mg/ml) and used without further purification (see exact protocol in (5)). The photolabeling was performed on two biological systems. On primary cultured rat neurons, autoradiography of the electrophoresis of membrane proteins revealed specific labeling of three main components with Mr = 86,000, 30,000 and 23,000. On purified synaptic membranes from the rat brain, the labeling pattern was with major bands corresponding to Mr = 86,000 and 59,000. It appeared plausible to propose that the largest band with an apparent molecular weight of 86,000 carries the binding site of apamin and that the other bands, differing between the two biological systems, represent proteins that are present in the immediate environment of the apamin receptor. Separately, another group reported that cross-linking of ^{125}I-apamin to rat synaptosomal membranes with disuccinimidyl suberate labeled a single protein of Mr = 28,000 - 33,000 (7, 8). Obviously, more work was needed in order to better characterize the chemical probes used in our experiments and try to explain the discrepencies.

Apamin possesses two amino groups with respective pK of 6.7 (α-amino group of Cys^1) and 10.3 (ε-amine of Lys^4) as established by NMR experiments (9). Ion exchange chromatography of the mixture obtained after reaction of ^{125}I-apamin with ANPAA-OSu on a SP-Sephadex C-25 column in 10 mM sodium phosphate, pH 7.0, 10 M BSA, with a linear gradient of sodium chloride from 0 to 100 mM, followed by a sharp gradient to 300 mM, permitted separation of four fractions as detected by radioactivity. Attribution was clear when comparing the relative importance of each peak with increasing amounts of photosensitive reagent added, from 0 to 10 excess, and the time of reaction extended from 1 to 3 hrs (Table 1). The first peak at the void volume representing 15 to 28 % of radioactivity is always present even when iodinated apamin is applied straight to the column without any modifi-

Table 1. Relative radioactivity in the four peaks eluted after SP-Sephadex C-25 chromatography of the reaction mixture of ^{125}I-apamin with ANPAA-OSu.

Excess of ANPAA-OSu	Time of reaction (hr)	% radioactivity			
		Peak 1	Peak 2	Peak 3	Peak 4
0		25			75
1	1	28	10	30	32
2	1	15	26	41	18
10	3	27	63	7	3

cation. It corresponds probably to some degradation product. The fourth peak eluting in the 300 mM NaCl buffer is the non-modified ^{125}I-apamin, the second and the third peaks eluting at 50 mM and 85 mM NaCl are likely to be respectively the dimodified and monomodified ^{125}I-apamin. Photolabeling on the cultured cells with derivatives 2 and 3, i.e. the dimodified and the monomodified ^{125}-apamin, gave distinct results. The dimodified ^{125}I-apamin labeled components with molecular weights of 86,000, 59,000, 33,000 and 22,000. The monomodified ^{125}I-apamin labeled only the bands at 33,000 and 22,000, meaning that probably the photosensitive derivative is specifically attached to one of the two amines. To assess which amino group is the most reactive a mono-biotinyl apamin derivative was prepared and then reacted with dansylchloride to block the remaining amine group. Amino acid analysis of the hydrolyzate showed recovery of one residue of lysine per mole of apamin derivative, meaning that the monomodification in phosphate buffer, pH 8.5, lead to the acylation of exclusively the ε-lysine amine, contrary to what would be expected by the respective pKs of the α- and ε-amines. Iodination of mono-biotinyl apamin and reaction with ANPAA-OSu gave us a radiolabeled apamin with the photosensitive group placed at the α-amine position. Photoactivation experiment with this probe in the cultured cells resulted in labeling of the components at 86 and 59 kilodaltons, meaning that, when apamin is bound to its receptor, the α-amine is near the 86,000 and 59,000 components and the ε-amine near the 33,000 and 22,000 components. These experiments would also explain why the chemical crosslinking of ^{125}-apamin on its receptor labeled only the low molecular weight components of the potassium channel. On a molecular block model (10) of apamin it is clear that the ε-amine of lysine is way out of the apamin core and that the α-amine is partially inaccessible. This model shows also that with a photosensitive arm, only the α-amine can reach the neighbourhood of the two arginines that are essential for activity. The 86 kilodaltons oligopeptide constituent of the apamin sensitive potassium channel seems to hold the essential binding site of apamin. The 59,000 component may be a degradation product of the 86,000 as its intensity varies from one experiment to the other.

REFERENCES

1 - Banks, B.E.C., Brown, C., Burgess, G.M., Burnstock, G., Claret, M., Cooks, T.M. and Jenkinson, D.H. (1979) Nature 282, 415-417.

2 - Hugues, M., Schmid, H., Romey, G., Duval, D., Frelin, C. and Lazdunski, M. (1982) EMBO J. 1, 1039-1042.

3 - Hugues, M., Romey, G., Duval, D., Vincent, J.P. and Lazdunski, M. (1982) Proc. Natl. Acad. Sci. U.S.A. 79, 1308-1312.

4 - Seagar, M.J., Granier, C. and Couraud, F. (1984) J. Biol. Chem. 259, 1491-1495.

5 - Seagar, M.J., Labbé-Jullié, C., Granier, C., Van Rietschoten, J. and Couraud, F. (1985) J. Biol. Chem. 260, 3895-3898.

6 - Darbon, H., Jover, E., Couraud, F. and Rochat, H. (1983) Biochem. Biophys. Res. Commun. 115, 415-422.

7 - Hugues, M., Schmid, H. and Lazdunski, M. (1982) Biochem. Biophys. Res. Commun. 107, 1577-1582.

8 - Schmid-Antomarchi, H., Hugues, M., Norman, R., Ellory, C., Borsotto, M; and Lazdunski, M. (1984) Eur. J. Biochem. 142, 1-6.

9 - Okhanov, V.V., Afanasiev, V.A., Gurevich, A.Z., Elyakova, E.G., Miroshnikov, A.I., Bystrov, V.F. and Ovchinnikov, Yu. A. (1980) Bioorg. Khim. 6, 840-864.

10 - Bystrov, V.F., Okhanov, V.V., Miroshnikov, A.I. and Ovchinnikov, Yu. A. (1980) FEBS Letters 119, 113-117.

THE DESIGN AND CONSTRUCTION OF PEPTIDES AND PROTEINS WITH AMPHIPHILIC SECONDARY STRUCTURES

E. T. Kaiser, B. Lynch and B. Rajashekhar

Laboratory of Bioorganic Chemistry and Biochemistry
The Rockefeller University, New York NY 10021

Introduction

Peptides and proteins possessing diverse biological activity and binding at biological interfaces have been shown to contain substantial segments which have the potential to form amphiphilic secondary structures in amphiphilic environments.[1,2] Through the design, construction and testing of the biological and physical properties of model peptides, we have been able to demonstrate the importance of amphiphilic secondary structural regions to the fundamental properties of apolipoproteins, peptide toxins, chemotactic agents and peptide hormones. We shall focus in this brief article on a particular class of compounds, the peptide hormones, for which we have tested the importance of amphiphilic secondary structural regions by the design and construction of appropriate models.

Several years ago we proposed[2-4] that peptide hormones which contain between 10 and 50 amino acids and which are not multiply crosslinked (that is, peptide hormones that contain no more than one disulfide linkage) might contain regions which could be induced to form amphiphilic secondary structures when bound in the vicinity of or at the appropriate receptors. We envisaged at least three advantages to the existence of amphiphilic secondary structures in the hormones. First, such regions could facilitate the diffusion of the peptide hormone to its receptor. In particular, when the hormone encounters an appropriate surface the amphiphilic structural region which is induced may aid diffusion of the peptide along the surface, reducing the problem of diffusion to the receptor from a three dimensional search to a two dimensional one.[5,6] A second feature of the induction of amphiphilic secondary structures which may be important is that such regions may act to hold the "specific recognition site" of the peptide hormone in the right geometry to produce a suitable response from the receptor. Finally, the peptide hormones are likely to utilize amphiphilic secondary structures in intramolecular interactions with other regions of the peptide molecule, leading to stabilization of the hormone to enzymatic degradation.

To test the importance of amphiphilic secondary structures in the peptide hormones, in our design of models we have taken account of the following factors in constructing amino acid sequences:

a) We have determined through model building and often through data on various analogs which have been published in the literature where and how long the amphiphilic structural regions in the natural system are.

b) We have constructed our models using amino acids having a high tendency to form the particular type of amphiphilic structure (α-helix, β-strand, β-turn, etc.) that we believe to be crucial to the activity of the particular hormone under study.

c) We have tried to maintain the overall charge balance of the amphiphilic region.

d) We have maintained the ratio of hydrophobic/hydrophilic residues of the natural hormone as closely as possible in our synthetic models in the amphiphilic region.

e) Most importantly, we have prepared models in which the sequence homology between the natural system and the designed system is minimized in the amphiphilic secondary structural region.

With respect to point e, if one had the objective principally of producing as potent a designed peptide hormone as possible, one might not focus strongly on the reduction of sequence homology. However, if the emphasis is on showing the importance of the secondary structural region for the properties of a peptide hormone, then the similar characteristics of a model peptide having in common with the natural system primarily the potential to form a related secondary structure, while having a rather different amino acid sequence, would be a rather strong argument.

Results and Discussion

Among the peptide hormones which we have studied most thoroughly from the standpoint of modelling are calcitonin and β-endorphin. In the present article the modelling approach we have used for these hormones will be reviewed briefly and new results on the proposed structure for yet another peptide hormone, calcitonin gene-related peptide, will be presented.

According to the structural model for the 32 amino acid calcium regulating hormone calcitonin which we have proposed,[7,8] there are three structural regions that are important to calcitonin's biological activity. These are a seven residue cyclic segment at the N-terminus which contains a disulfide bridge between the Cys-1 and Cys-7 residues, an amphiphilic α-helix starting from residue 8 and proceeding to the proline residue at position 23, and then a hydrophilic "spacer" unit which spans the region from residue 23 all the way to the carboxyl-terminal proline amide.

Two peptide models for the hormone which we designed have been described.[7,8] In both of these models we have optimized the amphiphilic α-helical structure in the region from residues 8 to 22 while reducing substantially the sequence homology to the corresponding region of any natural calcitonin.

Indeed, for the models the sequence homology with natural calcitonins in the hydrophilic face of the amphiphilic α-helix is minimal. In the less effective of the model peptides we designed, there are several changes in the hydrophobic face of the 8-22 amphiphilic α-helix from the sequence of a very active naturally-occurring calcitonin such as the salmon species. There are fewer such changes in the hydrophobic face for the more effective model. In both models, instead of the Glu residue at position 15 which was located in what appeared to be otherwise the hydrophobic face of the 8-22 amphiphilic α-helical region we utilized a Leu residue. Although in the 8-22 region the model calcitonin has only 40% sequence homology to salmon calcitonin, its biological activity is comparable. In particular, it binds with similar tightness to brain and kidney membranes, shows approximately the same behavior in activating adenylate cyclase, and exhibits approximately the same hypocalcemic potency in vivo. From such studies we have argued that the amphiphilic α-helical structure in the region from residues 8-22 is important for binding to calcitonin receptors. Furthermore, the results with our model peptides show that although a hydrophilic residue (Asp or Glu) usually occurs on the hydrophobic face (position 15) of the amphiphilic α-helical region in the natural calcitonins, it is not necessary for high biological activity.

Very recently, we have started to investigate the surface properties of calcitonin gene-related peptide (CGRP). This 37 residue peptide hormone, derived from alternative RNA processing of the calcitonin CGRP gene in both rats and humans,[9-12] has been localized by immunocytochemistry to specific areas throughout the central and peripheral nervous system.[10,13] The exons coding for calcitonin and CGRP are thought to have arisen from the same primordial gene via gene duplication and rearrangement events. This postulate is supported by reports that calcitonin and CGRP cross react with each other's receptors[14] and by our structural analysis presented here.

As a working hypothesis, we propose here in analogy to calcitonin that CGRP consists of three structural regions. These correspond to the region from the N-terminus through residue 7, including the disulfide linkage between residues 2 and 7, a possible amphiphilic helix from residues 8 through 25 and a spacer from residues 26 to the C-terminal residue. Additionally, the structure may involve a β-turn from residues 28 through 31 including the Pro residue at position 29. If such a β-turn were present, it could allow the spacer region to fold over the putative amphiphilic helix in such a way that the Phe amide at the C-terminus (present in the human and rat species) is brought in close proximity to the disulfide loop connecting positions 2 and 7. A picture of the proposed amphiphilic helical region of rat CGRP is drawn in Figure 1 together with a representation of the corresponding region of salmon calcitonin. The separation of the hydrophobic and hydrophilic faces in the rat-CGRP amphiphilic helix is quite striking although, in contrast to the amphiphilic helix of the salmon species where there are relatively few residues which do not have a reasonably high helix forming potential, the amphiphilic helix drawn for the rat CGRP does contain a number of Gly residues which are not generally preferred in α-helical structures. Also, it should be noted that the amphiphilic helix drawn for the rat CGRP is somewhat longer than that proposed for salmon calcitonin. Inspection of the proposed amphiphilic helical regions of various calcitonins shows that the salmon form has the most clearly amphiphilic α-helical structure[7,8]. The forms obtained from various other species tend to have either a less clear separation of hydrophobic and hydrophilic residues or contain residues with a lesser degree of potential for forming α-helical structures than the residues in the salmon species. Therefore, the comparison between the amphiphilic helix of the rat CGRP with the structure in salmon calcitonin probably gives a less favorable picture of the importance of the amphiphilic helix in the CGRP structure than is justified.

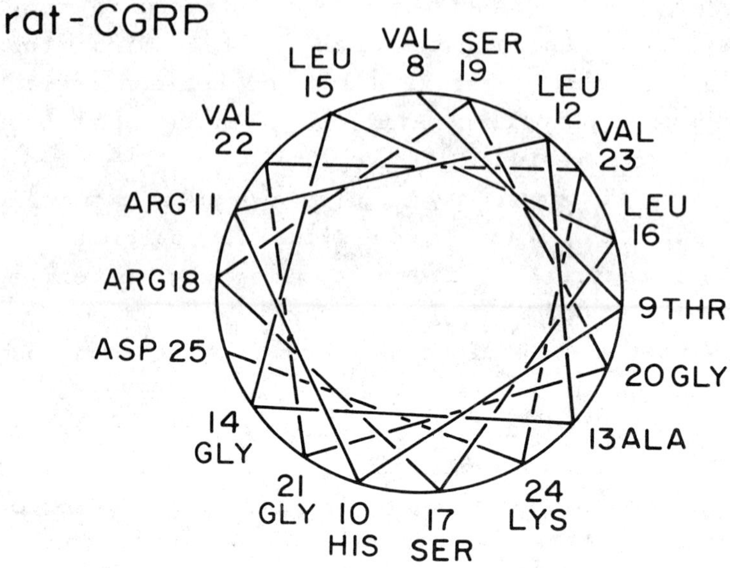

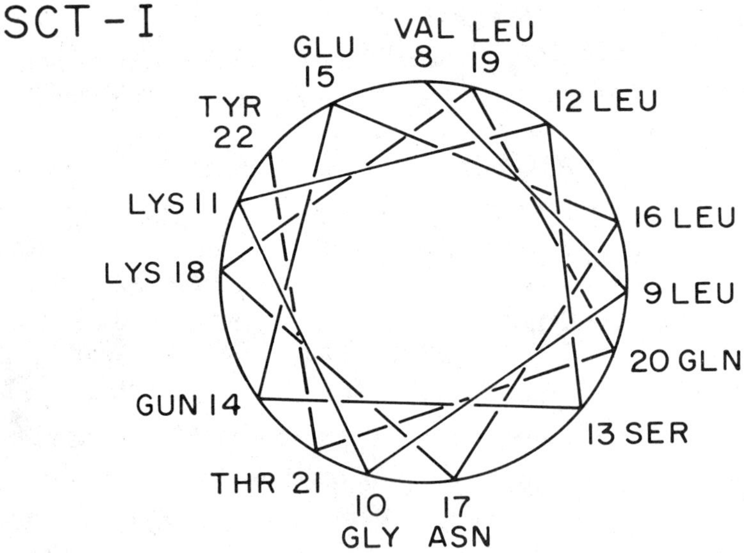

Fig. 1. top) Helical wheel representation of region from residues 8-25 in rat CGRP. bottom) Helical wheel representation of region from residues 8-22 in salmon calcitonin.

Our physical studies on rat CGRP support the proposed relationship between the structures of CGRP and calcitonin in amphiphilic environments. Thus, in a structure forming solvent such as 50% trifluoroethanol we have estimated from the value of $[\Theta]_{222}$ that approximately 46% α-helix is formed (CGRP concentration 4.3×10^{-6} M), behavior which is reminiscent of that shown by salmon calcitonin. Furthermore, in examination of the properties of a monolayer of CGRP formed at the air-water interface the collapse pressure measured (11 dyn/cm) was comparable to that seen earlier for salmon calcitonin (12 dyn/cm). We have been examining the competition of the rat CGRP with salmon calcitonin for receptors in rat brain homogenates and the competition may reflect at least in part the potential for both peptides to form amphiphilic α-helical structures.

Another hormone for which we have proposed the importance of an amphiphilic helical structure is β-endorphin.[15] In the case of this hormone, again three basic structural units can be distinguished: a highly specific opiate recognition site at the N-terminus (residues 1-5) which is connected through a hydrophilic link (residues 6-12) to a potential amphiphilic helix which includes the C-terminal residues 13-31.[15-19] Among the models for β-endorphin containing amphiphilic helical regions in their C-termini that we have constructed is a peptide where besides the amphiphilic α-helical region from residues 13-31 which contains a sequence having minimal homology to that of β-endorphin there was also a hydrophilic spacer region from residue 6-12 with minimal homology to the equivalent region of the natural hormone.[16] Furthermore, we have designed, constructed and tested the properties of a peptide which is a negative model for β-endorphin.[18] While many of the important feature of β-endorphin are retained in this peptide, there are important differences from the natural system in that in a helical conformation of the region from residues 13-31, the amphiphilic character present in our other model peptides is absent. In the light of our results with the negative model as well as with the various positive models for the C-terminal amphiphilic helical region we have been able to make very strong arguments for the importance of an amphiphilic helical structure in the region from residues 13-31 with respect to properties such as the resistance to proteolysis of the natural hormone. Indeed, the amphiphilicity of the C-terminal helical structural region was found to be essential for the high opiate activity on the rat vas deferens. On the other hand, there did not appear to be a similar requirement for interaction with the opiate receptors on the guinea pig ileum. We have also recently constructed a peptide model in which the whole C-terminal segment from residues 13-31 has been built using only D-amino acids. From circular dichroism measurements we have obtained spectral evidence that the C-terminal region of the model peptide assumes a left-handed helical arrangement. Our studies of the receptor binding properties, stability to enzymatic attack and analgesic activity of this model peptide showed that it behaves very similarly to β-endorphin itself. Since there is a drastic difference in the chirality of the amphiphilic helical region of β-endorphin and that of the model peptide we have concluded that the receptor binding and biological activity depend on the surface characteristics of the C-terminal helical region of β-endorphin rather than on the peptide backbone structure. Following this work we have begun the preparation of β-endorphin models in which whole regions will be replaced systematically by non-peptide structural units. We have focused, at the moment, on the hydrophilic link (residue 6-12) connecting the specific opiate recognition site at the N-terminus to the amphiphilic helix in the C-terminal region. We have prepared a model peptide containing a hydrophilic link consisting entirely of a repeating sequence involving a single amino acid, the S-isomer of γ-amino-γ-hydroxymethylbutyric acid. While the hydrophilic link still contains amide bonds, their distribution is different from that corresponding to what is seen for the usual peptide. We have found that using guinea pig brain whole membrane preparation the binding of this β-endorphin model to μ- and δ-receptors is actually slightly enhanced relative to observations made with β-endorphin. The relative potency of the new β-endorphin model is comparable to that of a related model differing only in that the hydrophilic linker has the natural human sequence. Most importantly, the potency of the new model relative to human β-endorphin in inhibiting the twitching of an electrically stimulated rat vas deferens is only about 7-fold less. Thus, the hydrophilic spacer region can be prepared from a structural unit which does not contain the usual peptide bonds, and we are encouraged to replace other regions of the molecule as well by non-peptide analogs. Finally, it should be noted that in the amphiphilic environment of the air-water interface the latest β-endorphin model showed a collapse pressure of 22 dyn/cm and a limiting area of 16.1 $Å^2$/residue, values very reminiscent of the corresponding β-endorphin model in which the hydrophilic spacer, however, has the human β-endorphin amino acid sequence.

Acknowledgement

The partial support of this research by NIH Program Project HL-18577, a grant from the Dow Chemical Co. Foundation and an NSF Predoctoral Fellowship (B.L.) is gratefully acknowledged.

References

1. Kaiser, E.T. and Kézdy, F.J. (1983) Proc. Natl. Acad. Sci. USA 80, 1137-1143.
2. Kaiser, E.T. and Kézdy, F.J. (1984) Science 223, 249-255.

3. DeGrado, W.F., Kézdy, F.J., and Kaiser, E.T. (1981) J. Am. Chem. Soc. 103, 679-681.

4. Taylor, J.W., Osterman, D.G., Miller, R.J., and Kaiser, E.T. (1981) J. Am. Chem. Soc. 103, 6965-6966.

5. Katchalski-Katzir, E., Rishpon, J., Sahar, E., Lamed, R., and Henis, Y.I. (1985) Biopolymers 24, 257-277.

6. Kaiser, E.T. (1985) Proc. New York Acad. Sci., submitted.

7. Moe, G.R., Miller, R.J., and Kaiser, E.T. (1983) J. Am. Chem. Soc. 105, 4100-4103.

8. Moe. G.R. and Kaiser, E.T. (1985) Biochemistry 24, 1971-1976.

9. Amara, S.G., Jonas, V., Rosenfeld, M.G., Ong, E.S., and Evans, R.M. (1982) Nature 298, 240-244.

10. Rosenfeld, M.G., Mermod, J., Amara, S.G., Swanson, L.W., Sawchenko, P.E., Rivier, J., Vale, W., and Evans, R.M. (1983) Nature 304, 129-135.

11. Morris, H.R., Panico, M., Etienne, T., Tippins, J., Girgis, S.I., and MacIntyre, I. (1984) Nature 308, 746-748.

12. Jonas, V., Lin, C.R., Kawashima, E., Sernon, D., Swanson, L.W., Mermod, J., Evans, R.M., and Rosenfeld, M.G. (1985) Proc. Natl. Acad. Sci. USA 82, 1994-1998.

13. Mulderry, P.K., Ghatei, M.A., Rodrigo, J., Allen, J.M., Rosenfeld, M.G., Polak, J.M., and Bloom, S.R. (1985) Neuroscience 14, 954-957.

14. Goltzman, D. and Mitchell, J. (1985) Science 227, 1343-1345.

15. Taylor, J.W., Miller, R.J., and Kaiser, E.T. (1981) J. Am. Chem. Soc. 103, 6965-6966.

16. Taylor, J.W., Miller, R.J., and Kaiser, E.T. (1982) Mol. Pharmacol. 22, 657-666.

17. Taylor, J.W., Miller, R.J., and Kaiser, E.T. (1983) J. Biol. Chem. 258, 4464-4471.

18. Blanc, J.P., Taylor, J.W., Miller, R.J., and Kaiser, E.T. (1983) J. Biol. Chem. 258, 8277-8284.

19. Blanc, J.P. and Kaiser, E.T. (1984) J. Biol. Chem. 259, 9549-9556.

THE AMPHIPATHIC HELIX IN PHOSPHOLIPID-PROTEIN INTERACTIONS: THE ROLE OF HYDROPHOBICITY AND HELICAL POTENTIAL

James T. Sparrow, Doris A. Sparrow, Henry J. Pownall, Gabriel Ponsin, Pradip K. Bhatnagar, Giroom G. Chang and Antonio M. Gotto, Jr.

Department of Medicine, Baylor College of Medicine and The Methodist Hospital, One Baylor Plaza, Houston, Texas 77030

Introduction

The amphipathic or amphiphilic helix was first hypothesized as an important structural feature of lipid associating proteins by Segrest et al. in 1974[1]. The unique features of this helical structure are an amino acid sequence such that two distinct faces are formed, one polar and the other nonpolar. These are illustrated in Figure 1 by residues 33 to 53 of apolipoprotein C-I[2]. In addition, the polar face contains pairings of charged residues, usually with the acidic residues near the center of the polar face paired with basic residues at the edge separating the polar and nonpolar faces. The nonpolar face frequently contains paired aromatic amino acids. With this arrangement of amino acids, the polar surface that is generated can orient towards the aqueous compartment near the polar head region of cellular membranes or plasma lipoprotein phospholipids and the nonpolar surface can associate with the hydrocarbon chains of these phospholipids. Using synthetic peptides, we have defined the lipid-binding region of the serum apolipoproteins as amphipathic helices[3]. The amphipathic helix has also been identified as a major structural feature in an increasing number of biologically important peptides and proteins.

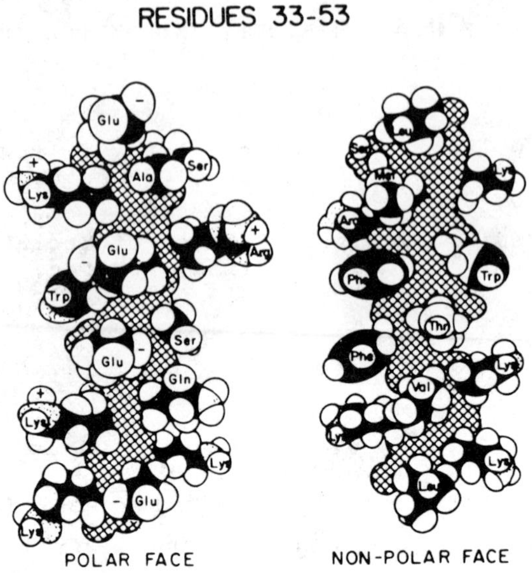

Fig. 1. An illustration of the amphipathic helix predicted for residues 33 to 53 of apolipoprotein C-I.

To investigate the amphipathic helix as a unique protein structural feature, it was important to have a model system in order to demonstrate the role of charge distribution, hydrophobicity, peptide chain length, and helix potential. Therefore, we began to design and synthesize peptides which incorporate these structural features. The first of these were presented at the Fourth American Peptide Symposium in 1975[4]. These peptides were sixteen and twenty residues in length and only one was weakly bound to phospholipid. Since that time we have continued our investigations and have synthesized a particularly useful twenty residue peptide, LAP-20, (Figure 2) which displays all the properties of the serum apolipoproteins, i.e., a high helical content in the presence of phospholipids, a blue-shift in the intrinsic tryptophan fluorescence spectrum and formation of stable lipid-peptide complexes[5]. In addition, this peptide greatly enhances the activity of the plasma enzyme, lecithin:cholesterol acyl transferase (LCAT) responsible for plasma cholesteryl ester synthesis[6].

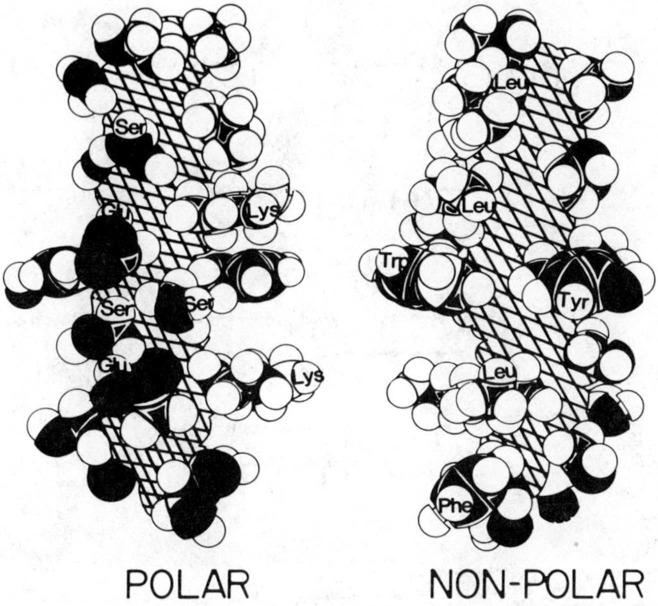

Val-Ser-Ser-Leu-Leu-Ser-Ser-Leu-Lys-Glu-
Tyr-Trp-Ser-Ser-Leu-Lys-Glu-Ser-Phe-Ser

Fig. 2. Amino acid sequence and polar and nonpolar faces of the model lipid-associating peptide, LAP-20.

To investigate the role of hydrophobicity we have replaced leucine residues in LAP-20 with alanine and studied the interaction with dimyristoylphosphatidylcholine (DMPC) vesicles by density gradient ultracentrifugation. We have also shortened the peptide by deleting the first five amino acids and acylated the amino terminus with aliphatic carboxylic acids of increasing length.

To investigate the importance of helical potential we have synthesized a series of LAP-20 peptides with proline substituted at various positions. Figure 3 shows the Chou-Fasman predicted structures for these six peptides and their

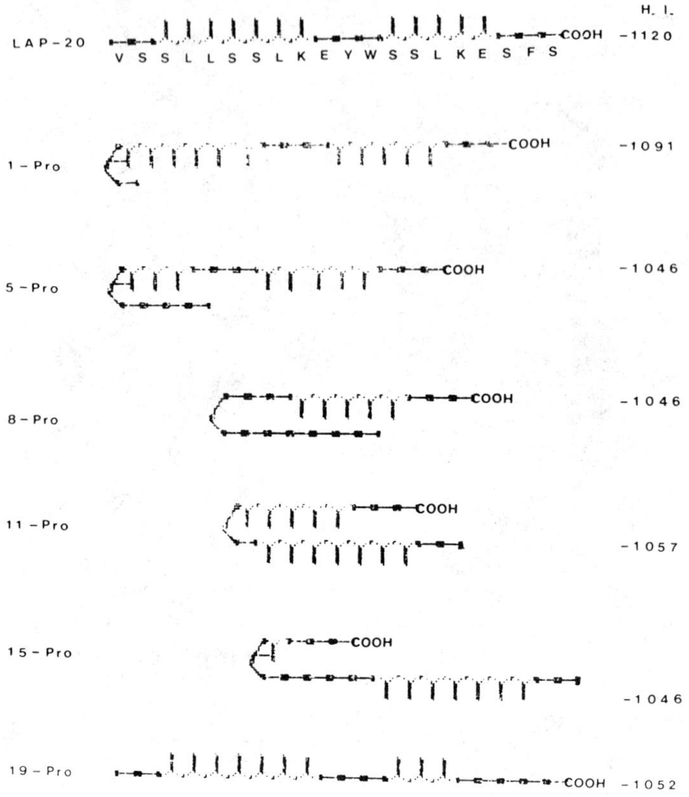

Fig. 3. A computer representation of the Chou-Fasman predicted structure of Pro-LAP-20 peptides. The average hydrophobicity (H.I.) in cal/residue was calculated using the scale of Bull and Breese[7].

average hydrophobicities calculated using the hydrophobicity scale of Bull and Breese[7]. The substitution of proline changes the average hydrophobicity very little but, as expected, has a dramatic effect on the predicted structure. Below we present the results of our studies on the interaction of these peptides with phospholipids. These results demonstrate the importance of helical potential, peptide chain length and hydrophobicity to protein-phospholipid interactions.

Results and Discussion:

The peptides were synthesized by solid phase techniques, cleaved with anhydrous HF and purified by reversed phase HPLC. After purification amino acid compositions were as expected. The yield of purified peptide varied from 15% to 30% based on the initial loading of the resin.

Table I. Fluorescence Maximum of Ala-LAP-20 Peptides

	Buffer	+ DMPC	Complex	L:P
5-Ala	350 nm	332 nm	335 nm	52
4,5-Ala_2	351 nm	348 nm	---	--
4,5,11-Ala_3	351 nm	350 nm	---	--
5,8,11-Ala_3	350 nm	350 nm	---	--

The peptides were studied for their ability to bind to vesicles of DMPC using shifts in the intrinsic tryptophan fluorescence spectrum and ellipticity changes at 222 nm as an indication of lipid-peptide interaction. As seen in Table I for the Ala-LAP-20 peptides, 5-Ala-LAP-20 showed a significant shift (~18 nm) in the fluorescence maximum. Density gradient ultracentrifugation of these mixtures permitted the isolation of a lipid-peptide complex of 52:1 only with 5-Ala-LAP-20 (Figure 3). By substituting two or three alanines for leucines in the hydrophobic face, the hydrophobic area is decreased and the peptide no longer binds to DMPC. These results support the importance of hydrophobicity in lipid-protein interaction.

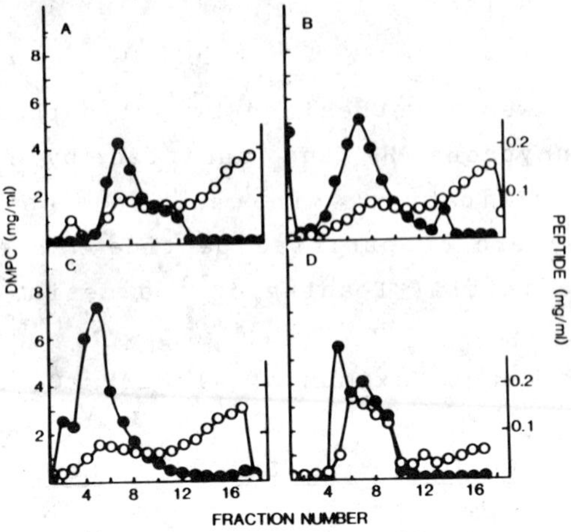

Fig. 4. Density gradients of Ala-LAP-20 peptide:DMPC mixtures. Peptide (o--o); DMPC (●——●). A, 5,8,11-Ala_3-LAP-20; B, 4,5,11-Ala_3-LAP-20; C, 4,5-Ala_2-LAP-20; and D, 5-Ala-LAP-20.

When the acylated LAP-15 peptides were studied using the same techniques we found that C_8-LAP-15 and peptides with longer acyl chains showed blue shifts of ~15 nm in their maxima (Table II). In addition, the interaction of these

Table II. Fluorescence and Circular Dichroic Properties of Acylated LAP-15 Peptides

No. of Carbons	Buffer	+ DMPC	$[\theta]_{222}$ Buffer	$[\theta]_{222}$ +DMPC
0	356 nm	356 nm	nil	nil
4	356 nm	356 nm	nil	-1000
8	356 nm	343 nm	nil	-10900
12	350 nm	341 nm	-6000	-10900
16	345 nm	341 nm	-6000	-11000

peptides with a C_4 reversed-phase increases in a regular fashion. There is a log-linear relationship between the retention time determined by reversed-phase HPLC or the equilibrium constant determined by equilibrium dialysis and the number of methylene units[8]. Therefore, by acylating the shorter peptide, LAP-15, with aliphatic carboxylic acids of increasing carbon number, the hydrophobicity can be increased incrementally to produce a peptide with sufficient hydrophobicity to interact with phospholipid. Those peptides which bind also display increases in their ellipticity at 222 nm indicative of α-helix formation (Table II). We estimate a helical content of ~35% from the C.D. spectra of those peptides which bind.

To determine the importance of helix potential the interaction of Pro-LAP-20 peptides was studied in a similar manner. C.D. spectroscopy in 33% hexafluoro-isopropanol indicated that 1- and 5-Pro-LAP-20 formed 59% and 43% α-helix, respectively, whereas 15- and 19-Pro-LAP-20 were 36% and 40% helical. As expected, 8- and 11-Pro-LAP-20

Table III. Fluorescence Maximum of Pro-LAP-20 Peptides

	Buffer	+ DMPC	Complex	L:P
1-Pro	349 nm	332 nm	333 nm	40:1
5-Pro	350 nm	333 nm	334 nm	53:1
8-Pro	352 nm	351 nm	--	--
11-Pro	352 nm	352 nm	--	--
15-Pro	351 nm	344 nm	345 nm(Unstable)	33:1
19-Pro	350 nm	341 nm	340 nm	42:1

displayed the least helical content (Table IV). We found in the presence of DMPC that 1-, 5- and 19-Pro-LAP-20 showed a shift of 17, 11, and 9 nm, respectively, in the intrinsic fluorescence spectrum (Table III); the 15-Pro peptide had a smaller shift of 7 nm. Density gradient ultracentrifugation of the mixtures permitted isolation of complexes with these same peptides (Figure 5). However, the 15-Pro-LAP-20

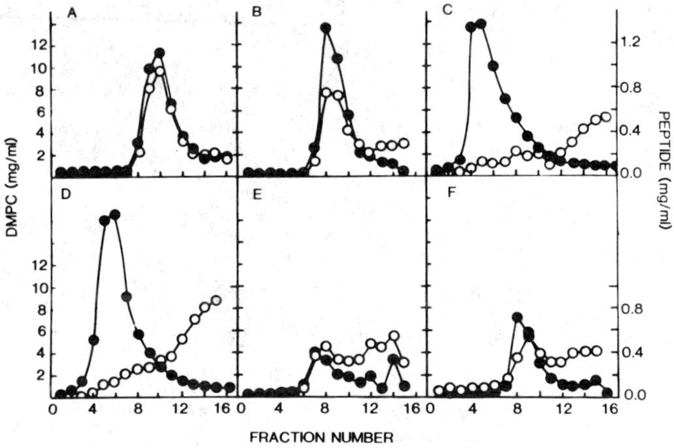

Fig. 5. Density gradients of Pro-LAP-20 peptide:DMPC mixtures. Peptide (o—o); DMPC (●—●). A, 1-Pro-LAP-20, B, 5-Pro-LAP-20; C, 8-Pro-LAP-20; D, 11-Pro-LAP-20; E, 15-Pro-LAP-20; and F, 19-Pro-LAP-20.

lipid:peptide complex was unstable to re-centrifugation. In the complex with DMPC, the 1- and 5-Pro-LAP-20 formed the complexes with the highest helical content; 63% and 54% respectively, 19-Pro-LAP-20 had 47%, and 15-Pro formed an unstable complex with 24-35% α-helix in the mixture, but was unstable to recentrifugation. The 8- and 11-Pro-LAP-20 did not form a complex or show changes in their helical content. Therefore, the restriction of helix propagation in a hydrophobic peptide imposed by the introduction of a proline residue prevents the formation of a stable phospholipid-peptide complex when the helical segment is less than 15 residues. We believe these results demonstrate the importance of helical potential and amphipathic helical length to lipid-protein interaction.

Table IV. Ellipticity of Pro-LAP-20 Peptides in the Presence and Absence of DMPC

	$[\theta]_{222}$			
Peptide	Hexafluoropropanol	Buffer	+DMPC	Complex
1-Pro	-19,800	-6,675	-16,812	-21,413
5-Pro	-14,600	-2,363	-12,668	-18,183
8-Pro	-9,800	-2,347	-3,903	--
11-Pro	-1,791	-2,019	-9,250	--
15-Pro	1,040	-11,613	-11,200	-6,256
19-Pro	-8,642	13,299	12,250	-15,431

Acknowledgments

The authors wish to thank Drs. William Mantulin, Germain Fernando and Martha Mims for obtaining some of the C.D. spectra, Mrs. Marjorie Sampel for preparing the manuscript, and Mrs. Susan Kelly for preparing the figures. This work was supported by the National Institutes of Health Specialized Center of Research in Atherosclerosis (HL 27341).

References

1. Segrest, J.P., Jackson, R.L., Morrisett, J.D., and Gotto, A.M.,Jr. (1974) FEBS Lett. **38**, 247-253.

2. Jackson, R.L., Sparrow, J.T., Baker, H.N., Morrisett, J.D., Taunton, O.D., and Gotto, A.M.,Jr. (1974) J. Biol. Chem. **249**, 5308-5313.

3. Sparrow, J.T. and Gotto, A.M.,Jr. (1982) Crit. Rev. Biochem. **13**, 87-107.

4. Sparrow, J.T., Morrisett, J.D., Pownall, H.J., Jackson, R.L., and Gotto, A.M.,Jr. (1975) Peptides: Chemistry, Structure, and Biology, Proceedings of the Fourth American Peptide Symposium, R. Walter and J. Meienhofer, eds., Ann Arbor Science Publishers, Ann Arbor, MI pp. 597-602.

5. Morrisett, J.D., Jackson, R.L., and Gotto, A.M.,Jr. (1975) Ann. Rev. Biochem. **44**, 183.

6. Pownall, H.J., Hu, A., Gotto, A.M.,Jr., Albers, J.J., and Sparrow, J.T. (1980) Proc. Natl. Acad. Sci. U.S.A. **77**, 3154-3158.

7. Bull, H. R. and Breese, K. (1974) Arch. Biochem. Biophys **161**, 665-670.

8. Ponsin, G., Strong, K., Gotto, A.M.,Jr., Sparrow, J.T., and Pownall, H.J. (1984) Biochemistry **23**, 5337-5342.

CONFORMATIONAL PROPERTIES AFFECTING THE BIOLOGICAL ACTIVITY OF SALMON CALCITONIN

Richard M. Epand and Raquel F. Epand
Department of Biochemistry, McMaster University
Hamilton, Ontario, Canada, L8N 3Z5

R.C. Orlowski and J.K. Seyler
Armour Pharmaceutical Company
Kankakee, Illinois 60901

Introduction

Polypeptides whose hydrophobic residues are regularly spaced at every third or fourth position in the sequence can fold into an amphipathic helix. This conformational feature allows for the spacial segregation of hydrophobic and hydrophilic residues. The presence of an amphipathic helix is thought to be important for enabling serum apolipoproteins to associate with phospholipids[1]. A number of polypeptide hormones have regularly spaced hydrophobic amino acid residues in their sequence[2] and surprisingly several of them are capable of solublizing phospholipids[2]. This conformational feature has led to the successful design of biologically active analogs of calcitonin[3]. Although calcitonins do not have to fold into a highly helical conformation in order to interact with phospholipids[4], analogs which do attain structures of higher helical content in the presence of phospholipid are generally more potent[5]. In the present work we describe several additional conformational features which determine the activity of the peptide.

Results and Discussion

Location of the amphipathic helix. The sequence of salmon calcitonin (sCT) is: CSNLSTCVLGKLSQELHKLQTYPRTNTGSGTP. A disulfide bridge links residues 1 and 7 and the terminal proline is amidated. We have calculated the helix probabilities for sCT using the statistical mechanical theory of Mattice[6]. This method correctly predicts that the peptide attains little helical structure in water and that the helical content is greater in the presence of anionic lipids than zwitterionic lipids. The section of the molecule which has the highest probability of folding into a helical conformation comprises residues 11-19. Residues 23-32 have virtually no tendency to form a helical structure either in the presence or in the absence of lipid. The Chou-Fasman rules predict that the carboxyl-terminal section of the peptide forms β-turns. This section of the molecule exhibits a definite periodicity in the distribution of hydrophobic residues. If this region of the molecule were to fold into a β-structure it would tend to segregate hydrophobic and hydrophilic residues on opposite sides of the peptide chain.

Hydrophobic moment of amphipathic helix. The hydrophobic moment is a quantitative estimate of the segregation of hydrophilic and hydrophobic residues in an α-helical conformation[7]. We have synthesized a series of sCT analogs with the deletion of one or more amino acids from the segment that can form an amphipathic helix. Our results demonstrate that there is no direct relationship between the hydrophobic moment and hypocalcemic activity. For example, des-Ser^{13}-sCT has as much activity as the native hormone despite a marked reduction in the hydrophobic moment of the peptide. The des-Ser^{13}-sCT has a greater tendency to fold into a helical conformation which may compensate for the loss of the amphipathic character of the helix which is formed. Thus the formation of an amphipathic helix, although important, is not an essential requirement of high biological potency. Other conformational properties such as helix-forming probability, can compensate for the loss of amphipathic character.

Conformation flexibility. Other conformational features also affect biological activity. The analogs [Gly^{8}]-sCT and [Ala^{16}]-sCT have slightly lower hydrophobic moments and lower tendency to form helical structures than sCT, yet these analogs have a higher biological potency than the native hormone. The only property which could lead to enhanced biological potency in these two derivatives is their increased conformational flexibility resulting from the smaller size of the substituted residues (Gly vs Val and Ala vs Leu) and their lower tendency to form structures of higher helical content. This increased conformational flexibility may facilitate the folding of the peptide into the conformation required for receptor binding.

Conformational coupling. A mechanism by which increased conformational flexibility can stabilize the folding of sCT is by facilitating interactions among residues which are separated in the primary structure. That such "long-range" interactions take place is suggested by results from a study of sCT segments and analogs. Although sCT-(12-32)-peptide amide and sCT-(11-23)-peptide amide contain the segment of sCT which forms an amphipathic helix, these peptides have weak helix-forming capabilities. In contrast, sCT-(1-23)-peptide amide forms an amount of helix comparable to sCT. This suggests that the amino terminus of the peptide facilitates helix formation in residues 12-20. Furthermore, although substitution of D-Arg at position 24 slightly enhances the activity of sCT, this same substitution decreases the activity of [Gly^{8}]-sCT. These results demonstrate that effects on activity caused by changes in one section of the molecule are dependent on the properties of other regions of the molecule. It is possible that hormone-receptor binding is coupled to cooperative conformational interactions among different regions of the peptide in an analogous manner to that which has been proposed for the binding of substrates to enzymes[8].

Conclusions

The presence of an amphipathic helix enhances the potency of calcitonin analogs.

The hydrophobic moment of the helix can be significantly reduced without loss of biological activity.

The enhanced activity of certain analogs can best be explained in terms of increased conformational flexibility.

There is conformational coupling among various regions of the calcitonin molecule.

References

1. Segrest, J.P., Jackson, R.L., Morrisett, J.D. and Gotto, A.M. (1974) Fed. European Biochem. Soc. Lett.38, 247-253.
2. Epand, R.M. (1983) Mol. Cell. Biochem. 57, 41-47.
3. Moe, G.R. and Kaiser, E.T. (1985) Biochemistry 24, 1971-1976.
4. Epand, R.M., Epand, R.F., Orlowski, R.C., Schleuter, R.J. Boni, L.T. and Hui, S.W. (1983) Biochemistry 22,5074-5084.
5. Epand, R.M., Epand, R.F. and Orlowski, R.C. (1985) Int. J. Pept. and Protein Res. 25, 105-111.
6. Hamed, M.M., Robinson, R.M. and Mattice, W.L. (1983) Biopolymers 22, 1003-1021.
7. Eisenberg, D., Weiss, R.M. and Terwilliger, T.C. (1982) Nature 299, 371-374.
8. Taniuchi, H. and Bohnert, J.L. (1975) J. Biol. Chem. 250, 2388-2394.

SYNTHESIS OF THE SIGNAL PEPTIDE OF BACTERIORHODOPSIN

Paul Wrede and Georg Büldt
Freie Universität Berlin, FB Physik, Abt. Biophysik, Arnimallee 14, 1000 Berlin 33

Martin Engelhard and Benno Hess
Max-Planck-Institut für Ernährungsphysiologie, Rheinlanddamm 201, 4600 Dortmund

Introduction

The extended amino-terminal sequence, the signal sequence of membrane proteins plays an important role in their translocation to, and integration into membranes. Several genetical and biochemical studies revealed that most signal sequences with a length of 20 to 40 amino acids start with a positive charged residue, further a stretch of hydrophobic amino acids and an amino acid with a small side chain close to the cleavage site for the leader peptidase[1]. An exception to this rule is the signal sequence of bacteriorhodopsin (bR), an integral membrane protein of the archaebacteria Halobacterium halobium, which serves as a light driven proton pump[2]. The signal sequence of bacteriorhodopsin encompasses only 13 amino acids, none of which is positively charged.

In order to elucidate the role of this signal sequence, the precursor of bR was obtained by solid phase peptide synthesis (SPPS): NH_2-Met-Leu-Glu-Leu-Leu-Pro-Thr-Ala-Val-Glu-Gly-Val-Ser-Gln-Ala. Two amino acids Glu and Ala were included to clarify also the role and specificity of the leader peptidase. For further biochemical studies polyclonal antibodies have been raised against this oligopeptide.

Materials and Methods

The peptide was synthesized, using 2g of BocAla-OCH_2-PAM-resin (0.35 mmol Ala/g resin) [3]. The Boc protecting group was removed by 50% TFA/CH_2Cl_2 for 30 min. Each protected amino acid was coupled to the neutralized (with 5% DIEA in CH_2Cl_2) growing peptide chain by symmetrical anhydrides (1 mM anhydride in DMF) for one hour. After washing with DMF, DIEA and DMF a second coupling step (1 hour) was added, during which the extent of coupling was monitored by the ninhydrin method [4]. The yield was usually between 99.2 to 99.8%. After the final amino acid was coupled, the peptidyl-resin was washed with DMF, CH_2Cl_2/Ac, Ac and CH_2Cl_2 and dried in vaccuo. The peptide was then cleaved from the resin, using dimethylsulfide to reduce methionine sulfoxide. The peptide was extracted from the resin by 3 x 20 ml 20% acetic acid. The combined solvents were removed by freeze drying. The crude product was purified via an automatic HPLC-system (Beckmann Instr.) which was set up for repetitive injection (Wisp, Waters) and fractionation (LKB). The sequence of the pure peptide and of the side product was established by micro-sequencing[6] and amino acid analysis.

The purified oligopeptide was covalently linked by glutaraldehyde to the keyhole limpet hemocyanin (KLH) protein. The KLH-protein (5 mg) was dissolved in 125 μl of phosphate buffer (0.4 M pH 7.5), and the oligopeptide (4 mg) was added in 375 μl H_2O, adjusted to pH 6 and the mixture was shaken for 15 min at room temperature. Then, 250 μl glutaraldehyde was added dropwise. The reaction was stopped after 30 min by the addition of 87 μl 10% glycine in water and stirred another 30 min. The reaction mixture was dialyzed against 20 mM $NaH_2PO_4 \cdot H_2O$, 15 mM NaCl, pH 7.5 (PBS buffer) at 4^oC overnight. 55 μl of this solution was diluted with 1995 μl PBS and 2500 μl of complete Freund' adjuvans. Four New Zealand rabbits were injected subcutaneously with 1 ml each of this suspension. Booster injections with the same protein concentration but with incomplete Freund' adjuvans were given after 8 and 18 days. After the 25th day about 20 ml blood was removed, giving about 10ml antiserum. Further, antibodies against purified purple membrane have been raised. The antibody titer was determined by an enzyme linked immunoabsorbent assay (ELISA) using horseradish peroxidase conjugated antirabbit IgG antibodies[7].

Results and Discussion

The cleavage and deprotection of the protected peptide from 1.15 g of peptidyl-resin gave 0.22 g (68%) of crude product. The purification of the signal-peptide was done on an analytical HPLC-system where injection and fractionation was automated. This method is quite efficient since it combines the resolution of the analylical reverse phase column with the fast purification of preparative quantities. The chromatogram revealed beside the main product (78%) three peptide containing peaks which are most probable artifacts from the HF-cleavage.

Having established the purity and the correct sequence of the main product by amino acid and sequence analysis, the peptide was coupled by glutaraldehyde to the KHL-protein. After immunizing rabbits with this protein-peptide conjugate and with bacteriorhodopsin as a control the corresponding antisera were collected after 25 days. The sera had a high titer (1:10000) for their corresponding antigen but did not crossreact as determined by ELISA. This finding is a necessary prerequisite for further studies on the processing of bR. Crossreactions are not unusual for polyclonal antibodies and since antibodies recognize sequence and folding domains even monoclonal antibodies could crossreact with different sequences (M. Engelhard and N. Uwira, unpublished). Antibodies are only recognizing the exposed sequences of membrane proteins. The signal sequence of bR is bound to the N-terminus of bacteriorhodopsin. It is known that 4-6 amino acids of the N-terminus are succeptible to proteases. Therefore, it is probable that the signal peptide with its relatively high polarity is not buried in the membrane. Hence, we would expect that the precursor protein of bR is recognized by the antibodies against the signalpeptide. Indeed, in preliminary experiments with spheroplasts the signal sequence is accessible to the antibodies. This result makes further studies on the fate of the signal peptide once it is synthesized on the ribosomes possible.

Acknowledgements

We would like to thank Dr. R.-M. Kamp for sequencing the peptides and Dr. L. Ehrlicher for discussion. Part of this work was supported by the Deutsche Forschungsgemeinschaft (SFB 312).

References

1. Michaelis, S. and Beckwith, J. (1982) Ann. Rev. Microbiol. 36, 435-465.

2. Hess, B., Kuschmitz, D. and Engelhard, M. (1984) in Information and Energy Transduction in Biological Membranes, Bolis, C.L., Helmreich, E.J.M. and Passow, H., Eds., Alan R. Liss Inc., New York, 164, pp. 81-91.

3. Mitchell, A.R., Kent, S.B.H., Engelhard, M. and Merrifield, R.B. (1978) J. Org. Chem. 43, 2845-2852.

4. Sarin, V.K., Kent, S.B.H., Tam, J.P. and Merrifield, R.B. (1981) Anal. Biochem. 117, 147-157.

5. Tam, J.P., Heath, W.F., and Merrifield, R.B. (1982) Tetrahedron Lett. 23, 2939-2942.

6. Wittmann-Liebold, B. and Kimura, M. (1984) Meth. Mol. Biol. 1, 221-242.

7. Dillner, J., Sternås, L., Kallin, B., Alexander, H., Ehlin-Henriksson, B., Jörnvall, H., Klein, G., and Lerner, R. (1984) Proc. Natl. Acad. Sci. USA 81, 4652-4656.

BIOPHYSICAL STUDIES OF GENETICALLY-DEFINED SYNTHETIC SIGNAL SEQUENCES

Martha S. Briggs* and Lila M. Gierasch

Department of Chemistry, University of Delaware, Newark, DE 19716 and *Department of Molecular Biophysics and Biochemistry Yale University, New Haven, CT 06511

Donald G. Cornell

USDA, Eastern Regional Research Center, 600 Mermaid Lane, Philadelphia, PA 19118

Nearly every protein destined to function outside of a cell is synthesized with a signal sequence, a 15- to 30-residue long, largely hydrophobic N-terminal extension, which is subsequently cleaved by a signal peptidase residing on the opposite side of the membrane (either the endoplasmic reticulum membrane in eukaryotes or the cellular membrane in prokaryotes).[1,2] While details of the process of protein translocation are poorly understood, the requirement of a signal sequence is well established. We have been examining the properties of synthetic signal peptides with a goal of clarifying the roles of the signal sequence in protein translocation.[3,4]

The signal peptides synthesized (details of synthesis to be reported elsewhere) were derived from an Escherichia coli outer membrane protein, called LamB. Emr and Silhavy[5] isolated strains of E. coli in which mutations had occurred in the signal region of the LamB gene (Figure 1). One mutant strain lost four residues from the signal region and no longer exported the LamB protein. Two strains containing this four-residue deletion and either of two additional point mutations showed restored capacity to export the LamB protein. As originally noted by Emr and Silhavy,[5]

WILD TYPE

Met Met Ile Thr Leu Arg Lys Leu Pro Leu Ala Val Ala Val Ala Ala Gly Val Met Ser Ala Gln Ala Met Ala/Val
-25 -20 -15 -10 -5 -1 1

DELETION MUTANT

Met Met Ile Thr Leu Arg Lys Leu Pro —— Val Ala Ala Gly Val Met Ser Ala Gln Ala Met Ala/Val

Gly → Cys PSEUDOREVERTANT

Met Met Ile Thr Leu Arg Lys Leu Pro —— Val Ala Ala Cys Val Met Ser Ala Gln Ala Met Ala/Val

Pro → Leu PSEUDOREVERTANT

Met Met Ile Thr Leu Arg Lys Leu Leu —— Val Ala Ala Gly Val Met Ser Ala Gln Ala Met Ala/Val

Fig. 1. The amino acid sequences of LamB wild type and mutant signal peptides. A slash indicates the site of cleavage by signal peptidase.

these secondary mutations appear to make more probable the adoption of an α-helical conformation by the signal sequence, since each replaces a helix destabilizing residue (Pro or Gly) by a helix-favoring residue (Leu or Cys).

Circular dichroism spectra (data not shown) indicate that the functional wild type and revertant peptides are predominantly α-helical in aqueous sodium dodecyl sulfate (SDS) solution while the nonfunctional deletion mutant signal peptide is unstructured. All of the signal peptides show CD spectra typical of "random" or unstructured conformations in buffer: Table I summarizes the secondary structural characterization of these peptides based on fitting their CD

Table I. Signal Peptide Conformations from CD[a]

Peptide	Solvent	% α Helix	% β Structure	% "Random"
Wild Type	Buffer[b]	7	17	76
	40 mM SDS[c]	80	0	20
Deletion Mutant	Buffer	8	13	79
	40 mM SDS	30	1	69
Gly → Cys Revertant	Buffer	9	10	81
	40 mM SDS	52	0	48
Pro → Leu Revertant	Buffer	20	1	79
	40 mM SDS	76	0	24

[a] Observed CD data were fitted to reference spectra.[6]
[b] 5 mM Tris, pH 7.3. [c] In 5 mM Tris, pH 7.3.

spectra to reference curves.[6]

Affinities of these signal peptides for lipid-water interfaces also correlate with the ability of the corresponding E. coli strain to export the LamB protein. Addition of signal peptide to the aqueous subphase beneath a phospholipid monolayer leads to a substantially larger net change in surface pressure (monitored by tensiometry) for the functional sequences than for the nonfunctional sequences, and the change occurs at a lower total concentration of peptide (data not shown).

These two properties of the signal peptides, tendency to take up an α-helical conformation and affinity for a phospholipid monolayer, have been correlated to each other by measuring the CD spectra of the peptides in phospholipid monolayers (details of the method described in Cornell).[7] Eight quartz plates were submerged in buffer in a Langmuir trough. A lipid monolayer was spread at the desired

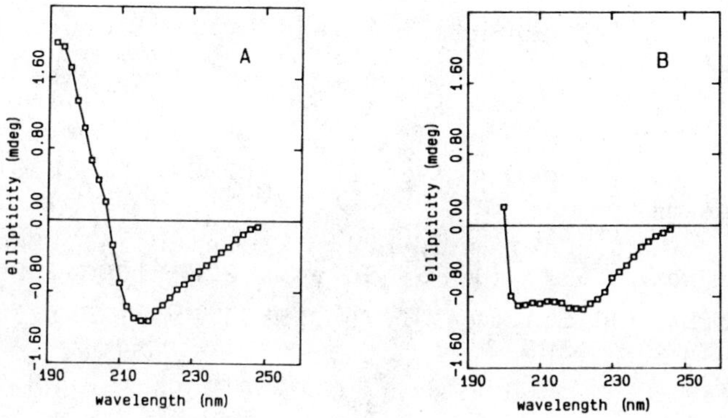

Fig. 2. CD spectra of 16 phospholipid/wild type signal peptide monolayers coated onto quartz plates, A. at high surface pressure, such that the signal peptide adsorbs onto, but does not insert into, the lipid monolayer, and B. at low surface pressure, such that the signal peptide inserts into the lipid monolayer. (The spectrum in B. has been corrected for electrostatically adsorbed, but not inserted, peptide.)

surface pressure, and signal peptide was injected into the subphase. Withdrawing the quartz plates while maintaining the surface pressure transferred a portion of the monolayer onto the plates. As shown in Figure 2B, the wild type signal peptide is α-helical when inserted into a phospholipid monolayer. Interestingly, when this peptide adsorbs to a high-pressure monolayer (formed at a surface pressure above the peptide's critical pressure of insertion) with no insertion, it adopts a β-like conformation (Figure 2A) distinct from its bulk aqueous conformation.

While several other components are certainly active participants in protein secretion, our observations argue strongly that direct interaction of signal sequences with membrane lipids plays a role in protein translocation.

This work was supported in part by a grant from the National Institutes of Health (GM 27616).

References

1. Silhavy, T.J., Benson, S.A. and Emr, S.D. (1983) Microbio. Rev. 47, 313-344.

2. Walter, P., Gilmore, R. and Blobel, G. (1984) Cell 38, 5-8.

3. Briggs, M.S. and Gierasch, L.M. (1984) Biochemistry 23, 3111-3114.

4. Briggs, M.S., Gierasch, L.M., Zlotnick, A., Lear, J.D. and DeGrado, W.F. (1985) Science 228, 1096-1099.

5. Emr, S.D. and Silhavy, T.J. (1983) Proc. Natl. Acad. Sci. USA 80, 4599-4603.

6. Greenfield, N. and Fasman, G.D. (1969) Biochemistry 8, 4108-4116.

7. Cornell, D.G. (1979) J. Colloid Interface Sci. 70, 167-180.

^{1}H AND ^{31}P NMR STUDIES OF MORPHINE, MET-ENKEPHALIN AND β-ENDORPHIN BINDING TO DODECYLPHOSPHORYLCHOLINE.

Antonio De Marco and Lucia Zetta
Istituto di Chimica delle Macromolecole, CNR, Milano, Italy

Pierluigi Gariboldi
Dipartimento di Chimica Organica, Università di Milano, Italy

Enea Menegatti
Istituto di Chimica Farmaceutica, Università di Ferrara, Italy

Introduction

Opioid peptides (β-endorphin and derivatives) share with opiates (morphine and analogs) the membrane receptors in the brain. The interaction of Met-enkephalin, human and camel β-endorphin and morphine with dodecylphosphorylcholine ($C_{12}PN$) is investigated by ^{1}H and ^{31}P NMR. The linewidth of the phosphoryl resonance appears as a suitable probe for determining the physico-chemical state of $C_{12}PN$ micelles and their interaction with the opiate molecules.

Results and Discussion

$C_{12}PN$ has been synthesized starting from n-dodecanol and 2-chloro-2-oxo-1,3,2-dioxaphospholane, according to Thuong and Chabrier.[1] Reagents and yields are indicated in the following scheme:

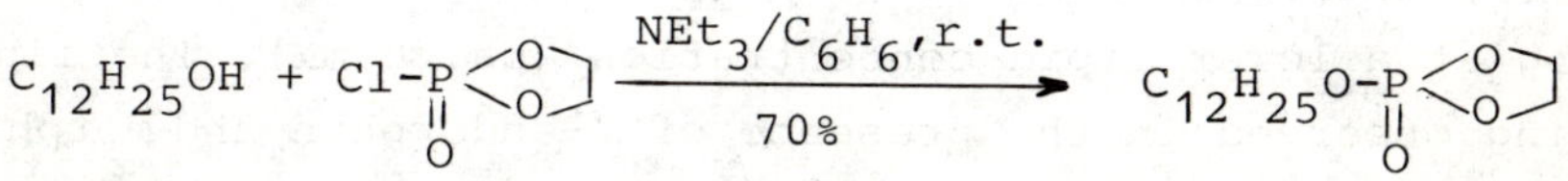

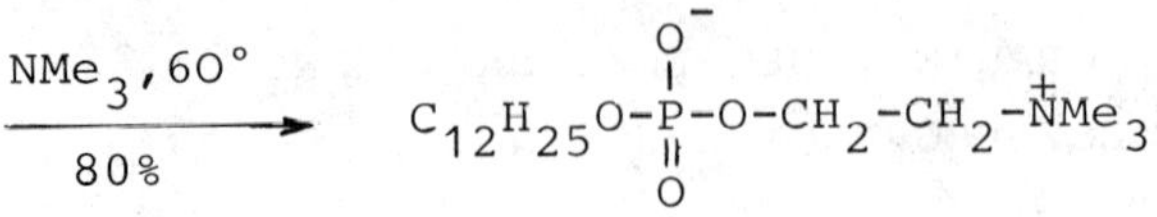

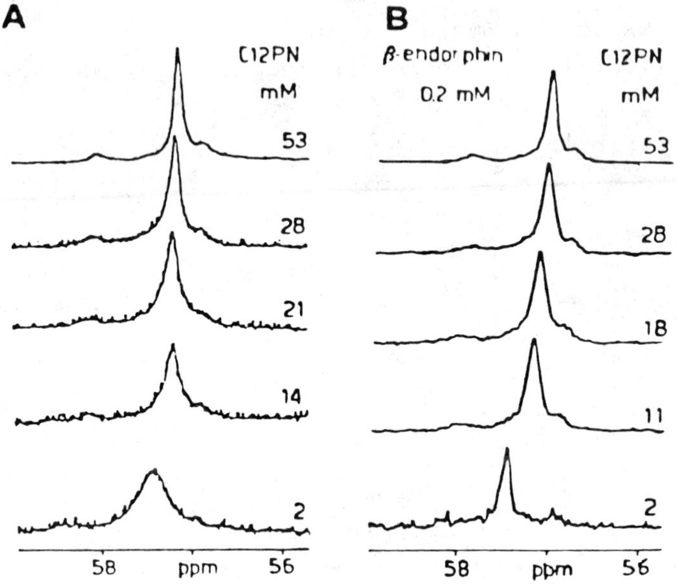

Fig.1. 80.96 MHz ^{31}P NMR spectra of $C_{12}PN$ at various concentrations. (A), alone; (B), in the presence of 0.2 mM camel β-endorphin. The ppm scale refers to external H_3PO_4; pH=6.2, T=25°C.

Figure 1. shows ^{31}P spectra of $C_{12}PN$ in $^{2}H_2O$. A progressive line sharpening is observed by increasing the lipid concentration. An inflexion near 13 mM (Fig.2A, □) most likely corresponds to the critical micelle concentration (CMC). Addition of Met-enkephalin does not sensibly affect the above trend (Fig. 2A, o). In the presence of camel β-endorphin (Fig.1B) the ^{31}P signal is already sharp at the initial $C_{12}PN$ concentration, then it broadens, reaching a maximum linewidth near 15 mM, and eventually resharpens (Fig.2A, ✱). In the presence of morphine (Fig.2A, △) the same trend is observed, the maximum linewidth occurring at a lower lipid concentration (ca. 8 mM). The line sharpening observed in the presence of β-endorphin and morphine suggests a lowering of the CMC, owing to formation of mixed micelles. It is known that in the presence of surfactants surface-active drugs may form aggregates containing molecules of both species.[2] The lack of such effect with Met-enkephalin has to be ascribed to the lower affinity of the shorter peptide for the micelles. Met-enkephalin seems not to form mixed aggregates, its interaction with the lipid occurring probably at the micelle surface.

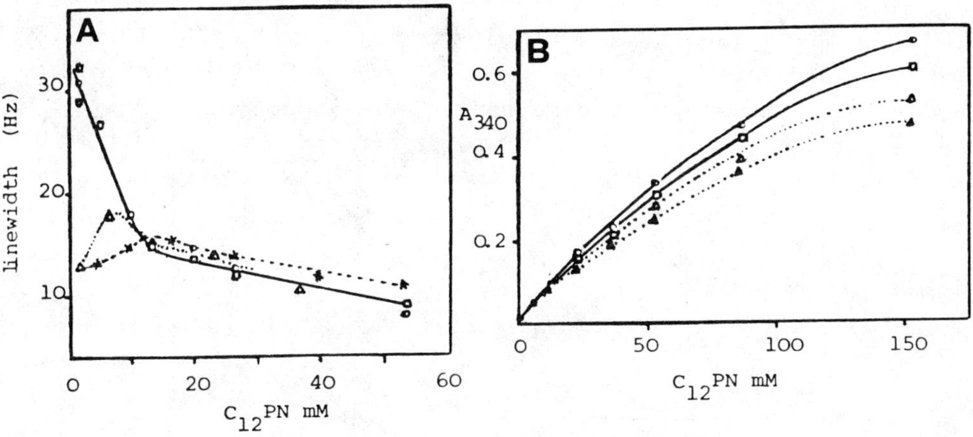

Fig.2. (A), linewidth *vs.* lipid concentration as measured in ^{31}P NMR spectra; (B), UV absorption at 340 nm *vs.* lipid concentration for C_{12}PN (□, ——), C_{12}PN plus Met-enkephalin 2.6 mM (o, ——), C_{12}PN plus camel β-endorphin 0.2 mM (✱, - - -), C_{12}PN plus morphine 2.1 mM (△, ····), C_{12}PN plus morphine 43.5 mM (▲, ····).

The formation of mixed micelles affects also the optical properties of the solution. Figure 2B shows that the maximum transparency occurs in the presence of morphine (△, ▲), whereas Met-enkephalin causes the opposite effect (o), the solution absorbing more than that containing C_{12}PN alone (□). Although the phenomenon is still not too clear, it can be hypothesized that mixed micelles are different in shape or size from those formed by C_{12}PN alone.

The higher affinity for C_{12}PN micelles exhibited by β-endorphin in comparison with Met-enkephalin is also manifested in

^{1}H NMR photo-CIDNP experiments.[3] From the intensities of the Tyr-1 resonances in the dark-minus-light difference spectra the correlation time of the aromatic side chain can be calculated for the peptides at various lipid concentrations.

Table I. Correlation Time of Tyr-1 in Met-Enkephalin and Camel β-Endorphin at Various C_{12}PN Concentrations.

C_{12}PN (mM)	τ_c (ns) Met-enkephalin 0.5 mM	β-endorphin 0.5 mM
0	2.0	2.4
2		2.4
7	2.0	3.5
14		3.6
20	2.2	
50	2.2	

From the τ_c values quoted in Table I it appears that in the pentapeptide Tyr-1 is little affected in mobility by the addition of a large excess of lipid. Conversely, for camel β-endorphin τ_c is sensibly affected by small amounts of C_{12}PN.

References

1. Thuong, N.T., and Chabrier, P. (1974) Bull.Soc.Chim.Fr. 667-671.
2. Florence, A.T. (1968) Adv.Colloid Interface Sci. 2, 117-121.
3. De Marco, A., Zetta, L., and Kaptein, R. (1985) Eur. Biophys.J. 11, 187-193.

This work was supported by the SP 8 Controllo del Dolore, Progetto Finalizzato Medicina Preventiva e Riabilitativa.

INTERACTION OF OPIOID PEPTIDES WITH MODEL MEMBRANES. ^{1}H-NMR EVIDENCE FOR THE FORMATION OF A FOLDED STRUCTURE OF β-ENDORPHIN IN MICELLAR SYSTEMS.

Lucia Zetta and Antonio De Marco
Istituto di Chimica delle Macromolecole, CNR, Milano, Italy

Marise Bourdenneau and Christian Brevard
Bruker Spectrospin SA, Wissembourg, France

Introduction

Photo-CIDNP and NOE experiments have shown that, in the presence of micelles, the accessibility and mobility of Tyr-1 and Tyr-27 in human β-endorphin are strongly reduced, irrespective of the ionic properties of the micelles.[1-3] It has been assumed that the change in accessibility and the immobilization arise from the secondary structure formed upon binding and stabilized by the micelle apolar chains. The presence of an α-helix in the C-terminal half of the peptide has been suggested by CD studies performed on β-endorphin in phospholipid micelles.[4] Further NMR studies on camel β-endorphin in micelles of sodiumdodecylsulfate confirmed the existence of a secondary structure based on the chemical shift dispersion and line broadening observed for the aromatic and methyl resonances, and on the slow exchange rate of some amide protons with the solvent deuterons.[3] In order to characterize the folded structure, pH titration and 2D NMR experiments have been performed on camel and human β-endorphin, in the presence of perdeuterated SDS micelles.

Tyr-Gly-Gly-Phe-Met-Thr-Ser-Glu-Lys-Ser-Gln-Thr-Pro-Leu-Val-
-Thr-Leu-Phe-Lys-Asn-Ala-Ile-Ile-Lys-Asn-Ala-Tyr-Lys-Lys-Gly-
-Glu human β-endorphin

Results and discussion

In the 2D COSY spectrum (not shown) the J-connectivities for all the methyl protons have been assigned to the corresponding amino acids. In Figures 1A and 1B the 2D COSY and NOESY spectra of the aromatic protons are reported. In Figure 1B two cross peaks connecting Phe-4 with the δ and ε protons of Tyr-27 suggest the presence of a folded form in which the N-terminal fragment interacts with the C-terminal portion of β-endorphin. In Figure 2 the NOE connectivities are reported for the aromatic and methyl protons. Cross peaks are observed connecting Tyr-1 δ-protons with the Met-5 and Val-15 methyl groups, Phe-4 with the Ala-21 and Ala-26 methyl protons. In addition, the methyl protons of Leu-14 and Leu-17, Ile-22 and Ile-23 show connectivities with the aromatic rings. These results indicate the existence of a hydrophobic domain involving residues with both aromatic and aliphatic side chains, the latter mainly located in the fragment 14-27.

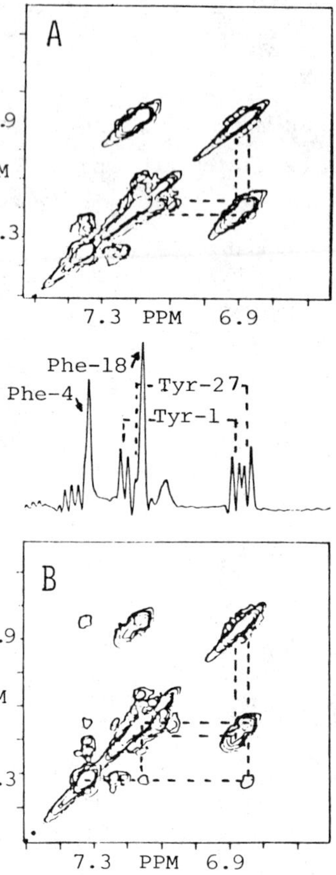

Fig.1. 400 MHz NMR spectra of human β-endorphin (10mM) in SDS (150mM): aromatic region. A. COSY B. NOESY.

In order to characterize this apolar domain, pH titrations have been performed on camel β-endorphin in aqueous solution and in the presence of SDS, in the pH range 5-13. The pK_a values obtained in the micellar system are considerably higher than in aqueous solution (Table 1). In addition, several methyl

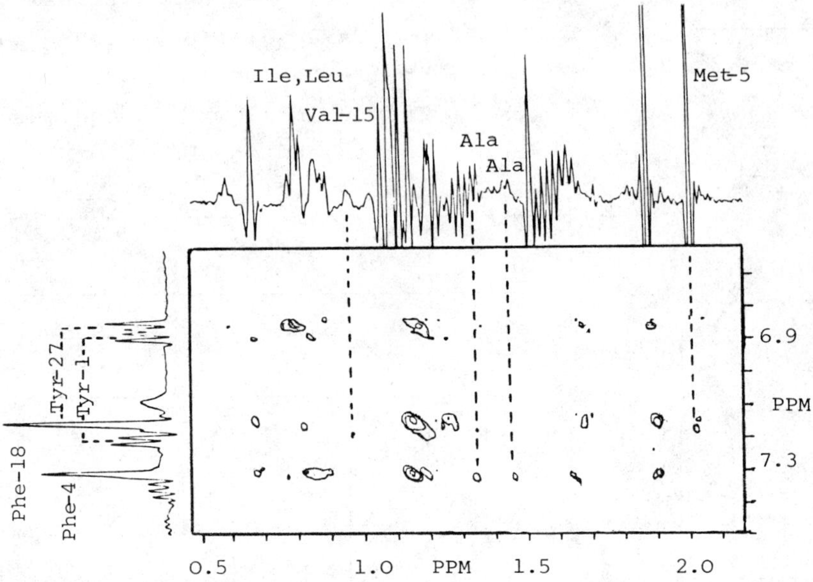

Fig.2. 400 MHz NOESY spectrum of human β-endorphin (10mM) in SDS (150mM).

groups show extrinsic pK_a values (Figure 3), not observed in aqueous solution, indicative of some non bonded contacts between the central apolar segment and the N-terminal residue of the molecule.

Combined 2D NOE results and pH titration indicate that in the presence of SDS micelles, β-endorphin adopts a tertiary structure in which the N-terminal fragment is buried inside a hydrophobic pocket. Very likely, this apolar domain is localized in the hydrocarbon core of the micelles where low accessibility is allowed to Tyr-1.[1-3] This is consistent with the hipothesys that, in the presence of membranes, Tyr-1 is concealed in a folded structure and hence protected against proteolytic attack.[5] The tertiary structure is stable over the pH range 5-13, although above 12, during the titration of the lysyl groups, the micelles partly break and β-endorphin is released in its random form. It can be concluded that the hydrophobic interactions with the micelles are the major factor which con

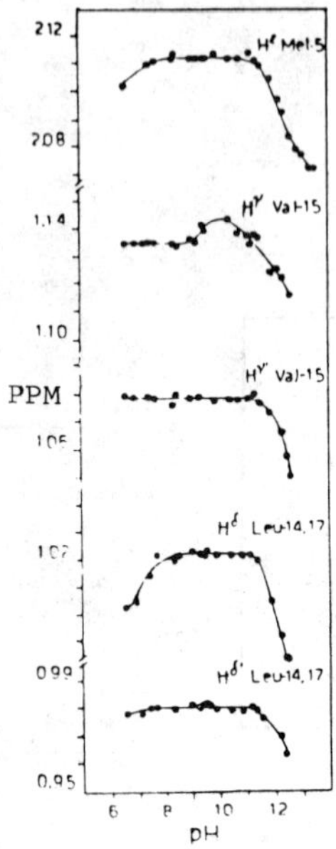

Fig. 3. ^{1}H-NMR of camel β-endorphin in SDS: pH dependence of chemical shifts for selected methyl groups.

Table I. Intrinsic (A) and Extrinsic (B) pK_a Values for Selected Groups in ^{1}H-NMR Spectra of Camel β-Endorphin (10mM) in Aqueous Solution and in SDS (150mM).

	pK_a		
	water	SDS	
	A	A	B
Tyr-1 OH	10.3	12	-
" NH_2	7.3	9	-
His-27 NH	6	8	12
Lys ε-NH_2	11	> 12	-
Met-5 CH_3	-	-	12
Val-15 CH_3	-	-	12
Leu-14,17 CH_3	-	-	12

tributes to the maintenance of the tertiary structure.

References

1. Zetta,L. Kaptein,R. and Hore,J.P. (1982) FEBS Lett. 145, 277-280.

2. Zetta,L., Hore,J.P. and Kaptein,R. (1983) Eur. J. Biochem. 134, 371-376.

3. Zetta,L. and Kaptein,R. (1984) Eur. J. Biochem. 145, 181-186.

4. Wu,C.-S.C., Lee,N.M., Loh,H.H., Yang,J.T. and Li,C.H. (1979) Proc. Natl. Acad. Sci. USA 76, 3656-3659.

5. Austen, B.M. and Sm th, D.G. (1977) Biochem. Biophys. Res. Commun. 76, 477-482.

This work was supported by the SP 8 Controllo del Dolore, Progetto Finalizzato Medicina Preventiva e Riabilitativa.

SPECTRAL STUDIES OF BACTERIORHODOPSIN FRAGMENTS AND THEIR RELEVANCE TO ITS SECONDARY STRUCTURE

Alexander T. Kozhich, Sergey V. Sychev, Galina V. Ovechkina
Shemyakin Institute of Bioorganic Chemistry USSR Academy of Sciences, ul. Miklukho-Maklaya 16/10, 117871, Moscow, USSR

Bacteriorhodopsin (BRh), an integral protein of the purple membrane of Halobacterium halobium is responsible for direct transformation of solar energy into pH gradient[1]. Important bioenergetic function and relative simplicity of this membrane protein have induced intensive structural and mechanistic investigations.

Using a method of electron microscopy Henderson and Unwin have developed a three-dimensional model of BRh, consisting of seven presumably α-helical rods, roughly perpendicular to the membrane plane[2]. A model has been proposed in which particular segments of the amino acid sequence is assigned to each helical rod[3-5]. An alternative model containing five α-helices and four antiparallel β-sheet strands has been also proposed[6,7] and indirectly supported by CD and IR measurements[6].

To shed more light on the problem we have attempted preparation and conformational study of a series of partial oligopeptide BRh sequences. Isolation and synthesis of peptides were carried out according to procedures published elsewhere[8,9].

We succeeded in isolation of segments 72-108(C), 163-208 (F), 1-36(A), 156-248(FG), 1-71(AB) from enzymatic and acidic hydrolyses of BRh (Letters A-G designate the fragments in accordance with[4]). Peptides Z-34-65-OH, Z-59-65-OH, Z-34-44-OMe were obtained by synthesis. CD and IR spectra of these peptides in trifluoroethanol (TFE) and SDS micelles were measured. Analysis of secondary structure was carried out from CD spectra by the Greenfield-Fasman method.

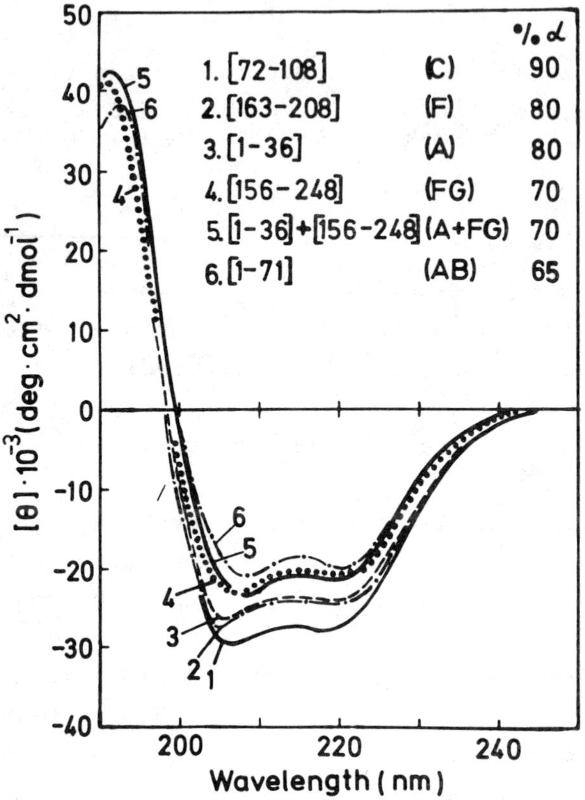

Fig. 1. CD spectra of the long BRh fragments in TFE. Peptide concentration is $(0,7-3,7)\cdot10^{-4}$ M.

As seen from Figure 1 the long transmembrane BRh fragments in TFE have high content (>60%) of α-helix. The remaining part is a random coil. Similar results were obtained for SDS micelles (SDS concentration 0,07M). Almost complete lack of β-sheet structure was deduced from the absence of β-structural component at 1620-1630 sm^{-1} in IR spectra in both TFE and SDS micelles.

According to the recent proposal[10] segments A-G are assigned to α-helical strands 1-7 in the native bacteriorhodopsin as shown below.

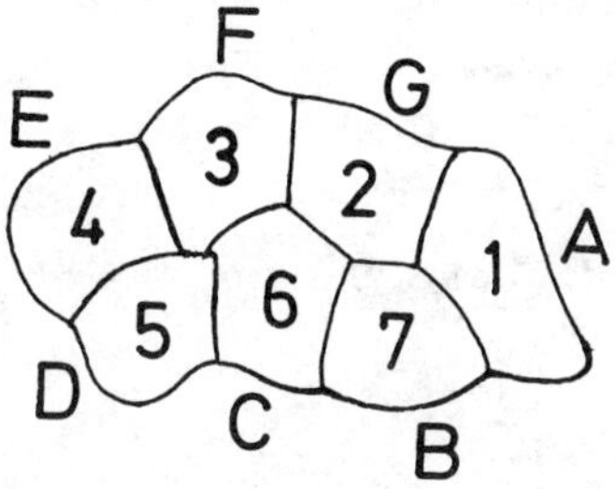

In the alternative model strands 1 and 2, i.e. segment A and G form a β-sheet. To probe this possibility we measured the CD spectra of equimolar mixture of segments A and F-G. As seen from the Figure 1 no indication of their association and β-sheet formation is observed.

CD spectra of segment B (amino acid sequence 34-65) (Fig. 2) reveal pronounced tendency of this segment to form α-helix even in the absence of supporting α-helical neighbours. The adjacent sequences 34-44 and 59-65 which in the model[4,5] serve as links connecting the α-helical rods as expected show no trend to helix formation. Instead, they prefer elongated (β) or random structures.

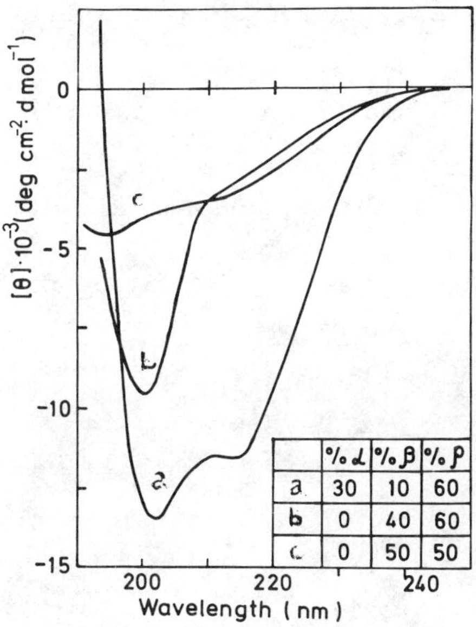

Fig. 2. CD spectra of a - Z-34-65-OH; b - Z-59-65-OH, c - Z-34-44-OMe in TFE. Peptide concentration $1{,}5\cdot10^{-3}$ M.

References

1. Stoeckenius, W., and Bogomolni, R.A. (1982) Annu. Rev. Biochem. v. 51, 587-616.

2. Henderson, R., and Unwin, P.N.T. (1975) Nature v. 257, 28-32.

3. Ovchinnikov, Yu.A., Abdulaev, N.G., Feigina, M.Yu., Kiselev, A.V., and Lobanov N.A. (1979) FEBS Lett. 100, 219-224.

4. Trewhella, J., Anderson, S., Fox, R., Gogol, E., Khan, S., Engelman, D., and Zaccai, G. (1983) Biophys. J. v. 42, 233-241.

5. Vturina, I.Yu., Kuryatov, A.B., Kiselev, A.V., Khoroshilova, N.I., Ovechkina, G.V., Abdulaev, N.G., Tsetlin,V.I., and Vasilov, R.G. (1985) Biol. Membr. (USSR) v. 1, 1661-1170.

6. Jap, B.K., Maestre, M.F., Hayward, S.B., and Glaeser, R.M. (1983) Biophys. J. v. 43, 81-89.

7. Glaeser, R.M., and Jap, B.K. (1984) Biophys. J. v. 45, 95-97.

8. Tsetlin, V.I., Zakis, V.I., Aldashev, A.A., Kuryatov,A.B., Ovechkina, G.V., and Shnyrov, V.L. (1983) Bioorgan. Khimia v. 9, 1589-1605.

9. Kozhich, A.T., and Vaskovsky, B.V. (1984) in PEPTIDES, Structure and Function, Proc. 8th Amer. Pept. Symp. Hruby, V.J., and Rich, D.H., Eds., Pierce Chemical Company, Rockford, IL, pp. 191-194.

10. Jubb, J.C., Worcester, D.L., Crespi, H.L., and Zaccai, G. (1984) EMBO J. v. 3, 1455-1461.

SOLID PHASE SYNTHESIS OF HUMAN PLASMA APOLIPOPROTEIN C-II AND SOME OF ITS FRAGMENT PEPTIDES.

T. Fairwell*, A.V. Hospattankar, G. M. Powell and J.C. Osborne, Jr.
Molecular Disease Branch, NHLBI, National Institutes of Health
Bethesda, Maryland 20205

S.A. Khan

The Rockfeller University,
New York, New York 10021

Introduction

Human plasma apolipoprotein (apo) C-II plays an important role in lipid metabolism as an activator of lipoprotein lipase, the enzyme that hydrolyses the triglycerides in plasma chylomicrons and very low density lipoproteins. An inherited deficiency of plasma apoC-II has been reported in patients with severe hypertriglyceridemia and type I hyperlipoproteinemia. Infusion of normal blood or plasma containing apoC-II results in a temporary reduction of plasma triglycerides and an elevation of very low density and high density lipoproteins. ApoC-II is a single polypeptide chain of 79 amino acids[1] and two separate functional domains have been tenatively identified in the sequence. It is reported that the lipid binding region is located within residues 43-51 and the carboxy-terminal 56-79 residues are essential for complete activation of lipoprotein lipase. Recently we have reported some preliminary data on the total synthesis of apoC-II[2]. A major chain termination occurred during this synthesis after incorporation of the lysine at position 19, probably due to some contaminants present in the Boc-Lys derivative used. However, this terminated peptide could be easily separated from the intact protein by ion-exchange chromatography. In the present synthesis, this chain termination is avoided and a number of peptides representing progressively longer segments from the carboxy-terminal of the molecule were also synthesized in order to study in detail, the activation of lipase by various fragment peptides and to explore their physiological functions.

Results and Discussion

The total synthesis of apoC-II and the fragment peptides representing amino acid residues 7-79, 40-79, 51-79 and 61-79 were accomplished by the Merrifield solid phase method[3] using the phenylacetamidomethyl (PAM) resin as the solid support. Boc-Glu(-OBZl)-O-Resin used in this synthesis was prepared according to the modified procedure of Khan (unpublished work), with a low substitution level of 0.21 mmol of Glu/g of resin. Successive amino acids were added to the Glu(-OBzl) resin with some modifications of the procedures described earlier[2,4]. The BrZ-derivative of Boc-Tyr was used instead of the Cl_2Bzl derivative and 1-hydroxy-benzotriazole was used as a coupling agent in every cycle. The amino acid sequence of apo-C-II showing the side chain protecting groups used during the synthesis is shown in Fig. 1. Portions of the fully protected peptide-resin were removed as needed, in order to obtain the fragment peptides. The peptides were deprotected and cleaved from the resin by treatment with anhydrous hydrogen fluoride according to the modified procedure of Sakakibara et al[5]. The cleavage was done at -3 to -4°C for 60 minutes in presence

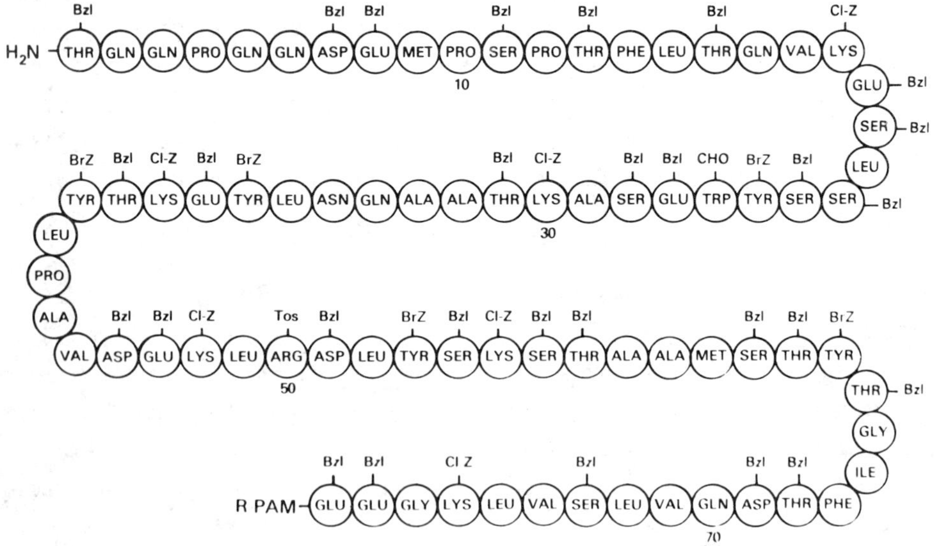

Fig. 1 Amino acid sequence of apoC-II with the side chain protecting groups used during the synthesis.

of anisole, dimethylsulfide and 1,2-ethanedithiol. The relative yields after HF cleavage were found to be only 53% for the intact molecule and 50-58% for the various peptides. Some peptide was found to be still attached to the resin, with the side chain protecting groups removed, by solid phase sequence analysis, suggesting that the low temperature used is not adequate to completely cleave the peptide-resin bond. However, the crude peptides obtained were fairly homogeneous and easy to purify.

After desalting on a Bio-Gel P6 column the crude apoC-II was initially purified by ion-exchange chromatography on a DE-52 column, with a linear gradient of Tris-HCl in 6 M urea (0.01-0.165 M, pH 8.0)[2]. The fraction eluting at the position of the native apoC-II under identical chromatography conditions was collected and further purified by high performance liquid chromatography using a Zorbax PEP-RP1 column. Peptides 7-79 and 40-79 were also purified with the same buffer gradient on a the DE-52 column, while the other peptides were purified by HPLC only. Considerable aggregation of the peptides was occuring, which resulted in these peptides becoming more and more insoluble. The purified synthetic apoC-II eluted at the same position as the native protein on HPLC analysis.

Amino acid composition of the synthetic protein, after acid hydrolysis at 110°C for 24 hours in presence of mercaptoethanol, is in excellent agreement with that of the native apoC-II and the various peptides also showed the predicted composition. Complete structural analysis of the synthetic apoC-II was achieved by the Edman degradation of the intact molecule as well as the various fragment peptides. No deletion sequences or rearranged amino acid derivatives were detected. Polyacrylamide gel electrophoresis (10% acrylamide) in Tris buffer (0.12 M, pH 8.9) containing 8M urea produced only a single band for the intact molecule and its mobility was identical to that of the native apoC-II. The immunopotency of the synthetic protein was also found to be identical to that of the native protein by the Immuno-Blot analysis using rabbit anti-apoC-II antibody.

Lipoprotein lipase assays were carried out according to the procedure described by Jahn et al[6]. The reaction mixture contained 0.046 mmol (^{3}H) triolein, 0.1 M Tris-HCl, 1.5% BSA and 0.1 M sodium chloride (final pH 8.0). Synthetic apoC-II and the fragment peptide 7-79 produced complete activation of lipoprotein lipase, identical to that produced by the native protein. C-II fragment 40- 79 produced only slight activation while peptides 51-79 and 61-79 produced no significant activation at the peptide concentration levels used in these assays. These results indicate that detailed studies using other fragment peptides of apoC-II may be needed to learn more about the mechanism of activation of lipoprotein lipase.

Thus the synthetic apoC-II has been found to be highly homogeneous and is indistinguishable from the native protein with respect to its amino acid sequence and composition, polyacrylamide gel electrophoretic mobility, and immunological and biological properties. The final yield of 19.5% could be improved by improving the HF cleavage yield. However, the HF cleavage conditions used in this synthesis[5] reduced some of the side reactions associated with the cleavage at 0°C. The successful synthesis of the total molecule and the 7-79 fragment now opens the way for further physiological studies and eventually the apoC-II 7-79 peptide may prove to be therapeutically useful in the treatment of apoC-II deficiency.

References

1. Hospattankar, A.V., Fairwell, T., Ronan, R. and Brewer, H.B. Jr. (1984) J. Biol. Chem. 259, 318-322.

2. Fairwell, T., Hospattankar, A.V., Brewer, H.B. Jr. and Khan, S.A. (1984) Proceedings Forum International sur les peptides, Cap D' Agde, France (in press).

3. Merrifield, R.B. (1963) J. Am. Chem. Soc. 85, 2149-2154.

4. Fairwell, T., Hospattankar, A.V., Ronan R., Brewer, H.B. Jr., Chang, J.K., Shimizu, M., Zilner, L and Arnand, C.D. (1983) Biochem 22, 2691-2697.

5. Chino, N., Kumagaye, S., Masni, S., Kimura, T and Sakakibara, S. (1984) Proceedings, Forum International sur les peptides, Cap D' Agde, France (in press).

6. Jahn, C.E., Osborne, J.C. Jr., Schaefer, E.J. and Brewer, H.B. Jr. (1981) FEBS Lett. 131, 366-368.

SYNTHETIC PEPTIDE ANALOGS OF APOLIPOPROTEINS

G. M. Anantharamaiah, Christie G. Brouillette, Byung H. Chung, Charlie F. Schmidt, Ali Gawish and Jere P. Segrest

Departments of Pathology, Biochemistry and Medicine and the Atherosclerosis Research Unit, University of Alabama at Birmingham, Birmingham, Alabama 35294

Introduction

The amphipathic helix has been generally presumed to be the structural form of the lipid-associating domains of the exchangeable apolipoprotein classes A, C and E from plasma lipoproteins.[1] The amphipathic helix model defines a general α-helical domain containing opposing polar and non-polar "faces". Further, the polar face contains a specific distribution of the charged residues with the positive occurring along the interface between the polar and non-polar faces, and the negative along the center of the polar face. The present study discusses the structural properties of the complexes formed between dimyristoylphosphatidylcholine (DMPC) and six synthetic peptide analogs compared to the complexes formed by apolipoprotein A-I (Apo A-I), the ability of these peptides to displace native apolipoproteins A-I and A-II from high density lipoprotein (HDL), and their ability to activate the enzyme, lecithin: cholesterol acyl transferase (LCAT).

Five out of six peptide analogs (Table I) have been designed based on model building studies and have no homology to the natural sequences of apolipoproteins. Peptide 18A is an idealized amphipathic helix and, although not homologous, resembles the consensus domain of apo A-I. Peptide reverse18A has the charged amino acids positions reversed to test the importance of the position of charges. Peptide desVal^{10}18A is the same as 18A but val at the 10th position deleted which produces a rotation of the polar/non-polar interface by 100°, resulting in an analog which resembles the apo A-IV consensus domain.

18A-pro-18A has a pro at the center of two 18A peptides, to test the co-operativity, if any, in amphipathic helical domains. Peptide 22dmrA is designed to test the theory that an 11-mer sequence (represently 33 base pairs) duplication produces an amphipathic helix with precise centered polar and non-polar residues. Such a peptide would be expected to be more active than duplicated 10-mer or 12-mer sequences. Peptide Apo A-I (8-33) is a highly conserved region of apo A-I that represents one of the 22-mer repetative domains.

Table I. Peptides Synthesized

1. 18A[2]: D-W-L-K-A-F-X-D-K-V-A-E-K-L-K-E-A-F
2. desVal^{10}18A[2]: K-W-L-K-A-F-Y-D-K-A-E-K-L-K-E-A-F
3. Reverse18A[2]: K-W-L-D-A-F-Y-K-D-V-A-K-E-L-E-K-A-F
4. 18A-pro 18A[2]
5. 22dmrA: L-W-D-K-L-K-D-L-F-S-S-L-W-D-K-L-K-D-L-F-S-S
6. ApoA-1(8-33): W-D-R-V-K-D-L-A-T-V-Y-V-D-V-L-K-D-S-G-R-D-Y-V-S-Q-F

Results

Density gradient ultracentrifugation, negative stain electron microscopy (EM) and non-denaturing gradient gel electrophoresis (GGE) studies indicate that 18A, 18A-pro-18A and 22dmrA produce stable discoidal complexes with DMPC. At an equal weight/weight ratio apo A-I and 18A formed discoidal complexes of essentially identical dimensions. DesVal^{10}18A formed quite larger discs of approximately 200A^{o} diameter and 18R formed even larger discs. In GGE the complexes formed by reverse18A did not enter the gel at all. Peptide apoA-I(8-33) did not form complexes with DMPC as seen by the EM and as studied by GGE. ^{1}H-NMR spectroscopy (400MHZ) studies are compatible with a protein annulus-bilayer disc structure for the 18A/DMPC complex, the same structure indicated by similar studies for apoA-I DMPC complexes[2,3]. Differential scanning calorimetry studies of the peptide:DMPC complexes (studied with peptides 18A, 18A-pro-18A, desVal^{10}18A, and reverse18A) are also supportive of the peptide-annulusbilayer disc model. The CD studies indicate that the rank order percent α -helicity of the peptide in its DMPC complex is 22dmrA > 18A-pro-18A>18A > desVal^{10}18A > reverse18A > 18R, an order inverse from the complex diameter.

Peptides were compared for their ability to displace native apolipoproteins from HDL and in their ability to activate the enzyme LCAT[4]. Peptide 22dmrA was not subjected to these studies due to its poor solibility (and probably aggregation) in buffer systems used for these studies.

Peptides (0.5 to 4 mg) were incubated with isolated HDL (2 mg) overnight at 7^{0}C, and the displaced apolipoproteins were characterized. The amount of apo A-I displaced from HDL was quantitated by the radial immunodiffusion method[4]. The major changes noted were a decrease in the apo A-I/A-II ratio for HDL modified by incubation with 18A-pro-18A (to a lesser extent, in 18A). The absolute level of apo A-II does not appear to change significantly after incubation with any of the four peptides. Peptide 18A-pro-18A, with two amphipathic helical domains linked through a pro, displaced 80o/o of the total of apo A-I at a 1:2 peptide to HDL weight ratio, while even at a 1:1 ratio peptide 18A displaced only 70o/o of apo A-I. Peptides reverse18A and A-I(8-33) displaced only a nominal amount of apo A-I from HDL, while desVal^{10}18A produced 20o/o displacement of apo A-I. LCAT activation studies indicate 18A-pro-18A is very active and in fact under the assay system this peptide showed 140o/o of the activity of apo A-I. Peptide 18A has about 20o/o of the LCAT activating activity of apo A-I, whereas desVal^{10}18A and reverse18A have no detectable activity. The unusual results obtained with the peptide 18A-pro-18A have been attributed to its capacity to spontaneously form discoidal complexes with eggPC which are better substrates for LCAT than vesicles. The other peptides and even apoA-I failed to form discoidal complexes with egg PC. Peptides 22dmrA and apoA-I(8-33) have not yet been subjected to these studies.

Conclusions

1. Analogs 18A, 18A-Pro-18A and 22dmrA whose sequences strongly mimic native amphipathic sequences also strongly mimic apoA-I in their interaction with DMPC. Analog desVal^{10}18A, which is a

reasonable analog for the consensus 22 residue repeate in apo A-IV, shows considerably less interaction.

2. Peptide 18A-Pro-18A has greater lipid-associating affinity than 18A, which presumably is the result of cooperativity provided by two covalently linked lipid-associating domains. Displacement studies also support this conclusion.

3. The rank order lipid-associating abilities of peptides can be represented as follows: 22dmrA > 18A-Pro-18A > 18A > $desVal^{10}$ > 18A > Reverse18A.

4. Reverse18A has the lowest lipid associating ability suggesting the importance of the presence of positively charged amino acids at the polar/non-polar interface.

5. The reason for 18A-Pro-18A to exceed apo A-I in its LCAT activating activity is its ability to spontaneously convert preformed egg pc vesicles to small protein-annulus bilayer discs which are better substrates for LCAT than vesicles.

6. These studies support the concept that protein annulus bilayer disc complexes, whether formed by analog peptides or by apo A-I and A-II, appear to be intrinsically good LCAT substrates.

7. Surprisingly apo A-I (8-33), which is an amphipathic and the highly conserved region of apo A-I sequences from various species, does not possess lipid binding ability. The probable role of this region in apo A-I is being investigated.

References

1. Segrest, J. P., Jackson, R. L., Morrisett, J. D., and Gotto, A. M. Jr. (1973) FEBS Lett. 28, 247-253.

2. Anantharamaiah, G. M., Jones, J. L., Brouillette, C. G., Schmidt, C. F., Chung, B. H., Hughes, T. A., Bhown, A. S., and Segrest, J. P., (1985) J. Biol. Chem., 260, 10248-10255.

3. Brouillette, C. G., Jones, J. L., Ng, T. C., Kercret, H., Chung, B. H., and Segrest, J. P., (1984) Biochemistry, 23, 359

4. Chung, B. H., Anantharamaiah, G. M., Nishida, T., Brouillette, C. G., and Segrest, J. P. (1985) J. Biol. Chem., 260, 10256-10262.

ATRIAL NATRIURETIC FACTORS. INTRODUCTORY REMARKS AND PRELIMINARY CLINICAL RESULTS.

Ralph Hirschmann
Merck Sharp & Dohme Research Laboratories
P.O. Box 2000
Rahway, NJ 07065

Michael Rosenblatt
Merck Sharp & Dohme Research Laboratories
West Point, PA 19486

Several related polypeptides of varying lengths have been isolated from rat and from human atria by laboratories in Canada, in the U.S., and in Japan during the recent past, and their structures have been elucidated. These have been shown to contain a common sequence, they have been synthesized chemically and by recombinant DNA techniques; sensitive radioimmuno-assays have been developed.

The importance of this family of peptides derives, in part, from the fact that they possess potent diuretic and vasodilatory activities in animals, and that they can block the release of aldosterone. Furthermore, it has been shown that these peptides are likely to be true hormones in that they have been detected in plasma by RIA. Further, specific membrane receptors have been found in renal, vascular, and in other tissues. Thus these compounds may be considered to be counter-regulatory to the renin-angiotensin-aldosterone system, and the heart should no longer be regarded as serving only the role of a mechanical pump. Rather, it is also an endocrine organ that can control its work load. It has also been observed that more than one receptor type exists for these compounds, and structure activity relationships have been established for related compounds both in vitro and in vivo, and separation of activities has been observed. The advances that have been made by workers in this field during the past two years are truly astonishing, and have been made possible through the productive interaction of biologists and chemists. It is not surprising that this class of peptides has already generated much enthusiasm in perhaps every discipline represented at this conference, with the possible exception of the X-ray crystallographers. Particularly satisfying is the possibility that one or more of these compounds may find a place in therapy in spite of the fact that the compounds isolated and synthesized to date are not active after oral administration.

We are aware of three clinical studies which have been carried out, all of them in normal volunteers. The results summarized below are, therefore, of a very preliminary nature. One study was carried out in New Zealand, with alpha-h-ANP given as a single bolus by i.v. administration. A second study was carried out in Japan at three centers using the same compound. In these latter trials, the peptide was given both as an i.v. bolus and by i.v. infusion. In a third study carried out in Switzerland with rat ANF, the peptide was given to salt-loaded volunteers in a 4-hour infusion. Taken together, these studies suggest that the atrial peptides are highly effective in lowering blood pressure and in producing diuresis and natriuresis without producing significant tachycardia, that they have a rapid onset of action, and that they have adequate duration of action when given parenterally. Only future clinical studies will establish to what extent these compounds have unique clinical potential for the treatment of congestive heart failure, hypertensive crises, renal failure, and in post-operative intensive care.

ATRIAL NATRIURETIC FACTOR. HISTORICAL INTRODUCTION AND OVERVIEW

Adolfo J. de Bold
Department of Pathology, Queen's University and Hotel-Dieu Hospital, Kingston, Ontario K7L 3H6

Just four years after atrial muscle extracts were reported to contain a factor with natriuretic and hypotensive properties[1], [atrial natriuretic factor (ANF)], a large amount of information has been gathered regarding the significance of this discovery at several levels. This rapid development attests to the impressive technical advancements in peptide chemistry and molecular biology of the last two decades. The discovery of ANF has its earliest basis on the introduction of the then relatively new technique of electron microscopy. Using this technique Kisch[2] in 1956, first described a difference in the cytological make up between atrial and ventricular muscle cells (cardiocytes) in the heart of the guinea pig. This difference consisted in that atrial cardiocytes, unlike ventricular cardiocytes, contained "microbodies". Subsequent work by several investigators led to a better definition of this finding and by the mid 60's it was clear that in all mammals, atrial cardiocytes were morphologically differentiated as both secretory and contractile cells[3]. It became apparent that the microbodies present in the atrial cardiocytes resembled storage granules found in polypeptide-hormone producing cells. These granules - currently known as specific atrial granules - could be found as part of a complex formed by the granules themselves, a highly developed Golgi complex and elements of the rough endoplasmic reticulum (Fig. 1).

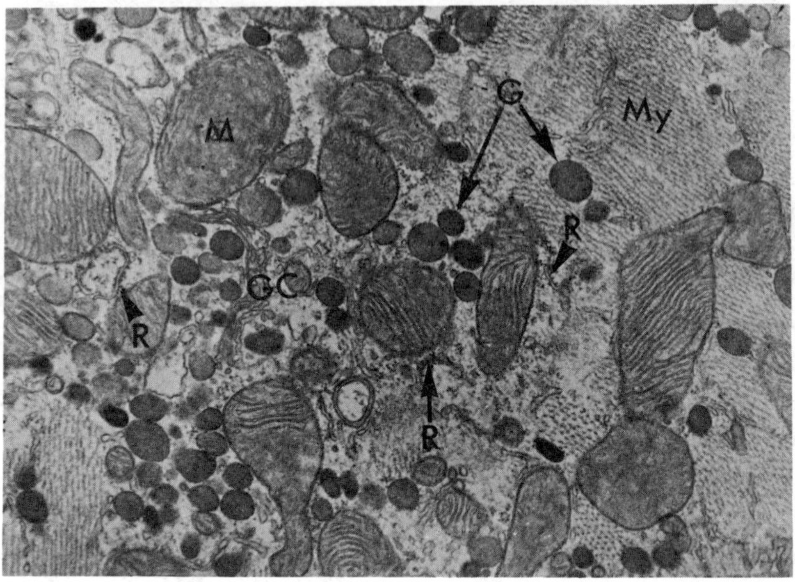

Fig. 1. Electron microscopic view of a portion of the central sarcoplasmic core of a rat atrial cardiocyte. G= specific atrial granules. GC= Golgi complex. R= rough endoplasmic reticulum. M= mitochondria. My= myofibrils.

From a functional point of view the above findings suggested the presence of an endocrine function for mammalian atrial cardiocytes although conceptually, this view had to be reconciled with the general consensus that polypeptide hormone-producing cells are of neural crest origin and cardiac muscle cells are of mesodermal origin. In addition, little if any morphological evidence existed regarding the release of the atrial granules content into the extracellular space. Interesting but somewhat confusing findings[4] were the description of specific granules in both the atria and ventricles of non-mammalian vertebrates (e.g. frog, fish, chicken) and the fact that the number of atrial granules is higher in small mammals (i.e. rats, mice) than it is in larger ones such as in dogs and cattle. To date there exist no satisfactory explanation for the latter two findings.

Through the 60's and 70's several efforts were made to define the chemical nature of the specific atrial granules. These investigations were carried out using mainly histochemical approaches at the electron microscopic level and confirmed the fact that the atrial granules did not share properties in common with organelles such as lysosomes and microbodies. Our histochemical work[5] was carried out at the light microscopic level because there exists a greater variety of techniques to study tissue components at this level than there is using electron microscopic techniques. However, unlike other secretory granules, the specific atrial granules are difficult to visualize at the light microscopic level because of their size and the high content of myofibrils and mitochondria which often interfere with their visualization. Nevertheless we were encouraged to try light microscopy histochemistry after it was found[6] that the stain lead-haematoxylin tartrazine was a highly specific and sensitive technique to demonstrate atrial granules by light microscopy. This technique was used to compare the tissue distribution of atrial granules with that of the histochemical reactions. Screenings were carried out for general tinctorial properties, carbohydrates, lipids and specific amino acids (Table I).

Of several stains tried one of interest was Gomori's aldehyde fuchsin because it clearly demonstrated the granules after oxidation in acidic permanganate or after thiosulfation by alkaline sulfite in the presence of cupric ions. The latter procedure suggested that sulfocysteine was being generated. If the sections were first reduced by thioglycolate then blocked with N-ethylmaleimide followed by thiosulfation, aldehyde fuchsin did not stain the granules.

Negative results were obtained also if thiosulfation was followed with cyanide treatment which blocks thiosulfate groups through the formation of thiocyanates. These procedures suggested the presence of sulfur-containing amino acids.

Techniques used for the histochemical detection of indole such as dimethylaminobenzaldehyde-nitrite and the formaldehyde-HCl gas phase reaction were positive for the atrial granules. In experiments in which the granules were stained with toluidine blue it was observed that the staining is quenched below pH 5.0. This staining characteristic is different from that observed for phospholipids, sialomucins and heparin which show staining at lower pH. In addition, it was observed that the granules stained orthochromatically suggesting the absence of significant amounts of innate sulfate. This was supported by experiments showing that the reactivity of the granules towards lead-haematoxylin and aldehyde fuchsin was restored by demethylation of methylated sections which suggests a desterification process as opposed to methanolic cleavage as it would occur for sulfate groups. Finally, tests for carbohydrates and for lipids proved negative in the granules.

Together, these findings suggested that the atrial granules store a basic polypeptide containing tryptophan and sulfur amino acids. The elucidation of the amino acid sequence of preproANF has corroborated these finding and neither past histochemical studies nor present biochemical investigations on ANF suggest that species other than ANF peptides are present in the atrial granules.

During the above described histological and histochemical investigations it was noticed that prolonged fixation of atrial tissues in fixatives containing acetic acid, the

Table I. Main Staining and Histochemical Properties of Rat Atrial Specific Granules

Procedure	Reacting species	Granule reactivity
Lead-haematoxylin	Side chain -COOH	+++
Aldehyde fuchsin (after thiosulfation)	-SH, -S-S-	+++
DMAB-nitrite	Indole	++
Formaldehyde-HCl in gas phase	Indole	++
Periodic acid-Schiff	Carbohydrate	-
Acid hematin-oil red	Phospholipids, neutral lipids	-

+++= intense reaction, ++= moderate reaction, -= nonreactive

reactivity of the atrial granules was diminished. This observation was used to develop the first isolation procedure for rat ANF[7] which relied on an acetic acid extraction. After fractionation of these extracts by gel chromatography most of the biological activity was recovered in fractions eluting at a fractionation range of 5,000 daltons or less. However, variations in the elution pattern between experiments suggested proteolytic breakdown of the peptides during isolation. For this reason we modified our extraction medium by incorporating 0.1 to 1.0 N HCl[7]. Using this approach it is possible to consistently isolate four main natriuretic peptides from rat atria which we have designated cardionatrins I - IV. A number of different peptides have been isolated by other laboratories[8-17]. All these peptides are derived from a precursor molecule, preproANF, which in the rat is 152 amino acids long. All biologically active peptides share an C-terminal region of the precursor containing a disulfide loop which is essential for biological activity. It is not clear at present which of the isolated peptides are naturally occurring forms. Under strongly acidic extraction conditions used in our laboratory the predominant immunoreactive form in tissue as well as in isolated granule extracts is cardionatrin IV. This peptide comprises residues 25 to 150 of preproANF. The predominant immunoreactive form in plasma elutes after RP-HPLC in the same position as cardionatrin I. This peptide comprises residues 123 to 150 of preproANF. The remainder of the plasma immunoreactivity appears to be cardionatrin IV[18].

The finding of natriuretic peptides in the mammalian atrial muscle confirms a view advanced many years ago regarding the possibility that the mammalian atrial cardiocyte is differentiated in part as a secretory cell. In a preliminary integration of the details of this new endocrine system it appears that its actions are both acute and long term regulation of extracellular fluid volume and blood pressure. This effect is accomplished by modulating the renin-angiotensin-aldosterone system, cardiac performance, vessel tonicity and renal handling of water and electrolytes.

References

1. de Bold, A.J., Borenstein, H.B., Veress, A.T., and Sonnenberg, H. (1981) Life Sci. **28**, 89-94.

2. Kisch, B. (1956) Exp. Med. Surg. **14**, 99-112.

3. Jamieson, J.D. and Palade, G.E. (1964) J. Cell Biol. **23**, 151-172.

4. Bencosme, S.A. and Berger, J.M. (1971) In: Methods and Achievements in Experimental Pathology, Bajusz E. and Jasmin, G. Eds., S. Karger, Basel, **5**, 173-213.

5. de Bold, A.J., Raymond, J., and Benscome, S.A. (1978) J.

Histochem. Cytochem. **26**, 1094-1102.

6. de Bold, A.J. and Bencosme, S.A. (1975) Stain Technol. **50**, 203-205.

7. Flynn, T.G., Davies, P.L., Kennedy, B.P., de Bold, M.L., and de Bold, A.J. (1985) Science **228**, 323-325.

8. Flynn, T.G., de bold, M.L., and de Bold, A.J. (1983) Biochem. Biophys. Res. Commun. **117**, 859-865.

9. Currie, M.G., Geller, D.M., Cole, B.R., Seigel, N.R., Fok, K.F., Adams, S.P., Eubanks, S.R., Galluppi, C.R., and Needleman, P. (1984) Science **223**, 67-69.

10. Geller, D.M., Currie, M.G., Siegel, N.R., Fok, K.F., Adams, S.P., Needleman, P. (1984) Biochem. Biophys. Res. Commun. **121**, 802-807.

11. Atlas, S.A., Kleinert, H.D., Camargo, M.J., Januszewicz, A., Sealey, J.E., Laragh, J.H., Schilling, J.W., Lewicki, J.A., Johnson, L.K., and Maack, T. (1984) Nature **309**, 717-719.

12. Kangawa, K., Tawaragi, Y., Oikawa, S., Mizuno, A., Sakuragawa, Y., Nakazato, H., Fukuda, A., Minamino, N., and Matsuo, H. (1984) Nature **312**, 152-155.

13. Kangawa, K., Fukuda, A., Minamino, N., and Matsuo, H. (1984) Biochem. Biophys. Res. Commun. **119**, 933-940.

14. Napier, M.A., Dewey, R.S., Albers-Schonberg, G., Bennett, C.E., Rodkey, J.A., Marsh, E.A., Whinnery, M., Seymour, A.A., and Blaine, E.H. (1984) Biochem. Biophys. Res. Commun. **120**, 981-988.

15. Seidah, N.G., Lazure, C., Chretien, M., Thibault, G., Carcia, R., Cantin, M., Genest, J., Nutt, R.F., Brady, S.F., Lyle, T.A., Paleveda, W.J., Colton, C.D., Ciccarone, T.M., and Veber, D.F. (1984) Proc. Nat. Acad. Sci. **81**, 2640-2644.

16. Misono, K.S., Fukumi, H., Grammer, R.T., and Inagami, T. (1984) Biochem Biophys. Res. Commun. **119**, 524-529.

17. Misono, K.S., Grammer, R.T., Fukumi, H., and Inagami, T. (1984) Biochem. Biophys. Res. Commun. **123**, 444-451.

18. Davies, P.L., de Bold, A.J., de Bold, M.L., Flynn, T.G., and Kennedy, B.P. (1985) Proc. Cdn. Fed. Biol. Soc. **28**, 164.

IDENTIFICATION OF RECEPTORS FOR ATRIAL NATRIURETIC FACTOR BY AFFINITY CROSSLINKING

M. A. Napier+*, K. E. Arcuri, M. E. Keegan, L. Hupe, and R. L. Vandlen+
Merck Institute for Therapeutic Research, P.O. Box 2000, Rahway, NJ 07065

Introduction

Specific, high affinity receptors for atrial natriuretic factor (ANF) have been identified in membranes prepared from vascular tissues[1], the presence of which is consistent with the observation that ANF is a potent vasorelaxant polypeptide. The synthetic peptide used for these studies corresponds to the C-terminal 26 amino acid sequence of rat ANF and exhibits the full biological activity of the natural ANF peptides[2]. The HPLC purified radioiodinated analog of ANF used to identify the receptors contains monoiodotyrosine with a specific activity of ca. 600 μCi/μg[1]. ANF binds to receptors in membranes prepared from rabbit aorta with a K_d of ca. 100 pM with 100 fmoles binding sites per mg protein. Similar high affinity receptors are present in porcine and bovine aorta. An excellent quantitative correlation was found between the affinity of ANF and analogs of intermediate activity to aorta membranes and the half-maximal concentration needed for relaxation of rabbit

---

+ Current Address:
Genentech, Inc., 460 Pt. San Bruno Blvd., So. San Francisco, CA 94080

* To whom correspondence should be addressed

aorta rings contracted by serotonin. The specificity for the ANF peptide was shown by the absence of interaction by other peptide hormones including angiotensin II and vasopressin. Similar binding has been found in other rabbit vascular tissues [3,4]. These data suggest that the receptors for ANF are responsible for mediating the physiological actions of ANF in vascular tissues.

In this communication we describe the identification of polypeptides which comprise the ANF receptor in rabbit aorta and other target tissues. The subunit structure of the ANF binding polypeptide was determined by crosslinking of ^{125}I-ANF with bifunctional crosslinking agents and by labeling with a radioiodinated photoaffinity analog of ANF. The apparent size of the solubilized receptor complex was determined by gel exclusion chromatography.

Results and Discussion

When membranes from rabbit aorta are labeled with ^{125}I-ANF and subsequently treated with bifunctional crosslinking agents, several polypeptide bands show specific incorporation of radioactivity (Figure 1). Nonspecific labeling is determined by binding and crosslinking in the presence of 100 nM unlabeled ANF. Autoradiography after SDS polyacrylamide gel electrophoresis (SDS-PAGE) reveals predominant labeling of a polypeptide of 60,000 daltons. Polypeptides of 70,000 and 120,000 daltons are also labeled to a lesser extent. Similar peptides are labeled by three different crosslinking agents. The most efficient labeling is with the rigid 1,5-difluoro-2,4-dinitrobenzene (DFDNB).

A radiolabeled photoaffinity analog of ANF (^{125}I-AzB-ANF) was prepared by coupling N-hydroxysuccinimidyl-4-azido-

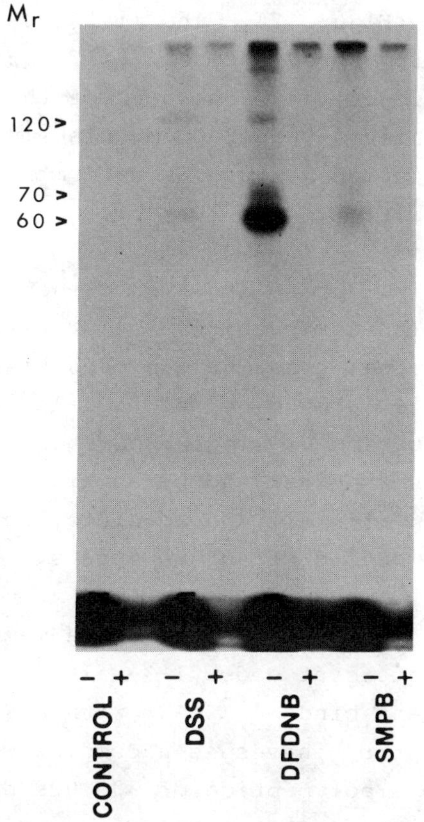

Fig. 1 Autoradiography of SDS-PAGE (11.5% acrylamide) separation of rabbit aorta membranes labeled with ^{125}I-ANF. Left to right, uncrosslinked control and samples crosslinked with DSS (disuccinimidyl suberate), DFDNB and SMPB (succinimidyl-4- (p-maleimido-phenyl) butyrate. Paired lanes are without (-) and with (+) 100 nM unlabeled ANF during the binding for total and nonspecific binding respectively. Molecular weight markers, M_r, are indicated.

benzoate (HSAB) to ANF followed by chloramine-T radioiodination of the tyrosine residue of ANF and purification by reversed phase HPLC. In rabbit aorta membranes, the same three polypeptides, 60,000, 70,000 and 120,000 daltons, are specifically labeled by the photoaffinity analog (Figure 2). The most intensely labeled band is a 120,000 dalton polypeptide. Half-maximal incorporation of radioactivity into each of the three peptide bands was achieved when 200 pM unlabeled ANF was included in the incubation mixture as determined by densitometric quantitation.

High affinity ANF receptors are also present in membranes prepared from several other rabbit vascular tissues and in rabbit kidney and adrenal tissues. In each tissue the same three polypeptides are labeled by the ^{125}I-AzB-ANF (manuscript in preparation). Similar polypeptides are also identified in tissues from rat and human kidney and from rat, bovine and human adrenal. No specific labeling is seen in tissues which are devoid of ANF receptors.

The ANF binding activity in rabbit aorta membranes is solubilized by the detergent dodecyl-β-maltoside. The 100,000 x g supernatant binds ^{125}I-ANF with a $K_d \sim 200$ pM. The solubilized receptor behaves as a complex with a Stokes radius equivalent to a polypeptide of 300,000 daltons on gel exclusion chromatography. ^{125}I-AzB-ANF specifically labels the same three polypeptides in the solubilized receptor preparation as in the membranes.

Several research groups have reported the stimulation of guanylate cyclase by ANF in target tissues including aorta[5] and kidney slices[6]. To identify cultured cells which might be useful for studies of the mechanism of action several lines were tested for their interaction with ANF. A10

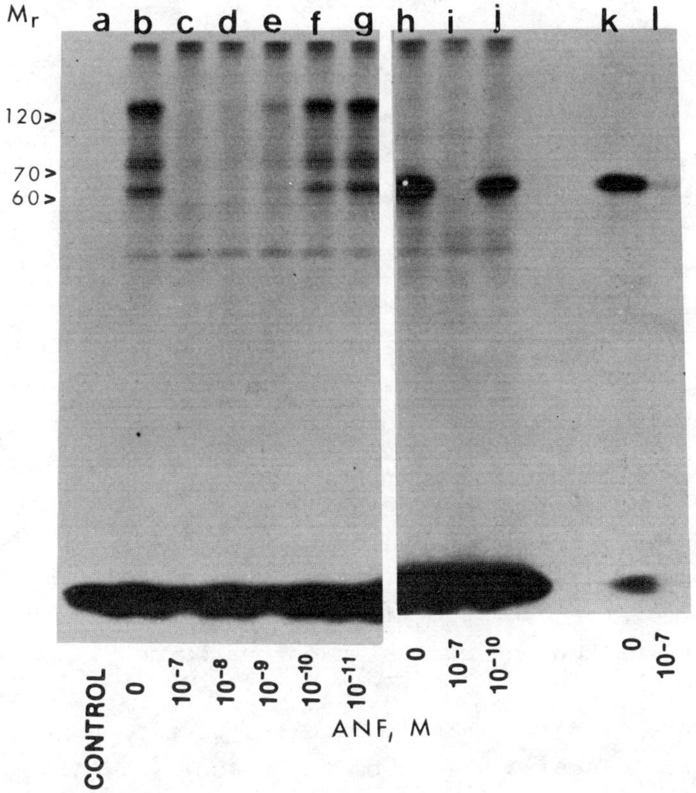

Fig. 2 Autoradiogram of SDS-PAGE separation of rabbit aorta membranes (lanes a-g), A10 cell membranes (h-j) and intact A10 cells (k,l) labeled with ^{125}I-AzB-ANF. Control (a) without photolysis. Unlabeled ANF (M) added during incubation as indicated below respective lane.

cells, derived from rat thoracic aorta and which possess properties characteristic of smooth muscle[7], respond to ANF with a dose dependent increase in cGMP. In the presence of the phosphodiesterase inhibitor isobutylmethylxanthine, incubation with ANF results in up to a 100-fold increase in cGMP as determined by radioimmunoassay.

In intact living A10 cells, specific high affinity (K_d 50 pM) binding is measured. Nonrelated peptide hormones, including angiotensin II, insulin, bradykinin and vasopressin, do not compete with the ^{125}I-ANF binding. In these cells only a 60,000 dalton peptide is labeled with either the photolabel analog of ANF (Fig. 2) or by the crosslinking agent DFDNB, with no evidence of the higher molecular weight peptides which are present in the rabbit aorta. Similar results are obtained for membranes prepared from these cells although the nonspecific background is somewhat higher. This is evidence that the receptor is localized in the plasma membrane of the A10 cells and that the 60,000 dalton peptide is probably not due to proteolytic degradation of higher molecular weight species. The solubilized receptor from the A10 cells behaves as a complex of ca. 250,000 daltons on exclusion chromatography.

The presence of only a single labeled subunit and the lower apparent molecular weight of the solubilized A10 receptor suggest that the ANF receptor in A10 cells is structurally different from that in the rabbit aorta membranes. The presence of different ANF receptors was first suggested by Currie *et al.*[8] based on the relative sensitivity of rabbit aorta and chick rectum smooth muscle to the ANF peptide atriopeptin-I (AP-I). AP-I has little or no activity in rabbit aorta but is active on the chick rectum smooth muscle. The low affinity of AP-I for rabbit aorta membranes (IC_{50} > 90 nM) is consistent with the low biological activity in this tissue. In A10 cells, however, AP-I is only slightly less active than ANF (IC_{50} for competition with ^{125}I-ANF binding is 330 pM and 31 pM, respectively). This observation further differentiates the ANF receptor in A10 cells and rabbit aorta smooth muscle.

In summary, we have shown the presence of specific high affinity receptors for ANF on membranes from vascular smooth muscle, kidney and adrenal tissues and on intact cultured cells. The affinity of ANF in rabbit aorta membranes correlates with the biological activity of ANF in this tissue. The cultured A10 smooth muscle cells possess high affinity receptors for ANF and respond to ANF with increased levels of cGMP as has been observed for other target tissues. By photoaffinity labeling techniques, a common 60,000 dalton polypeptide is labeled in both A10 and rabbit aorta membranes; other polypeptides are also specifically labeled in aorta membranes. Further studies including receptor purification are required to determine the exact stoichiometry and other molecular properties of the ANF receptor.

References

1. Napier, M.A., Vandlen, R.L., Albers-Schönberg, G., Nutt, R.F., Brady S., Lyle, T., Winquist, R., Faison, E.P., Heinel, L.A. and Blaine, E.H. (1984) Proc. Natl. Acad. Sci. USA 8:5946-5950.

2. Nutt, R.F., Brady, S.F., Lyle, T.A., Dylion-Colton, C., Paleveda, W.J., Ciccarone, T.M., Blaine, E.H., Winquist, R.J., Bennett, C.D., Hirschmann, R. and Veber, D.F. (1984) in "Peptides 1984. Proceedings of the 18th European Peptide Symposium"; Ragnarsson, U., Ed.; Almquist and Wiksell Intl., Sweden, 513-516.

3. Napier, M.A., Keegan M.E., Faison, E.P., Winquist, R., Arcuri, K.E. and Vandlen, R.L. (1985) Federation Proceedings 44:1237, Abst.

4. Winquist, R.J., Napier, M.A., Vandlen, R.L., Arcuri, K.E., Keegan, M.E., Faison, E.P. and Baskin, E.P. (1985), Clinical and Experimental Hypertension, in press.

5. Winquist, R.J., Faison, E.P., Waldman, S.A., Schwartz, K., Murad, F., Rapoport, R.M. (1984) Proc. Natl. Acad. Sci. USA 81:7661-7664.

6. Hamet, P., Tremblay, J., Thibault, G., Garcia, R., Cantin, M. and Genest, J. (1983) 65th Ann. Mtg. of the Endocrine Soc., San Antonio, TX, p. 289, Abst.

7. Kimes, B.W. and Brandt, B.L. (1976) Experimental Cell Research 98:349-366.

8. Currie, M.G., Geller, D.M., Cole, B.R., Siegel, N.R., Fok, K.F., Adams, S.P., Eubanks, S.R., Gallupie, G.R. and Needleman, P.N. (1984) Science 223:67-69.

ATRIAL NATRIURETIC POLYPEPTIDES (ANP): MOLECULAR FORMS AND DISTRIBUTION IN MAMMALIAN TISSUES AND PLASMA

Hisayuki Matsuo, Kenji Kangawa and Atsuro Miyata
Department of Biochemistry, Miyazaki Medical College,
Kiyotake, Miyazaki 889-16, Japan

Atrial natriuretic polypeptides (ANP), eliciting potent natriuretic-diuretic activity, have been isolated from human[1-3] and rat[4-11] atrial tissues and implicated in the control of fluid volume and vascular function. From human atria, three distinct forms(α;28, β;56, γ;126 amino acid residues) have been isolated[1-3]. α-hANP has been first identified as a 28-residue peptide containing a disulfide linkage[1] and β-hANP is known to be an antiparallel dimeric form of α-hANP[2]. Furthermore, γ-hANP of 126 residues has been revealed to be an N-terminally extended peptide of α-hANP, which is the largest form processed out of a 151-residue precursor[2,3]. In rat atria, a 28-residue peptide, designated α-rANP[4] or cardionatrin[5] has been identified as the peptide identical with α-hANP having a single amino acid replacement at position 12 of Ile for Met. In addition, γ-rANP of 126 residues, which is highly homologous to γ-hANP, has also been identified as the largest form of rANPs[11]. However, a number of N-terminally extended or deleted forms of α-rANP, varying in the chain length, have been isolated by various research groups, though in very low yield and named variously[6-10]. All of them have been found to be truncated peptides of γ-rANP. Identification of cDNAs encoding the human[3] and rat[11-14] ANP precursors revealed that sequences of all the atrial peptides isolated to date are contained within the precursor molecules, though their processing features remain uncertain. It is essential to solve the question as to which molecular forms of ANPs are present in and secreted from the cardiocytes and circulating in the blood. In this context, we have established a highly sensitive radioimmunoassay(RIA) for ANPs to directly measure the ANP content, as described in the present paper. By using the RIA, combined with chromatographic separation, we identified molecular forms of ANP present in heart and in blood of mammals.

MATERIALS AND METHODS

Radioimmunoassay for ANP: Performed by double-antibody method. The anti-α-hANP antibody was raised in rabbit by immunizing with α-hANP-bovine thyroglobulin conjugate emulsified with Freund's complete adjuvant. The incubation buffer for RIA was 0.05M sodium phosphate buffer (pH 7.4), containing 1% BSA, 0.1% Triton X-100, 0.08M NaCl, 0.025M EDTA 2Na, 0.05% NaN_3, and Trasylol 500 KIU/ml. The sample (100 μl) was incubated with anti-α-hANP antiserum diluent (500 μl) for 12 hrs. Then the tracer solution (purified monoiodinated α-hANP: 18,000-20,000 cpm in 200 μl) was added. All assay procedures were performed at 4°C. After the incubation for 36 hrs, anti-rabbit IgG goat serum diluent (200 μl) was added. After kept standing for 40 hrs, the tubes were centrifuged at 2,000g x 30 min at 4°C and radioactivity of the precipitate was measured.

Preparation of tissue and plasma samples: i) Tissue sample: The tissues were homogenized with a polytron mixer for 60 sec in 20 volume (v/w) of 0.1M AcOH containing 1% Triton X-100 and immediately boiled for 5 min to inactivate proteases. After chilling, the homogenate was centrifuged at 25,000g x 20 min. The supernatant was pooled and stored at -20°C before use. ii) Plasma sample: Plasma samples were taken with EDTA-2Na (1 mg/ml) and Trasylol (500 KIU/ml), and applied on a Sep-Pak C-18 cartridge(Waters) in the presence of saline. After washing the column with saline, the adsorbed material was eluted with 60% acetonitrile in 0.1% TFA. The eluate was evaporated and subjected to RP-HPLC.

RESULTS AND DISCUSSION

Characterization of the antiserum: The antiserum was usable at a final dilution of 1:400,000, utilizing ^{125}I-labeled ligand. Half maximum inhibition by α-hANP was observed at 60 pg/tube and α-hANP was detectable as low as 10 pg/tube. The antiserum mainly directs to the subsequence flanked by two cysteine residues(positions 7 and 23) in α-hANP molecule. Accordingly, the antiserum recognizes equally α-, β- and γ-hANPs. The antiserum shows 40% crossreactivity with α-rANP, which is a 28-amino acid peptide identical to α-hANP, having an only single residue replacement (Met^{12}- Ile^{12}). In addition, γ-rANP(126 residues), which is an N-terminally extended form of α-rANP, and all other rANPs so far isolated also crossreact appreciably with the antiserum (40%). Thus, hANPs and rANPs were found to be precisely measured by this RIA system. However, special care must be taken in the assay: i) ^{125}I-hANP must be purified immediately after iodination by RP-HPLC and stored -80°C before use. ii) BSA used for RIA buffer must be treated with N-ethylmaleimide before use. Otherwise, free thiol group in BSA may participate in opening and rejoining of disulfide bond of ANPs to form BSA-ANP conjugates, which interfere the assay. iii) The C-terminal Tyr carrying ^{125}I is so susceptible to proteolytic cleavage, that addition of Trasylol in the assay buffer is essential for obtaining reproducible results.

ANP in atrial tissue: The present procedure for tissue extraction is quite efficient to prevent nonspecific degradation of ANP by intrinsic proteases. Dilutions of the acid extracts from human, rat, bovine and porcine atrial tissues yielded competition curves which were parallel to the standard curve of α-hANP. Thus, the present RIA can precisely detect all the known forms of both human and rat ANPs so far identified. Furthermore, bovine and porcine[15] ANPs, which remain to be identified, can also be detected by this RIA. Fig.1a shows a typical gel-filtration pattern on Sephadex G-75 of rat atrial extract, monitored by the RIA, where only one major immunoreactive peak, with more than 90% of the total immunoreactivity, was observed at the position corresponding to γ-rANP. For unambiguous identification, we have performed further chromatographic comparison with authentic γ-rANP on RP-HPLC. As shown in Fig.2d, nearly all of the ANP-immunoreactivity of rat atria comigrated exactly with γ-rANP. Thus, it is concluded that γ-rANP was a predominant form of rANP in rat atrial tissue. Although a variety of truncated peptides of γ-rANP with low or intermediate molecular size (Mr 3,000-5,000) has been isolated from rat atrial extracts[4-9], no significant quantities of immunoreactive species corresponding to these peptides were observed in the present experiments, despite the fact that the antiserum employed recognizes all of these peptides. Accordingly, it is concluded that γ-rANP is the only major component present in rat atrial tissues. Considering past observations that isolation yield of each smaller peptide was very low, representing not more than 1% of that of γ-rANP, the present data indicate that the low- and intermediate-sized peptides may be not endogenous ANPs, but secondary products derived from γ-rANP by non-specific proteolysis during isolation. As shown in Fig.1b,c,e,f, essentially similar molecular distributions of ANP-immunoreactivity were also observed in atrial tissues of porcine and bovine species.

However, the molecular patterns in human atria showed some complexity, representing three distinct types(I,II,III), with 12 autopsied human hearts thus far tested. Figure 2a shows a gel-filtration chromatography on Sephadex G-75 of the extract of human autopsied atrial tissue(type I). There were observed two major immunoreactive peaks of hANP, of which the first smaller peak eluted at a molecular weight of γ-hANP and the larger peak at that of β-hANP. In Fig.2b, reverse phase HPLC of the same extract clearly indicates that a major immunoreactivity peak is identical with the authentic β-hANP, while a minor peak identical with the authentic γ-hANP. Thus, the human atrium of type I contains two components comprising β- and γ-hANP. However, another type of autopsied heart (Type II) shows a different pattern(Fig. 2c), indicating the presence of three components; α-, β- and γ-hANP, of which γ-hANP was found to be the major part. Type III, as shown in Fig.3d, exhibits a pattern very similar to those observed in rat, porcine and bovine tissues, where γ-ANP is present as a single major component. It should be noted that β-form of ANP has been found only in human, but never in other mammals. Incidentally, the content of ir-hANP widely varied from 6.11 to 47.8 ng/mg wet tissue. It should be also noted that type I, where β-hANP is major component, showed a relatively higher level of ANP content in the tissue, compared with other types in which γ-hANP was predominant. In addition, among the hearts of type I, remarkable differences in molar ratio of β- vs γ-hANPs were observed ranging from 1.2/1 to 7.1/1. These data may imply that formation of β-form could be implicated with overproduction or accumulation of hANP in the cell through so far unknown mechanisms.

ANP in plasma: In gel-filtration of rat plasma, we observed one major immunoreactive peak appeared in the range of lower molecular weight (Mr 3,000), along with minor peaks in the higher molecular weight region. Fig.3a shows HPLC pattern of rat plasma, where a predominant peak possessing 90% of the total immunoreactivity emerged at the position exactly identical with the authentic rat α-ANP(28 residues), positively indicating that a major component of the ANP in the circulation is α-form.Different from rat plasma, human plasma showed some complex feature, as observed in the atrial tissue. As shown in

Fig.3b, the HPLC of human plasma indicates that the ANP immunoreactivity comprises α- and β-hANP, along with other minor components unidentified. Incidentally, the content in rat plasma (Wistar rats, n=5) was found to be 389.3 ± 53.0 pg/ml for α-rANP eq. The content of ir-α-hANP in healthy humans was determined to be 231 ± 37.1 pg/ml, by RIA using 100 μl of plasma specimens (n=6, average age 24.8y).

Regional Distribution in Rat: As summarized in Table 1, the highest concentration of ANP was observed in the atrium, especially much in the right atrium. The small but appreciable ANP-immunoreactivity was found in brain, thyroid, submaxillary gland and adrenal gland, suggesting that ANP might have a specialized function in these regions.

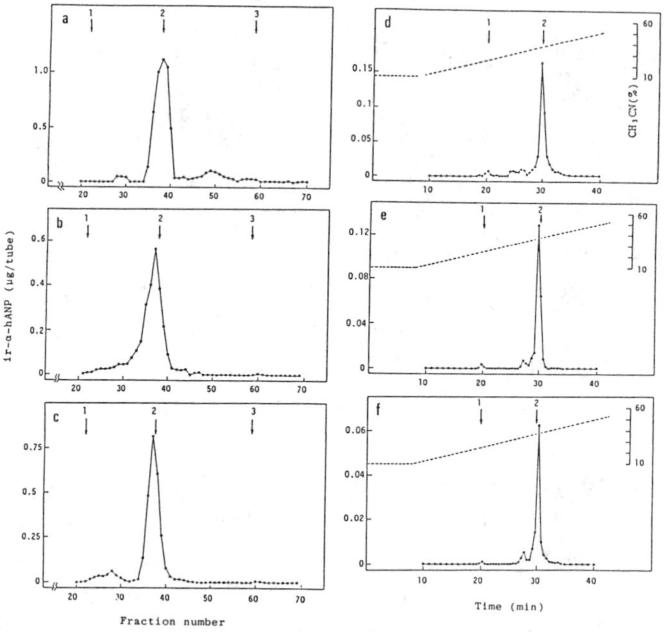

Fig.1 (a)(b)(c): Gel-filtration of atrial extract : (a) rat, (b) porcine (c) bovine. (d)(e)(f): Reverse phase HPLC of atrial extract: (d) rat, (e) porcine, (f) bovine. (1) α-hANP; (2) γ-hANP.

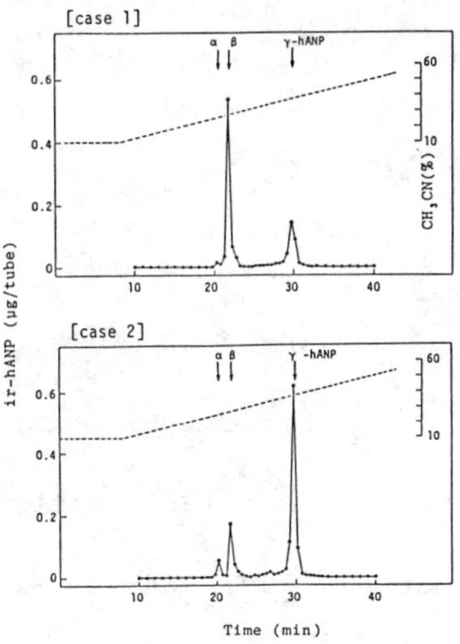

Fig.2 (a): Gel-filtration of human atrial extract. Type I (from a 78y male). (1) void volume, (2) γ-hANP, (3) β-hANP, (4) α-hANP.
(b)(c)(d): Reverse phase HPLC of human atrial extract (b) Type I, (c) Type II (from a 63y male), (d) Type III (from a 35y male). (1) α-hANP, (2) β-hANP, (3) γ-hANP.

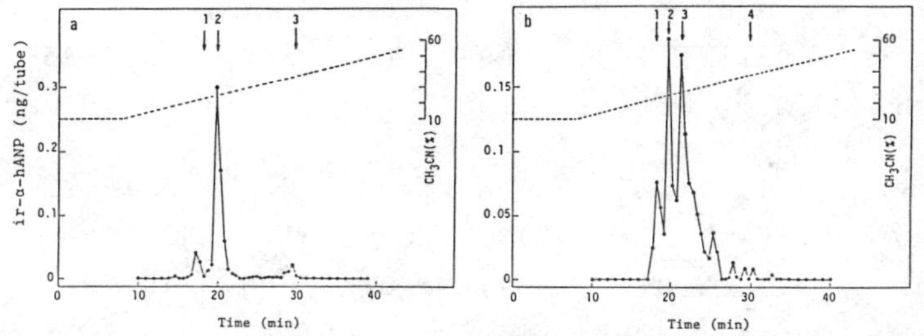

Fig.3 Reverse phase HPLC of rat (a) and human (b) plasma. Arrows: in (a): 1, α-hANP[1-26]; 2, α-rANP; 3, γ-rANP.
in (b): 1, α-hANP[1-26]; 2, α-hANP; 3, β-hANP; 4, γ-hANP.

Table 1

Regional Distribution of ANP in Rat
(pg/mg wet tissue)
(*pg/ml plasma)

Whole brain	6.03 ± 0.45	Lung	< 5
Pituitary gl.	< 5	Kidney	< 5
Thyroid gl.	94.9 ± 8.57	Pancreas	< 5
Submaxillary gl.	241 ± 38.8	Liver	< 5
Sublingual gl.	< 5	Spleen	< 5
Parotid gl.	< 5	Adrenal gl.	5.56 ± 0.58
Atrium right	99720 ± 14270	Stomach	< 5
left	56140 ± 4240	Small intestine	< 5
Ventricle	313 ± 110	Colon	< 5
Plasma	389 ± 53*		

REFERENCES

1. Kangawa, K. and Matsuo, H. (1984) Biochem. Biophys. Res. Commun., 118, 131-139.
2. Kangawa, K., Fukuda, A. and Matsuo, H. (1985) Nature, 313, 397-400.
3. Oikawa, S., et al. (1984) Nature, 309, 724-726.
4. Kangawa, K., et al. (1984) Biochem. Biophys. Res. Commun., 121, 585-592.
5. Flynn, T.G., Debold, M.L. and Debold, A.J. (1983) Biochem. Biophys. Res. Commun., 118, 131-139.
6. Kangawa, K., Fukuda, A., Minamino, N. and Matsuo, H. (1984) Biochem. Biophys. Res. Commun., 119, 933-940.
7. Currie, M.G., et al. (1984) Science, 223, 67-69.
8. Thibault, et al. (1984) FEBS Lett., 167, 352-356.
9. Seidah, N.G., et al. (1984) Proc. Natl. Acad. Sci. USA, 81, 2640-2644.
10. Atlas, S.A, et al. (1984) Nature, 309, 717-719.
11. Kangawa, K., et al. (1984) Nature, 312, 152-155.
12. Yamanaka, M., et al. (1984) Nature, 309, 719-722.
13. Maki, M., et al. (1984) Nature, 309, 722-724.
14. Seidman, C.E., et al. (1984) Science, 225, 324-326.
15. Forssmann, W.G., et al. (1983) Anat. Embryol., 168, 307-313.
16. Miyata, A., Mizuno, K., Minamino, N., and Matsuo, H. (1984) Biochem. Biophys. Res. Commun., 120, 1030-1036.
17. Nakao, K., et al. (1984) Biochem. Biophys. Res. Commun., 124, 815-821.
18. Tanaka, I., et al. (1984) Biochem. Biophys. Res. Commun. 124, 663-668.
19. Miyata, A., et al. Biochem. Biophys. Res. Commun. in press.

ATRIOPEPTIN RELEASE

David Schwartz*, Pamela T. Manning *† and Philip Needleman*
Department of Pharmacology, Washington University* and Monsanto Co†, St. Louis, MO 63110

deBold et al.[1] made the pioneering observation that extracts of rat atrial tissue produce a marked natriuretic and diuretic response when injected into anesthetized rats. Shortly thereafter, Currie et al.[2] demonstrated that the extracts also contain potent smooth muscle relaxant activity. Several laboratories subsequently isolated and characterized the biologically active species which proved to be a family of low molecular weight peptides all derived from a common 128 amino acid precursor molecule (see reference 3 for review). Because the atria were known to be a monitoring site for intravascular volume[4] and because of the potent biological effects of the atrial extracts, it was proposed from the onset that the atrial peptides may represent a novel endocrine system regulating fluid and electrolyte homeostasis.[1,2] However, purification of the atrial peptides or atriopeptins (AP) did not provide sufficient evidence to demonstrate the existence of a new hormonal system. Critical support for this claim requires the demonstration and validation of AP release from the heart into the circulation in response to physiological or pharmacological manipulations.

Initial efforts used biological assays to demonstrate atriopeptin release. Isolated hearts were perfused with either buffer or a fluorocarbon emulsion and isolated atria were incubated in vitro in standard tissue culture preparations. The perfusate and culture medium were assayed for biological activity, measuring either their natriuretic/diuretic potency in anesthetized rats or spasmolytic activity on isolated smooth muscle strips. Currie et al.[5] demonstrated that the isolated perfused rabbit heart releases a potent trypsin-sensitive vasorelaxant substance. Renal venous effluent in an analogous preparation had no apparent relaxant activity. When the cardiac effluent was purified over a Sephadex G-75 column and subsequently by HPLC only a single peak of natriuretic/spasmolytic activity was detected. This activity migrated in the vicinity of, but did not exactly comigrate with the rat atriopeptins (I, II, or III.) These experiments suggested that the high molecular weight precursor atriopeptigen is the storage form of the released active low molecular weight atriopeptin. Prohormone processing enzymes associated with the membranes of secretory granules have been identified for a number of peptides.[6,7] The existence of such a specific proteolytic enzyme acting on atriopeptigen is suggested by these data but remains to be directly demonstrated.

Evidence of stimulated atriopeptin release was provided by Dietz[8] using an isolated rat heart-lung preparation. Working with the hypothesis that an increase in atrial volume or stretch should release atriopeptin, he compared the natriuretic activity present in the perfusate of hearts subjected to low and high central venous pressure (CVP). One sample of perfusate was collected with the reservoir set at a height of 1-2 cm (low CVP) above the right atrium; a second sample was obtained with the reservoir 7-12 cm (high CVP) above the heart. When comparable aliquots of the samples were tested for natriuretic activity in anesthetized assay rats, the high CVP perfusate contained more natriuretic substance than did the low CVP perfusate. These results are consistent with the hypothesis that atrial stretch causes release of atriopeptin into the perfusate.

Studies with in vitro atrial cultures by Sonnenberg and Veress[9,10] also suggest stimulated atriopeptin release.

Several agonists, including acetylcholine, epinephrine, and arginine vasopressin (AVP) appear to induce the release of atriopeptin into the culture medium. Media from incubations of atria with these agonists caused greater natriuresis and diuresis when injected into assay rats that did media with agonists incubated in the absence of atria. However, the important control of media from an incubation of tissue without agonists was not presented. This experiment would have assessed the non-specific release (or leakage) of atriopeptin into the medium. Regardless, release of the natriuretic substance by acetylcholine and epinephrine was blocked by an appropriate competitive antagonist (atropine and phentolamine, respectively.) The effect of AVP could not be accounted for by activation of adenlyate cyclase since deamino-D-arg^{8}-vasopressin (which also activates this enzyme) did not stimulate release.

With the development of specific radioimmunoassays came more direct evidence of atriopeptin release. Atriopeptin immunoreactivity (APir) has been detected in rat plasma by several groups.[11-13] Tang et al.[14] demonstrated that ^{125}I-APIII has a circulating half-life of 2.5 min. in the rat. If the naturally occurring form(s) of circulating atriopeptin has a half-life comparable to that of labelled APIII, then there must be appreciable release to maintain the 0.5-1.0 ng/ml plasma concentrations detected. Since the heart contains the vast majority of AP present in the rat, it is likely that the atria are the source of the AP found in the circulation. Tanaka et al.[11] placed rats on a high salt diet and found the plasma APir to be twice that of control plasma, consistent with the hypothesis that increased vascular volume (resulting from the high NaCl treatment,) would stimulate greater AP release from the heart needed to meet the greater fluid challenge.

Lang et al.[15] used the immunoassay to confirm and advance the findings of Currie et al.[5] and Dietz.[8] The APir released by an isolated perfused rat heart preparation comigrated on HPLC with the APir present in normal rat plasma. In both cases, the immunoreactivity migrated in the vicinity of, but not coincident with standards of APII and APIII. Interestingly though, when blood was collected in the absence of protease inhibitors, the APir comigrated with APIII. These results are consistent with those obtained by biological assay of the perfusate from an isolated rabbit heart. In addition, in vivo changes in right atrial pressure (produced by volume expansion,) caused increases in APir in the plasma. Release was transient (peak at 1-2 min. and over by 10 min.,) and proportional to the magnitude of change in right atrial pressure. These data concur with those from the isolated heart-lung preparation mentioned above.

Manning et al.[13] demonstrated that vasopressin can produce a profound and sustained release of APir into the circulation. Arginine vasopressin (AVP) and its 1-deamino analog (dAVP) produced dose-dependent increases in plasma APir. dAVP at doses of 0.1 to 10 ug/rat raised the plasma APir 2.5 to 8-fold. In contrast to the time course of release in the volume expansion experiments, dAVP-stimulated release peaked at 5-15 min and, with the higher doses, was still evident at 60 min In these experiments, release was also shown to be pressure related. 1-deamino-D-arg^8-vasopressin, an analog devoid of the pressor activity (but retaining antidiuretic activity,) did not induce APir release. dAVP in the presence of a specific pressor antagonist likewise did not cause release. A number of vasopressin-related compounds and pressor agents were also tested. Oxytocin stimulated release as did angiotensin II (AII) and phenylephrine. Indeed, where it was measured, the time course of APir release closely paralleled changes in arterial pressure. These data, though, also questioned the complete dependence of APir release on changes in pressure suggested by the earlier work. When AVP, dAVP, AII, and phenlyephrine were administered at doses producing matched changes in mean arterial pressure, there were marked differences in the magnitude of APir release (dAVP = AVP >> AII = phenylephrine.)

Whether or not release proves to be solely dependent on atrial pressure, the mechanisms of signal-transduction responsible for release remain to be determined. Regardless of the mechanisms, these data represent the first in vivo demonstration of hormone-induced AP release. A recent study reported that infusion of APIII lowers elevated AVP levels in dehydrated or hemorrhaged animals.[16] The possibility of a negative feedback loop exists whereby AVP can stimulate AP release which in turn supresses AVP release. The works suggest an interaction between these two opposing hormones in the regulation of fluid and electrolyte homeostasis.

A consideration which must be kept in mind when interpreting all of the release data presented above is the use of biological or radioimmunological assay to demonstrate release. None of the experiments directly demonstrated the presence or release of AP into the circulation. Natriuretic/diuretic activity or immunoreactivity of the samples was assessed, which, although strongly suggestive, is still indirect evidence. The first direct demonstration of atriopeptin in the circulation was provided by Schwartz et al.[17] Using dAVP to greatly elevate circulating APir levels, the immunoreactive material was purified for characterization and sequence identification. When normal rat plasma was extracted on octadecylsilane columns, and the eluate applied to reversed phase HPLC only a single 2.5 min. (1 ml/min) fraction containing APir was detected. This fraction migrated in the vicinity of the low molecular weight (LMW) AP standards. When dAVP-stimulated plasma was purified by the same methods, the predominate (>90%) peak of immunoreactivity from the stimulated plasma co-migrated with the peak obtained from normal plasma. The single large peak was resolved into two distinct peaks (I and II) upon HPLC separation into 0.5 min. fractions. Peaks I and II contained

APir in a ratio of 10:1 and migrated with standards of ser-leu-arg-arg-APIII (SLRR-APIII) and APIII, respectively. These peaks were tested for biological activity and both showed potent dose-dependent vasorelaxant activity on precontracted rabbit aorta strips. Absolute identification was obtained by gas phase sequencing. As predicted by the HPLC migration patterns, peak I was identified as the 28 amino acid peptide, SLRR-APIII and peak II as APIII. The minor peak of APIII probably represents an artifact of proteolytic cleavage of SLRR-APIII. Lang et al.[15] showed that the migration pattern of the APir shifted from a position between standards of APII and APIII to a position co-migrating with APIII if blood was collected in the absence of protease inhibitors.

Although SLRR-APIII was the major circulating atrial peptide following dAVP stimulation, the question of which peptide is released remains unanswered. Prohormone processing enzymes for a number of peptides are associated with the membranes of secretory granules. If this is the case for the atrial peptide, the SLRR-APIII would represent the released form. However, since the blood used to purify the circulating peptide was collected 5 minutes following dAVP stimulation, the data are also consistent with release of the high molecular weight peptide which is then subjected, in the plasma, to specific cleavage yielding SLRR-APIII. Further experiments are required to resolve this question.

Regardless of the form of atrial peptide released, it is clear now that release does occur and biologically active atriopeptin is present in the circulation. The in vivo experiments discussed above suggest that volume loading and hormonal stimulation may be important factors regulating atriopeptin release.

The recent demonstration of atriopeptin release may help to resolve some observations in the literature that, to date, have lacked a satisfactory explanation. Left atrial distension and intravascular volume expansion in conscious dogs produce natriuresis and diuresis.[18] In addition, a paradoxical natriuresis has been shown to occur following administration of vasopressin.[19] Evidence of stimulated release of atriopeptin presented in this review now provides a viable explanation for these observations. The demonstration of atriopeptin release and the direct demonstration of the peptide in the circulation lend strong support to the claim that atriopeptin is a novel hormone involved in the regulation of fluid and electrolyte homeostasis.

REFERENCES

1. deBold, A.J., Borenstein, H.B., Veress, A.T., Sonnenberg, H.. A Rapid Potent Natriuretic Response to Intravenous Injection of Atrial Myocardial Extract in Rats. Life Sci. (1981) 28:89-94.

2. Currie, M.G., Geller, D.M., Cole, B.R., Boylan, J.G., YuSheng, W., Holmberg, S.W., Needleman, P. Bioactive Cardiac Substances: Potent Vasorelaxant Activity in Mammalian Atria. Science (1983) 221:71-73.

3. Needleman, P., Adams, S.P., Cole, B.R., Currie, M.G., Geller, D.M., Michener, M.L., Saper, C.B., Schwartz, D., Standaert, D.G.. Atriopeptins: Potential Cardiac Hormones. Hypertension. In Press.

4. Gauer, O.H., Henry, J.P., Sieker. Cardiac Receptors and Fluid Volume Control. Prog. Cardiovasc. Dis. (1961) 4:1-26.

5. Currie, M.G., Sukin, D., Geller, D.M., Cole, B.R., Needleman, P. Atriopeptin Release from the Isolated Perfused Rabbit Heart. Biochem. Biophys. Res. Comm. (1984) 124:711-717.

6. Austen, B.M., Smyth, D.G. in Molecular Endocrinology, Elsevier North-Holland Biomedical Press (1979) 99-110.

7. Zingg, H.H., Patel, Y.C. Processing of Somatostatin-28 to Somatostatin-14 by Rat Hypothalamic Synaptosomal Membranes. Life Sci. (1983) 33:1241-1247.

8. Dietz, J.R. Release of natriuretic factor from rat

heart-lung preparation by atrial distension. Am. J. Physiol. (1984) 247:R1093-R1096.

9. Sonnenberg, H., Krebs, R.F., Veress, A.T. Release of atrial natriuretic factor from incubated rat heart atria. IRCS Med. Sci. (1984) 12:783-784.

10. Sonnenberg, H., Veress, A.T. Cellular Mechanism of Release of Atrial Natriuretic Factor. Biochem. Biophys. Res. Comm. (1984) 124:443-449.

11. Tanaka, I, Misono, K.S., Inagami, T. Atrial Natriuretic Factor in Rat Hypothalamus, Atria, and Plasma: Determination by Specific Radioimmunoassay. Biochem. Biophys. Res. Comm. (1984) 124:663-668.

12. Gutkowska, J., Horky, K., Thibault, G., Januszewicz, P., Cantin, M., Genest, J. Atrial Natriuretic Factor is a Circulating Hormone. Biochem. Biophys Res. Comm. (1984) 125:315-323.

13. Manning, P.T., Schwartz, D., Katsube, N.C., Holmberg, S.W., Needleman, P. Vasopressin Stimulated Atriopeptin Release: Endocrine Antagonists in Fluid Homeostasis. Science. In Press.

14. Tang, J., Webber, R.J., Chang, D., Chang, J.K., Kiang, J., Wei, E.T. Depressor and natriuretic activities of several atrial peptides. Regulatory Peptides (1984) 9:53-59.

15. Lang, R.E., Tholken, H., Ganten, D., Luft, F.C., Rushpaho, Unger, Th. Atrial natriuretic factor - a circulating hormone stimulated by volume loading. Nature (1985) 314:264-266.

16. Samson, W.K. Atrial Natriuretic Factor Inhibits Dehydration and Hemorrhage-Induced Vasopressin Release. Neuroendocrinology (1985) 40:277-279.

17. Schwartz, D., Geller, D.M., Manning, P.T., Siegel, N.R., Fok, K.F., Smith, C.E., Needleman, P. Ser-Leu-Arg-Arg-APIII is the Major Circulating Form of Atrial Peptide. Science. In Press.

18. Fater, D.C., Schultz, W.D., Sundet, J.S., Mapes, K.L., Goetz, K.L. Effects of left atrial stretch in cardiac-denervated and intact conscious dogs. Am. J. Physiol. (1982) 242:H1056-H1064.

19. Smith, Jr., M.J., Cowley, Jr., A.W., Guyton, A.C., Manning, Jr., R.D. Acute and chronic effects of vasopressin on blood pressure, electrolytes, and fluid volumes. Am. J. Physiol. (1979) 237:F232-F240.

ISOLATION OF CARDIODILATIN (CDD 1-126), THE COMPLETE GENE PRODUCT OF AN ATRIAL PEPTIDE.

Wolf-Georg Forssmann, Dieter Hock, Erich Fey, Christa Kuhn, Manfred Reinecke, Christian Birr, Mats Carlquist, and Viktor Mutt

Department of Anatomy III, University of Heidelberg (FRG),
ORGANOGEN GmbH, Heidelberg (FRG),
Department of Biochemistry II, Karolinska University, Stockholm/Sweden

Introduction

Heart atria contain three main cell types which exhibit specific functions: (1) initiation and conduction of the electrical impulse for heart action, (2) the production of a pumping force through contractility of the myofibrils, and (3) the hormonal regulation of blood volume and blood pressure. The specialized cell for hormonal function, however, is the "myoendocrine cell". It contains, in addition to the organelles of the normal working myocardial cells, a well-developed rough endoplasmic reticulum, an extensive Golgi apparatus, and specific secretory granules which are located mainly within the perinuclear pole[1-3]. The endocrine apparatus is greatly influenced by changes in water and electrolyte homeostasis[4]. Corresponding extracts of those regions of the atria containing the myoendocrine cells, that is the atrial appendages, induce a strong diuresis and natriuresis[5] and relaxation of vascular smooth muscle[6-8].

Results and Discussion

From atrial appendages, we have purified and chemically characterized a peptide hormone containing 126 amino acids which is called cardiodilatin (CDD)[8]. Cardiodilatin induces both vascular smooth muscle relaxation and diuresis[8-10]. Within its primary structure this peptide contains all smaller molecules of other cardiac hormones previously described. Remarkable sequence homologies exist between rat and human cardiac peptides[11,12]. Thus, the variety of peptides published are all fragments of this larger peptide and cDNA and gene studies[13,14] revealed that the peptide, expressed as the complete hormone, is the 126-amino-acid-containing form. A large precursor of 151 amino acids, including the signal sequence, is also postulated from these gene studies. Our present investigation is directed towards identification of the various CDD-molecules functionally expressed in myoendocrine cells and later released into the blood stream.

The analysis of the naturally occurring CDD-molecules was carried out by several methods: (1) isolation and sequencing of the biologically active CDD using three batches of approximately 100 kg porcine atria. The initial steps of isolation were identical to the procedures described in our earlier publications[8-10]. However, the material obtained by the aqueous-ethanol precipitation was submitted to different chromatographic steps. We developed a two-step-purification procedure from the ethanol-water cold precipitate with a high yield of cardiodilatin. This variation of the method includes an ion-exchange chromatography with an ammonium-bicarbonate buffer and one step of reverse-phase HPLC. Thus, a highly purified CDD containing less than 10 % contamination was obtained. (2) Antibodies were raised against small synthetic segments of the CDD-molecule and in combination with SDS-gel electrophoresis and immunoblotting, the immunochemical properties and molecular size of the CDD-molecules were determined. (3) Immunohistochemistry was used to show that the correlated molecules are stored in the cytoplasmic structures of myoendocrine cells.

The results of our study indicate the existence of three main CDD-molecules: the data shows that the largest molecule occurring in myoendocrine cells is the extended CDD-126 which was substantiated as a 151-amino-acid-containing precursor by use of genome studies. This molecule has been identified by immunoblotting and was partially sequenced. Furthermore, two molecules, CDD-126 and CDD-88, seem to naturally occur in porcine and bovine atrial myoendocrine cells. This has been repeatedly confirmed by sequence studies. The amount of CDD-126 and CDD-88 is slightly variable in different batches of extracted atria. The occurrence of smaller CDD-molecules in well-preserved atrial extracts is negligible. However, from the blood serum, no such large molecules can be extracted. The characteristics of the CDD processed as a circulating hormone indicate a biologically active molecule in the range of 3,500 Daltons which is released from the atrial tissue into the blood stream. We are continuing our present work to obtain the complete structure of the circulating cardiodilatin molecule. In conclusion, the present study demonstrates the molecular pathway of cardiodilatin synthesized in the myoendocrine cells as a hormone precursor of 151 amino acids. After intracellular cleavage, CDD-126 and CDD-88 are stored as a main bulk of prohormones. From these molecules, a smaller C-terminal fragment is released into the blood stream which is considered the proper CDD-hormone. However, functional studies indicate that all large molecules which contain the complete C-terminal structure exert identical biological activities. The many smaller molecular forms of cardiac hormones described by other working groups must represent species variations or breakdown products during the isolation procedure.

References

1. Kisch, B. (1956) Exp. Med. Surg. 14, 99-112.
2. Bompiani, G.D., Rouiller, Ch., Hatt, P.Y. (1959) Arch. Maladies Coeur et Vaisseaux 52, 1257-1274.
3. Jamieson, J.D., Palade, G.E. (1964) J. Cell Biol. 23, 151-172.
4. Marie, J.P., Guillemot, H., Hatt, P.Y. (1976) Path. Biol. 24, 549-554.
5. De Bold, A.J., Borenstein, H.B., Veress, A.T., Sonnenberg, H. (1981) Life Sciences 28, 89-94.
6. Deth, R.C., Wong, K., Fukozawa, S., Rocco, R., Smart, J.L., Lynch, C.J., Awad, R. (1982) Federation Proc. 41, 983.
7. Currie, M.G., Geller, D.M., Cole, B.R., Boylan, J.G., Yusheng, W., Holmberg, S.W., Needleman, P. (1983) Science 221, 71-73.
8. Forssmann, W.G., Hock, D., Lottspeich, F., Henschen, A., Kreye, V., Christmann, M., Reinecke, M., Metz, J., Carlquist, M., Mutt, V. (1983) Anat. Embryol. 168, 307-313.
9. Forssmann, W.G., Birr, C., Carlquist, M., Christmann, M., Finke, R., Henschen, A., Hock, D., Kirchheim, H., Kreye, V., Lottspeich, F., Metz, J., Mutt, V., Reinecke, M. (1984) Cell Tissue Res. 238, 425-430.
10. Forssmann, W.G., Hock, D., Kirchheim, F., Metz, J., Mutt, V., Reinecke, M. (1984) Clin. and Exper.-Theory and Practice A6(10&11), 1873-1878.
11. Flynn, T.G., De Bold, M.L., De Bold, A.J. (1983) Biochem. Biophys. Res. Comm. 117, 859-865.
12. Kangawa, K., Matsuo, H. (1984) Biochem. Biophys. Res. Comm. 118, 131-139.
13. Yamanaka, M., Greenberg, B., Johnson, L., Seilhamer, J., Brewer, M., Friedemann, T., Miller, J., Atlas, S., Laragh, J., Lewicki, J., Fiddes, J. (1984) Nature 309, 719-722.
14. Greenberg, B.D., Bencen, G.H., Seilhamer, J.J., Lewicki, J.A., Fiddes, J.C. (1984) Nature 312, 656-658.

STRUCTURE-ACTIVITY STUDY OF α-HUMAN ATRIAL NATRIURETIC POLYPEPTIDE(α-hANP)

Naoyoshi Chino, Yuji Nishiuchi, Yuko Noda, Takushi X. Watanabe, Terutoshi Kimura, and Shumpei Sakakibara

Peptide Institute, Inc., Protein Research Foundation, Minoh-shi, Osaka 562, Japan

Introduction

In early 1984, the isolation and structure determination of several natriuretic and spasmolytic peptides were reported simultaneously by two different groups. The Matsuo group isolated three different components(α, β and γ) from human atrial cardiocytes, and the structure of the α-component (α-hANP) was determined to be a 28-peptide with a cystine cyclic structure as follows:[1]

```
1               5                 10                   15
Ser-Leu-Arg-Arg-Ser-Ser-Cys-Phe-Gly-Gly-Arg-Met-Asp-Arg-Ile-
                20                25          28
Gly-Ala-Gln-Ser-Gly-Leu-Gly-Cys-Asn-Ser-Phe-Arg-Tyr
```

(Cys-7 and Cys-23 linked by a disulfide bridge)

The Needleman group isolated two similar peptides from rat atrial extracts and determined the structures to be 21- and 23-peptides, atriopeptin I and II, respectively.[2] Atriopeptin I has a structure similar to α-hANP(5-25), and II to α-hANP(5-27), but these rat peptides contain Ile at position 12, which is Met in α-hANP.

To elucidate the structure-activity relationship, we synthesized α-hANP(1-28) and a series of its analogs by the solution procedure, and compared their natriuretic and spasmolytic activities.

Synthesis of Peptides

α-hANP and its analogs with the natural type (SS-type) cyclic structure were synthesized as reported in our previous paper.[3] Analogs with an ethylene linkage in place of the disulfide bond, deamino-dicarba-analogs, were synthesized as shown in Fig. 1 as a typical example. This route for cyclizing *via* the branched intermediates effectively eliminated the problem of insolubility of the fully protected linear intermediates in ordinary organic solvents. The cyclization reaction was carried out in DMF using a water-soluble carbodiimide(WSCI), 1-ethyl-3-(3-dimethylaminopropyl)-carbodiimide, in the presence of HOBT, the fully protected peptides were deprotected by HF, and the crude products were purified as in the case of the natural-type peptides.

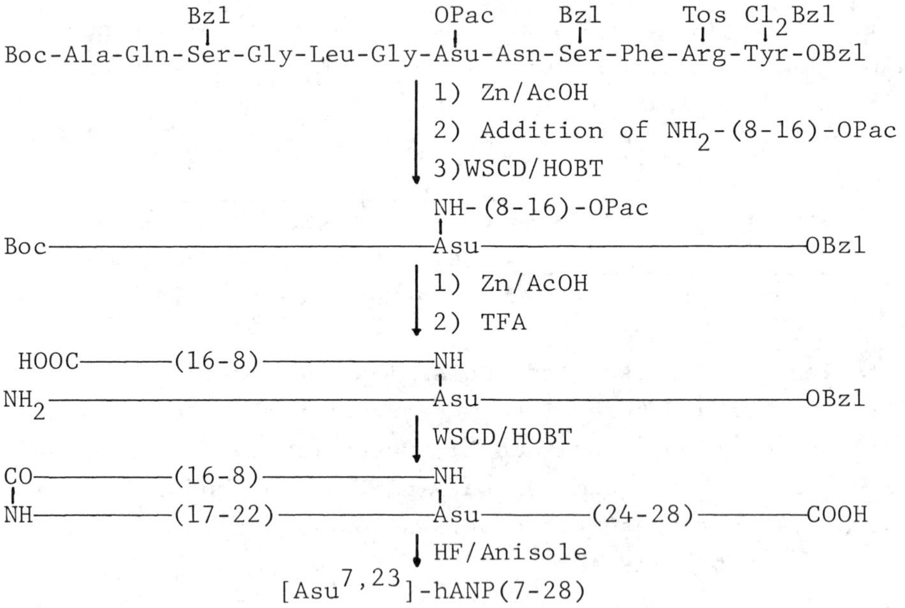

Fig. 1. Procedure for the synthesis of [Asu^{7,23}]-hANP(7-28).

Table I. Relative Biological Potencies of Synthetic Peptides using hANP(1-28) as the standard.

Compound	Spasmolytic Activity		Natriuretic Activity	
	Chick Rectum	Rat Aorta	1 nmol/Kg	10 nmol/Kg
hANP(1-28) derivatives				
hANP(1-28)	100[a]	100[b]	100[c]	100[d]
[Met(O)-12]	2.3 (6)	3.6 (6)	0	5.2 (1)
[Ile-12]	122 (6)	226 (6)	301	219 (3)
[Nle-12]	45.9 (6)	69.1 (6)	99.2	97.2 (2)
hANP N- and C-terminal truncated derivatives				
(4-28)	64.9 (6)	79.3 (6)	29.0	71.7 (2)
(5-28)	66.4 (6)	95.5 (6)	5.2	94.4 (3)
(7-28)	365 (6)	154 (6)	218	177 (3)
(5-27)	53.3 (6)	65.8 (6)	37.5	75.0 (3)
(5-25)	30.7 (6)	5.1 (6)	41.3	77.9 (2)
(7-27)	200 (5)	73.7 (2)	117	97.2 (1)
(7-23)	139 (6)	3.0 (4)	0	33.2 (1)
hANP(7-28) derivatives				
[Nle-12]	169 (6)	61.1 (6)	17.5	162 (3)
des(FGG)	<0.02 (3)	<0.01 (3)	ND	ND
des(AQSGLG)	<0.01 (6)	<0.01 (5)	ND	ND
des(Gly-9)	1.8 (6)	0.2 (2)	0	0 (2)
[D-Ala-9]	356 (5)	121 (4)	186	125 (2)
[D-Asp-13]	1.7 (5)	0.4 (2)	0	0 (2)
$[Asu^{7,23}]$-hANP derivatives				
(7-23)	76.9 (7)	0.13 (6)	0	4.7 (2)
(7-28)	77.9 (5)	10.9 (6)	0	14.0 (3)
[Ile-12](7-28)	90.9 (6)	124 (2)	43.8	155 (2)
[Nle-12](7-28)	37.0 (3)	10.0 (2)	0	4.1 (2)

Values are mean. Number of experiments are in parentheses. a) $ED_{50} = 2.28 \pm 0.16 \times 10^{-9}$ M(n = 61). b) $ED_{50} = 1.82 \pm 0.17 \times 10^{-9}$ M (n = 60). c) $\Delta U_{Na}V = 1.76 \pm 0.2$ μEq/min (n = 31). d) $\Delta U_{Na}V = 3.58 \pm 0.44$ μEq/min (n = 31). ND: not determined.

Bioassay and Structure-Activity Relationship

The pharmacological properties of these synthetic peptides were evaluated by the smooth muscle spasmolytic activities in chick rectum and rat thoracic aorta, and the natriuretic activity in anesthetized rat. The results, summarized in Table I, clearly show the role of intact Met at position 12 for the human-type hormone and Ile for the rat-type hormone. Interestingly, (7-28) showed the highest potency in the all assay systems among peptides examined. This agreed well with the results reported by Thibault et al. using rat atrial peptides.[4,5] Replacement of the disulfide bond with an ethylene linkage did not eliminate all of the biological activities. However, removal of several residues, or even one Gly residue, from the ring structure resulted in almost no activity in every assay system. Substitution of Gly^9 for D-Ala did not alter the biological activities, but substitution of Asp^{13} for D-Asp resulted in almost no activity in every assay system. These findings led to the conclusion that the original ring size of the 17-amino acid structure may be important for expressing the biological activities, and Asp^{13} may be one of the essential amino acid residues for maintaining the significant conformation.

References

1. Kangawa, K. and Matsuo, H. (1984) Biochem. Biophys. Res. Commun. 118, 131-139.
2. Currie, M.G., Geller, D.M., Cole, B.R., Siegel, N.R., Fok, K.F., Adams, S.P., Eubanks, S.R., Galluppi, G.R., and Needleman, P. (1984) Science 223, 67-69.
3. Chino, N., Nishiuchi, Y., Masui, Y., Noda, Y., Watanabe, T.X., Kimura, T., and Sakakibara, S. (1985) Peptide Chemistry 1984, 241-246.
4. Thibault, G., Garcia, R., Carrier, F., Seidah, N.G., Lazure, C., Chretien, M., Cantin, M., and Genest, J. (1984) Biochem. Biophys. Res. Commun. 125, 938-946.
5. Garcia, R., Thibault, G., Seidah, N.G., Lazure, C., Cantin, M., Genest, J., and Chretien, M. (1985) Biochem. Biophys. Res. Commun. 126, 178-184.

SYNTHESES AND PROPERTIES OF PEPTIDES RELATED TO ATRIAL NATRIURETIC HORMONE

Yoshiaki Kiso, Masanori Shimokura, Satoshi Hosoi, Toshio Fujisaki, Yoichi Fujiwara and Makoto Yoshida

Department of Medicinal Chemistry, Kyoto Pharmaceutical University, Yamashina-ku, Kyoto 607 JAPAN

Recently several peptides with the potency of natriuresis and diuresis were isolated from human and rat atrial extract, and the precursors of the peptides were sequenced. Of the peptides, α-human and rat atrial natriuretic polypeptide (α-hANP, α-rANP) consisting of 28 amino acids are thought to be essential to the potency and to play an important role in the blood pressure regulation system. Amino acid sequence of α-hANP (SLRRSSCFGGRMDRIGAQSGLGCNSFRY) is different from α-rANP only at the position 12; isoleucine in α-rANP.[1,2]

In the present study, we synthesized ANPs and their analogs using a new deprotection procedure based on the concept of push-pull mechanism.[3] Using the synthetic ANP analog, we also developed a radioimmunoassay for α-ANP[4] and examined the structure and activity relationship.

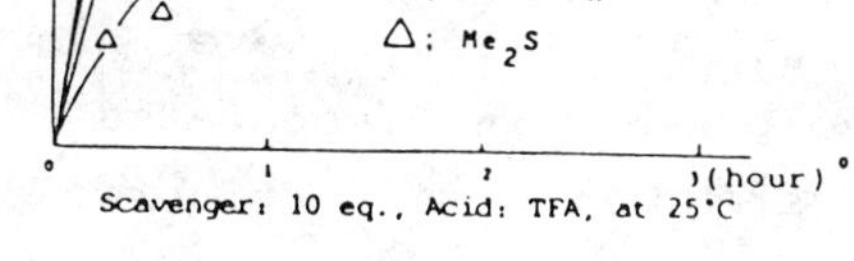

Fig.1 Deprotection of Lys(Z) by selenide-acid system.

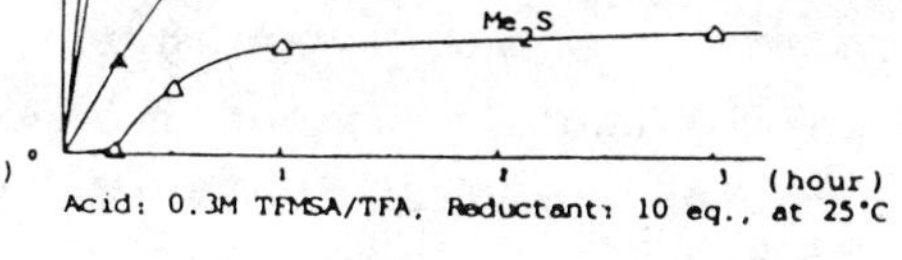

Fig.2 Reduction of Met(O) by selenide-acid system.

We synthesized α-hANP and α-rANP using a new substituted hydrazine, 2,2,2-trichloro-t-butoxycarbonyl hydrazine (Tcboc-$NHNH_2$) and a cysteine derivative, S-(2,4,6-trimethylbenzyl)cysteine (H-Cys(Tmb)-OH).[5] At the final deprotection we also used dimethylselenide as a soft base with TFMSA.[6] We confirmed the soft-basicity of selenides using dimethyl selenide and selenoanisole in the deprotection of Lys(Z) with TFA (Figure 1). Selenides with TFA completely deprotected Z grous in 2 hours, while sulfides in more than 3 hours. Furthermore, selenides with acids had a strong reductivity (Figure 2). We examined the stability of Cys(Tmb) which was first introduced by Brtnik et al.[7] We concluded that Cys(Tmb) was useful for peptide syntheses in which α-amino protecting groups were removed by TFA. Tcboc-$NHNH_2$ was prepared by the reaction of Tcboc chloride and 2 eq. of hydrazine hydrate in THF at 0°C. Tcboc group can be removed by Zn in AcOH more easily than 2,2,2-trichloroethoxycarbonyl (Troc) group. Synthetic scheme for α-hANP is shown in Figure 3. α-rANP was synthesized by a similar route.

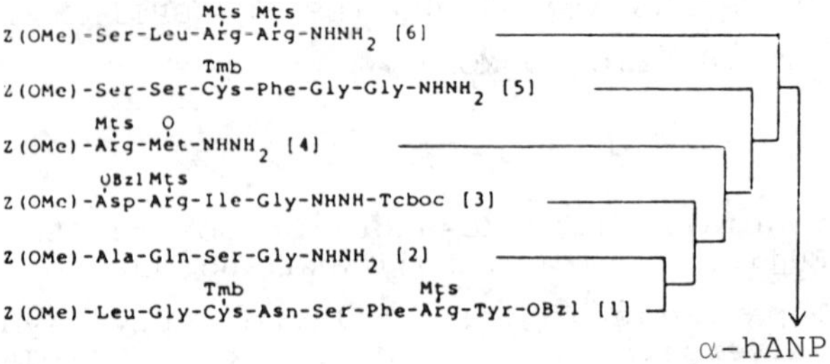

Fig.3 Synthetic scheme for α-hANP.

α-ANP(17-28) and α-ANP(24-28) are intermediates of the above synthesis. Synthetic α-ANP(17-28) was conjugated to bovine thyroglobulin using the carbodiimide coupling procedure. Conjugated α-ANP(17-28) was emulsified with a Freund's complete adjuvant and used for immunization of rabbits. With this anti-ANP antiserum, the distribution of α-ANP like immunoreactivity in human atria and rat hearts were examined. Three and two immunoreactive components were separated by gel filtration HPLC of the extracts from human right auricle and rat atrium respectively.[4]

Synthetic α-hANP caused potent, rapid and short-acting

increases in Na^+ and Cl^- excretion and also an increase in urine flow and K^+ excretion with lesser magnitude when injected into rat.

Also we synthesized a cyclic part of α-hANP, α-hANP(7-23)-NH_2 (Figure 4). Since this peptide had a little potency of diuresis and natriuresis (Table), we attempted to synthesize a chemically stable α-hANP analog. We considered disulfide bond would be equivalent to propylene on the point of interatomic distance and employed 8-aminocaprylic acid instead of cystine (Figure 5). This cyclic peptide, named, cyclonatrin-54, was a little more potent than α-hANP(7-23)-NH_2 in diuresis and natriuresis as expected (Table)[8].

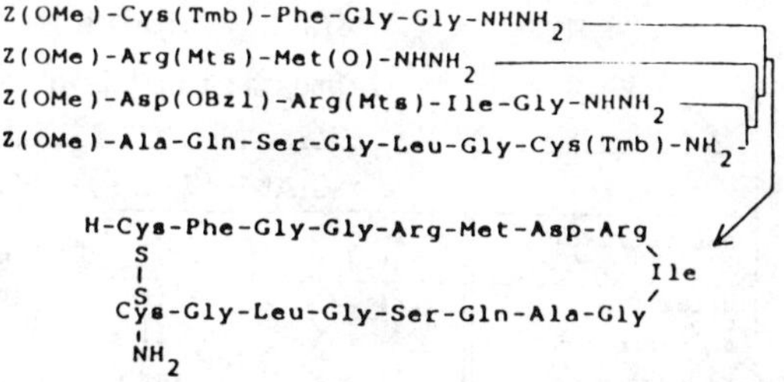

Fig.4 Synthetic Scheme for α-hANP(7-23)-NH_2.

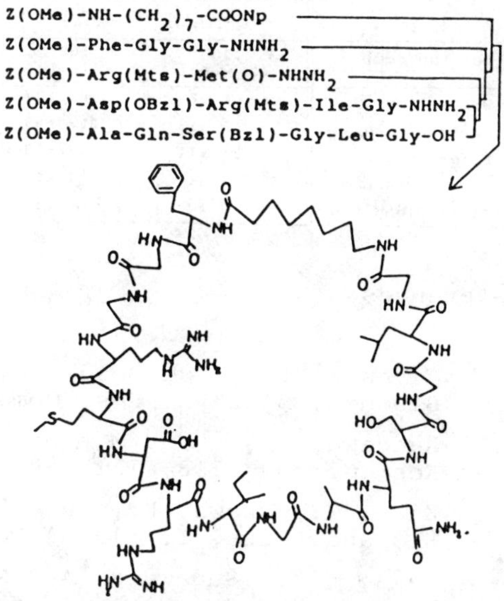

Fig.5 Synthetic scheme for cyclonatrin-54.

Furthermore, using a synthetic intermediate of cyclonatrin-54, we prepared a linear ANP analog, α-hANP(8-22), Phe-Gly-Gly-Arg-Met-Asp-Arg-Ile-Gly-Ala-Gln-Ser-Gly-Leu-Gly. This linear 15-amino acid peptide had a dose dependent natriuretic and diuretic activity, but no hypotensive effect (Table).

It was surprising that a linear peptide exhibited a potent natriuretic acitvity. For the first time, a linear peptide has been prepared which has substantial potency of natriuresis and diuresis.

Table. Renal Responses in Rat to Intravenous Administration of ANP Analogs.

Sample	α-hANP(7-23)-NH_2			Cyclonatrin-54		
Dose(μg/rat)	10 (n=3)	30 (n=3)	100 (n=3)	10 (n=6)	30 (n=6)	100 (n=6)
Urine Flow	21 ±2	42 ±30	98 ±57	37 ±8	58 ±14	68 ±19
Na^+excretion	54 ±36	138 ±116	409 ±267	199 ±58	278 ±105	378 ±122
K^+ excretion	39 ±25	37 ±9	65 ±50	54 ±14	57 ±15	39 ±9
Blood Pressure	0 ±0	- 3 ±2	- 7 ±2	0 ±0	- 5 ±1	- 7 ±1
Renal Blood Flow	18 ±15	16 ±7	22 ±9	13 ±6	17 ±3	18 ±7

Sample	α-hANP(8-22)	
Dose(μg/rat)	30 (n=6)	100 (n=6)
Urine Flow	28 ±8	82 ±27
Na^+excretion	42 ±13	269 ±89
K^+ excretion	31 ±11	43 ±13
Blood Pressure	2 ±1	1 ±1
Renal Blood Flow	0 ±0	-1 ±3

The increase (%) in each parameter was assesed by the difference between the values of control period ("before") and the experimental period ("after", maximum response). Data are expressed as the mean ± S.E.

References

1. Flynn, T.G., de Bold, M.L., and de Bold, A.J. (1983) Biochem. Biophys. Res. Commun. 117, 859-865.

2. Kangawa, K., and Matsuo, H. (1984) Biochem. Biophys. Res. Commun. 118, 131-139.

3. Kiso, Y., Ukawa, K., Satomi, M., Nakamura, S., Kitagawa, K., and Akita, T. (1980) Peptide Chemistry 1979, 193-198.

4. Nakao, K., Sugawara, A., Morii, N., Sakamoto, M., Suda, M., Soeda, J., Ban, T., Kihara, M., Yamori, Y., Shimokura, M., Kiso, Y., and Imura, H. (1984) Biochem. Biophys. Res. Commun. 124, 815-821.

5. Shimokura, M., Hosoi, S., Okamoto, K., Fujiwara, Y., Yoshida, M., and Kiso, Y. (1985) Peptide Chemistry 1984, 235-240.

6. Kiso, Y., Fujisaki, T., Shimokura, M., Okamoto, K., Kaimoto, M., and Uemura, S. (1985) Peptide Chemistry 1984, 289-294.

7. Brtnik, F., Krojidlo, M., Barth, T., and Jost, K. (1981) Coll. Czech. Chem.Commun. 46, 286-299.

8. Kiso, Y., Shimokura, M., Hosoi, S., Fujisaki, T., Fujiwara, Y., Yoshida, M., Nakao, K., Sugawara, A., Morii, N., Sakamoto, M., Suda, M., Imura, H., Kihara, M., Nakayama, K., and Yamori, Y. (1985) Peptide Chemistry 1984, 305-310.

STRUCTURE-ACTIVITY STUDIES OF ATRIOPEPTINS

Kam F. Fok, Foe S. Tjoeng, John A. Houbion, Kerry L. Spear, Sean T. Nugent, Shad R. Eubanks, Mark E. Zupec, Gillian M. Olins, Delores J. Blehm, and Steven P. Adams

Monsanto Company, St. Louis, MO 63198

K. Wakatani and Philip Needleman

Department of Pharmacology, Washington University Medical School, St. Louis, Missouri 63110

Introduction

Atriopeptins(AP) are a family of peptides isolated from rat atria which possess diuretic and natriuretic activity and relax vascular and intestinal smooth muscle[1,2].

1	2	3	4	5	6	7	8	9	10	11	12	13	14	15
Ser-	Ser-	Cys-	Phe-	Gly-	Gly-	Arg-	Ile-	Asp-	Arg-	Ile-	Gly-	Ala-	Gln-	Ser-

16	17	18	19	20	21	22	23	24
Gly-	Leu-	Gly-	Cys-	Asn-	Ser-	Phe-	Arg-	Tyr

Atriopeptin III

In order to investigate the structure-activity relationships of atriopeptins, a series of synthetic atrial peptides were prepared by the solid-phase method and their biological activities were compared. In this report we have focused on the NH_2-terminal and the COOH-terminal portions adjoining the Cys_3 to Cys_{19} disulfide bridge of the peptide.

Table I. Relative Biological Activity of the Atrial Peptides

	Rabbit aorta strip assay		Chick rectum smooth muscle assay	
	ED_{25}* (pmole)	R.P.@ %	ED_{50}# (pmole)	R.P.@ %
AP_{3-19}	21,000	0.35	84	59
AP I	5,400	1.4	220	23
AP_{1-22} (Des-Arg^{23}-AP II)	200	38	--	--
AP II	33	200	760	6.6
AP III	75	100	50	100
Ser-Leu-Arg-Arg-AP III	14	540	66	77

* Effective dose relative to 30 ng of nitroglycerin
# Effective dose relative to 3 ng of isoproterenol
@ Relative potency

Results and Discussion

Table I shows the relative biological potency of atrial peptides in relaxing vascular smooth muscle (rabbit aorta strips) and intestinal smooth muscle (chick rectum). The core structure, the Cys_3 to Cys_{19} peptide loop, has little activity in the rabbit aorta assay but remains quite active in the chick rectum assay. In the rabbit aorta assay, Ser-Leu-Arg-Arg-AP III is the most active of the atrial peptides, followed by AP II and AP III; Des-Arg^{23}-AP II remains substantially active, while AP I is much less active. The intestinal smooth muscle is less responsive to structural changes, indeed nearly all analogs exhibit similar activity. The results indicate that there is some sequence and tissue specificity with atrial peptides. Since the results of rabbit aorta assays are more in parallel with those obtained from the *in vivo* natriuretic activity and renal vasodilation assays in dog[3], we selected the rabbit aorta assay as a tool to study the structure-function relationship of atrial peptides *in vivo*.

Table II. Relative Activity of Synthetic Atrial Peptide Analogs Evaluated by Rabbit Aorta Assay

A. Effect of NH_2-terminal changes

Name	Structure (1 2)	Activity[a]
AP III	NH_2-Ser-Ser-	(100)
(Des-Ser[1],des-Ser[2])AP III	NH_2-	77
Arg-AP III	NH_2-Arg-Ser-Ser-	250
Arg-Arg-AP III	NH_2-Arg-Arg-Ser-Ser-	500
Ser-Leu-Arg-Arg-AP III	NH_2-Ser-Leu-Arg-Arg-Ser-Ser-	500
AP II	NH_2-Ser-Ser-	200
(Des-Ser[1])AP II	NH_2-Ser-	200
(Des-Ser[1],Des-Ser[2])AP II	NH_2-	56
Arg-Arg-AP II	NH_2-Arg-Arg-Ser-Ser-	300
Ser-Leu-Arg-Arg-AP II	NH_2-Ser-Leu-Arg-Arg-Ser-Ser-	150

B. Effect of COOH-terminal changes

Name	Structure (20 21 22 23 24)	Activity[a]
AP III	Asn-Ser-Phe-Arg-Tyr-OH	(100)
AP II	Asn-Ser-Phe-Arg-OH	210
(Des-Arg[23])AP II	Asn-Ser-Phe-OH	38
AP I	Asn-Ser-OH	1.4
(Des-Ser[21])AP I	Asn-OH	1.0
(Des-Phe[22])AP II	Asn-Ser- Arg-Tyr-OH	7.0
(Lys[23])AP III	Asn-Ser-Phe-Lys-Tyr-OH	17

[a]Relative to AP III

Table II shows the effects of NH_2-terminal changes and COOH-terminal changes on the activity of atrial peptides. Deletion of Ser^1 or Ser^1-Ser^2 from AP II or AP III does not severely decrease the bioactivity, while addition of Arg-, Arg-Arg-, or Ser-Leu-Arg-Arg- to either AP II or AP III results in enhancement of bioactivity. These results indicate that the NH_2-terminal region of atrial peptides is not crucial for expression of bioactivity, however this region may serve to modulate the activity. The presence of arginine residue(s) at the NH_2-terminus enhances the bioactivity. The carboxy terminal Tyr^{24} residue in AP III is not essential for bioactivity, while Arg^{23} and Phe^{22} make significant contributions to bioactivity. Deletion of Asn^{20}, Ser^{21} or Phe^{22} further decreases the activity. The results indicate that except for Tyr^{24}, all the amino acid residues of the COOH-terminal region of AP III are critical for bioactivity. Furthermore, replacement of Arg^{23} with lysine decreases the activity to less than 20% of that of AP III, suggesting that a guanidinium group at the COOH-terminal region of atrial peptides, with proper spatial orientation relative to the disulfide ring, may be required for bioactivity.

References

1. Currie, M.G., Geller, G.M., Cole, B.R., Siegel, N.R., Fok, K.F., Adams, S.P., Eubanks, S.R., Galluppi, G.R. and Needleman, P. (1984) Science **223**, 67-69.

2. Geller, D.M., Currie, M.G., Wakitani, K., Cole, B.R., Adams, S.P., Fok, K.F., Siegel, N.R., Eubanks, S.R., Galluppi, G.R. and Needleman, P. (1984) Biochem. Biophys. Res. Commun. **120**, 333-338.

3. Katsube, N., Wakitani, K., Fok, K.F., Tjoeng, F.S., Zupec, M.E., Eubanks, S.R., Adams, S.P., and Needleman, P. (1985) Biochem. Biophys. Res. Commun. **128**, 325-330.

Index to Contributors

INDEX TO CONTRIBUTORS

SUBJECT INDEX

A

I

K

L

M

P

R

T